Radiopharmaceuticals II:

PROCEEDINGS
2nd International Symposium on Radiopharmaceuticals
March 19–22, 1979
Seattle, Washington

Program Committee

Chairman
Vincent J. Sodd, PhD
Professor of Radiology
Director of Nuclear Medicine
Bureau of Radiological Health, FDA
University of Cincinnati
Cincinnati, Ohio 45267

David R. Allen, PhD
Assistant Professor
Pharmaceutical Sciences and Radiology
Director, University of Washington Nuclear Pharmacy
University of Washington
Seattle, Washington 98195

Dennis R. Hoogland, PhD
Assistant Professor of Radiology and Pharmacy
Head Nuclear Pharmacy
University of Minnesota Hospital
Minneapolis, Minnesota 55455

Rodney D. Ice, PhD
Dean and Professor
College of Pharmacy
University of Oklahoma
Oklahoma City, Oklahoma 71190

Book Coordinator

James A. Sorenson, PhD
University of Utah Medical Center
Salt Lake City, Utah 84132

Published by The Society of Nuclear Medicine, Inc.
475 Park Avenue South, New York, NY 10016.

ISBN 0-932004-05-9

Library of Congress Number 79-67730
Printed in the United States of America

TABLE OF CONTENTS

BOXED TITLE DENOTES INVITED PAPER

PANCREAS, PROSTATE, AND ADRENALS

RADIONUCLIDE PRODUCTION

PREFACE

Proceedings of the Second International Symposium on Radiopharmaceuticals

This volume is a compilation of the papers presented at the Second International Symposium on Radiopharmaceuticals which was held in Seattle, Washington, March 19-22, 1979. The Symposium was organized under the auspices of the Radiopharmaceutical Science Council of the Society of Nuclear Medicine (SNM) and was financially supported by the Bureau of Drugs and the Bureau of Radiological Health of the U. S. Food and Drug Administration, the U. S. Department of Energy, and the Health Protection Bureau of Health and Welfare, Canada.

The First International Symposium of Radiopharmaceuticals was held in Atlanta, Georgia, February 12-15, 1974. That meeting was considered highly successful, and its published proceedings have been considered a primer by those practicing this science.

Recognizing the need for a timely update of the information and expertise available in the field, a proposal for this Second International Symposium was submitted in 1976 by Drs. Buck A. Rhodes and Vincent J. Sodd through Dr. John A. Burdine, Chairman, Scientific Affairs Committee, to the Board of Trustees of the SNM. With the formation of the Radiopharmaceutical Science Council in the fall of 1976, this endeavor was transferred under the jurisdiction of the Council.

In 1977, the Board of Directors of the SNM appointed Drs. Vincent J. Sodd (Chairman), David R. Allen, Dennis R. Hoogland, and Rodney D. Ice as the Program Planning Committee for this meeting. During the planning for this Symposium, this committee worked closely and diligently solely to provide the most stimulating scientific program available at that time.

The objectives of the meeting were to present a comprehensive review of those topics considered important in the production and uses of radioactive tracers in diagnostic nuclear medicine, to introduce current developments in radiopharmaceutical technology, to discuss the present status of the regulatory process governing the use of these agents both in the United States and abroad, to provide an assembly for the exchange of ideas, and to publish this book as a comprehensive documentation of the present state of the art.

This publication contains 16 manuscripts by authors invited to write review articles on those general subjects considered appropriate to the present state of interest, namely, Production of Radionuclides, Quality Control, Organic and Inorganic Radiopharmaceuticals, Radioimmunoassay, Oncology, Hematology, Pharmacokinetics, The Clinical Needs in Functional Imaging, Cardiopulmonary Physiology, and the Endocrine, Skeletal, Renal, RES/Biliary, and Central Nervous Systems. Initially, the interest in Pharmacokinetics was not anticipated and, hence, this subject was not assigned as a specific category. However, because a significant number of proffered abstracts were received in this area, the Planning Committee decided to invite an expert in this field to introduce the subject.

Also, 53 manuscripts are included by authors who proffered abstracts relating to the subjects mentioned above. Furthermore, remarks by the Keynote

Speaker, Dr. Dixy Lee Ray, Governor of the State of Washington, and those of the invited panelists relating to the regulation of the use of radiopharmaceuticals are also included.

The symposium was initially highlighted by the keynote address of Governor Ray and the opening panel discussions, all of which dealt with current regulations governing the use of radioactive tracers in medicine. While the introductory remarks of the panelists are published herein, it was decided that for the sake of expediting the publication of this volume and to keep its size within reasonable limits, the vigorous discussion generated during this session could not be included.

The editors thank the many contributing authors who submitted wheir manuscripts promptly and in order. They gratefully acknowledge the assistance of Dr. James A. Sorenson, Book Editor of the Society and Mr. David Kellner of the Society's Central Office, for their concentrated efforts in the rapid publication of this document, and Ms. Cathy Tulloch and her staff for their continued assistance.

Vincent J. Sodd, Ph. D.
David R. Allen, Ph. D.
Dennis R. Hoogland, Ph. D.
Rodney D. Ice, Ph. D.

Introduction: President Society of Nuclear Medicine

The Radiopharmaceutical Science Council is to be congratulated for the program it has put together for this meeting. It is truly an international symposium.

The Radiopharmaceutical Science Council is setting an outstanding example for other councils within the Society of Nuclear Medicine. The past few years of its development have not been without difficulty, but due to the perseverance of individuals such as David Allen, Dennis Hoogland, Jerry Bruno, and others, the Council has become a very meaningful part of the Society of Nuclear Medicine. How the Council continues to function within the parent organization will have a significant effect on the future on the Society itself.

We are truly a multi-disciplinary institution and must provide the framework for the growth and development of the various components such as the Radiopharmaceutical Section. Allowing enough autonomy and independence for such groups to grow and be productive within their own capabilities, while at the same time maintaining the integrity of the umbrella organization, has not been and will not be easy. The leadership of the Radiopharmaceutical Science Council has demonstrated the patience and understanding to make this happen and has certainly set the example for others to follow.

The growth of nuclear medicine over the past 15 years has been astronomical. It is in large part due to the development of new and better radiopharmaceuticals, many by individuals at this symposium. Great strides have been made in obtaining clinically useful information with less radiation exposure to the patient, but our work is not complete. I challenge you to continue in our search for better diagnostic tools so that we can continue to improve the specificity of our diagnosis. New radiopharmaceuticals are essential in the continued growth of our discipline.

There has been much talk during the past few years about the so-called "attack" on nuclear medicine by newer disciplines such as computerized tomography and ultrasound. The strength of nuclear medicine, however, rests in the unique combination of physiology and anatomy, not in structure alone. We must remember the origin of our strength.

Since last June, I have attended many nuclear medicine meetings, including the World Congress meeting in Taiwan, smaller meetings in Peking and Shanghai, and have had innumerable contacts with our colleagues in the United States and around the world. From these experiences, I am convinced that nuclear medicine is <u>alive</u> and <u>well</u>!!

Again, my congratulations to the Council and my very best wishes for a successful meeting.

Douglas Maynard, M.D.
President, Society of Nuclear Medicine

EXCERPTS FROM A KEYNOTE ADDRESS:

POTENTIAL USES OF ATOMIC ENERGY IN LIGHT OF INCREASED GOVERNMENT REGULATIONS

DELIVERED BY GOVERNOR DIXY LEE RAY

SECOND INTERNATIONAL SYMPOSIUM ON RADIOPHARMACEUTICALS

SEATTLE, WASHINGTON, MARCH 19, 1979

Thank you very much ladies and gentlemen. I appreciate that warm welcome, and I am happy to tell you it is nice to be, however briefly, back among a group of scientists and educators.

It is a little different from the political arena. I used to tell audiences when I was Chairman of AEC that I may not know everything about nuclear science, but I wager that I would know more than any other marine biologist in my acquaintance. And, now I feel in the political arena that I might not know everything about science and public life and being Governor and all of the rest, but I guess I still know more about marine biology and radiopharmaceuticals than any other Governor of my acquaintance.

Anyhow, it is a great pleasure to be with you this morning and to be able on behalf of the citizens of this state to welcome you to our beautiful corner of the Northwest. We hope you enjoy your stay here. We hope you take advantage of the many amenities--get out and see our mountains, lakes, streams, rivers, forests, sound what have you--if, of course, they let you out of this room long enough to be able to enjoy being a human being in a wonderful setting. We are glad you are here in Seattle. It is our largest and most outstanding city, and again, welcome to the Evergreen State.

I am pleased that Seattle has been selected as the site for the Second International Symposium on Radiopharmaceuticals. Our state has had a very prominent spot in the history of the development and knowledge of nuclear science in general. The Richland Operation Office, United State's Department of Energy located on the Hanford Reservation in the Eastern part of our state has been involved both in the utilization and development of the perfection of radioisotopes and in nuclear energy since the very early days and continues to be the home of many respected scientists, engineers and technicians highly skilled in all aspects of nuclear science.

The largest single project, in this nation, with construction and investigation of more ways to capture the energy of the atom is located in the State of Washington; that enterprise is known as the FFTF, Fast Flux Test Facility. It has been under construction for a long time and is now in the process of getting its sodium fill and will be operating in a short time. We also have the largest operation, in terms of electrical power production, in the Washington Public Power Supply System of five nuclear plants under construction in various parts of the state. It is a project involving 32 municipal governments, 26 public utility districts, 5 investor-owned utilities and 52 cooperatives located in 7 states. That there is argument about it is obvious from its organization.

Despite all of the comments that one hears in many ways about nuclear matters today, the international cooperation in the peaceful uses of nuclear science has had a fine background, and had a broad scale beginning in December 1953 in President Eisenhower's talk and his declaration of the use of atoms for peace.

This milestone proposal over 25 years ago by the President of the United States announced an Atoms for Peace Program and urged the establishment of an international agency to promote peaceful applications of the atom. Less than a year later, President Eisenhower signed the Atomic Energy Act of 1954. As amended, this act has made possible greater participation by United States' private industry and by United States' scientists in cooperation with other nations in developing all aspects of nuclear knowledge.

Also, in 1954, the Society of Nuclear Medicine was preparing for its first annual meeting. And its First Annual Meeting was held at the Ben Franklin Hotel, located not far from this Olympic Hotel, May 29 and 30, 1954, in Seattle. During that historical meeting, wider use of the peaceful medical atom was initiated, giving birth to nuclear medicine as we know it today. Discussions during those scientific meetings included estimates of cardiac output,in vivo iodine I-131 labeled HSA, a panel of thyroid tracers, the examination and training of residents in radioisotope techniques, evaluation of the role of radioactive gold, and the use of radioactive phosphorus to note just a few of the subjects that were discussed.

Nuclear medicine, on an international scale, as well as a wide variety of other applications of nuclear science were all part of that historic beginning in 1954. When the First United Nations International Conference on the Peaceful uses of Atomic Energy took place in Geneva, less than one year later, a Russian physician presented a paper on the effects on blood of microcurie amounts of radioiodine.

The Atoms for Peace Program also provided for the training of international scientists and technicians from all over the world in nuclear science. The United States Atomic Energy Commission organized training programs at its facilities here in the United States as well as through its Atoms in Action Program designed to travel abroad. In the 10-year period between 1959 and 1969, the Atoms in Action Program visited 35 countries and provided training in the use of scientific reactors, research reactors, industrial applications and techniques of nuclear medicine. It sponsored programs in education ranging from secondary school level through post-doctoral training.

During the period 1955 to 1974, over 11,000 non-Soviet Bloc Scientists, educators and technicians participated in atomic research programs at facilities here in the United States. There were also 110 scientists and technicians from the Soviet Bloc countries who also participated in these programs.

The history of the international development of the peaceful atom will probably show that nuclear science has made important contributions in at least four areas: basic research and education, medicine, electrical energy and industrial applications. These contributions have become so much a part of our everyday lives and the benefits of them are so well known that it seems that it has long been forgotten how life was like in the days before there was nuclear knowledge. Indeed, we have so far forgotten the origins of our understanding of the atom, its versatility, and how, through knowledge, through education, through application, through engineering, that knowledge can be turned to the benefit of mankind. We have so far forgotten those origins that today it seems people everywhere are launching attacks on the very source of all of our benefits. Of these four major areas, basic research in education, medicine, electrical energy and industrial application, I want to take a few moments and compare, in contrast, the advances and the application of two of them--medicine and the production of electrical power.

My introduction to the radioisotope scanning techniques came during a visit to the Medical School at the University of California in 1973, where Benjamin Cassen had developed new scanning techniques by moving a gamma-ray detector mechanically back and forth across the thyroid gland. In England, about the same time, there was developed a system of a series of point measurements over the thyroid glands using a template marked off in centimeter squares that would fit over the neck. Just think how far we have come in a brief five years in radio scanning techniques.

Today, nuclear medicine is responsible for a long list of outstanding contributions in medical science. In certain areas of the United States, thyroid scans are a routine part of an annual physical examination. Other data indicate that radioisotope techniques are used in 9 out of 10 diagnostic procedures and a department of nuclear medicine has become a requirement for hospital accreditation. Nuclear medicine, through the nuclear atom, is providing a broad spectrum of diagnostic procedures. The medical information provided by radionuclides is important, routine, safe and certain and it is today difficult for the layman and the professional alike to appreciate the kind of dilemma that must have existed in diagnosis previous to the introduction of these important techniques.

About 10 million United States' hospital patients, that is one out of 10, plus millions of outpatients depend upon nuclear medicine for diagnosis of their illness. Radiopharmaceutical scientists have provided valuable contributions to diagnostic and therapeutic medical applications for nuclear energy.

The prospect for future development and application of radioactive drugs and techniques in medicine is even greater as our knowledge continues to grow. Such prospects become the interests of people like Dr. Wil Nelp, Mr. Babb and Arnold Stamm in developing the theory and techniques for using a neutron source, such as Californium-252, in neutron activation analysis procedures for detecting cystic fibrosis in new born babies by using a fraction of the infants' fingernail as the material for diagnostic activities.

But, despite the contribution that nuclear medicine has made through innovative therapeutic and diagnostic techniques, despite the prolongation of life, despite the knowledge that has developed about the process of aging, despite all that has been learned about many activities, despite all of the pharmaceuticals and the nuclear medicine application, like so many other noteworthy scientific and technical innovations, they appear now to be headed toward their time of trial, over regulation. There are those, indeed, who believe that the drug regulation in the form of Senate Bill 2755 marks the beginning of restrictive regulation on the field of nuclear medicine.

Those of us who have also been involved in nuclear science as an efficient and effective source of electrical energy, are all too familiar with the end results of excessive regulation and do not think that your science is one that can escape unscathed. In 1957, commercial nuclear electrical energy was first generated by the Shippingport nuclear power station in the State of Pennsylvania. This initial success created national and international interest in the potential of producing commercial, economical, electrical energy by capturing the energy of the atom. This interest and the Atomic Energy Act of 1954 as amended, gave birth to the nuclear electric industry both in the United States and abroad.

In the United States today, we are fast approaching a four generation society. National predictions show that by the end of the 1980's, 12 percent of the population of the United States will be beyond the age of 65 and, by the year 2000 about 9 percent of our population will be above the age of 70. As Governor of the State of Washington, programs for senior citizens are becoming very much more important and are high among the executive priorities. Plainly, people's life spans have been noticeably extended in the recent past. Scientists now know much more about the aging process, about nutrition, cellular functions and tissue functions, intricate chemical changes in the body and the timing mechanisms within the body. I cannot help but believe that the new frontier of this knowledge has been made available in part because of the radioisotopes that have been used as tracers, as probing tools and have enabled the scientific community to learn so much of the intricacy of life processes and of the nature of disease.

Biological science has advanced to levels never before thought possible. For example, bioengineering now provides instrumentation to monitor and to read out all manner of complex biological functions. Synthetic pharmaceuticals and other chemicals have been introduced for treating tension and diseases. When administered, these compounds can alter, combat, speedup, or coincide with normal body functions--some are even developed with time-release accuracy. In large measure, these scientific probing tools are in the form of radioisotopes, and they have been responsible for much of our new biological knowledge.

Although my own experience and training has been in biology, my real introduction to the biological value of radioisotopes as tracers occurred during my time as Chairman of the Atomic Energy Commission. During that period, I was impressed, overwhelmed, and indeed proud, to realize the depth of scientific knowledge that nuclear science made possible; and, the way in which it was developed through the supportive research in our nation, the willingness of American scientists to work in cooperation with scientists all over the world. Major publications of the Atomic Energy Commission developed, involving nuclear science abstracts, now receive worldwide distribution.

I had the opportunity during those years at the Atomic Energy Commission to visit the various laboratories including the Atomic Energy Commission's Isotopes Division at Oak Ridge, Tennessee, where the first sales of Radioisotopes were made during the early stages of the development of nuclear medicine. The Isotopes Division became an equally important partner with the Abbott Laboratories which developed the first radiopharmaceutical. And, may I say that it is the history of radioisotopes, radiopharmaceuticals, that more than anything else show that it is possible, though rare, that there can be developed originally under government oversight, government sponsorship, government grant that can, when properly handled, move into the private sector without strain and without tension. Indeed, radioisotopes stand out as the most remarkable example of that transfer of technology out of the government's sphere and into the private company.

To see the applications of radioisotopes in nuclear medicine, the Medical Division of the Oak Ridge Associated University, where work with cancer patients is being carried on, done through radioisotope tracers and Gallium through long term, whole body, low dose exposure techniques and so on, is a rewarding experience for everyone. And, I say the same for the remarkable Hutchinson Cancer Center located here in our own city. The Medical Division at Oak Ridge was the home of Dr. Marshal Brucer who must have been really the father of nuclear medicine at Oak Ridge. Dr. Brucer's Vignettes in Nuclear Medicine have played a major role in advancing the cause of nuclear medicine.

Today, reliance on nuclear power varies from nation to nation and is based on the available supplies of fossil fuels and hydroelectric sites, as well as access to uranium resources. The oil embargo of 1973 and the more recent price increases by the Organization of Petroleum Export Countries have had a profound effect on the energy policies and the energy supplies of the developing and industrialized nations. Many governments are now having to re-evaluate their energy plans in order to be able to be assured of adequate energy supplies for now and for the future. Realizing that nations cannot depend indefinitely on fossil fuels, energy planners are working to substitute other power sources; these include geothermal and solar energy and other sources as yet undeveloped such as nuclear fusion. But there exists one source in the production of energy, which is under so much attack, and that is nuclear. Even so, nuclear power plants now produce in the United States about 12 percent of our electrical energy, and they have shown to be effective, safe, economical and environmentally the most benign method of producing electricity that we know. Here are actual figures for the year 1978, not speculation, not accusation, not projection, but actual operating figures for 1978 provided by Commonwealth Edison of Illinois, which operates more nuclear power plants than any other utility in our country.

This particular utility produced in 1978, 31 billion kilo-watt hours of electricity, of which 12.5 million kilo-watt hours were nuclear. The cost of production for the fossil fuel plants was 24.9 mills per kilo-watt hour. The cost for producing the nuclear electricity was 12.5 mills per kilo-watt hour, or about fifty percent. Nuclear power plants are twice as effective, twice as economical, based on today's costs, as producing electricity by a fossil fuel plant, whether coal or oil. And, their figures by the way in mills per kilo-watt hour include the fuel, its cost, the operating and maintenance of the plants and carrying charges. Forty-five percent of Commonwealth's electricity in 1978 was produced by nuclear power plants. Had that 45 percent been produced by burning coal, it would have required 17.1 million tons of coal. Had the 45 percent of electricity been produced by petroleum, it would have required 54.7 million barrels of oil. What sort of costs are involved? If those kilo-watt hours had been produced by coal, the cost to customers of Commonwealth Edison would have been $366 million more than what the rate payers paid. Had they burned oil to produce the electricity, it would have cost the rate payer $798 million more than they paid. And yet, in the United States, though we are continuing to import oil to a point where we are now totally dependent upon the largess of other countries, the amount that we are importing this year will cost about $8 million just for the generation of electricity. That electricity could safely, economically, dependably be produced by nuclear power plants. As of

April last year, Atomic Industrial Forum conducted a survey of this nation. They find that the nuclear energy component of our total electrical supply is about 12 percent and that results in the saving of about 120 million tons of coal or 202.6 trillion cubic feet of natural gas or 425 million barrels of oil. Nuclear power plant operating results show that every million kilo-watt hours produced at the plant during a year saves 2.3 million tons of coal or 10 million barrels of oil. These fuel savings are significant when considering the prices of imported oil and the dependence upon whims of the international market and nations located far away.

Even so, nuclear power plants are the source of a bitter national controversy. Today we have sit-ins at nuclear power plant construction sites, picket lines at operating power plants, demonstrations in the nation's capitol and so on. So deep is the emotional feeling, and so widespread is the opposition that on the basis of the climate today it is difficult to consider nuclear power as a serious contender among the United States' future sources of energy. And it is difficult at the present time in the climate of today to consider whether in fact the nuclear industry in this nation can survive. If the nuclear industry in this nation does not survive, where will you look then for radioisotopes? If nuclear industry in this nation is closed down, as the opponents of nuclear electricity would like to do, do you really think that will not effect nuclear medicine in the future of this important part of the application of our nuclear knowledge? Do you really feel that that will not affect the production of nuclear scientists and the climate for working with radioactive compounds?

Nuclear power has been challenged on a number of grounds. These have shifted from health effects, to low-level radiation, to the integrity of the reactor vessel, to the emergency core cooling systems, to terrorism and diversion, to Price-Anderson Insurance, to the proliferation of nuclear weapons constructed from materials within the spent cores of nuclear reactors to the handling of radioactive wastes and so on and so on. In every instance, there is not any guarantee, as there is no guarantee of anything in life, of 100 percent freedom from risk but, the fact is that nuclear industry has about the best record, handling this, and understanding what it is doing, of any industry in our complex world today.

I hope that the many benefits of nuclear medicine will never be subjected to that kind of regulation, but it is a climate in which we now live and unless changed, you are headed for the same road. It could happen. Acceptance of nuclear medicine has lead to a new frontier in providing advanced medical and health services. The contribution of nuclear medicine through diagnostic and therapeutic techniques should continue to play a vital role in improving the health care not only of our nation but of all nations, and it is among those things for which the sick, the ill, the infirm hope and believe it is going to happen.

Radioisotopes, and utilization of nuclear knowledge is falling into the kind of question and opposition which has already overcome nuclear industry. I challenge you, as participants in this Second International Symposium on Radiopharmaceuticals, to dedicate your expertise to the task of ensuring that these medical and health services will always be available. But, to do that, you will have to tell the story of the benefits. You will have to tell the story of greater benefit than risk, because the risk is always there as you well know. You cannot utilize any kind of radiopharmaceutical without some kind of risk, but it is up to you to tell the story to the public that the benefits are better and outweigh the risks. Because if you do not, the institutions which produce the radioisotopes and all of the other institutions involved in the production of energy in which all of our activities are involved will be under attack. Regulation, in the end over regulation, unless it itself is carefully regulated, leads to prevention. And, that leads to stopping. I hope that in your conference today, in your symposium, you will rededicate yourselves to the challenge of keeping your profession alive, and you will not allow it to be subjected to over regulation. Thank you very much.

REGULATION OF RADIOPHARMACEUTICALS: AN ANALYSIS OF RADIOPHARMACEUTICALS APPROVED FOR MARKETING IN THE UNITED STATES 1951 - 1978

Jerome A. Halperin

Deputy Director
Bureau of Drugs

Stanley A. Stringer
Roger D. Eastep

Product Coordination Staff
Office of New Drug Evaluation
Bureau of Drugs
Food and Drug Administration
Rockville, Maryland 20857

Since termination in 1975 of the exemption from the investigational new drug provisions granted to the Atomic Energy Commission in 1963, radiopharmaceuticals have been regulated in the same way as all other drugs. One of the 17 drug review groups in the six new drug evaluation divisions is responsible for the evaluation of applications for marketing (new drug applications) and investigational new drug exemptions (IND's) for radioactive drugs.

The first new drug application for a radiopharmaceutical was approved on November 23, 1951, when Abbott Laboratories was granted approval for radioactive Sodium Iodide-131 for diagnosis and treatment of thyroid carcinoma. Seven months later, in June of 1952, Abbott's Sodium Phosphate P-32 oral solution and injection for treatment of polycythemia vera, certain leukemias, and diagnosis of certain occular and cerebral tumors was approved. Between 1951 and 1960, 11 radiopharmaceutical new drug applications were approved. Seventeen more were approved in the decade of the 1960's, and since 1970, another 101 have been approved, for a total of 129 NDA's. Table 1 displays the number of NDA's for radioactive drugs approved in each of the years from 1951 through 1978.

As of the end of 1978, eighteen sponsors account for the 129 NDA's (Table 2)— Mallinckrodt and Squibb each have more than 20; Abbott, NEN, and CIS having between 10 and 20 each, while another thirteen companies have from 1 to 9 approved applications.

Radioactive drugs having approved NDA's represent only 21 radionuclides. Table 3 lists each radionuclide used in an approved radiopharmaceutical drug product; the year the first new drug application containing that radionuclide was approved; the number of carrier molecules to which the radionuclide is "tagged"; and the number of new drug applications approved for products using that radionuclide as of December 31, 1978. Since approval of the first radiopharmaceutical, 21 radionuclides attached to 41 carrier molecules in a total of 129 applications have been approved. In recent years, FDA policy has been to approve a separate new drug application for each dosage form produced by a given manufacturer. Previously, several dosage forms of a particular drug were included in a single application; thus, it is not possible to make a one to one relationship between applications approved and individual products appearing in the marketplace. Moreover, several NDA's authorize distribution of approved products by other manufacturers, as well as the application holder. These are not included in this tabulation. Finally, FDA has recently withdrawn 13 new drug applications for radiopharmaceuticals because the sponsoring firm informed us that they were no longer interested in marketing the drugs and requested that the approval of the respective NDA's be withdrawn (Table 4). This action resulted in one radionuclide (Mercury-203) no longer being available as an approved new drug.

In 1974, FDA adopted a system for internal use to classify new drugs by chemical type and therapeutic potential. The purpose of classification is to identify potentially important new drugs to facilitate their evaluation so that useful new drugs can reach the market promptly. The classification system has two major components: chemical type and therapeutic potential.

The chemical component classifies drugs into six types: 1) new molecular entities, i.e., molecules never before marketed as approved drugs in the United States; 2) new salts or esters; 3) new formulations; 4) new combina-

TABLE 1. RADIOPHARMACEUTICAL NDA'S APPROVED BY YEAR

1951 - 1978

Year	No. of NDA's Approved
1978	6
1977	10
1976	30
1975	1
1974	28
1973	15
1972	3
1971	5
1970	3
1969	-
1968	5
1967	2
1966	3
1965	1
1964	2
1963	4
1962	-
1961	-
1960	-
1959	2
1958	-
1957	4
1956	-
1955	1
1954	2
1953	-
1952	1
1951	1
	129

TABLE 2. TOTAL NUMBER OF APPROVED RADIOPHARMACEUTICAL NDA'S BY SPONSOR 1951 - 1978

Mallinckrodt	28
Squibb	21
Abbott	17
New England Nuclear	12
CIS	10
Medi Physics	9
Union Carbide	8
Diagnostic Isotopes	6
Amersham/Searle	4
General Radioisotopes	3
Gamma Diagnostics	2
Miles	2
3M Company	2
AN Radiopharmaceuticals	1
General Electric	1
Imaj International	1
Merck Frost Labs	1
Proctor & Gamble	1
	129

TABLE 3. RADIONUCLIDES USED IN AN APPROVED RADIOPHARMACEUTICAL DRUG PRODUCT

Radionuclide	Year 1st NDA Approved	No. of "Tagged" Carrier Molecules	No. of Approved NDA's*
^{198}Au	1954	1	2
^{57}Co	1963	1	6
^{58}Co	1974	1	1
^{60}Co	1959	1	3
^{51}Cr	1955	2	4
^{18}F	1972	1	1
^{59}Fe	1973	1	1
^{67}Ga	1976	1	3
^{197}Hg	1972	1	2
^{203}Hg	1973	1	1
^{123}I	1976	1	1
^{125}I	1963	5	10
^{131}I	1951	6	22
^{32}P	1952	2	5
^{75}Se	1973	1	5
^{85}Sr	1973	1	2
^{87m}Sr	1974	1	1
^{99m}Tc	1972	11	52
^{201}Tl	1977	1	1
^{133}Xe	1973	1	6
^{169}Yb	1976	1	1
21 Radionuclides		41	129

*As of 12/31/78

tions; 5) drugs already marketed by other sponsors; or 6) drugs for which approval is being sought for new indications. The application of the chemical classification system to radioactive drugs differs somewhat from its application to stable drugs. For non-radioactive drugs, the chemical classification refers to the chemical moiety of the molecule. For radioactive drugs, however, the moiety of interest is the radionuclide tagged onto a particular carrier molecule. Type 2 drugs are new salts or esters of already approved moieties. For radiopharmaceuticals, Type 2 is used to represent different forms of the same carrier molecule, e.g., human serum albumin, macroaggregated HSA, and microaggregated HSA. Type 3 (new formulations) refers to new dosage forms of already approved radioactive drugs.

In the therapeutic potential classification, the important breakdowns are for drugs offering potentially important therapeutic gain (Type A), modest therapeutic gain (Type B), and little or no therapeutic gain over products already in the marketplace (Type C). While the assignment of the chemical classification is objective and relatively easy, assignment of therapeutic potential is subjective and compares the new drug against drugs already approved for that indication at the time the new NDA is approved. Once established, the classification does not change, i.e., downgrade, if a better new drug is approved for the same indication.

As every drug is required to demonstrate its safety and effectiveness before it can be approved for marketing, classification of a drug as having little or no therapeutic gain does not imply that it is not as good as drugs currently approved for that indication. Rather, it means that these drugs are as good as, but probably no better than, currently available diagnostic or therapeutic modalities. However, as different persons may react differently to different drugs, a "C" classification drug may offer advantages in terms of sensitivity or acceptability for individual patients in whom other drugs are not well tolerated or have not been effective.

The 129 approved NDA's for radioactive drug products have been arrayed in a matrix comparing the chemical type against the therapeutic potential at the time each was approved (Table 5). The classifications A, B, and C represent the subjective view of the FDA staff and are not based on a survey of the profession, or the view of drug manufacturers. The cells of interest are those showing therapeutic gain and new molecular entities. The number of new molecular entities is a measure of innovative strength in radioactive drug development. There have been 38 new molecular entities (30%) approved as radioactive drugs, 3 (2%) Type 2 drugs (primarily different types of human serum albumin); and 9 (7%) different formulations. Seventy-nine (61%) are drug products already manufactured by another firm.

Looking at the other axis of the matrix, only 8 radioactive drug products (6%) have been rated as important therapeutic gains; 25 (19%) have been rated as modest therapeutic gains, while the majority of the radiopharmaceuticals approved, 96 (74%), have been drugs which offered little or no therapeutic gain at the time of their approval.

To compare radiopharmaceuticals against all other drugs, the 413 NDA's approved in the 5 year period of 1974 through 1978 were analyzed (Table 6), of which 112 (27%) represented new molecular entities, 30 (7%) represented important therapeutic gains, and 65 (16%) represented modest therapeutic

TABLE 4. RADIOPHARMACEUTICAL NDA'S WITHDRAWN FEBRUARY 1979*

Radionuclide	"Tagged" Carrier Molecule(s)	Sponsor
^{198}Au	colloidal gold	Abbott
^{57}Co	cyanocobalamin	Abbott
^{51}Cr	sodium chromate	Abbott
^{203}Hg†	chlormerodrin	Mallinckrodt
^{131}I	iodohippurate	Abbott
	macroaggregated human serum albumin	Abbott
	human serum albumin (RISA)	Abbott
^{32}P	chromic phosphate	Abbott
	sodium phosphate	Abbott
^{85}Sr	strontium nitrate	Abbott
^{99m}Tc	sodium pertechnetate	Abbott
	sulfur colloid	Mallinckrodt
	sodium pertechnetate	Mallinckrodt

*Date of Bureau of Drugs memo initiating withdrawal action

†Only approved NDA for this radionuclide

TABLE 5. ALL RADIOPHARMACEUTICAL NDA'S EVER APPROVED
(Table Shows Distribution of Approved NDA's Based on Classification)

	Therapeutic Potential Type					
	A	B	C	D	E	
	Important Therapeutic Gain	Modest Therapeutic Gain	Little or No Therapeutic Gain	Special Situation	DESI/OTC Claim	TOTAL
1. New molecular entity (active moiety not yet marketed in U.S.)	8 (6.2%) 21.1%	21 (16.3%) 55.3%	9 (7.0%) 23.7%	--	--	38 (29.5%) 100%
2. New salt, ester or derivative (active moiety is marketed in U.S.)	--	2 (1.6%)	1 (0.8%)	--	--	3 (2.3%)
3. New formulation (compound marketed in U.S. but dosage form or formulation not)	--	1 (0.8%)	8 (6.2%)	--	--	9 (7.0%)
4. New combination (contains two or more components not previously marketed together)	--	--	--	--	--	--
5. Already marketed drug product (product duplicates a drug marketed in U.S. by another firm)	--	1 (0.8%)	78 (60.5%)	--	--	79 (61.2%)
6. Already marketed drug product by the same firm (used for new indications for marketed drugs)	--	--	--	--	--	--
TOTAL	8 (6.2%)	25 (19.4%)	96 (74.4%)	--	--	129 (100%)

TABLE 6. NDA'S APPROVED DURING THE PERIOD, JANUARY 1, 1974 THROUGH DECEMBER 31, 1978
(Table Shows Distribution of Approved NDA's Based on Drug Classification)

	Therapeutic Potential Type					
	A	B	C	D	E	
	Important Therapeutic Gain	Modest Therapeutic Gain	Little or No Therapeutic Gain	Special Situation	DESI/OTC Claim	TOTAL
1. New molecular entity (active moiety not yet marketed in U.S.)	17 (4.1%) 15.2%	46 (11.1%) 41.1%	49 (11.9%) 43.8%	--	--	112 (27.1%) 100%
2. New salt, ester or derivative (active moiety is marketed in U.S.)	1 (0.2%)	4 (1.0%)	8 (1.9%)	--	--	13 (3.1%)
3. New formulation (compound marketed in U.S. but dosage form or formulation not)	11 (2.7%)	10 (2.4%)	71 (17.2%)	--	--	92 (22.3%)
4. New combination (contains two or more components not previously marketed together)	--	1 (0.2%)	7 (1.7%)	1 (0.2%)		9 (2.2%)
5. Already marketed drug product (product duplicates a drug marketed in U.S. by another firm)	--	3 (0.7%)	180 (43.6%)	--	--	183 (44.3%)
6. Already marketed drug product by the same firm (used for new indications for marketed drugs)	1 (0.2%)	1 (0.2%)	2 (0.5%)	--	--	4 (1.0%)
TOTAL	30 (7.3%)	65 (15.7%)	317 (76.8%)	1 (0.2%)	--	413 (100%)

gains. Thus, a total of approximately 23% of the applications approved represented drugs which offered some gain over drugs on the market at the time they were approved.

Included in the total of 413 NDA's were 75 NDA's for radiopharmaceuticals (Table 7). The comparison is startling. Only 5 new radionuclides were introduced during that 5 year period (Gallium-67, Iodine-123, Strontium-87m, Thallium-201, and Ytterbium-169). The FDA judged only one of these as an important therapeutic gain (Sodium Iodide-123). The other four were judged as modest therapeutic gains. The overwhelming majority, 69 (91%), represented drugs already marketed by other firms which offered little or no therapeutic gain. As a caveat, however, it is important to recognize that this was the period just prior to and following termination of the exemption to the Atomic Energy Commission, during which time most of these drugs were widely distributed under loosely controlled investigational applications and under the AEC's "well established" rubric.

A subject which enjoys significant interest and for which FDA is often criticized is the time required for NDA's to be approved. The mean time from first receipt to final approval of the 413 new drug applications approved in the years 1974 through 1978 was 26.2 months. By comparison, the mean time of the 71 radiopharmaceutical NDA's approved in that same time (5 radioactive drugs were originally approved as licensed biologic products and are not included in this part of the analysis) was 31.5 months, but the mean of 9 "new" radiopharmaceutical NDA's, i.e., those which represent drugs which were not on the well established use list, is 25.2 months, close to the mean for all drugs, and slightly lower.

There are several reasons which account for longer review times for radiopharmaceuticals than non-radioactive drugs: 1) many of the radiopharmaceutical NDA's languished during the interval between their submission and final approval because of a lack of appreciation of urgency in resolving deficiencies in those applications by manufacturers who had been and were continuing to distribute the drugs under the AEC exemption; 2) enormous volume overload for FDA reviewers; 3) a number of radiopharmaceutical manufacturers are new to the drug business and have had difficulty in understanding and complying with FDA regulations, especially the new drug provisions and Good Manufacturing Practices Regulations. In this regard, radiopharmaceutical development and manufacturing are still a "cottage industry" including a number of small firms and nuclear pharmacies; the large firms getting into it only when a radioactive drug has been developed to the point where its economic potential is reasonable. Finally, there were, in a few instances, technological problems relating primarily to analytical methodology which could not be agreed upon between the sponsor and the agency. A histogram (Figure 1) displays the months required for approval of the 71 radiopharmaceutical NDA's approved during the 5 year period. The shaded blocks represent the "new" radiopharmaceutical approvals. The range for the 71 NDA's spanned 74 months, from 5 months to 79 months. The range for the "new" radiopharmaceuticals was only 34 months, one being approved within 9 months and one in the 43rd month. Figure 2 displays comparable data for the total 413 NDA's approved during that period.

Sales of technetium labeled products for 1977 accounted for approximately 75% of the total sales of radiopharmaceuticals ($67 million). The total of

TABLE 7. ALL RADIOPHARMACEUTICAL NDA'S APPROVED DURING THE PERIOD, JANUARY 1, 1974 THROUGH DECEMBER 31, 1978
(Table Shows Distribution of Approved NDA's Based on Drug Classification)

	Therapeutic Potential Type					
	A	B	C	D	E	
	Important Therapeutic Gain	Modest Therapeutic Gain	Little or No Therapeutic Gain	Special Situation	DESI/OTC Claim	TOTAL
1. New molecular entity (active moiety not yet marketed in U.S.)	1 (1.3%) 7.1%	7 (9.3%) 50.0%	6 (8.0%) 42.9%	--	--	14 (18.7%) 100%
2. New salt, ester or derivative (active moiety is marketed in U.S.)	--	1 (1.3%)	1 (1.3%)	--	--	2 (2.7%)
3. New formulation (compound marketed in U.S. but dosage form or formulation not)	--	1 (1.3%)	2 (2.7%)	--	--	3 (4.0%)
4. New combination (contains two or more components not previously marketed together)	--	--	--	--	--	--
5. Already marketed drug product (product duplicates a drug marketed in U.S. by another firm)	--	--	56 (74.7%)	--	--	56 (74.7%)
6. Already marketed drug product by the same firm (used for new indications for marketed drugs)	--	--	--	--	--	--
TOTAL	1 (1.3%)	9 (12.0%)	65 (86.7%)	--	--	75 (100%)

Fig. 1. Radiopharmaceutical NDA's Approved January 1, 1974 thru December 31, 1978

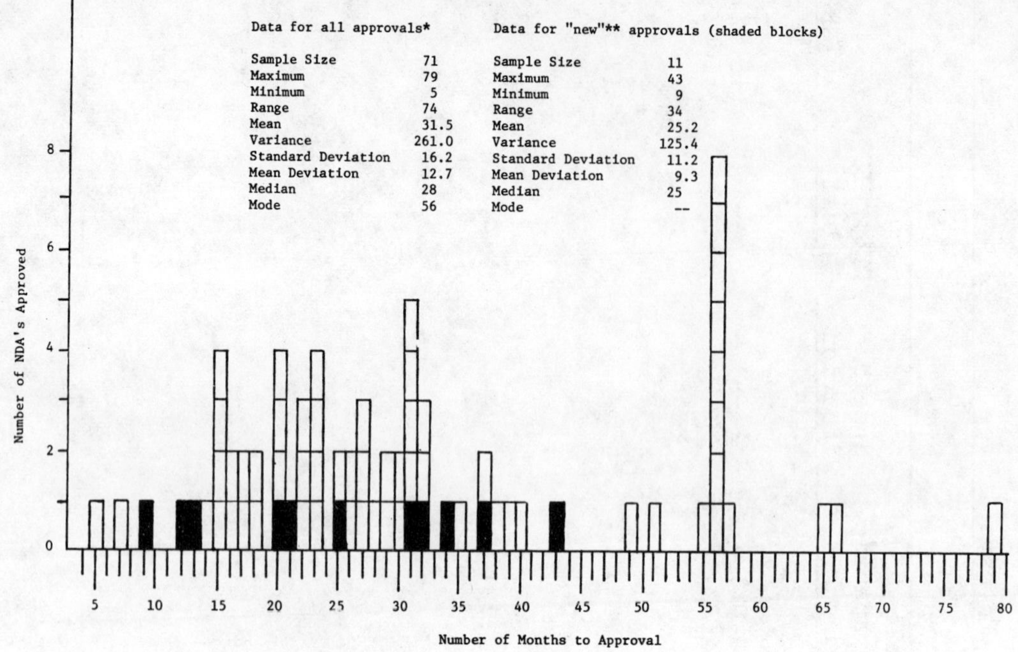

*excludes those radiopharmaceutical products originally reviewed and/or licensed by the Bureau of Biologics

**not previously regulated by AEC or NRC

Fig. 2

Graph Representing the Number of Months and the Number of NDAs Approved by Period of Months from NDA Receipt fo Final NDA Approval covering NDAs Approved from 1/1/74 thru 12/31/78

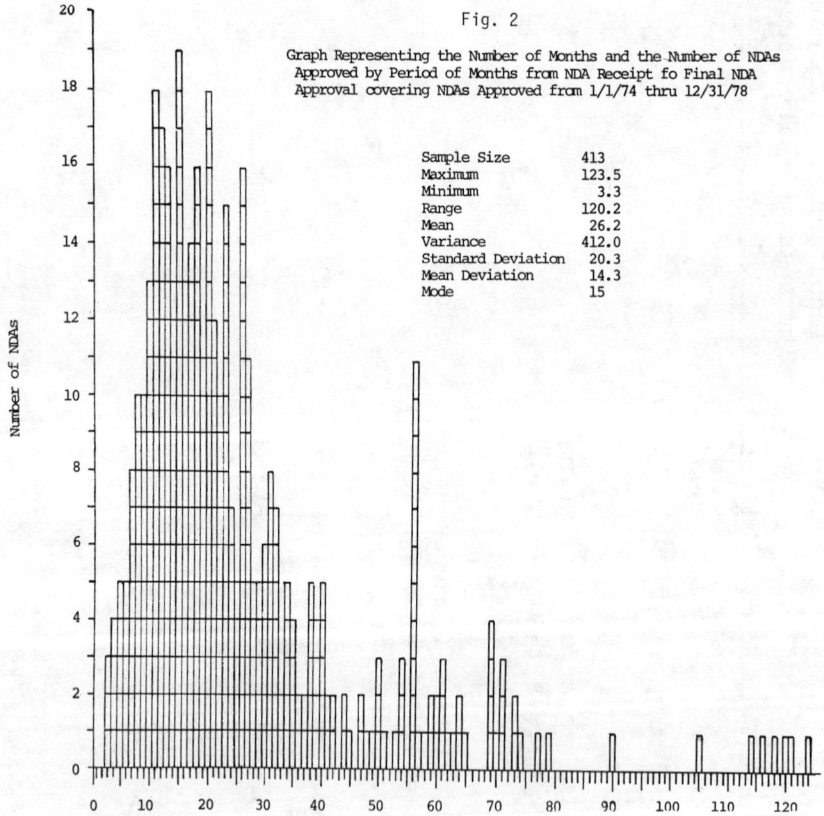

all other radiopharmaceutical drug products represented $22 million in sales with products of isotopes of iodine, gallium, and xenon representing $5 million each, products of selenium and thallium $2 million each, and products of all other radiopharmaceuticals $1 million or less (Table 8) (1).

In terms of current workload, the bolus of applications resulting from the promulgation of the radioactive drug regulations has passed, but there are still almost 20 NDA's for radiopharmaceuticals pending, including NDA's for three new radionuclides (Indium-111 and 113m and Krypton-81m). Most, however, are of class 5-C type, where a competitor firm wants to market a radioactive drug already marketed by other firms.

In the investigational stage, there are about 100 active IND's for radiopharmaceuticals. Most are for radionuclides and carrier molecules which are already approved for marketing, but for which some new investigational work, e.g., new indications, dosages, etc., is in progress. More exciting, however, is the realization that there are a number of IND's for new carrier molecules and at least nine new radionuclides.

Innovation then in a sense of developing new radioactive drugs for nuclear medicine is continuing. In addition, innovation is increasing in terms of application of new technologies to obtain more diagnostic information, especially in the area of dynamic studies, with radioactive drugs already available.

To summarize, regulation of radioactive drugs has evolved from a system of control outside the new drug provisions to where they are fully integrated into FDA's new drug evaluation process. Radiopharmaceuticals are now subject to the new drug classification system and, through use of that system, the percentages of new radioactive molecular entities and important therapeutic gains are noted to be lower than the percentages of new molecular entities and important gains for non-radioactive drugs. Finally, the average time for approval of radiopharmaceuticals is longer than for non-radioactive drugs, although the time for recently developed radiopharmaceuticals is less than the mean for all NDA's. Innovative strength in radiopharmaceutical development is strong as a number of new radionuclide-tagged carrier molecule combinations are either pending approval or in clinical trials.

REFERENCE

1. Hospital and Private Laboratories, IMS, 1977

TABLE 8. SALES OF RADIOPHARMACEUTICALS

	1977 $ (millions)	% Change From '76
Technetium Products	67	+22
Non-Technetium Products	22	+40
Iodine-125 + 131	5	+18
Gallium-67	5	+75
Xenon	5	+50
Selenomethionine	2	-4
Thallium	2	+183
Cobalt	1	-14
Phosphorus-32	0.5	+48
Chromium-51	0.3	+84
Indium*	0.1	-65
Iron-59	0.1	-12
Chlormerodrin	0.1	-10
Gold-198	<0.1	
Strontium-85	<0.1	
Fluorine-18	<0.1	
Other	2	+118
Labeled Compounds	1.4	-3
C-14	0.4	+6
H-3	0.2	+9
P-32	0.1	-13
I-125	<0.1	
Other	0.6	-15

*Not Approved

REGULATORY CONTROL OF RADIOPHARMACEUTICALS IN CANADA

E. G. Létourneau and W.J. Welsh

Radiation Protection Bureau
Department of National Health and Welfare
Ottawa, Ontario K1A 1C1
Canada

ABSTRACT

The Radiation Protection Bureau of the Department of National Health and Welfare is responsible for the regulatory control in Canada of radiopharmaceuticals, generators, and "cold" kits for the preparation of radiopharmaceuticals. A manufacturer must possess a Food and Drug licence before offering his products for sale and, if applicable, be in compliance with the new drug regulations. Continuing control is maintained through periodic plant inspections, a quality assessment program of licenced products, and in some instances by promulgation of regulations.

INTRODUCTION

The Health Protection Branch of the Department of National Health and Welfare is responsible for ensuring that only safe, efficacious radiopharmaceuticals are supplied to the Canadian nuclear medicine community. The authority to control the distribution of these products is contained in the Food and Drugs Act and Regulations. The drugs controlled include final dose form radiopharmaceuticals, as for example, gallium citrate Ga 67, thallous chloride Tl 201, radioiodinated fibrinogen; in vivo kits for the preparation of radiopharmaceuticals, such as macroaggregated albumin, sulphur colloid, methylene diphosphonate and radionuclide generators, the most common, of course being the Technetium Tc 99m generator. In vitro radiodiagnostics, for example radioimmunoassay kits, are classed as medical devices and therefore subject to the requirements of the Medical Devices Regulations.

Our authority does not encompass hospital radiopharmacies who compound radioactive drugs for use within their own institutions. The regulation of these actions is dependent on the hospital administrative organization and provincial college of pharmacy requirements. A person may compound whatever drugs he desires, and so long as he does not attempt to distribute these products, he is not subject to the dictates of the Food and DrugsAct. However radiopharmacies which compound and distribute radiopharmaceuticals to non-affiliated institutions are manufacturing for the purpose of distribution, and must therefore comply with all of the regulatory requirements which will be subsequently discussed. Manufacturing includes such operations as aliquoting from one container to another, preparation of radiopharmaceuticals using commercially procured kits and synthesis of radiopharmaceuticals from basic raw materials.

Radiopharmaceuticals, generators and in vivo kits are controlled by licencing the manufacturing premises. If any of the manufacturer's product line is considered to be in new drug status, then compliance with the regulations respecting new drugs is also required. The balance of this presentation will be devoted to detailed descriptions of each of these control methods, and concluding with a brief description of the regulatory control of in vitro radiodiagnostics.

CANADIAN FOOD AND DRUG SCHEDULE C,D LICENCE

A manufacturer must be in possession of a Canadian Food and Drug licence before offering radiopharmaceuticals, generators or in vivo kits for sale. The in vivo kits have just recently been added to the licencing schedule. An applicant for a licence must furnish adequate details concerning:

(i) The facility description including an area acceptable for the production of sterile, pyrogen free products, if these are to be produced.

(ii) A complete description of the manufacturing processes including procedures to ensure sterile, pyrogen free products if injectables are to be produced.

(iii) Quality control procedures to assess the radioactivity content, radiochemical, radionuclidic and chemical purities, pH, colour, clarity, extraneous material content, sterility, apyrogenicity, etc. (as appropriate) and the specifications for each of these parameters.

(iv) Instrumentation used for the above procedures including the relevant primary and secondary calibrations.

(v) Receipt, quarantine and clearance procedures for raw materials.

(vi) Quarantine and release procedures for final products.

(vii) A description of the lot number assignment and the actions required to institute a recall should it become necessary.

(viii) A description of the associated record keeping systems.

(ix) Stability studies to justify the recommended expiry dates for various products to be manufactured.

(x) Curricula vitae of staff with expertise in radiochemistry, radiopharmacy and preparation of injectables.

(xi) Samples of labels, package inserts, advertising literature, etc.

A pre-licencing inspection visit is carried out to verify the claims made on the application, and to assess the suitability of the candidate to manufacture safe, efficacious products. As previously indicated, it is the manufacturing premises which are licenced, not the distributor, agent or other middle man. Licences are issued for a one year period and may be renewed upon application.

Continuing regulatory control is maintained through periodic plant inspections, and a quality assessment program in which samples are analyzed in the laboratories of the Radiation Protection Bureau in Ottawa. Domestic manufacturers are inspected annually, American manufacturers every second year and European facilities every third year. Personnel performing inspections possess expertise in the disciplines of radiochemistry, radiopharmacy, radiation detection instrumentation, radiation dosimetry, microbiology, etc. During an inspection, all aspects of a manufacturer's operations are reviewed, including: raw materials qualification procedures, label inventory preparation and control, product synthesis, final product testing, instrument calibrations, adequacy of procedures with respect to injectable products, accuracy and completeness of lot records, etc.

The laboratory program concentrates on in vivo kits and generators because of the present dominance of these items in nuclear medicine usage. Final dose form radiopharmaceuticals are sampled on a random basis, the products of interest at the moment include gallium citrate Ga 67, thallous chloride Tl 201, fibrinogen I 125, and to a lesser extent, some of the older well established radiopharmaceuticals. Samples are evaluated with respect to radionuclidic purity, radiochemical purity, radioactivity concentration, and in some instances, biological distribution in mice and rabbits. Tc 99m generators are being obtained on a regular basis and all of the usual items of interest are investigated throughout the manufacturer's recommended lifetime of the product. These items include: Tc 99m yield, aluminum breakthrough, Mo 99 breakthrough, other radionuclidic impurities (alpha, beta and gamma) and pH. Macroaggregated albumin and normal albumin kits have been sampled quarterly for the past couple of years and the results of some of the analyses have recently been published. Bone imaging kits are now being added to the program and will probably be sampled at the same frequency. With kits, emphasis is placed on monitoring tagging efficiency using both in vitro and mouse biodistribution studies.

NEW DRUG REGULATIONS

Should any of the products which a manufacturer wishes to sell be classified as a new drug then compliance with the new drug regulations is also required. To paraphrase the definition in the Food and Drug Regulations, a new drug is one which contains a substance which has not been sold as a drug for sufficient time and in sufficient quantity to establish the safety and effectiveness of that substance for use as a drug. A new drug is also defined as one which has not been sold for the clinical indications recommended by the manufacturer for sufficient time and in sufficient quantity to establish the safety and effectiveness of those recommended conditions of use. Hence a new drug is one which contains a substance whose properties or actions are not fully established or which is promoted with other than the accepted clinical indications.

A person wishing to know the status of a product which he proposes to market should request a drug status decision from the Radiation Protection Bureau. By way of examples, albumin kits (except microaggregated albumin for RES studies), bone imaging kits and sulphur colloid kits are all considered to be established drugs. Kits in the new drug category include the various imidodiacetic acid compounds, the previously mentioned microaggregated albumin and macroaggregated albumin for venography studies. Radiopharmaceuticals considered to be new drugs include thallous chloride Tl 201, gallium citrate Ga 67 and the various radioiodinated adrenal scanning agents. These are examples only and should not be taken as a present listing of new drugs. Since nuclear medicine is a rapidly changing field, the drugs employed tend to remain in the new drug category for a shorter period of time than for other disciplines.

A manufacturer cannot sell or advertise a new drug unless he has filed a new drug submission (referred to as an NDS) and received a notice of compliance. The requirements for a new drug submission are specified in the Food and Drug Regulations. A manufacturer may, however, sell a new drug to qualified investigators for the sole purpose of clinical testing if certain conditions are met, including the filing of a preclinical submission (referred to as an IND). Pre-marketing clearance is required for IND's as well as NDS's. Much of the information required for either type of submission is the same and includes manufacturing and quality control procedures, pharmacology, toxicology and dosimetry calculations. Items required for the new drug submission but not the pre-clinical submission include demonstration of product stability, perhaps further evidence of efficacy and safety and of course supporting clinical data. Samples of the product are generally requested in conjunction with the NDS. An investigator's brochure is required in conjunction with an IND while an acceptable monograph must accompany an NDS. The brochure supplies the investigating physician with information concerning the proposed clinical testing, contraindications, precautions associated with the use of the new drug, etc. A product monograph gives information concerning the chemistry of the drug, indications and dosages, contraindications, precautions, adverse reactions, pharmacology, toxicology, dosimetry, literature references, etc. A cleared IND permits the distribution of a new drug for clinical trials in order to obtain information with respect to the safety, dosage and effectiveness of that new drug. The distribution is restricted to qualified medical practitioners who have agreed in writing to be investigators on behalf of the manufacturer.

The Regulations stipulate that we must respond to a submission within 120 days of its receipt. We generally have a reply back to a manufacturer within about 5 - 6 weeks for an investigational submission and 8 - 10 weeks for an NDS. By comparison, manufacturers may take anywhere from 3 months to 2 years to answer our generally straight forward queries.

Much of the information contained in a drug submission is also necessary for the subsequent licencing procedure. Needless to say, manufacturers are not required to duplicate information which has already been submitted. As this program is administered by a small group of people, the bureaucracy is minimized.

REGULATIONS RESPECTING RADIOPHARMACEUTICALS

There are no specific regulations controlling the new drug and licencing aspects of radiopharmaceuticals, generators, or kits. They are covered by the logical application of the same regulations which are used for all licenced drugs and new drugs. Because of their unique nature, it has been found necessary to promulgate separate labelling regulations, as the general regulations are not applicable to these products. In drafting these requirements we attempted to ensure compatibility with the dictates of other regulatory agencies, both Canadian and foreign, and thus minimize the need for separate labelling for materials imported into Canada. We have been reasonably successful in this endeavour with the exception of nomenclature for in vivo kits. The USAN and hence USFDA designations follow the format "Technetium Tc 99m Aggregated Albumin Kit". This designation is misleading and incorrect as it infers the kit contents are radioactive which, of course, is not the case. The previous USAN designations were much more logical and of the format which we prefer, for example "Kit for the Preparation of Technetium Tc 99m Aggregated Albumin", or alternately "Aggregated Albumin Kit" or something similar.

With one exception there are no regulations covering the purity (radiochemical purity, radionuclidic purity, etc.) of radiopharmaceuticals, generators, or kits. For radiopharmaceuticals appearing in a recognized pharmacopeia (generally the USP or BP), the monograph is accepted as the standard.

While there are no monographs for in vivo kits, we are of the opinion that radiopharmaceuticals prepared from kits should comply with the relevant pharmacopeia monograph. For example Tc 99m labelled macroaggregated albumin prepared from a commercially available kit should be able to satisfy the requirements of the USP monograph for the radioactive product, and manufacturers are now being asked to voluntarily accept these requirements. Should there be reluctance, regulations can, of course, be promulgated.

For products which are not listed in any recognized pharmacopeia the concept of professed standards is used, that is a manufacturer is allowed to establish his own standard. The standard must, of course, be realistic. In assessing the acceptability of a professed standard we refer to any analytical data which have been generated in our laboratories. If the same product is made by more than one manufacturer, we attempt to ensure that the same standard is adopted by all concerned.

Stipulation of radionuclidic impurity limits for Tc 99m is the only instance where it has been deemed necessary to introduce regulatory requirements. Our requirements, which apply to both generator eluates and final dose form radiopharmaceuticals, are somewhat more restrictive than the widely recognized USP limits, especially for Mo 99 (0.03μCi/mCi Tc 99m vs. 1.0μCi/mCi). Our specifications were derived as the result of consideration of patient radiation dose, which must be minimized, and state-of-the-art manufacturing capabilities. With respect to the latter, our fully involved laboratory support program was a requisite in order to formulate what we feel are realistic specifications. Regulations can only be written by persons who are fully cognizant of the practical problems as the result of their day-to-day working experiences.

IN VITRO RADIODIAGNOSTICS

While this Symposium is concerned primarily with radiopharmaceuticals, a few words on the regulatory control of in vitro radiodiagnostics may be of interest. The term "in vitro radiodiagnostics" includes all radioimmunoassay (RIA), competitive protein binding (CPB) and similar test kits, for example T3, T4, hepatitis, digoxin, TSH, etc. These kits are all classed as medical devices.

The major mechanisms for regulating medical devices are through adequate labelling and a device notification system. In addition there is authority to require pre-marketing clearance for certain classes of devices. This course of action is reserved for devices which may present a significant safety or efficacy question. No pre-marketing clearance is required for RIA or CPB kits.

The labelling requirements are common sense items, for example: name of the device, name of the manufacturer, lot number, model designation, nature of benefits claimed to be obtainable through the use of the device, whether or not the device is sterile, expiration date, etc.

Device notification requires a manufacturer or his authorized representative to file with the Bureau of Medical Devices of the Health Protection Branch information which adequately describes and identifies the device including the purpose for which the device is recommended; copies of the directions for use, labelling and package insert; names and addresses of the manufacturer, importer, Canadian distributor, as applicable; whether or not the device is sold in a sterile condition and the method of sterilization; whether or not it may be resterilized and the recommended method of resterilization; and several other notification requirements which are not applicable to in vitro radiodiagnostics.

Device notifications must be filed within 10 days from the day of first sale and be amended as required to keep the filed information up-to-date. Additional information relating to the design, construction and performance characteristics of the device, promotional literature, number of units imported or manufactured in the preceding 12 months may be requested from the manufacturer.

SUMMARY

In summary, our regulatory program is logical and reasonable. We promulgate official regulations only when absolutely necessary and rely on good working relationships with the industry for the balance of the time. To operate on this philosophy requires personnel who are fully knowledgeable of the field which they are regulating.

RADIOPHARMACEUTICAL MANUFACTURERS' INTERACTION WITH REGULATORY AGENCIES

James F. Finn

Diagnostic Isotopes Incorporated
Bloomfield, N.J. 07003

Good morning, Mr. Chairman, Governor Ray, ladies and gentlemen. I appreciate the invitation to speak at this session and appear here on my own behalf as a manufacturer. My remarks do not necessarily reflect those of the industry.

While looking for inspiration on how best to present the view of a small company's interaction with the regulatory agencies, I consulted an area apparently not yet regulated.

A random selection of a passage reads as follows .... 'And Judas went out and hanged himself ....' A second random selection .... 'Go forth and do likewise ....' and one more try! .... 'Do what must be done quickly' .... The obvious message is that I be brief.

Diagnostic Isotopes is a privately owned company, incorporated in New Jersey, February 1, 1971. At the present time, the organization includes Radiopharmacy services in New Jersey, New York and Florida. Diagnostic Isotopes International is a subsidiary responsible for international distribution.

While we believe the company has been responsive to the needs of the nuclear medicine community in products and service, we appreciate your continued input and support.

Manufacturing and distribution of radiopharmaceuticals are the company's only business. Government interaction is an increasing intergral part of this and every business, including apparently religion: Render to Caesar the things that are Caesar's .... .

At last count approximately 30 Federal departments, agencies, boards, committees or commissions interact with the Radiopharmaceutical industry. Add to this the regulations of Pharmacy Boards, the 28 Agreement States and the distinction between cyclotron and reactor produced nuclides, which may or may not be regulated on a state by state basis, the separate and distinct requirements of the SCHARM Committee of New York City, one can see the high probability of duplication of effort and double jeopardy. Most certainly, the opportunity to hang up ourself in regulations.

Interaction with individuals in most of the regulatory agencies is, in general, intercourse with normal, helpful, hardworking, reasonable people. An inspector recently asked ....Who is easier to deal with .... NRC or the Agreement States? .... a well intentioned question but not really relevant since it is not a matter of choice.

Interaction with impersonal agencies is another matter indeed. It is a constantly changing environment, sometimes open and flexible, sometimes rigid and by the book. One is never comfortable with the ever present threat of a violation or infraction of some obscure regulation or proposed regulation from NRC, EPA, OSHA, FAA, DOT, FDA, MTB, etc. Overlapping authority and regulations of one agency may be in direct conflict with another agency; thus creating the opportunity to be hung between agencies. Amid the conflict there also exists inconsistencies. For example, the Nuclear Regulatory Commission's jurisdiction for radiation protection extends only to byproduct material. Are we to conclude no hazard exists or no protection is required for accelerator produced nuclides? Obviously not, that's just the way the regulations are written! For example, the Department of Transportation - or, I should say, the Materials Transportation Bureau - and the Nuclear Regulatory Commission share responsibility for insuring use of safe methods of preparing and transporting radiopharmaceuticals. Of course, why your properly placarded package is sitting on the loading dock in an airport across the country is not the responsibility of the Federal Government. Could it be the hazardous classification has an adverse reaction on freight forwarders or airline pilots? Once injected, on an outpatient basis for example, this restricted, hazardous threat is no longer subject to puncture proof, drop resistant, leak-tight packaging requirements and labeling. Are we to conclude a hazard no longer exists?

Recently, a new agency has been created called the IRLG for Interagency Regulatory Liaison Group. One function of this group is to develop plans to eliminate duplication of regulatory efforts of FDA, CPSC, EPA and OSHA. Hopefully, a new gallows has not been created.

Fortunately, the Sunshine Act calls for agency reevaluation and hopefully elimination. For example, the National Center for Productivity and Quality of Work Life self destructed in September 1978.

The message of this agency from its Council of Economic Advisors is .... the most significant problem of recent years is the deterioration of production efficiency. It recommended reversal of this economic slowdown by enhancing technological innovation, stimulation of investment and improving government-business cooperation. Technological innovation in the field of Nuclear Medicine and instrumentation is obvious. The field probably had achieved its greatest growth in the latter part of the 1960's through the early 1970's. A sure stimulation of capital investment.

Enter, the Food and Drug Administration. Our protector from conception to grave.

Revocation of the exemption of byproduct radiodiagnostics from FDA jurisdiction has brought all radiopharmaceuticals under FDA scrutiny.

In an industry dealing with short product life expectancy and even shorter half-lives, one develops a sense of urgency and appreciation for time; an appreciation not shared by the regulatory agencies.

Since there is no distinction between therapeutic drugs or contrast media and radiodiagnostics, the FDA regulatory review process is time-consuming. The 180-day review period is seldom, if ever, met.

The New Drug Application review process .... an accusatory document if you will .... appears to operate under the premise of one's being quilty .... now prove yourself innocent. Hardly conducive to government-business entente.

And the Lord said .... 'Go forth and multiply ....'

The Federal Food, Drug and Cosmetic Act, as amended in 1976, brought us the Bureau of Medical Drug Devices.

The Drug Regulation Reform Act of 1978 will involve us more deeply in generation of paperwork.

What does all this government regulation mean to you, the practitioner or consumer. First, the increased costs of compliance for products and services will be passed on to the consumer. Industry is forced to take a second and third look at the return on investment when considering new radiopharmaceuticals. Competition is discouraged, if not prevented, from entering the marketplace. Any new agents will be delayed in entering the marketplace. Limited use products may be withdrawn from distribution.

How might we resolve some of these problems?

I would like to offer a few suggestions.

... Exempt properly packaged radiopharmaceuticals from the hazardous classification to help expedite shipments and eliminate surcharges.

... Establish abbreviated New Drug Application procedures for radiopharmaceuticals.

... Allow distribution after the 180-day review period since the risk to benefit ratio for radiopharmaceuticals is recognized as extremely low.

... Improvise a uniform format for all countries which regulate radiopharmaceuticals and maintain reciprocal agreements of approval.

My plea to the regulators is .... We will render to Caesar what we must, but as we walk through life, don't be afraid, with your hand in my pocket, be careful, you have worn the lining threadbare; so, do what must be done quickly!

NOTES FOR A PANEL ON REGULATORY AFFAIRS

Knud Kristensen

The Isotope-Pharmacy
The National Health Service of Denmark
DK-2700 Copenhagen-Brønshøj

Apart from the legal aspects any regulatory action must be based on a solid scientific and professional background. The purpose of regulations are not only to protect the patient, the workers, the environment and other third party interest, but it is also to make sure, that radiopharmaceuticals of the highest quality can be used in the best possible way. It shall also make sure that new developments may take place, if it is beneficial to the individual patient and to the society. Regulations are needed but according to my philosophy, they would normally not be used in daily practice. They are considered a sort of safety net or minimum standards. The basis for this statement is a high moral and scientific quality level among the professions involved. There may be an economic pressure, that may be working against this. It is therefore important to have a high level of professional competance, which can only be obtained by training and the continous provision of information. It is not expected, that anybody would normally act against solid scientific evidence in his own profession. It may be so, that to fulfill this ideals becomes more and more difficult as one moves away from direct contact with the patients.

Requirements for quality and safety should be under continous development and improvements as the technical and economical background developes. Detailled regulations are therefore not ideal as they tend to give improvements in steps, instead of a continous evolution. This again calls for a high professional standard, where nobody would be interested in using products, that do not have a high level of quality and safety. In order to get such a system to work, it is very important to be able to create a high level of confidence between the authorithies and the industry and between authorithies and the pharmaceutical and medical profession.

The Isotope-Pharmacy, that act as the official governmental control laboratory for radiopharmaceuticals in Denmark, has to the best of my knowledge been able to obtain reasonable standard of radiopharmaceuticals used in Denmark without having to take any legal actions. The regulations in Denmark are general in structure and details can be put in by the National Health Service but are normally provided as advice and recommendation by the Isotope-Pharmacy.

Due to the structure of supply of radiopharmaceuticals the Isotope-Pharmacy was given the duty to supply radiopharmaceuticals, kits and generators to Danish hospitals. It has been shown, that this has placed the Isotope-Pharmacy in an ideal position, where the background information for the necessary steps are provided by the experience gained in the actual handling of a large proportion of the products used in Denmark. Besides the function as a pharmacy most resources are used for studying products and collecting and providing information to hospitals. The evaluation and continously improvements of product specifications is another important duty. By the availibility of the most up to date information on product quality in a comparable form are the use at hospitals of the best quality encouraged.

It has turned out to be of extremely great importance, that the Isotope-Pharmacy both in the formal organisation and physically is placed between the State Institute for Radiation Hygiene and the Drug Control Laboratory, which gives very good possibilities for avoiding conflicting advice or requirements. It also provides a very important scientific background.

QUALITY CONTROL ANALYSIS AT THE HOSPITAL

Knud Kristensen

The Isotope-Pharmacy
The National Health Service of Denmark
DK-27oo Copenhagen - Brønshøj

ABSTRACT

Quality control analysis is an integral part of quality assurance. In a system as with radiopharmaceuticals where part of the finishing of the product takes place at individual hospitals, the need for quality control analysis at the hospital can be discussed. Data are presented that stresses the importance of quality control by the manufacturer as a basis for limitation of such work at hospitals. A simplified programme is proposed.

INTRODUCTION

The requirements of modern nuclear medicine are radiopharmaceuticals of uniform properties, that are effective with regard to the intended use and where the risk of use is minimal compared to the benefit obtained. The risks discussed should be taken to involve both the risk to the worker and to the patient. Principles used for safeguarding non-radioactive pharmaceuticals could be applied with certain modifications. However an evaluation of quality control measures needed will have to be done seperately for each area. It is the intention of this paper to discus quality control measures for radiopharmaceuticals, and how these measures should be placed in the total system, particularly which part should take place at the hospital.

QUALITY ASSURANCE

High quality does mean radiopharmaceuticals of a uniform quality in relation to established standards. Important elements in a production system whereby this may be obtained are the operators, premises, equipment, documentation system, working procedures and quality control measures. Quality assurance is the sum of all such measures taken to obtain the required quality based upon development work, clinical trials, and good manufacturing practice (GMP) (1-2). GMP is that part of quality assurance ensuring, that products are prepared to a quality appropriate to their intended use (specifications). Quality control is part of GMP and is more limited in its scope than quality assurance. It is mainly concerned with sampling and analytical testing and with the establishment of specifications and documentation which form the basis for release of raw material and finished products. The purpose of quality assurance is thus to give finished products in accordance with specifications, that contains the correct ingredients of the desired purity in correct amounts, which has been processed as defined appears correctly labelled in its proper containers and is transported and stored according to specifications.

In principle there may be no difference between radiopharmaceuticals and other drugs. Industrial produced radiopharmaceuticals delivered ready-for-use or products prepared from raw materials at an Isotope-Pharmacy or at an individual hospital e.g. cyclotronproduced radiopharmaceuticals must all be produced in a system as just described. There may be a need for certain modifications of rules and regulations because of radioactivity, because of short-half-lives or because of the need for special handling and production procedures but a system based on well-developed, well-controlled and documented procedures should be applied (3). Specifications should be laid down based on development work, clinical trials and investigations of possible impurities, side reactions, and adverse reactions (4).Good Radiopharmacy Practice (GRP) is defined as all the procedures and handling of radiopharmaceuticals, that take place at the hospital. It is thus parallel to Good Manufacturing Practice (GMP) in the industrial production of pharmaceuticals including radiopharmaceuticals (5-1o).

HANDLING AT HOSPITALS

There is one main difference between radiopharmaceuticals and other drugs: A larger part of the handling takes place at the hospitals. Even with most ready-for-use radiopharmaceuticals it may be necessary to make a calculation of the volume of a patientdose followed by a dispensing of the dose. A large group of radiopharmaceuticals covering more than 5o% of the use in most countries is the Tc-99m labelled radiopharmaceuticals prepared from Tc-99m generators and kits. In this case even a larger part of the work take place at the hospital. Such materials may be prepared with all the necessary quality assurance build in, but the producer can never garantee for the working procedures carried out at the hospital. A special problem may be created by the fact, that the generator and the kit may come from different manufacturers.

All this leads to the conclusion, that part of the total quality assurance program possibly including quality control analysis will have to be carried out at the hospital.

It may be useful to consider 3 different types of radiopharmaceuticals as they are handled at the hospitals:

Type I: Radiopharmaceuticals from raw materials, from isotope generators, from home-produced preparation kits, complicated labelling procedures with patient samples and sometimes including the production of radionuclides by a cyclotron.

Type II: Radiopharmaceuticals from generators and preparation kits and/or labelling of patient samples with such material.

TypeIII: Finished radiopharmaceuticals in the form of individual patient doses or/and radiopharmaceuticals in multidose form.

The large proportion of nuclear medicine procedures in most countries rely on the use of products from the last two groups. The first type is mainly found at places, where research and development of radiopharmaceuticals take place or where cyclotronproduced shortlived materials are prepared. There is no doubt, that requirements very similar to the GMP for industrial production of radiopharmaceuticals will develop for this type of work. However, for types II and III, there may be a case for a much simpler system at the hospital because of the possibility for building in quality assurance in the semi-manufactured products delivered to the hospital. Tc-99m radiopharmaceuticals from generators and kits is at present the main product among type II-preparations.

STANDARDS

Evaluation of a radiopharmaceutical by quality control methods requires a set of specifications (a standard). Such standards are based on development work including clinical trials. The specifications describes individual quality parameters such as radionuclidic and radiochemical purity, specific activity, physico-chemical properties etc. and the analytical methods to be used in controlling these. The setting of standards are performed through a pharmacopoeae-monograph (11-12) through a licensing procedure or by individual user when prescribing a product. In Denmark standards for individual products not generally licensed are set by the Isotope-Pharmacy, a govermental quality assurance laboratory.

The experimental background for the present day standards are often very limited. Many parameters are set at limits, that could be obtained in practice rather than by systematic studies of the needs. It has to be admitted, that such studies often may be difficult to perform due to ethical reasons or due to the lack of relevant animal models.

The relationship between specifications of the radiopharmaceutical used and the clinical results obtained is an area, that is just started to be developed for example by groups, that do more systematic development of new radiopharmaceuticals. It is very difficult to find information in the litterature because it is not practice to give detailed information on the product used.

It is outside the scope of this paper to do a detailed evaluation of the specifications for radiopharmaceuticals, but a few general comments are offered: (a) The requirements for radionuclidic purity are in general well worked out, based most often on radiation dose consideration such as for Mo-99 in Tc-99m or both on radiation dosimetry and instrument requirements such as I-124 in I-123. Good analytical methods are most often available allthough methods for control of α and β impurities may be very tedious.(b) Radiochemical purity is a more complicated subject, as it is more difficult to determine, and as it also is more difficult to establish the need for a certain level. The methods in several standards for Tc-99m radiopharmaceuticals are not sufficient for a full determination of the radiochemical purity. For manyTc-99m radiopharmaceuticals only free pertechnetate is determined. The problems with free pertechnetate is mostly, that it gives a higher general background, while other radiochemical impurities such as free iodide in hippuran also may add substantially to the radiation dose. (c) Specific activity is seldom a problem. Methods for control are in several cases not available. (d) Physical-chemical properties such as particle size in colloidal or particulate preparations may still need a better characterisation. (e) General pharmaceutical properties such as pH, isotonicity, sterility, pyrogenes may be a problem for some products but with the acceptance of quality assurance as a total system not only based on quality control analyses of the actual batch but also on control of the method, it should be possible to find sensible solutions in most cases.

In general it is the impression, that most specifications may be adequate for the present day use of the products, but the analytical methods are an area for improvements. A second generation of specifications will certainly be improved also with regard to radiation hygiene and according to the use of products for dynamic studies.

QUALITY CONTROL

Several review papers have been presented in recent years on quality control methods (13-17) and several symposia (18-19) were devoted to quality control in Nuclear Medicine. Other publications relate to comparative evaluation of individual Tc-99m radiopharmaceuticals product types. (2o-28)

Data on more systematic studies of the performance of the radiopharmaceutical quality assurance systems are however very scarce. In United States a reporting system on Radiopharmaceutical Drug defects were established in 1976. During the first year 43 defect products were reported. 28 of these were concerned with Tc-99m labelled radiopharmaceuticals. (29)

In Canada a report on quality control analysis of 42 batches of lo different products of Tc-99m macroaggregated albumin and Tc-99m human serum albumin gives results from studies to the end of 1977. At least 3 products did not fulfil the specifications. (3o)

In order to be able to decide on the necessary official quality programme and to be able to advice hospitals of any need for in-house quality control 3 different studies were initiated, when the Isotope-Pharmacy in Denmark were established. The evaluation of products has been done against allready existing standards but have included a variety of analytical methods. In the following a summary description will be given of the results of an evaluation of all radiopharmaceuticals used in Denmark in relation to established standards.

The background for these studies is the supply of radiopharmaceuticals to all Danish hospitals. Table 1 gives some statistical information on the number of nuclear medicine procedures carried out. The population is 5 million people but the numbers of nuclear medicine procedures is probably significantly lower than in the United States, which means that the volume from which this samples were taken would be less than 1% of the use in the United States.

1.The Isotope-Pharmacy supply control

The national govermental Isotope-Pharmacy in Denmark supply about 6o% of all radiopharmaceuticals used in Denmark as ready-for-use products or as Tc-99m-generators and kits. All other products must have a license, and the Isotope-Pharmacy take samples for analysis on a random basis. As a limited but more direct approach all products received at the Isotope-Pharmacy are inspected, papers and labels compared to specifications, total radioactivity determined and radionuclidic identity measured along with a simple radionuclidic purity test in the form of measurement in a brass and a plastic shielding besides the unshielded measurement in an ionization chamber. Finally external contamination is checked by smear testing of suspected products. The results are presented in figure 1. There has been a decrease in the frequence of defect product as companies responded well to complaints. It should however also be remembered, that the selection of product used is done according to the quality obtainable. It is noted that only very few defects were due to transport. Most could be related directly back to the manufacturers and therefore could be considered as a lack of quality assurance at the manufacturers.

2.Compliance testing

All radioactive products apart from Tc-99m labelled radiopharmaceuticals are sampled on a random basis for a complete analysis of all parameters in the specification. The results from 1977 and 1978 are presented in table 2. Apart from a product that was a mixture of I-125 and I-131, were all deviations from specifications minor in nature.

3.Tc-99m preparation kits

Tc-99m preparation kits have during a 2 1/2 year period been subject to

TABLE 1. STATISTICAL INFORMATION ON NUCLEAR MEDICINE IN DENMARK 1977.

All procedures :	62 departments
	683oo procedures (1366o/mill.)
	Yearly increase : 16%
Tc-99m :	25 departments
	383oo procedures (56% of total)
	13oo Tc-99m generators (287 Ci Mo-99)
	23oo preparation kits (1o5oo preparation)
	324 Ci Tc-99m injected.

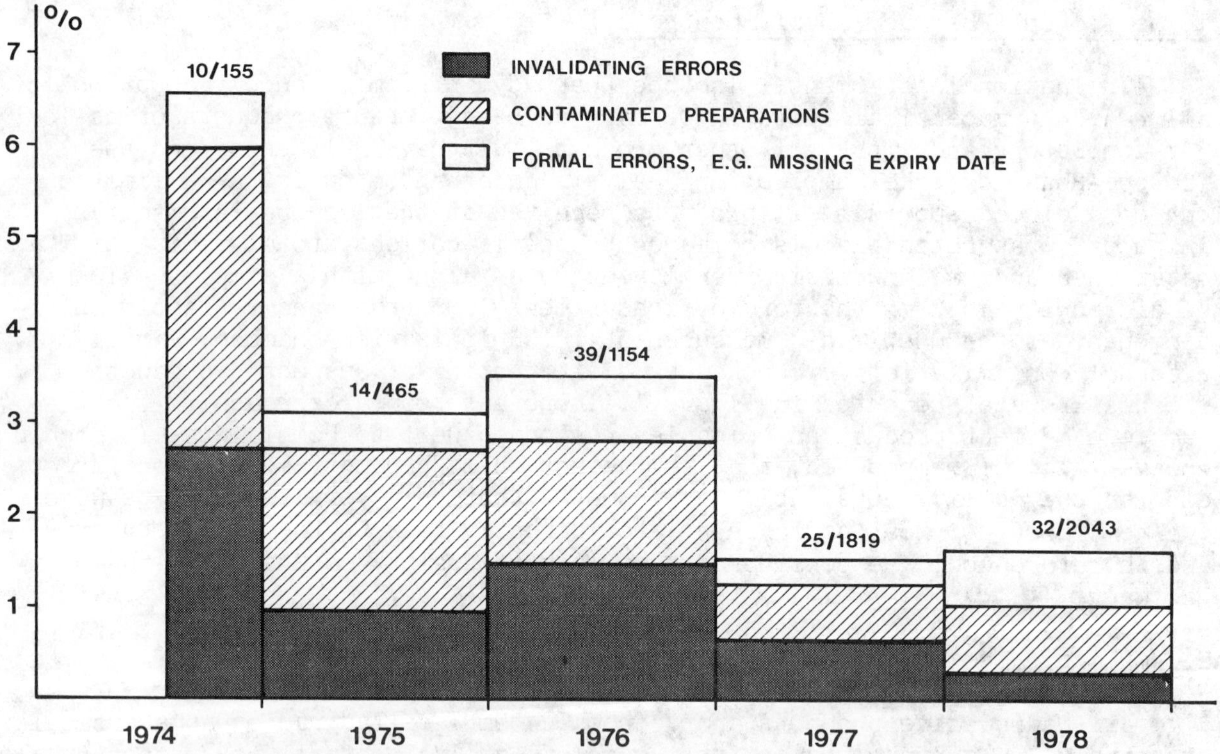

Figure 1.

Percentage of radiopharmaceuticals received at the Isotope-Pharmacy that were found defective by inspection and simple measurements.

a systematic compliance study. For each product used in Denmark 3 different batches have been analyzed. For some products 3 batches have not yet been studied because so many batches of products have not been used. For each batch labelling has been carried out on 3 different days and the resulting products analyzed immediately afterwards with at least 6 different methods for evaluating the radiochemical purity.

The methods were chosen such that they should be able to show free pertechnetate, reduced/colloidal Tc-99m, and if possible labelling efficiency directly. Parameters such as pH, isotonicity, particle size etc. were also included where relevant. All parameters were determined by double determinations. The methods used are outlined in table 3 (27,31,32). All labelling was carried out with Tc-99m from the same type loo mCi generator. All labelling and all analysis were carried out by the same technecian. The results obtained by this comprehensive study whereby more than 4ooo analysis have been carried out are presented in summarized form in table 4.

The results show that 16 batches (9%) were defect on the basis, that at least 2 out of 3 preparations did deviate from specifications confirmed by more than one analytical method. Further 6 batches would be considered defect, if it was required that all three preparations should fulfil the specifications. If the product were evaluated by just one analytical method, 95 preparations would have been considered defect. The most frequent single analytic parameter, that would have lead to rejection of the product, if no other method was used to provide evidence of a false result, is the widely used methanol ITLC system (ITLC Silicagel SG-thin layer chromatography with 85% methanol). About 65% was of this type.

A detailed analysis shows, that the number of defect are distributed rather even all over the product spectra and among the supplying companies.

No obvious correlation was observed between the day number in the generators life and the resulting products in the comprehensive batch study but a detailed analysis might reveal such differences.

In order to see if any changes did take place during the period covered the material was divided into two parts covering first and second part of the period. The results are given in table 4. It can be seen that while the analytical problems are distributed evenly in the two periods, the number of defect products are significantly higher in the first period. Which may of course be due to the selection of products by hospitals in accordance with the experience gained, but could also indicate improvements in product quality.

Analysis were made immediately after preparation. The programme has recently been extended to look at stability. 5 different product (4 MDP - 1 Phytate) were analyzed again after 6 hours. All products fulfilled specifications immediately after preparation while one MDP product after 6 hours gave 3 preparations, which all were significantly below specifications.

The limitations of this study are, that it is performed under more idealized conditions than under hospital routine, and that the Tc-99m source were a small size generator of just one type. The results are therefore considered to be minimum and only an indication of how the system may work.

TABLE 2. RESULTS OF COMPLIANCE TESTING AT THE ISOTOPE-PHARMACY 1977 AND 1978.

	Number of products tested	Number of defect products	Types of defects			
			Rad./conc. total radio-activity	Specific activity	Pos.limulus test	Other
1977	46	6	2	1	1	1
1978	64	lo	4	2	2	2

TABLE 3. METHODS USED IN BATCH CONTROL OF TC-99M PREPARATION KITS

Physical :	Appearance particle size - numbers pH - isotonicity filtration through membranefilter (Nuclepore)
Chemical :	Thinlayer or paperchromatography (2-4 diff. systems) high voltage electrophoresis Sephadex column scanning
Biological :	Organ distribution (mice or rats).

TABLE 4. RESULTS OF QUALITY CONTROL OF TC-99M LABELLED RADIOPHARMACEUTICALS.

Type	Products	Batches	Prep.	Batches defect	Single prep. defect	Single analytical parameter deviating	Single analytical result deviating
Antimoni sulfide coll.	1	5	15	2	o	1	o
Albumin	5	11	33	2	o	6	2
Citrate	1	3	9	o	o	o	o
DMSA	3	1o	3o	1	o	3	3
DTPA	6	11	33	1	o	o	o
EHDP	3	9	27	o	o	18	2
Glucoheptonate	2	5	15	o	o	2	1
HIDA	3	8	24	1	o	4	o
MAA	7	17	51	4	o	7	3
MDP	9	27	81	o	1	3o	13
Microcoll.	2	5	15	o	1	2	1
Microspheres	6	15	45	2	1	8	8
Phytate	4	11	33	o	o	4	o
Polyphosphate	1	3	9	o	o	o	o
Pyrophosphate	4	9	27	1	o	6	1
RBC labelling	1	2	6	o	o	o	o
Sulphur coll.	6	16	48	2	2	4	3
Tin colloid	2	4	12	o	1	o	1
Total	66	171	513	16	6	95	38
Total first half of study	-	94	282	12	3	41	15

Danish hospitals have been asked to report product problems to the Isotope-Pharmacy. 7 such reports on Tc-99m radiopharmaceuticals were received during 1978. The reports were all on unexpected organ distribution. Most reports related to bone scintigraphy agents. In most cases it was possible to analyse either the actual product or a vial from the same batch. Only in one case was it possible by the quality control system mentioned to verify product defects. In that particular case all Tc-99m in both the pertechnetate-vial as in the prepared kits were in the colloidal form. The reason for this is unexplained.

At present it is our impression that most defect products in our study relate to the kits rather than to the generator used, but further studies are needed to prove this. An earlier study of all available Tc-99m sulphur colloid kits labelled with Tc-99m from all available Tc-99m generators did indicate a slight correlation between free pertechnetate and oxidizing/reducing properties of the Tc-99m eluate (23).

Report has been published on the effect of the amount of Tc-99m in the eluate (33) and of the possible effect of oxidizing substances(34).At an earlier time also aluminium ions were considered a problem (35).The only case we have been able to document was a case of a MAA preparation, that gave a lower labelling yield with an elute from a generator, that was not eluted for a week, than with an eluate from a generator eluted 24 hours before. The results are given in table 5. It should be noted, that the biological distribution were all normal, which may indicate, that the preparation is unstable under the conditions of analysis. The stannous chloride content of this kit was as low as 9o µg per vial. Such more casuistical data are included here because they illustrate the problems encomtered in doing quality control analyses and evaluating the results.

The influence of the generator eluate on the number of preparations that fall outside specifications may be judged from the number of cases, where only one out of three preparations were defect. This was only the case 6 times among 513 preparations indicating that in this study, it may only be a small problem.

EVALUATION OF QUALITY CONTROL RESULTS

From the available information the following conclusion may be drawn: The level of quality of radiopharmaceuticals judged in relation to present specifications seems to be comparable to other drugs. The total percentage of defect products from this rather large sample of products varied from 2-15 % for different groups of products found by different methods, however many of the defects were minor in nature. There has been a decrease in the number of defects found in products used in Denmark during the period studied. Most defects seems to be related to products rather than to transport or handling at the final preparation stage, which means, that they may be corrected by manufacturers. For Tc-99m labelled radiopharmaceuticals the rate of defect products may have been lower in 1978 as compared to 1977. There seems to be no reason why the rate of defect products in general would not continue to be at this or a lower level. However, other testing procedures may reveal other deficiencies. New more stringent specifications may not be fulfilled without improvements in the system.

Quality control methods for Tc-99m radiopharmaceuticals in pharmacopoeaes and other specifications do require modifications particularly with regard to determination of radiochemical purity.

SUGGESTIONS FOR A HOSPITAL QUALITY CONTROL PROGRAMME FOR RADIOPHARMACEUTICALS

It has been shown that most defects revealed by a quality control study might be corrected already by the manufacturer. For the hospitals that only use ready-for-use radiopharmaceuticals or generators and kits (type II and III radiopharmaceuticals), it is therefore in Denmark recommended, that only a limited quality assurance programme be set up. Handling at hospitals shall take place in accordance with Good Radiopharmacy Practice (5,9), which will include requirements for premises, hygienic standards, written instructions, documentation systems, training of personel etc.

The content of quality programmes would then be build according to the types of products. Such programmes require as basis a specification and for each products used a document should be found where the specifications needed from the clinical point of view were fixed at the time, when the clinical procedure was set-up.

1.Ready-for-use radiopharmaceuticals (type III products)

A quality control programme could be restricted to a comparison of information on labels, package insert etc. to the specification. (or as simple as comparison to a previous received label etc.) Visual inspection for transport damage and finally to measurement of the radioactivity of the dispensed patient dose.

This programme will make sure, that it is the same product received as it was the intention, when the prescription was made, and it gives a quality control measurement of the dispensed dose. It does presuppose, that the quality control measures by the manufacturers are sufficient to pick up any mistakes made during the production process. The results so far obtained by official quality control testing or by the defect product reporting programme cannot substantiate further quality control analysis by the user.

2.Tc-99m labelled radiopharmaceuticals from generators and kits (type II)

Quality control should be carried out with regard to the part of the work done at hospitals. For Tc-99m pertechnetate the following parameters may be relevant: The elution yield from the generator should be calculated. This will indicate whether the system works properly. Mo-99 is one possible impurity, that has been noted a few times primarily due to wrong mounting of the generator either by the manufacturer or by the operator but also as a possibility after transport damage. It is very easily controlled by a very simple method (36).Testing of the first eluate is considered most relevant. Finally control of the aseptic technique should be carried out. One way of doing this is by testing the last eluate for microbial contamination, but it should be stressed that such testing can only reveal a rather grossly faulty system. Sterility of the product must controlled by a continous evaluation of the whole system.

For Tc-99m radiopharmaceuticals prepared with preparation kits the recommendations are that no analytical quality control testing of each preparation should be necessary in the daily routine. Compatibility between generators and kits may be a problem, but this should be dealt with, when setting up the procedure. At that stage a radiochemical purity test may be needed. Further two cases have been identified, where a radiochemical purity test may be useful: When the type or size of generator is changed, and when a new technecian is introduced to the procedure. It may further be useful to the hospital to be able to perform such a test in cases where the clinician, when evaluation a scintigram, comes up with the question, if a particular finding could be due to the radiopharmaceutical. This can then often be ruled out rather fast.

For such purposes a simple chromatographic test may be useful. The experience is however, that for a complete evaluation of a Tc-99m-radiopharmaceutical a variety of different methods may be needed. It may therefore be considered more useful, if quality control analysis by the hospital is needed, to perform this as a control of the production method rather than as a condition for release of individual preparations.

A simplified quality control kit may be useful (37) but the experience with commercially available systems has not allways been good. In our study several products would have been rejected, if based solely on a particular system (38). Quality control analysis at hospitals of each individual Tc-99m preparations are only expected to have rather small implications on the clinical work, if the results found in our studies could be taken to be representative. Further systematic studies from other places are however needed to prove this.

It should be possible to produce Tc-99m kits, that with a great margin of safety, should give products that fulfil specifications, whereby resources at hospitals could be used in a better way than to a large number of daily quality control procedures. This do not necessary apply, if generators and kits are used in a special way e.g. in an Isotope Pharmacy for dispensing and distribution of a large number of doses from each vial.

It is considered useful for a group of hospitals to have acces to one laboratory, that is equipped for more extensive studies and which then may help in setting up procedures and in trouble-shorting at other hospitals. This laboratory could also do routine testing for methods control. In Denmark with 25 hospitals using Tc-99m radiopharmaceuticals we consider one such place sufficient.

CONCLUSION

This first systematic study of the effect of quality control analysis seems to indicate, that it is possible to produce radiopharmaceuticals that fulfil present specifications. Analytical methods in these specifications are in need for improvements. It is expected to be most costeffective to centralise quality control analysis as much as possible either to the manufacturers or, if needed to a central or regional laboratories.

REFERENCES

1. Guide to Good Pharmaceutical Manufacturing Practice, 1977 Department of

Health and Social Security, HMSO, London, 1977.
2. Food and Drud Administration, U.S.A.:Code of Federal Regulations Tittle 21 Part 21o-211.
3. Good Practice in the Manufacture of Radiopharmaceuticals.A proposal put forward by a group of Pharmacists from the Nordic countries.Arch Pharm Chem 78:1oo2-1oo9, 1971.
4. Food and Drug Administration U.S.A: Guidelines for the clinical evaluation of Radiopharmaceutical Drugs 1977.HEW publication No 77-3o44.
5. Kristensen K: Preparation and Control of Radiopharmaceuticals in Hospitals. A Guidebook published by the Int.Atomic Energy Agency (In press).
6. Kristensen K, Müller T, Rhodes B A: Good Radiopharmacy Practice in Quality Control in Nuclear Medicine p.268-275 (editor B A Rhodes)Mosby,St.Louis (1977).
7. British Institute of Radiology: Guidelines for the Preparation of Radiopharmaceuticals in Hospitals Special Report no.11, London,1975.
8. The Hospital Preparation of Radiopharmaceuticals, Radionuclidic Topic Group, The Hospital Physicists Association, Scientific Report Series 16,1976.
9. On the preparation of Tc-99m labelled Radiopharmaceuticals,Report IA-5, The Isotope-Pharmacy, Copenhagen, 1976.
1o. Trott N G: The Preparation of Radiopharmaceuticals in Hospitals.Proc. of the Bad Gastein symp. 1978. Radioactive Isotope in Klinik und Forchung band 13, 517-28 (Editor R.Höfer) Egemann Verlag, Vienna 1978.
11. European Pharmacopoeia. Vol.3,1975, suppl. 1973,1977. Maisonneuve S A, France, 1977.
12. The United States Pharmacopoeia XIX edition, 1975.
13. Stelmack H A, Qinn III J L: Radiopharmaceuticals Quality Control. Sem Nucl Med 4: 295-3o3, 1974.
14. Cohen Y, Besnard M: Analytical Methods of Radiopharmaceutical Quality Control in Radiopharmaceuticals (editor G Subramanian et al.) The Soc Nucl Med New York, 1975.
15. Srivastava S C, Meinken G, Smith T D, Richards P: Problems Associated with Stannous Tc-99m Radiopharmaceuticals. Int J Appl Rad Isotopes 28: 83-95, 1977.
16. Krohn K A, Jansholt A: Radiochemical Quality Control of Shortlived Radiopharmaceuticals. Int J Appl Rad Isotopes 28: 213-227, 1977.
17. Pauwles E K J, Feitsma R: Radiochemical Quality Control of Tc-99m labelled Radiopharmaceuticals. Some Daily Practice Guidelines. Eur J Nucl Med 2: 97-1o3, 1977.
18. Quality Factors in Nuclear Medicine. Procedings of 13.Int. Meeting Soc of Nucl Med, Copenhagen, 1975.
19. Rhodes B A: Quality Control in Nuclear Medicine, Radiopharmaceuticals, instrumentation, and in vitro assays. The C.V.Mosby Company, St.Louis, 1977.
2o. Vinberg N, Kristensen K: Comparative evaluation of Tc-99m-generators. Eur J Nucl Med 1: 219-233, 1976.
21. Sood V J, Fortman D L: Analysis of the Sr-89 and Sr-9o content in eluates of fissionproduced Mo-99 Tc-99m-generators. Health Physics 3o: 179-182, 1976.
22. Sood V J, Grand R J, Montgomery D M: An investigation of the Plutomium content in fission product Mo-99 Tc-99m-generators. Health Physics 29: 425-427, 1975.
23. Petersen B, Kristensen K: A comparative study of Tc-99m labelled radiopharmaceuticals for liver scintigraphy. Proc. of XIII Int.Ann Meeting of Soc Nucl Med, Copenhagen, 1975.

24. Krogsgaard O W: Radiochemical purity of various Tc-99m labelled bone scanning agents. Eur J Nucl Med 1: 15-17, 1976.

25. Krogsgaard O W: Tc-99m sulphur colloids. Eur J Nucl Med 1: 31-35, 1976

26. Stetten Ov, Thiel E: Radiochemical purity of a Tc-99m labelled Technescan PYP. Eur J Nucl Med 1: 267-68, 1976.

27. Warbick A, Ege G N, Henkelman R M, Maier G, Lyster D M: An Evaluation of Radiocolllid Sizing Techniques. J Nucl Med 18: 827-34, 1977.

28. Petit W A, Deland F H, Pepper G H, Blanton L: Characterization of Tin-Technetium Colloid in Technetium-Labelled Albumin Preparations. J Nucl Med 19: 387-392, 1978.

29. Ford L,Shroff A, Benson W, Atkins H,Rhodes B A: SNM Drug Problem Reporting System. J Nucl Med 19: 116-117, 1976.

3o. Results of Quality Control Studies of Technetium-99m Labelled Radiopharmaceuticals prepared from Kits. Report 78 EHD-24. Health and Welfere, Canada, 1978.

31. Vivian A,Ice R D, Shen V, Hetzel K R: Radiochemical Purity of Radiopharmaceuticals using Gelman Seprachrom TM (ITLC TM) Chromatography. Tech Bull 32, Gelman Instrument, Michigan.

32. Persson B R R: Gel Chromatography Column Scanning: A Method for Identification and Quality Control of Tc-99m Radiopharmaceuticals in Radiopharmaceuticals (editor G Subramanian et al.) Soc Nucl Med, New York, 1975.

33. Lamson M L, Kirschner A S, Hotte C E, Lipsitz E L, Ice R D: Generator-produced $^{99m}TcO_4^-$: Carrier Free ?. J Nucl Med 16: 639-41,1975.

34. Colombetti L G, Barnes W E: Effect of Chemical and Radiochemical Impurities from Eluents on Tc-99m-labelling Efficiency. J Nucl Med 16: 271-74, 1977.

35. Webber M L, Cragin M D, Victery W K: Albuminium Content in Eluents from Commercial Technetium Generators. J Nucl Med 12: 7oo, 1971.

36. Richards P, O´Brien M J: Rapid Determination of Mo-99 in separated Tc-99m. J Nucl Med lo: 517, 1969.

37. Zimmer A, Pavel DG: Rapid Miniaturized Chromatographic Quality Control Procedures for Tc-99m Radiopharmaceuticals.

38. On Testing of an Analytical Kit for Quality Control of Tc-99m-labelled Radiopharmaceuticals, Report IA-6, 1978. The Isotope-Pharmacy, Copenhagen-Brønshøj.

TABLE 5. IN VIVO AND IN VITRO PROPERTIES OF MACROAGGREGATED SERUM ALBUMIN, WHEN LABELLED WITH TC-99M ELUATES FROM GENERATORS NOT ELUTED FOR DIFFERENT PERIODS OF TIMES.

Lot no.	22o	22o	Blool	Blool
Days between Tc-99m eluates	1	7	1	7
Lung uptake (rats)	97%	97%	98%	96%
ITLC (Rf 1) (% TcO_4^-)	o.98%	48%	1.7%	62.2%
% TcO_4^- (Sephadex)	o.5%	8.8%	o.5%	12.3%
8 μ Nuclepore filtration (% retained on filter)	98.3%	71%	98.3%	75%

^{99m}Tc-SULFUR COLLOID - EVALUATION OF PREPARATION PARAMETERS FOR KITS FROM FOUR COMMERCIAL MANUFACTURERS

Donald L. Fortman
Vincent J. Sodd

FDA, Nuclear Medicine Laboratory
Bureau of Radiological Health
Cincinnati, Ohio 45267

ABSTRACT

This study investigated the adequacy of the preparation instructions listed in the package inserts of four commercial suppliers of ^{99m}Tc-sulfur colloid kits. All of these kits utilize the reaction of an acid on sodium thiosulfate to produce the sulfur colloid. The labeling efficiency vs. time of reaction, temperature of reaction, ^{99m}Tc quantity, and effect of complexing agents and stabilizers were studied. With the exception of one manufacturer that listed water bath temperature specifications that could result in erratic labeling efficiencies, the instructions found in the package inserts were found to be adequate, and should be carefully followed.

INTRODUCTION

In this era of radiopharmaceutical preparation utilizing technetium-tin reactions in one-component kits, ^{99m}Tc-labeled sulfur colloid remains a hold-out as a multicomponent kit that requires both chemical and physical manipulation for its production.

The ^{99m}Tc-sulfur colloid is formed in these kits by the reaction of an acid with sodium thiosulfate in the presence of $Na^{99m}TcO_4$, forming colloidal particles that contain the ^{99m}Tc. The reaction sequence goes through the formation of thiosulfuric acid, which is extremely unstable and quickly breaks down into elemental sulfur, sulfur dioxide and water. This reaction is slow at room temperature; therefore, the reaction mixture is subjected to boiling water temperature for various lengths of time to increase the reaction rate. Somewhere in the reaction sequence, ^{99m}Tc and sulfur unite to form technetium-sulfur colloid. The exact structure of the final product is not known since the level of technetium in these kits is so low (nanograms and lower) that conventional methods of instrumental and chemical analysis cannot be used to determine molecular configurations. Suggested configurations come from either extrapolated data obtained using microgram or milligram levels of ^{99}Tc, or by analogy with the chemistry of rhenium which is in the same group (VIIB) as technetium in the periodic table. These configurations range from some reduced form of technetium physically absorbed on elemental sulfur particles (1), to some type of formal bonding between technetium and sulfur (2).

The purpose of this study was to determine the effects of temperature and time of reaction on the labeling efficiency of sulfur colloid kits, from four commercial manufacturers (USA), and to determine if the instructions found in the package insert are adequate for formulation of a final product that is within the 92% labeling efficiency specified by the U.S. Pharmacopeia (USP).

MATERIALS AND METHODS

The kits used in this study were obtained from four different commercial manufacturers and represented at least two different lots from each manufacturer. The contents of the kits are listed in Table 1. The $^{99m}TcO_4^-$ solutions were obtained from two different generator suppliers and were tested for both ^{99}Mo and Al^{+3} content prior to use. The ^{99}Mo was assayed using the dose calibrator method where the generator eluant was placed in a 4-mm thick lead container and the ^{99}Mo activity was read on a dose calibrator at the appropriate ^{99}Mo setting. All $^{99m}TcO_4^-$ solutions used were within the USP and United States Nuclear Regulatory Commission (USNRC) specifications of not more than 1 μCi ^{99}Mo per mCi of ^{99m}Tc. In most cases no ^{99}Mo could be detected by this method. The aluminum concentration was determined by placing a drop of generator eluant on aluminum test paper (Macherey, Nagel & Co., Duren, Germany). The paper was then treated with HCl-acetate solution and exposed to NH_3 vapor for 10 sec. A red coloration against a yellow background indicates the presence of aluminum. The limit of sensitivity of the test is 10 μg aluminum per ml. The specifications of the USP and USNRC for aluminum concentration are not more than 10 μg/ml of generator eluate. None of the generator eluants used in this study gave a positive test for aluminum.

The labeling efficiency was determined via Instant Thin Layer Chromatography (ITLC) using silica gel ITLC media (Gelman Inst. Co., Ann Arbor, MI.) and methanol: water (85:15) as the solvent. The chromatograms were allowed to develop to 15 cm, dried, cut into 1-cm strips and counted in a well-type NaI(Tl) crystal. The R_f values for this system are 0 to 0.1 for ^{99m}Tc-sulfur colloid and 0.9 to 1.0 for free $^{99m}TcO_4^-$ and ^{99m}Tc-EDTA. While this method deviates slightly from the USP method, in that ITLC is used instead of "chromatographic paper," the use of ITLC silica gel media has been widely used as an alternate method. The gel filtration work in the study was performed using a 0.7 cm x 25 cm column of Sephadex G-25 (medium) (Pharmacia Fine Chemicals Inc., Piscataway, N.J.). The column was eluted with saline containing 0.9% benzyl alcohol. The eluate was collected in 1-ml fractions and counted in a well-type NaI(Tl) crystal. The elution profile of the samples was compared to the profile of samples containing known concentrations of various ^{99m}Tc radiochemical species. In this system ^{99m}Tc-labeled sulfur colloid and ^{99m}Tc-EDTA are eluted from the fourth to seventh ml and free $^{99m}TcO_4^-$ is eluted from the eleventh to seventeenth ml.

Temperature studies were carried out in a water bath that held temperature to within ± 0.5°C of that desired.

RESULTS

As mentioned above, the reaction between thiosulfate and acid is slow at room temperature. The data in Fig. 1 show that the reaction proceeds quite rapidly at 97°C. The curve represents composite data from all four manufacturers. It was found that the labeling efficiency reached the 92% level between 4-5 min. The curve then reaches a plateau at about 5 min and remains essentially unchanged at times greater than 5 min. The time required for the reaction vials to be in the boiling water bath specified by the kit manufacturers ranged from "5 min" to "10 ± 2 min." Some samples were purposely heated for up to 30 min to see if the vial would crack, or the top would blow off, or if this extended heating time would have a detrimental physical or chemical effect on the sulfur colloid. The vials remained intact and no physical or chemical problems were observed in the final product.

Fig. 2 shows the effect of temperature on labeling efficiency. Again, these data represent a composite of the four manufacturers, and the curve shows that a temperature of 95°C or greater is necessary to ensure a 92% yield. At 90°C, the labeling efficiency becomes quite erratic, with yields ranging from 70% to 94%. While instructions for three manufacturers call for "vigorously" boiling or a "rolling" boil (which can be loosely interpreted to mean temperatures of 96°C and above), one manufacturer, (C), simply specifies a "boiling water bath (95°C ± 5°C)." One can, therefore, have a water bath within the specifications of the package insert at 90°C and be in the temperature range where it is felt that labeling is erratic. Mention should be made here that all of these samples had a "cloudy" appearance. So that, while the colloid is forming, visual observance of the colloid is no indication of the labeling progress.

While it was not originally planned to study the effects of ^{99m}Tc quantity on labeling efficiency, it was found that in these kits ^{99m}Tc quantity can affect the labeling yields. In observing ^{99m}Tc quantity vs. labeling efficiency, represented in Fig. 3, it was found that instead of a single curve

TABLE 1. CONTENTS OF COMMERCIAL ^{99m}Tc-SULFUR COLLOID KITS

Manufacturer:	A	B	C	D
Reagents:	$Na_2S_2O_3$ (6 mg)	$Na_2S_2O_3$ (1.7 mg)	$Na_2S_2O_3$ (12 mg)	$Na_2S_2O_3$ (2.0 mg)
	H_3PO_4 (100 mg)	HCl (15 mg)	HCl (15 mg)	HCl (7 mg)
	Gelatin		Gelatin	Gelatin
	NaCl		Na_2EDTA (3.6 mg)	Na_2EDTA (2.3 mg)
			K_3PO_4	
Buffer:	Acetate	Phosphate-Hydroxide	Phosphate-Hydroxide	Phosphate-Hydroxide
	Na_2EDTA (4 mg)	Providone		

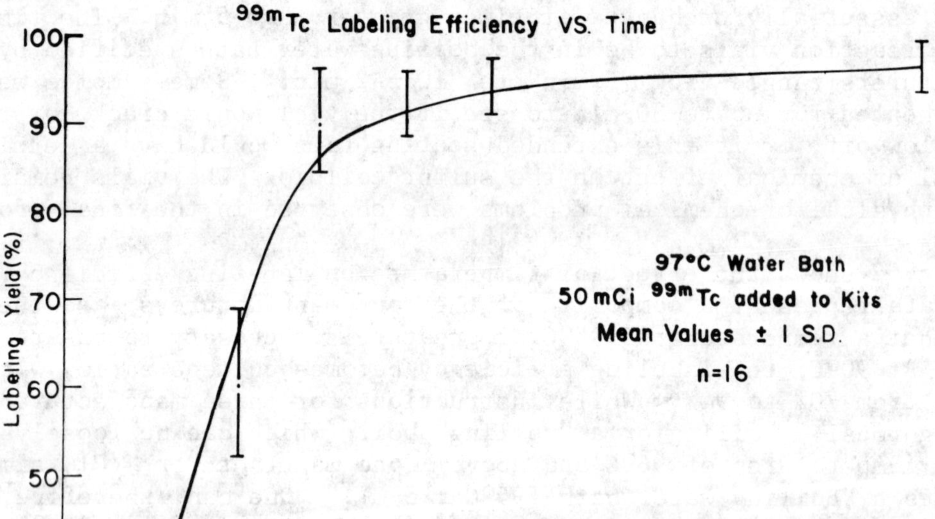

Fig. 1. Effect of time in 97°C water bath on labeling efficiency of Tc-99m-sulfur colloid. Curve represents composite data for all four manufacturers.

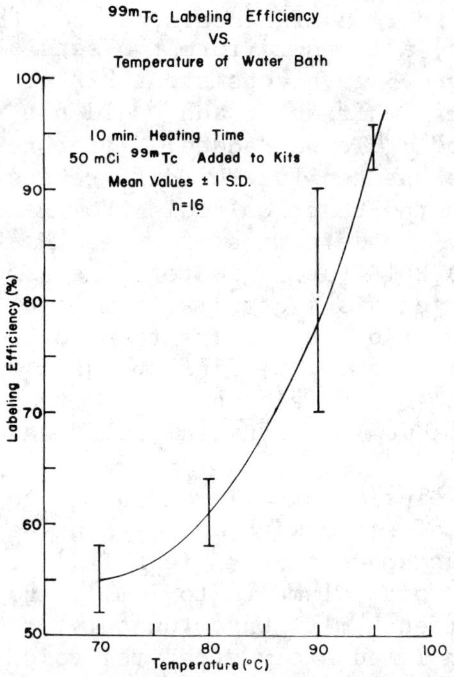

Fig. 2. Effect of temperature of water bath on the labeling efficiency of Tc-99m-sulfur colloid. Curve represents composite data for all four manufacturers.

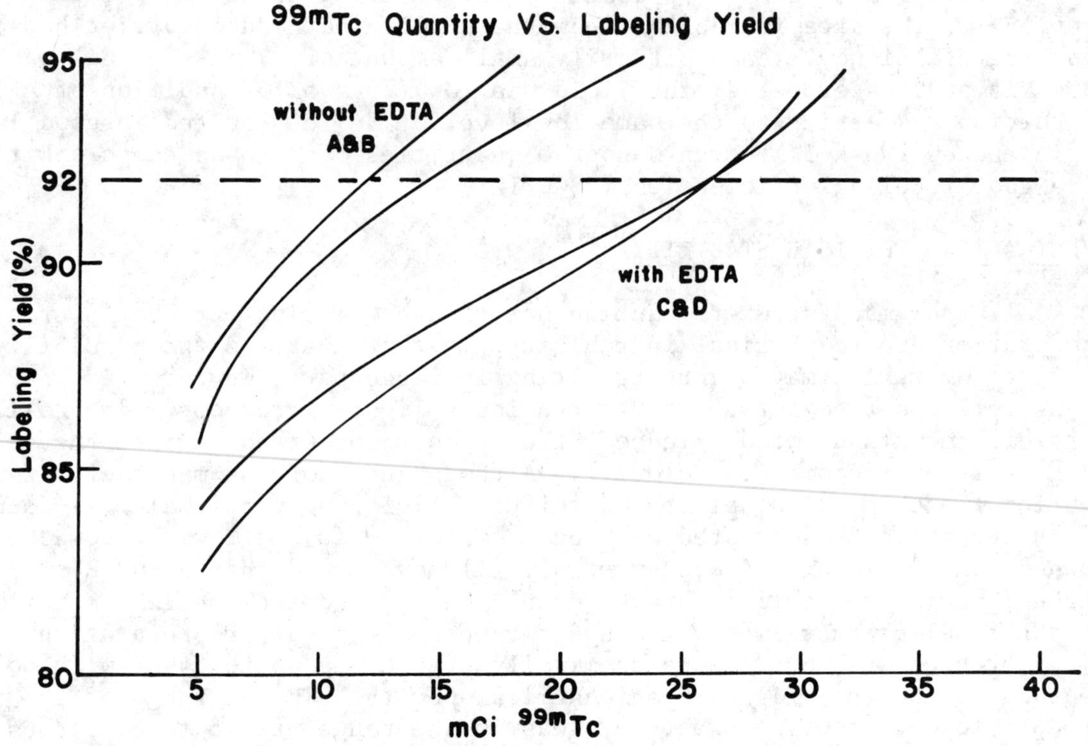

Fig. 3. Tc-99m quantity added to kits with and without Na_2EDTA vs. labeling efficiency. Reactions were subjected to 97°C water bath for 10 min.

representative of all four manufacturers, as was the case in the time and temperature parameters, a trend to form two sets of curves was observed. Kits from Manufacturers A and B gave consistent 92% yields when the ^{99m}Tc quantity was 10-12 mCi or higher; whereas, C and D did not reach the 92% level until approximately 25 mCi of ^{99m}Tc was added. After re-examining the contents of the kits, the reason suspected for the differences in the data was that Na_2EDTA was present in the reaction vials from Manufacturers C and D and not in vials from A and B. To determine if indeed Na_2EDTA was interfering in the reaction, 2.5 mg of Na_2EDTA (approximately the same amount found in the kits) was added to the reaction vials of kits from Manufacturers A and B. The labeling efficiency curve of these kits then followed the pattern generated by kits with Na_2EDTA in them. The final product from both types of kits, with and without Na_2EDTA, were also subjected to gel filtration chromatography using Sephadex G-25 (medium) with normal saline as the eluant.

Fig. 4 shows the elution profile of sulfur colloid on G-25 (medium) Sephadex. Both types of kits, with and without Na_2EDTA, show the same profile. Previous experience has shown that ^{99m}Tc-EDTA is eluted from the column immediately after the void volume (4 to 7 ml), and free $^{99m}TcO_4^-$ is eluted at the eleventh to seventeenth ml. Unfortunately, colloidal material is also eluted from the column immediately after the void volume, as Pettit et al. (3) in their work with colloids on Sephadex have reported. However, as shown in Fig. 5, ITLC of the #5 fraction showed two peaks for kits with EDTA (^{99m}Tc-S and ^{99m}Tc-EDTA) and only one peak (^{99m}Tc-S) for kits without EDTA. The overall amount of technetium that was bound to EDTA in the entire product was calculated to be between 2 to 3%.

In relation to the ^{99m}Tc quantity vs. labeling efficiency mentioned previously, the data in Table 2 show the effect that added carrier has on the labeling efficiency. An arbitrarily small amount of ^{99m}Tc (30 μCi) was added to a kit and, as expected, the yield was low (75 ± 5%). Addition of a few μg of rhenium as carrier to the same low level of ^{99m}Tc resulted in good labeling efficiencies (96 ± 3%). Even nanogram quantities of ^{99}Tc added to a kit will give good labeling efficiencies (98 ± 1%).

CHEMICAL AND PHYSICAL STABILITY

All four manufacturers caution not to use the final product 6 hr after preparation due to physical instability, in that there is the possibility that agglomeration may occur, resulting in large particle formation. Preparations were observed for up to 24 hr using a light microscope, with no large particle formation noted. Indeed, the opposite occurred in that they tended to "clear-up" (become less cloudy). Although none of the manufacturers mention the chemical stability of sulfur colloid, Harper et al. (2) found that if sulfur colloid is treated with base, free $^{99m}TcO_4^-$ is formed from the breakdown of the colloid, and Kelley et al. (1) have found that at a pH of 7 or higher, the sulfur colloid breaks down. The pH and the chemical stability of the final products were examined for up to 24 hr after preparation. The final product was found to be chemically stable for up to 24 hr with no degradation of the original radiochemical purity. The pH range of the final product for all four manufacturers was in the range of 5.5 to 6.5, thus, avoiding the problem caused by neutral or basic pH.

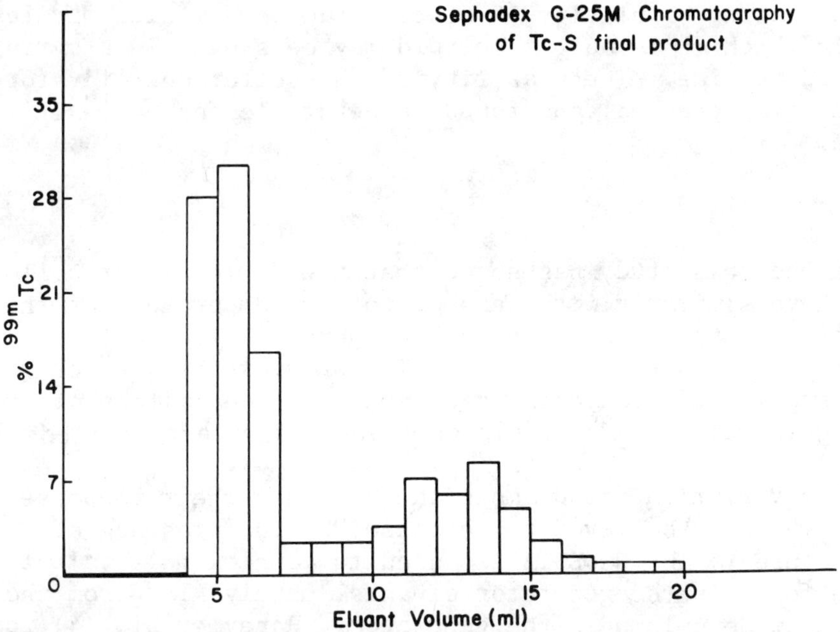

Fig. 4. Elution profile of ^{99m}Tc-sulfur colloid on Sephadex G-25m column (0.7 x 25 cm) eluted with normal saline.

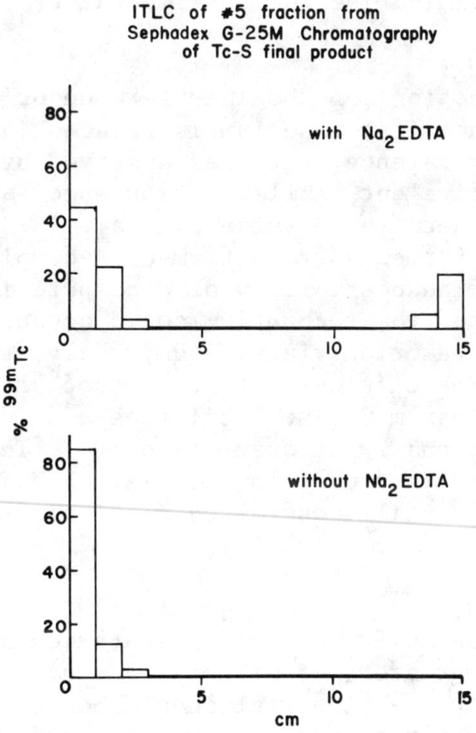

Fig. 5. Instant Thin Layer Chromatography of the 5th ml eluted from a Sephadex G-25m column of ^{99m}Tc-sulfur colloid preparations with and without Na_2EDTA. Silica gel ITLC media was used with methanol: water (85:15) as solvent.

The increasing popularity of central radiopharmacies may lead to a situation in which ^{99m}Tc-sulfur colloid may be stored in a syringe for a long period of time. The stability of the sulfur colloid stored in 1 cc syringes was also checked, and found to be stable for at least 24 hr after preparation.

DISCUSSION

In essence this study has shown that the ^{99m}Tc-sulfur colloid kits examined behave similarly with respect to the temperature of the water bath, and time in the water bath, even though there is a variation in the amount of reactants found in the kits. This is not unexpected since these kits contain mg quantities of reactants, and produce mg quantities of product to react with quantities of ^{99m}Tc six or seven orders of magnitude lower.

The only variation found among the kits was their response to ^{99m}Tc quantities, which is believed to be caused by the presence of Na_2EDTA in the reaction mixtures. The Na_2EDTA is added to complex multivalent cation species that may be found in the generator eluants, mainly Al^{+3} from the breakdown of the aluminum oxide column in the generator. Haney et al. (4) found that concentrations of Al^{+3} as low as 5 µg/ml (which is below the USP maximum limit of 10 µg/ml) caused a white, flocculent precipitate to form in the sulfur colloid preparation within 5 min after preparation. These large particles of aluminum hydroxide or aluminum phosphate physically trap the sulfur colloid and result in high lung uptake of the preparation when injected. Addition of Na_2EDTA promotes the formation of a soluble aluminum-EDTA complex, thus, eliminating the detrimental effect of insoluble aluminum salts.

A possible explanation for the interference of Na_2EDTA on the labeling efficiency could be that if technetium is reduced in the thiosulfate-acid reaction to some lower valence state, as observed by Kelley et al. (1), and Cifka et al. (5), multivalent, labile, cation species of technetium could form an EDTA complex. Furthermore, Galateanu et al. (6) observed that reduced species of technetium formed a technetium-EDTA complex in only a few seconds at room temperature. The possibility of a competition between the sulfur colloid and the EDTA for the technetium could account for the behavior of kits that have EDTA in the reaction vials. Admittedly, these data are a fine point of the study, and in the clinical setting do not present a problem for two reasons: 1) most nuclear medicine facilities add ^{99m}Tc quantities in excess of 25 mCi to the kits; and 2) if quantities of ^{99m}Tc around the 10-15 mCi level are added to kits that contain Na_2EDTA, the yield should be in the area of 88-90% which is still considered clinically useful.

CONCLUSIONS

In general, a good sulfur colloid preparation should be subjected to a water bath at 95°C or greater for at least 4 min, and ideally, 5 min. Ideally, the contents of the reaction vial should be kept as simple as possible, with a minimum number of reagents (e.g., acid, sodium thiosulfate and, of course, the ^{99m}Tc). The Na_2EDTA should be put in the buffer component of the kit, along with the stabilizing agents such as gelatin or povidone. In this manner, they will be just as effective, and any interference with the main reaction will be eliminated. With this in mind, the instructions for preparation of sulfur colloid kits as they appear in the package insert have been found to be adequate, with the one exception noted in the text, (the temperature specification of manufacturer C, where stating 95°C or greater would be better than "95 ± 5°C") and should, as noted by the manufacturers, be closely followed.

REFERENCES

1. KELLEY WN, ICE RD: Pharmaceutical quality of technetium-99m sulfur colloid. Amer J Hosp Pharm 30: 817-820, 1973

2. HARPER P, LATHROP K, GOTTSCHALK A: Pharmaco-dynamics of some technetium-99m preparations. In Radioactive Pharmaceuticals, Andrews GA, Kniseley RM, Wagner HN, eds. CONF-651111, Springfield, VA, 1965, pp 335-358

3. PETTIT WA, DELAND FH, PEPPER GH, et al: Characterization of tin-technetium colloid in technetium-labeled albumin preparations. J Nucl Med 19: 387-392, 1978

4. HANEY TA, ASCANIO I, GIGLIOTTI JA, et al: Physical and biological properties of a ^{99m}Tc-sulfur colloid preparation containing disodium edetate. J Nucl Med 12: 64-68, 1971

5. CIFKA J, VESELY P: Non-pertechnetate radiochemical impurity in sulfur colloid labeled with ^{99m}Tc. In Radiopharmaceuticals and Labeled Compounds, Vol. I, IAEA, Vienna, 1973, pp 53-63

6. GALATEANU I, BRATU C, PETRIDE A: The reduction of ^{99m}Tc pertechnetate by hydrazine. Radiochem Radioanal Letters 28: 95-104, 1977

TABLE 2. CARRIER EFFECTS ON LABELING EFFICIENCY

	Yield (%) $\pm$ 1 S.D.
30 μCi ^{99m}Tc added to kit	75 $\pm$ 5
30 μCi ^{99m}Tc added to kit plus 25 μg $KReO_4$	96 $\pm$ 3
30 μCi ^{99m}Tc added to kit plus 25 ng ^{99}Tc	98 $\pm$ 1

COMPARISON OF THE RAT AND MOUSE MODEL FOR MONITORING THE RADIOCHEMICAL PURITY OF Tc-99m HUMAN SERUM ALBUMIN

J.R. McLean, L.J. Rockwell, and W.J. Welsh

Radiation Protection Bureau
Department of National Health and Welfare
Ottawa, Ontario K1A 1C1
Canada

ABSTRACT

Kits from 4 manufacturers were reconstituted according to package insert instructions. The radiochemical purity of each preparation was assessed by in vitro techniques prior to biological distribution studies in rats and mice. Tissues of interest were blood and liver. Analysis of variance (at $p = 0.05$) indicated that there was a significant difference between the tissue distribution results for rats and mice. Also, there was a significant difference between preparations based on the blood results for either rats or mice but not the liver results. Only the blood results reflected the in vitro test values.

INTRODUCTION

The Radiation Protection Bureau of the Canadian Department of National Health and Welfare conducts an ongoing program to monitor the purity and indirectly the efficacy of licenced radiopharmaceutical products (final dose form radiopharmaceuticals, generators, "cold" kits for the preparation of radiopharmaceuticals). Final dose form radiopharmaceuticals and Tc-99m generators are obtained from manufacturers on an "on request" basis while kits are sent to us according to an established sampling program. At the moment we are routinely sampling normal albumin, macroaggregated albumin and bone imaging kits. Manufacturers' lot release testing results accompany all samples which are submitted.

Some manufacturers use rats for their biodistribution studies while we prefer mice (a Swiss Webster strain) as they are more economical to house and feed, require less space and are easy to handle. The only potential drawback with mice might be the adequacy of injections. However this is not a problem for experienced technicians. Over a period of time we noted what appeared to be a consistent difference between our mouse tissue distribution results for Tc-99m labeled human serum albumin (^{99m}Tc-HSA) and manufacturers' results obtained in rats. Hence we decided to compare the rat and mouse as models for monitoring the radiochemical purity of ^{99m}Tc-HSA. At the same time we investigated the validity of using blood clearance and liver uptake as criteria for assessing the adequacy (and indirectly the efficacy) of this kit prepared radiopharmaceutical. A product which is rapidly cleared from blood is of limited clinical usefulness while appreciable liver uptake is indicative of "reduced hydrolysed" Tc-99m ($^{99m}TcO_2$) and/or denatured albumin.

We are now requiring that kit prepared ^{99m}Tc-HSA should contain less than 10% $^{99m}TcO_4^-$ and not more than 15% $^{99m}TcO_2$, but with a total impurity content not exceeding 20% (i.e. labeling efficiency greater than 80%). At 30 minutes post-injection the blood content in mice, assuming a blood volume of 6.5%, should be greater than 35% and the liver uptake less than 15%. Stomach activity should be less than 1%.

MATERIALS AND METHODS

Human serum albumin kits from four manufacturers were reconstituted according to package insert instructions, generally 5 ml of $Na^{99m}TcO_4$ containing 10-15 mCi of radioactivity from a New England Nuclear generator. The prepared radiopharmaceuticals were allowed to stand for 30 minutes to ensure the labeling reaction was complete.

In vitro tests were conducted to monitor the radiochemical purity of the preparations. These tests, which were all performed in quadruplicate, included paper chromatography with 85% methanol, paper chromatography with albumin loaded paper and nitrogen purged saline (1) and trichloroacetic acid (TCA) protein precipitation (2,3). The paper chromatography-methanol system does not require elaboration; the $^{99m}TcO_4^-$ could be easily quantitated as it migrated with an $R_f > 0.5$ while everything else remained at the origin. With the albumin loaded paper chromatography-saline system $^{99m}TcO_4^-$ and ^{99m}Tc-HSA migrated with an $R_f > 0.7$, and the radiochromatogram peaks were not readily separable. However the $^{99m}TcO_2$ did not migrate and the origin activity was indicative of this impurity. Figures 1 and 2 show typical radiochromatograms for this solvent system. The activity between the origin and the combined $^{99m}TcO_4^-$ plus ^{99m}Tc-HSA peaks varied from manufacturer to manufacturer as well as between lots and could represent up to 30% of the spotted activity. It appeared to behave like albumin in vivo and could be a dimeric or polymeric form of albumin. Since only the origin activity was relevant in this particular case, this additional component did not have to be considered. The TCA protein precipitation assay was rapid and easy to perform. The precipitate was separated by centrifugation with the $^{99m}TcO_4^-$ and $^{99m}TcO_2$ remaining in the supernatant. Any colloid entrapped in the precipitate was readily removed with one or two TCA washes. Hence the supernatant activity represented the total radiochemical impurity content and provided confirmation of the chromatographic assays.

Tissue distribution studies were performed in Sprague-Dawley rats and in our own Health Protection Branch strain of Swiss Webster mice. Six mice and six rats were used for each preparation, 0.1 ml aliquots being injected into the tail vein of mice and femoral vein of rats.

At 30 minutes post-injection the mice were sacrificed by decapitation. Triplicate 0.025 ml blood samples were removed from the heart and assayed for radioactivity. Livers were quickly excised, washed well with phosphate buffered saline then cut into small (1-2 mm) pieces. The pieces were placed in centrifuge tubes and the red blood cells lysed with 0.83% ammonium chloride solution. After standing at room temperature for 2-3 minutes the suspension was centrifuged at 500G for 5 minutes, the supernatant discarded and the liver tissues assayed for radioactivity. Finally the percent of injected dose, corrected for tail activity, was calculated for both liver and blood. Blood volume of the mouse was assumed to be 6.5% of body weight. This value was derived from blood volume determinations in our laboratory.

The rats were anesthetized with halothane. Triplicate 1 ml blood samples were collected from the vena cava and assayed for radioactivity. Livers were perfused through the portal vein for 1.5 minutes with phosphate buffered saline while the animals were expiring. The livers were removed, assayed for radioactivity and the percent of injected dose calculated for both liver and blood. Blood volume of the rat was assumed to be 7% of body weight (4).

The data were subjected to a one way analysis of variance at the $p = 0.05$ level of significance (5). A Duncan's multiple range test was applied when significant differences were detected by analysis of variance (6).

RESULTS AND DISCUSSION

The in vitro test results are summarized in Table 1. With the exception of Company A, the $^{99m}TcO_4^-$ and $^{99m}TcO_2$ levels are quite acceptable. The sums of the individual chromatographic assays are in reasonable agreement with the TCA protein precipitation results.

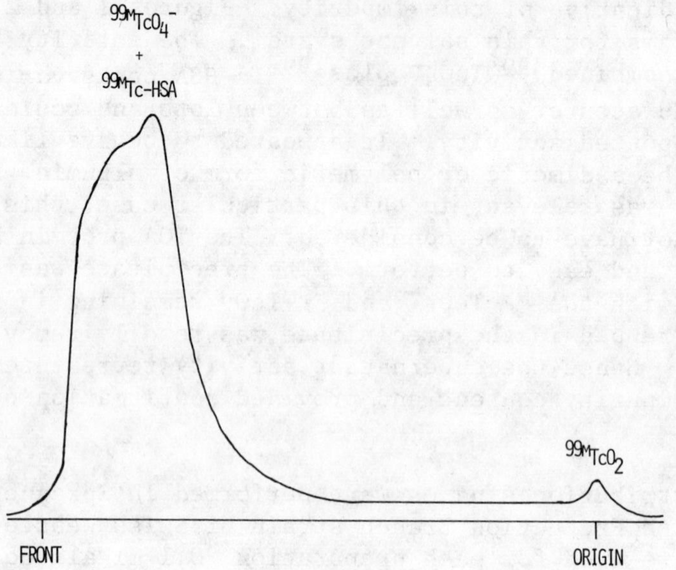

Figure 1.

Paper chromatography of ^{99m}Tc-HSA using albumin loaded paper (Whatman No. 1) and nitrogen purged saline; $^{99m}TcO_2$ content is approximately 3% and negligible activity between the peaks.

Figure 2.

Paper chromatography of ^{99m}Tc-HSA using albumin loaded paper (Whatman No. 1) and nitrogen purged saline; $^{99m}TcO_2$ content is approximately 5% and about 29% of the activity between the peaks.

The tissue distribution results are presented in Tables 2, 3 and 4. For ease of comparison, blood results for both animal species are itemized in Tables 2 and 3 while the liver results are in Table 4. Two way analysis of variance, at a level of significance of $p = 0.05$, indicated that there was a significant difference between the blood and liver results for rats as compared to those for mice. In addition, one way analysis of variance ($p = 0.05$) indicated that there was a significant difference between companies based on the blood results for either mice or rats. When the means were compared by the Duncan's multiple range test (Table 3), Company A's product was found to give blood values which were significantly lower than the rest of the products tested. The blood levels at 30 minutes after the administration of ^{99m}Tc-HSA to rats or mice therefore tend to reflect the in vitro test results, in that Company A's product contained the highest level of radiochemical impurities and produced the lowest level of radioactivity in the blood.

The uptake of radioactivity in the liver however did not reflect the results for the $^{99m}TcO_2$ fraction obtained from the radiochemical purity assays. In addition, the liver values did not relate in any meaningful manner to the activities found in the blood. Therefore, radioactivity in the livers of rats or mice at 30 minutes after the injection of ^{99m}Tc-HSA appears to have little utility as a measure of product quality.

ACKNOWLEDGEMENTS

The authors wish to express their appreciation to Dr. R.S. McCullough and Mr. R.A. Tate for assistance with the statistical manipulations.

REFERENCES

1. Lin MS, Kruse SL, Goodwin DA, et al: Albumin-loading effect: A pitfall in saline paper analysis of ^{99m}Tc-albumin. J Nucl Med 15: 1018-1020, 1974.

2. Lamson M, Callahan RJ, Castronovo FP, et al: A rapid index of free activity in preparations of ^{99m}Tc-albumin. J Nucl Med 15: 1061-1062, 1974.

3. McLean JR, Welsh WJ: Measurement of unbound ^{99m}Tc in ^{99m}Tc-labeled human serum albumin. J Nucl Med 17: 758-759, 1976.

4. Diet Manual. Cleveland, ICN Pharmaceuticals Inc., Life Sciences Group, 1976, p.5.

5. Guenther WC: Analysis of Variance. Englewood Cliffs, Prentice Hall Inc., 1964.

6. Dunnett CW: Statistics in Endocrinology. Cambridge, MIT Press, 1967, p.79-104.

TABLE 1. IN VITRO ASSESSMENT OF RADIOCHEMICAL PURITY OF Tc-99m HUMAN SERUM ALBUMIN (n=4)

Company	Sodium Pertechnetate Tc-99m Content[(a)]	"Reduced Hydrolysed" Tc-99m Content[(b)]	Total Unbound Tc-99m Content[(c)]
A	18.1±2.8%	18.0±4.3%	36.4±3.4%
B	0.5±0.1%	6.7±1.2%	5.9±2.6%
C	7.7±0.4%	3.0±0.3%	9.1±0.1%
D	6.5±0.2%	5.5±0.1%	7.3±0.2%

(a) paper chromatography with 85% methanol

(b) paper chromatography with albumin loaded paper and nitrogen purged saline

(c) protein precipitation with 10% trichloroacetic acid solution

TABLE 2. Tc-99m HUMAN SERUM ALBUMIN IN BLOOD (RATS AND MICE) AT 30 MINUTES POST INJECTION

Company	Percent of Injected Dose in Total Blood* Rat (n=6)	Mouse (n=6)
	(Coefficient of Variation < 10%)	
A	37.4	22.3
B	48.4	26.5
C	45.1	36.4
D	52.2	36.0

*Rat blood volume assumed to be 7.0% of body weight

Mouse blood volume assumed to be 6.5% of body weight

TABLE 3. Tc-99m HUMAN SERUM ALBUMIN IN BLOOD (RATS AND MICE) AT 30 MINUTES POST INJECTION. NON-SIGNIFICANT RANGES ARE UNDERSCORED (DUNCAN'S MULTIPLE RANGE TEST, P=0.05)

Percent (Mean) of Injected Dose in Total Blood*
(Coefficient of Variation < 10%)

Company	A	B	C	D
Rat (n=6)	37.4	48.4	45.1	52.2
Mouse (n=6)	22.3	26.5	36.4	36.0

*Rat blood volume assumed to be 7% of body weight

Mouse blood volume assumed to be 6.5% of body weight

TABLE 4. Tc-99m HUMAN SERUM ALBUMIN IN LIVER (RATS AND MICE) AT 30 MINUTES POST INJECTION

Company	Percent of Injected Dose in Whole Organ (Coefficient of Variation < 10%) Rat (n=6)	Mouse (n=6)
A	6.5	9.3
B	6.4	10.2
C	9.5	10.4
D	8.6	10.5

LABELED PROTEINS: CRITERIA FOR QUALITY

William A. Pettit, Frank H. Deland
and Linda Blanton

Nuclear Medicine Service
Veterans Administration Medical Center
Lexington, Kentucky 40507

ABSTRACT

Several labeled protein preparations have been evaluated in regard to 1) contamination with unbound radionuclide, 2) high molecular weight radioactive contaminants, 3) stability of the radionuclide-protein bond, and 4) preservation of biological activity. Regardless of which radionuclide and labeling procedure is used, evaluation of the labeled protein can be effectively achieved by the assay procedures set forth. Iodine provides the most desirable products, followed by technetium and mercury in that order.

INTRODUCTION

In nuclear medicine biologically significant proteins labeled with radionuclides have potential for a variety of diagnostic uses. Iodine radionuclides have been used for labeling tumor specific antibodies for tumor localization (1,2), fibrinogen for thrombus detection (3), albumin for blood volume (4) and heart studies (5); technetium-99m has found application in the labeling of albumin (4), immunoglobulin G (IgG) (6), and fibrinogen (7); mercury (II) ion labeling of fibrinogen has been reported also (8). Studies dealing with radionuclide labeled proteins frequently lack accurate or adequate assay procedures. Obtaining a high degree of radiochemical purity in labeled protein preparations is often frustrated by several elusive problems of analyses. These problems and their detection are discussed below.

ANALYTICAL PROBLEMS

1. Separation of bound and unbound radionuclide.

An initial problem encountered with the preparation of labeled proteins of radiopharmaceutical quality is the separation of protein bound from the unbound activity. Thin-layer chromotography (9), paper chromotography (10), dialysis (11), trichloroacetic acid precipitation (12,13), electrophoresis (7), and column chromatography (14,15,16,17) have been used to determine the percentages of bound and unbound activity in labeled protein preparations. Most labeled protein preparations require a separation procedure simply for removal of unbound activity.

2. Contamination of labeled protein with high molecular weight artifacts.

In some labeling preparations high molecular weight materials may also be produced. For example, in technetium labeled protein preparations formation of colloid may also result. These colloids may be produced from both the stannous chloride reduction (16) or electrolytic reduction procedures (18). Since these materials are of high molecular weight they may mimic the labeled protein in some analytical procedures. These colloids are most likely composed of technetium and aggregates of tin ions or zirconium ions that have undergone hydrolysis.

Other high molecular weight artifacts such as protein dimers or polymers may also be produced (7). High molecular weight artifacts present a special problem since it is often difficult to separate them from the desired labeled protein.

3. A. Lability of the protein-radionuclide bond.

Stability of the radionuclide linkage to the protein also determines the usefulness of the preparation. In general the diagnostic or therapeutic purpose of the preparation of the labeled proteins will determine the degree of stability required. Iodinated ones demonstrate the greatest stability since covalent bonds are formed between the protein and the iodine, nevertheless, the iodine bond is subject to hydrolysis, albeit relatively slowly(19). Although technetium labeled proteins have a lower degree of stability than iodinated ones, preparations of Tc-albumin have been found to be comparable to iodinated albumin for blood volume determinations if used soon after preparation(20).

B. An additional evaluation of label stability entails the degree of transfer of the radionuclide from the labeled protein to other unlabeled proteins such that the unlabeled protein becomes labeled. This phenomenon is an important consideration if the labeled material is to be used <u>in vivo</u> or in the presence of other proteins. The extent to which such transfers occur appears to depend largely on the nature of the chemical bond between the protein and the radionuclide. Covalently bound iodine does not undergo the transfer. On the other hand mercury ion apparently bonds in an ionic fashion to sulfhydryl groups on the protein and transfer of the label is rapid. Technetium which also binds to sulfhydryl groups of proteins also undergo the transfer process, but to a much lesser extent than mercury.

4. Retention of biological activity.

Retention of native biological properties and activity following the labeling procedure is the final consideration before use. Biological activity can be partially destroyed due to either the conditions of the labeling procedure, the presence to the radionuclide at or near the active site of the protein, or radiolysis. Further the mere presence of the radionuclide in the protein may alter its normal metabolism and distribution.

ANAYLTICAL PROCEDURES

1. Separation of high and low molecular weight components.

Separation of labeled high molecular weight materials (greater than 5000 M.W.) from radioactive salt components may be accomplished on a short (6.0 cm) Sephadex G-25* or Sephacryl S-200* column. The labeled protein is eluted in the void volume of the column (See Figure 1). Ideally, this should be the only procedure necessary prior to use of the material. The following assay procedures are performed on material obtained in the void volume of the column.

2. Detection of high molecular weight artifacts.

Analysis for high molecular weight contaminants (colloidal material, dimeric or polymeric protein) can be effectively achieved on a larger (80-90 cm) column of Sephadex G-200 or Sephacryl S-200. Figure 2 shows a comparison of a uv (280 nm) trace and the relative counts obtained from each fraction obtained from the column. The presence of radioactivity with no uv absorbance indicates colloidal material. These results were obtained from labeling albumin with technetium by the stannous chloride reduction procedure. Dimeric or polymeric protein, of course, would also show uv absorbance, but would be readily distinguishable from the labeled monomer because of its different elution volume.

*Pharmacia Fine Chemicals

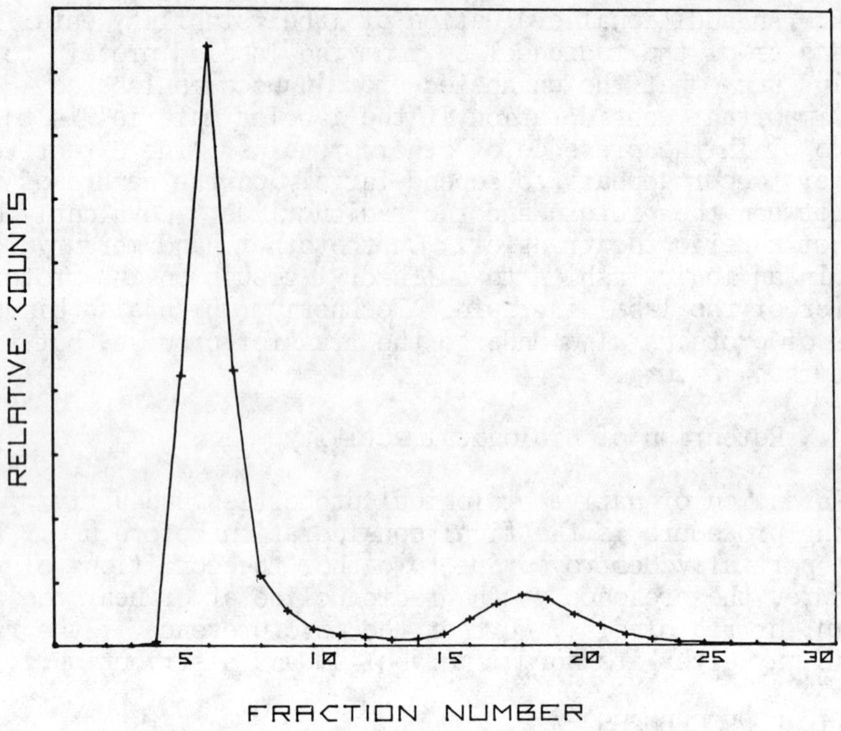

Figure 1. Separation of I-125 labeled goat IgG and unbound I-125 on 6.0 cm Sephadex G-25. Fraction size is 0.5 ml.

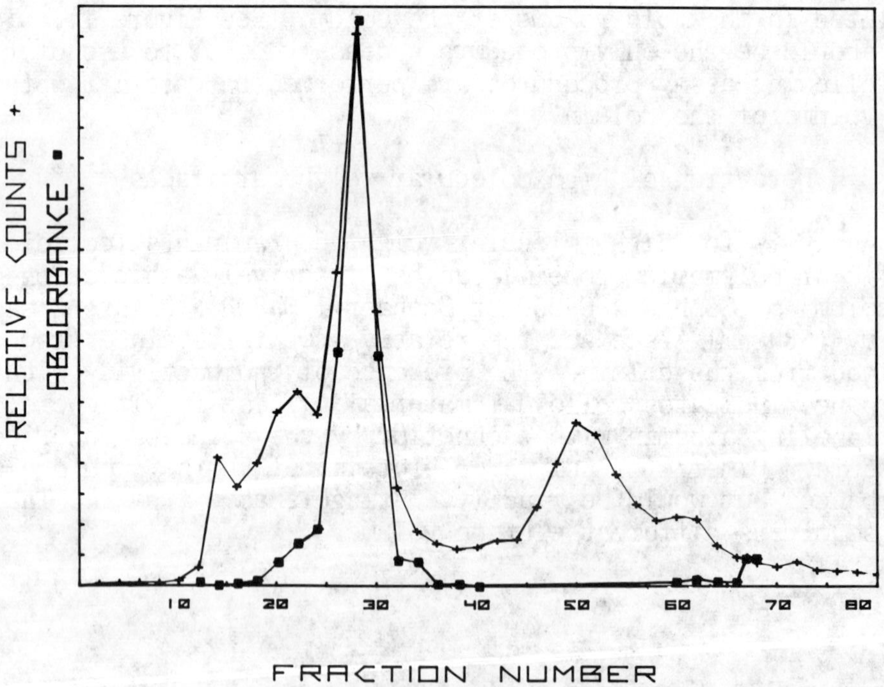

Figure 2. Comparison of activity and protein absorbance from Sephacryl S-200 separation of Tc-99m albumin prepared by stannous shloride reduction illustrating colloid formation.

3. Evaluation of label stability.

Stability of the radionuclide-protein complex can be ascertained by the following methods:

A. Gel filtration of the sample in temporal sequence will demonstrate loss of label from the protein monomer as illustrated in Figure 3. In this example aliquots from a sample of technetium labeled IgG were chromatographed at 0,4, and 22 hours after labeling. Following decay correction of the 4 and 22 hour samples, plots of the results indicate a decrease in activity in the monomer peak, i.e., a loss of label from the protein.

B. Transfer of the label from one protein to another is demonstrated by incubation of the labeled protein (from the short column separations above, part 1) with an unlabeled protein of different molecular weight. Figure 4 shows the results obtained from Sephacryl S-200 separation following incubation of Hg-203 labeled goat IgG with albumin for 1 hour. The IgG peak is centered at fraction 30 and the albumin peak is at fraction 37. Although the IgG was the original labeled protein essentially quantitative transfer of the mercury to the albumin has occurred.

Figure 5 shows the results obtained from gel filtration of goat IgG was labeled with technetium by an electrolytic procedure (5) and subsequent incubated with albumin. Both colloid formation and label transfer are indicated.

4. Retention of biological activity.

A. In vitro

Each protein requires a specific assay procedure to determine if and how much biological activity is lost due to the labeling procedure or due to the presence of the radionuclide on the protein. Labeled and unlabeled materials should be compared simultaneously. Assuming the unlabeled material to be 100% active, any difference in this comparison represents the per cent loss of biological activity. Radioimmunoassay procedures are probably the most sensitive and accurate for this purpose although other procedures may be used. For example, substrate turnover rates could be used in the evaluation of labeled enzymes.

B. In vivo

The most efficacious test for a labeled protein preparation is its *in vivo* behavior.

1) Although a labeled protein may appear to be essentially identical to the unlabeled material by physical and chemical procedures, the presence of the radionuclide in the protein may significantly alter the normal biological disposition or metabolism. Unfortunately standards for normal distribution are not easily obtained.

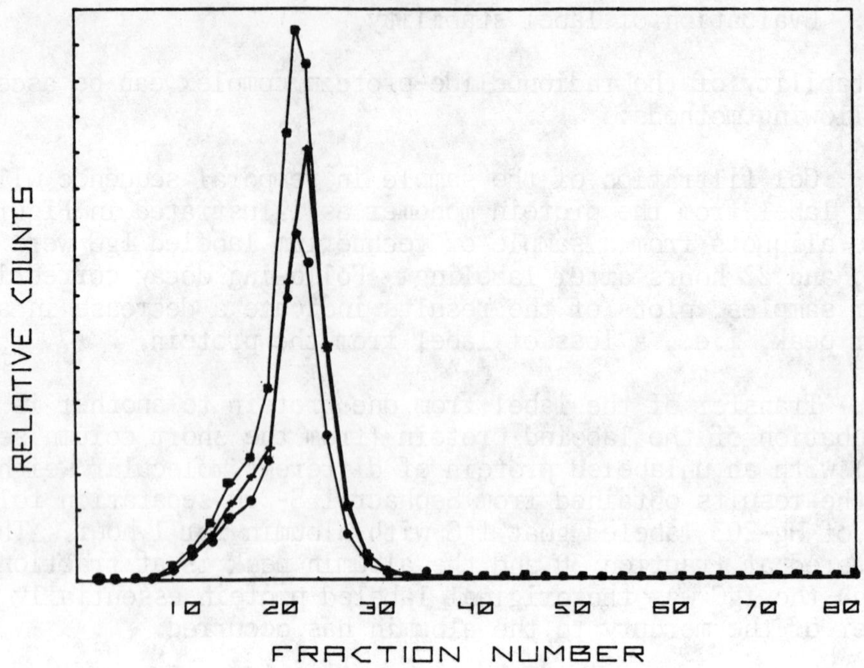

Figure 3. Sequential gel filtration of aliquots from a Tc-99m albumin preparation. Samples were chromatographed at 0 (■), 4 (+), and 22 (●) hours. Curves are decay corrected to illustrate the loss of the label from the protein.

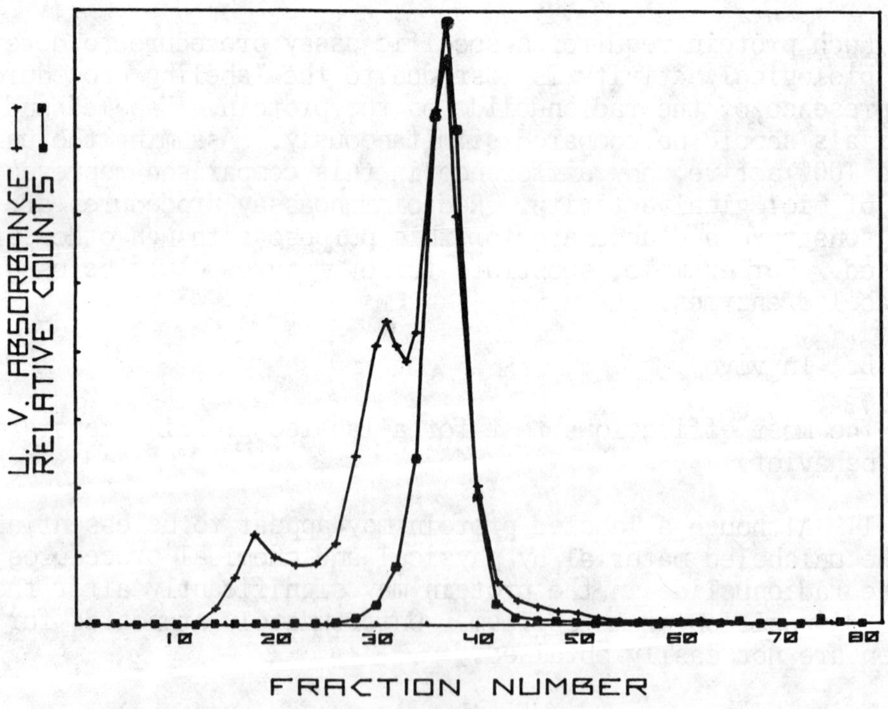

Figure 4. Gel filtration results following labeling of goat IgG with Hg-203. Prior to gel filtration the sample was incubated with excess albumin. The label has been almost quantitatively transferred from the IgG (fraction 30) to albumin (fraction 37).

2) The final criterion that must be evaluated is the performance of the expected *in vivo* function. For example, does a particular preparation of labeled fibrinogen indeed delineate thrombi *in vivo* or does a labeled antitumor antibody localize in a tumor? These procedures are best carried out in appropriate animal models where the results can be obtained from tissue assay as well as scintillation imaging.

Discussion

The above procedures, although time consuming, provide a framework for the evaluation of proteins labeled with radionuclides. The pitfalls encountered in protein labeling can be detected by these methodologies. Radionuclide interaction with the protein can be represented by the following equations where R* represents the radionuclide and the k's are the rate constants for the processes shown.

$$\text{protein} + R^* \underset{k_2}{\overset{k_1}{\rightleftharpoons}} \text{protein-}R^*$$

$$\downarrow k_3$$

hydrolysis, oxidation, etc. (i.e., competing reactions)

Although the scheme presented may be oversimplified, it offers explanations for the observed problems associated with protein labeling.

For iodinations, k_1 is very large compared to k_2, and k_3 would be significant compared to k_1 only if an agent was present which inhibited labeling. For example, a reducing agent would convert I_2 to I^-. In the case of mercury (II) labeled proteins, k_3 is not detectable when compared to k_1, (if no other proteins or sulfhydryl compounds are present), but k_2 is significant. If another protein is present that can react with the unbound mercury k_3 becomes evident due to the presence of another protein. Typically, k_2 is best observed by this transfer process which mercury labeled IgG undergoes in the presence of albumin. With technetium labeled proteins, unless large amounts of protein are used or unless the tin reductant is protected from oxidation and hydrolysis (e.g., by a chelating agents such as tartrate or DTPA) k_3 may be observable and comparable to k_1. The reverse process, k_2, is small compared to k_2 when mercury is considered, but is nonetheless detectable as either transfer or direct loss of label from the protein.

By far the most acceptable tracer proteins are those labeled with iodine. Generally once unreacted iodide is removed from the preparation, retention of biological activity is the remaining concern. A high substitution level appears to have a deleterious effect on biological activity (21). Technetium labeling of proteins potentially offers several advantages over iodination. Among these are superior imaging characteristics and lower patient radiation burden. In some cases the half-life of technetium may be too short for some procedures.

Difficulties associated with technetium labeling include colloid formation, label transfer and lower stability of the preparation in comparison to iodine labeled materials. Mercury labeled proteins are the most unsatisfactory since the radionuclide is rapidly lost from the protein when other receptors for the mercury are present.

Acknowledgement

This work was supported by the Medical Research Service of the Veterans Administration.

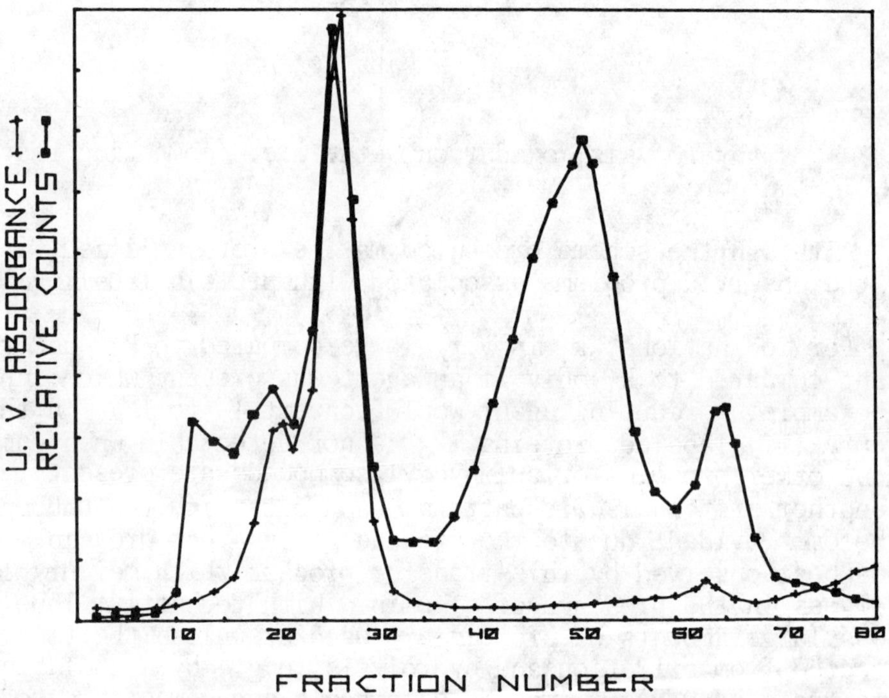

Figure 5. Gel filtration results of Tc-99m IgG labeled by electrolytic reduction. Following the labeling procedure the product was incubated with albumin. Both colloid formation and label transfer have occurred.

REFERENCES

1. Goldenberg DM, DeLand FH, Kim E, et al: Use of radiolabeled antibodies to carcinoembryonic antigen for the detection and localization of diverse cancers by external photoscanning. New Engl J Med 298:1384, 1978

2. Belitsky P, Ghose J, Agireno J, et al: Radionuclide imaging of metastases from renal-cell carcinoma by ^{131}I-labeled antitumor antibody. Radiology 126:515, 1978

3. Knight LC, Primeau JL, Siegel BA, et al: Comparison of In-111-labeled platelets and iodinated fibrinogen for detection of deep vein thrombosis. J Nucl Med 19:891, 1978

4. Yang SSL, Nicholoff EL, McIntyre PA, et al: Tc-99m human serum albumin: a suitable agent for plasma volume measurements in man. J Nucl Med 19:804, 1978

5. Wang DW, Huang JT: Labeling of human immune gamma globulin with ^{99m}Tc. Int J Appl Radiat Isot 28:719-722, 1977

6. Pettit WA, DeLand FH: An improved procedure for labeling proteins with Tc-99m. J Nucl Med 19:694, 1978 (abst.)

7. Meinken G, Srivastava SC, Smith TD, et al: Is there is "good" Tc-99m-albumin? J Nucl Med 17:537, 1976 (abst.)

8. Saha GB, Girschek PK, Tarrer PA: Labeling of human fibrinogen with Hg-197. J Nucl Med 17:542-543, 1976 (abst.)

9. Billinghurst MW: Chromatographic quality control of ^{99m}Tc-labeled compounds. J Nucl Med 14:793-797, 1973

10. Lin MS, Kruse SL, Goodwin DA et al: Albumin-loading effect: A pitfall in saline paper analysis of ^{99m}Tc-albumin. J Nucl Med 15:1018-1020, 1974

11. Williams MJ, Deegan T: ^{99m}Tc-albumin: An assessment of the physical methods for measuring labelling efficiency. Int J Appl Radiat Isot 22:775-777, 1971

12. McLean JR, Welch WJ: Measurement of unbound ^{99m}Tc in ^{99m}Tc-labeled human serum albumin. J Nucl Med 17:758-759, 1976

13. Lamson M III, Callahan RJ, Castronovo FP, et al: A rapid index of free activity in preparations of ^{99m}Tc-albumin. J Nucl Med 15:1061-1062, 1974

14. Eckelman WC, Richards P: Analytical pitfalls with ^{99m}Tc-labeled compounds. J Nucl Med 13:202-204, 1972

15. DeLigny CL, Gelsema WJ, Beums MH: The influence of experimental conditions on the efficiency of labeling human serum albumin with ^{99m}Tc, using SN (II) as the reductant. Int J Appl Radiat Isot 20: 241-248, 1976

16. Pettit WA, DeLand FH, Pepper GH, et al: Characterization of tin-technetium colloid in technetium-labeled albumin preparations. J Nucl Med 19:387-392, 1978

17. Persson RBR, Strand SE, Knoos T: Radioanalytical studies of ^{99m}Tc-labeled colloids and macro-molecules with gel chromatography column scanning technique. J Radioanal Chem 43:275-286, 1978

18. Pettit WA, DeLand FH, Blanton L: unpublished data

19. Knight LC, Harwig YSL, Welch MJ: In vitro stability and in vivo clearance of fibrinogen or serum albumin labeled with ^{77}Br, ^{131}I, or ^{125}I by direct or indirect synthetic methods. J Nucl Med 18: 282-288, 1977

20. Yang SSL, Nicholoff EL, McIntyre PA, et al: Tc-99m human serum albumin: A suitable agent for plasma volume measurements in man. J Nucl Med 19:804-807, 1978

21. Nord S, Weissman IL: Radiolabeled antitumor antibodies. III Highly iodinated and highly radioiodinated antibodies. J Natl Cancer Inst 53:959-965, 1974

ACCEPTANCE TESTING OF ^{123}I-SODIUM IODIDE SOLUTION PRODUCED BY THE ^{127}I(p,5n) ^{123}Xe$\rightarrow$^{123}I REACTION

Kenneth Breslow, Jeffrey Barnes,
Shea-Kei Mak, Kenneth T. Study, and Buck A. Rhodes

University of New Mexico
College of Pharmacy, Radiopharmacy
Albuquerque, New Mexico 87131

ABSTRACT

Methods for rapidly assessing the pharmaceutical quality of ^{123}I-NaI solution produced by the (p,5n) reaction are discussed. A test to determine, within a short time, the radionuclidic acceptability of ^{123}I-NaI solution is described. This test, which is based on the initial ratios of the 159 KeV gamma rays to the X-rays determined at the time of calibration, precludes dispensing radionuclidically impure ^{123}I-NaI solution. Techniques for quickly determining the radiochemical purity of radioiodine solutions are also discussed.

INTRODUCTION

To date, pharmaceutical quality ^{123}I-NaI solution approved by the United States FDA for routine use is available from only one commercial supplier.* This product, produced by the direct ^{122}Te(d,n)^{123}I reaction (hereafter referred to as the (d,n) reaction) contains as much as 1.1% ^{124}I and ^{126}I 3.2% ^{130}I, and 0.8% ^{131}I at time of calibration (1). A significant relative increase in radionuclidic contaminants occurs 24 hours after calibrattion. As much as 4% ^{124}I, ^{130}I, ^{131}I and 10% ^{126}I may be present at this time as contaminants, which can result in a higher radiation dose to the thyroid and a decrease in the quality of the scintiscan.

The relatively pure ^{123}I-NaI solution prepared by the indirect method, the ^{127}I(p,5n)^{123}Xe$\rightarrow$^{123}I reaction (hereafter referred to as the (p,5n) reaction) is commercially available as a radiochemical.+ Producing ^{123}I directly by the (d,n) reaction has the disadvantage of creating significant quantities of radioiodine contaminants, which cannot be chemically separated from ^{123}I. These contaminants probably arise from impure target materials and competing nuclear reactions. By generating ^{123}I through the ^{123}Xe intermediate, monoatomic ^{123}I species are formed which can be easily separated from the two hour ^{123}Xe precursor. In addition, the ^{127}I target material, which is 100% naturally occurring, is inexpensive compared to the enriched tellurium targets required for direct production methods.

The only significant radionuclidic contaminant produced by the (p,5n) reaction is ^{125}I. DeNardo et al. (2) reports less than 0.1% ^{125}I contamination at 6.6 hours after bombardment, and 0.4% ^{125}I thirty hours after bombardment. Baker et al. (3) found the ^{125}I contamination to be 2% at thirty-eight hours after bombardment. Both investigators analyzed samples of ^{123}I obtained from the same commercial source.+

In order to assure the pharmaceutical quality of a preparation of ^{123}I NaI solution produced by the (p,5n) reaction, we need to verify two parameters: 1) the radionuclidic purity, and 2) the radiochemical purity. This report describes a method for the rapid analysis for small amounts of ^{125}I in a solution of ^{123}I-NaI, and presents results from radiochemical analysis on radioiodine solutions produced by the (p,5n) reaction.

RADIONUCLIDIC ANALYSIS

Because of the short half-life of ^{123}I (13.3 hours) and the relatively long half-life of ^{125}I (60.2 days), the percentage of radionuclidic contamination of ^{125}I in the ^{123}I solution increases rapidly. Iodine-125 has a higher absorbed radiation dose per μCi than does ^{123}I; therefore, as the proportion of ^{125}I contamination increases, so does the absorbed radiation dose to the patient. In order to insure the radionuclidic quality of the product being dispensed and to adequately determine its expiration time after calibration, a rapid method was developed for quantifying the degree of ^{125}I contamination in ^{123}I-NaI solution.

Methods: A method was devised to determine the proportion of ^{125}I contamination in ^{123}I-NaI solution using a 1024 channel multichannel analyzer (MCA) and NaI(Tl) detector. The method involves comparing the relative abundance of X-rays in a fixed aliquot of solution at two (or more) different times and calculating the percent ^{125}I contamination from formulae developed for this purpose. This is an absolute method in that no standard is required, and the results are independent of the detector and the counting geometry used.

Five to ten µCi of ^{123}I solution is placed on a small sheet of thin transparent plastic wrap and is precipitated with a drop of $AgNO_3$ solution (as was done to collect the data reported), or evaporated to dryness. The plastic wrap is folded in half and placed between two layers of cardboard with circular cut-outs which allow the plastic area containing the sample to be directly exposed to the detector. This planchet serves as the source for the entire period. At time t_1, the planchet is placed over the NaI(Tl) well detector, thus insuring low deadtime and avoiding loss of counts in the X-ray peak resulting from coincidence counting. Counts are accumulated in the MCA until the X-ray peak reaches 100,000 counts. The integral of counts in the X-ray peak is determined along with the time required to accumulate the counts (in seconds). At a second time, t_2, the above process is repeated, making sure the source has the same geometric relationship to the crystal as it had during the first count.

Analysis: The derivation of the formulae to calculate the percentage of ^{125}I contamination is as follows:

At time t_1, $C_1 = E_{123}MA_{123} + E_{125}NA_{125}$, (1)

where C_1 = cps in X-ray peak (with deadtime correction);

E_{123} = the efficiency of counting ^{123}I X-rays.

E_{125} = the efficiency of counting ^{125}I X-rays.

A_{123} = activity of ^{123}I in sample at calibration time.

A_{125} = activity of ^{125}I in the sample at calibration time.

M and N = decay factors (see Table 1).

At time t_2, $C_2 = E_{123}OA_{123} + E_{125}PA_{125}$, (2)

where C_2 = cps in the X-ray peak (with deadtime correction), and

E_{123}, E_{125}, A_{125}, A_{123} are as above;

O and P = decay factors (see Table 1).

Since the energy range and geometry of counting are the same for the ^{123}I X-rays and the ^{125}I X-rays, and since the mean number of X-rays in ^{125}I is 1.64 times that in ^{123}I (Table 2), then

$$1.64\ E_{123} = E_{125}. \quad (3)$$

TABLE 1 DECAY FACTORS

$$M = e^{-\lambda_{123} t_{1,123}} = e^{-0.05211 t_{1,123}}$$

$$N = e^{-\lambda_{123} t_{1,125}} = e^{-0.00048 t_{1,125}}$$

$$O = e^{-\lambda_{123} t_{2,123}} = e^{-0.05211 t_{2,123}}$$

$$P = e^{-\lambda_{125} t_{2,125}} = e^{-0.00048 t_{2,125}}$$

(λ in hours^{-1})

TABLE 2 RELATIVE ABUNDANCE OF X-RAYS IN ^{123}I AND ^{125}I*

^{123}I		^{125}I	
Mean #/Disint.	Mean Energy (KeV)	Mean #/Disint.	Mean Energy (KeV)
0.472	27.5	0.067	35.5
0.242	27.2	0.738	27.5
0.127	31.2	0.378	27.2
0.026	31.8	0.199	31.0
$\Sigma_x = 0.867$		0.041	31.8
		$\Sigma = 1.423$	

$$\text{Relative abundance of } ^{125}\text{I X-rays} = \frac{^{125}\text{I X-rays}}{^{123}\text{I X-rays}} = \frac{1.423}{0.867} = 1.64$$

*Source: MIRD/Dose Estimate Report No. 5, Summary of Current Radiation Dose Estimates to Humans From ^{123}I, ^{124}I, ^{125}I, ^{126}I, ^{130}I, ^{131}I, and ^{132}I as Sodium Iodide. J. Nucl. Med., Vol. 16, No. 9, p. 857 -860

Substituting equation 3 into equations 1 and 2, they become:

$$C_1 = E_{123}MA_{123} + 1.64\ E_{123}NA_{125} \tag{1a}$$

and

$$C_2 = E_{123}OA_{123} + 1.64\ E_{123}PA_{125}\ . \tag{2a}$$

By rearranging and manipulating equations 1a and 2a they become:

$$\frac{A_{125}}{A_{123}} = \frac{MC_2 - OC_1}{1.64PC_1 - 1.64NC_2} = x. \tag{4}$$

From equation 4,

$$\frac{A_{125}}{A_{123}} = x, \quad \text{and} \tag{4a}$$

$$A_{123} + A_{125} = 100 \text{ (total activity) or}$$

$$A_{123} = 100 - A_{125}\ . \tag{4b}$$

Substituting equation 4b into 4a:

$$A_{125} = x(100 - A_{125}), \text{ and}$$

$$(1 + x)A_{125} = 100x,$$

$$A_{125} = \frac{100x}{1 + x} \tag{5}$$

<u>Determination of t_1 and t_2</u>: Since the time (t) required to accumulate the data used to calculate A_{125} and A_{123} (in equation 4) can range from a few minutes to several hours, a redefinition of t is needed. What is needed is a time (t_x in hours) somewhere between the time from calibration at the beginning of the count (t_b in hours) and the time from calibration at the end of the count (t_e in hours), such that the number of disintegrations between t_b and t_x equals the number of disintegrations between t_x and t_e (Figure 1). In other words, what is needed is an instantaneous t_x which will represent the hypothetical situation in which the number of radioactive atoms (N_x) remains constant throughout the counting period t_b and t_e.

Integrating and solving for areas 1 and 2:

$$\text{Area 1} = \frac{-N_o e^{-\lambda t_x}}{\lambda} - N_x t_x + \frac{N_o e^{-\lambda t_b}}{\lambda} + N_x t_b, \text{ and} \tag{6}$$

$$\text{Area 2} = N_x t_e + \frac{N_o e^{-\lambda t_e}}{\lambda} - N_x t_x - \frac{N_o e^{-\lambda t_x}}{\lambda}\ . \tag{7}$$

Solving for t_x such that area 1 equals area 2 (equation 6 = equation 7)

$$\frac{N_x}{N_o} = \frac{e^{-\lambda t_e} - e^{-\lambda t_b}}{\lambda(t_b - t_e)} . \tag{8}$$

Since $N_x = N_o e^{-\lambda t_x}$, $\quad \frac{N_x}{N_o} = e^{-\lambda t_x}$. (9)

From equations 8 and 9,

$$e^{-\lambda t_x} = \frac{e^{-\lambda t_e} - e^{-\lambda t_b}}{\lambda(t_b - t_e)} .$$

Taking the natural logarithm of both sides,

$$t_x = \frac{-1}{\lambda} \ln \left[\frac{e^{-\lambda t_e} - e^{-\lambda t_b}}{\lambda(t_b - t_e)} \right] . \tag{10}$$

t_x is calculated for each of the isotopes ^{125}I and ^{123}I at each of the counting times (t_1 and t_2), giving t values to be used in calculating decay factors M, N, O, and P (Table 1), i.e., $t_{1,123}$, $t_{1,125}$, $t_{2,123}$, and $t_{2,125}$.

Since these calculations are lengthy, a computer program has been written to facilitate the determination of the percentage of ^{125}I contamination at calibration time. An Altair computer model 8800-B was used for this purpose; it also calculates the standard deviation when three or more counting periods have been accumulated.

Results: Table 3 compares the measured percentage of ^{125}I contamination at time of calibration (approximately 40 hours post-bombardment), as measured by the University of New Mexico Radiopharmacy (using the methods described above), and by the National Bureau of Standards. Sample 1370 was analyzed at only two counting intervals, and therefore cannot be expressed with a standard deviation. Standard deviations were calculated when a comparison was made among computed percentages of ^{125}I at various times. For example, if four counting times are used, results obtained for each counting period are compared, and the percentage of ^{125}I is calculated for each comparison (i.e., t_1 vs. t_2, t_1 vs. t_3, t_1 vs. t_4, t_2 vs. t_3, etc.). The final calculated percentage of ^{125}I contamination is the mean of all comparisons, plus or minus the standard deviation.

The mean percentage of ^{125}I contamination determined in fifteen samples of ^{123}I-NaI solution analyzed (as produced by the (p,5n) reaction) was 1.52% $\pm$ 0.56, with the range being from 0.58% to 2.3% ^{125}I at time of calibration. This is in close agreement with both DeNardo, et al. and Baker, et al.

Discussion: Since the above method for calculating the percentage of ^{125}I contamination in ^{123}I-NaI solution requires counting times spanning at least one half-life (13.3 hours), a more rapid criterion for acceptance was necessary in order to determine whether or not to release the material for patient use.

TABLE 3 MEASURED PERCENT OF ^{125}I IMPURITY IN ^{123}I-NaI SOLUTION AT CALIBRATION TIME AS DETERMINED BY TWO INDEPENDENT OBSERVERS

Sample	U.N.M.*	N.B.S.**
1370	1.52	1.37 ± 5%
1380	2.86 ± 0.32	2.77 ± 10%
1382	3.26 ± 0.43	3.39 ± 10%

* University of New Mexico

** National Bureau of Standards

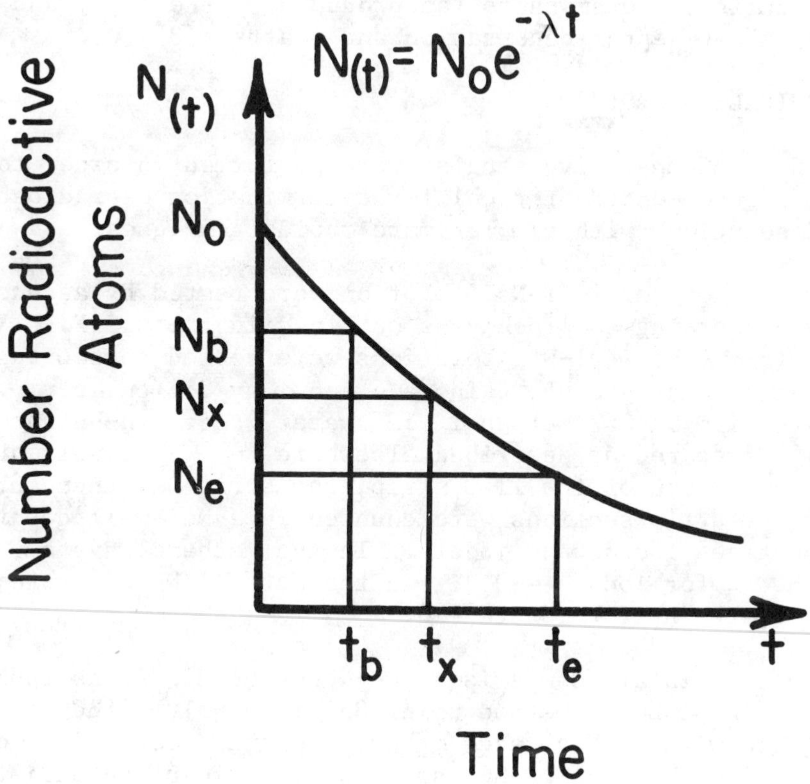

Figure 1.

Exponential decay of ^{123}I containing ^{125}I impurity. Area 1 = $[(N_b t_b), (N_x t_x)]$ Area 2 = $[(N_x t_x), (N_e t_e)]$.

A comparison of the 159 KeV gamma ray/X-ray ratio with the percentage of ^{125}I impurity as calculated above was made for several samples (Figure 2). This is a relative method, and the calibration curve shown in Figure 2 is applicable only to an individual detector and a fixed counting geometry.

Weekly supplies of ^{123}I-NaI solution arrive at the laboratory on or close to calibration time. An initial accumulation is collected on the MCA, and the peak height ratio (159 KeV gamma rays/X-rays) is determined. For example, an initial peak height ratio of 0.8 will, from Figure 2, give an estimate of 1.5% ^{125}I at calibration time. The time after calibration which results in a given percentage of ^{125}I can be determined, using a simple computer program for calculating the relative decay of both isotopes. The expiration time after calibration is defined as being the time at which 10% ^{125}I is present.

Assuming a 5% thyroidal uptake of an administered dose of 100 μCi ^{123}I solution with an impurity of 10% ^{125}I (from Figure 2), the dose to the thyroid will be 1.75 Rads, in contrast to 26 Rads from the same dose of isotopically pure ^{131}I (4).

Dosimetry parameters were not strikingly different for ^{123}I produced by the (p,5n) reaction, when compared with ^{123}I produced by the (d,n) reaction (Figure 3). The (p,5n) ^{123}I can be imaged acceptably with a scintillation camera using a low-energy collimator, whereas the ^{123}I containing higher levels of radionuclidic contaminants (as produced by the (d,n) and $^{124}Te(p,2n)^{123}I$ reaction) shows septal penetration and scatter (3).

RADIOCHEMICAL ANALYSIS

Method: Comparative studies were performed in order to correlate the U.S.P. XIX recommended protocol for determination of radiochemical purity of ^{131}I-NaI solution, with simpler more rapid techniques.

Several lots of ^{123}I-NaI solution were tested by ascending paper chromatography, according to procedures outlined in the U.S.P. XIX for ^{131}I-NaI solution (5). The ^{123}I-NaI solutions were tested concomitantly by ascending thin layer chromatography, using a 6.5 cm. by 0.8 cm. strip of ITLC-SG** chromatography medium in 85% methanol, 15% water, and in n-butanol saturated with 3N NH_4OH(1:1) prepared as described elsewhere (6). The solvents were allowed to migrate to the top of the ITLC strip; the strip was then cut 2 cm. from the origin. The strip sections were counted in a NaI well counter, and analyzed for percentages iodide and iodate. In the methanol system, the R_f for iodide = 1, and the R_f for iodate = 0.2. In the butanol/NH_4OH system, the R_f for iodide = 1; the R_f for iodate = 0.

Results: In all comparisons between the U.S.P. ascending paper chromatography and the ITLC-SG method using 85% methanol as the solvent, the two methods were within 1% of each other. However, comparison of the two solvent systems used with the ITLC-SG, shows a significant inconsistency in the results when n-butanol saturated with 3N NH_4OH was used. For eight samples, the mean percentage iodide determined in the methanol solvent system was 95.6% ± 4.5, while the same samples tested in n-butanol/3N NH_4OH resulted in a mean percentage iodide of 88.1% ± 17.7. A further advantage of the methanol solvent is that its average development time is five minutes, as compared to twelve minutes for the n-butanol/ammonium hydroxide solvent.

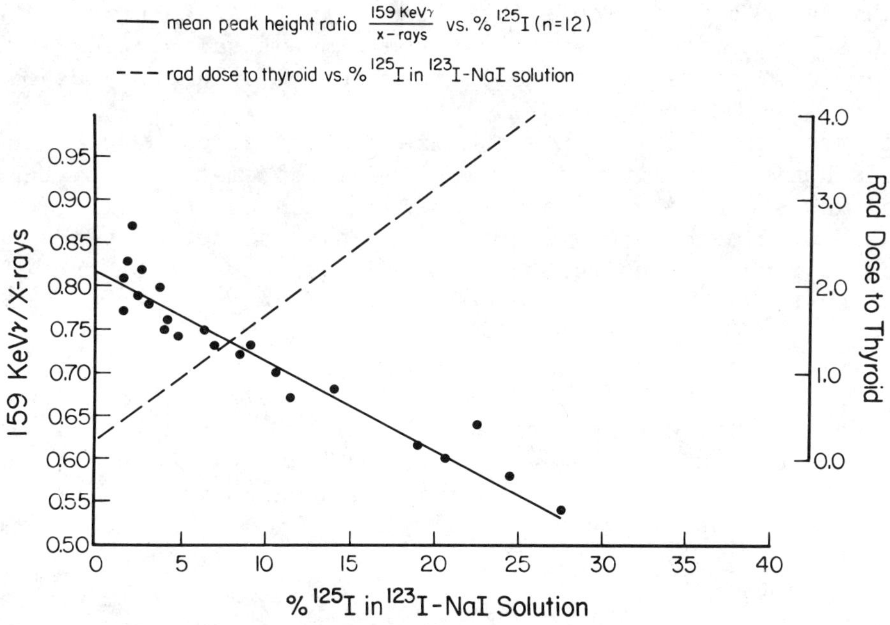

Figure 2.

Relationship of the % ^{125}I in ^{123}I-NaI solution to the initial 159 KeV gamma ray/X-ray peak height ratio and the radiation dose to the thyroid.

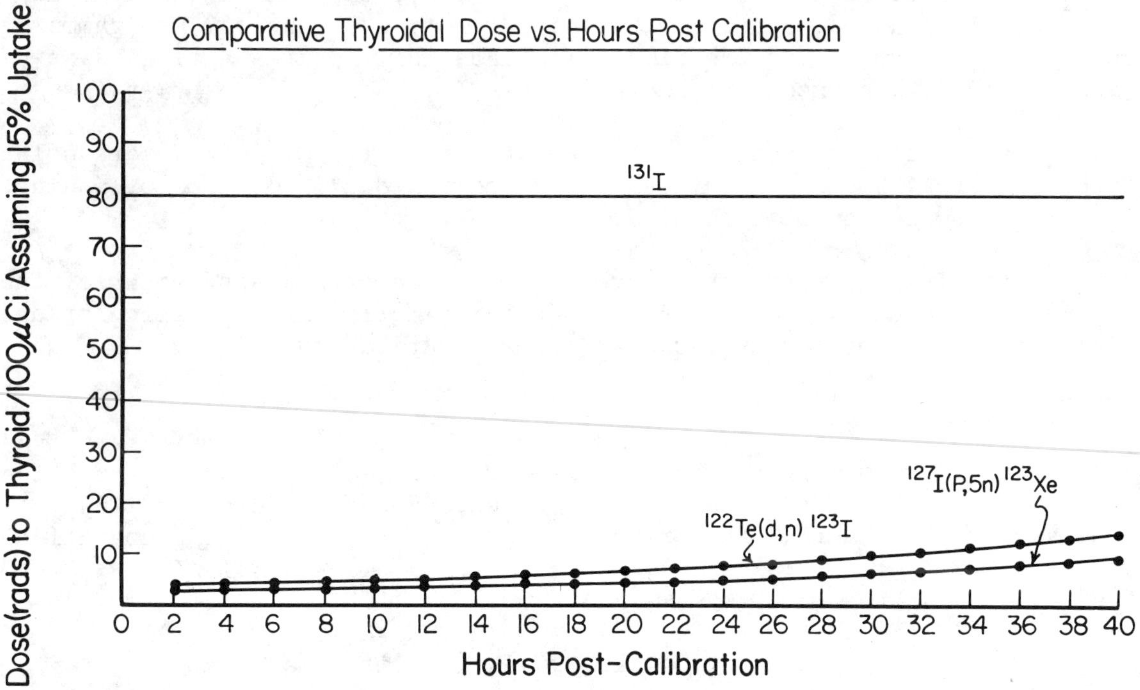

Figure 3.

Comparison of the radiation dose to the thyroid for three radioiodine preparations with increasing time after calibration.

In a large series (Table 4), lots of ^{123}I-NaI solution were tested using the ITLC-SG, 85% methanol system at the time of calibration (n=18), and were tested again 24 hours later (n=5). Negligible oxidation was observed.

Discussion: Results show excellent correlation between the U.S.P. test for radiochemical purity and the ITLC-SG in 85% methanol system. Poor correlation was shown with the n-butanol/3N NH_4OH system. In view of the fact that the U.S.P. methodology requires about four hours development time, as compared to five minutes for the ITLC method, we endorse the ITLC-SG/85% methanol system.

The significance of a small percentage of iodate in the ^{123}I-NaI solution intended for oral use is questionable. DeNardo et al. found no significant difference in thyroidal uptake between orally administered ^{123}I iodide and ^{131}I iodide, even when the ^{123}I was contaminated with as much as 70% ^{123}I iodate. It is presumed that iodate is reduced to iodide (by the gastric juices) prior to thyroid uptake.

SUMMARY

Methods to rapidly assess the pharmaceutical quality of ^{123}I-NaI solution produced by the (p,5n) reaction were discussed.

The determination of radionuclidic purity (i.e., ^{125}I in ^{123}I) is complicated because the low energy gamma and X-rays of ^{125}I are "hidden" in the X-ray spectrum of the ^{123}I. Equations were derived for determining the ^{125}I contamination of ^{123}I as a function of the change in the relative count rates in the X-ray region of the spectrum as compared to the count rates in the photopeak region. Accurate analysis requires measurements over a period of a half-life. A test to determine, within a short time, the radionuclidic acceptability of ^{123}I-NaI solution, based on the initial ratios of the 159 KeV gamma rays to the X-rays determined at the time of calibration, insures against dispensing a batch of ^{123}I with excessive amounts of ^{125}I.

Determination of radiochemical purity using T.L.C. in 85% methanol proved to be both reliable and consistent in detecting oxidized forms of iodine. Good correlation was seen with the U.S.P. methods.

In conclusion, the methods presented for the rapid assessment of the purity of ^{123}I-NaI solution produced by the (p,5n) reaction are suitable for determining final acceptance criteria and pharmaceutical quality.

ACKNOWLEDGMENTS

This work was supported by a grant from the Gammaceutics Corporation. The authors are indebted to Carol McColm for her expert technical editing. We are also grateful to Laura Fleharty and Max Trejo for manuscript preparation.

FOOTNOTES

* Medi-Physics Corporation, Emeryville, California

\+ Crocker Nuclear Laboratories, Davis, California

** Gelman Instrument Company, Ann Arbor, Michigan

TABLE 4 RADIOCHEMICAL PURITY OF ^{123}I-NaI SOLUTION AS DETERMINED ON ITLC-SG IN 85% METHANOL, 15% WATER

Time	% Iodide	% Iodate
At Calibration (n=18)	97.38 ± 1.80	2.63 ± 1.80
24 Hours Post Calibration (n=5)	97.28 ± 1.39	2.72 ± 1.39

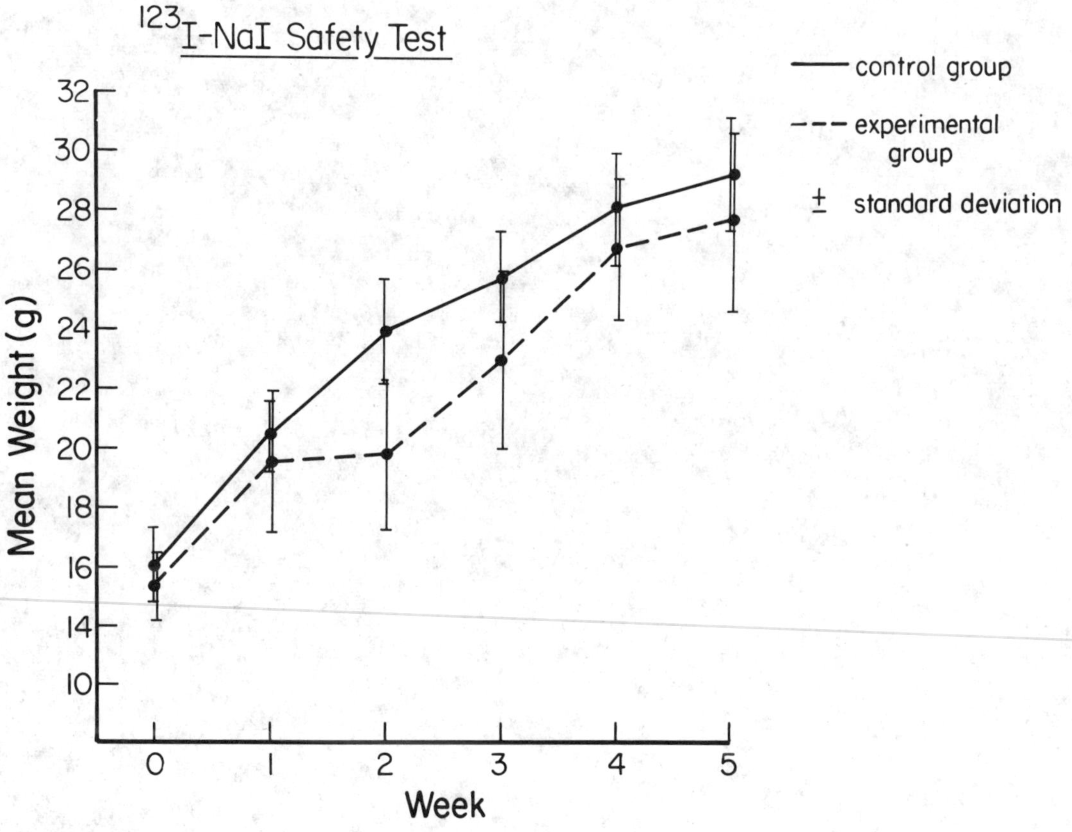

REFERENCES

1. Sodium Iodide ^{123}I Capsules and Solution for Oral Administration Diagnostic. Product information insert, Medi-Physics Corp., Emeryville, California, December, 1975

2. DeNardo GL, Krohn KA, Jansholt A-L, et al.: Present and Future Applications of ^{123}I. In Medical Radionuclide Imaging, Vol. II Vienna, IAEA-SM-210/308, 1977, pp.3-20

3. Baker GA, Lum DJ, Smith EM, et al.: Significance of Radiocontaminants in ^{123}I for Dosimetry and Scintillation Camera Imaging. J Nucl Med 17:740-743, 1976

4. MIRD/Dose Estimate Report No. 5: Summary of Current Radiation Dose Estimates to Humans From ^{123}I, ^{124}I, ^{125}I, ^{126}I, ^{130}I, ^{131}I, and ^{132}I as Sodium Iodide. J Nucl Med 16:857-860, 1975

5. The United States Pharmacopeia, Nineteenth Revision. p. 462, Mack Printing Company, Easton, Pa., 1974

6. Procedure Manual: Radiochemical Purity of Radiopharmaceuticals Using Gelman ITLC Chromatography. Technical Bulletin No. 32, Gelman Instrument Co., Ann Arbor, Michigan, June, 1977

EFFECT OF SOLUBILITY ON THE BIODISTRIBUTION AND EVALUATION OF 16-[^{123}I]-9-HEXADECENOIC ACID AS A MYOCARDIAL IMAGING AGENT

Dennis R. Hoogland, Jamey P. Weichert, Steven A. Sirr, Mathis P. Frick, Lee A. Forstrom, and Merle K. Loken

University of Minnesota Hospitals
Minneapolis, Minnesota 55455

A method for synthesizing 16-[^{123}I]-9-hexadecenoic acid (IHDA) free of particulates and macro-colloids with a 75% final yield has been developed. Our results indicate biodistribution and product quality of myocardial imaging agents such as IHDA with short effective biological half times in blood and myocardial tissue can be evaluated in the mouse. The importance of ascertaining biodisposition in non-cardiac tissues when evaluating fatty acids as myocardial imaging agents was demonstrated. Initial dog studies demonstrated that IHDA has potential for imaging sites of decreased perfusion in the heart, and may be suitable for sequential rest and stress studies because of its short effective half time in myocardial tissue (15-20 minutes) and acceptable non-target activity levels.

INTRODUCTION

Evans and Beierwaltes (1,2) used radioiodinated oleic acid prepared by addition of iodine-131 monochloride to the double bond of the fatty acid. This results in large iodine and chlorine atoms on the fatty acid molecule which cause the labeled molecules to be biologically distinguishable from the parent compound. Labeling fatty acids in the carboxyl group with isotopes of carbon has no effect on the physiological properties of these agents. By labeling fatty acids by interhalogen replacement on the terminal atom, Robinson and Lee have demonstrated that these molecules retain the biological extraction characteristics of the parent fatty acids (3) as the iodine atom occupies a volume and position similar to that of a terminal methyl group.

Poe and coworkers (4) also demonstrated that 16-[^{123}I]-hexadecenoic acid (IHDA) when administered intravenously to dogs concentrates in myocardial tissue in proportion to relative coronary blood flow, has a short clearance half time for blood or myocardium (1.7 and 20 minutes, respectively), has distribution patterns similar to that of potassium-43 in acute and chronic infarction, and has high specific activity such that good quality images were obtained in as little as two to three minutes per view using a pin hole collimator. The iodine-123 labeling of the fatty acid also was desirable because of the high flux of 159 keV photons which are optimal for imaging with scintillation cameras, and result in a low radiation exposure per mCi administered.

This study was initiated to develop a reliable synthetic method for terminal labeled IHDA, and to demonstrate its potential clinical efficacy in various animal models.

MATERIALS AND METHODS

Chemistry: The synthesis of 16-[^{123}I]-hexadecenoic acid (IHDA) was developed from the method of Robinson and coworkers (3,5). Using 16-bromo-9-hexadecenoic acid* as starting material, IHDA is produced by interhalogen replacement of the terminal bromine atom. In a dark environment, less than 10 mg of starting material is refluxed in a 10 ml pear-shaped flask for three hours in 3 ml of acetone and 5 mCi of carrier free [^{123}I]-sodium iodide. The [^{123}I]-sodium iodide** is obtained in 1 ml of acetone. When the reaction is complete, the solvent is evaporated with a stream of nitrogen.

Then 2 ml of 25 percent HSA is added, and the solubility of the residue is facilitated by placing the reaction flask into a 50°C ultrasonic bath for 30 minutes. The HSA-fatty acid solution is passed through a 1x3 cm anion exchange column (Dowex 1-8x) to remove any unreacted iodine-123. The resultant solution is then diluted to six percent HSA with physiologic saline and passed through a filtering stack consisting of 0.45 and 0.22 micron Millipore filters. The filter stack, which sterilizes the final product, can also be assayed to determine quantities of labeled particulates and macro-colloids. Current yields are 65-75 percent as determined by thin layer and anion exchange chromatography.

Proton nuclear magnetic resonance (NMR) and mass spectroscopy were used to determine relative purity and structure of the final reaction product. The

* Aldrich Chemical Company, Milwaukee, Wisconsin.

** Crocker Nuclear Lab, Davis, California.

NMR results indicated the starting material (16-bromo-9-hexadecenoic acid) and final product were equally pure, but no significant shift due to iodine replacement of bromine was detected. Mass spectroscopy showed the final reaction product was essentially the bromo compound with very little iodo compound. This reflects the small amount of iodine in the high specific activity and carrier free [^{123}I]-sodium iodide used in the synthesis.

Animal Studies: The distribution of IHDA was determined in female white mice following intravenous administration of 15 μCi in less than 0.2 ml. The animals were sacrificed by decapitation at intervals up to eight minutes. Blood was collected on a wax block to prevent coagulation, and selected organs (heart, lung, liver, spleen and kidneys) were excised and weighed. Aliquots of blood and the entire organs were assayed with an iodine-123 standard. Total injected dose was determined by assaying the empty syringe and the tail of each mouse (site of injection), and subtracting the results from the original contents of the syringe. The percent dose in each gram of tissue was calculated, and ratios to heart tissue determined at each sacrifice interval.

Normal mongrel dogs were administered IHDA intravenously either unfiltered (1.0 mCi) or subjected to ultrasound and filtration (1.6 mCi), and left anterior oblique (LAO) images (200 K counts) were obtained every two minutes for the first 20 minutes, and five minute intervals over the next 25 minutes.

The effects of myocardial ischemia on the biodistribution of IHDA were then studied in the dog. A suture snare was tied on the circumflex artery, 1 cm distally from its bifurcation with the LAD, in a 20 kg mongrel dog. A Doppler flow probe was placed just distally to the snare. Then 15 days later, IHDA (2.0 mCi) was administered intravenously with normal blood flow in the circumflex artery. Data was collected for 30 minutes with a low energy mobile camera and low energy all purpose collimator. Data were stored on disc, and displayed on digital cathode ray tube.

Then seven days later, blood flow in the circumflex artery was reduced 50 percent, and IHDA (2.0 mCi) administered intravenously. Data was collected and stored for 30 minutes as in the control study above.

RESULTS

Iodine-123 labeled hexadecenoic acid prepared in our laboratories using the method of Robinson (3,5) resulted in a similar percent yield of final product. However, intravenous administration of unfiltered product in mice resulted in high lung and liver activity in comparison to myocardial tissue (Table 1) which indicated the presence of labeled particulates and colloidal material. Administration of unfiltered product in a dog yielded similar results 9-13 and 17-21 minutes after injection (Fig. 1A).

Review of the synthesis procedures indicated insolubility of IHDA in the 25 percent human serum albumin (HSA) as a possible source of labeled particulates and colloid. Filtration of this solution through a 0.45 micron Millipore filter resulted in 42 percent of initial activity remaining on the filter. However, heating the 25 percent HSA solution slightly to 50°C while subjecting the product to ultrasound for 30 minutes increased the solubility without denaturating the albumin, and lowered the quantity of product on the filter to less than one percent of initial activity.

Mice intravenously administered IHDA prepared with the above modifications indicated a dramatic decrease in lung and liver activity with a corresponding increase in myocardial activity (Tables 2 and 3). Maximum myocardial activity was observed in 1.5-2.5 minutes, and a decrease in activity of approximately

TABLE 1.

TISSUE CONCENTRATIONS OF 16-[^{123}I]-9-HEXADECENOIC ACID ADMINISTERED INTRAVENOUSLY* TO MICE WITHOUT PRIOR ULTRASOUNDING OR FILTRATION

	Percent Dose per Gram Tissue			Ratio to Heart Concentration		
	1.5-2.0 min.	2.0-2.5 min.	2.5-3.0 min.	1.5-2.0 min.	2.0-2.5 min.	2.5-3.0 min.
Heart	4.5	3.9	5.1	-	-	-
(n)**	(3)	(3)	(3)			
Lung	19.9	17.9	24.0	0.2	0.2	0.2
(n)	(3)	(3)	(3)	(3)	(3)	(3)
Liver	20.0	14.8	19.3	0.2	0.3	0.2
(n)	(3)	(3)	(3)	(3)	(3)	(3)
Blood	3.0	3.4	3.4	1.5	1.1	1.5
(n)	(3)	(3)	(3)	(3)	(3)	(3)

* IHDA (15 µCi in less than 0.2 ml) administered via tail vein to female white mice in 6% HSA solution.

** n = number of animals studied.

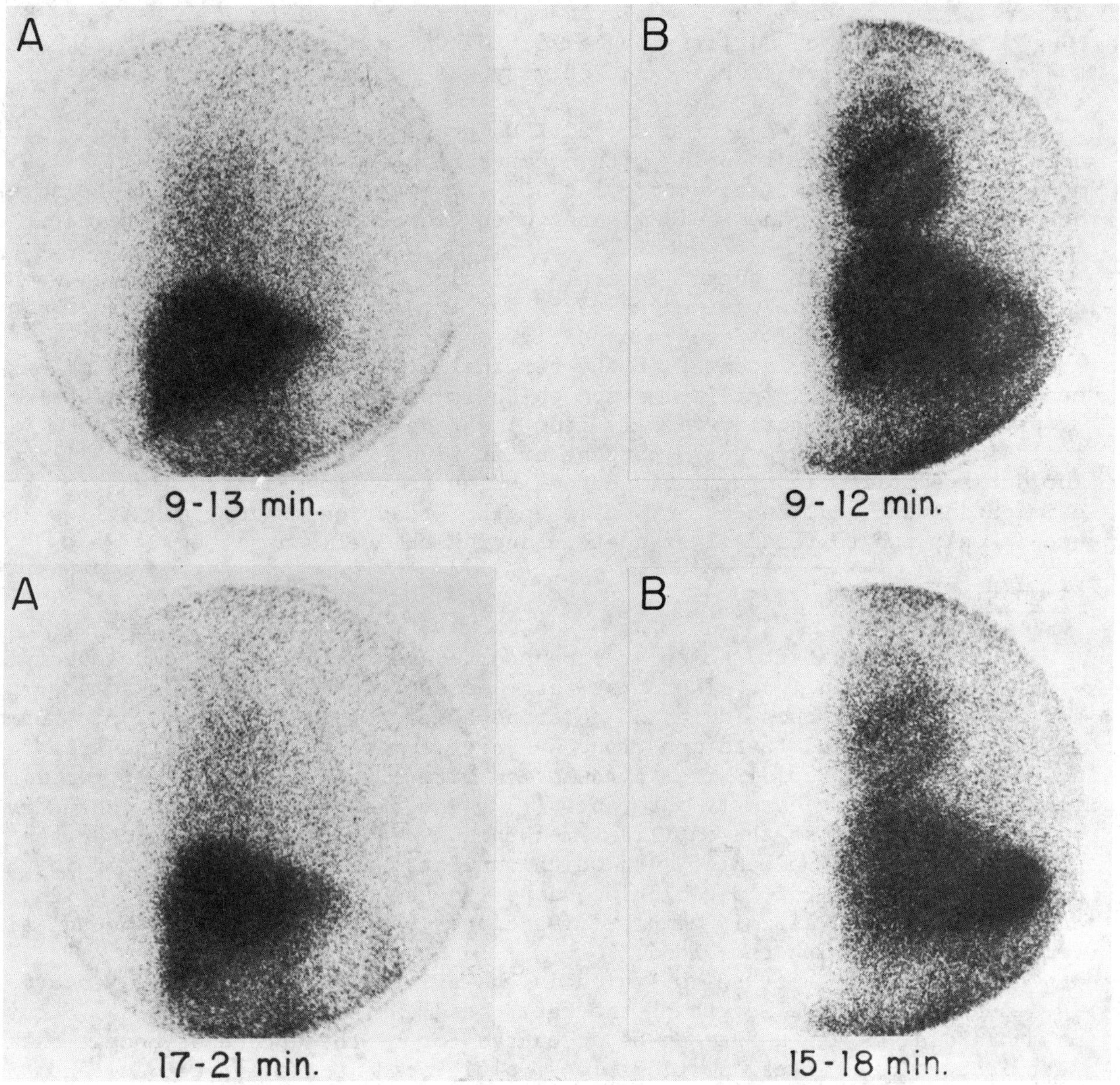

Figure 1A.

Sequential LAO images of dog administered 1.0 mCi ^{123}I-Hexadecenoic Acid not subjected to ultrasound or filtration prior to intravenous administration. Note lack of heart activity.

Figure 1B.

Sequential LAO images of dog administered 1.6 mCi ^{123}I-Hexadecenoic Acid subjected to ultrasound and filtration prior to intravenous administration. Note right ventricular activity (A,B).

33 percent was observed by 4-6 minutes after administration (Tables 2 and 3). Lung, liver and blood activity peaked during the same time interval as heart tissue, but a similar decrease in activity was not observed 4-6 minutes after administration.

A dog intravenously administered IHDA prepared with the above modifications demonstrated myocardial uptake with acceptable background activity in 9-12 minutes (Fig. 1B). Right ventricular activity was observed in the left anterior oblique view in 9-12 and 15-18 minutes with decreased activity at the later interval.

A well demarcated posterilateral perfusion defect was demonstrated when circumflex artery flow was decreased 50 percent (Fig. 2C, Arrow). By 20 minutes, activity in myocardial tissue was nearly gone (Fig. 2D).

Interhalogen replacement of the terminal bromine with iodine-123 on hexadecenoic acid is a Finkelstein reaction which is normally sensitive to light (6). By protecting the reaction vessels from light during the reflux reaction, final yield increased from 65 to 75 percent after final filtration. The [^{123}I]-sodium iodide is supplied in one ml of acetone which helps to maintain high yields by eliminating the addition of any water to the reaction medium, and allows the use of only two to three ml of acetone during the refluxing step.

DISCUSSION

Nonesterified fatty acids are transported in the blood as soluble complexes with plasma albumin (7). That free fatty acids are mobilized from adipose tissue by several hormones including norepinephrine, adrenal cortocotropic hormone (ACTH), insulin, and thyroid hormone has been demonstrated (8). The free fatty acids are removed rapidly from circulation with a half life of about two minutes when injected intravenously into dogs (7). The rate of passage of the fatty acid molecules across the capillary membrane is many times greater than the passage of albumin molecules with which the fatty acids are associated in the blood. Therefore, this association serves to keep fatty acids in solution, and the fatty acid-albumin complex dissociates at the surface of endothelial cells in contact with the blood.

Oxidation of fatty acids, which is the primary myocardial energy source, is dependent upon the length of the fatty acid chain (9,13). From 25-75 percent of myocardial oxygen usage can be accounted for by the uptake of nonesterified fatty acid (10). In circulating blood, oleic acid is the largest component of the free fatty acids and for these free fatty acids, the myocardial uptake rate is the highest for oleic acid in both man and dog (11, 12). Long chain fatty acids (16 carbons or greater in length) are dependent upon the enzyme carnitine acyl-COA transferase for their translocation from the extramitocondrial to the intramitocondrial sites. Under aerobic conditions myocardial fatty acid beta oxidation yields intracellular accumulation of citrate and isocitrate. Weiss and coworkers demonstrated that beta oxidation of fatty acid is dependent upon the length of fatty acid when palmitate was metabolized at a rate greater than octanoate (9).

Evans and coworkers observed in the isolated heart that nonesterified fatty acids are readily extracted by heart muscle and that a portion of the uptake is temporarily stored as tissue triglyceride and phospholipid (1). Infusion of glucose and insulin with radioiodinated oleic acid increased the ratio of radioactivity in heart muscle to blood by enhancing storage of the fatty acid as a myocardial triglyceride and decreasing oxidation of the fatty acid in heart and other tissues. Concentration of lipid bound radio-

TABLE 2.

CONCENTRATION OF 16-[^{123}I]-9-HEXADECENOIC ACID (IHDA)
IN HEART, LUNG, LIVER AND BLOOD FOLLOWING INTRAVENOUS ADMINISTRATION* TO MICE

	Percent Administered Dose Per Gram of Tissue						
	1.0-1.5 min.	1.5-2.0 min.	2.0-2.5 min.	2.5-3.0 min.	3.0-3.5 min.	3.5-4.0 min.	4.0-6.0 min.
Heart	17.3	20.1	22.2	12.3	15.9	15.7	14.0
(n)**	(2)	(2)	(2)	(3)	(4)	(3)	(2)
Lung	5.2	8.9	8.3	5.5	7.5	8.4	7.5
(n)	(2)	(2)	(2)	(3)	(4)	(3)	(2)
Liver	9.1	11.1	17.6	13.2	12.9	14.5	16.2
(n)	(2)	(2)	(2)	(3)	(4)	(3)	(2)
Blood	2.3	3.5	3.9	2.2	2.9	4.4	4.3
(n)	(2)	(2)	(2)	(3)	(4)	(3)	(2)

* IHDA (15 μCi in less than 0.2 ml) administered via tail vein to female white mice in 6% HSA solution.

** n = number of animals studied with product filtered (0.45 and 0.22 micron Millipore filters) prior to injection.

TABLE 3.

RATIO* OF 16-[^{123}I]-9-HEXADECENOIC ACID CONCENTRATION IN HEART VS. LUNG, LIVER AND BLOOD FOLLOWING INTRAVENOUS ADMINISTRATION** TO MICE

	1.0-1.5 min.	1.5-2.0 min.	2.0-2.5 min.	2.5-3.0 min.	3.0-3.5 min.	3.5-4.0 min.	4.0-6.0 min.
Heart/Lung	3.6	2.3	2.8	2.3	2.2	2.0	1.9
(n)***	(2)	(2)	(2)	(3)	(4)	(3)	(3)
Heart/Liver	2.0	2.0	1.3	1.0	1.2	1.1	0.9
(n)	(2)	(2)	(2)	(3)	(4)	(3)	(3)
Heart/Blood	7.9	5.8	5.8	6.0	5.9	3.7	4.0
(n)	(2)	(2)	(2)	(3)	(4)	(3)	(3)

* Ratio of % Dose/gm Tissue.

** IHDA (15 µCi in less than 0.2 ml) administered via tail vein to female white mice in 6% HSA solution.

*** n = number of animals studied with product filtered (0.45 and 0.22 micron Millipore filters) prior to injection.

activity per unit weight of tissue three to four hours after injection of radioiodinated oleic acid is higher in heart muscle than in other organs with comparable metabolic activity or vascularity. They also observed high concentration of radioactivity in gastric contents which was attributed to rapid excretion by gastric mucosa of water soluble iodine-131 metabolized from the radioiodinated fatty acid.

The studies by Evans and coworkers were not optimal as only a low specific activity radioiodinated oleic acid was available (8 to 12 μCi per ml), and the images obtained were of poor quality due primarily to the relatively poor physical characteristics of iodine-131 as an imaging agent. Since the rate of fatty acid extraction by the myocardium of the isolated heart was shown to depend upon the concentration of the fatty acids in the blood (12), Beierwaltes and coworkers attempted to increase the myocardial uptake of fatty acids labeled with higher specific activities of iodine-131 and small volume injections (2). However, blood in the ventricular chamber contained sufficient iodine-131 to prevent separate imaging of ventricular myocardial ischemia in the dog. Thus it was impossible to produce curves which accurately reflected the uptake and disappearance of radioactive fatty acids from the myocardium as an indication of abnormalities in fatty acid metabolism by the heart muscle. They also found radioactive concentration of carbon-14 labeled oleic acid in the myocardium was significantly higher than that of iodine-131 labeled oleic acid. This difference suggests that the mechanism of myocardial uptake of iodinated oleic acid may be different from that of unlabeled oleic acid.

Evans and Beierwaltes (1, 2) used radioiodinated oleic acid prepared by addition of iodine-131 monochloride to the double bond of the fatty acid. This results in large iodine and chlorine atoms on the fatty acid molecule which cause the labeled molecules to be biologically distinguishable from the parent compound. Labeling fatty acids in the carboxyl group with isotopes of carbon has no effect on the physiological properties of these agents. By labeling fatty acids by interhalogen replacement on the terminal atom, Robinson and Poe have demonstrated that these molecules retain the biological extraction characteristics of the parent fatty acids (3), as the iodine atom occupies a volume and position similar to that of a terminal methyl group. Metabolic and distribution studies with these fatty acids will better reflect true physiological processes.

Studies by Robinson and Poe (3-5, 14-18) have demonstrated hexadecenoic acid labeled in the terminal position with iodine-123 potentially provides a radiopharmaceutical which is readily imaged following intravenous administration, can be prepared and purified within a short period of time, is distributed proportionate to a reasonable myocardial blood flow, is rapidly cleared by normal cells (approximate biological half time is 1.7 and 20 minutes for blood and myocardium respectively), and allows repeat studies within a few hours of the first dose.

This study demonstrates that methods which increase the solubility of fatty acids in HSA solutions, will also increase the final yield, specific activity, and biological effectiveness of radiopharmaceuticals which are bound to proteins.

The evaluation of IHDA and other fatty acids as potential myocardial imaging agents may be misleading if a solution containing labeled particulates and macro-colloids is administered and biodisposition in non-cardiac organs is not examined.

Additional improvements in the synthesis of IHDA have been proposed by Robinson (5). He obtained a 90 percent yield prior to filtration after heating ten minutes at 120°C with methyl ethyl ketone in a flame sealed three ml glass ampule. This method offers a high yield rapid labeling procedure in a closed system, which could prove to be the method of choice.

The mouse has been demonstrated to be capable of predicting biodistribution of fatty acids as myocardial imaging agents despite their short effective half-times in blood and myocardial tissue.

A dog study where flow through the circumflex artery was decreased 50 percent and a posteriolateral defect was demonstrated with IHDA supports the conclusion that IHDA can be used to evaluate coronary artery disease which results in significant reduction of coronary flow.

These data also suggest that IHDA may be administered sequentially within reasonable intervals which will allow same-day rest and stress myocardial studies.

ACKNOWLEDGEMENTS

This work was partially supported by Grant #A0-77-78 of the University of Minnesota Medical Foundation.

We would like to acknowledge the contributions of Richard A. Ponto and Dr. Jeffrey Schwartz in the dog studies, and Doris Lunden for preparation of the manuscript.

REFERENCES

1. Evans JR, Gunton RW, Baker RG, et al.: Use of radioiodinated fatty acid for photoscans of the heart. Cir c Res 16: 1-10, 1965
2. Beierwaltes WH, Ice RD, Shaw MJ, et al.: Myocardial uptake of labeled oleic and linoleic acids. J Nucl Med 16: 842-845, 1975
3. Robinson GD, Lee AW: Radioiodinated fatty acids for heart imaging: Iodine monochloride addition compared with iodide replacement labeling. J Nucl Med 16: 17-21, 1975
4. Poe ND, Robinson GD, Graham LS, et al.: Experimental basis for myocardial imaging with ^{123}I-labeled hexadecenoic acid. J Nucl Med 17: 1077-1082, 1976
5. Robinson GD: Synthesis of ^{123}I-16-iodo-9-hexadecenoic acid and derivatives for use as myocardial perfusion imaging agents. Int J Appl Rad Iso 28: 149-156, 1977
6. Ingold CK, Structure and Mechanism in Organic Chemistry, London, Cornell University Press, 2nd ed., 1969, p.435
7. French JE, Morris B, Robinson DS: Removal of lipids from the blood stream. Br Med Bull 14: 234-238, 1958
8. Oram JF, Bennetch SL, Neely JR: Regulation of fatty acid utilization in isolated perfused rat hearts. J Biol Chem 248: 5299, 5309, 1973
9. Weiss ES, Hoffman EJ, Phelps ME, et al.: External detection and visualization of myocardial ischemia with ^{11}C-substrates in vitro and in vivo. Circ Res 39: 24-32, 1976
10. Gordon RS, Cherkes A: Unesterified fatty acids in human blood plasma. J Clin Invest 35: 206-212, 1956
11. Rothlin ME, Bing RJ: Extraction and release of individual free fatty acids by the heart and fat deposits. J Clin Invest 40: 1380-1386, 1961
12. Bing RJ: Metabolic activity of the intact heart. Am Med 30: 679-691, 1961
13. Carlsten A, Hallgren B, Jagenburg R, et al.: Myocardial metabolism of glucose, lactic acid, amino acids and fatty acids in healthy human individuals at rest and at different work loads. Scan J Clin Lab Inv 13: 418-428, 1961
14. Poe ND, Robinson GD, MacDonald NS: Myocardial extraction of labeled long-chain fatty acid analogs. Proc Soc Exp Med 148: 215-218, 1975

15. Poe ND, Robinson GD, MacDonald NS: Myocardial extraction of variously labeled fatty acids and carboxylates. J Nucl Med 14: 440, 1973
16. Poe ND, Eber LM, Norman AS, et al.: Myocardial images in nonacute coronary and noncoronary heart diseases. J Nucl Med 18: 18-23, 1977
17. Poe ND: Rationale and radiopharmaceuticals for myocardial imaging. Semin Nucl Med 7: 7-28, 1977
18. Poe ND: A critical evaluation of myocardial imaging. In Radiopharmaceuticals, New York, Society of Nuclear Medicine, Inc., 1975, pp.359-369

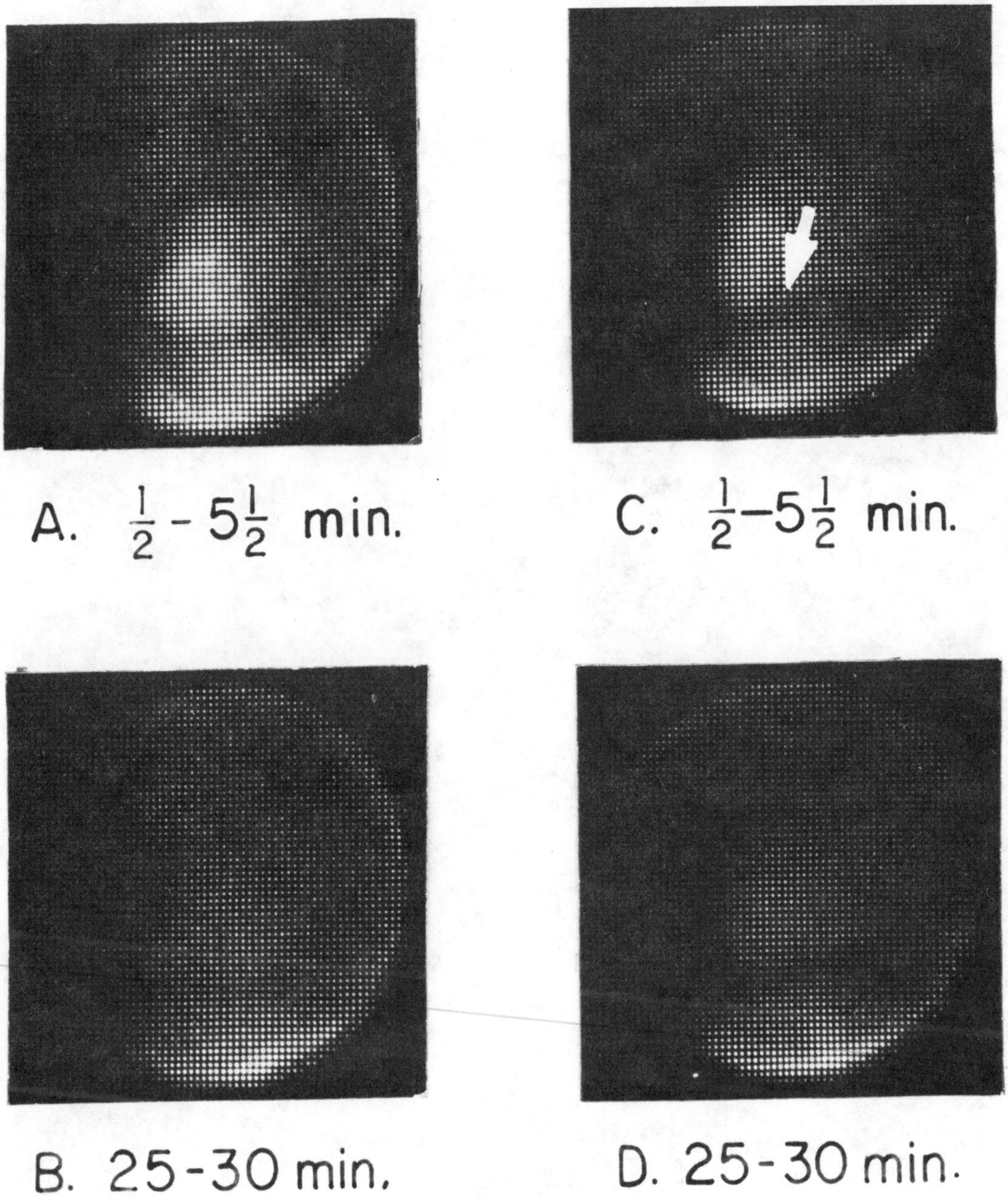

Figure 2.
Sequential left lateral images following administration of 2.0 mCi ^{123}I-Hexadecenoic Acid with a snare in place at bifurcation of circumflex artery with LAD (A,B) and with 50 percent decreased flow in circumflex artery (C,D) one week later. Note posteriolateral defect (Arrow).

REEVALUATION OF THE CONTAMINANTS IN THE 99-Mo 99m-Tc GENERATOR ELUANTS

Ivor L. Preiss and Adrienne S. Frank

Rensselaer Polytechnic Institute
Department of Chemistry
Troy, New York 12181

Raj Kishore

Veterans Administration Medical Center
Department of Nuclear Medicine
Albany, New York 12208

There have been many studies to determine the level of contamination in generator produced 99m-Tc. Such studies are required so that reasonably accurate dose calculations, which include all radiations from a sample, can be made. We have conducted a set of experiments designed to establish the origin of the as yet unassigned 322 Kev gamma-ray often observed as the major contaminant in generator-produced 99m-Tc eluant fractions: Classical radiochemical milking of daughter from parent + grandparent was employed along with high resolution gamma-ray spectroscopy. These experiments, in which energy and half-life were determined, indicate that the 322 Kev gamma-ray is indeed the result of beta-decay of 99m-Tc. Our results do not preclude other beta or gamma decays from this parent.

INTRODUCTION

The use of 99mTc in diagnostic nuclear medicine has grown considerably over the past decade. Like all radiopharmaceuticals the quality control aspects are critical. Various measurements[1] have been made to determine the origin of gamma-ray contaminants, especially in 99 Mo 99mTc generator produced 99mTc. Most of the gamma-rays have been assigned to "impurities" such as 99 Mo, 131 Te, 132I, and 97Ru. One of the gamma-rays which has eluded positive assignment is the 322 Kev line which appears in the spectrum of 99mTc regardless of chemical and physical purification procedures employed.

The dose calculations have been based on the assumption that the presently accepted decay scheme of 99mTc is correct. Thus the existence of any extraneous radiation that could accompany the decay of 99mTc will have an effect on patient dose. Jones and Griffin[2] and Décombaz et al[3] each report that there could be a beta-decay branch associated with the decay of the 99 mTc isomeric state (labelled B-1 in Figure 1). The generalized decay scheme shown in Figure 1 indicate those beta-decays and gamma transition that are theoretically probable in the decay of 99mTc. The solid lines represent well established transitions. The dashed lines marked B-2 and B-3 represent likely beta-decay branches that could compete with the beta-decay labelled B-1. We have conducted a series of classical parent daughter milking experiments utilizing a standard 99Mo99mTc generator (fission produced 99 Mo) which conclusively proves the existence of a beta-decay from 99 mTc to a level in 99 Ru which gives rise to a 322 Kev gamma-ray.

The chemical separation of daughter activity from parent nuclide at regular time intervals allows one to establish the growth rate of the daughter and hence the decay constant of the parent. The classical presentation of such data shows the activity of the daughter at the time of milking plotted versus the time between successive milkings (or chemical separations). The slope of such a plot is the half-life of the parent. This technique is ideally suited for establishing any generic relationships which might be present in the Mo-Tc generator. The intensity of the 322 Kev gamma, at the time of separation from the column plotted versus the time interval between separations gives the T 1/2 of the parent associated with this gamma-ray. At the same time, the decay rate of the gamma-ray itself can be followed in each milked sample, thus yielding a series of independent measurements of the T 1/2 for the daughter. These data are shown in Figure 2.

PROTOCOL

Measurements were performed using standard Squibb Minitec(R) generators. The elution of the generator was carried out following the manufacturer's suggested procedures. The age of the several generators used in each of the separate experiments

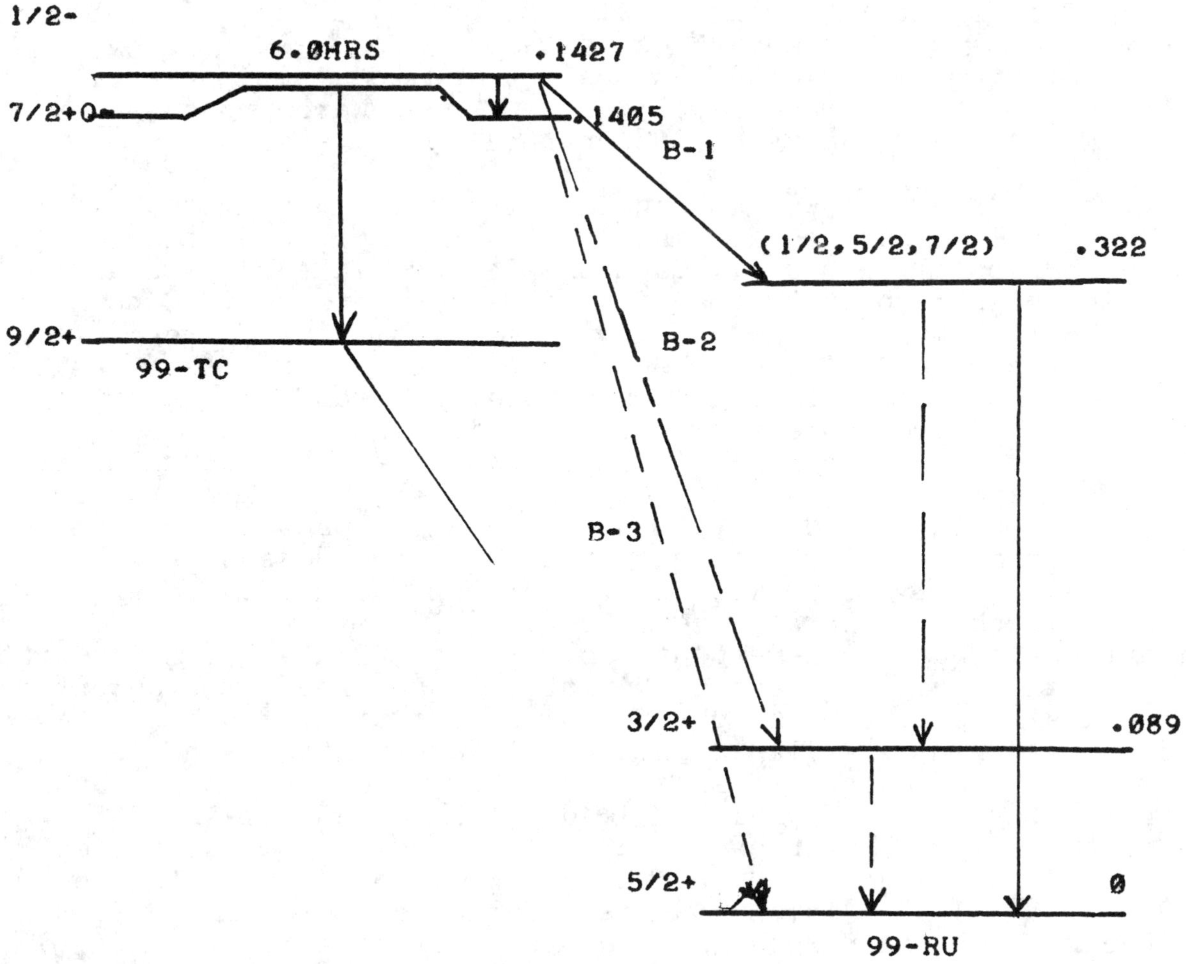

Figure 1.

GENERALIZED DECAY SCHEME OF Tc-99m. THOSE TRANSITIONS SHOWN AS SOLID LINES ARE CONFIRMED. THOSE THAT ARE DASHED ARE PROBABLE.

varied from one to several weeks. The Ge(Li) detectors utilized had volumes of 35 cm^3 and 45 cm^3 detector was used exclusively in the milking experiments and was coupled to an FET preamplifier and a spectroscopy grade amplifier. Signals were processed by a 100 MHZ ADC coupled to a 4096 channel memory system.

The Tc fraction was milked from the generator column with 5 ml of saline solution at exactly 12 hour time intervals over a period of 72 hours. Each sample was placed at a standard position with respect to the 45 cm^3 Ge(Li) detector and counted through a calibrated lead shield. The position was determined such that the dead time of the analyzer system was less than 10 percent. The decay of each sample was followed for a 24 hour period. This procedure was followed utilizing two different generators, one giving an initial activity (at the beginning of the measurement) of 90 mCi and the other 88 mCi. A third generator was utilized to establish gross gamma spectra and half-lives of the several gammas reported as being present by others[1,2,3]. A four week old generator was milked to determine if the presence of very long lived activities in the eluant could have been responsible for the 322 Kev gamma-ray. No 322 Kev gamma-ray peak above statistical background was observed from this generator. The photo peak intensities of the 322 Kev and 140 Kev gamma-ray lines in each milked fraction were determined. The integrated areas were corrected for background, absorption in the glass vial and lead shield, and normalized to the detector efficiency for each energy. The relative intensities of the 322 Kev and 140 Kev gamma-rays were found to be in the order of 6×10^{-6}. This value is in agreement with the results of Décombez[3] which were based on the separation Tc from (n,γ) products in a molybdenum target. It is important to note that this relative gamma-ray intensity corresponds to the minimum relative beta decay intensity.

CONCLUSIONS

The milking experimental data are shown in Figure 2 B and C. These represent the decay of the 322 Kev gamma in the Tc fraction separated from the generator at fixed time intervals. The agregate half-life is found to be 6.05± .03 hours. The activities extrapolated to the times of separation are plotted in Figure 2A. The decrease in activity with respect to the time between the separations shown in Figures 2B and C is a measure of the half-life of the parent radionuclide. This was found to be 67 ± 2 hours. This unique combination of half-lives, 67 hours feeding 6 hours, confirms the Mo-Tc-Ru decay chain at mass 99.

Detailed consideration of the nuclear structure physics involved and the decay systematics leads us to the conclusion that there is at least one other beta-decay branch from 99m-Tc, Labelled B-2 on Figure 1. This decay is to the first excited state of 99-Ru at 89 Kev. This branch could be an order of magnitude more intense than that to the 322 Kev level and should be at least equal in population. A third beta decay branch, labelled B-3, is

Figure 2

RESULTS OF THE RADIOCHEMICAL MILKING OF A Mo-Tc GENERATOR. 'A' INDICATES THE DECAY OF THE PARENT NUCLIDE OF 322 KEV GAMMA-RAYS 'B' AND 'C' SHOW THE TYPICAL DECAY DATA OF SEQUENTIALY MILKED Tc FRACTIONS; FROM TWO DIFFERENT GENERATORS.

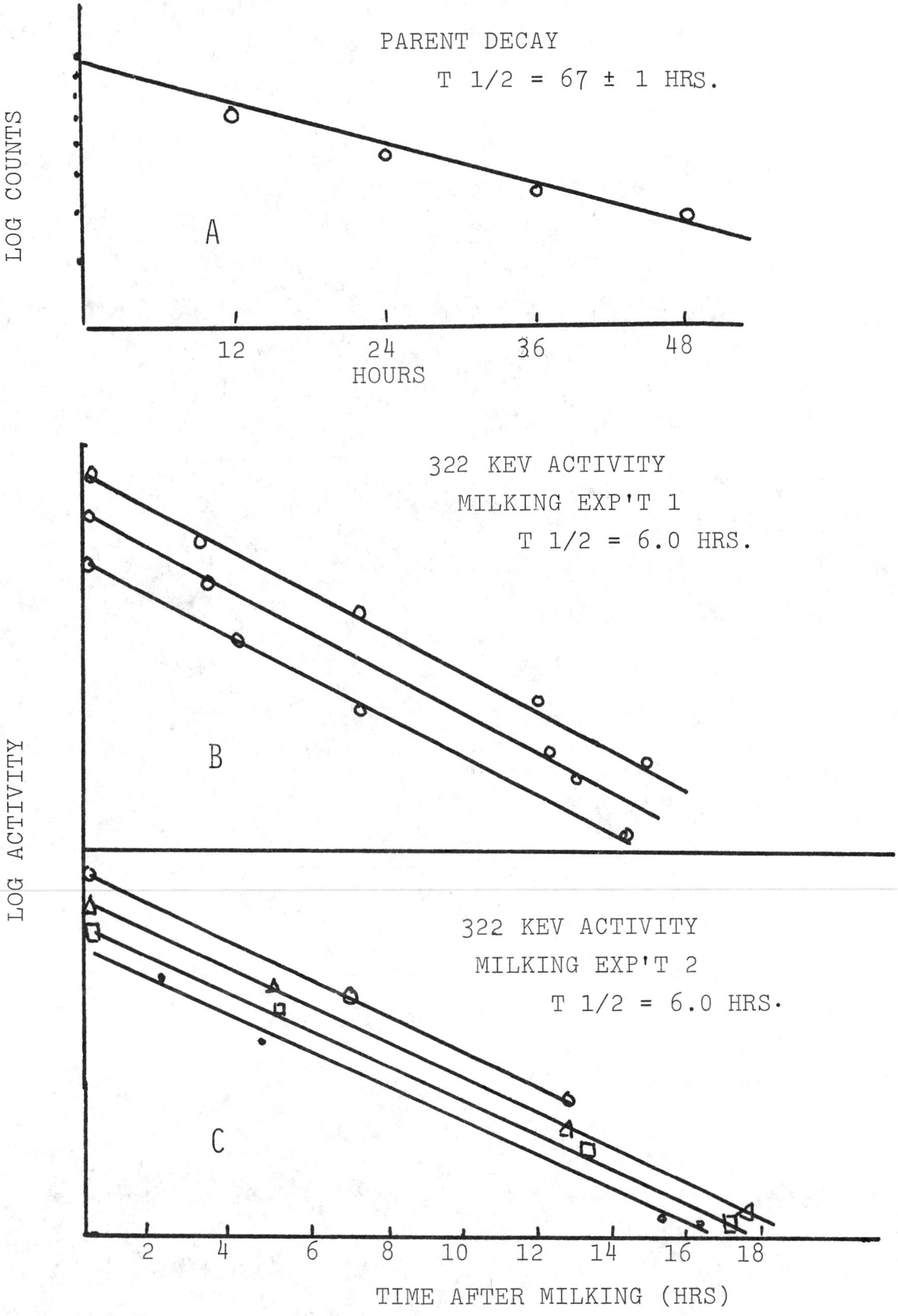

possible with its intensity dependent on the spin assigned to the state at 322 Kev. Two as yet unobserved gamma rays should therefore exist. The 89 Kev ground state transition and a 233 Kev gamma between the 2nd and 1st excited states. The extent to which the gamma-rays are converted is dependent on the details of the nuclear structure but the decay rates of the beta branches and gamma-ray intensities can be approximated based on systematics. Therefore, a reevaluation of the detailed dose calculations of 99 mTc should be undertaken even at this juncture, and certainly when the decay characteristics are fully elucidated by more complete studies.

REFERENCES

1. For example see: Anderson, D.W., Raeside, D.E. and Ficken, V.J., J. Nuc. Med. 15 889(1974).

   Blaha, V and Krivsky, Z. Ceskoslov. Farn. 22, 176 (1973).

   LeBlanc, A.D., Int. J. App. Rad. and Iso. 26, 785 (1975).

2. Jones, J.D. and Griffin, H.C., Radiochem. Radioanal. Let. 4 381 (1970).

3. Décombaz, M., Gostley, J.J. and Lerch, P. Ibid 10, 119 (1972).

ORGANIC RADIOPHARMACEUTICALS

RECENT ADVANCES

Alfred P. Wolf and Joanna S. Fowler

Department of Chemistry
Brookhaven National Laboratory
Upton, New York 11973

ABSTRACT

Organic radiopharmaceuticals are considered in light of accelerator and nuclide production requirements, special problems relating to the carrier-free state, including terminology,of the special technology required to prepare and manipulate these compounds and new trends in compound design and synthesis. The emphasis is on medical cyclotrons and the positron emitting radionuclides, carbon-11, nitrogen-13, oxygen-15, and fluorine-18. New routes to synthetic precursors and organic compounds of high specific activity labeled with carbon-11, fluorine-18 and iodine-123 including monosaccharides, aromatic amines, neuroleptics, fatty acids, steroids, and other classes of compounds are discussed. Some compounds are considered in terms of the development and evaluation of structure-activity relationships and including some newer concepts such as metabolic trapping.

The list of organic radiopharmaceuticals was modest and rather short as recently as 1973, especially if taken in light of known pharmaceuticals, to say nothing of physiologically active compounds and organic compounds in general. There has been a rapid increase in research and in publication in this area since 1973 (1) due to several factors, principal among which are the interest in compounds labeled with positron emitters, the development of organic compounds which probe function and metabolism in the brain, heart, liver, and adrenals, and the development of compounds tied to chelating agents which attach to technetium and other "non-organic" radionuclides. However, another factor has had a profound stimulus on synthetic work and that has been the increasing sophistication in the design of labeled compounds for studying function and metabolism in vivo. So, for example, structure activity relationships, the principle of metabolic trapping, the utilization of the flow-metabolism couple, and macromolecular pulse labeling among other approaches are being developed (2) to probe biochemistry and physiology in normal and pathological states.

New areas of research and application using labeled organic compounds which draw heavily on the expertise developed in radiopharmaceutical work and which may or may not relate to nuclear medicine include the quantification of drug regimen, especially in the area of chemotherapeutic agents, drug testing with relation to the enormously increased data base required by federal regulation, application to pollution problems, and studies on photosynthesis and in other living systems. It is not, however, the purpose of this paper to give an overview of application, but rather to focus on advances in recent years, and in particular to cover the period since the First International Symposium on Radiopharmaceuticals sponsored by the Society of Nuclear Medicine and held in Atlanta in 1974 on the broad aspects and problems associated with compound preparation. The labeling of organic compounds has many facets ranging from nuclide production, precursor preparation, radiopharmaceutical design, and organic synthesis to quality control, the "carrier-free" problem, and the technological side of compound production, e.g. laboratory design, assay procedures, personnel monitoring, etc. These overall questions in one form or another have been addressed since 1974 in a number of symposia (3) and in the literature (4,5). Indeed, the expansion in this field has been so extensive that an in depth review of the whole organic area is outside the scope of this paper. The topics to be covered in this paper can be identified in the section listing given below. It must be emphasized that this review is not intended to be exhaustive in nature but rather to focus on specific examples which illustrate principles and practice. The choice is based in some measure on the informal training carried on at Brookhaven to help new groups in being exposed to the significant aspects of radionuclide incorporation with special emphasis on the positron emitting radionuclides. Many elegant procedures which have appeared in the literature since 1974 could not be covered here. However, the reader is directed to the extensive bibliographies on accelerator produced nuclides for use in biology and medicine (6,7) which are complete through June, 1976 and which contain listings of organic syntheses. The third volume in this series is in preparation. In addition, other areas such as chelates, metallo-organics, labeled macromolecules, and iodination procedures are not covered.

The desirable properties and developments in compounds labeled with positron emitting radionuclides could not, however, have been brought to its present level of application without parallel development and advancement of not only positron imaging devices, but the successful design of positron emission transaxial tomography (PETT) which allows the quantitative determination of radioactivity in a volume element of tissue *in vivo*. The history, theory, and capabilities of these instruments have been reviewed in depth (8). Unique characteristics of positron radiation, namely, the annihilation of the positron at the end of its track to give two photons of 511 keV each emitted in opposite direction at an ∢ of 180° ± 0.3° results in two advantages over single photon imaging. Sensitivity is increased and the imaging system is isosensitive to positron emitters along a coincidence line in the containing system. Thus, since one had labeled biomolecules whose transport, localization, metabolism, and catabolism could first be studied in animal systems in order to study basic biochemistry and physiology, two further steps could be taken; the development of compounds labeled with positron emitters labeled at levels of concentration, "carrier-free" vide infra, evoking no macro-physiological response (thus acting as "true" tracers), and the determination of dynamic properties of these compounds in a living system non-invasively and quantitatively by means of PETT. It should, however, be noted that advances in single photon tomography may ultimately lead to a similar facility in quantitative measurement of radioactivity in a volume element of tissue. Although the range of organic radiopharmaceuticals is somewhat more limited if only single photon emitters such as iodine are considered with the concomitant disadvantages of possible alteration of the bio-active compounds' functionality *in vivo* due to major structural perturbation, a major attraction is the possible wider dissemination of approach in clinical practice since the compound need not be prepared on site or in the near vicinity of a "medical cyclotron".

Topics to be covered in sections in this overview include nuclide production, precursor preparation, the "carrier-free" problem, the design and synthesis of radiopharmaceuticals, and a short comment on technological aspects.

NUCLIDE PRODUCTION

With the increasing interest in radiopharmaceuticals labeled with positron emitters, making a proper choice in the routes used in preparing the desired nuclide become essential. This is especially true of those institutions which have not as yet put together a facility (accelerator and associated laboratories) for preparing the required nuclides. In considering the positron emitters carbon-11, nitrogen-13, oxygen-15, and fluorine-18 the nuclear reactions of choice have been clearly delineated. Table 1 lists the choices for each nuclide based on consideration of yield and target availability. Yield depends on the particle energies and currents available from the accelerator. There is at present some controversy as to which accelerator is ideal but there are some considerations which must be taken into account before a particular set of accelerator conditions is championed. The ultimate choice depends on the use to which the radionuclides are put. There is a wide gulf between an application where the main thrust is animal research and perhaps one or two human subjects/day, and an application where clinical research involving camera and/or tomograph capacity to its fullest on a daily basis is required. A further consideration is whether or not the program is based on simple gases and compounds exclusively, i.e. ^{11}CO, $^{11}CO_2$, $^{13}NH_3$, $^{13}N\text{-}N_2$, $^{15}O\text{-}O_2$, $H_2{}^{15}O$, $C^{15}O$, $^{15}O\text{-}CO_2$, etc. or is a more complex mixture of gases and radiopharmaceuticals. Going on the assumption that a broad clinical program and a program of basic research involving animals and human subjects is contemplated, the question that must be asked is whether or not sufficient quantities of radiopharmaceuticals are available at delivery for such a program. Thus, stated yields of radionuclides for specific medical cyclotron types can be misleading in that they are usually based on a calculated maximum derived from excitation functions which may or may not be accurate. Furthermore, it is often the case that a yield is based on a stated maximum particle energy ignoring the fact that the material to be bombarded is in some form of container during bombardment and therefore the actual particle energy the target sees is lower than the energy of the particle impinging on the container wall. In Table 2 a list of some yields is given at energies and currents chosen to match approximately some of the specifications of currently available "low" energy machines. Factors such as optimum time and cost of enriched targets is not considered here. It must again be emphasized, however, that these are 100% recovery yields at saturation, an ideal rarely achieved in practice. Nevertheless, yields of some precursors approaching these limits are possible, especially with carbon-11. There is hardly any question that as more sophisticated engineering skills are brought to bear on production problems yields will more closely approximate theoretical maxima. Regretably to date,accelerator manufacturers have focussed on physics and not on chemistry and chemical engineering.

It is the purpose of this paper to update what is possible and to delineate new trends in organic radiopharmaceuticals. In the foregoing discussion no mention has been made of the many other accelerator produced nuclides that may be of interest. If one wishes to include such nuclides as ^{123}I, ^{76}Br, ^{77}Br, ^{68}Ga, ^{67}Ga, ^{38}K, or ^{201}Tl which can be incorporated into organic radiopharmaceuticals or used in conjunction with chelating agents, the considerations as to a suitable accelerator becomes considerably more complex. Thus, a more detailed analysis of radionuclide production problems is not germane (9). Nevertheless, careful consideration of these problems should be given by all who wish to embark on a program based on accelerator produced nuclides.

The next step in synthesis is the preparation of appropriate precursors labeled with the desired radionuclides.

PRECURSOR PREPARATION

The literature (4,10) contains many descriptions for target and precursor preparation. It is safe to say that in almost no case with the possible exceptions of ^{11}CO and $^{11}CO_2$ has the yield of precursor been optimized in the production sense. In general, methods have been described which are usually optimized for the machine parameters and other factors peculiar to the institution of origin. In any event, methods for the preparation of a wide variety of simple precursors can be found in the literature (1,3,4,5,6,7,10). A list of these precursors is given in Table 3. A number of revised, improved or new methods continue to appear in the literature.

In the area of carbon-11 precursors,post irradiation chemical methods for the preparation of ^{11}C-cyanate (11) and ^{11}C-phosgene (12) from ^{11}C-cyanide and ^{11}C carbon monoxide respectively have been described. More recently, on-line recoil syntheses of carbon-11 precursors are again being developed. ^{11}C-methyl iodide from the on-line bombardment of nitrogen-HI mixtures with protons can give radiochemical yields up to 25% (13). ^{11}C-methylamine in 45% radiochemical yield, and ^{11}C-guanidine in 49% radiochemical yield can be produced from the proton bombardment of crystalline NH_4Cl and NH_4I, respectively (14). Rapid and facile separation of these products from the target material could provide a new route to these precursors.

TABLE 1. POSITRON EMITTERS FOR LABELED ORGANIC RADIOPHARMACEUTICALS

Nuclide	Reactions of Choice	Threshold (MeV)	Energy at Cross Section Max (MeV)	Maximum Cross Section (millibarns)
Carbon-11	$^{10}B(d,n)^{11}C$	0	2.3	260
	$^{11}B(p,n)^{11}C$	3.0	6.5	400
	$^{14}N(p,\alpha)^{11}C$	3.1	7.6	253
Nitrogen-13	$^{13}C(p,n)^{13}N$	3.2	6.6	276
	$^{12}C(d,n)^{13}N$	0.3	4.7	150
Oxygen-15	$^{15}N(p,n)^{15}O$	3.7	6.3	380
	$^{14}N(d,n)^{15}O$	0	2.5	200
Fluorine-18	$^{18}O(p,n)^{18}F$	2.6	5.1	700
	$^{20}Ne(d,\alpha)^{18}F$	0	-	194

TABLE 2. CARBON-11 AND FLUORINE-18 YIELDS

Nuclide	Reaction	Energy of Particle (p or d) on Target			
		8	10	12	14
		Yield in mCi/μA at Saturation			
Carbon-11	$^{14}N(p,\alpha)^{11}C$	29	45	66	93
Fluorine-18	$^{18}O(p,n)^{18}F$	110	147	187	216
	$^{20}Ne(d,\alpha)^{18}F$	46	63	75	84

In the area of nitrogen-13 chemistry a method has been described for the bombardment of high pressure (∿ 15 atm) O_2 targets with protons to give $^{13}N-N_2$, $^{13}N_2O$, and $^{13}NO_2$ directly in a ratio of 8:1:3.5 (15). A target system involving the utilization of the $^{16}O(p,\alpha)^{13}N$ reaction in water for preparing $^{13}NH_3$, $^{13}N-N_2$, $^{13}NO_2-$, and $^{13}NO_3-$ by subsequent steps has also been described (16). The preparation of $^{13}NH_3$ via reduction of the oxides of nitrogen by De Varda's alloy (17) is a convenient method for $^{13}NH_3$ preparation in good yield. Nitrous oxide ($^{13}N-N_2O$) can be prepared starting with $^{13}NO_3-$ from the proton bombardment of water (18).

Oxygen-15 water is usually prepared by exchange of $^{15}O-CO_2$ with water. A direct recoil synthesis has been recently described involving the deuteron bombardment of H_2-N_2 mixtures which yields a very high specific activity product one minute after end of bombardment (19).

A patent (20) issued on the preparation of anhydrous fluorine-18 gives no experimental detail on the actual gas composition. New targetry and methodology for the preparation of on-line high specific activity anhydrous $^{18}F-F_2$ has been developed (21) utilizing the $^{20}Ne(d,\alpha)^{18}F$ reaction in high pressure neon targets containing scavenger F_2. A method for "carrier-free" $H^{18}F$ utilizing the $^{20}Ne(^{3}He,\alpha n)^{18}Ne$; $^{18}Ne \xrightarrow[\text{sec}]{1.5} {}^{18}F + \beta^{+}$ reaction in a $Ne-H_2$ mixture similar to previously described deuteron bombardments of $Ne-H_2$ mixtures has been published (22). The preparation of ^{18}F-DAST (diethylaminosulfur trifluoride) (23), ^{18}F-trifluoromethyl hypofluorite (24), and ^{18}F labeled $Cl^{18}F$ (25) have also been described.

There is little question that with the increasing interest in positron emitters for labeling organic compounds more attention will have to be paid to developing on-line production methods for the needed synthetic precursors to say nothing of wholly automated methods for compound preparation. We can expect to see more work on the refinement of procedures now in use and the development of new approaches, especially those yielding precursor at EOB or very shortly thereafter, be they by recoil or radiation synthesis, excitation labeling, or by fast chemical synthesis directly following EOB.

THE CARRIER-FREE PROBLEM

One of the important advantages of the short lived nuclides is that if the specific activity used is high enough they will act as true tracers in that their administration to whatever living system will not elicit a measurable physiological response (drug action) which may or may not perturb the purpose to which the tracer is put. In the case of carbon-14 or tritium this condition of lack of concomitant physiological response cannot always be met even if it were possible to prepare the labeled compound in the true carrier free state. Table 4 lists pertinent facts that focus on the relationships between specific activities and other physical parameters in an intercomparison of the nuclides of carbon and fluorine. If one considers the ratio of half lives the greatly superior sensitivity possible with the short lived tracer becomes quickly apparent. If one considers that many biochemical, biological, and medical experiments are carried out at the millicurie level it can be seen that for carbon-14 weighable amounts of the nuclide are involved, but that this is not the case for carbon-11 and fluorine-18. When one now considers that 1 millicurie of carbon-11 in the carrier free state is only 6.52×10^{10} atoms it

TABLE 3. PRECURSORS FOR SYNTHESIS OF COMPOUNDS LABELED WITH POSITRON EMITTERS

Carbon-11

CO CO_2 CN^- CH_2O CH_3I CH_3OH

C_2H_2 C_2H_4 C_2H_5OH CH_3-COOH $NH_2-\overset{\displaystyle O}{\overset{\|}{C}}-NH_2$

Nitrogen-13

NO NO_2 N_2O N_2 HCN NH_3

Oxygen-15

O_2 H_2O CO CO_2

Fluorine-18

F_2 F^-

Salts of Fluorine

$(C_2H_5)_4N^+F^-$ or Crown Ether salts

NOF ClF CF_3OF $(C_2H_5)_2NSF_3$

SbF_3 $LiBF_4$ $FClO_3$ BiF_3 etc.

TABLE 4. CARBON, FLUORINE, AND TRITIUM AS TRACERS

	Carbon-11	Fluorine-18	Tritium	Carbon-14
Half-life	20.4 min	110 min	12.3 yrs	5700 yrs
Ratio of Specific Activities Carbon-14 = 1	1.5×10^8	2.7×10^7	464	1
Curies/mol (atomic state)	9.22×10^9	1.71×10^9	2.92×10^4	62.8
Atoms/Curie	6.52×10^{13}	3.52×10^{14}	2.06×10^{19}	9.6×10^{21}

becomes very quickly apparent that the meaning of carrier free implies a tiny amount of material in an environment where the same material in stable (i.e. carbon-12) form may exist at levels many orders of magnitude higher than the radioactive material and still be undetectable by the usual analytical methods. To cite two examples of the problems in producing carrier free substances we can consider the preparation of $H^{11}CN$ (26) and the preparation of $H^{11}CHO$ (27) where it was possible to set upper limits on the available specific activity. It should be mentioned that no one to date has devised an experimental method that will define the proximity to the true carrier free state for carbon-11, nitrogen-13, oxygen-15, or fluorine-18 in quantitative terms when the dilution of the neat nuclide does not exceed one or two orders of magnitude. In the case of $H^{11}CN$ from $^{11}CH_4$ it was possible to obtain a maximum specific activity of 2000 Ci/mmol or a dilution of ca 5000 (26) and in the case of $H^{11}CHO$ from $^{11}CO_2$ it was possible to obtain a specific activity of 600 Ci/mmol or a dilution of ca 15,000 (27). In each case the stable carbon-12 came from traces of carbon-12 in the bombardment gas (consider for example the problem of removal of 1 ppm of $^{12}CO_2$ or $^{12}CH_4$ from commercial N_2; or the removal of 1 ppm $^{12}CO_3^=$ from boron oxide) and in the equipment and chemicals used in the subsequent steps needed to prepare the $H^{11}CN$ or $H^{11}CHO$. It can be argued that these specific activities are sufficiently high to satisfy the requirements of any trace level study. However, with the new interest in receptor site research it can be shown that near carrier free compounds are required in order to carry on meaningful experiments in many cases.

It is our contention that the question of the relation of the specific activity of the compound to that in the true carrier free state be considered in each case where it has bearing. Indeed the use, or more properly, misuse of the term carrier free is all too evident in the literature. What is almost universally meant when carrier free is used is actually <u>no carrier intentionally added</u>. Since it is clearly difficult to quantitate the amount of carrier in many cases and since no facile method has as yet been devised for measuring actual specific activity of the final product when no carrier has been intentionally added, we propose that a uniform nomenclature be used which will indicate to the reader of the scientific literature precisely what the condition of the product is. Thus, we propose:

1. <u>CF</u> to mean <u>carrier free</u> with evidence provided that its use is justifiable.

2. <u>NCA</u> to mean <u>no carrier added</u> and to be used for the vast majority of syntheses reported today. This term should basically replace CF since it is a more accurate designation.

3. <u>CA</u> to mean <u>carrier added</u>. This should be applied to any product where carrier is intentionally added whether or not the specific activity has been determined.

While this point may seem highly technical it will be an important issue as work with these short lived nuclides proliferates and experiments become increasingly sophisticated. We believe the use of CF, NCA, and CA will help to clarify the status of the compound for the reader of a paper and for those who depend on the accuracy of published work.

DESIGN OF RADIOPHARMACEUTICALS

Two major and related approaches have been applied in compound design, namely, (1) the targeting of a radiopharmaceutical to a particular organ by taking advantage of its unique metabolism or function, and more recently (2) to the use of labeled substrates to map metabolic activity or primary events which lead to a biological response in the case of hormones or drugs. Although the distinction here is subtle both areas of endeavor, when carried out in a scientifically thorough way contribute not only to direct application but also to forming a logical basis upon which new ideas and applications may be launched.

This section will deal with examples of some of the concepts used in radiopharmaceutical design and descriptions of the studies which support them. In many cases rather thorough studies on structure:activity relationships and metabolism have been described.

Recently a number of organ specific radiopharmaceuticals have been reported. Their design was based on their ability to elicit some physiological or biochemical response in a specific organ and they include enzyme inhibitors, receptor antagonists, adrenergic neuron blockers, and metabolic substrates or their analogs. Many of these studies include _in vitro_ assays for determining factors responsible for the biodistribution of the various classes of compounds.

The successful therapeutic use of enzyme inhibitors to inhibit human adrenocortical function formed the basis for an investigation of these labeled inhibitors as agents for adrenal scanning. One of these (±)-2-(3-iodo-4-aminophenyl)-2-aminophenyl)-2-phenethylamine (an inhibitor of steroid 11 β-hydroxylase) labeled with ^{123}I (^{123}I-(±)-1) has been used to image dog adrenals (28).

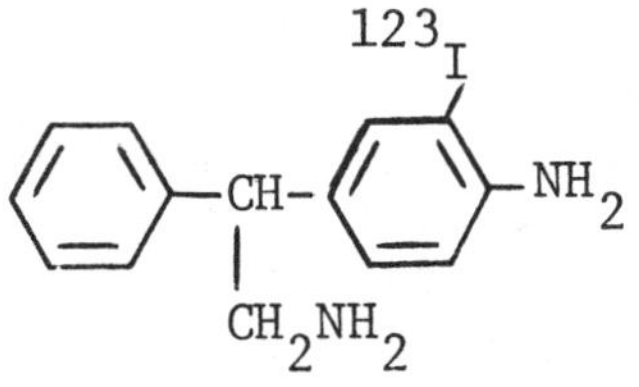

^{123}I-(±)-1

Resolution of (±)-1 into its optical isomers followed by radioiodination allowed a study of the stereoselectivity of adrenocortical enzymes. It was found that the (-)-1 enantiomer gave a 100% greater uptake in the adrenal cortex and a 3-fold increase in both the adrenal cortex/liver and adrenal cortex/medulla compared to the (+)-1 enantiomer demonstrating the possible clinical utility of a chiral radiopharmaceutical (29).

The possible use of radiolabeled cardioselective β-adrenergic antagonists as agents for imaging the myocardium has recently been investigated (30,31). These β-blockers would be predicted to localize in the myocardium by virtue of their specific and high affinity interaction with β-adrenergic receptors. Thus far the compounds studied are derivatives of the β-blockers, propranolol, practolol, and acetbutylol. Although a radioiodinated derivative of practolol showed favorable distribution characteristics for myocardial imaging, it was species dependent and in vivo displacement studies with propranolol showed the binding to be non-specific (32). A simple two-compartment equilibrium model has been developed in order to calculate the maximum target to blood ratio (33).

The observation that the presence of estrogen receptors in malignancy correlates with tumor remission after endocrine ablation suggested the importance of developing labeled ligands to bind to the estrogen-receptor and led to the investigation of several radioiodinated estrogen derivatives as candidates (34). These compounds were studied to determine their affinity for the estrogen receptor and it was found that one of these derivatives, iodohexestrol, binds to the cytosol estrogen-receptor protein. Its in vitro behavior was related to the in vivo distribution demonstrating the usefulness of the in vitro receptor assay system in predicting in vivo distribution.

Dog adrenal medullae have been imaged using ^{131}I-o-iodobenzyldimethyl-2-hydroxyethyl-ammonium (2) a radioiodinated analog of the adrenergic neuronal blocking agent bretylium (35). This study demonstrates the importance of the

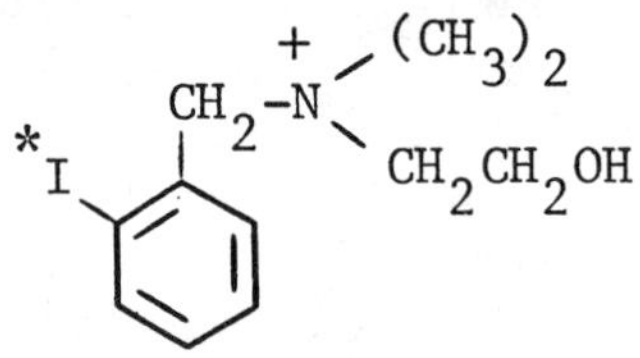

2

chemical structure of the cationic head of the molecule as well as the position of the iodine atom.

Since long chain fatty acids are a major energy source for the myocardium, their use for imaging the myocardium has received considerable attention (36). The successful use of ^{11}C-palmitic acid for this purpose has encouraged the development of analogs labeled with other radionuclides to extend the useful shelf life of the radiopharmaceutical. A recent study compared various long-chain fatty acids labeled with ^{11}C, ^{34m}Cl, ^{77}Br, and ^{123}I and evaluated their potential application to the measurement of myocardial metabolism in vivo. Comparative studies of the kinetics of accumulation and clearance from the heart muscle of mice showed that the extraction of ω-haloacids is more efficient than α-haloacids and that the uptake of ^{123}I-17-iodoheptadecanoic acid is nearly identical to that of ^{11}C-palmitic acid (37). A study of the chemical form of the radioactivity in the heart and blood after injection of this ω-iodo fatty acid showed that its metabolism is via β-oxidation with dehalogenation taking place only at the end (38).

The efficient incorporation of 5-fluorouridine into RNA forms the basis of another study in which ^{18}F-5-fluorouridine is used as a probe for measuring RNA synthesis in tumors. ^{18}F-5-fluorouridylate was found to be an effective precursor for RNA synthesis in mouse spleen and intestine as well as intracranial tumors (39).

In turning to the probing of metabolism in vivo ^{18}F-2-deoxy-2-fluoro-D-glucose has been shown to be an effective tracer for quantitatively mapping the first step of glycolysis in brain (40) and in heart (41,42). Factors responsible for the biodistribution of ^{18}FDG have been determined. Organs which show the highest uptake of activity (brain and heart) have the highest rates of phosphorylation *in vitro* and show all of the radioactivity present is ^{18}FDG-6-phosphate even at one minute post-injection. The rapid clearance of ^{18}FDG from liver, lungs, and kidneys and the retention by heart and brain is the result of metabolic trapping and reflective of glucose utilization (43).

An exciting new area of research in the design and synthesis of labeled tracers has been initiated by the discovery of the existence of discrete populations of neurotransmitter receptors in brain and realization of their importance in the understanding and therapy of mental disorders. It is apparent in examining the various problems encountered in studying receptor populations (44) that the use of receptor ligands labeled with positron emitters, because of their high specific activity and decay by body penetrating radiation,can offer unique possibilities for probing the dynamics of receptor populations in normal and disease states.

The dopamine/neuroleptic receptor has received considerable attention in recent years because its blockade is the one common denominator among the many drugs which show antipsychotic activity. Therefore, it is of considerable interest to develop labeled antagonists with a high specificity for the dopamine receptor which would allow mapping these receptor populations in the brain.

Many of the most potent antagonists of the dopamine receptor are derivatives of butyrophenones and have the following general structure. The majority of these also have a halogen in the aromatic ring. Variation in the $-NR_1R_2$ substituent changes the properties of the antagonist.

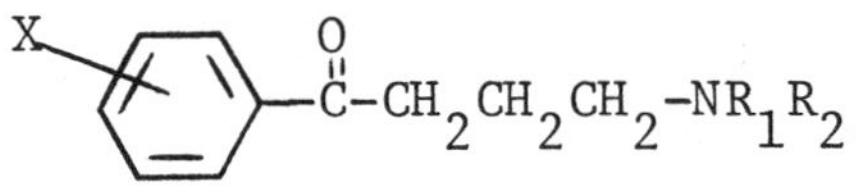

BUTYROPHENONE SKELETON

A general approach to the synthesis of ^{11}C-butyrophenone derivatives has been developed (45). It involves two steps using ^{11}C-cyanide as the isotopic precursor and allows the flexibility of varying the NR_1NR_2 substituent. The

$$R_1R_2NCH_2CH_2CH_2Cl + {}^{11}CN^- \longrightarrow R_1R_2NCH_2CH_2CH_2{}^{11}CN$$

$$\xrightarrow[2)\ H^+]{1)\ ArMgBr} R_1R_2NCH_2CH_2CH_2{}^{11}\overset{O}{\overset{\|}{C}}Ar$$

carbon-11 resides in the carbonyl moiety. Since the metabolism of molecules of this type has been shown to proceed via N-dealkylation followed by oxidation with ultimate excretion of the benzoyl moiety (46), the metabolized ^{11}C-butyrophenones (^{11}C-carbonyl) would be predicted to be rapidly excreted lowering the body background and radiation dosimetry associated with the use of this tracer.

This general synthesis of butyrophenones has been applied to the synthesis of ^{11}C-spiroperidol, a ligand with a high affinity for the dopamine receptor

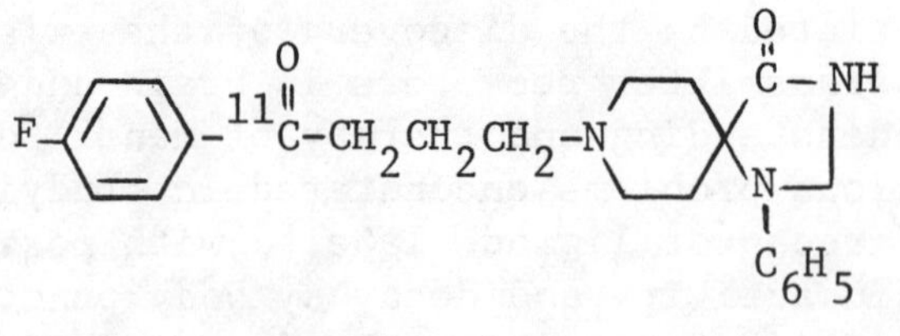

^{11}C-SPIROPERIDOL

(47). The application of this synthesis to other butyrophenones with different antagonist properties is also under investigation along with the parameters affecting <u>no carrier added</u> syntheses with multifunctional substrates.

NEW SYNTHESES OF RADIOPHARMACEUTICALS

Carbon-11: A new route to 2-deoxy-D-glucose has been developed (48) and applied to the synthesis of ^{11}C-2-deoxy-D-glucose (^{11}C-2-DG) (49). A particularly noteworthy feature of this sequence is the reduction of a nitrile

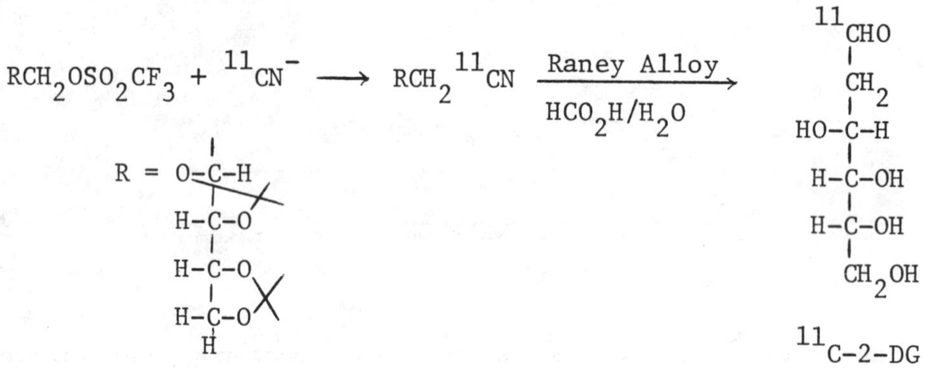

^{11}C-2-DG

to an aldehyde under no carrier added (NCA) conditions.

[1-^{11}C]-DL-valine has been prepared by a high temperature high pressure modification of the Bucherer-Strecker amino acid synthesis and evaluated for pancreas imaging (50). This material is produced at intervals of 1 hr or less in 100-200 mCi batches. Another amino acid, ^{11}C-DOPA (β-(3,4-dihydroxyphenyl)-D,L-α-alanine-1-^{11}C) has been synthesized by carboxylation of an α-lithio-isocyanide (51).

The preparation of ^{11}C-methyl labeled 1,1'-dimethyl-4,4'-dipyridinium diiodide (^{11}C-paraquat) has been accomplished using ^{11}C-iodomethane for use in *in vivo* studies of pulmonary toxicology (52).

In an unusual example of biosynthetic application ^{11}C-hippuric acid was prepared from ^{11}C-benzoic acid and glycine using rat liver mitochondria (53). The incorporation of carbon-11 into pyruvic acid, lactic acid, and L-alanine via enzymatic synthesis from $^{11}CO_2$ has been reported (54). Lactic acid and lactonitrile have also been prepared synthetically (55).

Fluorine-18: The formation of carbon-fluorine bonds is a difficult synthetic problem which is magnified when one attempts to apply known methods of carbon-fluorine bond formation to ^{18}F-labeling where it is desirable to use little or no carrier fluorine. The chemical forms of ^{18}F available for labeling can be classified as nucleophilic fluorine or electrophilic fluorine and the problems encountered with each of these forms are predictable based on the organic chemistry of fluorine.

With nucleophilic fluorine, displacement reactions are usually envisioned

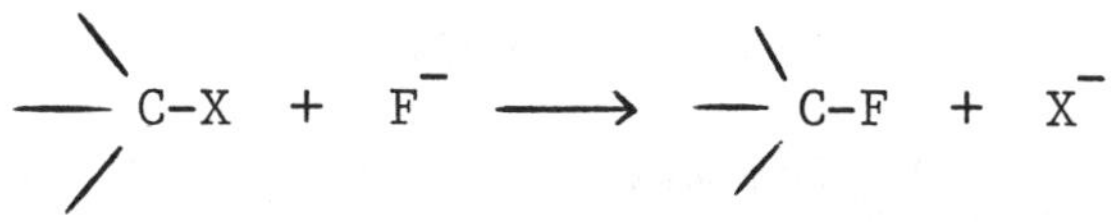

for C-F- bond formation. The success of a displacement reaction is strongly affected by the presence or absence of water in the reaction mixture because the high hydration energy of fluoride causes the energetics of C-F bond formation to be very unfavorable in protic media (56).

Several approaches to increasing the reactivity of fluoride ion have been applied to labeling with ^{18}F. The use of anhydrous $H^{18}F$ from the Ne + H_2 target is the most successful and has been applied to the synthesis of ^{18}F-3-deoxy-3-fluoro-D-glucose (57). In other studies where $^{18}F^-$ from the water target is

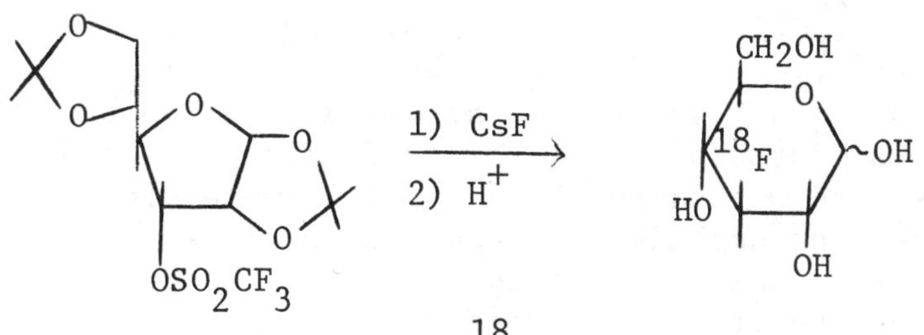

^{18}F-3-Deoxy-3-Fluoro-D-Glucose

used, the reactivity of the ^{18}F-fluoride (after careful drying) is increased by using hindered cations or by using crown ethers. For example, ^{18}F-6-deoxy-6-fluoro-α-D-galactopyranose has been prepared using tetraethylammonium fluoride-^{18}F in acetonitrile (58). The use of $K^{18}F$ and 18-crown-6 has been applied to the synthesis of ^{18}F-21-fluoroprogesterone (59,60), as well as to the ^{18}F for F exchange on benzotrifluoride (61). ^{18}F-β-D-glucosyl fluoride has been prepared by the reaction of tetra-O-acetyl-α-D-glucosyl bromide with $Ag^{18}F$ in acetonitrile (62).

Electrophilic fluorination using $^{18}F_2$ has found continued application in the synthesis of ^{18}F-2-deoxy-2-fluoro-D-glucose (63). The use of $^{18}F_2$ is considerably simplified by purchasing a premixed cylinder of 1% F_2 in neon as the source of F_2 carrier in the target gas. This is far safer than handling pure fluorine and eliminates the need for a high vacuum line for measuring small quantities of pure F_2.

Nitrogen-13: A procedure has been described for the preparation of millicurie quantities of ^{13}N-bis(2-chloroethyl)nitrosourea labeled in the nitroso group (64).

TECHNOLOGY

The application of β^+-emitters to many areas of research has generated considerable interest in the physical set-up of cyclotrons and adjacent chemistry laboratories which are capable of safely handling Curie levels of short-lived nuclides. Therefore, some of the important aspects of laboratory design, shielded work areas, radioactivity monitoring and experimental set-ups which have proved workable from our experience and which often use commercially obtainable or easily constructed components are summarized below.

The spatial relationship of the cyclotron vault to the chemistry laboratory is an important one, in that the delivery of target gases or liquids is generally most easily accomplished when the distance between the target and the area where the radionuclide is to be used is as short as possible. When the chemistry lab is adjacent to the vault, there is, of course, a heavy shielding wall between these areas with strategically placed ports to contain delivery lines.

Target gases as well as liquid or solid forms of labeled precursors are most safely and conveniently handled in a shielded hood or some form of hot cell. The simplest shielded work area can be obtained by modifying a commercially available laboratory hood. After providing structural support of the floor of the hood,lead plate is placed on the floor of the hood. Along the entire front face of the hood a short wall of lead bricks is built up. On the top of the wall a high density (6.2 gm/cc) lead glass window (65) (16.5 in high x 16.5 in wide) is installed on a track so that it slides across the front face of the hood. This provides a reasonably well shielded work area, especially when modular shielding is used within the hood. For example, lead shielding of various traps in which the nuclide is trapped as well as small lead shields for various reaction vessels, when used inside the modified commerical hood provides reasonably inexpensive and effective protection for manipulations which need to be carried out during synthesis. The use of remotely controlled equipment during synthesis as well as extension tools for performing various manipulations is also very effective. The automation of a radiopharmaceutical synthesis has recently been reported and includes all steps of precursor preparation, compound preparation, chromatography, and sterilization. The facility is enclosed within a hot cell (66).

Monitoring the amount of radioactive precursor delivered from a target to the synthesis laboratory is conveniently carried out using an ionization chamber with digital readout (67) in a shielded fixed geometry. This allows one to measure the accumulated radioactivity at any time during target gas delivery. Such a system is in use in our $^{18}F_2$ production. Monitoring the distribution of radioactivity during synthesis is conveniently done using a shielded ionization chamber with a large ($\sim$ 2 7/8 inch) diameter well. This unit is installed in the shielded hood already described and allows direct measurement of radioactivity of different vessel sizes and shapes (67).

CONCLUSION

This short paper was intended to update current trends and directions in the design and preparation of organic radiopharmaceuticals, particularly those labeled with positron emitters. Associated concerns such as nuclide production, specializedtechnology, and a presentation of problems engendered by the increasing need for carrier free or near carrier free materials were also addressed. The debate over whether or not accelerator produced radionuclides will impact in a major way on health care will no doubt continue but it is now becoming clear that the discussion is no longer based on utility but rather on economics. The coming years will surely bring major advances in cyclotron technology, focussing mostly on simplicity and reliability of operation, automated (computerized) control, and hopefully reduced cost in the sense of cost per watt of beam. Automation of nuclide production and ultimately radiopharmaceutical production (black box approach) according to sound engineering practice is a concurrent goal. While it is not feasible to suggest an accelerator in every hospital it is worth considering regional centers based on levels of activity and types of nuclides required. Oxygen-15, nitrogen-13, potassium-38, and probably carbon-11, etc. i.e. nuclides with half lives of less than one hour and the compounds in which they are incorporated will have to be produced at or near the site of use. However, nuclides such as fluorine-18 and its compounds could serve a wider geographic area. We have not addressed the utility of generator produced nuclides since they are covered in detail in other papers at this symposium.

Some accelerator produced radionuclides, especially ^{67}Ga, ^{201}Tl, and ^{81}Rb-^{81m}Kr in the U.S. (and ^{123}I in Europe) have already impacted in a major way on health care. The utility of the simple gases containing ^{13}N, ^{15}O, and ^{11}C and labeled organic compounds, metallo-organics, and organic complexes are now receiving intensive study worldwide with notable breakthroughs in the area of *in vivo* metabolic studies and the flow-metabolism couple. Thus, one can expect to see major advances in the coming years with accelerator nuclides taking their place in the pursuit of basic biochemical and biological research and in the armamentarium of nuclear medicine practice.

ACKNOWLEDGEMENTS

Research carried out at Brookhaven National Laboratory under contract with the U.S. Department of Energy and supported by its Office of Basic Energy Sciences and Office of Biomedical and Environmental Research.

REFERENCES

1. Symposium, Copenhagen, "Radiopharmaceuticals and Labelled Compounds", Vol I and II, IAEA, Vienna, 1973, cf list in A. P. Wolf, D. R. Christman, J. S. Fowler, and R. M. Lambrecht, Synthesis of Radiopharmaceuticals and Labelled Compounds Using Short-lived Isotopes, p. 345, Vol I.

2. Symposium, "Radiopharmacology", CRC Press publication, in press, 1979, cf B. M. Gallagher, J. S. Fowler, R. R. MacGregor, R. M. Lambrecht, A. P. Wolf, E. J. Crawford, and M. Friedkin, "The In Vivo Measurement of Transport and Metabolic Processes Using Radiotracers" and other articles in this publication.

3. First (Brookhaven, 1976) and Second (Oxford, 1978) International Symposium on Radiopharmaceutical Chemistry, Abstracts of these Symposia can be found in J. Lab. Comp. and Radiopharm. 13, 1976 and 16, 1979.

4. Radiopharmaceuticals and Other Compounds Labelled with Short Lived Nuclides M. Welch, ed., Oxford, Pergamon Press, 1977.

5. Tubis, M. and Wolf, W.: Radiopharmaceuticals, New York, Wiley Interscience, 1976.

6. "Accelerator Produced Nuclides for Use in Biology and Medicine: A Bibliography", 1939-1973, TID4500-BNL-50448, Vol I., compiled by D. R. Christman and K. Karlstrom, available from National Technical Information Service, U.S. Department of Commerce, 5285 Port Royal Road, Springfield, Virginia 22161.

7. Ibid, 1974-June, 1976, TID4500-BNL-50448, Vol. II., UC-23.

8. (a) Brownell, G. L., Correa, J. A., and Zamenhof, R. G.: Recent Advances in Nuclear Medicine, Lawrence, J. H. and Budinger, T. F., eds., Vol. 5, Grune and Stratton, 1977; (b) Ter-Pogossian, M. M.: Sem. in Nucl. Med. 7: 109-128 (1977); (c) Budinger, T. F.: Sem. in Nucl. Med. 7: 285-298 (1977), Phelps, M., Hoffman, E. J., and Kuhl, D. E.: ref. 9, Vol. I., 233-253.

9. cf Wolf, A. P. in Symp. Los Angeles, "Medical Radionuclide Imaging", p. 343-353, Vol. I., IAEA, Vienna, 1977 for some other facets of this problem.

10. Clark, J. C. and Buckingham, P. D." Short Lived Radioactive Gases for Clinical Use, Boston, Butterworth, 1975.

11. Winstead, M. B., Chern, C. I., Lin, T. H., et al.: "Synthesis and preliminary scintigraphic evaluation of in vivo distribution of ^{11}C-hydroxyurea/isohydorxyurea and ^{11}C-cyanate", J. Appl. Radiat. Isot. 29: 443-447 (1978).

12. Brinkman, G. A., Hass-Lisewska, I., Veenboer, J. T. et al.: "Preparation of $^{11}COCl_2$", J. Appl. Radiat. Isot. 29: 701-702 (1978).

13. Stöcklin, G. and Wagner, R., Kernforschungsanlage, Julich, Germany, Private Communication, G. Stöcklin, 1979.

14. Rössler, K., Stöcklin, G., and Wagner, R., Ibid, Private Communication, K. Rössler, 1979.

15. Parks, N. J., Peek, N. F., and Goldstein, E.: "The synthesis of ^{13}N labeled atmospheric gases via proton irradiation of a high pressure oxygen target", J. Appl. Radiat. Isot. 26: 603-687 (1975).

16. Parks, N. J. and Krohn, K. A.: "The synthesis of ^{13}N labeled ammonia, dinitrogen, nitrile, and nitrate using a single cyclotron target system", J. Appl. Radiat. Isot. 29: 754-756 (1978).

17. Vaalburg, W., Kamplinis, J. A. A., Beerling, D. et al.: "An improved method for the cyclotron production of ^{13}N-labeled ammonia", Int. J. Appl. Radiat. Isot. 26, 316-318 (1975).

18. Nickles, R. J., Gatley, S. J., Hichwa, R. D. et al.: "The synthesis of ^{13}N labeled nitrous oxide", J. Appl. Radiat. Isot. 29: 225-227 (1978).

19. Ruiz, H. V. and Wolf, A. P.: "Direct synthesis of oxygen-15 labeled water at high specific activities", J. Label. Comp. and Radiopharm. 15, 185-189 (1978).

20. Winchell, H. S., Wells, D. K., Lamb, J. F. et al.: "Process for preparing fluorine-18", U.S. Patent 3,981,769, Sept. 21, 1976.

21. Casella, V. R., Ido, T., Ruth, T., et al.: "Anhydrous $^{18}F-F_2$ for Radiopharmaceutical Preparation" paper in preparation.

22. Crouzel, C. and Comar, D.: "Production of carrier-free ^{18}F-hydrofluoric Acid", J. Appl. Radiat. Isot. 29: 407-408 (1978).

23. Straatman, M. G. and Welch, M. J.: "Fluorine-18 labeled diethylaminosulfur trifluoride (DAST): An F-for-OH fluorinating agent", J. Nucl. Med. 18: 151-158 (1977).

24. Neirinckx, R. P., Lambrecht. R. M.. and Wolf. A. P.: :Cyclotron Isotopes and Radiopharmaceuticals. XXV. An anhydrous ^{18}F-fluorinating intermediate: Trifluoromethyl hypofluorite", J. Appl. Radiat. Isot. 29: 323-327 (1978).

25. Lambrecht, R. M., Neirinckx, R. P., and Wolf, A. P.: "Cyclotron Isotopes and Radiopharmaceuticals. XXIII. A Novel Anhydrous ^{18}F-fluorinating Intermediate", J. Appl. Radiat. Isot. 29, 175-183 (1978).

26. Christman, D. R., Finn, R. D., Karlstrom, K. I., et al.: "The Production of ultra high activity ^{11}C-labeled hydrogen cyanide, carbon dioxide, carbon monoxide, and methane via the $^{14}N(p,\alpha)^{11}C$ reaction. XV." Int. J. Appl. Radiat. Isot. 26: 435-442 (1975).

27. Berger, G., Maziere, M., Sastre, J. and Comar, D.: Service Hospitalier Frederic Joliot, Orsay, France, Private Communication, D. Comar, 1978.

28. Beierwaltes, W. H., Wieland, D. M., Mosley, S. T. et al.: "Imaging the adrenal glands with radiolabeled inhibitors of enzymes: Concise communication", J. Nucl. Med. 19: 200-203 (1978).

29. Wu, J., Wieland, D. M., Beierwaltes, W. H. et al.: "Radiolabeled enzyme inhibitors -- enhanced localization following enantiomeric purification", J. Label Comp and Radiopharm 16: 6 (1979).

30. Jiang, V. W., Gibson, R. E., Rzeszotarski, W. J. et al.: "Radioiodinated derivatives of adrenoceptor blockers for myocardial imaging", J. Nucl. Med. 19: 918-924 (1978).

31. Hanson, R. N., Holman, B. L., and Davis, M. A.: "Synthesis and Biologic Distribution of Radioiodinated β-Adrenergic Antagonists", J. Med. Chem. 21: 830-833 (1978).

32. Rzesotarski, W. J., Eckelman, W. C., Gibson, R. E. et al.: "Beta Adrenergic Blocking Agents for Myocardial Imaging", J. Label Comp. and Radiopharm. 16: 72-73 (1979).

33. Eckelman, W. C., Gibson, R. E., Rzeszotarski, W. J. et al.: "The design of receptor binding radiopharmaceuticals", J. Label Comp and Radiopharm 16, 4-5 (1979).

34. Komai, T., Eckelman, W. C., Johnsonbaugh, R. E. et al.: "Estrogen derivatives for the external localization of estrogen dependent malignancy" J. Nucl. Med. 18: 360-366 (1977).

35. Wieland, D. M., Swanson, D. P., Brown, L. E. et al.: "Imaging the adrenal medulla with an I-131 labeled antiadrenergic agent", J. Nucl. Med. 20: 155-158 (1979).

36. Weiss, E. S., Hoffman, E. J., Phelps, M. E. et al.: "External detection and visualization of myocardial ischemia with ^{11}C-substrates in vitro and in vivo", Circ. Res. 39: 24-32 (1976).

37. Machulla, H. J., Stöcklin, G., Kupfernagel et al.: "Comparative evaluation of fatty acids labeled with C-11, Cl-34m, Br-77, and I-123 for metabolic studies of the myocardium: Concise communication", J. Nucl. Med. 19: 298-302 (1978).

38. Machulla, H. J., Kupfernagel, Ch., Stöcklin, G. et al.: "Comparative studies of ^{123}I-fatty acids in myocardial metabolism; development of ω-^{123}I-heptadecanoic acid to clinical application for in vivo diagnosis", J. Label Comp. and Radiopharm 16: 75-76 (1979)

39. Crawford, E. J., Friedkin, M., Fowler, J. S. et al.: "5-[F-18]fluoro-uridylate as a probe for measuring RNA synthesis and tumor growth rates in vivo", J. Nucl. Med. 19: 702 (1978).

40. Reivich, M., Kuhl, D., Wolf, A. P. et al.: "The [^{18}F]fluorodeoxyglucose Method for the measurement of local cerebral glucose utilization in man", Circ. Res. 44: 127-137 (1979).

41. Phelps, M. E., Hoffman, E. J., Selin, C. et al.: "Investigation of [^{18}F] 2-fluoro-2-deoxyglucose for the measure of myocardial glucose metabolism", J. Nucl. Med. 19: 1311-1319 (1978).

42. Gallagher, B. M., Ansari, A., Atkins, H. L. et al.: "Radiopharmaceuticals XXVII: ^{18}F-Labeled 2-deoxy-2-fluoro-D-glucose as a radiopharmaceutical for measuring regional myocardial glucose metabolism in vivo: Tissue distribution and imaging studies in animals", J. Nucl. Med. 18: 990-996 (1977).

43. Gallagher, B. M., Fowler, J. S., Gutterson, N. I. et al.: "Metabolic trapping as a principle of radiopharmaceutical design: Some factors responsible for the biodistribution of [^{18}F] 2-deoxy-2-fluoro-D-glucose", J. Nucl. Med. 19: 1154-1161 (1978).

44. Snyder, S. H., U'Prichard, D. C., and Greenberg, D. A.: "Neurotransmitter receptor binding in brain" in Psychopharmacology: A Generation of Progress, Lipton, M. A., DiMascio, A., and Killam, K. F. eds., New York, Raven Press, 1978, 361-370.

45. Fowler, J. S. and Wolf, A. P.: work in progress.

46. Soudijn, W.,Van Wijngaarden, W. and Allenwijn, F.: "Distribution, excretion, and metabolism of neuroleptics of the butyrophenone type", Eur. J. Pharmacol. 1: 47-57 (1967).

47. Kuhar, M. J., Murrin, L. C., Malouf, A. T. et al.: "Dopamine receptor binding in vivo: The feasibility of autoradiographic studies", Life Sci. 22: 203-210 (1978).

48. Shiue, C.-Y., MacGregor, R. R., Lade, R. E. et al.: "A new synthesis of 2-deoxy-D-glucose", Carbohydrate Res., in press.

49. Shiue, C.-Y., MacGregor, R. R., Lade, R. E. et al.: "The synthesis of 1-C-11-2-deoxy-D-glucose for measuring regional brain glucose metabolism in vivo", J. Nucl. Med. 19: 676-677 (1978).

50. Washburn, L. C., Wieland, B. W., Sun, T. T. et al.: "[1-^{11}C]-DL-valine, A potential pancreas-imaging agent", J. Nucl. Med. 19: 77-83 (1978).

51. Reiffers, S., Beerling-Van der Molen, H. D., Vaalburg, W. et al.: "Rapid synthesis and purification of carbon-11 labeled DOPA. A potential agent for brain studies", Int. J. Appl. Radiat. Isot. 28: 955-958 (1977).

52. Palmer, A. J.: "The preparation of ^{11}C-methyl labelled 1,1'-dimethyl-4,4'-dipyridinium diiodide", J. Label Comp and Radiopharm 14: 27-33 (1978).

53. Gatley, S. J., Crawford, J. S., Halama, J. R. et al.: "Synthesis of ^{11}C-hippuric acid from ^{11}C-benzoic acid and glycine using rat liver mitochondria" Oxford abstr. # A21, 1978.

54. Cohen, M. B., Spolter, L., Chang, C. C. et al.: "Enzymatic synthesis of C-11 pyrivic acid, lactic acid, and L-alanine", J. Nucl. Med. 19: 701 (1978).

55. Winstead, M. B., Chern, C., Tz-Hong, L. et al.: "Synthesis and Preliminary scintigraphic evaluation of in vivo distribution of ^{11}C-lactic acid and ^{11}C-lactonitrile", Int. J. Appl. Radiat. Isot. 29: 69-73 (1978).

56. Sheppard, W. A. and Sharts, C. M.: Organic Fluorine Chemistry, New York W. A. Benjamin, Inc., 1969, p. 16.

57. Tewson, T. J., Welch, M. J., Raichle, M. E.: "[^{18}F]-labeled 3-deoxy-d-fluoro-D-glucose: Synthesis and preliminary biodistribution data", J. Nucl. Med. 19: 1339-1352 (1978).

58. Christman, D. R., Orhanovik, Z., Shreeve, W. W. et al.: "The synthesis of 6-deoxy-6-^{18}F-α-D-galactopyranose", J. Label. Comp. and Radiopharm. 8: 555-559 (1977).

59. Irie, T., Fukushi, K., Ido, T.: "^{18}F-fluorination by K^{18}F-crown ether system", Oxford Abstr. #8.

60. Spitznagel, L. A., Marino, C. A., and Erg, R. R.: "A Study on the use of a crown-ether to synthesize a fluorine-18 labeled steroid", WFNMB, 1978, p.71.

61. Ido, T., Irie, T., and Kasida, Y.: "Isotope Exchange with ^{18}F on a Superconjugate system", Oxford Abstr. #A5.

62. Lemire, A. and Reed, M. F.: "Preparation of ^{18}F-β-D-Glucosyl Fluoride", J. Label Comp. and Radiopharm. 15: 105-109 (1979).

63. Ido, T., Wan, C.-N., Casella, V. et al.: "Labeled 2-deoxy-D-glucose analogs. ^{18}F-labeled 2-deoxy-2-fluoro-D-glucose, 2-deoxy-2-fluoro-D-mannose, and ^{14}C-2-deoxy-2-fluoro-D-glucose", J. Label Comp. and Radiopharm. 14: 175 (1978).

64. Pettit, W. A., Tilbury, R. S., Digenis, G. A. et al.: "A convenient synthesis of ^{13}N-BCNU", J. Label Comp. and Radiopharm. 13: 119-122 (1977).

65. High density lead glass windows are available from Penberthy, Division of Nuclear Pacific Corp., Seattle, Washington.

66. Berger, G., Maziere, M., and Comar, D.: "An example of a completely automatic method to synthesize a radiopharmaceutical molecules: ^{11}C-imipramine", J. Label Comp. and Radiopharm. 16: 97-99 (1979).

67. Available from Capintec, Inc., Mountvale, New Jersey.

DESIGN AND DEVELOPMENT OF Te-123m RADIOPHARMACUETICALS

Garo P. Basmadjian, Rodney D. Ice, Robert A. Magarian,
George R. Parker, Stanley L. Mills, and Alan S. Kirschner

College of Pharmacy, The University of Oklahoma at the Health
Sciences Center, Oklahoma City, OK. 73190

Tellurium-123m has favorable physical characteristics, (159 keV γ isomeric transition photons, 84% abundance, $t_{1/2}$ 119 d) and forms in many instances stable C-Te bonds, allowing for the synthesis of numerous compounds containing a γ-emitting radionuclide without affecting the parent molecule's physiological characteristics.

The development of a quick method for solubilizing Te-123m metal into two inorganic nucleophiles in aqueous solution, i.e. NaHTe and Na_2Te_2, has facilitated the handling of Te-123m metal and the synthesis of new Te-123m radiopharmaceuticals.

Nucleophiles with the structure $R\text{-}Te^-Na^+$ were prepared by the reaction of alkyl halides with NaHTe. Alkyl halides reacted with Na_2Te_2 produced stable, colored dialkyl ditellurides which could be reduced *in situ* to the desired sodium alkyl telluride nucleophiles. The reaction of these Te-123m labeled organic nucleophiles with short and long chain alkyl halides enables the synthesis of Te-123m fatty acid analogs.

INTRODUCTION

The synthesis of radiopharmaceuticals emitting gamma radiation has utilized both inorganic and organic synthetic schema. The inorganic procedures used have been either chelation, complexation or coordinate compound formation of the pharmaceutical with a gamma-emitting radionuclide, e.g. Tc-99m. The organic synthetic procedures have involved the synthesis of a radiopharmaceutical containing a covalently bonded gamma-emitting radionuclide, e.g. I-131 or Se-75.

The labeling of a potential pharmaceutical with a radionuclide has been called "foreign labeling" and can be attempted in two ways: a) the exchange of an existing isotope in the molecule with its radioactive counterpart or b) the covalent bonding of the radionuclide within the synthetic schema, preferably at the latest possible synthetic step to prevent undue radioactive product handling, thus decreasing radiation exposure and minimizing product decay.

In 1975, we reported (1) a simple method for introducing Se-75 into an organic molecule and applied this procedure to the synthesis of Se-75-6B-benzyl seleno methyl-19 norcholest-5(10)-en-3B-ol (2). Although Se-75 has better physical properties than I-131 and is a "versatile" radionuclide, an even better radionuclide is tellurium-123m. Both Se-75 and Te-123m have similar half lives ($t_{1/2}$ 120 and 119 days, respectively), but Te-123m decays by isomeric transition, emitting a single 159 keV gamma photon (84% abundance), and does not emit beta particles. Thus on a per disintegration basis, Te-123m will give a smaller absorbed radiation dose and a greater number of useful photons than either I-131 or Se-75. In addition, current nuclear medicine instrumentation has been optimized for 100-200 keV photons and thus would not need any special adaptation for imaging Te-123m radiopharmaceuticals.

Tellurium being isosteric with sulfur, is capable of giving stable C-Te bonds, and allows the radiochemist to synthesize numerous compounds containing the Te atom without affecting the parent compound's physical characteristics as is usually the case with radiopharmaceuticals chelated with technetium.

Tellurium chemistry applicable to the synthesis of radiopharmaceuticals is sparse. In addition to presentations by our group (3,4), scientists at Oak Ridge Laboratories (5), Amersham (6) and a patent issued in the Netherlands (no actual Te-123m work performed) (7) are the only known works on Te-123m radiopharmaceuticals.

The design and development of new Te-123m labeled radiopharmaceuticals is an extension of our knowledge acquired with Se-75. As with selenium, an easy method was needed for the solubilization of Te-123m metal into inorganic nucleophiles before Te-123m labeled organic nucleophiles could be synthesized, followed by the introduction of Te-123m into the desired radiopharmaceuticals. We have developed such a facile method for tellurium (8) and used the method to synthesize Te-123m labeled fatty acids and steroids.

MATERIALS AND METHODS

Stable metallic (grayish-black powder) tellurium and metallic Te-123m (maximum specific activity 1 mCi/mg) were used in the synthesis outlined below. All reactions were performed in an inert atmosphere of N_2 gas to avoid the decomposition of reactive inorganic and organic intermediate nucleophiles.

Solubilization of Te-123m in water

Production of NaHTe-(Te-123m): 12.7 mg (0.1 mmole) of metallic tellurium-123m was weighed into a screw capped vial, assayed in a dose calibrator and transferred quantitatively into a 25 ml 3-necked flask with about 2 ml of distilled water. The flask was fitted with a water cooled condenser. Nitrogen gas was introduced through a side arm. The exiting nitrogen gas was allowed to bubble through an I_2-KI aqueous solution to trap any released volatile Te-123m compounds. The other side arm was capped with a rubber septum. Heating of the flask was started and 10 mg of $NaBH_4$ (0.25 mmole) in 1 ml of distilled water was added via a syringe through the rubber septum. Heating to reflux was continued until the stirred suspension of the Te acquired a violet tinge which, within 15 minutes, turned to a colorless, clear solution. The heat was then removed and the aqueous pH 8 solution of NaHTe-Te-123m was ready for the next reaction.

Production of Na_2Te_2-(Te-123m): To 6.35 mg. of metallic Te-123m (0.05 mmole), weighed, radioassayed and transferred as described above, was added 4 mg of $NaBH_4$ (0.1 mmole) in 1 ml of distilled water. This stirred slurry was heated at reflux until a clear, colorless solution was obtained. To this was added a suspension of another 0.05 mmole of Te-123m (6.35 mg) in 1 ml of 1% NaOH via a syringe. Immediately, the solution acquires a violet-purple tinge as the added Te metal starts to dissolve, and within 5 minutes with continuous stirring and heating to reflux, a clear dark purple solution is obtained. The species produced is disodium ditelluride, in a pH 12 aqueous solution.

Synthesis of Te-123m Sodium Alkyl Tellurides

This procedure applies to aliphatic sodium tellurides. An alkyl halide, primary or secondary, when added (in alcohol or THF) in equimolar amount to the radioactive solution of the NaHTe in water, followed by basifying the solution to pH 12 with NaOH, produces the radioactive sodium alkyl telluride. The sodium alkyl tellurides were not isolated as such but were allowed by air oxidation to form the more stable dialkyl ditelluride or derivatized to a stable dialkyl telluride.

Two examples, the preparation of isopropyl sodium telluride isolated as diisopropyl ditelluride and the preparation of hexadecanyl sodium telluride isolated as dihexadecanyl telluride are outlined below.

Synthesis of Isopropyl-Te^-Na^+-(Te-123m): The Te-123m metal (6.3 mg, 0.05 mmole, 1.2 mCi) was solubilized to NaHTe-(Te-123m) as previously described. To the room temperature solution of NaHTe-(Te-123m) was added 0.5 ml of isopropyl bromide in peroxide-free tetrahydrofuran (6.8 mg, 0.055 mmole). The solution was stirred for 15 minutes at room temperature, and 10 ml of ethyl ether added. This mixture was stirred (exposed to air) to extract the yellow diisopropyl ditelluride. The ether layer was collected, dried over anhydrous $MgSO_4$, reduced in volume with a stream of N_2 and layered on top of a 1 x 10 cm column of dry silica gel for chromatography and eluted with hexane. Over 90% of the radioactivity was contained in a yellow band that was collected and chromatographed on silica gel (Eastman) with hexane as solvent to show a single radioactive band yellow in color with an Rf of 0.5 corresponding to "cold" diisopropyl ditelluride. The radioactive yield was 935 μCi.

The isopropyl-Te^-Na^+, needed for further reactions, was prepared from the ditelluride by reduction in alcohol with $NaBH_4$ under a nitrogen atmosphere.

Preparation of Hexadecanyl-Te^-Na^+-(Te-123m): The Te-123m metal (12mg, 0.1 mmole, 0.8 mCi) was solubilized to NaHTe-(Te-123m) as previously described. To the room temperature solution of NaHTe-(Te-123m) was added 0.11 mmole of 1-iodohexadecane (39 mg) in 0.5 ml of an ethanol-THF (50:50) solution. This solution was stirred for 15 minutes and 0.15 mmole of NaOH in water (0.1 ml) was added followed by 0.25 mmole of 1-iodohexadecane. The clear solution was stirred for 30 minutes at room temperature followed by the addition of 10 ml of ethyl ether. This mixture was stirred and the clear colorless ether layer removed and dried over anhydrous $MgSO_4$. Greater than 95% of the radioactivity was accounted for in the ether layer.

An aliquot chromatographed on silica gel (Eastman) with $CHCl_3$ as solvent showed a single radioactive band with an Rf of 0.8.

Upon evaporation the ether solution deposited white needles of dihexadecanyl telluride-Te-123m. The radioactive yield in the crystals was 705 μCi. No attempt was made to prove the structure by the synthesis of the "cold" compound as all other possible compounds were ruled out by TLC chromatography using known "cold" compounds.

Synthesis of Dialkyl Ditelluride-Te-123m via Na_2Te_2

Using the Na_2Te_2-(Te-123m) aqueous solution, short chain aliphatic dialkyl ditellurides, e.g., dimethyl ditelluride (CH_3-Te-Te-CH_3), diisopropyl ditelluride $(CH_3)_2CH$-Te-Te-$CH(CH_3)_2$ as well as long chain dialkyl ditellurides, e.g., dihexadecyl ditelluride ($C_{16}H_{33}$-Te-Te-$C_{16}H_{33}$) and dihexadecenoic acid ditelluride (HOOC-CH_2-$(CH_2)_6$CH=CH-$(CH_2)_5$-CH_2-Te$)_2$ were synthesized in near quantitative yields. The "cold" ditelluride was synthesized, isolated and characterized. In the radioactive synthesis, the synthesized ditelluride was immediately reduced with $NaBH_4$ to the sodium alkyl telluride which was further reacted with other alkyl halides *in-situ*.

Synthesis of dihexadecyl ditelluride $-C_{16}H_{33}$-Te-Te-$C_{16}H_{33}$: Stable tellurium metal (1.27 gm, 0.01 M) was solubilized (as described under "Production of Na_2Te_2-(Te-123m)" to a colorless solution. The solution was made basic to pH 12 with NaOH. Additional 12.7 gm (0.01 M) of tellurium metal was added and the solution stirred and refluxed to cause total dissolution of the Te to give a dark purple clear solution.

The solution of Na_2Te_2 was allowed to cool to room temperature, and to it was slowly added 6.7 gm (0.022 M) of hexadecanyl bromide (1-bromohexadecane) in 40 ml of peroxide-free THF via a syringe. Heat was applied to the solution whereby a color change from dark purple to brownish-red was observed. After 2 hours of gentle heating and stirring, the ditelluride formed was extracted into CCl_4. The organic layer was washed with water and dried over anhydrous $MgSO_4$. The CCl_4 solution was evaporated under vacuum to yield 7 gms of a red oil (90% yield) which solidified upon standing. The red solid has a m.p. of 40-42°. TLC on silica gel (Eastman) with $CHCl_3$ indicated a single compound with an Rf of 0.9.

The NMR spectrum of the CCl_4 solution indicated the desire triplet at 2.6 ppm corresponding to the $-C\underline{H_2}$-Te-Te-$C\underline{H_2}$-hydrogens.

RESULTS

The solubilization of Te-123m by $NaBH_4$ in water requires heat and an inert atmosphere. A 2:1 equivalent of sodium borohyride to tellurium metal in water, refluxed for 10-15 minutes will produce a colorless, pH 8 solution of sodium hydrogen telluride (NaHTe) according to the following reaction:

$$2Te + NaBH_4 + H_2O \xrightarrow{\text{Heat, } N_2} 2\ NaHTe + Na_2B_4O_7 + 14\ H_2$$

No volatile tellurium products are emmitted. The by-product of the reaction, sodium borate ($Na_2B_4O_7$), does not interfere with subsequent reactions. The sodium hydrogen telluride is very sensitive to air oxidation, and when exposed to air acquires a violet-reddish tinge followed by the complete deposition of a black precipitate of Te metal.

Upon the addition of another molar equivalent of Te metal in 1% NaOH suspension to the refluxing solution of the NaHTe, a violet to purplish solution results as the Te metal dissolves to form the ditelluride anion:

$$NaHTe + Te + NaOH \xrightarrow{\text{Heat, } N_2} Na_2Te_2 + H_2O$$

This species in water is colored deep violet-purple, is very sensitive to air oxidation like NaHTe, but stable in an inert atmosphere. As in the preparation of sodium hydrogen telluride, no volatile radioactive products were detected in the I_2-KI solution trap.

The addition of alkyl halides to the aqueous sodium hydrogen telluride resulted in near quantitative yields of the sodium alkyl tellurides within 5 minutes to 2 hours reaction times. The colorless sodium alkyl tellurides dimerize to yellow-red dialkyl ditellurides very easily when exposed to air, but when maintained in a N_2 atmosphere are strong nucleophiles displacing primary or secondary alkyl halides. As seen from our synthesis, yields of sodium alkyl tellurides isolated as the stable colored ditellurides or colorless dialkyl tellurides were in the 90-95% range.

Dialkyl ditellurides were also synthesized in yields of over 90% by the addition of alkyl halides, primary or secondary, to an aqueous solution of Na_2Te_2. Reaction times and temperatures range from a few minutes at room temperature to 1-2 hours at 100°C.

DISCUSSION

Until now, the development of synthetic chemical methods which can be used to introduce tellurium into organic molecules in a facile manner was lacking. The current literature available (5,6) only describes a cumbersome and hazardous procedure using a solubilization technique involving sodium metal in liquid ammonia developed for stable tellurium metal (9). Our experiences with selenium chemistry appeared applicable to tellurium with minor modification. As in our work with Se-75 solubilization (1), sodium borohydride also has the ability to reduce tellurium metal to Na_2Te_2 and subsequently to NaHTe in aqueous media. Reports by Klayman and Griffin (10) which describe the unreactivity of $NaBH_4$ on Te metal in alcohol or aqueous solution have been proven incorrect in our laboratories.

Compared to the Se metal reduction with $NaBH_4$, which is an exothermic reaction, the dissolution of Te by sodium borohyddide requires heat and longer reaction times in both aqueous and alcoholic solutions (8). In addition, the dissolution of Te in absolute ethanol with an excess of $NaBH_4$ results not only in the formation of NaHTe but also H_2Te, a volatile gas. Therefore, this method of solubilization is not suitable for Te-123m because of the loss of about 50% of the radioactivity. In comparison, the dissolution of Te in water with $NaBH_4$ results in NaHTe or Na_2Te_2 depending on the stoichiometry of $NaBH_4$ to the Te metal (8). In aqueous media the reaction may be carried out with no loss of volatile tellurium products, thus making this reaction ideal for Te-123m dissolution. This process of tellurium dissolution with $NaBH_4$ has facilitated considerably the synthesis of sodium alkyl tellurides, dialkyl ditellurides and dialkyl tellurides in very high yields and in "single pot" reactions.

At present, we have been successful in the synthesis of various Te-123m sodium alkyl tellurides, which, as strong nucleophiles can be introduced into organic compounds to produce stable tellurides; Te-123m dialkyl ditellurides (stable intermediates that can be easily reduced with $NaBH_4$ back to sodium alklyl tellurides); and long chain fatty acid tellurides.

Although the developed solubilization procedures will lead to the easy introduction of Te-123m via organic nucleophiles into new radiopharmaceuticals there are two major drawbacks associated with the synthesis and evaluation of Te-123m i.e., 1) the high cost of Te-123m metal and 2) the low specific activity of the presently available Te-123m.

As additional new Te-123m radiopharmaceuticals are developed and their potential evaluated, the above disadvantages might be minimized.

REFERENCES

1. Basmadjian GP, Hetzel KR, Ice RD: New methods for introducing Se-75 into radiopharmaceuticals. Int J of Appl Rad and Isotopes 26: 695-696, 1975

2. Basmadjian GP, Ice RD, Hetzel KR: Synthesis and Biodistribution of 75-SE-6 β-Benzylseleno methyl-19-nor-cholest-5(10)-en-3 β-ol(BSeC). 1st Annual Meeting on General Radiopharmaceutical Science, Atlanta, Georgia, Jan. 22, 1978

3. Basmadjian GP, Parker GR, Magarian RA, Kirschner, AS, Ice RD: Potential Te-123m radiopharmaceuticals: A simple method for the introduction of Te-123m into organic compounds. Proceedings, Second International Symposium on Radiopharmaceutical Chemistry. Oxford, England. July 3, 1978, pp 33-34

4. Basmadjian GP, Mills SL, Parker GR, Kirschner AS, Ice RD, Magarian RA: 16-Methyl telluro-Te-123m-9-Hexadecenoic acid: A potential myocardial agent. J Nucl Med 19: 718, 1978

5. Knapp Jr FF, Callahan AP: The synthesis of tellurium-123m labeled steroids. J Nucl Med 18: 610, 1977

6. Boyd GS, Merrick MV,Monks R, Thomas IL: Bile acid analogs labeled with Selenium-75 or Tellurium-123m. Proceedings, Second International Symposium on Radiopharmaceutical Chemistry. Oxford, England. July 3, 1978, pp 38-40

7. van Popta EW: Selenyl - en telluryl derivaten van steroiden, alsmede van deze verbindingen in diagnostische preparaten. Netherlands Patent No. 7415526, 1976

8. Basmadjian GP, Parker GR, Magarian RA, Ice RD, Mills SL, Kirschner AS: Synthesis of Te-123m nucleophiles: Potential use in the synthesis of new Te-123m radiopharmaceuticals. Int J of Appl Rad and Isotopes (submitted)

9. Irgolic KG: The Organic Chemistry of Tellurium, Gordon and Breach Science Publication, N.Y., N.Y. 1974, pp 121-137

10. Klayman, DL, Griffin JS: Reaction of selenium with sodium borohydride in protic solvents. Method for the introduction of selenium into organic molecules. J Amer Chem Soc 95: 197, 1973

TELLURIUM-123m-LABELED ISOSTERES OF PALMITOLEIC AND OLEIC ACIDS SHOW HIGH MYOCARDIAL UPTAKE

Furn F. Knapp, Jr., Kathleen R. Ambrose, and Alvin P. Callahan

Nuclear Medicine Technology Group
Health and Safety Research Division
Oak Ridge National Laboratory
Oak Ridge, Tennessee 37830

and

Robert A. Grigsby and Kurt J. Irgolic
Chemistry Department
Texas A & M University
College Station, Texas 77840

ABSTRACT

These studies were directed at determining if the telluro fatty acids prepared by the isosteric replacement of the Δ^9-double bonds of oleic and palmitoleic acids with ^{123m}Te would show heart uptake in rats. The isostere of palmitoleic acid, 9-tellurapentadecanoic acid(II), was prepared by basic hydrolysis of the product formed by the coupling of ^{123m}Te-sodium hexyl tellurol with methyl-8-bromooctadecanoate. Similarly, the isostere of oleic acid, 9-telluraheptadecanoic acid(IV), was prepared by the same route beginning with the reaction of ^{123m}Te-sodium octyl tellurol with methyl-8-bromooctadecanoate. Both ^{123m}Te-(II) and ^{123m}Te-(IV) showed remarkably high heart uptake in rats (2-3% dose/gm) ten minutes after intravenous administration, and the heart/blood ratios were high (20-30/1). Finally, the hearts of rats injected with ^{123m}Te-(IV) have been clearly imaged with a rectilinear scanner.

INTRODUCTION

Since it is well established that the biological properties of many organic compounds are retained when selected double bonds are isosterically substituted with S, Se or Te (1, 2), we wished to determine if this isosteric behavior was also exhibited by aliphatic tellurium isosteres of olefinic compounds. Long-chain unsaturated fatty acids were obvious choices for these investigations since oleic acid (9-octadecenoic acid) is one of the major serum fatty acids and long-chain fatty acids are utilized for energy production by the normal myocardium (3). The preparation and testing of the ^{123m}Te-labeled isosteres of unsaturated fatty acids were therefore prime candidates for determining the potential usefulness of the concept of isosteric replacement of olefinic linkages with a chalcogen radionuclide for the development of new and potentially useful radiopharmaceuticals.

In the present investigation we have prepared ^{123m}Te-labeled 9-tellurapentadecanoic acid [$CH_3(CH_2)_5$-^{123m}Te-$(CH_2)_7$-COOH] and 9-telluraheptadecanoic acid [$CH_3(CH_2)_7$-^{123m}Te-$(CH_2)_7$-COOH] as isosteres of palmitoleic acid and oleic acid, respectively. Tissue distribution experiments in rats have demonstrated the pronounced heart uptake of radioactivity following intravenous administration of the ^{123m}Te-labeled fatty acids. In addition, rat hearts have been clearly imaged following injection of ^{123m}Te-labeled 9-telluraheptadecanoic acid.

PREPARATION OF TELLURO FATTY ACIDS

The general method that has been developed for the synthesis of telluro fatty acids involves the following transformations:

$$\text{Te} + \text{Na} \xrightarrow[\text{Step 1}]{NH_3} Na_2Te_2 + \text{RX} \xrightarrow[\text{Step 2}]{} R_2Te_2 + NaBH_4 \xrightarrow[\text{Step 3}]{\text{MeOH}} \text{Na-Te-R} +$$

$$\text{Br-R'-COOMe} \xrightarrow[\text{Step 4}]{\text{MeOH}} \text{R-Te-R'COOMe} + \text{NaOH} \xrightarrow[\text{Step 5}]{\text{EtOH-}H_2O} \text{R-Te-R'-COOH}.$$

For the preparation of 9-tellurapentadecanoic acid (II), iodohexane was used in Step 2 with subsequent conversion to sodium hexyl tellurol which was coupled with the methyl-8-bromooctanoate in Step 4 followed by basic hydrolysis to give (II). The 8-bromooctanoic acid was obtained commercially and converted to the methyl ester by treatment with diazomethane. Sodium octyl tellurol was prepared in the same manner and coupled with methyl-8-bromooctanoate to give methyl-9-telluraheptadecanoate(III) which was treated with base to yield 9-telluraheptadecanoic acid (IV). Compounds (I), (II), (III), and (IV) were fully characterized by thin-layer chromatographic, ultraviolet, infrared, mass-spectral and nuclear magnetic resonance spectral methods and these data are consistent with the proposed structures (Table 1).

PREPARATION OF TELLURIUM-123m-LABELED FATTY ACIDS

Tellurium-123m-labeled dioctyl ditelluride was prepared by iodooctane alkylation of ^{123m}Te-sodium ditelluride (31.3 mCi) by the general procedure described earlier (4). For convenience, these reactions were performed on the 1 mmole scale. Following sodium borohydride reduction (Step 3) of the orange-colored ditelluride solution (19.62 mCi) in benzene-methanol, the resulting colorless solution of ^{123m}Te-sodium octyl tellurol was reacted in Step 4 with methyl-8-bromooctanoate (59 mg, 250 μmoles) by refluxing for 30 minutes. The tellurols in Step 4 were used in excess to insure complete consumption of the haloacid methyl ester. The reaction mixture was cooled, poured into water and the benzene layer washed several times with water, dried over anhydrous Na_2SO_4 and the solvent evaporated *in vacuo* to give a yellow oil which was dissolved in petroleum ether and applied to a 60-200 mesh silicic acid (acidic) column (2 x 20 cm). The column was eluted with petroleum ether (25 ml fractions) which removed a radioactive peak that contained ^{123m}Te-dioctyl ditelluride and ^{123m}Te-octyl ditelluride. Further elution with 2% ether in petroleum ether removed a radioactive component with the same mobility as authentic methyl-9-telluraheptadecanoate (Figure 1). The peak fractions were combined to give ^{123m}Te-methyl-9-telluraheptadecanoate, 5.41 mCi (26% from methyl-8-bromooctanoate). Only one radioactive component was detected upon thin-layer radiochromatographic analyses on silica gel G (solvent, benzene). This material had the same mobility as authentic methyl-9-telluraheptadecanoate (R_f 0.6). The purified material had a specific acticity of ∿ 31 mCi/mmole. For long-term storage the sample was transferred to break-seal tubes, the solvent evaporated under argon and the tubes sealed. Samples were stored in the dark at 8°C.

The ^{123m}Te-labeled 9-telluraheptadecanoic acid (IV) was prepared by basic hydrolysis of the methyl ester (III) in refluxing aqueous ethanol under argon. Aliquots of (III) were dissolved in 20 ml of absolute ethanol and 0.50 ml of a 1 N NaOH solution was added and the reaction system flushed thoroughly with argon. The mixture was then refluxed in the dark for 30 minutes. Following cooling, the mixture was poured into water the pH adjusted to pH 2-3 by addition of 1 N HCl and the resulting cloudy solution extracted with ethyl ether. The organic layer was washed well with water, dried over anhydrous sodium sulfate and the solvent evaporated under argon. The methyl ester (III) was relatively stable and could be stored as a neat oil under argon in the dark at 0°C for several weeks with only marginal decomposition. In contrast, the free acid (IV) was relatively unstable and was formed by hydrolysis of (III) immediately prior to the biological experiments.

The ^{123m}Te-labeled methyl-9-tellurapentadecanoate (I) was prepared exactly in the same manner as described earlier for ^{123m}Te-labeled methyl-9-telluraheptadecanoate (III). The ^{123m}Te metal (24.8 mCi) was converted to ^{123m}Te-

Table 1. Physical Properties of Telluro Fatty Acids and Telluro Fatty Acid Methyl Esters

R_1-Te-$(CH_2)_7COOR_2$

I. Methyl-9-Tellurapentadecanoate; $R_1 = n - C_6H_{13}$, $R_2 = CH_3$
II. 9-Tellurapentadecanoic Acid; $R_1 = n - C_6H_{13}$, $R_2 = H$
III. Methyl-9-Telluraheptadecanoate; $R_1 = n - C_8H_{17}$, $R_2 = CH_3$
IV. 9-Telluraheptadecanoic Acid; $R_1 = n - C_8H_{17}$ $R_2 = H$

Compound	TLC, R_f	Ultraviolet	Infrared	Mass Spectrum	Nuclear Magnetic Resonance Spectrum
I	0.53*	λ_{max} 234 nm	ν_{max}^{neat} 1750 cm^{-1} (carbonyl)	m/e 372, M m/e 341, M-31	0.91, s, 3H, terminal methyl 2.29, t, 2H, α-CH_2 2.61, t, 4H, CH_2's flanking Te 3.66, s, 3H, -OCH_3
II	0.55**	λ_{max} 235 nm	. . .	. . .	0.76, s, 3H, terminal methyl 1.99, t, 2H, α-CH_2 2.49, t, 4H, CH_2's flanking Te
III	0.53*	λ_{max} 234 nm	ν_{max}^{neat} 1745 cm^{-1} (carbonyl)	m/e 400, M m/e 369, M-31	0.96, s, 3H, terminal methyl 2.31, t, 2H, α-CH_2 2.59, t, 4H, CH_2's flanking Te 3.66, s, 3H, -OCH_3
IV	0.55**	λ_{max} 236 nm	. . .	. . .	0.80, s, 3H, terminal methyl 1.90, t, 2H, α-CH_2 2.40, t, 4H, CH_2's flanking Te

*SiO_2, benzene; **SiO_2, Petroleum ether-ether-acetic acid, 70:30:1

sodium ditelluride (Step 1) and alkylated with iodohexane to give, after work-up, ^{123m}Te-labeled dihexyl ditelluride (11.17 mCi). Following sodium borohydride reduction (Step 3), coupling with methyl-8-bromooctanoate (Step 4) and column purification, ^{123m}Te-labeled methyl-9-tellurapentadecanoate (II) was obtained (1.35 mCi). The methyl ester (II) was hydrolyzed with base to yield the ^{123m}Te-labeled 9-tellurapentadecanoic acid.

BIOLOGICAL STUDIES

Following basic hydrolysis, the ^{123m}Te-labeled fatty acids were dissolved in a small volume of absolute ethanol and added dropwise to a stirred solution of 6% bovine serum albumin at 40°C. Female Fischer strain rats were injected via a tail vein with 1.0 ml. of the bovine serum albumin solution containing 5-10 μCi of the ^{123m}Te-labeled fatty acid. After various time intervals the animals were sacrificed by decapitation after being anesthesized with ether. The various organs were removed, washed with saline, weighed and the radioactive contents determined by counting the organs directly in a autogamma counter. Three animals were used for each time point. For rectilinear scans, the animals were injected with 30-50 μCi of the radiolabeled fatty acid in 1.0 ml. of the 6% bovine serum albumin. The animals were anesthesized by intraperitoneal injection of pentabarbital for the rectilinear scans which were performed with a small animal scanner equipped with a 64-hole gold collimator. The rats were scanned at a rate of 0.25 in/min.

RESULTS AND CONCLUSIONS

The preparation of ^{123m}Te-labeled fatty acids (II) and (IV) is a simple method and the radiolabeled compounds can be synthesized and purified easily in a single day. The use of the ω-halogenated fatty acid methyl ester in Step 4 was chosen as the preferred method after initial attempts to couple the free ω-halogenated fatty acid with the sodium alkyl tellurols resulted in low and non-reproducible yields of the telluro fatty acids. In addition, the methyl esters of the telluro fatty acids have been found to be relatively stable and can be stored in the dark under argon at 0°C for several weeks with very little decomposition. Unfortunately, the free acids (II) and (IV) are relatively unstable and must be generated by basic hydrolysis of the methyl esters immediately before use. Methods to stablize the free acids have not yet been investigated in detail but will be the subject of future studies.

Following intravenous administration of the ^{123m}Te-labeled acids (II) and (IV) complexed in bovine serum albumin very pronounced heart uptake was detected in rats as evidenced by the data illustrated in Figures 2 and 3. It is interesting to note that the % dose/gm tissue values for maximal heart uptake (3-4%) very closely paralleled similar data reported for ^{14}C-labeled oleic acid (5). These data suggest that the radioactivity detected in heart tissue may well represent myocardial uptake of the intact ^{123m}Te-labeled fatty acids. More importantly, these preliminary data may indicate that replacement of the Δ^9-double bonds of palmitoleic acid and oleic acid may be a

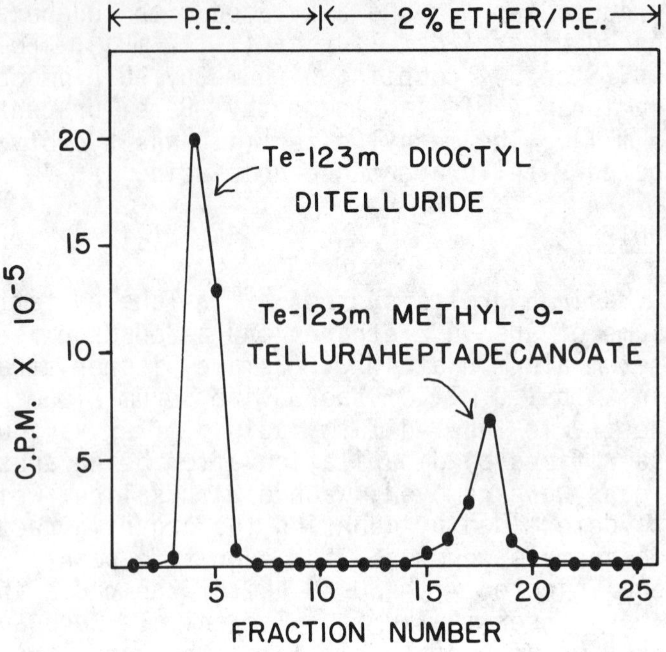

FIGURE 1

Silicic acid column chromatographic profile obtained upon purification of ^{123m}Te-methyl-9-telluraheptadecanoate (III). The radioactive peak for the radiolabeled methyl ester co-chromatographed with authetic unlabeled (III). The peak fractions were combined and aliquots hydrolyzed with base to give ^{123m}Te-9-telluraheptadecanoic acid which was used for the biological experiments described in Figures 3 and 4.

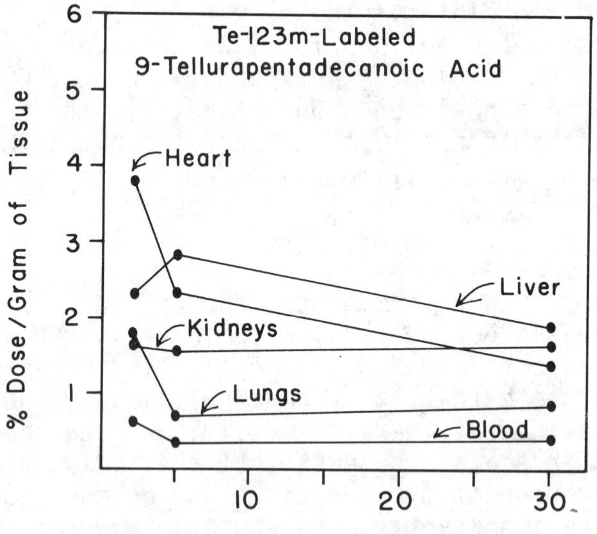

FIGURE 2

The tissue distribution of radioactivity in female Fischer rats following intravenous administration of ^{123m}Te-9-tellurapentadecanoic acid (II). The experimental points represent averaged data (% dose/gm tissue) from three animals. The animals were each injected via a tail vein with 5-10 μCi of radioactive (II) complexed in 1.0 ml of a 6% bovine serum albumin solution.

true isosteric conversion. The hearts of rats were clearly imaged with a rectilinear scanner after administration of ^{123m}Te-labeled 9-tellurahepta-decanoic acid (Figure 4). The areas of significant radioactivity accumulation were the liver and heart. Subsequent rectilinear scans of the excised, washed heart showed an identical pattern. These results confirm the tissue distribution data and indicate that the radioactivity was present primarily in the heart muscle and not in the blood pooled within the cardiac chambers.

Finally, these studies indicate that the isosteric replacement of selected double bonds with ^{123m}Te or ^{75}Se may be an efficient strategy for the design and development of new radiopharmaceuticals. We feel our results may have some application for the design of new heart imaging agents. Studies are now in progress to determine the effects of both total chain length and the position of the tellurium heteroatom on the heart uptake of a series of ^{123m}Te-labeled fatty acids.

ACKNOWLEDGEMENTS

Research sponsored by the Office of Health and Environmental Research, U. S. Department of Energy under contract W-7405-eng-26 with the Union Carbide Corporation. The partial support of the investigations at Texas A & M University by the Robert A. Welch Foundation of Houston, Texas is gratefully acknowledged. The technical assistance of L. A. Ferren is gratefully acknowledged, and we thank Dr. D. V. Woo for performing the rectilinear scans, Dr. B. M. Benjamin and L. L. Brown for the nuclear magnetic resonance studies, and C. W. Pritchard and Dr. W. D. Rainey, Jr. for the mass spectral measurements. We also thank C. A. Floyd for typing the manuscript.

REFERENCES

(1) KOROLKOVAS A, BURCKHALTER JH: *Essentials of Medicinal Chemistry*, New York, John Wiley & Sons, 1976.

(2) ZANATI G, GAARE G, WOLFF ME: Heterocyclic Steroids · 5 · Sulfur, selenium, and tellurium 5α-androstane derivatives and their 7α-methylated congeners. *J Med Chem* 17:561-563, 1974.

(3) POE ND, ROBINSON GD, MACDONALD NS: Myocardial extraction of labeled long-chain fatty acid analogs *Proc Soc Exp Biol Med* 148:215-218, 1975.

(4) KNAPP FF, CALLAHAN, AP: The Synthesis of Tellurium-123m-labeled steroids. *J Nucl Med* 18:610, 1977.

(5) BEIERWALTES WH, ICE RD, SHAW MJ, et al: Myocardial Uptake of labeled oleic and linoleic acids. *J Nucl Med* 16:842-845, 1975.

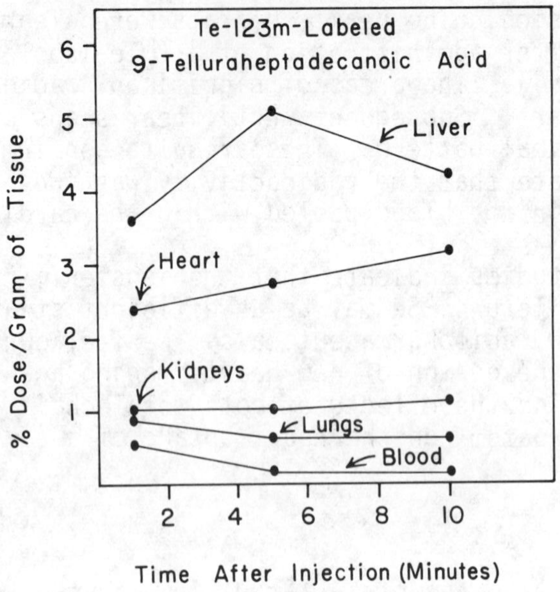

FIGURE 3

The tissue distribution of radioactivity in female Fischer rats following intravenous administration of ^{123m}Te-9-telluraheptadecanoic acid (IV). The experimental points represent averaged data (% dose/gm tissue) from three animals. The animals were each injected via a tail vein with 5-10 μCi of radioactive (IV) complexed in 1.0 ml of a 6% bovine serum albumin solution.

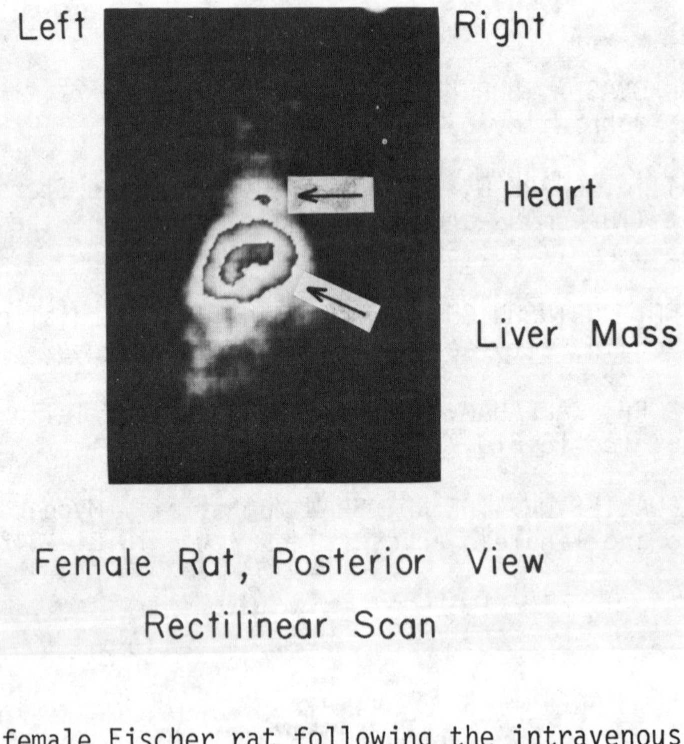

FIGURE 4

Rectilinear scan of a female Fischer rat following the intravenous administration of 50 μCi of ^{123m}Te-9-telluraheptadecanoic acid (IV). The animal was injected via a tail vein with radiolabeled (IV) complexed in 1.0 ml of a 6% bovine serum albumin solution.

SELENAPROLINE-Se-75: A RADIOLABELED IMINO ACID

Steven H. Wong, Richard P. Spencer,
and Abraham Weaver

Department of Nuclear Medicine
University of Connecticut Health Center
Farmington, Connecticut 06032

ABSTRACT

The radiolabeled imino acid Se-75 - selenaproline was synthesized from Se-75-L-selenocystine. The imino acid was transported against a concentration gradient by intestinal sacs in vitro. Following intravenous administration in mice and rats, the ratio of radioactivity in brain was 4 to 5 fold higher with Se-75 - selenaproline than with Se-75-L-selenocystine (but still less than the blood concentration). The compound shows some instability and has to be purified prior to use. Radiolabeled selenaproline may be of interest in following collagen synthesis up to the 4-hydroxylation stage, and in examining amino acid/imino acid penetration across the blood/brain and other biological barriers.

INTRODUCTION

While the distribution of amino acids labeled with gamma ray emitting radionuclides (such as Se-75-selenomethionine) has been studied in vivo, there are almost no data on the fate of similarly labeled imino acids. Selena-proline, an analogue of the imino acid proline, has been noted to have inhibitory effects on protein synthesis(1,2). Since selenaproline contains a selenium atom, we have studied the synthesis of the compound with Se-75 and the biological distribution of the resultant product.

MATERIALS AND METHODS

Synthesis of Se-75-selenaproline was modified from the production of the nonlabeled compound as reported by DeMarco and coworkers (3). The synthesis (Figure 1) involved condensation of Se-75-L-selenocystine with formaldehyde after reduction with sodium borohydride. Se-75-L-selenocystine (Amersham) had a specific activity of 8.7 mCi/17 mg. The HCI solution was neutralized with IN NaOH and made basic with 0.4N NaOH. Under continous bubbling of N_2, the Se-75-L-selenocystine was reduced by addition of 9mg of $NaBH_4$ over a period of one half hour. To destroy excess $NaBH_4$, concentrated acetic acid was added to the reaction mixture to bring the pH to 5-5.5. Ring closure was achieved by addition of 40 µl of 35% formaldehyde and the reaction was carried out for an additional 20 minutes. The mixture was then lypophilized and the residue was analyzed by preparative scale thin layer chromatography. This was accomplished on cellulose plates (Brinkmann, Westburg,N.Y.) using n-butanol/acetic acid/water, 3:1:3 (V:V:V). The Rf of the imino acid was 0.66. The labeled imino acid, eluted from the thin layer chromatographic plates, had a specific activity of 4 mCi/7 mg. The solid material was kept refrigerated. Subsequent studies were carried out with the compound dissolved in physiologic saline just prior to use. Purity of the Se-75 – selenaproline was also checked in two other chromatographic systems. The first was cellulose on aluminum plates, n-butanol/acetic acid/water 4:1:4(V:V:V), upper phase, with an R_f of 0.36. The second was paper chromatography with Whatman No. 1 paper,n-butanol/ethanol/water, 4:1:5 (V:V:V) upper phase with an R_f of 0.25. Samples were periodically removed to check on the stability of the compound. Syntheses were also carried out using nonlabeled D,L-selenocystine (CalBiochem) as the starting material in order to check on chromatographic behavior and the melting point.

Biodistribution studies were carried out by intravenous administration of the compound (in saline) to control A/J mice and those bearing a transplantable neuroblastoma, as well as to Sprague Dawley rats of both sexes. Tailvein injections were carried out in the mice, and the left saphenous vein was used in the rats. Volumes of 0.2 ml and activities of 3-30 microcuries were used. A comparison was made with the distribution of Se-75-L-selenocystine. Four mice from each group (selenocystine: normal, selenaproline: neuroblastoma, and selenaproline: normal) were counted at 1,2,4,24,48 and 72 hours in order to determine whole body retention. Other animals were sacrificed at 0.5, 1,2,4, 24 and 72 hours, and tissues weighed, counted for radioactivity and compared with standards. The percent dose per organ and percent dose per gram of tissue were calculated. Urine was analyzed for radioactivity by using the chromatographic systems previously described and counting each 0.5 cm of the plates.

Transport across the intestine in vitro was studied by using the everted sac technique (4,5). Mice were sacrificed, the small intestine removed, washed in lactated Ringer's solution, and everted. Segments of about 4 cm in length were formed and one end of each ligated. Each sac was filled with about 1 ml of a Se-75 – selenaproline solution (1.7×10^{-5}M) in Ringer's. The second end of each sac was then ligated after admitting a small bubble of oxygen. Each sac was then placed in a separate 25 ml flask which contained 5 ml of the same Se-75 – selenaproline solution. Each flask was gassed with 95% O_2 and 5% CO_2. After being stoppered, flasks were incubated in an oscillating water bath at 37°C for 1 hour. Sac contents were drained and the "serosal" fluid radioactivity was compared with that in the bathing "mucosal" fluid. Each solution was also chromatogramed to identify the radioactive species.

RESULTS

Radiochemical yield of Se-75 – selenaproline was 40% of the starting material. Radiochemical purity was 98% in the 3 chromatographic systems immediately after elution from the preparative phases. The non-radioactive D,L-selenaproline, synthesized from D,L-selenocystine, had a melting point of 158°-162°C (reported 160°-164°C)(3). Elemental analysis (Schwarzkopf Microanalytical, Woodside, N.Y.) for $C_4H_7NO_2Se$ yielded: calculated C 26.68, H 3.92, N 7.78, found C 26.60, H 3.91, N7.87. Purified Se-75 – selenaproline, upon storage in the refrigerator, showed slow decomposition by chromatography (30% after 9 days). The breakdown of the product was noted in the original synthesis of the nonlabeled material (3).

Whole body retention of Se-75 in mice, after intravenous administration of selenocystine or selenaproline, is shown in Table 1. Selenaproline, in both normal A/J mice and those with neuroblastoma, showed greater initial retention than selenocystine. The distribution of Se-75 per gram of tissue after injection of Se-75 – selenaproline in mice is shown in Table 2 (and the same data are shown in Table 3, expressed in terms of activity per organ). Corresponding data for Se-75-L-selenocystine are given in Tables 4 and 5. Results in mice bearing a neuroblastoma are shown in Tables 6 and 7. In Tables 8 and 9 are shown a comparison of distribution of Se-75 – selenaproline and Se-75-L-selenocystine in rats at one-half and one hour.

Chromatographic analysis of the urine of mice, after intravenous administration of Se-75 – selenaproline, showed the slight (15%) presence of other radioactive species (a principal spot corresponded to selenocystine).

In the intestinal transport study, each intestinal sac showed a decrease in radioactive counts in the "mucosal" solution, and an increase in the "serosal" solution. This indicated transport from the mucosal to serosal sides. The mean serosal/mucosal concentration of radioactivity after 1 hour was 1.30. Chromatography of mucosal and serosal solutions revealed approximately 10 percent degradation of the Se-75 – selenaproline, principally to the starting material (selenocystine).

DISCUSSION

A relative disadvantage of Se-75 – selenaproline is that the compound has to be repurified prior to use. The ring system may be labile, as shown by apparent slow reversion to the starting material, selenocystine. Selenaproline underwent transport, against a concentration gradient by the everted gut sac. It also entered the brain, so that the selenaproline/

selenocystine ratio in the mouse and rat brain was about 4 to 1 up to 5 to 1. However, the brain/blood ratio did not exceed unity. Selenaproline initially showed greater body retention than selenocystine. However, the material did not concentrate in a neuroblastoma.

Selenaproline is known to be an inhibitor of protein synthesis, likely by interfering with proline incorporation (1,2). However, since selenium is in the 4 position in the ring, the molecule can not undergo 4-hydroxylation (such as occurs in proline-4-hydroxyproline). Hence, selenaproline can be used as a probe of collagen production up to, but not beyond, the stage of 4-hydroxylation. There is also a potential role for selenaproline in the synthesis of peptides which contain proline.

ACKNOWLEDGEMENTS

Supported by Grant CA 17802 from the National Cancer Institute, USPHS

REFERENCES

1. Antonucci, A.; Foppoli, C.; Demarco, C. and Cavallini, D.: Inhibition of protein synthesis in rabbit reticulocytes by selenaproline. Bull.Mol. Biol. Med., 2:80-84, 1977.

2. DeMarco, C.; Busiello, V.; DiGirolamo,M. and Cavallini, D.: Selenaproline and protein synthesis. Biochim. Biophys. Acta., 478:156-166, 1977.

3. DeMarco, C.; Coccia, R.; Rinaldi, A. and Cavalli, D.: Synthesis and chromatographic properties of selenazolidine-4-carboxylic acid (selenaproline). Ital. J. Biochem., 26:51-58, 1977.

4. Wilson, T.H. and Wiseman, G.: The use of sacs of everted small intestine for the study of the transference of substances from the mucosal to the serosal surface. J. Physiol., 123:116-125, 1954.

5. Spencer, R.P. and Samiy, A.H.: Intestinal transport of L-tryptophan in vitro: inhibition by high concentrations. Am. J. Physiol.,199: 1033-1036, 1960.

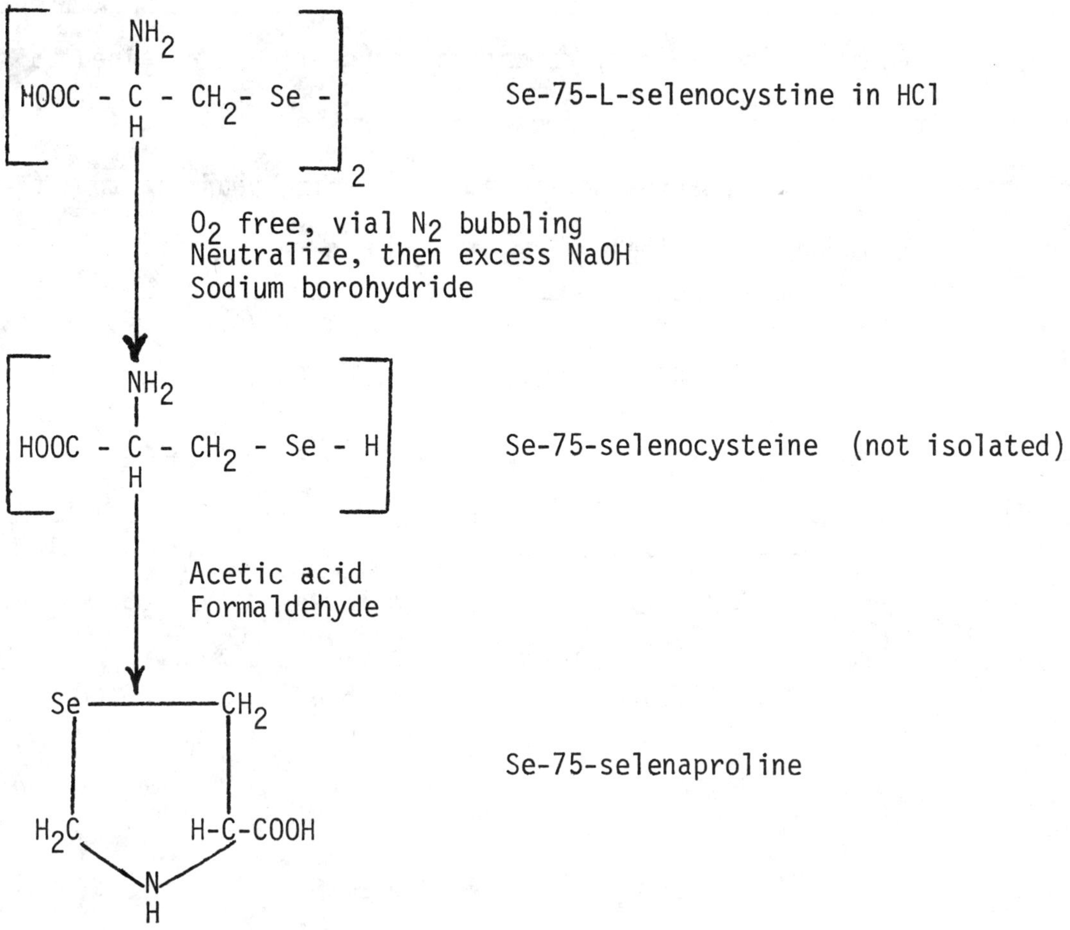

Figure 1. Synthesis of Se-75-selenaproline

TABLE 1. Comparison of whole body retention (percent dose retained) of Se-75 at various times after intravenous administration of Se-75-selenocystine or Se-75-selenaproline. N = mice bearing neuroblastoma (4 mice at each time, whole body counting).*

PERCENT DOSE RETAINED

Time In Hours	Selenocystine	Selenaproline	N selenaproline
1	52.38 ± 7.25	87.35 ± 13.44	85.11 ± 8.34
2	49.93 ± 6.78	82.82 ± 15.88	86.00 ± 2.86
4	24.53 ± 2.10	29.96 ± 3.69	45.57 ± 5.58
24	17.10 ± 1.09	14.18 ± 1.98	19.25 ± 1.46
48	16.16 ± 1.29	13.28 ± 1.01	15.67 ± 0.17
72	13.87 ± 1.44	13.88 ± 2.38	15.40 ± 0.78

* Mean ± std. dev.

TABLE 2. Distribution of radioactivity (per gram tissue) at different times after injection into 4-5 mice of Se-75-selenaproline.*

PERCENT DOSE/GRAM TISSUE

ORGAN	1 H	2 H	4 H	24 H
Blood	2.27 ± 0.10	2.51 ± 0.32	1.60 ± 0.32	1.18 ± 0.08
Liver	25.58 ± 4.87	16.56 ± 3.49	8.92 ± 2.17	4.05 ± 0.15
Spleen	2.46 ± 0.84	2.79 ± 0.66	1.79 ± 0.29	1.17 ± 0.11
Kidneys	13.96 ± 0.70	10.37 ± 0.15	7.91 ± 0.33	5.23 ± 0.09
Heart	2.17 ± 0.27	2.34 ± 0.81	1.89 ± 0.52	0.83 ± 0.03
Lungs	3.64 ± 0.36	3.68 ± 0.91	2.40 ± 0.75	1.17 ± 0.11
Brain	0.89 ± 0.04	0.75 ± 0.14	0.62 ± 0.12	0.34 ± 0.01
G.I.	4.85 ± 0.53	5.88 ± 0.73	3.83 ± 0.27	1.81 ± 0.57

TABLE 3. Distribution of radioactivity in organs at different times after injection into 4-5 A/J mice of Se-75-selenaproline.*

PERCENT DOSE/ORGAN

ORGAN	1 H	2 H	4 H	24 H
Blood	3.56 ± 0.32	3.36 ± 0.28	2.11 ± 0.27	1.57 ± 0.58
Liver	30.20 ± 0.03	15.42 ± 2.09	8.32 ± 1.47	4.16 ± 0.22
Spleen	0.20 ± 0.11	0.27 ± 0.09	0.12 ± 0.05	0.08 ± 0.01
Kidneys	4.60 ± 0.55	3.15 ± 0.47	2.11 ± 0.17	1.70 ± 0.02
Heart	0.24 ± 0.04	0.24 ± 0.07	0.17 ± 0.04	0.08 ± 0.01
Lungs	0.70 ± 0.15	0.58 ± 0.13	0.37 ± 0.10	0.21 ± 0.04
Brain	0.33 ± 0.03	0.30 ± 0.05	0.21 ± 0.07	0.13 ± 0.01
G.I.	13.54 ± 0.98	12.49 ± 2.58	8.49 ± 0.40	4.15 ± 1.35

* Mean ± std. dev.

TABLE 4. The distribution of Se-75 radioactivity per gram of tissue at different time intervals after intravenous injection of Se-75-selenocystine in A/J mice (4 or 5 mice at each point in time).*

PERCENT DOSE/GRAM TISSUE				
ORGAN	1 H	2 H	4 H	24 H
Blood	1.27 ± 0.07	0.97 ± 0.41	1.18 ± 0.48	0.94 ± 0.04
Liver	11.64 ± 0.88	8.19 ± 1.93	5.37 ± 1.78	3.25 ± 0.55
Spleen	2.78 ± 0.78	2.15 ± 0.21	1.52 ± 0.19	1.42 ± 0.18
Kidneys	14.90 ± 1.54	11.48 ± 0.41	6.84 ± 2.74	4.95 ± 0.72
Heart	0.95 ± 0.09	0.84 ± 0.01	0.72 ± 0.21	0.66 ± 0.08
Lungs	2.36 ± 0.29	1.83 ± 0.23	1.42 ± 0.41	1.27 ± 0.06
Brain	0.20 ± 0.06	0.17 ± 0.04	0.17 ± 0.04	0.13 ± 0.01
G.I.	2.70 ± 0.45	2.17 ± 0.51	1.93 ± 0.56	0.95 ± 0.27

TABLE 5. Distribution of Se-75 radioactivity per organ at different time intervals after injection of Se-75-L-selenocystine in A/J mice (4 or 5 mice at each time).*

PERCENT DOSE/ORGAN				
ORGAN	1 H	2 H	4 H	24 H
Blood	3.08 ± 0.20	2.36 ± 0.99	2.51 ± 0.95	1.99 ± 0.15
Liver	24.63 ± 0.76	17.83 ± 5.40	10.33 ± 2.32	5.82 ± 0.33
Spleen	0.31 ± 0.06	0.32 ± 0.07	0.32 ± 0.19	0.22 ± 0.04
Kidneys	9.52 ± 0.97	7.79 ± 0.70	3.89 ± 1.86	2.06 ± 0.06
Heart	0.16 ± 0.02	0.15 ± 0.03	0.11 ± 0.04	0.09 ± 0.01
Lungs	0.48 ± 0.08	0.44 ± 0.06	0.32 ± 0.11	0.31 ± 0.03
Brain	0.08 ± 0.02	0.07 ± 0.02	0.06 ± 0.02	0.05 ± 0.01
G.I.	10.56 ± 1.63	8.64 ± 1.98	7.40 ± 1.34	4.23 ± 0.28

* Mean ± std. dev.

TABLE 6. Distribution of Se-75 per gram of organ at different time intervals after injection of Se-75-selenaproline in A/J mice bearing a neuroblastoma (4 mice at each time).*

PERCENT DOSE/GRAM TISSUE

ORGAN	1 H	2 H	4 H	24 H
Blood	3.50 ± 0.21	2.45 ± 0.41	2.07 ± 0.24	1.47 ± 0.23
Liver	31.67 ± 5.37	9.18 ± 2.21	6.36 ± 0.81	3.11 ± 0.65
Spleen	3.07 ± 0.27	2.08 ± 0.55	1.39 ± 0.10	1.23 ± 0.17
Kidneys	16.05 ± 2.61	8.91 ± 1.37	7.72 ± 0.56	4.83 ± 0.35
Heart	3.07 ± 0.48	1.36 ± 0.13	1.85 ± 1.07	0.69 ± 0.20
Lungs	5.70 ± 1.57	2.55 ± 0.21	1.81 ± 0.23	1.39 ± 0.30
Tumor	2.29 ± 0.26	1.53 ± 0.11	1.51 ± 0.20	1.33 ± 0.21
Brain	1.24 ± 0.15	0.63 ± 0.06	0.64 ± 0.13	0.35 ± 0.05
G.I.	8.68 ± 3.38	4.07 ± 0.12	3.91 ± 0.13	1.91 ± 0.95

TABLE 7. Distribution of Se-75 radioactivity per organ at different time intervals after injection of Se-75-selenaproline into A/J mice bearing a neuroblastoma (4 mice at each time).*

PERCENT DOSE/ORGAN

ORGAN	1 H	2 H	4 H	24 H
Blood	3.99 ± 0.24	3.71 ± 0.51	3.12 ± 0.79	2.30 ± 0.19
Liver	27.61 ± 3.01	10.74 ± 1.18	7.16 ± 0.18	3.40 ± 0.68
Spleen	0.42 ± 0.15	0.42 ± 0.10	0.29 ± 0.10	0.20 ± 0.03
Kidneys	3.73 ± 0.52	3.00 ± 0.24	2.15 ± 0.32	1.55 ± 0.11
Heart	0.22 ± 0.01	0.15 ± 0.01	0.19 ± 0.11	0.08 ± 0.01
Lungs	0.77 ± 0.19	0.41 ± 0.12	0.28 ± 0.05	0.25 ± 0.05
Tumor	0.77 ± 0.20	0.61 ± 0.29	0.91 ± 0.20	1.21 ± 0.92
Brain	0.42 ± 0.08	0.23 ± 0.03	0.23 ± 0.05	0.13 ± 0.02
G.I.	10.53 ± 0.76	7.71 ± 0.61	7.31 ± 0.73	3.15 ± 0.73

* Mean ± std. dev.

TABLE 8. Distribution of Se-75 at two time intervals after intravenous administration of Se-75-selenaproline in Sprague Dawley rats (4 at each time).*

ORGAN	0.5 H		1 H	
	% D/Gram	% D/Organ	% D/Gram	% D/Organ
Blood	0.37 ± 0.07	6.24 ± 1.28	0.35 ± 0.08	5.50 ± 1.29
Liver	5.99 ± 1.21	55.98 ± 6.47	6.84 ± 1.02	58.24 ± 8.14
Spleen	0.32 ± 0.05	0.19 ± 0.02	0.27 ± 0.03	0.15 ± 0.02
Kidneys	3.09 ± 0.32	5.99 ± 0.28	3.05 ± 0.24	5.94 ± 0.04
Heart	0.44 ± 0.12	0.31 ± 0.08	0.39 ± 0.08	0.28 ± 0.06
Lungs	0.53 ± 0.06	0.89 ± 0.22	0.51 ± 0.07	0.78 ± 0.30
Pancreas	0.81 ± 0.08	0.79 ± 0.19	0.89 ± 0.13	0.60 ± 0.15
Brain	0.12 ± 0.01	0.20 ± 0.01	0.12 ± 0.02	0.21 ± 0.04
G.I.	0.63 ± 0.08	16.68 ± 1.68	0.61 ± 0.24	16.52 ± 5.51

TABLE 9. Distribution of Se-75 at two time intervals after intravenous administration of Se-75-L-selenocystine in Sprague Dawley rats (4 at each time).*

ORGAN	0.5 H		1 H	
	% D/Gram	% D/Organ	% D/Gram	% D/Organ
Blood	0.44 ± 0.14	7.92 ± 2.23	0.39 ± 0.07	6.79 ± 0.97
Liver	3.92 ± 0.33	45.84 ± 2.89	5.37 ± 0.82	53.37 ± 1.34
Spleen	0.53 ± 0.16	0.39 ± 0.09	0.46 ± 0.09	0.27 ± 0.05
Kidneys	4.42 ± 1.19	9.75 ± 1.49	4.81 ± 1.51	9.18 ± 1.78
Heart	0.20 ± 0.03	0.17 ± 0.01	0.19 ± 0.03	0.15 ± 0.01
Lungs	0.58 ± 0.29	0.53 ± 0.11	0.41 ± 0.08	0.63 ± 0.14
Pancreas	1.27 ± 0.27	1.63 ± 0.83	1.30 ± 0.25	1.96 ± 0.64
Brain	0.03 ± 0.01	0.04 ± 0.01	0.03 ± 0.01	0.05 ± 0.01
G.I.	0.37 ± 0.08	11.48 ± 1.72	0.43 ± 0.15	9.02 ± 1.52

* Mean ± std. dev.

SYNTHESIS OF 6-RADIOIODO-6-DEOXY-D-GALACTOSE, AND ITS TISSUE DISTRIBUTION AND EXCRETION IN RATS

Richard Flanagan, Brian C. Lentle and Leonard I. Wiebe

University of Alberta and Cross Cancer Institute
Edmonton, Alberta, Canada

and

Frank Helus and Wolfgang Maier-Borst
German Cancer Research Center
Heidelberg, Federal Republic of Germany

6-Radioiodo-6-deoxy-D-galactose was prepared by thermal neutron irradiation of the non-radioactive compound, and by iodine exchange on 6-iodo-1,2:3,4-diisopropylidene-6-deoxygalactose with either NaI-125, NaI-131 and $MgSO_4$, or I-123 on gold foil. The diisopropylidene compound was deblocked with aqueous acid to obtain the labeled sugar. Upon I.V. injection of I-125 or I-131 labeled 6-iodo-6-deoxygalactose into rats, rapid renal and heptic uptake resulted in elimination of 22 percent and 36 percent of the dose in 4 hour bile and 4 hour urine respectively. About 2 percent of a dose remained in the bodies of iodide pretreated rats, 24 hours after injection. After 4 hours, tissue: blood ratios ranged from 0.3 (Skeletal muscle) to 3.8(kidney). Although iodogalactose was found to be stable in saline at room temperature, label integrity *in vivo* was compromised after about 4 hours. Bile contained primarily free radioiodide, whereas urine contained mainly radioiodogalactose with varying proportions of radioiodide.

INTRODUCTION

The principles of isosterism, as they apply to functional groups or substituents of indigenous biochemicals, may be used to design xenobiotics with fairly predictable biological properties. As a result, it has been possible to introduce changes of molecular weight, polarity and chemical reactivity in an orderly fashion to bring about competitive and irreversible reactions with binding sites of specific enzymes, carrier proteins and membranes, or to alter elimination patterns. These studies have played an important role in the evolution of non-invasive techniques for the measurement of biochemical function *in vivo*.

Several saccharides, particularly hexoses, have been labeled with gamma-emitting radionuclides. Metabolic trapping agents such as fluorine-18 labeled deoxyglucoses(1,2) and C-11 labeled deoxyglucose(3) are currently used in limited clinical investigations of glucose metabolism. Others have been proposed as useful hepatic imaging agents(4) or as yardsticks of hexose metabolism in pathological conditions(5).

The transport of hexoses, specifically glucose, across cell membranes is thought to occur via a saturable process called facilitated diffusion(6). Studies have shown that binding of the hexose to the membrane carrier involves hydrogen bonding of the pyranose form at several sites, notably C-1, C-2, C-3, and C-6. Surprisingly, substitution of the C-6 hydroxyl by a lipophylic iodine atom resulted in virtually no change in the binding affinity of 6-iodo-6-deoxygalactose when compared with D-glucose(7). The rate of transport of the iodogalactose into human erythrocytes, however, was less than 40 percent of that observed for 6-deoxy-D-galactose(6).

This paper describes the synthesis of 6-radio-iodo-6-deoxygalactose, and its differential uptake and whole-body elimination by rats. The rationalization of a potential diagnostic value for this compound was based on its known binding and transport at the erythrocyte membrane carrier, and the assumptions of inter-cellular differences in binding affinity and/or the number of binding sites per cell in various tissues.

EXPERIMENTAL

All chemicals and solvents were of reagent grade quality. PMR spectra were obtained using a varian Ha-100 spectrometer. IR spectra were measured on a Nicolet 7199 FT-IR system. All chromatography was performed using camag silica gel for TLC.

1,2:3,4-Diisopropylidene-D-Galactose(A)

This was prepared after the fashion of Raymond and Schroeder(8). A mixture of anhydrous D-galactose (20 g, 0.11 moles), zinc chloride (24 g) and sulphuric acid (1 ml) were stirred for 4 hours at room temperature. The reaction mixture was added to a vigourously stirred solution of sodium carbonate (40 g) and water (100 ml). The supernatant was concentrated under reduced vacuum to give crude diisopropylidene galactose. This was further purified by dissolution in ether, washing with water and drying over magnesium sulphate. Removal of the ether gave 1,2:3,4,-diisopropylidene-D-galactose (27.5 g, 95% of theory).

6-Tosyl-1,2:3,4-diisopropylidene-D-galactose(B)

A mixture of (A) (10 g, 0.04 moles), acetone (15 ml) and pyridine (10 ml) was cooled to 0°C and treated with tosyl chloride (10 g, 0.05 moles). The latter was added in three portions over a period of one hour while maintaining the temperature of the mixture below 45°C. The mixture was allowed to stand overnight and then quenched by the addition of water (2 ml). The crude tosylate was precipitated by diluting the mixture with 200 ml of water. The precipitate was collected, dried, and recrystallized from ethanol (95%) to give 6-tosyl-1,2:3,4-diisopropylidene-D-galactose (yield 9.5 g, 60% of theory). M.P. 90-91°C. Mass spectrum: (414, 1%), (399, 100%). PMR ($CDCl_3$): δ7.8,δ7.3 (d, 2H, J=8), δ5.44 (d, 1H, $J_{1,2}$ =5) δ4.58 (d of d, 1H, $J_{2,3}$ =2.5, $J_{3,4}$ = 8), δ4.25 (d of d, 1H, $J_{1,2}$ = 5, $J_{2,3}$ = 2.5), δ4.0-4.2 (m, 4H), δ2.4 (s, 3H), δ1.5, δ1.32, 1.30, 1.28 (s, 3H).

6-Iodo-6-deoxy-1,2:3,4-diisopropylidene-D-galactose(C)

A solution of (B) (5 g, 0.012 moles) and sodium iodide (22 g, 0.12 moles) were heated in dimethyformamide at 120°C for 24 hours. The mixture was cooled and filtered to remove sodium p-toluenesulphonate. The filtrate was dissolved in a mixture of water (400 ml) and methylene chloride (100 ml). The organic layer was removed, washed with water, 5% sodium thiosulphate solution and dried over magnesium sulphate. Removal of the methylene chloride under reduced pressure gave crude 6-iodo-6-deoxy-1,2:3,4-diisopropylidene-D-galactose. This was recrystallized from ethanol (4.1 g, 90% of theory) M.P. 71°C. Mass spectrum: (380, 10%), (365, 100%). PMR ($CDCl_3$): δ5.5 (d, 1H, $J_{1,2}$ = 5), δ4.6 (d of d, 1H, $J_{2,3}$ = 2.5, $J_{3,4}$ = 7.75. δ4.37 (d of d, 1H, $J_{3,4}$ = 7.75, $J_{4,5}$ = 1.6) δ4.28 (d of d, 1H, $J_{1,2}$ = 5, $J_{2,3}$ = 2.5), δ3.95 (d of d, 1H, $J_{4,5}$ = 1.6, $J_{5,6}$+ $J_{5,6'}$ = 14.2), δ3.35 (M, 2H, $J_{5,6}$ = 7, $J_{5,6}$ = 7.2, $J_{6,6'}$ = 10), δ1.55,δ1.45,δ1.36 ,δ1.34 (s, 3H).

6-Iodo-6-deoxy-D-galactose(D)

A mixture of (C) (100 mg, 0.26 m moles) in acetic acid (5 ml) and water (5 ml) was refluxed for 2 hours while stirring vigourously. By this time the mixture had become clear. It was cooled and diluted to 50 ml with dioxane and concentrated under reduced pressure. The residue was redissolved in dioxane (20 ml) and reconcentrated. The crude 6-iodo-6-deoxy-D-galactose was recrystallized from methanol as colourless white needles M.P. 114°C (62 mg, 80% of theory). Mass spectrum (C.I./NH_3): (318, 100%).

^{125}I exchange labeling of (C) and preparation of ^{125}I-labeled(D)

A mixture of (C) (4 mg) and, $Na^{125}I$ (5 mCi, carrier free in 50 μl of 0.1N NAOH) in 1 ml of N, N-dimethylformamide was heated at 140°C for 48 hours. Analysis of the reaction mixture by TLC on silica gel / $CHCl_3$ showed 60% incorporation into the sugar. The DMF was removed by concentration of the reaction mixture under high vacuum. The residue was dissolved in methylene chloride (1 ml) and filtered. The filtrate was concentrated under high vacuum and the residue sublined at 80°C to give pure iodide-free(C). This was hydrolysed as described previously in a solution of acetic acid (0.5 ml) and water (0.05 ml) to give pure 6-[^{125}I]-iodo-6-deoxy-D-galactose (180 μCi, specific activity 0.23 Ci/m mole)

^{131}I exchange labeling of (C) and preparation of ^{131}I labeled(D)

A mixture of (C) (2 mg) and Na^{131}I (5 mCi in 50 μl of 0.1N NaOH, <20% radiochemical purity) and magnesium sulphate (10 mg) was heated in 1 ml of N,N-dimethylformamide at 120°C with stirring for 12 hours. Analysis by TLC on silica gel/$CHCl_3$ indicated >95% incorporation into the sugar moiety. Work-up of this mixture, by the procedures described above for ^{125}I-(C) gave ^{131}I-(D) in 88% yield (specific activity 0.75 Ci/m mole), containing <1% free iodide.

^{123}I exchange labeling of (C) and preparation of ^{123}I-(D)

A gold foil carrying sublimed ^{123}I (up to 500 μCi) (9) was submerged in a solution of (C) (2-3 mg) in DMF (1 ml) and heated (140°C). Upon removal of the solvent in vacuo after 2 hours, up to 50% of the radiolabel was incorporated ^{123}I-(D) was isolated by TLC in overall yields of about 20%.

Preparation of 6-[^{128}I]-iodo-6-deoxy-D-galactose(D)

A sample of (D) (1 mg) was irradiated at a flux of 10^{12} neutrons/cm^2/sec for 1 minute in the University of Alberta Slowpoke Reactor. Analysis by TLC on silica gel/$CHCl_3$:MeOH(4:1) showed that 60% of the resultant activity could be associated with the 6-iodo-6-deoxy-D-galactose. The rest of the activity corresponded mainly to free iodide with a smaller proportion running at the solvent front. A summary of labeling reactions depicted in Figure 1 is given in Table 1.

ANIMAL STUDIES

Male Wistar rats (200-250 g, Woodlyn Farms) were used; a limited number received Walker 256 tumor transplants into the right hind limb 7-10 days prior to study. Whole-body distribution patterns were observed by using several NaI scintillation detectors coupled to a multi-channel analyser operating in the time-sequence scaling mode (9) when I-125 labeled (D) was used, and by gamma scintillation camera with a pin-hole collimator when I-131 labeled (D) had been injected. The dose (1 μCi of I-125; 350-475 μCi I-131) was injected into the surgically-exposed femoral vein. Tissue distribution of radioactivity was quantitated by radioanalysis of organs dissected from rats which had been exsanguinated while under ether anesthesia, 4 hours after injection. Biliary elimination was estimated by radioanalysis of bile collected via a bile duct cannula while the rats were under urethane anesthesia(10). Limited chromatographic analysis of radioactive bile and urine were done by TLC, followed by radiochromatographic analysis using a chromatogram scanner of in-house design.

TABLE 1. REACTION CONDITIONS FOR THE INTRODUCTION OF RADIOIODINE INTO 6-IODO-1,2:3,4-DIISOPROPYLIDENE-6-DEOXYGALACTOSE AND INTO 6-IODO-6-DEOXY-GALACTOSE

Reaction Substrate	Reaction Product	Reaction Conditions	Radiochemical Yield
6-I-DIP-GAL	6-[^{123}I]-DIP-GAL	Sublimed ^{123}I on AU -DMF; 135°; 2 hrs. -MIk; reflux; 2 hrs.	 51% 76%
6-I-DIP-GAL	6-[^{125}I]-DIP-GAL	Na^{125}I - 0.1N NAOH DMF, 140°C, 48 hrs.	 60%
6-I-DIP-GAL	6-[^{131}I]-DIP-GAL	Na^{131}I - 0.1N NAOH DMF, 120°C, $MgSO_4$, 12 hrs.	 90%
6-iodo-6-deoxygalactose	6-[^{125}I]-deoxy-galactose	reactor ambient T(40°) 10^{12}n cm^{-2} sec^{-1}; 1 min.	60%

6-I-DIP-GAL: 6-iodo-1,2:3,4-diisopropylidene-6-deoxygalactose
DMF : dimethylformamide
MIk : methylisopropyl ketone

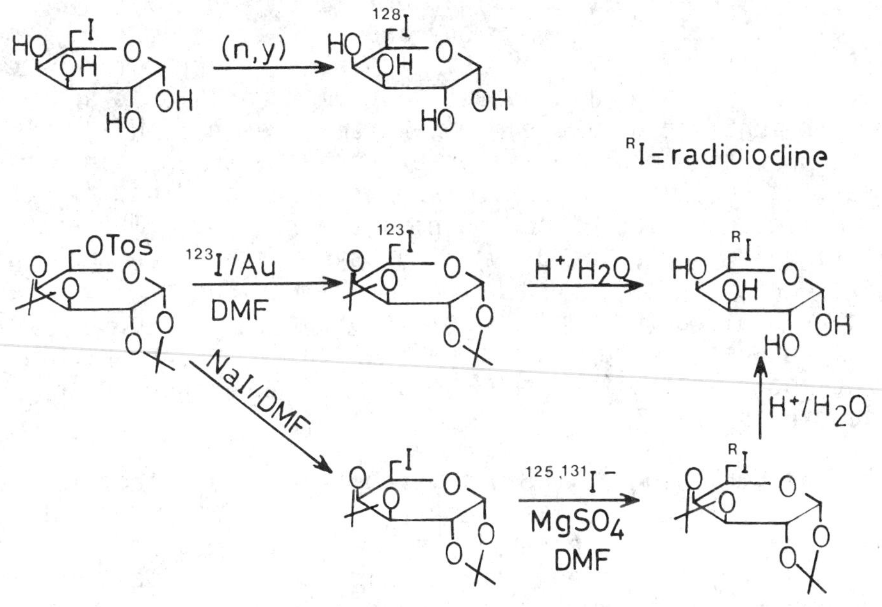

Figure 1.

Schemes for the introduction of radioiodine atoms into precursors of 6-radioiodo-6-deoxygalactose. See Table 1 for a summary of methods and yields.

RESULTS

Urinary elimination of radioactivity 4 hours after injection represented 35 percent of the dose to rats under long-acting urethane anesthesia (Table 2), whereas up to 50 percent of the dose was eliminated within 45 minutes when animals were conscious. Biliary elimination by cannulated rats accounted for 22 percent of the dose, but less than 1 percent of the dose was present in feces over a 6 day period. Pre-treatment of rats with iodide prevented accumulation of radioiodide in the thyroid, with the result that about 98 percent of the dose was eliminated within 24 hours. In rats not pre-dosed with iodide, a small slow-elimination compartment was observed.

Dynamic tissue radioactivity profiles (Figure 2) and quantitative distribution data showed rapidly declining blood levels, and accumulation of radioactivity in the urine after rapid concentration in the kidneys. Whole-body scintigrams (Figure 3) of non-cannulated rats were supportive of the quantitative data obtained upon dissection. The apparent accumulation of radioactivity in the abdominal region just below the diaphragm was taken to represent radioactivity in the duodenum, liver and stomach. Quantitative estimates of tissue : blood radioactivity ratios, and of the percent of the dose present in various tissues 4 hours after injection are presented in Table 2. TLC analysis of 4 hour urine showed the presence of 2 radioactive components judged to be radioiodide and radioiodo-(D), in various proportions. Similar analyses of bile indicated the presence of 1 or more components with chromatographic mobility near that of iodide. Unequivocable evidence for the chemical nature of radioiodine in either urine or bile was not obtained.

DISCUSSION

The observed rapid decline in blood radioactivity after injection of radioiodo-(D), and the prompt elimination via the urine, was in agreement with limited data published by Raymond and Schroeder(8)for (D) in dogs. The significant biliary excretion has also been reported for other radio-iodosugars(4). The insignificant uptake by membranes of tissues with high glucose requirements (e.g. brain, myocardium) does not support the assumptions made with respect to numbers of saturable transport sites in the cell membranes of various tissues. This rational was based on *in vitro* studies, however, and rigid conclusions concerning the role of membrane binding on tissue distribution cannot be drawn. The synthesis of 6-[^{18}F]-fluoro-6-deoxygalactose has been reported(5), and a comparison of its' distribution with that of radioiodo-(D) would be of interest in that both inhibit the transport of sorbose by the glucose carrier to a similar extent, although the fluoro compound exhibits strong polar binding at C-6 while (D) exhibits hydrophobic binding(7).

Hepatic accumulation of (D) was not observed, although a significant proportion (22%) of the dose was present in bile. As seen in Figure 3, radioiodo-(D) offers little potential as a tissue imaging agent in rats. However, the biliary excretion, coupled with the rapid decline in radio-activity in all tissues, suggests a potential role in the study of gall bladder morphology or hepatic function in animals which have a gall bladder.

TABLE 2. TISSUE DISTRIBUTION OF RADIOACTIVITY IN MALE WISTAR RATS, 4 HOURS, AFTER i.v. INJECTION OF 6-[^{125}I]-IODO-6-DEOXYGALACTOSE. A DOSE OF 1.0 μCi WAS GIVEN TO EACH OF 5 RATS

Tissue	Tissue:Blood (±S.D.)	% Dose g^{-1} (±S.D.)	% Dose (±S.D.)
Liver	1.7 ± 1.3	0.5 ± 0.2	5.4 ± 2.3
Kidney	3.8 ± 2.3	1.2 ± 1.0	3.6 ± 3.6
Lung	0.9 ± 0.3	0.2 ± 0.1	0.4 ± 0.1
Heart	0.6 ± 0.1	0.2 ± 0.1	0.2 ± 0.1
Spleen	0.5 ± 0.1	0.3 ± 0.1	0.1 ± 0.0
Thyroid	0.6 ± 0.3	0.3 ± 0.2	0.1 ± 0.1
Skeletal Muscle	0.3 ± 0.2	0.1 ± 0.1	a
Duodenum[b]	0.9 ± 0.2	0.3 ± 0.1	a
Blood	1.0	0.3 ± 0.1	3.8[c]
Bile	a	a	22.2 ± 6.0
Urine	a	a	36.1 ±12.6

a: not calculated
b: cannulated common bile duct
c: blood volume estimated to be 5% of total body weight

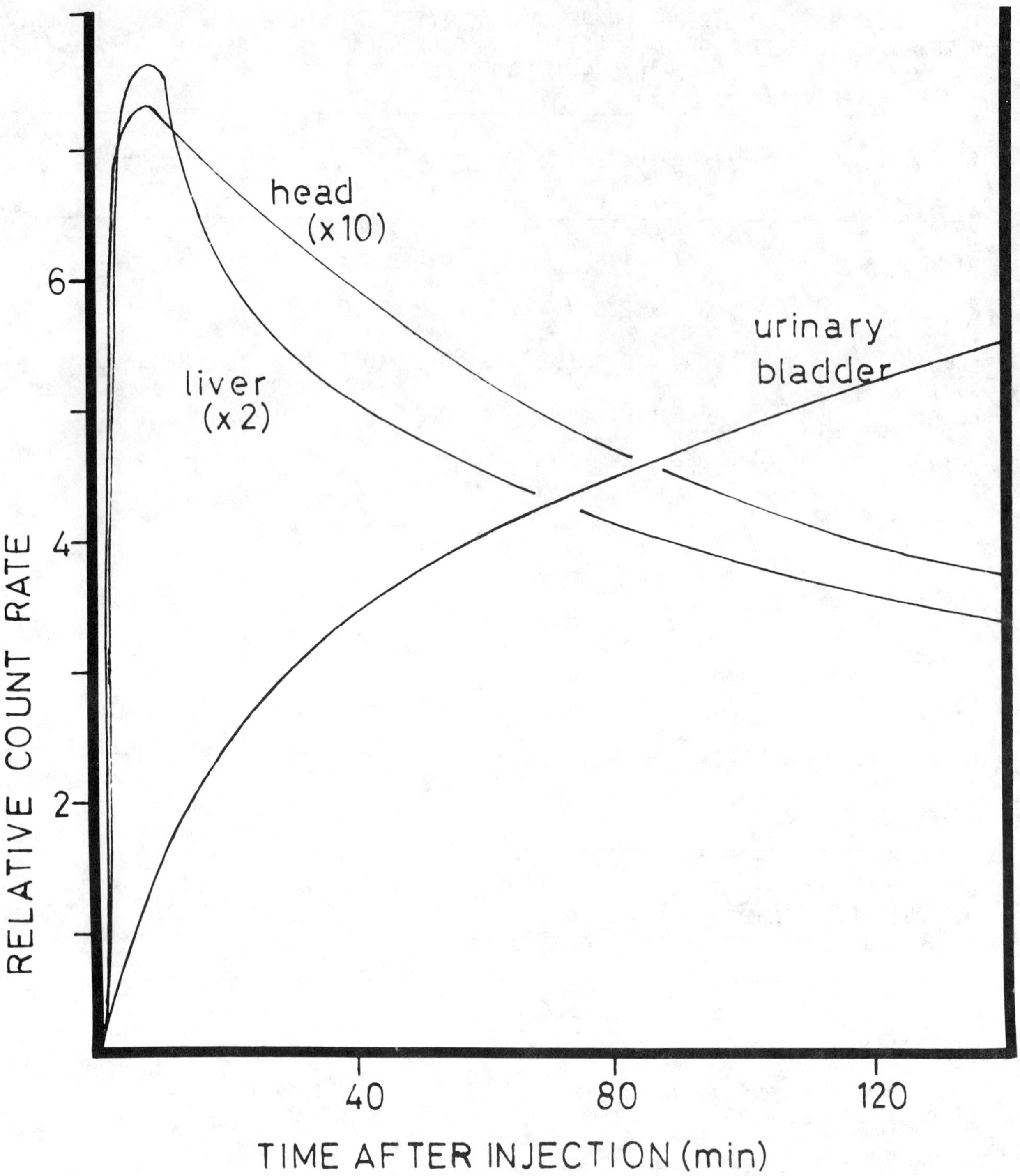

Figure 2.

Radioactivity levels in liver, blood(head) and urinary bladder, derived non-invasively over a period of 4 hours after i.v. injection of 6-[^{125}I]-iodo-6-deoxygalactose. These data are taken from one experiment, typical of four additional studies undertaken. The rats were immobilized by urethane anesthesia, and their bile ducts had been cannulated.

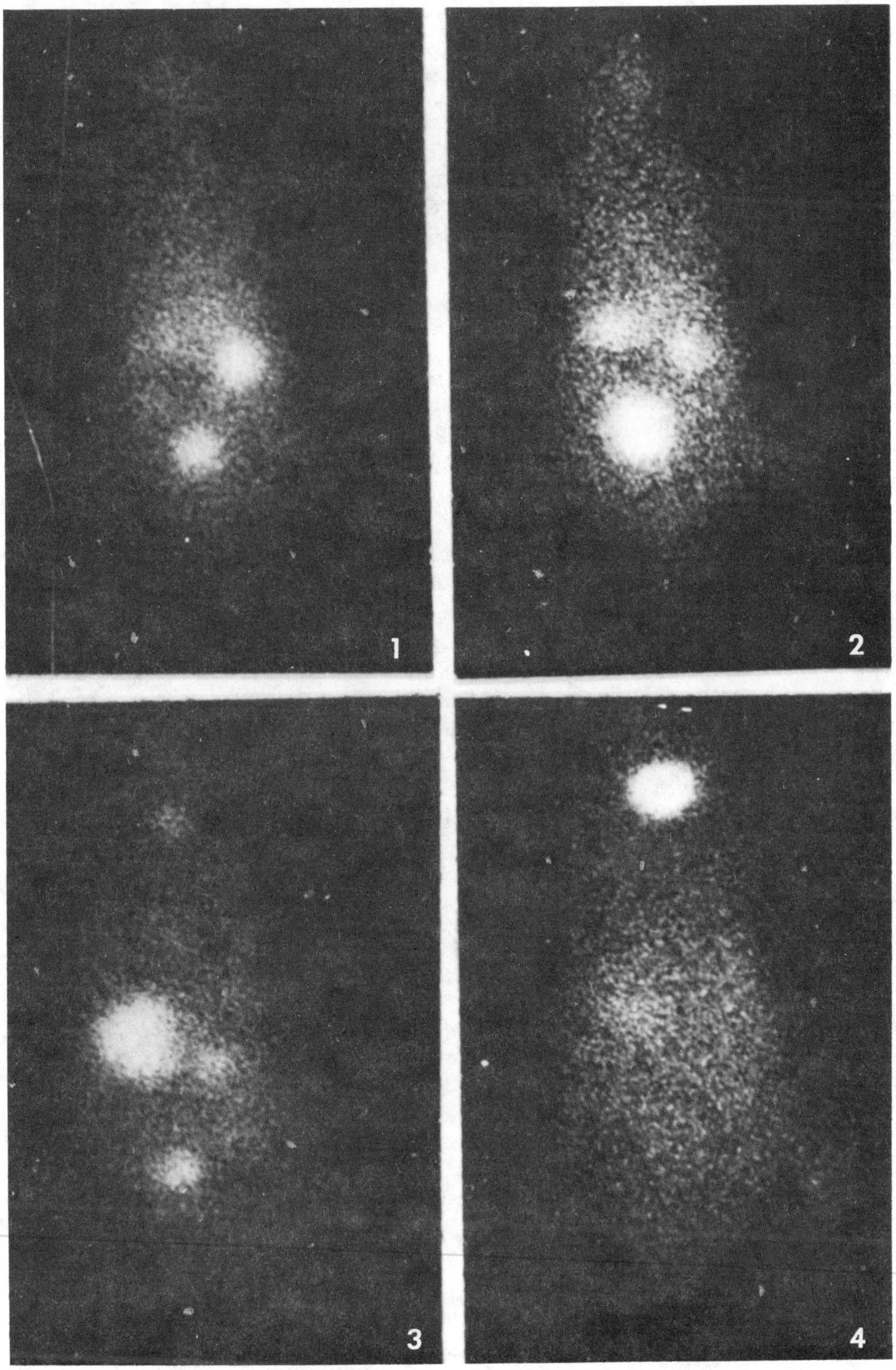

Figure 3.

Scintigrams of rats injected with 6-[^{131}I]-iodo-6-deoxygalactose (ca 400 μCi) taken after 1) 10 minutes, 2) 30 minutes, 3) 250 minutes, and 4) 24 hours. Rats were immobilized by light ether anesthesia during the imaging period. Scintigrams were obtained using a Searle pho/gamma TM camera, fitted with a pinhole (5mm) collimator; 50k data points were acquired for each image. Kidneys and bladder are evident in early scintigrams (1-3), whereas only the thyroid is clearly delineated after 24 hours.

ACKNOWLEDGEMENTS

The authors thank the von Humboldt Foundation and the Provincial Cancer Hospitals Board for the respective awards of a von Humboldt Fellowship(L.I.W.) and a Post-doctoral Fellowship(R.F.). The research was carried out with the financial assistance of the Heritage Trust Fund - Cancer Research grant number 9205. The assistance of Mr. Ken Campbell, Alberta Remote Sensing Center, for electronic enhancement of Scintigrams is gratefully acknowledged.

REFERENCES

1. Phelps ME, Hoffman EJ, Selin C, Huang SC, Robinson G, MacDonald N, Schelbert H and Kuhl DE: *Investigation of [^{18}F]-2-Fluoro-2-deoxy*-glucose for the measurement of myocardial glucose metabolism. J Nucl Med 19:1311-19, 1978.

2. Tewson TJ, Welch MJ and Raichle ME: [^{18}F]-labeled-3-deoxy-3-fluoro-D-glucose: Synthesis and preliminary biodistribution data. J Nucl Med 19:1339-45, 1978.

3. Shiue CY, MacGregor RR, Lade RE, Wan CN and Wolf AP: The synthesis of 1-C-11-2-deoxy-D-glucose(C-11-2-DG) for measuring regional brain glucose metabolism in vivo. J Nucl Med 19:676-7- 1978(abstract).

4. Tsuya A and Atikada S: 3'-Deoxy-3-iodoglucose as diagnostic agent. Japan Kokai 77 83959. Cited in Chem Abstr 87:206545m, 1977.

5. Christman DR, Orhanovic Z and Wolf AP: The synthesis of 6-deoxy-6-^{18}F-α-D-galactopyranose for metabolic studies. J Label Cmpds Radiopharmaceuticals XIII:283, 1977(abstract).

6. Barnett JEG, Holman GD, Chalkley RA and Munday KA: Evidence for two assymetric conformational states in the human erythrocyte sugar-transport system. Biochem J 145:417-29, 1975.

7. Barnett JEG, Holman GD and Munday KA: Structural requirements for binding to the sugar-transport system of the human erythrocyte. Biochem J 131:211-21, 1973.

8. Raymond AL and Schroeder EF: Synthesis of some iodo-sugar derivatives. J Amer Chem Soc 70:2785-91, 1948.

9. Helus F, Sinn H, Maierborst W and Sahm U: I-123 Trapping Mittels goldfolie. Radiochem Radioanal Letters 32:311-14, 1978.

10. Tse JW, Wiebe, LI, Ediss C and Shysh A: Clearance and elimination of ^{125}I-Rose Bengal in the sub-totally nephrectomized chronically uremic rat. Int J Nucl Med Biol 3:134-7, 1976.

11. Tse JW, Shysh A and Wiebe LI: Excretion patterns of ^{125}I-radioiodohippuric acid in normal and in chronically uremic rats. ibid 5:77-82, 1978.

INORGANIC RADIOPHARMACEUTICALS

Edward Deutsch

Department of Chemistry
University of Cincinnati
Cincinnati, Ohio 45221

ABSTRACT

The chemistries of gallium and technetium radiopharmaceuticals are discussed from the viewpoints of periodic relationships and the information to be gained *via* application of the techniques and methodologies of inorganic chemistry. Biodistribution of gallium radiopharmaceuticals appears to be determined by competetive equilibria, thermodynamically driven but possibly kinetically controlled. The chemistry of reduced technetium radiopharmaceuticals is complicated and subtle, but techniques such as single crystal x-ray structural analysis, electrochemistry, and high performance liquid chromatography, are being effectively used to elucidate this chemistry. Recent results from applications of these techniques are presented, emphasizing the growing synergism between inorganic chemistry and nuclear medicine.

INTRODUCTION

More than sixty of the 104 established elements have been used in, or have been investigated for use in, the preparation of radiopharmaceuticals. Figure 1 shows some of these elements in their periodic relationships. While this compilation is undoubtedly incomplete, it does dramatically reflect the wide ranging search that has been conducted for radionuclides which might satisfy the stringent requirements of nuclear medicine. It also emphasizes that the ultimate efficacy of a radiopharmaceutical depends upon the chemical properties of the element about which the pharmaceutical is constructed. It is the chemistry of the element which will determine the methods of production and purification of useful forms of a given radionuclide, the routes available for synthesis of radiopharmaceuticals, the properties of these radiopharmaceuticals, and in general the in vivo distributions of the radiopharmaceuticals.

Since it is not feasible to discuss the inorganic chemistry of sixty elements, the scope of this review has been limited to elements of the periodic Groups III and VII wherein reside some of the isotopes of greatest value to nuclear medicine (1). Within Group IIIA, the inorganic chemistry of gallium will be briefly discussed because of the current interest in Ga-68 for positron tomography, and also the widespread use of Ga-67 for tumor imaging. Within Group VIIB, the inorganic chemistry of technetium will be discussed at length since Tc-99m is clearly the preeminent radioisotope in clinical diagnostic imaging, and as such is currently the focus of the bulk of research aimed towards radiopharmaceutical development (1).

GALLIUM (GROUP IIIA)

Periodic Relationships. Gallium is a member of Group IIIA (Figure 1) which is composed of those elements having three valence electrons beyond a noble gas configuration. All of these elements readily lose three electrons, yielding the 3+ oxidation state, and the M(III) ions dominate the aqueous chemistry of Al, Ga and In. (The tendency to form the 1+ oxidation state increases as the Group is descended, and for Tl this is the more stable state in aqueous solution (2).) All the early elements of Group IIIB, from Sc through the lanthanides, exhibit the same predilection for the 3+ oxidation state, but the actinides are characterized by variable oxidation states (3). Thus within Groups IIIA and IIIB there are 20 elements that exhibit similar chemistry, existing in aqueous solution predominantly as M(III) ions with a d^0 electron configuration. From the viewpoint of biological relevance it is important to note that the chemistry of these ions mimics that of high spin iron(III) which has a d^5 electron configuration.

Chemistry. In aqueous solution only the 3+ oxidation state of gallium is viable. Complexes of gallium(III) are assumed to have octahedral stereochemistry, although to our knowledge this assumption has not yet been confirmed by single crystal x-ray structural analysis of a complex containing a ligand relevant to nuclear medicine. In the absence of coordinating ligands gallium(III) exists as the hexaaquo ion $[Ga(H_2O)_6]^{3+}$ (4). This species is quite acidic (pKa ca. 2.6) and on addition of base it is successively converted to several metastable hydroxide complexes and finally precipitates as the insoluble, crystalline GaOOH (5). The presence of a strong complexing ligand is therefore necessary to maintain gallium(III) in aqueous solution at neutral pH. Gallium(III) has a high affinity for fluoride and ligands binding through oxygen or nitrogen atoms, and a somewhat lesser affinity for chloride, but the half-life for exchange of monodentate ligands on gallium(III) is of the order of a millisecond (6). Thus, complexes of monodentate ligands (e.g. fluoride, chloride) rapidly decompose in aqueous solution unless a large excess of ligand is present (i.e., gallium(III) complexes with monodentate ligands are labile and consequently there is only a very small kinetic barrier to attainment of thermodynamic equilibrium). Stabilization of gallium(III) in neutral aqueous solution requires multidentate, chelating ligands which bind through oxygen and nitrogen atoms (e.g. EDTA, oxalate, 8-hydroxyquinoline, etc.), and which are relatively inert to displacement by water or other ligands (6). The applications of gallium chemistry to the development of radiopharmaceuticals revolve about the kinetic, thermodynamic and structural properties of these chelate complexes.

Radiopharmaceuticals. Gallium-67 radiopharmaceuticals have proven to be efficacious tumor imaging agents (7,8), and analogous gallium-68 agents are currently of interest for use in positron tomography (9,10). The preparation, quality control, and in vivo distribution of gallium (and other Group III element) radiopharmaceuticals are dominated by considerations of the kinetic and thermodynamic stability of chelate complexes (1,8-11).

Current Ga-68 generators produce the very stable Ga-EDTA complex (12) which is difficult to convert to less stable complexes (9,10). Ehrhardt and Welch (10) have recently reported the development of a Ga-68, solvent extraction, generator which produces a weak 8-hydroxyquinoline complex that in turn can be readily transformed to more stable complexes via simple ligand exchange reactions. Similarly, Arino, Skraba and Kramer (13) have developed a miniaturized, ion exchange, Ga-68 generator which delivers the weak Ga-oxalate complex. These generators should allow the synthesis of Ga-68 chelate complexes with a wide range of thermodynamic stabilities.

Iron(III) is transported in vivo by glycoproteins referred to as transferrins. Because of the similarity between high spin iron(III) and the M(III) ions of Group III (see above), transferrins also have a high affinity for Ga, In, and, to a lesser extent, the lanthanides. Therefore the in vivo distribution of Group III radiopharmaceuticals appears to be dominated by competition for the Group III metal between transferrin (including lactoferrin (14,15)) and the administered chelate ligand (1,7-9,11). Thus, similar tissue distributions are obtained when Ga-67 is administered as the chloride, citrate, nitrilotriacetate and salicylate complexes, indicating that all these relatively weak adducts decompose in vivo to generate Ga-67 labeled transferrin (1). Contrariwise, Ga-67 labeled EDTA is eliminated intact, indicating that in vivo this stable complex does not transfer gallium to transferrin. However, Ga-EDTA must acquire its in vivo stability from a kinetic phenomenon rather than from a lack of thermodynamic driving force to form Ga-transferrin. This conclusion (1) results from the observation that the tissue distribution of the thermodynamically more stable Ga-DTPA complex is similar to that of the weak Ga-citrate, and not to that of Ga-EDTA. Thus it appears that Ga-DTPA has available to it a kinetic path for gallium transfer that is not available to Ga-EDTA; this path probably involves participation by the dangling (i.e. not coordinated) chelate arms of Ga-DTPA which are not present in Ga-EDTA, but support for this hypothesis will have to await future kinetic and structural studies on this class of complexes.

TECHNETIUM (GROUP VIIB)

It has been noted many times, most recently in excellent reviews by Hayes (1) and Eckelman and Levenson (16), that the availability and optimum nuclear properties of technetium-99m have led to its wide spread use in diagnostic imaging. However, equally important in establishing the preeminence of Tc-99m among medically useful radionuclides is the fact that the several oxidation states of technetium exhibit diverse chemistries which allow Tc-99m to be incorporated into a variety of formulations that are specific for different organs (17). The potentialities for the development of new radiopharmaceuticals that are inherent in the rich, varied chemistry of technetium are only beginning to be realized, primarily because the inorganic chemistry of technetium has not yet been extensively developed. This is an historical situation, arising from both the nonoccurence of technetium in nature and the difficulties encountered in conducting macroscopic chemistry with radioactive materials. However, the pressing need (16-18) of nuclear medicine for extensive, detailed, fundamental research into the inorganic chemistry of technetium is providing a driving force to rectify this historical situation and generate the information which will be necessary for significant advances in the synthesis, use, and understanding of technetium radiopharmaceuticals.

Periodic Relationships. Technetium is a member of Group VIIB (Figure 1) which is composed of those transition metal elements having seven electrons beyond a noble gas configuration. These elements, Mn, Tc and Re, all readily lose seven electrons to yield the +7 oxidation state permetallate anions MnO_4^-, TcO_4^- and ReO_4^-, in analogy to the perhallate anions of Group VIIA (ClO_4^-, BrO_4^- and IO_4^-). While not as definitive as those noted for gallium, periodic relationships involving technetium have been very important in the early stages of developing the chemistry of this element (19-23). (a) From its position in the middle of the d-block transition metals technetium is expected to exhibit multiple oxidation states and an extensive redox chemistry, and indeed every oxidation state from -1 to +7 (19,23) has been reported. (b) The lanthanide contraction causes the chemistry of second and third row transition metal elements to be more alike than the chemistry of second and first row elements. This generalization holds true for the Group VIIB triad, the chemistries of Re and Tc being more similar than the chemistries of Tc and Mn. (c) From the aqueous chemistries of Mn and Re, it is expected that the aqueous chemistry of Tc will be dominated by the oxidizing power of soluble TcO_4^- and the great thermodynamic stability of insoluble TcO_2, and this is indeed the case. (d) Extrapolation from the chemistries of the neighboring elements Mo, Ru and Re, leads to the expectations that Tc complexes will often contain strongly bonded oxo ligands, can contain metal-metal bonds, and will exhibit variable coordination numbers and geometries. Data presented below show that these expectations are fully realized.

In general, the position of technetium in the periodic table ensures that technetium will exhibit the most complicated and subtle chemistry of any of the elements commonly used in nuclear medicine. While this intricate chemistry creates significant barriers to our current understanding of the behavior of technetium radiopharmaceuticals, it also creates the possibility of generating wide ranges of complexes with properties that will be advantageous to the design and synthesis of radiopharmaceuticals.

Chemistry. The systematic chemistry of technetium is relatively unexplored, but that which is established, or thought to be established, has been reviewed several times (1,16-25) from the viewpoints of both nuclear medicine and inorganic chemistry. Therefore only limited themes related to recent developments in technetium chemistry will be discussed here.

The aqueous chemistry of technetium is dominated by the driving force for reduction of pertechnetate to the very stable, insoluble TcO_2. This reduction can occur in vivo, although very slowly (1). In the presence of appropriate ligands, the lower oxidation states of technetium can be stabilized and TcO_2 precipitation prevented. Certain of these lower oxidation states, in certain coordination environments, can be expected to be inert to ligand substitution; this is a highly desirable property since it allows reduced technetium complexes to retain their integrity in vivo. Thus the two greatest challenges now confronting the development of technetium radiopharmaceuticals are (1) the controlled synthesis of substitution inert, reduced technetium complexes, and (2) the elucidation of the chemistry and structure of these complexes. Several inorganic chemistry research groups have recently joined the practitioners of nuclear medicine in taking up these challenges, and they have brought with them their own arsenal of tools, techniques and methodologies to add to the armamentarium already deployed. In the following sections, recent results obtained from single crystal x-ray structural analysis, electrochemical studies, and high performance liquid chromatography (HPLC) will be presented in some detail. Techniques involving electron spin resonance (ESR) (26,27), pulse radiolysis (28), mass spectroscopy, and ion probe analysis also appear to have the sensitivity required for technetium research and should prove valuable when further developed. In addition, ion exchange distribution (29) and electrophoretic (30-32) techniques have recently been used effectively to determine charges, hydrolysis constants, and ligand-to-metal ratios (32), of technetium complexes in solution. In an elegant series of studies, Loberg and Fields (33-35) have recently employed a variety of methodologies, including double-labeling, electrophoresis, ligand exchange kinetics, and HPLC, to determine the nature of IDA-technetium complexes in solution.

Inherent in the application of many of the techniques described above is the use of the long-lived Tc-99 isotope, and much concern has been voiced about the applicability of results obtained with Tc-99 to radiopharmaceutical problems involving Tc-99m. The chemical and biological properties of Tc-99 and Tc-99m have been shown to be identical (36,37), as indeed they must, and no example has yet been presented where Tc-99 and Tc-99m systems differ significantly (ignoring obvious cases such as insufficient tin(II) available for complete reduction of Tc-99 pertechnetate (38)). Reactions which are second-order in technetium concentration will, of course, be very sensitive to the total concentration of technetium (as both 99 and 99m), and this could engender differences between "carrier" and "carrier-free" systems. However, Tc-99m produced from Mo-99 generators is never "carrier-free" (39), and in fact clincially operated generators can produce eluants that are ca. 1 μM in total technetium (calculated from the data of reference 39 assuming 24 hr since last elution, and 700 mCi Tc-99m collected in 5 ml eluant). Several techniques, especially electrochemistry and ESR, are sensitive and effective at this, and even lower, concentration levels. Thus the range of technetium concentrations encountered in radiopharmaceutical studies overlaps considerably with the range of concentrations amenable to chemical investigation, making "carrier" and "carrier-free" Tc-99m/Tc-99 studies complementary rather than dichotomous.

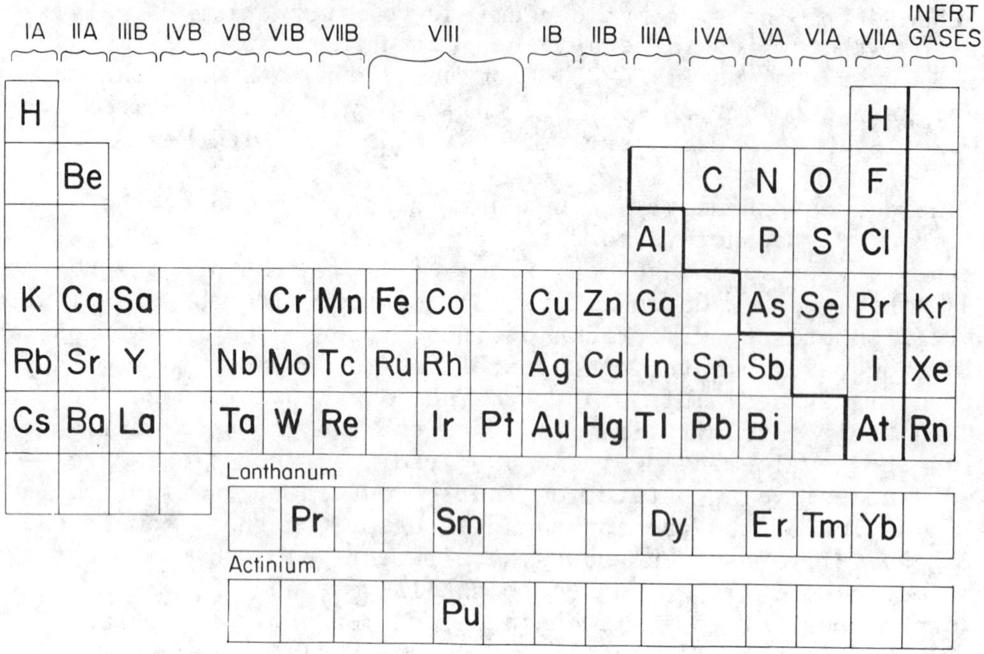

Figure 1.

Some elements used in, or proposed for use in, nuclear medicine.

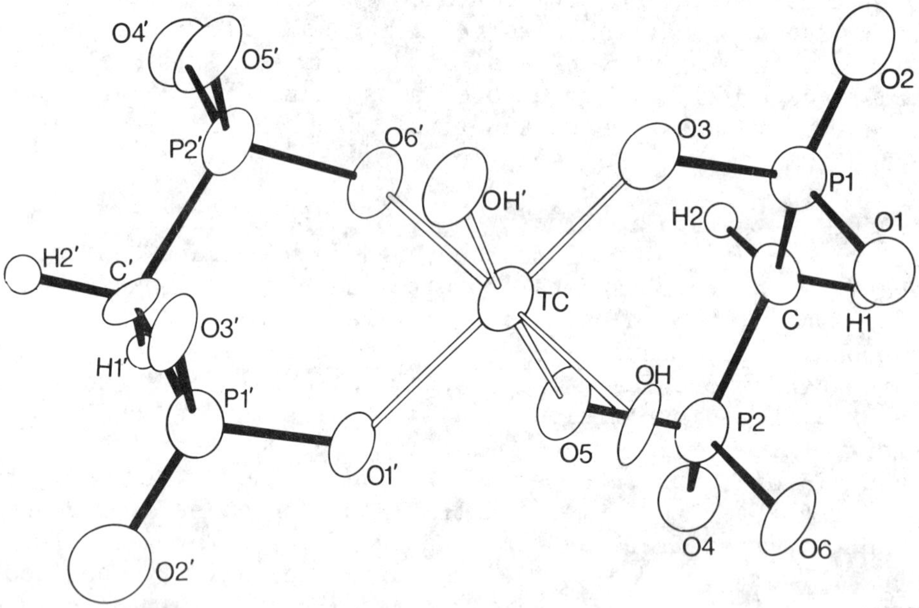

Figure 2.

Perspective view of a portion of the $[Tc(OH)(MDP)^-]$ infinite polymer where MDP represents methylene diphosphonate in an unknown protonation state. Hydrogen atoms bonded to oxygen, lithium counterions, and waters of hydration have been omitted for clarity.

Single Crystal X-ray Structural Analysis. Table 1 lists those technetium complexes known to us that have been characterized by single crystal x-ray methods. This compilation includes both published and unpublished, as of March 1979, data. Several results of considerable importance to radiopharmaceutical development, can be gleaned from these structures. (a) Reduced technetium complexes exhibit a wide range of coordination numbers and geometries. They are certainly not restricted to six-coordinate, octahedral structures as is often assumed in the radiopharmaceutical literature (16,36). There are five complexes in Table 2 that were prepared under conditions relevant to current radiopharmaceutical formulations: $Tc(dmg)_3(SnCl_3)(OH)$; $Tc(HBPz_3)Cl_2O$; $[Tc(OH)(MDP)^-]_n$, Figure 2; $TcO(SCH_2CH_2S)_2^-$; $TcO(SCH_2C(O)S)_2$. Of these, two are five-coordinate with square pyramidal geometry, two are six-coordinate with near octahedral geometry, and one is seven-coordinate with geometry approximating that of a monocapped octahedron. (b) Technetium(V) appears to be a prevalent oxidation state for complexes prepared under aqueous, aerobic conditions; at least four of the above five complexes exhibit the +5 oxidation state. (c) Three of the five complexes noted above, plus $TcOCl_4^-$, contain the $Tc^V{=}O$ center, and the other two contain bridging oxygen atoms which might reasonably be assumed to be derived from a Tc=O linkage. It is thus likely that technetium radiopharmaceuticals prepared in aqueous media (especially by the reduction of pertechnetate) will contain oxo ligands, and the properties associated with these ligands should be anticipated in the chemistry of technetium radiopharmaceuticals. The oxo group exerts a strong trans labilizing effect (54) which leads to a vacant trans coordination site in $TcO(SCH_2CH_2S)_2^-$, $TcO(SCH_2C(O)S)_2^-$ and $TcOCl_4^-$, although in solution this axial site presumably contains a loosely bonded solvent molecule. In $Tc(HBPz_3)Cl_2O$, the tridentate $HBPz_3^-$ ligand suppresses five-coordination but the influence of the oxo group leads to a weak trans ligand-metal bond as reflected in the trans nitrogen atom being 0.2 A° further from Tc than are the two cis nitrogen atoms. In contrast to the labile coordination site trans to the oxo ligand, the four equatorial coordination sites cis to the oxo ligand are resistant to ligand substitution, and thus should provide a suitable template about which to construct radiopharmaceuticals that will retain their integrity in vivo. The future design and synthesis of technetium radiopharmaceuticals should take into account the likely occurence of tightly bonded oxo centers such as $Tc^V{=}O$, and the markedly different bonding properties of the equatorial and axial ligation sites within these centers. (d) The structures of $Tc_2Cl_8^{3-}$ and $Tc_2(OOCC(Me)_3)_4Cl_2$ demonstrate that technetium can form complexes with metal-metal bonds, and this possibility should not be overlooked when formulating possible structures for reduced technetium radiopharmaceuticals. (e) tr-$Tc(NH_3)_4(NO)(OH_2)^{2+}$ represents the first member of what should be a large class of nitrosyl-technetium complexes analogous to the well known nitrosyl-ruthenium complexes (55). The NO ligand stabilizes low oxidation states (49) and Tc-NO centers should provide suitable templates for synthesis of substitution inert radiopharmaceuticals (27). (f) Figure 2 shows the results of the first structural analysis to be conducted on a technetium complex containing a clinically used radiopharmaceutical ligand, i.e. MDP. This very difficult analysis was conducted by Dr. Robert Barnett of the Procter & Gamble Miami Valley Laboratories. The solid state structure of $[Tc(OH)MDP^-]_n$ (prepared by substitution of MDP onto $TcBr_6^{2-}$) is polymeric with technetium centers bridged by MDP and OH (or OH_2) ligands, and MDP ligands in turn bridged by technetium centers. This structure, as well as other results mentioned later, indicate that diphosphonate (and presumably pyrophosphate, polyphosphate, etc.) technetium radiopharmaceuticals are most probably oligomeric or polymeric mix-

TABLE 1. TECHNETIUM COMPLEXES CHARACTERIZED BY SINGLE CRYSTAL X-RAY STRUCTURAL ANALYSIS

Complex*	Coordination Number	Oxidation State	Reference
$Tc(DIARS)_2Cl_4^+$	8	V	a
$Tc(DMG)_3(SnCl_3)(OH)\cdot 3H_2O$	7	V	40,a
$TcCl_3(CO)(Me_2PPh)_3\cdot EtOH$	7	III	41
$(TcOF_4)_3$	6	VI	42
$Tc(HBPz_3)Cl_2O$	6	V	a
$[Tc(OH)(MDP)^-]_n$	6	?	b
$TcCl_6^{2-}$	6	IV	43
$(Tc,Sn)Br_6^{2-}$	6	IV	a
$Tc(DIARS)_2Cl_2^+$	6	III	a
$TcCl_3(Me_2PPh)_3$	6	III	44
$TcCl(ACAC)_2(PPh_3)$	6	III	45
$Tc_2(OOCC(Me)_3)_4Cl_2$	6	III	46
$Tc_2Cl_8^{3-}$	6	II,III	47
$TcCl_2((EtO)_2PPh)_4$	6	II	48
<u>tr</u>-$Tc(NH_3)_4(NO)(H_2O)^{2+}$	6	I	49,c
$TPP[Tc(CO)_3]_2$	6	I	50
$TcO(SCH_2C(O)S)_2^-$	5	V	51
$TcO(SCH_2CH_2S)_2^-$	5	V	52
$TcOCl_4^-$	5	V	d
Tc_2O_7	4	VII	53

*Abbreviations used: DIARS = o-phenylenebisdimethylarsine, DMG = dimethylglyoxime in unknown protonation state, ACAC = acetylacetonate monoanion, $HBPz_3$ = hydrotris(1-pyrazolyl)borate monoanion, MDP = methylene diphosphonate in unknown protonation state, TPP = tetraphenylporphine, Me = CH_3, Et = C_2H_5, Ph = C_6H_5.

a. Deutsch E, Elder RC, et al: unpublished data.
b. Deutsch E, Barnett R, Libson K: unpublished data.
c. Hoard JL: unpublished data.
d. Davison A, Cotton FA, Day V: unpublished data

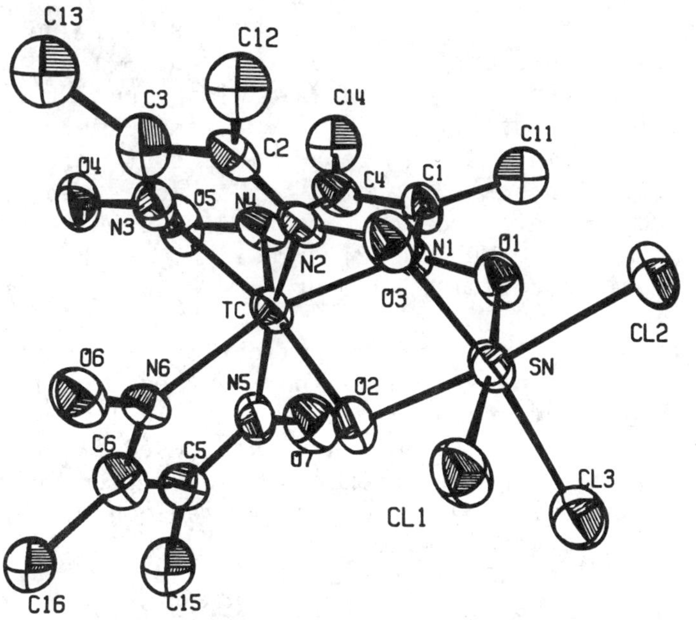

Figure 3.

Perspective view of the seven-coordinate technetium(V) complex Tc(DMG)$_3$-($SnCl_3$)(OH) where DMG represents dimethylglyoxime in an unknown protonation state. Hydrogen atoms and waters of hydration have been omitted for clarity. Reproduced from reference 40 with permission.

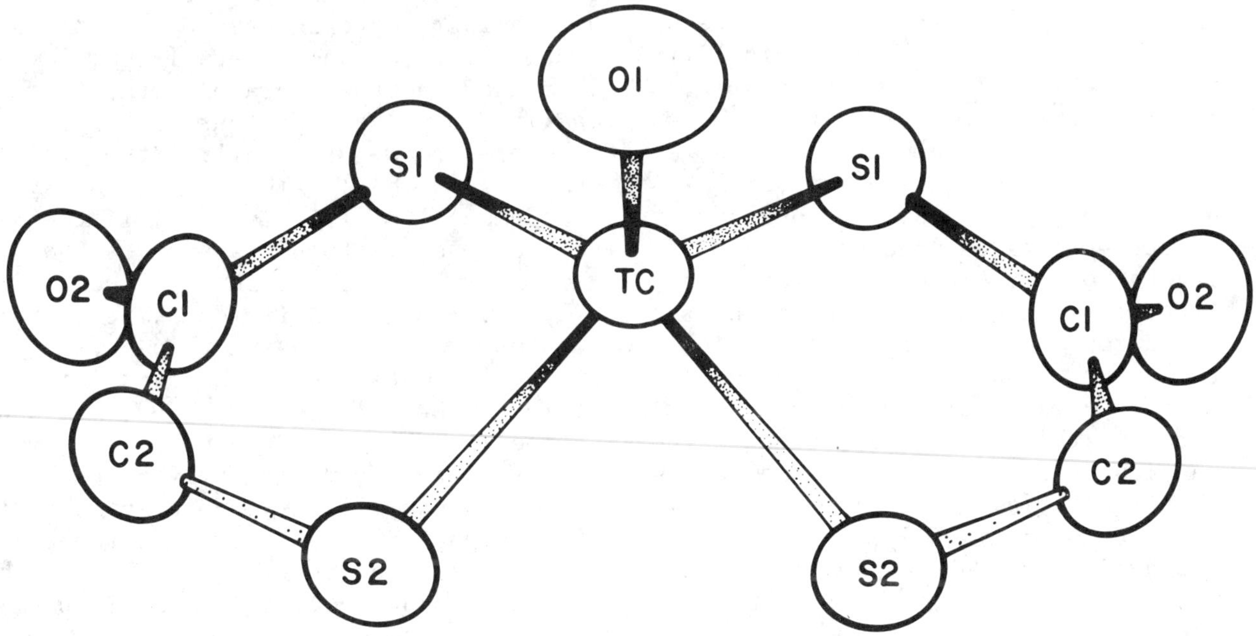

Figure 4.

Perspective view of the five-coordinate technetium(V) complex [TcO(SCH$_2$C(O)-S)$_2$]$^-$. Hydrogen atoms have been omitted for clarity. Reproduced from reference 51 with permission.

tures in aqueous solution. Other clincially used ligands such as EDTA and DTPA (but not IDA) are capable of forming dimeric (56,57) and oligomeric metal complexes, and the prevalent use of tin in radiopharmaceutical formulations (38) raises the additional possibility of mixed-metal dimers, oligomers, etc. In sum, aqueous radiopharmaceutical preparations containing ligands that have the possibility of simultaneously coordinating to more than one metal (e.g. DTPA), and especially those that have been designed to simultaneously coordinate to more than one metal (e.g. MDP, HEDP), are likely to be complex mixtures of oligomeric or polymeric materials, and not single, well defined, chemical species.

Electrochemistry. The first application of electrochemistry to radiopharmaceutical development was the actual synthesis of reduced technetium radiopharmaceuticals by electrochemical means. Chemically reactive electrodes such as tin (58,59) or zirconium (60-62) have been used extensively, but more recently Russell (63,64) has employed chemically inert electrodes such as platinum, gold, glassy carbon, or mercury, to prepare radiopharmaceuticals that are not contaminated by the electrode material.

In addition to the above synthetic applications, electrochemistry has been utilized in efforts to obtain fundamental information about the nature and mechanism of pertechnetate reduction in aqueous solution, and about the redox behavior of reduced technetium complexes. These electron transfer processes are crucial to the preparation of technetium radiopharmaceuticals, but the chemistry involved is labyrinthine and only poorly understood (38). The electrochemical reduction of pertechnetate in aqueous complexing media is complicated by the existence of several species of differing oxidation states, the great chemical and electrochemical reactivity of these reduced species, and the sensitivity of the system to pH, oxygen, electrolyte, nature of complexing ligand, etc. (65-68). The standard electrochemical techniques of polarography and cyclic voltammetry that have been applied to technetium chemistry have enjoyed only limited success, the results often being difficult to interpret and generally uninformative (69-71). This situation arises because the electro-generated technetium species often undergo one or more of several competetive reactions rather than the desired reverse electron transfer at the electrode surface (i.e. the electrochemical couples are irreversible or quasi-reversible). Three techniques have recently been used to overcome this fundamental difficulty. (a) If the electro-generated species is very reactive, reverse electron transfer can be favored over the competing reactions by very rapidly sweeping the applied potential through the region of interest. (b) A variety of different kinetic complications due to slow electron transfer can be avoided by using thin-layer techniques in which a small volume of solution is contiguous with a very high surface area electrode. Under these conditions the entire solution volume, and not just the solution adjacent to the electrode surface, is brought to electrochemical equilibrium. If the electrode is constructed from optically transparent materials, the approach to electrochemical equilibrium can be monitored spectrophotometrically and the technique is designated thin-layer spectroelectrochemistry (72-74). (c) If the electro-generated species reacts rapidly with water, the system may be simplified by the use of nonaqueous solvents. Some recent applications of these three techniques are outlined below:

(a) The logical first step in the reduction of pertechnetate by a variety of chemical agents is the generation of technetium(VI). Despite extensive studies (69-71), technetium(VI) had not been electrochemically detected owing to the rapid chemical and electrochemical reactions of this species in aqueous media. However, by the use of very fast scan cyclic voltammetry we have recently (28) been able to observe technetium(VI) in aqueous alkaline media and show that it has a lifetime on the order of milliseconds, more than sufficient to allow it to participate as an intermediate in redox reactions. (b) A recent polarographic study (68) of pertechnetate reduction in aqueous HEDP solution led to the usual ill-defined waves, and only estimates of $E_{1/2}$ and n values could be made. However, thin-layer coulometric analysis of the reduction of pertechnetate in 0.5 M HEDP (pH 7.4) shows that the first reduction step involves the transfer of exactly two electrons, converting Tc(VII) to Tc(V); the pH dependence of this process indicates that one proton is consumed for each electron transferred. Thin-layer spectroelectrochemical analysis of $TcCl_6^{2-}$ in aqueous chloride media shows that it is reversibly reduced to a Tc(III) species. (c) Astheimer and Schwochau (75) have used stringent anaerobic and nonaqueous conditions to electrochemically prepare salts of the metastable technetate anion, TcO_4^{2-}. Mazzochin and co-workers have electrochemically investigated some of the organometallic technetium complexes listed in Table 1 by utilizing nonaqueous media (76,77).

Recent reports (78,79) have noted that the in vivo distribution of reduced technetium radiopharmaceuticals depends upon the oxidation state of technetium in the final radiopharmaceutical formulation. Indirect chemical titrations have been employed several times in attempts to determine the oxidation state of technetium in radiopharmaceuticals, but interpretation of data obtained by this technique is difficult (16). Electrochemical procedures can much more directly probe this point, and should become both useful and effective once a satisfactory data base has been established. However, a more important question involves the oxidation state of technetium radiopharmaceuticals in vivo, i.e. after they have interacted with the biological redox buffer system. If various redox forms of a technetium radiopharmaceutical rapidly convert to the same oxidation state after injection, it will make no difference which form is injected. But if attainment of redox equilibrium is slow, then different initial oxidation states could have different in vivo distributions; this seems to be the case with some reduced technetium radiopharmaceuticals (78,79), and certainly is the case with pertechnetate (1). Modern electrochemical techniques have the sensitivity necessary for these in vivo determinations, and micro-electrodes suitable for in vivo placement have been developed (80). The future application of these techniques and methodologies to in vivo technetium analysis should provide crucial data on the chemistry and biological distribution mechanism(s) of technetium radiopharmaceuticals.

HPLC. High performance liquid chromatography (HPLC) promises to become the single most efficacious technique available for unraveling the complexities of reduced technetium radiopharmaceutical formulations. While conventional quality control chromatographic techniques can separate radiopharmaceuticals into TcO_2, pertechnetate and "complexed technetium" (81), HPLC has the additional capability of separating the "complexed technetium" into its component fractions. This raises the possibility of functionally specific quality control wherein only that component which is most specific for the target organ is isolated and used. Such an approach would resolve many of the irreproducibility problems (38) associated with reduced technetium radiopharmaceuticals. By coupling HPLC with extremely sensitive electrochemical detectors (82,83) it should be possible both to identify individual components and also to determine the amounts of these components in biological samples (80,82,83).

To our knowledge, only two applications of HPLC to technetium radiopharmaceutical development have yet been reported. Loberg and Fields (34) separated Tc(Sn)-HIDA into HIDA, Sn-HIDA and Tc-HIDA components, and in this system the Tc-HIDA fraction is presumably homogeneous.* Last year, Majerik and Russell (84) were able to separate pertechnetate from a variety of reduced technetium radiopharmaceuticals, but at that time had not yet been able to separate the reduced radiopharmaceuticals into their components. In very recent results we have been able to accomplish this separation on a Tc($NaBH_4$)-HEDP radiopharmaceutical analog; the HPLC analysis clearly shows that the "complexed technetium" in Tc($NaBH_4$)-HEDP is present in several different forms and that this radiopharmaceutical analog is not a single, well-defined, chemical species.

Radiopharmaceuticals.* Technetium radiopharmaceuticals have recently been reviewed several times (1,16-18,32) and therefore only limited themes of current interest will be discussed here.

Reduced technetium radiopharmaceuticals are usually formulated (a) with a small ligand (such as HEDP or DTPA) which provides the entire complex with the biological characteristics necessary for organ specificity, or (b) with a bifunctional ligand (such as HIDA (32,35) or EDTA-derivatized biomolecules (85)) which has both a moiety designed to bind to technetium, and a moiety designed to provide organ specificity. In either formulation, pertechnetate is reduced in the presence of the ligand, and stannous ion is the reductant most commonly used (38). A question fundamental to the design and development of technetium radiopharmaceuticals is whether or not tin is incorporated into the chemical structure of the final radiopharmaceutical. Double labeling experiments showing that radioactive tin localizes in the same tissues as does technetium, do not directly answer this point since independent tin-ligand and technetium-ligand complexes may very likely have the same biodistrobution. Likewise, experiments showing different biodistributions for radiopharmaceuticals produced with non-stannous reductants, also do not directly answer the point since different reductants may produce different technetium oxidation states or may themselves be incorporated into the technetium complex (e.g. thiols). Therefore chemical evidence must be used to answer this essentially chemical question, and the relevant information currently available is tabulated below:

(a) After HPLC separation of the original reaction mixture, Loberg and Fields (34) established that Tc(Sn)-HIDA does not contain tin. (b) Single crystal x-ray structural analysis of Tc(Sn)-dmg (40) establishes that this material does contain chemically bonded tin. (c) Macroscopically prepared and purified Tc(Sn)-HEDP is homogeneous and does contain tin, the Sn/Tc ratio being ca. 3 by x-ray fluorescence analysis. (d) Steigman, Meinken and Richards (68) have recently shown that Tc(Sn)-PO_4 in solution is a mixed-metal complex that does contain tin. (e) Single crystal x-ray structural analysis of Tc(Sn)-Br (Table 1) shows that this material is an isomorphous mixture of $TcBr_6^{2-}$ and $SnBr_6^{2-}$ octahedra. In addition, x-ray fluorescence analysis shows that the Sn/Tc ratio is 2 for a variety of samples; this implies some long-range order, and therefore Sn-Tc interaction, in the solid state. We have also seen indications of Sn-Tc interactions in solution for this system, reminiscent of better established Sn-Pt (86) and Sn-Tc (68) solution interactions.

---

* Radiopharmaceutical formulations prepared by the reduction of pertechnetate in the presence of excess ligand are designated as "Tc(reductant)-ligand".

Thus it appears that with certain ligands, especially those that are capable of simultaneously bonding to two metal atoms, tin can and does bond to the technetium complex. This is not unexpected considering that Sn(IV) and Tc(IV) centers have the same size and charge, and therefore should have an affinity for the same types of ligands. Also, in addition to being a good reductant, the $SnCl_3^-$ moiety is a good ligand in its own right and is known to interact strongly with low-valent transition metals (86-88); this provides a reasonable mechanism for the chemical incorporation of tin into technetium radiopharmaceuticals (40). Monomeric complexes of technetium do, of course, occur (see Table 1), and will be favored even in the presence of excess tin if the ligands do not readily bind simultaneously to two metal ions. This seems to be the case with the tri-dentate HIDA ligand (34).

Given that tin is likely to be incorporated into technetium radiopharmaceuticals containing good bridging ligands (exemplified by the phosphates and phosphonates) it is of interest to prepare analogous complexes without the use of tin in order to compare the resultant properties. This can be accomplished electrochemically at an inert electrode (63,64), or by use of an innocent chemical reductant which is unlikely to be incorporated into the final radiopharmaceutical (e.g. $NaBH_4$, formamidine sulfinic acid (89)). We have recently taken an alternate route, i.e. that of substituting a ligand onto a pre-reduced technetium center such as $TcBr_6^{2-}$ (37,90). Compared to currently used redox procedures this route allows much greater flexibility and subtlety in the design and synthesis of reduced radiopharmaceuticals -- in principle it allows control of the oxidation state, coordination number, and coordination geometry of the final radiopharmaceutical. Reduced Tc-99m complexes, such as $^{99m}TcCl_6^{2-}$, can be quantitatively prepared from $^{99m}TcO_4^-$ in "carrier-free" situations (91). Substitution of HEDP onto $TcBr_6^{2-}$ leads to a product (designated Tc-HEDP) that has the same in vivo distribution as Tc(Sn)-HEDP and Tc($NaBH_4$)-HEDP, indicating that it is the HEDP ligand which primarily controls biodistribution (37,90). Double labeling of Tc-HEDP with [^{3}H]HEDP shows that in solution this complex contains HEDP and Tc in a ratio of 1:1, in good agreement with the solid state structure of Tc-MDP (Figure 2) which is a 1:1 polymer. Therefore in solution Tc-HEDP must be at least a dimer, and more probably a polymer, since there is no way that a single HEDP ligand can satisfy the coordination requirements of reduced technetium within a monomeric complex. This observation provides independent support for the hypothesis presented earlier, i.e. that in solution diphosphonate technetium radiopharmaceuticals are most probably polymeric mixtures.

Summary. The results presented in this section demonstrate dramatically that the nascent application of the techniques and methodologies of inorganic chemistry have already been very effective in elucidating the chemistry of technetium. While there are still many fundamental questions to be asked and problems to be solved, within the next few years this synergism between inorganic chemistry and nuclear medicine should produce a quantum jump in our abilities to synthesize, understand, and effectively use, reduced technetium radiopharmaceuticals.

ACKNOWLEDGMENTS

Preparation of this manuscript was supported in part by the National Institutes of Health, Grant No. HL-21276. The author gratefully acknowledges receiving pre-publication results from Professors J. L. Hoard and A. Davison, and especially thanks his colleagues and students for their many significant contributions to the ideas and research presented in this paper.

REFERENCES

1. Hayes RL: Chemistry and radiochemistry of metal-ion nuclides commonly employed in radiopharmaceuticals. In The Chemistry of Radiopharmaceuticals, Heindel ND, Burns HD, Honda T, Brady LW, eds, New York, Masson, 1978, pp 155-168
2. Cotton FA, Wilkinson G: Advanced Inorganic Chemistry. New York, Interscience, 1972, p 280
3. Newton TW: The Kinetics of the Oxidation-Reduction Reactions of Uranium, Neptunium, Plutonium, and Americium in Aqueous Solutions. Springfield, Va. USERDA, 1975, pp 3-7
4. Fiat D, Connick RE: Magnetic resonance studies of ion solvation. The coordination number of gallium(III) ions in aqueous solution. J Amer Chem Soc 88: 4754, 1966
5. Craig HR, Tyree SY Jr: Hydrolytic behavior of gallium(III) perchlorate. Inorg Chem 8: 591-594, 1969
6. Margerum DW, Cayley GR, Weatherburn DC, Pagenkopf GK: Kinetics and mechanisms of complex formation and ligand exchange. In Coordination Chemistry, Martell AE ed, vol 2, ACS Monograph 174. Washington DC, ACS, 1978, pp 1-220
7. Larson SM, Milder MS, Johnston GS: Tumor-seeking radiopharmaceuticals: gallium-67. In Radiopharmaceuticals, Subramanian G, Rhodes BA, Cooper JF, Sodd VJ, eds, New York, SNM, 1975, pp 413-434
8. Thakur ML: Gallium-67 and indium-111 radiopharmaceuticals. Int J Appl Radiat Isot 28: 183-291, 1977
9. Hnatowich DJ: A review of radiopharmaceutical development with short-lived generator-produced radionuclides other than 99m-Tc. Int J Appl Radiat Isot 28: 169-181, 1977
10. Ehrhardt GJ, Welch MJ: A new germanium-68/gallium-68 generator. J Nucl Med 19: 925-929, 1978
11. Welch MJ, Welch TJ: Solution chemistry of carrier-free indium. In Radiopharmaceuticals, Subramanian G, Rhodes BA, Cooper JF, Sodd VJ, eds, New York, SNM, 1975, pp 73-79
12. Greene MW, Tucker WD: An improved gallium-68 cow. Int J Appl Radiat Isot 12: 62, 1961
13. Arino H, Skraba WJ, Kramer HH: A new 68-Ge/68-Ga radioisotope generator system. Int J Appl Radiat Isot 29: 1170120, 1978
14. Weiner RE, Hoffer PB, Thakur ML: Identification of Ga-67 binding component in human neutrophils. J Nucl Med 19: 732, 1978 (abst)
15. Hoffer PB, Huberty J, Samuel A: Use of an affinity chromatographic method to determine relative binding of 67-Ga to lactoferrin and transferrin in tear fluid. J Nucl Med 19: 732, 1978 (abst)
16. Eckelman WC, Levenson SM: Radiopharmaceuticals labelled with technetium. Int J App Radiat Isot 28: 67-82, 1977
17. Siegel JA, Deutsch E: Synthesis of radiopharmaceuticals by the reduction of 99m-TcO_4^-: 99m-Tc-labelled heart scanning agents. Ann Rep Inorg Gen Synth 1975: 311-326, 1976
18. Richards P, Steigman J: Chemistry of technetium as applied to radiopharmaceuticals. In Radiopharmaceuticals, Subramanian G, Rhodes BA, Cooper JF, Sodd VJ, eds, New York, SNM, 1975, pp 23-35
19. Colton R: The Chemistry of Rhenium and Technetium. New York, Interscience 1965
20. Peacock RD: The Chemistry of Technetium and Rhenium. Amsterdam, Elsevier, 1966.

21. Boyd GE: Technetium and promethium. J Chem Educ 36: 3-14, 1959
22. Kotegov KV, Pavlov ON, Shvedov VP: Technetium. Adv Inorg Chem Radiochem, 11: 1-90, 1968
23. Anders E: Technetium and astatine chemistry. Ann Rev Nucl Sci 9: 203-220, 1959
24. Schwochau K: Zur chemie des technetiums. Angew Chem 76: 9-19, 1964
25. Lavrukhina AK, Pozdnyakov AA: Technetium. In Analytical Chemistry of Technetium, Promethium, Astatine and Francium. Ann Arbor, Ann Arbor-Humphrey Sci Publ, 1970, pp 1-102
26. Heitzmann MW, Ford LA, Yang GC, Benson WR: ESR studies of 99-Tc in aqueous solutions II. Read before the American Chemical Society, Miami Beach, Fla., Sept. 11-14, 1978
27. Yang GC, Heitzmann MW, Ford LA, Benson WR: ESR nitrosyl halide technetium complexes. Read before the American Chemical Society, Miami Beach, Fla., Sept. 11-14, 1978
28. Deutsch E, Heineman WR, Hurst R, et al: Production, detection, and characterization of transient hexavalent technetium in aqueous alkaline media by pulse radiolysis and very fast scan cyclic voltammetry. J Chem Soc Chem Comm: 1038-1040, 1978
29. Owunwanne A, Marinsky J, Blau M: Charge and nature of technetium species produced in the reduction of pertechnetate by stannous ion. J Nucl Med 18: 1099-1105, 1977
30. Gorski B, Koch H: Zur chemie des technetium in wassriger losung -- I: J Inorg Nucl Chem 31: 3565-3571, 1969
31. Gorski B, Koch H: Uber die komplexbildung von technetium mit chelatbildenden ligenden -- II: J Inorg Nucl Chem 32: 3831-3836, 1970
32. Burns HD, Worley P, Wagner HN Jr, et al: Design of technetium radiopharmaceuticals. In The Chemistry of Radiopharmaceuticals, Heindel ND, Burns HD, Honda T, Brady LW, eds, New York, Masson, 1978, pp 269-289
33. Loberg MD, Fields AT: Stability of 99m-Tc-labeled N-substituted iminodiacetic acids: Ligand exchange reaction between 99m-Tc-HIDA and EDTA. Int J Appl Radiat Isot 28: 687-692, 1977
34. Loberg MD, Fields AT: Chemical structure of technetium-99m-labeled N-(2,6-dimethylphenylcarbamoylmethyl)iminodiacetic acid (Tc-HIDA). Int J Appl Radiat Isot 29: 167-173, 1978
35. Loberg MD: Radiolabeled drug analogs based on N-substituted iminodiacetic acid. In The Chemistry of Radiopharmaceuticals, Heindel ND, Burns HD, Honda T, Brady LW, eds, New York, Masson, 1978, pp 191-204
36. Hambright P, McRae J, Valk PE, et al: Chemistry of technetium radiopharmaceuticals I. Exploration of the tissue distribution and oxidation state consequences of technetium(IV) in Tc-Sn-gluconate and Tc-Sn-EHDP using carrier 99-Tc. J Nucl Med 16: 478-482, 1975
37. Deutsch EA, Libson KF, Becker CB, et al: Preparation, chemistry, and biological distribution of technetium phosphonate radiopharmaceuticals synthesized without stannous ion. Read before the Society of Nuclear Medicine 3rd Annual Western Regional Meeting, Vancouver, B.C., October 14, 1978
38. Srivastava SC, Meinken G, Smith TD, Richards P: Problems associated with stannous 99m-Tc-radiopharmaceuticals. Int J Appl Radiat Isot 28: 83-95, 1977
39. Lamson ML, Kirschner AS, Hotte CE, et al: Generator-produced 99m-TcO_4^-: Carrier-free?. J Nucl Med 16: 639-641, 1975

40. Deutsch E, Elder RC, Lange BA, et al: Structural characterization of a bridged 99-Tc-Sn-dimethylglyoxime complex: Implications for the chemistry of 99m-Tc-radiopharmaceuticals prepared by the Sn(II) reduction of pertechnetate. Proc Natl Acad Sci USA 73: 4287-4289, 1976
41. Bandoli G, Clemente DA, Mazzi U: Carbonyltrichlorotris(dimethylphenylphosphine)technetium - ethanol (1/1), the first seven coordinate complex of technetium; position of this molecule in the C_{3v} family. J Chem Soc Dalton Trans: 373-380, 1978
42. Edwards AJ, Jones GR, Sills RJC: Part XII. Trimeric technetium oxide tetrafluoride. J Chem Soc A: 2521-2523, 1970
43. Elder RC, Estes GW, Deutsch E: Ammonium hexachlorotechnetate(IV). Acta Cryst B35: in press, 1979
44. Bandoli G, Clemente DA, Mazzi U: Preparation and crystal and molecular structure of mer-trichlorotris(dimethylphenylphosphone)technetium(III). J Chem Soc Dalton Trans: 125-130, 1976
45. Bandoli G, Clemente DA, Mazzi U: Preparation and molecular structure of trans-chlorobis(pentane-2,4-dionato)(triphenylphosphine)technetium(III). J Chem Soc Dalton Trans: 1837-1844, 1977
46. Cotton FA, Gage LD: The preparation and molecular structure of tetrapivalatodichloroditechnetate(III) and its dirhenium congener: a Tc-Tc quadruple bond. Nouv J Chimie 1: 441-442, 1977
47. Cotton FA, Shive LW: Structure of tripotassium octachloroditechnetate hydrate. Inorg Chem 14: 2032-2035, 1975
48. Mazzi U, Clementa DA, Bandoli G, et al: Technetium-phosphonite complexes. Synthesis of hexacoordinate technetium(II) and technetium(III) complexes of diethyl phenylphosphonite and the crystal structure of dichlorotetrakis-(diethyl phenylphosphonite)technetium(II). Inorg Chem 16: 1042-1048, 1977
49. Armstrong RA, Taube H: Chemistry of trans-aquonitrosyltetraamine-technetium(I) and related studies. Inorg Chem 15: 1904-1909, 1976
50. Tsutsui M, Hrung CP, Ostfeld D, et al: Unusual metalloporphyrin complexes of rhenium and technetium. J Amer Chem Soc 97: 3952-3965, 1975
51. DePamphilis BV, Jones AG, Davis MA, Davison A: Preparation and crystal structure of oxotechnetiumbis(thiomercaptoacetate) and its relationship to radiopharmaceuticals labeled with 99m-Tc. J Amer Chem Soc 100: 5570-5571, 1978
52. Smith JE, Byrne EF, Cotton FA, Sekutowski JC: A thiol complex of technetium pertinent to radiopharmaceutical use of 99m-Tc. J Amer Chem Soc 100: 5571-5572, 1978
53. Krebs B: Technetium(VII)-oxid: Ein ubergansmetalloxid mit molekulstruktur im festen zustand. Z Anorg Allg Chem 380: 145-159, 1971
54. Shustorovich EM, Porai-Koshits MA, Buslaev YA: The mutual influence of ligands in transition metal coordination complexes with multiple metal-ligand bonds. Coord Chem Rev 17: 1-98, 1975
55. Cotton FA, Wilkinson G: Advanced Inorganic Chemistry. New York, Interscience, 1972, pp 1012-1013
56. Shibahara T, Sykes AG: Preparation and characterization of dimeric molybdenum(III)-ethylenediaminetetraacetate complexes. J Chem Soc Dalton Trans: 95-99, 1978
57. Kneale GG, Geddes AJ: The crystal structure of the potassium salt of the binuclear Mo(III) anion di-μ-hydroxo-μ-acetato-μ-ethylenediaminetetraacetato-bis[molybdenum(III)], $K[Mo_2(OH)_2(C_{12}H_{15}O_{10}N_2)]$. Acta Cryst B31: 1233-1236, 1975
58. Belkas EP, Varvarigou AD: A quick and reliable process for preparing 99m-Tc-1: hydroxy-ehtylidene-1,1-diphosphonic acid disodium salt (HEDSPA) using electrolysis. Int J Appl Radiat Isot 27: 590-591, 1976

59. Gil MC, Palma T, Radicella R: Electrolytical labeling of 99m-technetium radiopharmaceuticals. Int J Appl Radiat Isot 27: 69-72, 1976
60. Steigman J, Eckelman WC, Meinken G, et al: The chemistry of technetium labeling of radiopharmaceuticals by electrolysis. J Nucl Med 15: 75-80 1974
61. Chi SL, Hoag SG, Yanchick VA: Electrolytic complexing of glucoheptonate and technetium-99m. J Nucl Med 19: 520-524, 1978
62. Simon J, Ketring AR, Troutner DE: Tetraaza ligands as complexing agents for 99m-Tc. Read before the American Chemical Society, Miami Beach, Fla., Sept. 11-14, 1978
63. Russell CD: Carrier electrochemistry of pertechnetate: application to radiopharmaceutical labelling by controlled potential electrolysis at chemically inert electrodes. Int J Appl Radiat Isot 28: 241-249, 1977
64. Russell CD, Majerik J: Tracer electrochemistry of pertechnetate: chelation of 99m-Tc by EDTA after controlled-potential reduction at mercury and platinum cathodes. Int J Appl Radiat Isot 29: 109-114, 1978
65. Appelbaum GM, Kempf CR, Aidens DA, et al: Electrochemistry of 99-Tc-DTPA systems. Read before the American Chemical Society, Miami Beach, Fla., Sept. 11-14, 1978
66. Russell CD, Cash A: Products of pertechnetate reduction in complexing media. J Nucl Med 19: 694, 1978 (abst)
67. Kissel G, Feldberg SW: Disproportionation of the technetate ion in aqueous alkaline media. An electrochemical study. J Phys Chem 73: 3082-3088, 1969
68. Steigman J, Meinken G, Richards P: The reduction of pertechnetate-99 by stannous chloride - II. The stoichiometry of the reaction in aqueous solutions of several phosphorus(V) compounds. Int J Appl Radiat Isot 29: 653-660, 1978
69. Rulfs CL, Pacer R, Anderson A: The polarography of aqueous pertechnetate ion. J Electroanal Chem 15: 61-66, 1967
70. Russell CD, Cash AG: Polarographic reduction of pertechnetate. J Electroanal Chem 92: 85-99, 1978
71. Grassi J, Rogelet P, Devynck J, Tremillon B: Radiopolarography of technetium(VII) in acidic medium. J Electroanal Chem 88: 97-103, 1978
72. Kuwana T, Heineman WR: Study of electrogenerated reactants using optically transparent electrodes. Accts Chem Res 9: 241-248, 1976
73. DeAngelis TP, Heineman WR: An electrochemical experiment using an optically transparent thin layer electrode. J Chem Educ 53: 594-597, 1976
74. Heineman WR: Spectroelectrochemistry. Anal Chem 50: 390A-395A, 1978
75. Astheimer L, Schwochau K: Electrochemical reduction of MnO_4^-, TcO_4^- and ReO_4^- in organic solvents. J Inorg Nucl Chem 38: 1131-1134, 1976
76. Mazzochin GA, Seeber R, Mazzi U, Roncari E: Voltammetric behavior of technetium-99 complexes with π-acceptor ligands in aprotic medium. I. Oxidation of $TcCl_3(PMe_2Ph)_3$. Inorg Chim Acta 29: 1-4, 1978
77. Mazzochin GA, Seeber R, Mazzi U, Roncari E: Voltammetric behavior of technetium-99 complexes with π-acceptor ligands in aprotic medium. II. Reduction of $TcCl_3(PMe_2Ph)_3$ and of $TcCl_4(PMe_2Ph)_2$. Inorg Chim Acta 29: 5-9, 1978
78. Ikeda I, Osamu I, Kurata K: Chemical and biological studies on 99-Tc-DMS-II: Effect of Sn(II) on the formation of various Tc-DMS complexes. Int J Appl Radiat Isot 27: 681-688, 1976
79. Johannsen B, Syhre R, Spies H, Munze R: Chemical and biological characterization of different Tc complexes of cysteine and cysteine derivatives. J Nucl Med: 19: 816-824, 1978

80. Adams RN: Probing brain chemistry with electroanalytical techniques. Anal Chem 48: 1126A-1137A, 1976
81. Eckelman WC, Levenson SM: Chromatographic purity of 99m-Tc compounds. In Quality Control in Nuclear Medicine, Rhodes BA, ed, St. Louis, Mo., Mosby, 1977, pp 197-209
82. Kissinger PT, Refshauge CJ, Dreiling R, Adams R: An electrochemical detector for liquid chromatography with picogram sensitivity. Anal Lett 6: 465-477, 1973
83. Kissinger P: Amperometric and coulometric detectors for high performance liquid chromatography. Anal Chem 49: 447A-456A, 1977
84. Majerik JE, Russell CD: Determination of free pertechnetate in radiopharmaceuticals by high pressure, thin layer, and paper chromatography. Read before the Society of Nuclear Medicine, 19th Annual Meeting of the Southeastern Chapter, Birmingham, Ala. Nov. 1-4, 1978
85. Meares CF, Goodwin DA, Leung CSH, et al: Covalent attachment of metal chelates to proteins: the stability in vivo and in vitro of the conjugate of albumin with a chelate of 111-indium. Proc Natl Acad Sci USA 73: 3803-3806, 1976
86. Cheng CH, Eisenberg R: Homogeneous catalysis of the water gas shift reaction using a platinum chloride - tin chloride system. J Amer Chem Soc 100: 5986-5970, 1978
87. Olmstead MM, Benner LS, Hope H, Balch AL: Addition of tin(II) chloride to $Pd_2(Ph_2PCH_2PPh_2)_2Cl_2$. The crystal and molecular structure of $Pd_2(Ph_2PCH_2PPh_2)_2(SnCl_3)Cl$. Inorg Chim Acta 32: 193-198, 1979
88. Donaldson JD: The chemistry of bivalent tin. Prog Inorg Chem 8: 287-356, 1967
89. Fritzberg AR, Lyster DM, Dolphin DH: Evaluation of formamidine sulfinic acid and other reducing agents for use in the preparation of Tc-99m labeled radiopharmaceuticals. J Nucl Med 18: 553-557, 1977
90. Deutsch EA, Libson KF, Becker CB, et al: Application of inorganic chemistry to the preparation of technetium phosphonate bone scanning agents that do not contain tin. Read before the Society of Nuclear Medicine, 19th Annual Meeting of the Southeastern Chapter, Birmingham, Ala., Nov. 1-4, 1978
91. Shukla SK: Reduction of pertechnetate-99m ion by hydrochloric acid I. Preparation of chromatographically pure technetium-99m(IV). J. Chromatogr 151: 51-57, 1978

AN EFFICIENT MICROSCALE PREPARATION OF TIN-117m-TIN TETRACHLORIDE-A PIVOTAL INTERMEDIATE FOR THE SYNTHESIS OF TIN-117m-LABELED RADIOPHARMACEUTICALS

David V. Woo, Furn F. Knapp, Jr., Thomas A. Butler, and Alvin P. Callahan

Nuclear Medicine Technology Group
Health and Safety Research Division
Oak Ridge National Laboratory
Oak Ridge, Tennessee 37830

ABSTRACT

A microscale technique has been developed for the preparation of $^{117m}SnCl_4$. The radiolabeled tin tetrachloride is an important intermediate for the synthesis of ^{117m}Sn-labeled organic radiopharmaceuticals. The "flow-through" system involves high temperature chlorination of metallic ^{117m}Sn with subsequent cryogenic collection of the $^{117m}SnCl_4$ in a special reaction vessel. This vessel can then be used for the conversion of the tin tetrachloride to various useful intermediates. The reactions that have been studied include the formation of tetraalkyl and tetraaryltin compounds by reaction of $SnCl_4$ with Grignard or alkyl and aryl lithium reagents ($SnCl_4 \longrightarrow R_4Sn$ or Ar_4Sn). In addition, by stoichiometric control, the comproportionation of $SnCl_4$ with tetramethyltin can yield any one of the mixed methylchlorotin intermediates, $MeSnCl_3$, Me_2SnCl_2 or Me_3SnCl. These products are important intermediates for the preparation of ^{117m}Sn-labeled steroids, fatty acids, amino acids, barbituates and a variety of other potentially useful agents.

INTRODUCTION

The Sn-117m radionuclide decays by isomeric transition with the emission of a primary single γ-photon in 87% abundance with an energy of 158 keV. The efficient detection of the 158 keV photon with sodium iodide crystals and the moderate 14 day physical half-life suggests that Sn-117m-labeled agents would be an attractive new class of radiopharmaceuticals. In addition, these agents would have a reasonable shelf-life and the absorbed tissue dose would be lower than the dose experienced with agents labeled with longer-lived radionuclides such as Te-123m and Se-75. Organotin chemistry is very versatile and such compounds are often quite stable (1) and one could thus envision the preparation of a wide variety of tissue-specific Sn-117m-labeled radiopharmaceuticals. This combination of favorable radionuclidic properties, well-defined chemistry and synthetic versatility has prompted us to explore the preparation of organotin compounds of biological interest.

The synthesis of ^{117m}Sn-labeled organotin compounds requires the initial conversion of metallic ^{117m}Sn to a suitable intermediate. In our estimation, tin tetrachloride ($SnCl_4$) fills the requirements of a versatile and useful intermediate for the synthesis of a wide variety of organotin compounds. We now report a high-yield microscale method for the conversion of metallic ^{117m}Sn to the pivotal $SnCl_4$ intermediate. This flow-through method involves chlorination of the metallic tin with the subsequent distillation and trapping of the $SnCl_4$ in a specially designed reaction vessel. This reaction vessel can be subsequently used to prepare a variety of useful ^{117m}Sn-labeled intermediates.

APPARATUS FOR PREPARATION OF TIN TETRACHLORIDE

The apparatus shown in Figure 1 was specially designed for the in-line microscale preparation of $SnCl_4$. The receiver-reaction vessel can be used for the synthesis of various organotin intermediates. The Vycor reduction tube (48 cm x 2.5 cm) is connected to Pyrex tubing at either end with graded seals. A glass manifold fitted with Teflon gas-inlet stopcock valves is connected via a 35/20 ball joint to the inlet side of the reduction tube. The flow-rates of gases introduced through the manifold are monitored by precision flow meters. The reduction tube is heated with a split-tube furnace (20 cm x 3 cm; Hevi Duty Electric Co., Milwaukee, WI) with a maximum rating of 1000°C. A pyrex receiver-vessel is attached to the reduction tube to trap the $SnCl_4$. This vessel can subsequently be used to convert the $SnCl_4$ to various useful organotin intermediates by the introduction of reagents through a rubber or silicon septum. The conical shape of the reaction vessel enables the use of a conical magnetic stirring bar to agitate small volumes. Finally, the exit gases are passed through a tube filled with activated charcoal and then a NaOH scrubber.

PREPARATION OF TIN TETRACHLORIDE

The preparation of $^{117m}SnCl_4$ requires several steps rather than the direct chlorination of reactor produced metallic ^{117m}Sn as a result of the necessity to accurately determine the production yield of ^{117m}Sn. The ^{117m}Sn was formed by neutron irradiation (n,γ) of 94.74% enriched ^{116}Sn in the Oak Ridge High Flux Isotope Reactor (HFIR) at a thermal neutron flux of 2.5 x 10^{15} n/cm^2 · sec. The reactor target was dissolved in conc. HCl to give a solution of hydrated tin dichloride ($^{117m}SnCl_2 \cdot H_2O$) (Figure 2). Aliquots of this solution were assayed to accurately determine the radionuclidic purity, specific activity and production yield of the ^{117m}Sn. A preliminary assessment of the factors influencing the reactor production of ^{117m}Sn by the HFIR irradiation of enriched ^{116}Sn targets can be made on the basis of five irradiation experiments. All irradiations were conducted in the same physical location in the HFIR hydraulic tube facility. The ^{116}Sn targets varied in weight from 10 to 73 mg and the irradiation periods varied from 2 to 14 days. The ^{117m}Sn production yields are consistent with an effective cross section (σ) of 13.1 millibarns and follow the radioisotope production equation $A = N\phi\sigma\ (1\text{-}e^{\lambda t})$ After a seven day irradiation a ^{117m}Sn product with a specific activity of 1.3 mCi/mg was obtained and this increased to 2.2 mCi/mg after a fourteen day irradiation. The maximum specific activity of ^{117m}Sn which we can achieve is calculated to be ∿ 4.4 mCi/mg and an excess of 140 days irradiation would be required to reach this value. We can conclude that ^{117m}Sn does not have a significant neutron capture cross section and that target self-shielding is not a significant factor.

Tin(II) dichloride can be conveniently converted to tin(II) oxide by treatment with sodium carbonate (reaction 2). The hydrated $^{117m}SnCl_2$ solution is reduced to a minimal volume (∿ 5 ml) and a saturated Na_2CO_3 solution added until the phenolphthalein end-point is reached. The hydrated ^{117m}SnO is formed as a fine colloidal solid and is heated for 2-3 hours resulting in the quantitative conversion to a crystalline allotrophic form having a blue-black metallic luster. This product is washed several times with distilled water, centrifuged and dried in an oven at ∿ 140°C. Carrier $SnCl_2 \cdot H_2O$ can be added to the $^{117m}SnCl_2 \cdot H_2O$ acid solution if a product with a lower specific activity is required.

The high temperature hydrogen reduction of tin(II) oxide gives metallic tin (reaction 3). The ^{117m}SnO is transferred to a 76 mm x 16 mm x 10 mm silica boat (Amersil, Inc.; Hillside, NJ) which is placed in the center of the reduction tube. The apparatus is purged with argon and the reduction tube gradually heated to ∿ 200°C to remove all traces of moisture. Hydrogen gas is introduced through the system while the temperature of the reduction tube is slowly raised to 580°C over a one hour period. The water of hydration volatilizes at about 450°C and is swept out of the reaction system in the hydrogen stream. The ^{117m}SnO reduction is allowed to proceed for two hours and the system is then cooled to ambient temperature and flushed with argon. A minimal argon flow is maintained to prevent introduction of atmospheric oxygen and possible re-oxidation of tin.

Tin tetrachloride can be obtained in high yield by high temperature chlorination of tin metal (reaction 4). After cooling, the reaction system is

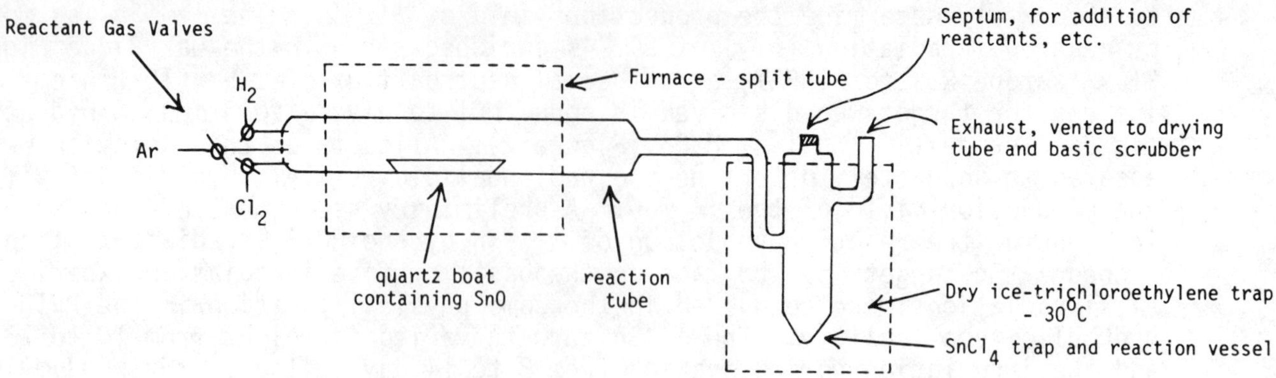

FIGURE 1

Diagram of the apparatus that was developed for the microscale preparation of tin tetrachloride. The apparatus consists of essentially two important components. The split-tube furnace heats the reaction tube in which the SnO ⟶ Sn and Sn ⟶ $SnCl_4$ conversions are conducted. The receiver-reaction vessel is used to cryogenically trap the $SnCl_4$ as it is distilled from the reaction tube. It can also be used for subsequent conversions, and the products listed in Table 1 were prepared in this receiver-reaction tube.

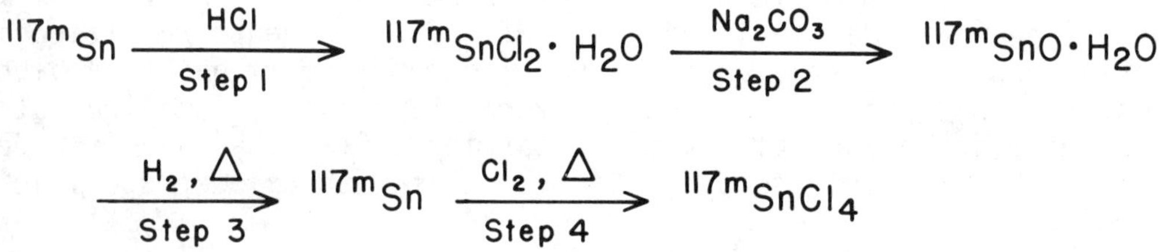

FIGURE 2

The sequence of reactions developed for the conversion of reactor produced ^{117m}Sn to $^{117m}SnCl_4$. The target is initially dissolved in HCl (reaction 1) so that aliquots can be counted to accurately determine the production yield and radionuclidic purity of ^{117m}Sn. Reaction 2 converts the tin chloride to the insoluble Tin(II) oxide in high yield. Reactions 3 and 4 are performed in the specially designed apparatus shown in Figure 1 flushed with dry chlorine gas at a flow rate of $\sim$ 20-30 ml/min. The furnace temperature is raised to $\sim$ 150°C and the $^{117m}SnCl_4$ forms as clear liquid droplets on the glass surface of the reduction tube. The receiver-reaction vessel is cooled to -30°C in a dry ice-trichloroethylene bath. The $^{117m}SnCl_4$ is then distilled cryogenically from the reduction tube into the receiver-reaction vessel by gentle heating of the reduction tube with a forced-air heat gun. A minimal flow of chlorine gas is required during this stage to sweep the $^{117m}SnCl_4$ vapor into the receiver-reaction vessel. The chlorination is terminated when there is no further evidence of tin metal in the silica boat. The furnace is removed and the system then flushed with a minimal flow of argon. The cooling bath is removed from the receiver-reaction vessel and the $^{117m}SnCl_4$ is allowed to warm to about 10°C. The reaction vessel is maintained at this temperature to minimize any potential evaporation and loss of the $^{117m}SnCl_4$ product. At this stage the $^{117m}SnCl_4$ appears a light yellow color due to trace amounts of dissolved chlorine. The color gradually fades as the dissolved chlorine is volatilized and swept away in the argon stream. The resulting colorless $^{117m}SnCl_4$ solution is ready at this stage for subsequent transformations.

In a parallel experiment, non-radioactive tin metal was converted to $SnCl_4$ using the apparatus shown in Figure 1 . Since $SnCl_4$ is a highly reactive and corrosive substance the identity of the tin tetrachloride was experimentally supported by freezing point and boiling point measurements. The experimentally determined freezing point (-35°C) and boiling point (114°C) values were identical to similar values determined with commercially available tin tetrachloride. Furthermore, the tin tetrachloride prepared with the apparatus shown in Figure 1 was converted to tetraphenyltin, a stable, crystalline substance having a well-defined melting point and which was readily characterized by a variety of physical methods (*vide infra*).

CONVERSION OF TIN TETRACHLORIDE TO ORGANOTIN INTERMEDIATES

The $^{117m}SnCl_4$ can be converted to a variety of ^{117m}Sn-labeled organotin intermediates by the direct introduction of reagents into the receiver-reaction vessel. The center neck of the vessel (Figure 1) is fitted with a septum through which the reagents are conveniently introduced with a hypodermic syringe. Reaction mixtures can be heated or cooled by immersing the conical tip of the reaction vessel in a water bath, and the reaction mixtures are stirred using a magnetic stirrer. The most important aspect of this system is of course that the highly reactive tin tetrachloride is distilled directly into the reaction vessel in an argon atmosphere eliminating exposure to atmospheric moisture.

Table 1 illustrates the variety of organotin intermediates that have been prepared using this system. The synthesis of tetraphenyl tin (ϕ_4Sn) will be used to illustrate the conversion of $SnCl_4$ to an organotin intermediate using this apparatus since ϕ_4Sn is a crystalline, well characterized substance.

Table 1. The apparatus shown in Figure 1 was used for the conversion of tin oxide to metallic tin by high temperature hydrogen reduction. The metallic tin was subsequently converted to tin tetrachloride by reaction with chlorine gas. The tin tetrachloride was distilled into the special reaction vessel and then reacted with a variety of alkyl metal reagents to form selected tin intermediates.

Experiment	Substrate	Initial Product	Reagent	Final Product	Overall Yield
I	$SnO \longrightarrow Sn$	$SnCl_4$	ϕLi	ϕ_4Sn	54%
II	$SnO \longrightarrow Sn$	$SnCl_4$	CH_3MgBr	Me_4Sn	59%
III	$SnO \longrightarrow Sn$	$SnCl_4$	Me_4Sn	Me_3SnCl	86%
IV	$*SnO \longrightarrow *Sn$	$*SnCl_4$	. . .	. . .	72%
V	$*SnO \longrightarrow *Sn$	$*SnCl_4$	$Me_4{*Sn}$	$Me_3{*SnCl}$	55%

(*Sn denotes the tin-117m radionuclide)

Tin metal (115 mg, 1 mmol) was chlorinated as described earlier and the $SnCl_4$ distilled into the special reaction vessel. Three milliliters of a 1.75 M phenyl lithium solution in benzene-ether, 70:30 (Aldrich Chemical Co.) were added dropwise over a 30 minute period to the $SnCl_4$ solution cooled to 0-4°C. The solution was gently stirred under argon and as the reaction proceeded a milky white precipitate of lithium salts appeared. After addition of the ϕLi solution the temperature was raised to 40°C and the reaction allowed to proceed for 30 minutes. After cooling to 0-4°C, 3-4 ml of water was added cautiously and the mixture was then agitated and heated to 70-80°C. The hydrolyzed mixture was then allowed to stand until the aqueous and organic layers separated. The benzene layer was decanted and the aqueous phase then extracted with warm chloroform. This procedure was repeated twice and the combined benzene and chloroform extracts washed well with water and evaporated to dryness under argon. The resulting white solid was re-crystallized from chloroform with cooling. The crystalline solid had an uncorrected melting point of 227°C (commercially available ϕ_4Sn melted at 227°C). In addition, the synthetic sample exhibited thin-layer chromatographic, ultraviolet, mass spectral and ^1H and ^{13}C nuclear magnetic resonance properties identical to similar values determined using an authentic sample of tetraphenyl tin (Aldrich Chemical Co.)

DISCUSSION

The apparatus shown in Figure 1 was specially designed for the preparation of a variety of useful intermediates from tin tetrachloride. As an example, reaction of $SnCl_4$ with excess Grignard reagents or alkyl lithium or aryl lithium reagents gives the tetraalkyl or tetraaryltin products in good yields (Table 1). Another interesting reaction is the non-statistical ratio of products (comproportionation) formed by treatment of $SnCl_4$ with Me_4Sn (1). As an example, reaction of $SnCl_4$ and Me_4Sn in a 1/3 molar ratio gives Me_3Sn-Cl as the sole end product. Similar reaction of these reactants in an equal molar ratio gives Me_2SnCl_2 and a 3/1 molar ratio yields $MeSnCl_3$. These products are formed in each case by reaction of the neat reactants without the use of a solvent. This is an obvious advantage since subsequent transformations on the products are therefore simplified. One of the most useful intermediates for the synthesis of trimethyltin substituted compounds is formed by simple reaction of Me_3Sn-Cl with lithium metal in tetrahydrofuran (2). A unique class of trimethyltin substituted steroids have been prepared from Me_3Sn-Li (3). Trimethylchloromethyltin (Me_3SnCH_2Cl) is another interesting and useful intermediate which can be prepared by the following route: $SnCl_4 + Me_4Sn \longrightarrow Me_3Sn\text{-}Cl + CH_2N_2 \longrightarrow Me_2(CH_2Cl)\text{-}Sn\text{-}Cl + CH_3MgBr \longrightarrow Me_3SnCH_2Cl$ (4). In this case the methylene group can serve as a "spacer" for introduction of the Me_3Sn- moiety into regions of molecules that are subject to steric hindrance. These are just a few examples of how the wealth of known organotin chemistry can be applied for the potential synthesis of ^{117m}Sn-labeled compounds of biological interest.

A microscale method for $^{117m}SnCl_4$ production had to be developed and optimized before the subsequent preparation of ^{117m}Sn-labeled organic radiopharmaceuticals could be initiated. Now that this efficient method has been developed for the preparation of the crucial $SnCl_4$ intermediate, we envision the synthesis of an exciting array of ^{117m}Sn-labeled agents, including various steriods, fatty acids, amino acids, barbituates and other unique compounds that are potentially useful for *in vivo* tissue imaging procedures.

ACKNOWLEDGEMENTS

Research sponsored by the Office of Health and Environmental Research, U. S. Department of Energy under contract W-7405-eng-26 with the Union Carbide Corporation. The authors thank R. A. Grigsby for his assistance in several of the tin tetrachloride conversions and C. A. Floyd for typing the manuscript.

REFERENCES

(1) NEUMANN WP: *The Organic Chemistry of Tin,* New York, John Wiley & Sons, 1970

(2) TAMBORSKI C, FORD FE, SOLOSKI EJ: Preparation and Reactions of Trialkyltinlithium. *J Org Chem* 28:237-239, 1963

(3) Unpublished experiments, 1979

(4) SEYFERTH D, ROCHOW EG: The Reaction of Organotin Halides with Diazomethane. *J Amer Chem Soc* 77:1302-1304, 1955

PREPARATION OF A NEW OXOTECHNETIUM-SULFUR COMPLEX

Bruno V. DePamphilis, Alun G. Jones, and Michael A. Davis

Department of Radiology, Joint Program in Nuclear Medicine,
Harvard Medical School, and Peter Bent Brigham Hospital
Boston, MA 02115

and

Alan Davison

Department of Chemistry
Massachusetts Institute of Technology
Cambridge, Massachusetts 02138

ABSTRACT

The reduction of $^{99}TcO_4^-$ in the presence of 3,4- toluenedithiol by dithionite in aqueous alkali produces the oxotechnetium (V) bis-(3,4- toluenedithiolate) anion. This has been isolated as the tetrabutylammonium salt. The similarity of this anion to previously characterized square pyramidal oxotechnetiumbis-(1,2-dithiolates) has been established from elemental micro-analysis, infrared spectroscopy, and field desorption mass spectrometry.

INTRODUCTION

Determining the chemical nature of clinically useful technetium-99m compounds is very difficult because of the minute concentrations of Tc involved in typical kit preparations. As a result, the ligation, stereochemistry, and even the oxidation state of the metal in most such radiopharmaceuticals are unknown.

We have established a research program to determine the preferred ligand environments of technetium in its various oxidation states and wherever possible the precise structures of its complexes. In order to employ the analytical methods routinely available to the inorganic chemist, it is necessary to isolate and purify macroscopic amounts (about 100 mg.) of these substances. This has necessitated using the long-lived isotope ^{99}Tc in our syntheses.

The first fruit of this research was the preparation of the oxotechnetium-bis-(thiomercaptoacetate) anion, $TcO(SCH_2COS)_2^-$, isolated as the tetrabutylammonium salt (1). This compound was subjected to single-crystal x-ray structural analysis, which revealed the anion to consist of a square pyramidally coordinated technetium atom, with the four sulfur atoms of the two chelating ligands defining the base of the pyramid and an oxygen atom its apex. The metal was calculated to be in the +5 oxidation state (Figure 1). A report of the preparation and characterization of the tetraphenylarsonium salt of oxotechnetiumbis-(1,2- ethanedithiolate) appeared simultaneously (2). This anion has the same square pyramidal structure as the thiomercaptoacetate complex, and is almost identical to it in terms of the bond lengths and angles within the $TcOS_4$ core. Both of these complexes show a strong absorption band in the same region of the infra-red, the thiomercaptoacetate at 950 cm^{-1}, and the 1,2-ethanedithiolate at 940 cm^{-1}(3). These bands have been assigned (1) to the Tc –O stretch by comparison with a series of isoelectronic oxomolybdenumbis-(dialkyldithiocarbamate) complexes (4), each of which has a strong absorption in the region 955 cm^{-1} –975 cm^{-1} that has been assigned as the Mo –O stretching frequency (4).

The commonly available ligand 3,4– toluenedithiol has been extensively used as a colorimetric reagent for the determination of transition metals (5,6). It has an aromatic backbone and, when complexed to transition metals, can lead to electronic delocalization across the entire complex via extended orbital overlap (7). We felt it would be of interest to see if this ligand would affect the pattern of behavior established by the two dithiolates already discussed (1,2).

MATERIALS, METHODS, AND RESULTS

Synthesis. 0.2474g of 3,4-toluenedithiol (1.59×10^{-3} mole) was dissolved in 50 ml of 0.6M NaOH solution, and 2.25 ml of 0.352M NH_4TcO_4 solution (7.92×10^{-4} mole) added. A freshly prepared solution of 0.555 g $Na_2S_2O_4$ (3.19×10^{-3} mole) in 10 ml of 0.6M NaOH was added to the reaction mixture rapidly, followed one minute later by 1.0 ml of 50% Bu_4NCl in H_2O. After five minutes of stirring the suspension was filtered, and the solids washed with a few milliliters of the hydroxide solution and distilled water.

The crude precipitate was extracted three times with 5 ml portions of acetone. The combined filtered extracts were mixed with 5 ml of water and stored for 24 hours at -23 degrees C. The resulting crystals were filtered, washed with water, and dried overnight on a vacuum line. The yield was 35%, based on the amount of TcO_4^- used initially.

Analysis. The elemental microanalysis results for the compound are summarized in Table 1. Carbon, hydrogen, nitrogen, and sulfur analysis were performed by Spang Microanalytical Laboratory, Eagle Harbor, Michigan. The technetium analysis was done by a liquid scintillation method as follows. A 0.0317g sample of the compound was dissolved in 50 ml of methanol, and the technetium oxidized to TcO_4^- with 5 ml of 30% H_2O_2, a few drops of 50% NaOH solution, and gentle heating. The digest was diluted to 1 liter with water, and 100 microliter aliquots of this solution analyzed by routine liquid scintillation using a commercial scintillation cocktail (RIAFLUOR, New England Nuclear). The absolute activity of the test solution was determined by the internal standardization method (8), using a standard pertechnetate solution quantitated by spectrophotometric analysis (λ_{max} = 244nm; ε = 6,220).

Physical Studies. The UV-visible spectrum of the product is shown in Figure 2. The spectrum is dominated by an absorption maximum at 361nm (ε = 13,000) and has, in addition, a peak at 313nm (ε = 7,400) and an inflection at about 405nm (ε = 6,300).

The infra-red spectrum of the compound is shown in Figure 3. As well as bands attributable to either the ligand or the cation, there is a strong band at 930 cm^{-1}.

The field desorption (9) mass spectrum is shown in Figure 4. The major peaks in this spectrum are readily identified (10) as the Bu_4N^+ cation (m/e = 242), and the oxotechnetium anion associated with two Bu_4N^+ cations (m/e = 907).

DISCUSSION

The reduction of pertechnetate in the presence of 1,2-ethanedithiol by dithionite in aqueous alkali has been shown to be an effective method of producing the oxotechnetiumbis-(1,2-ethanedithiolate) anion (11). Similarly, the dithionite reduction of pertechnetate in the presence of 3,4-toluenedithiol leads to the formation of the oxotechnetiumbis-(3,4-toluenedithiolate) anion. The positive identification of this species is supported by three observations. First, the microanalytical data corresponds closely to the expected formula $[Bu_4N]\{TcO[(CH_3)C_6H_3S_2]_2\}$ (or, $[Bu_4N][TcO(tdt)_2]$). Second, the presence of a strong band in the infra-red spectrum at 930 cm^{-1} suggests the presence of a Tc-O moiety. This assignment is based on a similar band in the spectra (1,3) of both $TcO(SCH_2CH_2S)_2^-$ and $TcO(SCH_2COS)_2^-$ and by comparison to the diagnostic Mo-O stretching frequencies in a number of oxomolybdenumbis-(dialkyldithiocarbamate) complexes (4). Finally, the field desorption mass spectrum of the isolated compound is most easily rationalized by the presence of Bu_4N^+ and $TcO(tdt)_2^-$ ions.

This study shows that the possibility of extensive electronic delocalization through a chelating dithiolate does not hamper the formation of the same type of oxotechnetium species as is obtained with simple 1,2-dithiolates under similar conditions.

TABLE 1. ELEMENTAL MICROANALYSIS OF PRODUCT

Calculated for $(Bu_4N)[TcO((CH_3)C_6H_3S_2)_2]$:					
	C 54.14%	H 7.22	N 2.11	S 19.25	Tc 14.89
Found:					
	C 54.30	H 7.23	N 2.11	S 20.56	Tc 14.51

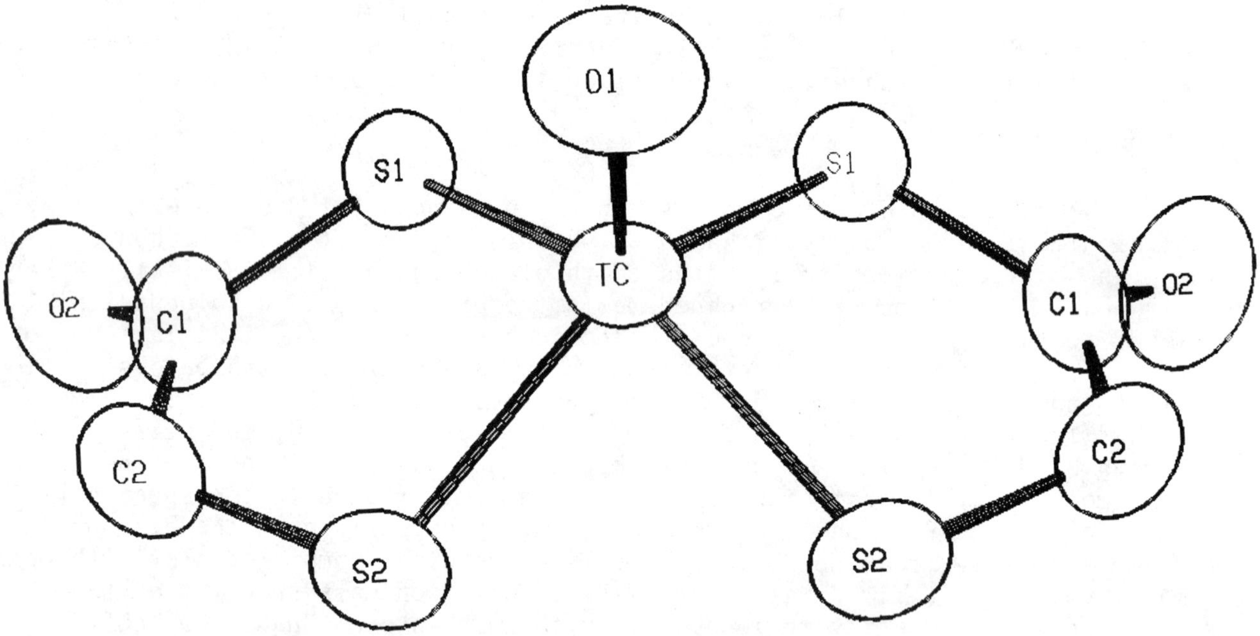

Figure 1

X-ray structure of $TcO(SCH_2COS)_2^-$ (1)

ACKNOWLEDGMENTS

This work was performed under USPHS Grant 5 R01 GM23270 and Department of Energy Contract EY-76-S-4115.

REFERENCES

(1) DePamphilis BV, Jones AG, Davis MA et al: Preparation and crystal structure of oxotechnetium bis(thiomercaptoacetate) and its relationship to radiopharmaceuticals labeled with ^{99m}Tc. J Am Chem Soc 100:5570-5571, 1978

(2) Smith JE, Byrne EF, Cotton FA et al: A thiol complex of technetium pertinent to radiopharmaceutical use of ^{99m}Tc. J Amer Chem Soc 100:5571-5572, 1978

(3) Byrne EF, Smith JE: Technetium complexes of aliphatic thiols. The synthesis and characterization of oxo-bis(1,2 and 1,3- dithiolato) technetate (V) anions. Inorg Chem, submitted for publication

(4) Stiefel EI: The coordination and bioinorganic chemistry of molybdenum. In Progress in Inorganic Chemistry, S.J. Lippard, ed, vol 22. New York, Wiley-Interscience, 1978, pp. 102-109

(5) Clark, RED: O-Dithiols in analysis.Part VII.Toluene-3,4-dithiol as a general analytical reagent in qualitative analysis. Analyst 83:396-402, 1958

(6) Clark RED, Neville RG: Toluene -3,4-dithiol and its derivatives as analytical reagents. J Chem Ed 36:390-393, 1959

(7) McCleverty JA: Metal 1,2- dithiolene and related complexes. In Progress in Inorganic Chemistry, S.J. Lippard, ed, vol.10, New York, Wiley-Interscience, 1968, p. 70

(8) Moss AJ Jr, Dalrymple GV, Boyd CM: Practical radioimmunoassay. St. Louis, Morby Company, 1976, pp. 72-73

(9) Beckey HD: Principles of field ionization and field desorption mass spectrometry, Oxford, Pergamon Press, 1977

(10) Wilson BW, Costello CE, Carr SA et al: Field desorption mass spectrometry of complex anions containing rhenium and technetium. Anal Lett, in press

(11) Orvig C, Trop HS, Davison A et al: A survey of reducing agents for the synthesis of a known technetium compound in aqueous solution. Submitted to J Nucl Med

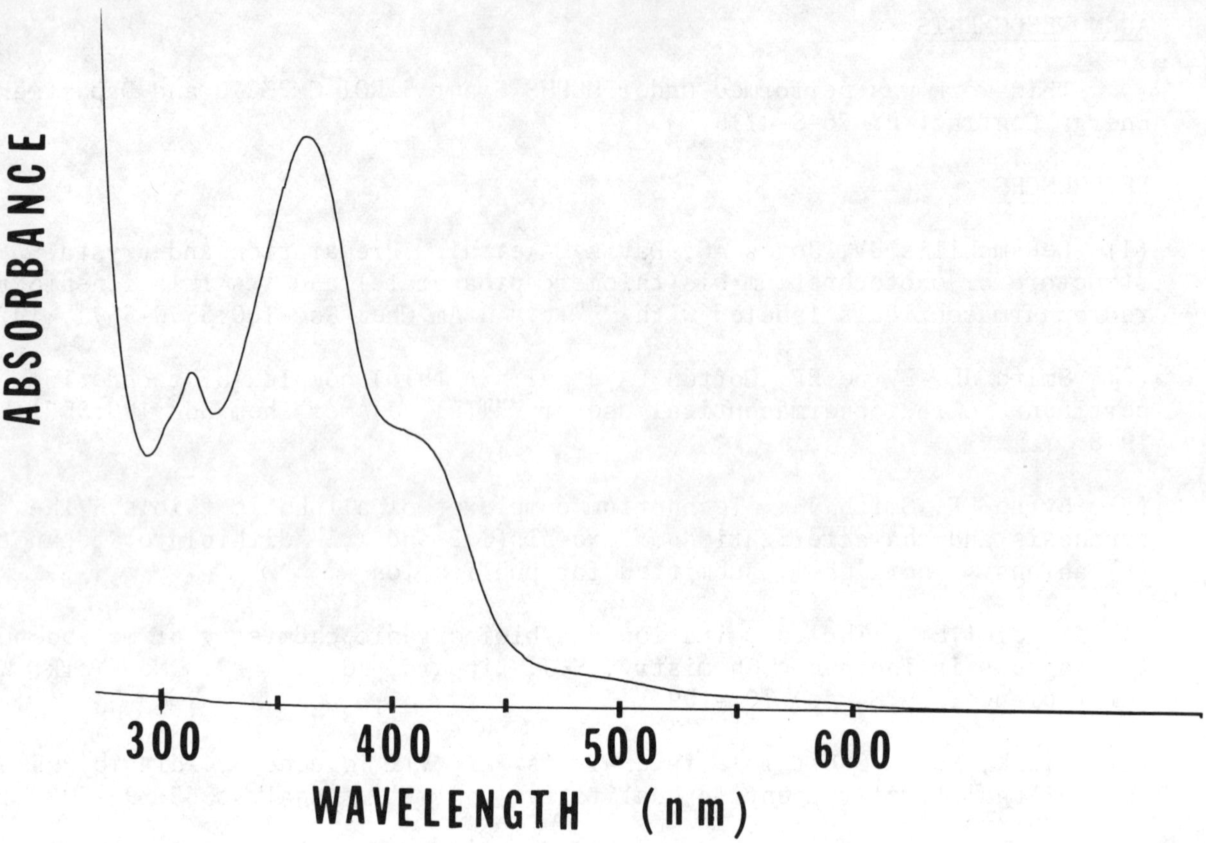

Figure 2

UV-visible spectrum of $(Bu_4N)[TcO(tdt)_2]$

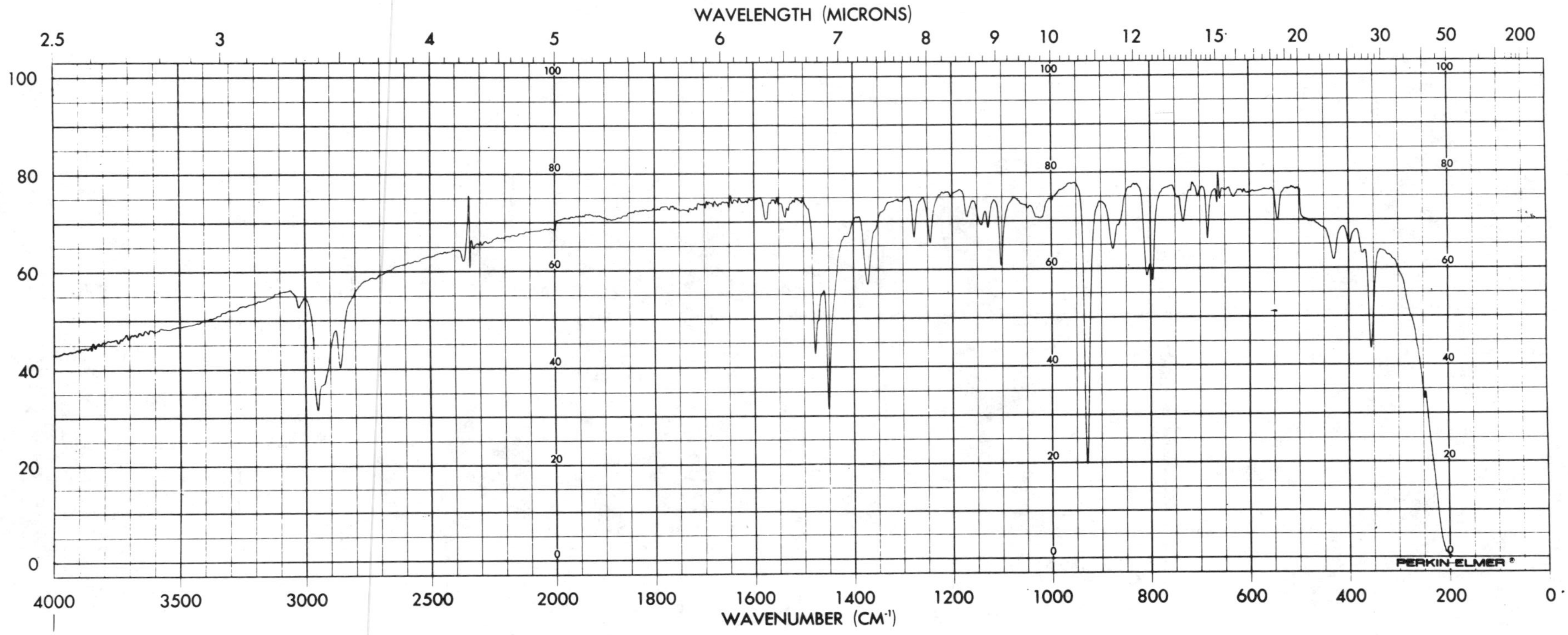

Figure 3

Infra-red spectrum of $(Bu_4N)[TcO(tdt)_2]$

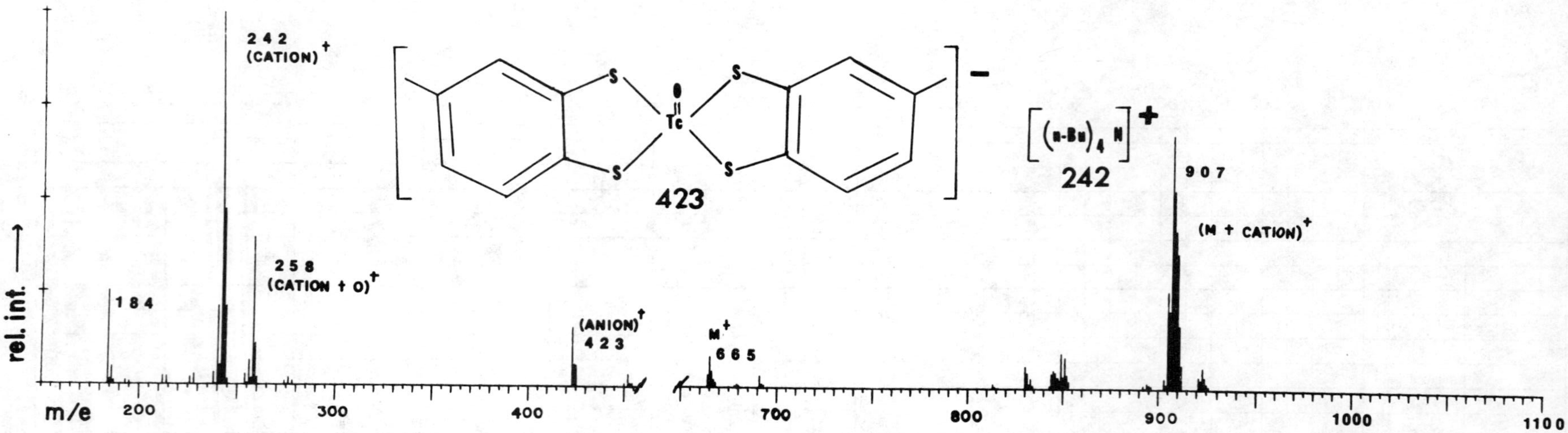

Figure 4

Field desorption mass spectrum of $(Bu_4N)[TcO(tdt)_2]$ (<u>10</u>)

ION EXCHANGE METHODS FOR THE DETERMINATION OF THE NET CHARGE ON SOME RADIOPHARMACEUTICALS

Azu Owunwanne, Robert E. O'Mara
Division of Nuclear Medicine
University of Rochester
School of Medicine and Dentistry
Rochester, NY 14642

and

Monte Blau, Jacob Marinsky
Departments of Nuclear Medicine and Chemistry
State University of New York/Buffalo
Buffalo, NY 14214

ABSTRACT

The ion exchange distribution method was used to determine the charge on the radioactive species of ^{67}Ga citrate, ^{201}TlCl, ^{169}YbDTPA, ^{99m}TcHEDP, $^{99m}TcO_4^-$, and the reactive ^{99m}Tc species produced when pertechnetate is reduced by stannous ion. The charge on ^{67}Ga citrate was +1.5, indicating a polymerized species; the charge of ^{201}TlCl was confirmed to be +1; ^{169}YbDTPA and $^{99m}TcO_4^-$ had charges of -2 and -1 respectively; ^{99m}TcHEDP had a curve with changing slopes indicating charges of -2 and -0.5.

INTRODUCTION

The theory and practice of the ion exchange distribution method have progressed to a level that its acceptance as one of the methods useful for the study of complexes in solution is fully established. The ion exchange equilibrium distribution technique was first introduced by Schubert (1). It was further developed by Marcus (2). Its major advantage is that the experiments can be done with very low concentrations or with carrier-free amounts of the unknown complex. Hence, this technique is quite appropriate for the study of solution chemistry of radiopharmaceuticals.

At present, limited studies have been done on the chemistry of some radiopharmaceuticals (3-5). Such important parameters as the charge, complex formation and stability constants are virtually unknown for most radiopharmaceuticals. In this paper we report the determination of the net charge on several ions and complexes of interest to nuclear medicine. These include gallium-67 citrate (^{67}Ga citrate), Tl-201 thallous chloride (^{201}TlCl), ytterbium-169 disodium diethylenetriaminepentate (^{169}YbDTPA), Tc-99m hydroxyethylidenediphosphonate (^{99m}TcHEDP), Tc-99m pertechnetate ($^{99m}TcO_4^-$) and the reactive Tc-99m intermediate produced when $^{99m}TcO_4^-$ is reduced by stannous ion.

Theory: With this method, the net charge on a cation, anion, or complex is determined by equilibrating the ion between an ion exchange resin (cation or anion) and different concentrations of bulk neutral electrolyte. The distribution of the trace ion, D_o is defined as follows:

$$D_o = \frac{\text{Activity in 1 gram dry resin}}{\text{Activity in 1 gram solution}} \tag{1}$$

In the examination of the valence of a cation by this method, the trace ion, M^{y+}, of charge (y) is exchanged with the bulk cation, C^{n+}, of known charge (n) and concentration as shown in equation 2 below:

$$nM^{y+} + y\bar{C}^{n+} \rightleftarrows n\bar{M}^{y+} + yC^{n+} \tag{2}$$

Bars are placed above a component to represent the resin phase. The distribution, D_o, is given by

$$D_o = \frac{[\bar{M}^{y+}]^n}{[M^{y+}]^n} \tag{3}$$

The equilibrium quotient (Q_o) is given by

$$Q_o = \frac{[\bar{M}^{y+}]^n[C^{n+}]^y}{[M^{y+}]^n[\bar{C}^{n+}]^y} \tag{4}$$

$$= (D_o)\ \frac{[C^{n+}]^y}{[\bar{C}^{n+}]^y} \tag{5}$$

On rearrangement of eq. 5 and taking log of both sides,

$$\log D_o = -\log Q_o + y/n \log [\bar{C}^{n+}] - y/n \log [C^{n+}] \quad (6)$$

Since Q_o and $\bar{C}$ are constants,

$$\log D_o = \text{constant} - y/n \log [C^{n+}] \quad (7)$$

The slope of a plot, $\log D_o$ vs. $\log [C^{n+}]$, is thus y/n, where y is the charge on the unknown cationic species as previoualy stated and n is the well-defined charge of the cation of the bulk electrolyte.

When the bulk electrolyte is monovalent, e.g. sodium perchlorate or perchloric acid, equation 7 reduces to

$$\log D_o = \text{constant} - y \log [C^{+}] \quad (8)$$

and the slope of $\log D_o$ vs. $\log [C^{+}]$ equals the charge, y, on the trace ion. When the bulk electrolyte is divalent, e.g. stannous perchlorate, equation 7 then becomes

$$\log D_o = \text{constant} - 1/2y \log [C^{2+}] \quad (9)$$

and the slope of a plot of $\log D_o$ vs. $\log [C^{2+}]$ equals one half the charge, y, on the unknown ion.

For anion exchange distribution studies, equation 7 becomes

$$\log D_o = \text{constant} - y \log [C^{n-}] \quad (10)$$

where C is bulk monovalent electrolyte with known concentration; y, the slope of the curve equals the charge on the unknown anion or negatively charged complex species.

MATERIALS AND METHODS

The radiopharmaceuticals studied were ^{67}Ga citrate*, ^{169}YbDTPA**, ^{201}TlCl*, ^{99m}TcHEDP† and $^{99m}TcO_4^-$‡, and the reactive species produced when $^{99m}TcO_4^-$ is reduced by stannous ion. ^{169}YbDTPA, ^{67}Ga citrate, ^{201}TlCl and $^{99m}TcO_4^-$ were used directly as supplied by the manufacturers. The ^{99m}TcHEDP was prepared by the addition of the appropriate volume of $^{99m}TcO_4^-$ to the "kit" according to the manufacturer's directions.

For ^{201}TlCl and ^{67}Ga citrate, the cation resin used was in the Na^+ form, and the bulk electrolyte was sodium perchlorate. The cation exchange resin AG-50W-X8, 50-100 mesh (H^+ form) from Bio-Rad was converted to the Na^+ form with 0.10M sodium perchlorate solution. For the reactive ^{99m}Tc species, the

*New England Nuclear
**3M Company
†Procter and Gamble Company
‡Central Radiopharmacy, School of Medicine, SUNY, Buffalo, NY

cation resin was in the Sn^{2+} form and the bulk electrolyte was stannous perchlorate. The above cation resin (H^+ form) was converted to the Sn^{2+} form with 0.05M stannous perchlorate. The $Sn(ClO_4)_2$ was prepared by the method of Noyes and Toabe (6). The $Sn(ClO_4)_2$ was stored in a glass-stoppered bottle over metallic tin in a nitrogen atmosphere. The concentration of the $Sn(ClO_4)_2$ was determined by titration with iodine, using starch indicator. For ^{169}YbDTPA, ^{99m}TcHEDP and $^{99m}TcO_4^-$ the anion resin was used and the resin was in ClO_4^- form. The anion exchange resin AG1-X8, 50-100 mesh (Cl^- form) from Bio-Rad was converted to ClO_4^- form using 0.10M sodium perchlorate. Both the cation and anion resins were washed with triply distilled water until the pH of the eluent equalled the pH of the elutriant. The resins air dried at room temperature were isopiestrically equilibrated with a saturated solution of ammonium chloride under nitrogen, to define their moisture content.

Approximately 0.7g of the resin (cation or anion) was weighed into a polyethylene vial previously flushed with nitrogen. About 10ml of the appropriate bulk electrolyte (previously bubbled with wet nitrogen) at the several different concentrations selected for study, were pipetted into each vial. A trace amount of the appropriate radiopharmaceutical was introduced into each vial, which was immediately recapped and reweighed. For the distribution studies on reactive technetium intermediate, tracer quantity of $^{99m}TcO_4^-$ was added to the vials containing $Sn(ClO_4)_2$ solution and stannous form resin. The $^{99m}TcO_4^-$ was reduced by $Sn(ClO_4)_2$ (the bulk electrolyte) to produce the reduced technetium species which distributed between the cation resin and solution phases. Each vial was equilibrated for 24 hrs. Activity measurements were made directly on the air-dried resin and solutions.

RESULTS AND DISCUSSION

Cation Exchange Distribution Studies: To demonstrate the validity of the cation distribution method, equilibration experiments were carried out with trace $^{22}Na^+$ ion of known charge with the $Sn(ClO_4)_2$ - Sn^{2+} resin system. The distribution, D_o of tracer concentrations of sodium ion and technetium species between the resin and solution phases was measured as functions of both stannous and hydrogen ion concentrations. These data are presented in graphical form in Figures 1 and 2. The molar concentrations of stannous perchlorate and hydrogen ion were relatively low and concentration has been taken as equal to activity.

With an ion of known charge, Na^+, the measured slopes are 1.2 and -.63 (Fig. 1a, 2a) when log D_o is plotted against pH and log $[Sn(ClO_4)_2]$ respectively. Within experimental error, these slopes correspond to the known charge of +1 on Na^+ ion, confirming the reasonable control of the value of Q_o by the experimental approach.

The measured slopes in Figures 1b and 2b are 2.18 and 0.98 respectively, indicating a technetium species with a charge of +2.0. The generally accepted redox reaction between TcO_4^- and Sn^{2+} at high concentrations is

$$3Sn^{2+} + 2TcO_4^- + 8H^+ \rightarrow 3Sn^{4+} + 2TcO_2 + 4H_2O \quad (11)$$

With Tc(VII) reduced to Tc(IV) by Sn^{2+}, the most likely technetium species carrying a charge of +2 is $Tc(OH)_2^{2+}$ or TcO^{2+}.

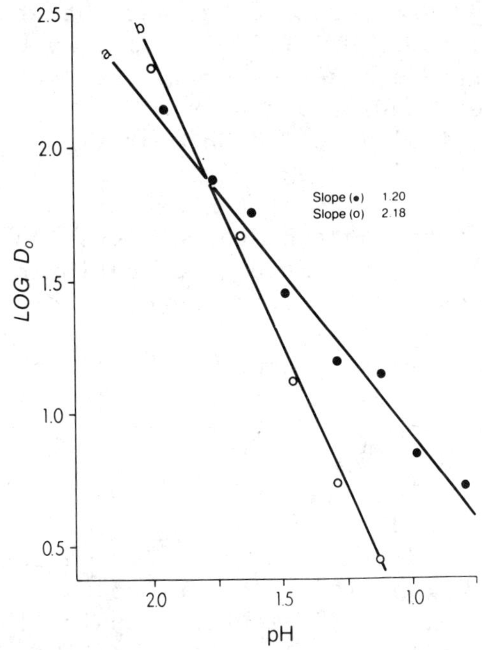

Figure 1

The distribution, D_o, of Na-22 (●) and reduced Tc-99m species (○) between cation resin and $HClO_4$-$Sn(ClO_4)_2$ solution phases plotted against pH.

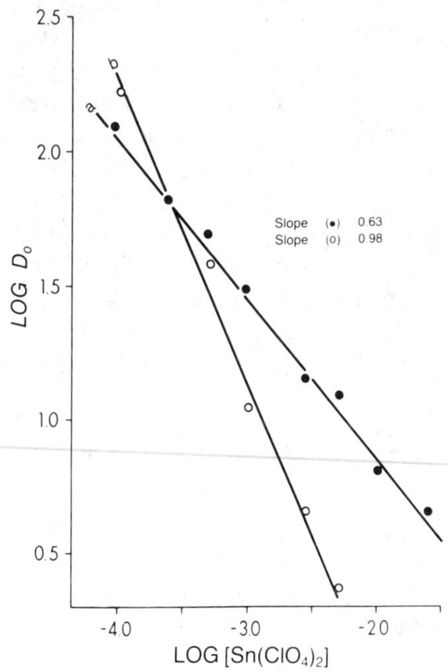

Figure 2

The distribution, D_o, of Na-22 (●) and reduced Tc-99m species (○) between cation resin and $HClO_4$-$Sn(ClO_4)_2$ solution phases plotted against $Sn(ClO_4)_2$ concentration.

Gorski and Koch (7) have studied the net charge of tetravalent technetium produced by the reduction of $NH_4{}^{99}TcO_4$ in $HClO_4$ by hydrazine, using the electrophoretic method. At pH 1, they determined the charge on the Tc(IV) ion in solution to be +2. They suggested that the Tc(IV) species is TcO^{2+} which hydrolyzes to $TcOOH^+$ and finally to $TcO(OH)_2$ or ($TcO_2 \cdot H_2O$). Guennec and Guillaumont (8), using a solvent extraction method, have proposed the existence of $TcOOH^+$ at pH 1. However, some of the extraction coefficients obtained in these experiments were very low and subject to sizeable error. At any rate, these two studies point to the existence of cationic species in solution resembling those proposed in the present study.

It is important to point out that physical-chemical techniques applied to solutions generally do not distinguish clearly between $M(OH)_2^{2+}$ and MO^{2+} ions. Ion exchange, electrophoretic and solvent extraction methods do not discriminate between $Tc(OH)_2^{2+}$ and TcO^{2+}.

^{201}Tl, Thallous Chloride and ^{67}Ga Citrate. The distribution, D_o, of tracer concentrations of $^{201}TlCl$ and ^{67}Ga citrate between resin and solution phases was measured as a function of sodium perchlorate concentration of the solution phase. These data are presented graphically in Figure 3. The measured slope for $^{201}TlCl$ is 0.95 (Fig. 3a), confirming the charge of +1.0 on thallous ion. When the equilibrium distribution experiment with ^{67}Ga citrate is performed in acidic medium (Fig. 3c) using $HClO_4$ as the bulk electrolyte, the measured slope of the radioactive species of ^{67}Ga citrate is 2.7. This slope indicates that in acidic medium the ^{67}Ga exists as Ga^{3+}. However, when the distribution experiment is performed at or near physiologic pH, using $NaClO_4$ as the bulk electrolyte and keeping the pH constant at pH = 6.8, the distribution graph (Fig. 3b) has a curve with changing slope. The measured slope of the straight line portion of the curve is 1.5. The changing slope indicates formation of more than one species in solution, whereas the fractional charge indicates probable polymerization. Previous studies (9,10) have indicated that gallium citrate polymerizes in solution at near neutral pH. There is no doubt that ^{67}Ga citrate species is positively charged because the species was not adsorbed by an anion exchange resin.

Anion Exchange Distribution Studies. The distribution, D_o, of tracer concentrations of $^{99m}TcO_4^-$, ^{99m}TcHEDP and ^{169}YbDTPA between resin and solution phases was measured as a function of the sodium perchlorate concentration of the solution phase. These data are presented in Figures 4 and 5. At these low concentrations, activity has been taken to equal the molar concentration.

^{99m}Tc Pertechnetate. The measured slope for $^{99m}TcO_4^-$ ion is 0.98 (Fig. 4a) and corresponds to the known charge of -1 on the TcO_4^- ion. The correct prediction of the charge on the TcO_4^- ion provides a test of the validity of the approach used to study the charge on the more complex species.

^{169}Yb Disodium Diethylenetriaminepentate. The measured slope for the distribution plot of ^{169}YbDTPA is 2.3 (Fig. 4b), indicating a charge of -2.0. The probable reaction is

$$Yb^{3+} + DTPA^{5-} \rightarrow YbDTPA^{2-} \quad (12)$$

The suggested charge in this study is in agreement with the charge for the structure of $^{169}YbDTPA^{2-}$ as given in Reference 11.

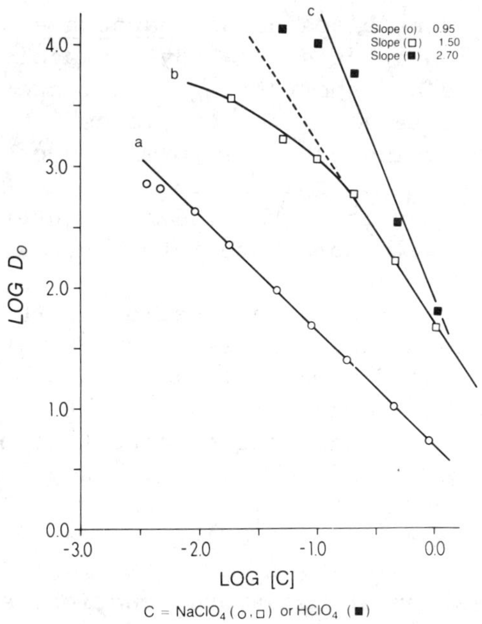

Figure 3

The distribution, D_o, of Tl-201 (○) and Ga-67 citrate (□,■) between cation resin and $NaClO_4$ or $HClO_4$ solutiom phases plotted against $NaClO_4$ or $HClO_4$ concentration.

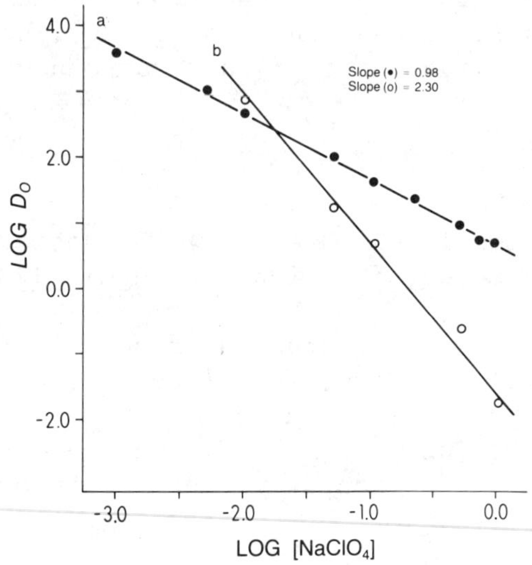

Figure 4

The distribution, D_o, of Tc-99m pertechnetate (●) and Yb-169 DTPA (○) between anion resin and $NaClO_4$ solution phases plotted against $NaClO_4$ concentration.

^{99m}Tc Hydroxyethylidene Disodium Diphosphonate. The ^{99m}TcHEDP (Fig. 5) plot yields a line with changing slope. By extrapolating the eventual straight line to the ordinate axis and subtracting these interpolated points from the observed D_o values, two straight lines of slopes 2.0 and 0.5 are resolved. This result is presumed to indicate the formation of complex technetium anions with charges of -2.0 and -0.5, respectively. Unlike the usual "curve-stripping" with simple exponential functions, the ordinate intercepts do not reflect the quantity of each complex. The ordinate intercept is the constant in equation 10.

In these studies, no direct determination of the oxidation state of technetium in the Tc radiopharmaceutical was made. However, other workers have reported the formation of Tc(IV)HEDP, Tc(IV) gluconate (12) and Tc(IV) phosphate 13,14). So we assume, that in the presence of HEDP, that Tc(VII) is reduced to Tc(IV) by Sn(II). We suggest that the component with a net charge of -2 is produced according to the following equation:

$$Tc(OH)_2^{2+} + 2H_2HEDP^{2-} \rightarrow (Tc(OH)_2(H_2HEDP)_2)^{2-} \quad (13)$$

The basis for the suggestion of the above equation is that at the pH of interest (pH 5-7), two of the four phosphonic acids are ionized (15) and therefore, the disodium and not the tetrasodium form of HEDP is used in the formulation of ^{99m}TcHEDP complex. There is also the possibility that tin is a part of the complex as suggested by Tofe and Francis (16) and implied by Deutsch (17) who studied the structural characterization of bridged ^{99m}Tc-Sn-dimethylgloxime complex.

The component with fractional net charge could be an aggregate of Tc(IV), Sn(II), Sn(IV), OH^- and HEDP ions in solution. It is not an uncommon phenomenon to have compounds with fractional charges (18-20), however, such a phenomenon is mostly observed in solids (defects) and compounds with metal-metal bonds (clusters).

REFERENCES

1. Schubert J, Russell ER, Myers LS: Dissociation constants of radium-organic acid complexes measured by ion exchange. J Biol Chem 185:387-398, 1950

2. Marcus Y: Ion exchange studies of complex formation. In Ion Exchange, Marinsky, JA, ed, vol I, New York, Mercel Dekker, Inc., 1966, pp 101-138

3. Owunwanne A, Marinsky J, Blau M: Charge and nature of technetium species produced in the reduction of pertechnetate by stannous ion. J Nucl Med 18:1099-1105, 1977

4. Welch MJ, Welch TJ: Solution chemistry of carrier-free indium. In Radiopharmaceuticals, Subramanian G, Rhodes BA, Cooper JF, Sodd VJ, eds, The Society of Nuclear Medicine, Inc., New York, 1975, pp 73-79

5. Loberg MD, Fields AT: Chemical structure of technetium-99m-labeled N-(2,6-diethylphenylcarbanonylmethyl) imino diacetic acid (Tc-HIDA). Int J Appl Radiat Isot 29:167-173, 1978

6. Noyes AA, Toabe R: The relative electrode potentials of tin and lead determined by equilibrium measurements with their perchlorates. J Am Chem Soc 39:1537-1545, 1917

7. Gorski B, Koch H: Zur chemie des technetium in wässeriger losung 1. Über den zustand des vierwertigen technetium in wässeriger losung. J Inorg Nucl Chem 31:3565-3571, 1969

8. Guennec JY, Guillaumont R: Comportement du Tc/IV d l'echelle des indicatteur en milieu non complexant. Radiochem Radioanal Let 13:33-38, 1973

9. Harris WR, Martell AE: Aqueous complexes of gallium (111). Inorg Chem 15:713-720, 1976

10. Glickson JD, Pitner TP, Webb J, et al: Hydrogen-1 and gallium-71 nuclear magnetic resonance study of gallium citrate in aqueous solution. J Am Chem Soc 97:1679-1683, 1975

11. Ytterbium Yb 169DTPA. Nuclear Products, 3M Company, St. Paul, 1976 p 1

12. Hambright P, McRae J, Valk PE, et al: Chemistry of technetium radiopharmaceuticals. 1. Exploration of the tissue distribution and oxidation state consequences of technetium (IV) in Tc-Sn-gluconate and Tc-Sn-HEDP using carrier ^{99}Tc. J Nucl Med 16:478-482, 1975

13. Thomason PF: Coulometric titration of potassium pertechnetate in a solution of potassium phosphate. Unclassifed Report, Oak Ridge National Lab, Tennessee, ORNL-2453, 1958, p 7

14. Terry AA, Zittel HE: Determination of technetium by controlled-potential coulometric titration in buffered sodium tripolyphosphate medium. Anal Chem 35:614-618, 1963

15. Castronovo FP: Method for the synthesis of 1-hydroxy-ethylidene-1, 1-disodium phosphonate (HEDSPA): A skeletal seeking radiopharmaceutical after labeling with ^{99m}Tc. J Nucl Med 15:127-130, 1974

16. Tofe AJ, Francis MD: Optimization of the ratio of stannous tin: ethane-1-hydroxy-1, 1-diphosphonate for bone scanning with ^{99m}Tc-pertechnetate. J Nucl Med 15:69-75, 1974

17. Deutsch E, Elder RC, Lange BA, et al: Structural characterization of a bridged ^{99}Tc-Sn-dimethylglyoxime complex: Implications for the chemistry of ^{99m}Tc-radiopharmaceuticals prepared by the Sn(II) reduction of pertechnetate. Pro Natl Acad Sci 73:4287-4289, 1976

18. Cotton FA, Wilkinson G: Advanced Inorganic Chemistry, 3rd ed, New York, Interscience Publishers, 1972, p 547

19. Baird MC: Metal-metal bonds in transition metal compounds. In Progress in Inorganic Chemistry, Cotton FA, ed, vol 9, New York, Interscience Publishers, 1968, pp 1-160

20. Penfold BR: Sterochemistry of metal cluster compounds. In Perspective in Structural Chemistry, Dunitz ID, Ibers JA, eds, vol II, New York, John Wiley & Sons, 1968, pp 71-150

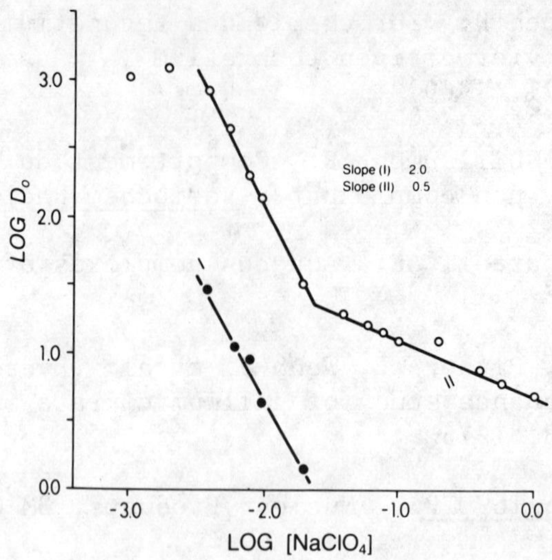

Figure 5

The distribution, D_O, of Tc-99m HEDP between anion resin and $NaClO_4$ solution phases plotted against $NaClO_4$ concentration. Curve 1 is produced by using curve stripping technique.

MICROSCALE SYNTHESIS AND BIODISTRIBUTION OF Pt-195m-LABELED *CIS*-DICHLORODIAMMINEPLATINUM(II), *CIS*-DDP

James D. Hoeschele, Thomas A. Butler, and John A. Roberts

Nuclear Medicine Technology Group
Health and Safety Research Division
Oak Ridge National Laboratory
Oak Ridge, Tennessee 37830

ABSTRACT

Cis-DDP is a unique metal-based antitumor drug which exhibits potent activity toward cancers of genitourinary origin. The major dose-limiting side-effect of *cis*-DDP is nephrotoxicity, which limits broader clinical use. We have initiated studies using Pt-195m-labeled *cis*-DDP to elucidate the critical pharmacological parameters affecting nephrotoxicity. The general details of the optimized microscale synthesis (50 μmole scale), the effects of varying the dose and concentration of *cis*-DDP, and the use of a diuretic on the biodistribution of Pt-195m-labeled *cis*-DDP are as follows: (1) yields of *cis*-DDP average 75%, (2) the product meets the well-defined purity criteria established for *cis*-DDP, (3) the % dose retained/gm tissue is independent of both dose and concentration (at a fixed dose) of *cis*-DDP, and (4) injection of the diuretic furosemide at 2 and 4 mg/kg leads to enhanced retention of Pt-195m in most organs at 24 hrs. The potential clinical implications of these results are discussed.

INTRODUCTION

Cis-Dichlorodiammineplatinum(II), *cis*-DDP, is a unique metal-based drug which was recently approved for human use by the FDA in the treatment of metastatic testicular and ovarian cancers. The major side-effect of *cis*-DDP is dose-limiting renal toxicity which limits both repetitive and broader clinical use of this agent. Various clinical procedures have been developed to minimize kidney toxicity; however, kidney toxicity can still occur despite the utilization of these procedures. The most successful procedure that has been developed thus far is the slow infusion of *cis*-DDP over a 6-8 hr period preceded by intravenous hydration and followed by mannitol diuresis (1, 2).

We report here recent results of a study designed to identify the critical pharmacological parameters which could lead to reduced kidney uptake (damage) by *cis*-DDP. Using Pt-195m-labeled *cis*-DDP, the parameters studied are the distribution/retention of *cis*-DDP as a function of (a) drug dose, (b) drug concentration, and (c) the administration of the diuretic, furosemide (LASIX), before, simultaneously with, and after *cis*-DDP administration.

We also report here the key features of an optimized microscale synthesis of Pt-195m-labeled *cis*-DDP. This synthesis scheme was refined and adapted to a scale of 0.1 mmole or less in order that labeled *cis*-DDP can be produced efficiently, in high purity, and be provided for collaborative laboratory and clinical studies. These studies include the pharmacokinetics, kidney toxicity, biodistribution, and *in vitro* binding studies aimed at elucidating the mechanism of action of this drug.

Platinum-195m ($T_{1/2}$ = 4.02d) is an ideal radiolabel for *cis*-DDP since it emits penetrating radiation (γ-photons of 99 and 129 keV) and thus provides the capability of sensitive detection of Pt at low concentrations in whole animals, tissues, and *in vitro* biochemical experiments. Transmigration of daughter radionuclides is not a problem since Pt-195m decays directly to stable Pt-195 by isomeric transition. Emission of approximately four electrons per disintegration also makes the application of this nuclide attractive for autoradiography in tissue localization studies (3).

OPTIMIZED MICROSCALE SYNTHESIS OF Pt-195m-LABELED *CIS*-DDP

The molecular structure of *cis*-DDP is illustrated in Figure 1 and a six-step reaction scheme for synthesizing Pt-195m-labeled *cis*-DDP is illustrated in Figure 2. This scheme is a highly refined version of the original method used to synthesize Pt-195m-*cis*-DDP (4). The reaction conditions (reactant concentrations, reaction time and temperature) have been optimized in order to obtain the highest yield and purity of *cis*-DDP. Pt-195m is produced by neutron irradiation (n, γ) of enriched Pt-194 (97.41%) in the High Flux Isotope Reactor (HFIR) at Oak Ridge National Laboratory at a flux of 2.5×10^{15} neutrons/$cm^2 \cdot$sec. The high burn-up cross-section of Pt-195m

cis-DICHLORODIAMMINEPLATINUM(II)
(cis-DDP)

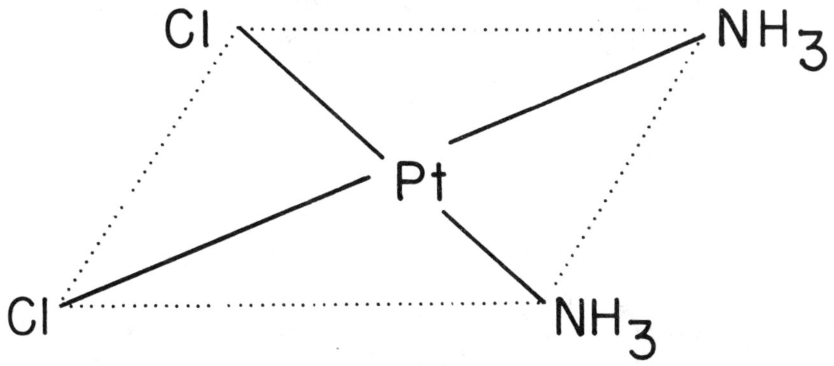

FIGURE 1

Molecular structure of *cis*-dichlorodiammineplatinum(II)

(1) Pt target + aqua regia $\xrightarrow[\text{heat to dryness}]{2\,NaCl}$ Na_2PtCl_6 (≤5)

(2) $Na_2PtCl_6 + \frac{1}{2}N_2H_4 \cdot 2HCl \xrightarrow{\sim 5°C} \xrightarrow[5\ min.]{85°C} Na_2PtCl_4$

(3) $Na_2PtCl_4 + 4KI \xrightarrow[\text{ambient temp.}]{2A}$ *cis*-$PtA_2I_2 \downarrow$

(steps 2–3: 10-15)

(4) *cis*-$PtA_2I_2 + 2AgNO_3 \xrightarrow{55°C}$ *cis*-$PtA_2(H_2O)_2^{2+}$

(5) *cis*-$PtA_2(H_2O)_2^{2+} + HCl_{(xs)} \xrightarrow{55°C}$ *cis*-$PtA_2Cl_2 \downarrow$

(6) Purification by Recrystallization

(a) 0·1N HCl, 0·15M NaCl, &/or DMF/HCl

(steps 4–6: 10-15)

A denotes NH_3 ; Pt denotes Pt-195m-labeled Pt

FIGURE 2

Microscale synthesis of Pt-195m-labeled *cis*-$[Pt(NH_3)_2Cl_2]$. Numbers in parentheses are average step-wise losses in %; nonessential reaction products have been omitted for the sake of brevity.

(13,000 barns) limits the specific activity to $\sim$ 1 mCi of Pt-195m/mg Pt metal. Major by-product radionuclides are Au-199, Pt-197, Ir-192, and Ir-194. Carrier-free Au-199 is removed by extraction into Methyl Isobutyl Ketone (MIBK) and elimination of iridium radionuclides occurs because Ir does not follow Pt in the synthesis scheme. Brief comments on each of the essential steps in the scheme are as follows:

(Step 1) The Pt metal target is dissolved in *aqua regia*, the concentration of HCl adjusted to 3M and extracted with MIBK to remove Au-199 and then evaporated just to dryness (in the presence of NaCl) to yield a residue of crystalline $Na_2PtCl_6 \cdot xH_2O$. The amount of Pt recovered from the target ($\sim$ 98%) is determined by spectrophotometric analysis of NO_3-free solutions of Na_2PtCl_6 in 0.5M HCl ($\varepsilon_{262\,nm} = 3.9 \times 10^4 M^{-1}\ cm^{-1}$). It is essential to know accurately the concentration of Pt at this stage because of the critical stoichiometry required in Step 2.

(Step 2) The Na_2PtCl_6 residue is dissolved in water and is rapidly reduced at 5°C to Na_2PtCl_4 on addition of a solution of $N_2H_4 \cdot 2HCl$ (0.4M). Stoichiometry is critical in this step since excess $N_2H_4 \cdot 2HCl$ could reduce Pt(IV) to Pt metal. Following the initial vigorous reaction, the solution is slowly warmed to 85°C to complete the reduction.

(Step 3) Excess KI (1.5 x stoichiometric amount) is added to the solution from Step 2 to form K_2PtI_4, followed by the addition of NH_4OH to produce the insoluble intermediate, *cis*-$Pt(NH_3)_2I_2$. The optimum I^-/Pt(II) mole ratio is 6. At a mole ratio of 4 or less, PtI_4^{2-} is hydrolytically unstable whereas at a mole ratio of greater than 6, some sacrifice in yield results.

(Step 4) *Cis*-$Pt(NH_3)_2I_2$ is washed once with H_2O and then converted (at 55°C) to the soluble diaquo species, *cis*-$[Pt(NH_3)_2(H_2O)_2]^{2+}$, on addition of $AgNO_3$ solution maintaining a temperature of 55°C for 15 min. A 10% excess of $AgNO_3$ (0.4M) is used to insure complete removal of iodide ion. Following separation of the AgI, the excess Ag^+ ion is precipitated as AgCl by addition of 45% of the amount of Cl^- (as 1M HCl) required. This amount of Cl^- insures removal of Ag^+ but does not lead to premature formation and co-precipitation of *cis*-DDP with AgCl.

(Step 5) The product *cis*-DDP is then formed on addition of excess HCl to react with the diaquo species at 55°C for 5 minutes. A relatively pure product is obtained on cooling to 0°C.

(Step 6) *Cis*-DDP is purified by recrystallization from either (a) 0.1N HCl or (b) 0.15M NaCl or, preferably, (c) by dissolution in N,N-dimethylformamide (DMF) at room temperature followed by addition of 2 volumes of 0.1N HCl and cooling to 0°C.

Prepared in this manner, Pt-195m-labeled *cis*-DDP meets the well-defined purity criteria established for unlabeled *cis*-DDP (5, 6). Preparations exhibit a major absorption maximum at ∿ 301 nm (ε = 131 $M^{-1}cm^{-1}$) and an absorbance ratio, [As 300nm/246nm], of 4.4 ± 0.2. The absorption at 301 nm is used to determine the yields of *cis*-DDP. Purity is also evaluated by employing paper chromatography. A pure compound exhibits a single homogeneous spot with an R_f value of 0.33 - 0.36, using Whatman 3MM paper when developed with 9:1 acetone: x(X = H_2O, 1M HCl) at 21°C. Approximately 2% of the applied compound (∿ 9 μg total) remains at the origin for the 9:1 acetone:1M HCl mixture, whereas 6-8% remains at the origin when using 9:1 acetone:H_2O. Greatest differentiation between *cis*-DDP and *trans*-DDP, a potential contaminant, is found using alumina (Bakerflex BF-1) and the above eluents.

The key features of the refined microscale procedure are as follows:

(1) High yields (∿ 75%) and a chemically and radiochemically pure product are obtained at a scale of 50 μmole or greater.

(2) After Step 1, all operations can be performed in convenient screw-capped 15 ml centrifuge tubes.

(3) Steps (1) and (2) of the synthesis can be used without modification for the synthesis of other Pt(II and IV) antitumor analogs.

DISTRIBUTION/RETENTION STUDIES

Single injections of Pt-195m-labeled *cis*-DDP only and in combination with furosemide (LASIX) were administered intravenously (lateral tail vein) to 10-13 week-old Fischer 344 strain rats. Pt-195m-*cis*-DDP was administered in 0.85% NaCl solution at a concentration of 1.0 mg/ml except where indicated otherwise. The amount of radioactivity administered per rat was adjusted to ∿ 50-75 μCi by dilution of the Pt-195m-labeled *cis*-DDP solution with stable *cis*-DDP in 0.8% NaCl at the same concentration. Furosemide, (courtesy of J. Coupal, VA Hospital, KY) was administered at a concentration of 10 mg/ml. Rats were sacrificed at 24 hrs post-injection of Pt by decapitation while under ether anesthesia. Organs were excised, rinsed with saline, weighed, and Pt-195m activity counted using a Packard Auto-Gamma. Three or four animals were used for each data point.

As a first step in assessing whether kidney retention (damage) of Pt might be altered by pharmacological parameters, we have examined the tissue distribution of Pt-195m activity as a function of the (a) dose of *cis*-DDP, mg/kg, (b) concentration of *cis*-DDP in the injected solution (at a fixed dose of *cis*-DDP) and (c) forced diuresis (without hydration) using furosemide (a potent diuretic which inhibits Na^+ ion reabsorption in the tubules). The results in Figure 3 and Table 1 indicate that tissue distribution (% dose retained/g tissue after 24 hrs) is essentially independent of both the injected dose and concentration of *cis*-DDP injected. This means that a constant fraction of the injected dose is retained in all tissues for a single i.v. injection over a three-fold range of concentrations (0.5 to 1.5

TABLE 1. THE DISTRIBUTION OF RADIOACTIVITY IN RAT TISSUES AFTER ADMINISTRATION (I.V.) OF SOLUTIONS OF Pt-LABELED-*CIS*-DDP VARYING IN THE CONCENTRATION OF *CIS*-DDP.

	Percent dose per gram tissue		
[*cis*-DDP], (mg/ml):	0.50	1.0[1]	1.5
Tissue			
Blood	0.17 ± 0.01	0.19	0.18 ± 0.01
Liver	0.29 ± 0.03	0.34	0.29 ± 0.01
Spleen	0.20 ± 0.01	0.23	0.21 ± 0.01
S. Intestine	0.11 ± 0.12	0.12	0.11 ± 0.01
Kidneys	1.32 ± 0.03	1.62	1.47 ± 0.10
Testes	0.036 ± 0.002	0.036	0.038 ± 0.002
Brain	0.010 ± 0.000	0.012	0.012 ± 0.001

Fischer 344 rat (male)

[1]single rat

mg/ml) and a six-fold range of doses (1.0 to 6.0 mg/kg). Hence, proportionally greater amounts of Pt would be retained by the kidney at higher doses. [Average values of the % dose/g tissue for other tissues not included in Figure 3 are as follows: spleen (0.25), intestine (0.14), ovaries (0.16), fallopian tubes plus uterus (0.25), and lungs (0.21)]. Comparative studies indicate that there are no dose-related differences in tissue distribution of male vs female rats, with the possible exception of the kidney which appears to be slightly higher in the male, eg., 1.66 ± 0.20 vs 1.26 ± 0.19 (female).

The influence of furosemide on the distribution of *cis*-DDP is illustrated in the histogram in Figure 4. Simultaneous injection of furosemide (at 10 mg/kg) and *cis*-DDP (at 2 or 4 mg/kg) exacerbates Pt retention in most tissues and blood after 24 hrs. The most pronounced effect is in the kidney, 30 and 27% for Pt doses of 2 and 4 mg/kg, respectively. These results agree qualitatively with those of DeSimone and coworkers (7). The preliminary results in the male rat showed increases in the kidney of approximately 60% over the *cis*-DDP controls. Administration of furosemide 30 minutes before or after *cis*-DDP treatment gives rise to approximately the same increase in tissue retention as in the simultaneous administration of both agents. The enhanced tissue retention caused by furosemide is slightly dose-dependent. Using a dose of 25 mg/kg of furosemide (at a *cis*-DDP dose of 2 mg/kg) leads to an increase in Pt retention of ∿ 12% over that shown in Figure 4 for the 10 mg/kg dose of furosemide.

DISTRIBUTION STUDIES-CLINICAL IMPLICATIONS

The significance of any clinical implications inferred from these studies depends on two major questions: (a) Is Pt retention in the rat kidney directly related to kidney damage or impairment? and (b) Is the rat a good pharmacokinetic model for man? Definitive studies are not available to answer the question in (a); however, recent work by Wolf (3) and Ward (8) indeed suggest that the rat is a good pharmacokinetic model for man.

The ideal drug regimen for *cis*-DDP is one which will maintain or potentiate its effectiveness and minimize dose-related kidney damage. If kidney damage is directly and proportionally related to kidney retention, there would appear to be no particular advantage (as far as kidney toxicity is concerned) in replacing a single-dose regimen by a fractionated dose schedule since the amount of Pt retained is likely to be the same in either case for the same total dose. This premise, of course, assumes that a constant fraction of Pt would be retained for each successive dose of the fractionated dose. From a pharmacokinetic and (cumulative) toxicological standpoint, it is exceedingly important to establish whether a constant fraction of Pt is retained for each successive dose administered.

Since rat kidney uptake is independent of both dose and drug concentration it would appear that kidney retention is not likely to be diminished by

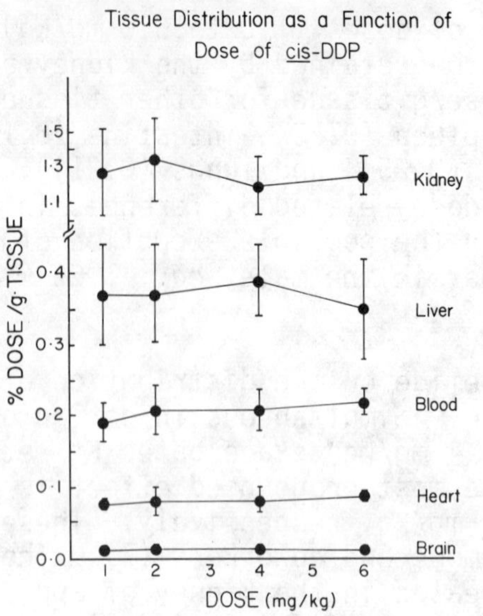

FIGURE 3

The distribution of radioactivity in female F344 rat tissues 24 hr after administration (i.v.) of various doses of Pt-195m-labeled *cis*-DDP. The distribution data are expressed in terms of the % of the injected dose/g. tissue; dose of *cis*-DDP is in terms of mg/kg body weight.

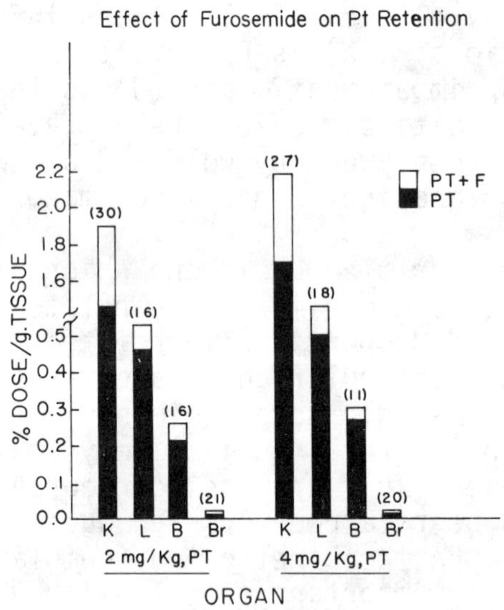

FIGURE 4

The enhancement of Pt uptake in various tissues of the female F344 rat 24 hr after simultaneous administration (i.v.) of Pt-195m-labeled *cis*-DDP (at 2 and 4 mg/kg) and furosemide (10 mg/kg). Enhancement is most pronounced in the kidney. Distribution data are expressed in the % of injected dose/g tissue. Numbers in parentheses represent the % increase in Pt uptake resulting from the use of furosemide. The symbols K, L, B and Br represent the organs, kidney, liver, blood and brain, respectively.

variation of drug concentration unless the pharmacokinetics of uptake are markedly different at very low drug concentrations. It is possible, therefore, that a very slow and therefore extended administration of *cis*-DDP could lead to reduced Pt uptake (damage) in the kidney. What effect an extended i.v. infusion will have on the efficacy of the drug is a big question mark.

It has been presumed that the use of diuretics (such as furosemide and mannitol) in combination with *cis*-DDP chemotherapy leads to reduced kidney retention. Our results suggest that this indeed may not be true. If the use of furosemide leads to enhanced retention but yet ameliorates kidney toxicity, either different sites of uptake are involved and/or a simple relationship between kidney retention and kidney damage (impairment) cannot exist. The enhanced tissue uptake raises still another question: Does the enhanced retention of Pt lead to reduced tolerance or hypersensitivity as regards allergic response to *cis*-DDP? Allergic reaction to selected platinum complexes is well documented.

Future studies will focus on the tissue retention of Pt as a function of long-term infusion and multiple-injection of Pt-195m-labeled *cis*-DDP.

ACKNOWLEDGEMENTS

Research sponsored by the Office of Health and Environmental Research, U. S. Department of Energy under contract W-7405-eng-26 with the Union Carbide Corporation. The technical assistance of L. A. Ferren and C. E. Guyer is gratefully acknowledged.

REFERENCES

1. MERRIN C: A New Method to Prevent Toxicity with High Doses of *Cis*-diammine Platinum: *Proc. Am. Assoc. Cancer Res. and ASCO,* 17:243,1976

2. HAYES D, CVITKOVIC E, GOLBEY R, SCHEINER E, and KRAKOFF IH: Amelioration of Renal Toxicity of High Dose *Cis*-platinum diammine dichloride (CPDD) by mannitol induced diuresis. *Proc Am Assoc Cancer Res and ASCO,* 17:169, 1976

3. WOLF, W: personal communication

4. HOESCHELE JD, and VAN CAMP L: Whole-Body Counting and Distribution of *cis*-$^{195m}Pt(NH_3)_2Cl_2$ in the Major Organs of Swiss White Mice, in *Advances in Antimicrobial and Antineoplastic Chemotherapy,* Vol. II, HEJZLAR M, SEMMSKY M, and MASAK S: pp. 241-242, University Park Press, Baltimore, 1972

5. HOESCHELE JD: The Preparation, Purification, and Analysis of *Cis*-$[Pt(NH_3)_2Cl_2]$ (manuscript in progress)

6. Clinical brochure for *Cis*-dichlorodiammineplatinum(II), NSC-119875 (August, 1974), National Cancer Institute.

7. DESIMONE PA, YANCEY RS, COUPAL JJ, BUTTS JJ, BUTLER TA, and HOESCHELE JD: The Effect of a Forced Diuresis on the Distribution and Excretion (Via Urine and Bile) of Pt-195m *Cis*-dichlorodiammineplatinum(II): submitted for publication

8. WARD JM, GRABIN ME, LEROY AF, and YOUNG DM: Modification of renal toxicity of *cis*-dichlorodiammineplatinum(II) with furosemide in male F344 rats. *Cancer Treat Rep.* 61:375-379, 1977

CLINICAL NEEDS IN FUNCTIONAL IMAGING

H. L. Atkins, R. G. Fairchild, P. Richards and H. Pate

Medical Department
Brookhaven National Laboratory
Upton, N.Y. 11973

Emission tomography, with positron or single photon emitting radionuclides, offers the possibility of quantitation, a necessary ingredient for functional studies. A figure of merit has been devised as an aid in selecting radionuclides that offer the greatest photon yield for the lowest radiation dose because emission tomography requires a greater information density than conventional imaging. Alterations in function generally precede anatomical alterations and functional imaging may therefore be a more sensitive method of investigating disease processes.

The work at BNL was done under Contract No. EY-76-C-02-0016 with the United States Department of Energy.

INTRODUCTION

The very foundation of nuclear medicine is the assessment of physiologic or metabolic function. However, for a period of time an emphasis was placed on morphological information available through such procedures as liver imaging, thyroid imaging, etc. Nevertheless, functional information was available through radionuclide procedures even though this aspect of the study was not the primary objective.

Rapid advances in the technology of computerized x-ray tomography and ultrasound have established the primacy of these modalities as providers of anatomical information. In some instances they can provide useful physiological information. A general reassessment in nuclear medicine as a consequence of these developments has resulted in greater emphasis on the functional information derived from tracer techniques.

Multiple types of functional studies using radionuclides are in routine use and applied to clinical care. These include rates of uptake and washout and degrees of concentration in an organ or region. Rapid dynamic studies and equilibrium studies are employed. The study of function implies quantitation. This has been partially possible through the now rather routine utilization of dedicated computer systems in Nuclear Medicine laboratories. Imaging studies can be used to evaluate regional function in organs and portions of the body thus permiting a more sensitive index of abnormalities compared to measurements of global function.

Except for special circumstances, true quantitation has not been readily achievable because of the difficulties in correcting for tissue attenuation of radioactivity and background activity in surrounding tissues. Not only do these factors defeat attempts at quantitation, but they also decrease sensitivity in evaluating abnormal regions because of the degradation of contrast. The recent application of tomographic techniques to nuclear imaging promises to correct this situation and greatly improve the prospect for true quantitative function imaging. These techniques have progressed somewhat further in the use of positron emitters but they are also being developed with regard to single photon emitters.

SELECTION OF RADIONUCLIDES

A) GENERAL CONSIDERATIONS

The considerations which enter into the selection of an appropriate radionuclide for in-vivo tracer studies have been extensively reviewed previously (1,2). A physical half-life matched to the time of observation, the absence of non-penetrating radiation and the emission of efficiently detected photons result in a high yield of detectable events for a low patient radiation dose. Also important is the need for a radionuclide that can be labeled firmly to the substance that is to be traced and that will not alter its biochemical behavior.

Especially important is a consideration of the specific activity when investigating metabolic pathways and receptor mechanisms. Low specific activities may result in flooding the pools, pathways, or receptors when a sufficient level of activity for adequate detection is administered. This may alter distribution considerably (3,4).

Other factors which govern the selection of a radionuclide include availability and radionuclidic purity. For the ultra short-lived positron emitters an accelerator must be in close proximity to the clinical site. Generator systems can provide very short-lived material in the absence of an accelerator and, with the development of instrumentation which extends the range of useful photon energies, a number of new parent-daughter relationships have been under investigation recently.

B) EMISSION COMPUTED TOMOGRAPHY (ECT)

The increasing interest in both single photon emitters and positron emitters for emission tomography has led to a consideration of radionuclide characteristics suitable for these modalities. While the general considerations discussed above also apply to radionuclides for conventional scintigraphy, the need for high photon yields is especially important for ECT. The reconstruction algorithms require a count density considerably higher than for conventional projection imaging. An improvement in detection efficiency of newer instrumentation will ease this problem to some extent.

In considering the nuclides available for imaging purposes we have reviewed the MIRD pamphlets 10 and 11 (5,6) and performed a calculation of a figure of merit for both single photon and positron emitting radionuclides. Nuclides with a half-life less than 193 hours (I-131) and with photon energies greater than 0.069 MeV (^{201}Tl) and equal to or less than the 0.511 MeV annhilation radiation of positron emitters were considered. Exceptions were ^{127}Xe with a 36.4 day physical half-life but a very short biological half-life, and ^{75}Se with a 120 day half-life, but used routinely in clinical imaging studies.

Assumptions were: a) whole body distribution; b) no excretion; c) attenuation by overlying tissues was not significantly different for the various energies; d) only photons present in more than 8% of decays were included unless the photon was the only one emitted; e) the radionuclide was administered as a single bolus; no equilibrium imaging was considered. Calculations were made for different time intervals from 15 minutes to 4 hours after administration.

Detector efficiency for single photon ECT was considered for two different physical situations relating to the two methods currently in use, for a 1/2 inch crystal thickness of sodium iodide employed in gamma cameras and a 1 inch thick crystal employed in a multicrystal focussed collimator system. Appropriate collimation for the photon energy and the ability to collect multiple photon energies were included in the calculations. Positron emitters were considered only for the 1 inch thick crystal.

Calculation utilized the following formulation:

$$Q = \frac{n\ e^{-\lambda t}\ GP}{S\tau}$$

where Q = figure of merit (photons/rad)
n = number of photons per μCi (calculated from No. of photons per disintegration = n_i)
λ = decay constant (h^{-1})
G = geometrical plane source efficiency (7)*
P = crystal photopeak efficiency (8)**
S = mean absorbed dose to total body per unit cumulated activity (rad/μCi·h)
τ = average life (T 1/2 x 1.44)

The results of these calculations are seen in Tables 1-6.

Technetium-99m is by far the preferred radionuclide for single photon emission tomography under these conditions at time intervals up to 4 hours. Next comes iodine-123. It must be remembered that the evaluation has ignored specific localization of the nuclide, especially when incorporated into certain compounds. If the highest concentration is in the organ of interest the figure of merit in the table is falsely low, whereas if significant concentration occurs in other sites the figure of merit is falsely high. The tables serve only as a general guide and cannot be applied to specific situations without some modification taking into consideration these other factors.

CLINICAL NEEDS

A) CARDIAC PHYSIOLOGY

First pass studies of myocardial contractility require only on the order of a minute to carry out, yet the most commonly used nuclide is technetium-99m with a half-life of 6 hours. It would be desirable to have a readily available nuclide with a half-life of a few minutes for these studies, particularly for children. Such an agent would permit the use of higher levels of activity with lower radiation dose and repeat studies at short intervals provided the counting or imaging equipment could handle the increased activity. An example of a promising radionuclide is tantalum-178 derived from the ^{178}W-^{178}Ta generator (9).

Myocardial perfusion studies are now routinely performed with thallium-201. The low energy photons of this radionuclide in combination with considerable background activity make it less than ideal. Emission tomography with ^{201}Tl does show some promise. A variety of positron emitting materials have been developed for myocardial imaging including ^{11}C-palmitate (10), ^{38}K (11) and ^{52m}Mn (12). Particularly interesting is ^{52m}Mn because it is the daughter of

* Footnote at end of paper
** Footnote at end of paper

iron-52 and therefore available in a generator system to users without a cyclotron. Its half-life of 21 minutes is well suited to its use---long enough for easy handling and detection, and short enough to permit serial studies at short intervals (Fig. 1).

B) BRAIN

Most of the recent interest in the brain has been related to the short-lived labels, particularly ^{18}F as 2-fluoro-2-deoxy-D-glucose for the study of regional cerebral glucose metabolism (13). The study of neurotransmitter and blocking agent receptor mechanisms is another area receiving considerable attention. There is possibly a role for labels other than ^{11}C and ^{18}F in the evaluation of brain metabolism.

We have been using ^{123}I-4, 3 (dimethylamino, propylamino)-7-iodoquinoline for melanoma localization (14). By chance, we have noted a high degree of localization of the compound in the brain, including the cerebellum (Fig. 2). The behavior is similar to another compound 4-iodo-2,5-dimethoxyphenylisopropylamine(4-I-DPIA) described by Sargent et al. (15). Other similar compounds which can be labeled with radioiodine probably exist and may provide a means for studying some aspect of cerebral metabolism.

C) TUMOR LOCALIZATION

Present methods for tumor localization lack consistency of action and sufficient specificity to be entirely satisfactory. More specific materials must be developed to identify a particular tumor and to separate inflammatory lesions from neoplasms. An example of one approach is the use of the iodinated quinolines for melanoma detection. Other materials which enter into metabolic processes of tumors should be sought. Perhaps the newer platinum compounds (16,17) or ruthenium labeled tracers (18) may offer possibilities for improved tumor detection.

D) POSITRON TOMOGRAPHY

Currently there is a rapid development of positron imaging systems which promises to provide high sensitivity and improved resolution in new instrumentation. These new systems are very expensive, particularly since the market for them is rather limited and largely confined to institutions with cyclotrons. While optimal utilization of positron tomographs for functional imaging requires a cyclotron for production of ^{11}C, ^{13}N, ^{15}O, and ^{18}F a number of generator systems for production of short-lived positron emitters would ensure a wider distribution of positron tomographs and extend utilization of the method to sites remote from accelerators.

A number of generator systems have been suggested in the past. The ^{68}Ge-^{68}Ga generator is potentially capable of providing a label for many of the agents now labeled with ^{99m}Tc (19). Recent developments have resulted in a generator in which the ^{68}Ga is not chelated to EDTA and is therefore more readily incorporated into other compounds (20,21). Unfortunately the radiation dose

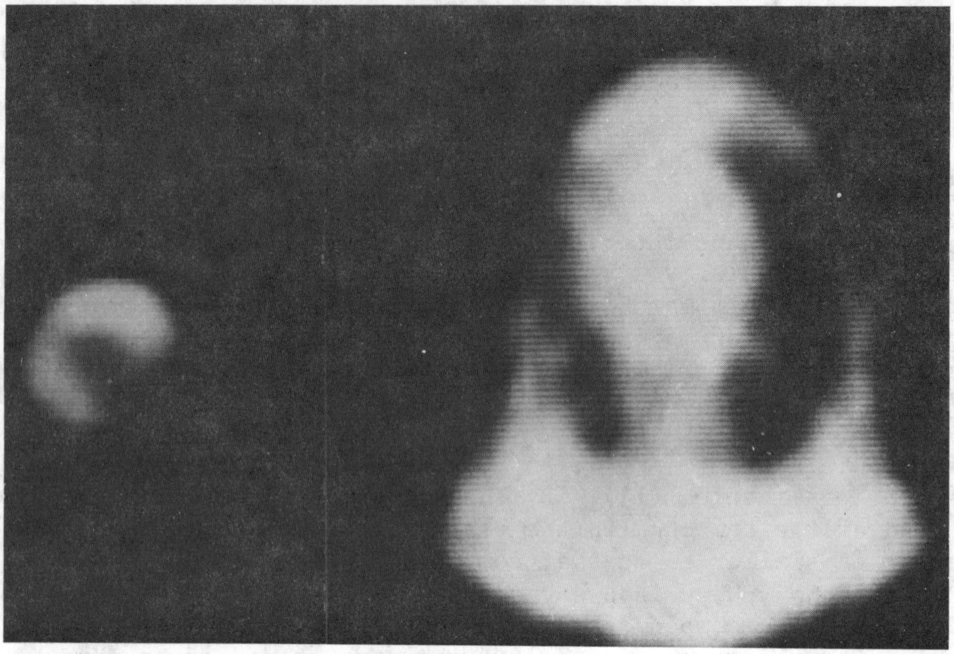

Figure 1. Emission scan performed with ^{52m}Mn, and transmission scan of a normal dog showing characteristic myocardial distribution of radioactivity.

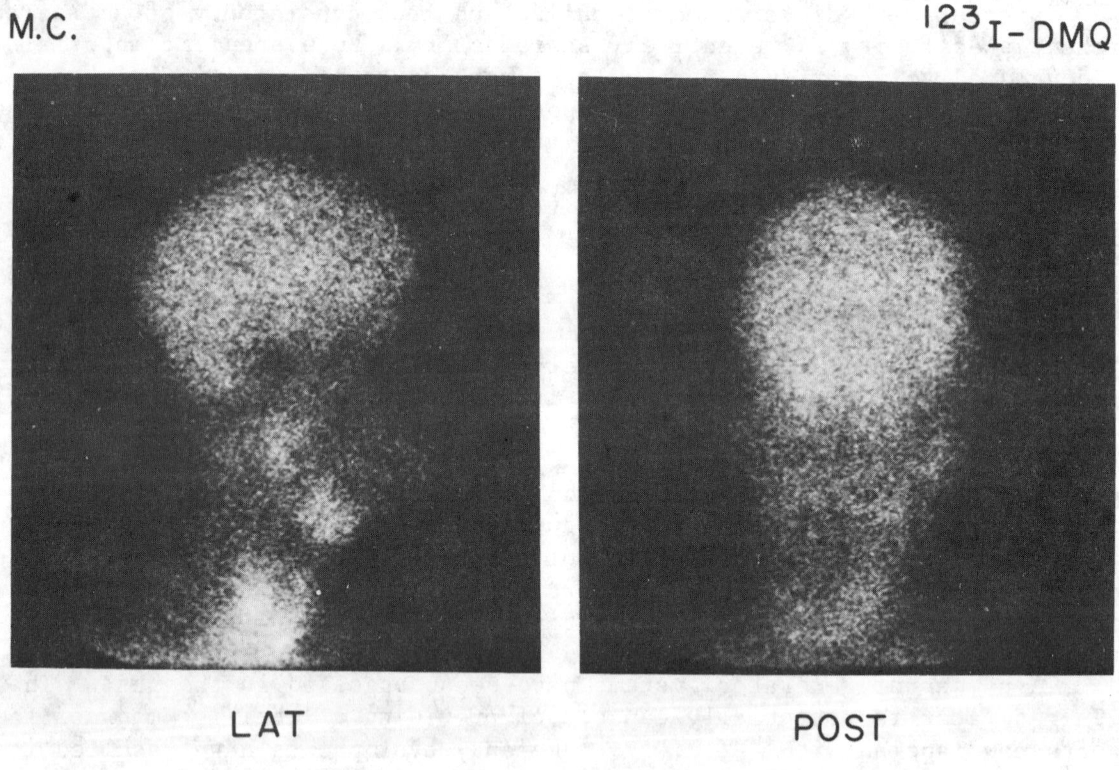

Figure 2. Anterior and right lateral scintiphotos of the head following administration of ^{123}I labeled 4,3(dimethylamino,propylamino)-7-iodoquinoline. Note uptake in cerebrum and cerebellum. There is some free iodine uptake in the salivary glands and thyroid.

from ^{68}Ga is considerably higher than from ^{99m}Tc and the utility of the compounds is limited in terms of the study of metabolic processes.

The ^{122}Xe-^{122}I generator system is of unique design, using gas phase separation (22). It provides a short-lived (3.5 min.) positron emitter which can be used, after the development of rapid iodination techniques, to label a variety of useful materials and which can permit repeat studies after short intervals.

Iron-52 is receiving renewed attention for two purposes. With appropriate instrumentation it is probably the best tracer for hemopoietic bone marrow distribution (23). It is also the parent of ^{52m}Mn which is being evaluated for myocardial perfusion imaging (12), but which might also be useful as a label for a number of compounds.

Several other generator systems have been suggested for positron emitting radionuclides. Generators are listed in Table 7.

CONCLUSIONS

The rapid development of emission computed tomography for both positron and single photon emitters promises to emphasize the quantitative and functional possibilities in radionuclide techniques. Advances in instrumentation are paralleling radiopharmaceutical development. A limitation of tomographic systems is the rapidity at which acquisition of data can occur. As acquisition times approach 1 second, time consumed in the reconstruction of multiple images can become a limiting factor. The use of equilibrium imaging with materials such as $^{15}O_2$ and ^{81m}Kr can be used instead of rapid dynamic procedures utilized with projection imaging of gamma emitters.

Improved methods for measuring brain metabolism and function, assessment of blood flow, delineating regional myocardial perfusion, and detection of tumors are greatly needed.

REFERENCES

1. Wagner HN Jr, Emmons H: Characteristics of an ideal radiopharmaceutical. In Radioactive Pharmaceuticals, G.A. Andrews, R.M. Kniseley, H.N. Wagner Jr, eds. U.S.A.E.C., CONF-651111, Oak Ridge, Tenn., 1966, pp 1-32.

2. McAfee JG, Subramanian G: Radioactive agents for imaging. In Clinhcal Scintillation Imaging, 2nd ed, L.M. Freeman, P.M. Johnson, eds, Grune & Stratton, N.Y., 1975, pp 13-114.

3. Fowler JS, Wolf AP, Christman DR, et al. Carrier-free "C-labeled catecholamines. In Radiopharmaceuticals, G. Subramanian, B.A. Rhodes, J.F. Cooper, V.J. Sodd, eds, Society of Nuclear Medicine, N.Y., 1975 pp 196-204.

4. Atkins HL, Christman DR, Fowler JS, et al: Organic radiopharmaceuticals labeled with isotopes of short half-life. V. ^{18}F-labeled 5- and 6-fluorotryptophan. J. Nucl. Med. 13: 713-719, 1972.

5. Dillman LT, Von der Lage FC: Radionuclide decay schemes and nuclear parameters for use in radiation-dose estimation. MIRD pamphlet No. 10, Society of Nuclear Medicine, Sept. 1975.

6. Snyder WS, Ford MR, Warner GG, Watson SB: "S" absorbed dose per unit cumulated activity for selected radionuclides and organs. MIRD pamphlet No. 11, Society of Nuclear Medicine, Oct. 1975.

7. Anger HO. Radioisotope Cameras, In Instrumentation in Nuclear Medicine, Vol. 1, G.J. Hine, ed., Academic Press, New York, 1967, pp 505-506.

8. Hine GJ, Erickson JJ: Advances in scintigraphic instruments, In Instrumentation in Nuclear Medicine, Vol. 2, G.J. Hine, J.A. Sorenson, eds. Academic Press, New York, 1974 pp 8-9.

9. Holman BL, Harris GI, Neirinckx RD, et al.: Tantalum-178 -- a short-lived nuclide for nuclear medicine: production of the parent W-178. J. Nucl. Med. 19: 510-513, 1978.

10. Sobel BE, Weiss ES, Welch MJ, et al: Detection of remote myocardial infarction in patients with positron emission transaxial tomography and intravenous ^{11}C-palmitate. Circulation 55: 853-857, 1977.

11. Lambrecht RM, Hara T, Gallagher BM, et al. Cyclotron isotopes and radiopharmaceuticals - XXVIII. Production of potassium-38 for myocardial perfusion studies. Int. J. Appl. Radiat. Isot. 29: 667-671, 1978.

12. Atkins HL, Som P, Fairchild RG, et al.: Myocardial emission tomography with manganese-52m (in preparation).

13. Reivich M, Kuhl D, Wolf A, et al.: The (^{18}F) fluorodeoxyglucose method for the measurement of local cerebral glucose utilization in man. Circ. Res. 44: 127-137, 1979.

14. Packer S, Redvanly C, Lambrecht RM, et al: Quinoline analog labeled with iodine 123 in melanoma detection. AMA Arch. Ophth. 93: 504-508, 1975.

15. Sargent T III, Budinger TF, Braun G, et al.: An iodinated catecholamine congener for brain imaging and metabolic studies. J. Nucl. Med. 19: 71-76, 1978.

16. Lange RC, Spencer RP, Harder HC: Synthesis and distribution of a radiolabeled antitumor agent: cis diamminedichloroplatinum (II). J. Nucl. Med. 13: 328-330, 1972.

17. Smith PHS, Taylor DM: Distribution and retentin of the antitumor agent ^{195m}Pt-cis-dichloroammine platinum II in man. J. Nucl. Med. 15: 349-351, 1974.

18. Srivastava S, Richards P, Meinken G, et al.: Evaluation of radiolabeled ruthenium compounds as tumor-localizing agents (this symposium).

19. Anger HO, Gottschalk A: Localization of brain tumors with the positron scintillation camera. J. Nucl. Med. 4: 326-330, 1963.

20. Ehrhardt GJ, Welch MJ: A new germanium-68/gallium-68 generator. J. Nucl. Med. 19: 925-929, 1978.

21. Neirinckx RD, Davis MA: Development of a chromatographic Ge-68/Ga-68 generator yielding ionic gallium (this symposium).

22. Richards P, Ku TH: The ^{122}Xe-^{122}I system: a generator for the 3.62 minute positron emitter, ^{122}I. J. Appl. Radiat. Isot. (in press).

23. Knospe WH, Rayuder VMS, Cardello M, et al: Bone marrow scanning with ^{52}iron (^{52}Fe). Regeneration and extension of marrow after ablative doses of radiotherapy. Cancer 37: 1432-1442, 1976.

FOOTNOTES

* Plane source efficiency is a measure of the fraction of γ-rays emitted by an extended plane source perpendicular to the collimator axis, that can pass through the collimator holes. This efficiency is given by (ref. 5)

$$G = \frac{\pi N do^4}{64L^2(1+L/F)^2}$$

where N = number of holes
do = hole diam. at large end of collimator (next to crystal face)
L = collimator length
F = focal distance (collimator face to point of focus)

For these calculations, a 5 inch diam. crystal was assumed, with Ohio-Nuclear collimators with F = 8.9 cm, L = 7.6 cm, and N = 151, 109, and 55, for 180, 370 and 550 keV (maximum energy) collimators. For the 1 inch crystal, $G = 200 \times 10^{-4}$ for $0<E<180$ keV; $G = 180 \times 10^{-4}$ for $180 \leq E \leq 370$ keV; $G = 91 \times 10^{-4}$ for $E>370$ keV. A value of $G = 91 \times 10^{-4}$ was assumed for positrons. For the ½ inch crystal, $G = 200 \times 10^{-4}$ for $0<E<180$ keV and $G = 180 \times 10^{-4}$ for $180 \leq E \leq 400$ keV.

** Photopeak detection efficiency defined as fraction of γ-rays striking the crystal which produce either direct photoelectric interaction or a multiple Compton-photoelectric interaction. Most multiple events occur within an inch (lateral direction) of the primary event, so that crystal diameter is not a significant contributing factor.

TABLE 1. FIGURE OF MERIT FOR SINGLE PHOTON EMITTERS (½ INCH CRYSTAL)

Nuclide	S (rad/μCi.h)	T½ (hours)	E (MeV)	n_i (phot/dis)	Q 0.25h	0.5h	1.0h	2.0h	4.0h
Cu-67	0.60E-05	61.7	0.185 0.093	0.471 0.171	0.66E+06	0.66E+06	0.66E+06	0.65E+06	0.64E+06
Ga-67	0.30E-05	78.1	0.093 0.185 0.300	0.380 0.239 0.161	0.13E+07	0.13E+07	0.13E+07	0.12E+07	0.12E+07
Se-75	0.46E-05	2880.0	0.121 0.136 0.265 0.280 0.401	0.156 0.539 0.571 0.239 0.116	0.40E+05	0.40E+05	0.40E+05	0.40E+05	0.40E+05
Se-77m	0.33E-05	0.005	0.162	0.508	0.53E-05	0.20E-20	0.29E-51		
Br-77+Se-77m	0.39E-05	56.0	0.239	0.258	0.30E+06	0.30E+06	0.30E+06	0.29E+06	0.29E+06
Kr-81m	0.33E-05	0.004	0.191	0.664	0.28E-10	0.40E-31	0.83E-73		
Sr-87m	0.54E-05	2.81	0.388	0.832	0.60E+07	0.56E+07	0.50E+07	0.39E+07	0.24E+07
Tc-99m	0.20E-05	6.03	0.140	0.879	0.35E+08	0.34E+08	0.32E+08	0.28E+08	0.22E+08
Ru-97	0.32E-05	69.6	0.216	0.853	0.11E+07	0.11E+07	0.11E+07	0.11E+07	0.11E+07
Ag-109m	0.26E-05	0.011	0.088	0.038	0.11E+03	0.18E-04	0.49E-18	0.38E-45	0.22E-99
In-111	0.58E-05	67.4	0.172 0.247	0.896 0.940	0.16E07	0.16E+07	0.16E+07	0.16E+07	0.15E+07
In-113m	0.71E-05	1.66	0.392	0.621	0.55E+07	0.49E+07	0.40E+07	0.26E+07	0.11E+07
I-123	0.31E-05	13.0	0.159	0.836	0.89E+07	0.88E+07	0.86E+07	0.81E+07	0.73E+07
I-131	0.98E-05	193.0	0.364	0.0820	0.54E+05	0.54E+05	0.54E+05	0.54E+05	0.53E+05
Xe-127	0.44E-05	874.0	0.172 0.203 0.375	0.250 0.676 0.180	0.87E+05	0.87E+05	0.87E+05	0.87E+05	0.87E+05

Xe-133	0.50E-05	128.0	0.081	0.360	0.27E+06	0.27E+06	0.27E+06	0.27E+06	0.27E+06
Cs-129	0.51E-05	32.1	0.372 0.411	0.419 0.292	0.52E+06	0.52E+06	0.51E+06	0.50E+06	0.48E+06
Cs-134m	0.38E-05	2.9	0.128	0.126	0.52E+07	0.49E+07	0.44E+07	0.35E+07	0.21E+07
Dy-157	0.44E-05	8.1	0.327	0.912	0.37E+07	0.36E+07	0.35E+07	0.32E+07	0.27E+07
Er-171	0.17E-04	7.52	0.112 0.124 0.296 0.308	0.207 0.091 0.289 0.644	0.24E+07	0.24E+07	0.23E+07	0.21E+07	0.17E+07
Re-188	0.25E-04	16.8	0.155	0.126	0.13E+06	0.13E+06	0.13E+06	0.12E+06	0.11E+06
Au-195m	0.61E-05	0.008	0.262	0.681	0.38E+01	0.50E-08	0.86E-26	0.25E-61	
Au-198	0.14E-04	64.7	0.412	0.956	0.12E+06	0.12E+06	0.12E+06	0.12E+06	0.11E+06
Au-199	0.55E-05	75.12	0.158 0.208	0.394 0.087	0.48E+06	0.48E+06	0.48E+06	0.47E+06	0.46E+06
Hg-197	0.33E-05	65.0	0.077	0.254	0.58E+06	0.57E+06	0.57E+06	0.56E+06	0.55E+06
Pb-212	0.70E-05	10.6	0.239	0.431	0.15E+07	0.14E+07	0.14E+07	0.13E+07	0.11E+07
Tl-201	0.26E-05	73.5	0.069 0.083	0.980 0.980	0.50E+07	0.50E+07	0.50E+07	0.49E+07	0.48E+07

TABLE 2. FIGURE OF MERIT FOR SINGLE PHOTON EMITTERS (1 INCH CRYSTAL)

Nuclide	S (rad/μCi.h)	$T_{1/2}$ (hours)	E (MeV)	n_i (phot/dis)	Q 0.25h	0.5h	1.0h	2.0h	4.0h
Cu-67	0.60E+-5	61.7	0.185 0.093	0.471 0.171	0.79E+06	0.79E+06	0.79E+06	0.78E+06	0.76E+06
Ga-67	0.30E-05	78.1	0.093 0.185 0.300	0.380 0.239 0.161	0.15E+07	0.15E+07	0.15E+07	0.15E+07	0.14E+07
Se-75	0.46E-05	2880.0	0.121 0.136 0.265 0.280 0.401	0.156 0.539 0.571 0.239 0.116	0.50E+05	0.50E+05	0.50E+05	0.50E+05	0.50E+05
Se-77m	0.33E-05	0.005	0.162	0.508	0.59E-05	0.22E-20	0.33E-51		
Br-77+Se-77m	0.39E-05	56.0	0.239	0.258	0.46E+06	0.46E+06	0.46E+06	0.45E+06	0.44E+06
Kr-81m	0.33E-05	0.004	0.191	0.664	0.35E-10	0.51E-31	0.10E-72		
Sr-87m	0.54E-05	2.81	0.388	0.832	0.60E+07	0.57E+07	0.50E+07	0.39E+07	0.24E+07
Tc-99m	0.20E-05	6.03	0.140	0.879	0.36E+08	0.35E+08	0.33E+08	0.30E+08	0.24E+08
Ru-97	0.32E-05	69.6	0.216	0.853	0.16E+07	0.16E+07	0.15E+07	0.15E+07	0.15E+07
Ag-109m	0.26E-05	0.011	0.088	0.038	0.11E+03	0.19E-04	0.51E-18	0.40E-45	0.23E-99
In-111	0.58E-05	67.4	0.172 0.247	0.896 0.940	0.21E+07	0.21E+07	0.20E+07	0.20E+07	0.20E+07
In-113m	0.71E-05	1.66	0.392	0.621	0.55E+07	0.50E+07	0.41E+07	0.27E+07	0.12E+07
I-123	0.31E-05	13.0	0.159	0.836	0.10E+08	0.99E+07	0.96E+07	0.91E+07	0.82E+07
I-131	0.98E-05	193.0	0.364	0.820	0.11E+06	0.11E+06	0.11E+06	0.11E+06	0.10E+06
Xe-127	0.44E-05	874.0	0.172 0.203 0.375	0.250 0.676 0.180	0.11E+06	0.11E+06	0.11E+06	0.11E+06	0.11E+06

Xe-133	0.50E-05	128.0	0.081	0.360	0.29E+06	0.29E+06	0.29E+06	0.29E+06	0.28E+06
Cs-129	0.51E-05	32.1	0.372	0.419	0.51E+06	0.51E+06	0.50E+06	0.49E+06	0.47E+06
			0.411	0.292					
Cs-134m	0.38E-05	2.9	0.128	0.126	0.55E+07	0.52E+07	0.46E+07	0.36E+07	0.23E+07
Dy-157	0.44E-05	8.1	0.327	0.912	0.67E+07	0.66E+07	0.63E+07	0.58E+07	0.49E+07
			0.112	0.207					
			0.124	0.091					
Er-171	0.17E-04	7.52	0.296	0.289	0.33E+07	0.32E+07	0.31E+07	0.28E+07	0.23E+07
			0.308	0.644					
Re-188	0.25E-04	16.8	0.155	0.126	0.14E+06	0.14E+06	0.14E+06	0.13E+06	0.12E+06
Au-195m	0.61E-05	0.008	0.262	0.681	0.64E+01	0.84E-08	0.14E-25	0.42E-61	
Au-198	0.14E-04	64.7	0.412	0.956	0.12E+06	0.12E+06	0.12E+06	0.11E+06	0.11E+06
Au-199	0.55E-05	75.12	0.158	0.394	0.55E+06	0.55E+06	0.55E+06	0.54E+06	0.53E+06
			0.208	0.087					
Hg-197	0.33E-05	65.0	0.077	0.254	0.61E+06	0.60E+06	0.60+06	0.59E+06	0.58E+06
Pb-212	0.70E-05	10.6	0.239	0.431	0.22E+07	0.22E+07	0.21E+07	0.20E+07	0.18E+07
Tl-201	0.26E-05	73.5	0.069	0.980	0.53E+07	0.52E+07	0.52E+07	0.52E+07	0.51E+07
			0.083	0.980					

TABLE 3. FIGURE OF MERIT FOR POSITRON-EMITTING RADIONUCLIDES

Nuclide	S (rad/μCi.h)	$T_{1/2}$ (hours)	n_i (phot./dis.)	Q 0.25h	Q 0.50h	Q 1.0h	Q 2.0h	Q 4.0h
C-11	0.23E-04	0.338	1.996	0.13E+08	0.80E+07	0.29E+07	0.37E+06	0.61E+04
N-13	0.26E-04	0.167	2.000	0.14E+08	0.50E+07	0.63E+06	0.99E+04	0.25E+01
O-15	0.33E-04	0.034	2.000	0.99E+06	0.64E+04	0.27E+00	0.48E-09	0.15E-26
F-18	0.18E-04	1.83	1.940	0.46E+07	0.42E+07	0.35E+07	0.24E+07	0.11E+07
Mn-52m	0.58E-04	0.355	1.931	0.50E+07	0.31E+07	0.12E+07	0.16E+06	0.33E+04
Fe-52	0.14E-04	8.30	1.156	0.84E+06	0.83E+06	0.79E+06	0.73E+06	0.62E+06
Cu-62	0.50E-04	0.162	1.957	0.72E+07	0.25E+07	0.29E+06	0.40E+04	0.77E+00
Cu-64	0.62E-05	12.8	0.386	0.42E+06	0.41E+06	0.40E+06	0.38E+06	0.34E+06
Zn-62	0.54E-05	9.15	0.158	0.27E+06	0.27E+06	0.26E+06	0.24E+06	0.20E+06
Ga-68	0.33E-04	1.14	1.784	0.35E+07	0.30E+07	0.22E+07	0.12E+07	0.36E+06
Se-73	0.24E-04	7.20	1.300	0.64E+06	0.62E+06	0.59E+06	0.54E+06	0.44E+06
Rb-81	0.96E-05	4.58	0.615	0.12E+07	0.11E+07	0.10E+07	0.89E+06	0.66E+06
Rb-81+Kr-81m	0.13E-04	4.58	0.615	0.86E+06	0.83E+06	0.77E+06	0.66E+06	0.49E+06
Rb-82	0.51E-04	0.022	1.893	0.50E+05	0.17E+02	0.20E-05	0.27E-19	0.49E-47
I-124	0.18E-04	101.0	0.512	0.24E+05	0.24E+05	0.24E+05	0.24E+05	0.24E+05

TABLE 4. SINGLE PHOTON EMITTERS LISTED IN DESCENDING ORDER OF FIGURE OF MERIT (½ INCH CRYSTAL)

0.25h	0.50h	1.0h	2.0h	4.0h
Tc-99m	Tc-99m	Tc-99m	Tc-99m	Tc-99m
I-123	I-123	I-123	I-123	I-123
Sr-87m	Sr-87m	Tl-201	Tl-201	Tl-201
In-113m	Tl-201	Sr-87m	Sr-87m	Dy-157
Cs-134m	Cs-134m	Cs-134m	Cs-134m	Sr-87m
Tl-201	In-113m	In-113m	Dy-157	Cs-134m
Dy-157	Dy-157	Dy-157	In-113m	Er-171
Er-171	Er-171	Er-171	Er-171	In-111
In-111	In-111	In-111	In-111	Ga-67
Pb-212	Pb-212	Pb-212	Pb-212	In-113m
Ga-67	Ga-67	Ga-67	Ga-67	Pb-212
Ru-97	Ru-97	Ru-97	Ru-97	Ru-97
Cu-67	Cu-67	Cu-67	Cu-67	Cu-67
Hg-197	Hg-197	Hg-197	Hg-197	Hg-197
Cs-129	Cs-129	Cs-129	Cs-129	Cs-129
Au-199	Au-199	Au-199	Au-199	Au-199
Br-77+Se-77m	Br-77+Se-77m	Br-77+Se-77m	Br-77+Se-77m	Br-77+Se-77m
Xe-133	Xe-133	Xe-133	Xe-133	Xe-133
Re-188	Re-188	Re-188	Re-188	Au-198
Au-198	Au-198	Au-198	Au-198	Re-188
Xe-127	Xe-127	Xe-127	Xe-127	Xe-127
I-131	I-131	I-131	I-131	I-131
Se-75	Se-75	Se-75	Se-75	Se-75
Ag-109m	Ag-109m	Ag-109m	Ag-109m	Ag-109m
Au-195m	Au-195m	Au-195m	Au-195m	Au-195m
Se-77m	Se-77m	Se-77m	Se-77m	Kr-81m
Kr-81m	Kr-81m	Kr-81m	Kr-81m	Se-77m

TABLE 5. SINGLE PHOTON EMITTERS LISTED IN DESCENDING ORDER OF FIGURE OF MERIT (1 INCH CRYSTAL)

0.25h	0.50h	1.0h	2.0h	4.0h
Tc-99m	Tc-99m	Tc-99m	Tc-99m	Tc-99m
I-123	I-123	I-123	I-123	I-123
Dy-157	Dy-157	Dy-157	Dy-157	Tl-201
Sr-87m	Sr-87m	Tl-201	Tl-201	Dy-157
In-113m	Tl-201	Sr-87m	Sr-87m	Sr-87m
Cs-134m	Cs-134m	Cs-134m	Cs-134m	Er-171
Tl-201	In-113m	In-113m	Er-171	Cs-134m
Er-171	Er-171	Er-171	In-113m	In-111
Pb-212	Pb-212	Pb-212	In-111	Pb-212
In-111	In-111	In-111	Pb-212	Ru-97
Ru-97	Ru-97	Ru-97	Ru-97	Ga-67
Ga-67	Ga-67	Ga-67	Ga-67	In-113m
Cu-67	Cu-67	Cu-67	Cu-67	Cu-67
Hg-197	Hg-197	Hg-197	Hg-197	Hg-197
Au-199	Au-199	Au-199	Au-199	Au-199
Cs-129	Cs-129	Cs-129	Cs-129	Cs-129
Br-77+Se-77m	Br-77+Se-77m	Br-77+Se-77m	Br-77+Se-77m	Br-77+Se-77m
Xe-133	Xe-133	Xe-133	Xe-133	Xe-133
Re-188	Re-188	Re-188	Re-188	Re-188
Au-198	Au-198	Au-198	Au-198	Au-198
Xe-127	Xe-127	Xe-127	Xe-127	Xe-127
I-131	I-131	I-131	I-131	I-131
Se-75	Se-75	Se-75	Se-75	Se-75
Ag-109m	Ag-109m	Ag-109m	Ag-109m	Ag-109m
Au-195m	Au-195m	Au-195m	Au-195m	Au-195m
Se-77m	Se-77m	Se-77m	Se-77m	Kr-81m
Kr-81m	Kr-81m	Kr-81m	Kr-81m	Se-77m

TABLE 6. POSITRON EMITTERS LISTED IN DESCENDING ORDER OF FIGURE OF MERIT

	0.25 h	0.5 h	1.0 h	2.0 h	4.0 h
1.	N-13	C-11	F-18	F-18	F-18
2.	C-11	N-13	C-11	Ga-68	Rb-81
3.	Cu-62	F-18	Ga-68	Rb-81	Fe-52
4.	Mn-52m	Mn-52m	Mn-52m	Fe-52	Rb-81+Kr-81m
5.	F-18	Ga-68	Rb-81	Rb-81+Kr-81m	Se-73
6.	Ga-68	Cu-62	Fe-52	Se-73	Ga-68
7.	Rb-81	Rb-81	Rb-81+Kr-81m	Cu-64	Cu-64
8.	O-15	Rb-81+Kr-81m	N-13	C-11	Zn-62
9.	Rb-81+Kr-81m	Fe-52	Se-73	Zn-62	I-124
10.	Fe-52	Se-73	Cu-64	Mn-52m	C-11
11.	Se-73	Cu-64	Cu-62	I-124	Mn-52m
12.	Cu-64	Zn-62	Zn-62	N-13	N-13
13.	Zn-62	I-124	I-124	Cu-62	Cu-62
14.	Rb-82	O-15	O-15	O-15	O-15
15.	I-124	Rb-82	Rb-82	Rb-82	Rb-82

TABLE 7. SOME GENERATOR SYSTEMS FOR POSITRON EMITTING RADIONUCLIDES

Parent ($T_{1/2}$)	Daughter ($T_{1/2}$)
^{44}Ti (48y)	^{44}Sc (3.9h)
^{52}Fe (8.2h)	^{52m}Mn (21.1m)
^{68}Ge (275d)	^{68}Ga (1.14h)
^{62}Zn (9.13h)	^{62}Cu (9.76m)
^{82}Sr (25d)	^{82}Rb (1.25m)
^{118}Te (6.0d)	^{118}Sb (3.5m)
^{122}Xe (20.1h)	^{122}I (3.5m)
^{134}Ce (3.0d)	^{134}La (6.8m)

RADIOPHARMACEUTICALS FOR NEUROLOGICAL STUDIES

Michael J. Welch and Timothy J. Tewson

The Edward Mallinckrodt Institute of Radiology
Washington University School of Medicine
510 South Kingshighway
St. Louis, Missouri 63110

ABSTRACT

Radiopharmaceuticals have been developed to measure several neurological parameters. These include, blood volume, brain tissue composition, blood flow, brain metabolism, brain receptor concentrations and the interactions of platelets with vessel walls. The production of specific radiopharmaceuticals to measure these functions is discussed, and some of the uses described. In general, these radiopharmaceuticals are labeled with short-lived positron emitting radionuclides although indium-111 can be used to label platelets to study platelet/vessel wall interactions.

INTRODUCTION

Until recently, radiopharmaceuticals used for neurological studies could be divided into three categories: 1) those demonstrating the breakdown of the blood brain barrier (1); 2) those which assess cerebrospinal fluid pathways (2); and 3) agents which measure regional cerebral blood flow (3). The first of these applications utilizes radiopharmaceuticals that do not penetrate the intact blood brain barrier. The second application usually utilizes labeled chelates, while the third commonly uses intra-arterial injections of dissolved ^{133}Xe. Over the past few years, new radiopharmaceuticals have been developed, and both single photon tomography (4) and positron emission tomography (5,6) have been utilized to measure other parameters related to neurological function and metabolism. These parameters include blood volume, chemical composition of tissue (i.e., local pH, local hematocrit, and lipid content), brain blood flow, metabolism, and platelet vessel interactions. There is also great interest in the development of agents to map brain receptor concentrations.

Some of these approaches were discussed earlier in a Society of Nuclear Medicine publication (7). In this review, we will emphasize radiopharmaceutical advances made since that publication; the tracer techniques utilized to measure parameters will not be discussed in detail as they are the subject of an extensive review by Raichle (8). We will discuss the radiopharmaceuticals grouped according to the particular neurological function being measured.

BLOOD VOLUME

Blood volume was one of the first parameters measured by emission tomography (9). In the initial study, ^{99m}Tc-labeled red cells were utilized in conjunction with the Kuhl Mark IV scanner. Any radiopharmaceutical that is contained in the vascular bed for the duration of the study can be utilized for the measurement. For use with a positron emission tomograph, carbon-11 labeled carbon monoxide, gallium-68 labeled red cells and gallium-68 labeled transferrin are all candidates. If carbon-11 labeled carbon monoxide is used, the subject simply inhales air containing the high specific activity tracer (10). The regional cerebral blood volume is calculated from data obtained from the equilibrium tomographic image of ^{11}C-carboxyhemoglobin in the brain and from venous blood samples drawn during the time of the emission scan. As mentioned above, labeled proteins and labeled red cells can also be utilized to measure cerebral blood volume; these tracers measure slightly different spaces. The measurement of the equilibrium distribution of a plasma tracer (^{11}C-labeled methyl albumin (11) or ^{68}Ga-labeled transferrin) combined with that of a cell tracer (^{11}CO-labeled red cells or ^{68}Ga-labeled red cells (12) allows the determination of tissue hematocrit (7).

TISSUE CHEMICAL COMPOSITION

Tomographic techniques can be used to measure the composition of tissue by selecting suitable compounds that rapidly distribute between the blood and the tissue. One can determine the partition coefficient of a material by utilizing the data from both equilibrium tomographic imaging and blood samples obtained during the time of the emission scan. If one knows the concentration of the compound of interest in the vascular compartment, the actual tissue composition can be calculated. Examples of this approach include carbon-11 labeled carbon dioxide, which allows an assessment of the tissue acid/base status (13), and the use of ^{11}C-octanol or another lipid soluble material to determine the tissue lipid content.

BLOOD FLOW

Several radiopharmaceuticals have been suggested for the measurement of blood flow, but to date no completely satisfactory technique has been developed. One such radiopharmaceutical is nitrogen-13 labeled ammonia. This compound was originally produced for medical use by bombarding aluminum carbide with deuterons (14,15). The aluminum carbide was dissolved in acid producing dissolved ^{13}N-ammonia in solution; this solution was then made basic, distilled, and the ammonia trapped in a small volume of acid. An alternative method of production utilizes deuteron bombardment of a gaseous methane target (16,17). These two methods suffer from the common disadvantage that significant amounts of carrier ammonia may be produced radiolytically because of nitrogen contamination of the target. The preferred method for ammonia production involves the bombardment of water to produce ^{13}N by the $^{16}O(p,\alpha)^{13}N$ nuclear reaction followed by the reduction of the ^{13}N compounds so formed to ammonia with titanous chloride (18) or some other reducing agent. Ammonia labeled with nitrogen-13 has been used to map perfusion in the brain (19) because it is efficiently trapped, presumably by incorporation into amino acids. Utilization of $^{13}NH_3$ to determine flow is unlikely to succeed since the tracer is not freely diffusible, and the amount of ammonia extracted from the blood decreases as blood flow increases. Thus as blood flow increases, the estimated flow based on the amount of ammonia extracted will be increasingly underestimated. The situation with ammonia is further complicated by the fact that the ratio of the diffusible species (NH_3) to the non-diffusible NH_4^+ varies as a function of blood pH; thus, pH differences will affect the ammonia uptake. Figure 1 shows the degree to which NH_3 underestimates blood flow at various flows, it is seen from this figure that even at normal flows (50 ml/min/100 g), the experimentally determined values are significantly lower than those predicted. In these experiments, the flow changes were accomplished by altering the pCO_2, which also alters the blood pH.

Another agent that has been postulated as a flow tracer is carbon-11 labeled nicotine (20) as its reported brain uptake index (BUI) is high. Carbon-11 labeled nicotine has been synthesized by methylating N-nornicotine by the addition of carbon-11 labeled formaldehyde followed by sodium borohydride reduction. Purification of the nicotine was accomplished by ether extraction from basic solution followed by acidification of the ether mixture with an ether/HCl solution. The precipitate, carbon-11 nicotine, was separated by centrifugation, dissolved in a physiologic buffer and millipore filtered prior to injection. Using 700 μg of N-nornicotine, labeling yields of ≃20% of carbon-11 labeled nicotine (>90% purity) were obtained. If nicotine can be used as a flow tracer, the single pass extraction in brain following carotid injection should approach 100% and the washout of the tracer from brain tissue should be negligible over the time required for measurement. We have compared the brain uptake of ^{11}C-nicotine and ^{15}O-labeled water by measuring the fraction of the activity extracted by the brain during a single capillary transit (21). Brain extraction fractions of 53% and 25% were obtained at flows of 50 ml/min/100g and 100 ml/min/100g, respectively. It appears that although nicotine is retained in brain tissue to a greater extent than water (Figure 2), the extraction depends on flow and therefore nicotine is unsatisfactory as a flow tracer.

Another compound that has been suggested as a flow tracer is ^{18}F-labeled fluoroethanol. This agent is highly extracted in the brain (22), and is expected to be trapped intracellularly because of its action as an inhibitor of the citric acid cycle (23). Reports in the literature on preparation of ^{18}F-labeled fluoroethanol (22,24) describe procedures leading to low specific activity material. Due to the high toxicity (23) of fluoroethanol, an agent for human use, this tracer would have to be prepared on a carrier free level. Our approach to this synthesis (25) is shown in the following reaction scheme:

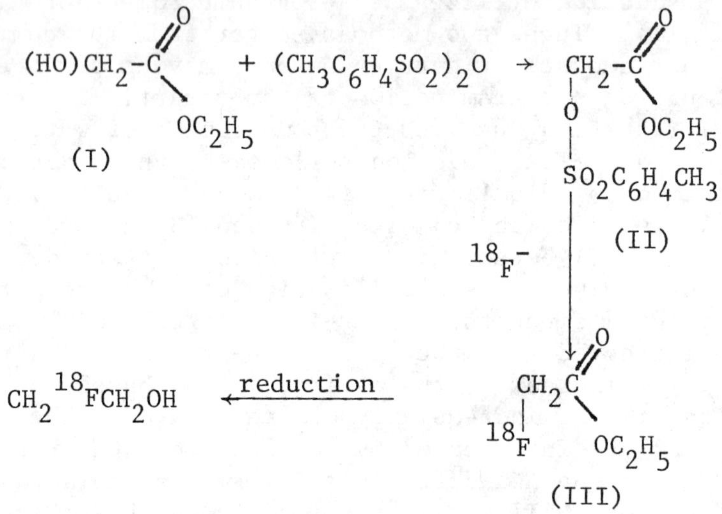

Carrier free cesium fluoride is used to displace the tosylate group from intermediate II to give III, which is reduced to fluoroethanol. Preliminary data on the single pass extraction of fluoroethanol following carotid injection in the monkey show that, although the compound is extracted by the brain, the tracer is not fixed in the tissue.

Another approach relates an equilibrium image of ^{15}O-labeled water, obtained following the continuous inhalation of ^{15}O-labeled water, with regional cerebral blood flow. Although several groups have shown that the amount of tracer in a particular part of the brain is related to blood flow (26,27), this relationship is not linear, thus accurate assessment of regional cerebral blood flow is not possible.

Another technique for the measurement of cerebral blood flow, which is used by the group at the Montreal Neurological Institute (28), utilizes the positron-emitting inert gas krypton-77 (29). Krypton-77 is produced by irradiation of sodium bromide with the 40 MeV proton beam of the McGill Synchrocyclotron. The krypton is separated from the target by distillation in vacuo and trapping at liquid nitrogen temperature. The patient inhales

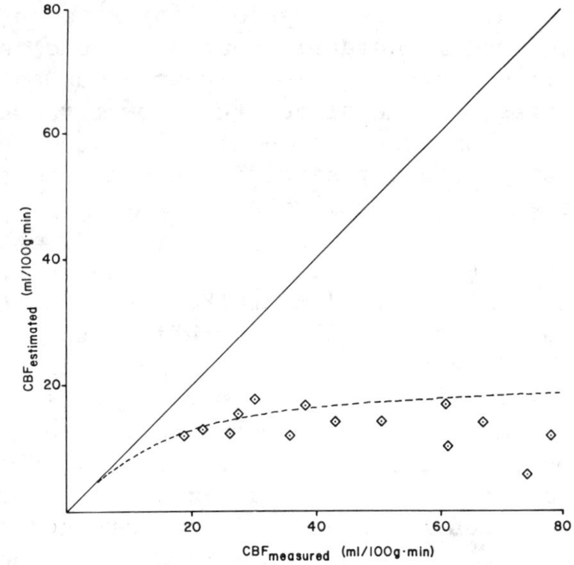

Figure 1.

This is a plot of the estimated blood flow utilizing $^{13}NH_3$ as a tracer. Data from a true flow indication would lie on the solid line while the dotted line is an indicator line; assume a P·S (capillary permeability x capillary surface area) of ≃22. The measured data at high flows are beneath this line; probably as flow was changed by altering the pCO_2 which causes pH effects.

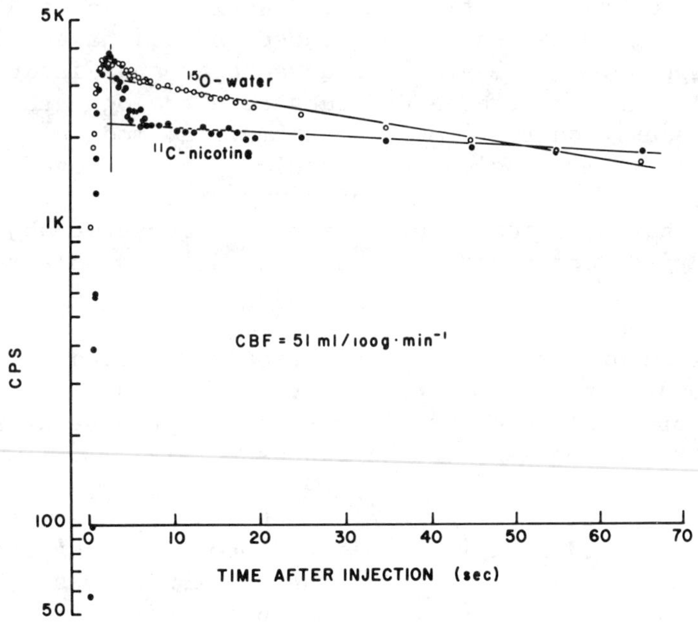

Figure 2.

The single pass extraction of ^{11}C-nicotine in the brain of a rhesus monkey compared to that of $H_2{}^{15}O$ at the same flow.

a bolus (4-10 mCi) of the carrier free krypton; the krypton distribution in a cross section of the head is determined at 30 second intervals for a period of 8 minutes. Time activity scans (krypton washout curves) are then generated for each pixel of the slice; the flow for each slice is generated by conventional techniques (3). This technique suffers from the problems of obtaining a statistically significant dynamic washout curve with instrumentation designed for static imaging.

One technique that can be used to measure regional cerebral blood flow with tomographic devices is an adaptation of the tissue autoradiography originally proposed by Landau et al (30), and refined by Reivich (31), Sokoloff (32) and their co-workers. In the original technique which was applied to animals, a freely diffusible tracer was infused intravenously for one minute followed by decapitation. Quantitative autoradiograms of brain slices and the time activity curve of arterial blood comprise the data from which rCBF is determined. A tomographic image could be substituted for the autoradiogram provided such an image could be obtained in a relatively short period of time (<10 sec). The development of fast tomographic devices such as PETT V (6) should allow the application of this technique. A number of radiopharmaceuticals could serve as the tracer for such studies; one candidate for the positron application is ^{11}C-labeled n-butanol, which has been shown to be a freely diffusible tracer even at high values of rCBF (7).

METABOLISM

There has been great interest over the past few years in the design of agents to measure local cerebral metabolism. Agents such as oxygen-15 labeled oxygen (33,34), carbon-11 labeled glucose (35), fluorine-18 labeled 2-fluoro-2-deoxy-D-glucose (36), fluorine-18 labeled 3-fluoro-3-deoxy-D-glucose (37), and carbon-11 labeled 2-deoxy-D-glucose (38,39) have all been synthesized for this purpose. There are both advantages and disadvantages associated with the use of labeled substrates and substrate analogs; the utilization of either type of tracer requires complex mathematical models to extract the metabolic rates from the emission tomographic data. However, it is not the purpose of this review to discuss the advantages of the various techniques. The various radiopharmaceuticals for measuring glucose metabolism have not been compared in vivo, and others have discussed the problems with the various techniques (8).

The preparation of the radiopharmaceuticals required for these measurements is more complex than those regularly carried out in a nuclear pharmacy. As examples, the preparation of two of the agents proposed for the measurement of regional glucose metabolism will be discussed.

The current preparation of carbon-11 labeled glucose used at Washington University is an adaptation of the published procedures of the biosynthetic method using swiss chard leaves (40,41) that have previously been light starved for over 24 hours and $^{11}CO_2$ that is produced directly in the cyclotron target. After extraction of the carbon-11 activity from the swiss chard leaves with 90% alcohol and the hydrolysis with hydrocholoric acid, the mixture of sugars is neutralized with sodium hydroxide and rotary evaporated to a volume of 2-3 ml. After millipore filtration, the sample is injected onto a 3' x 1" column packed with the calcium form of AG 50 W-X4 (Bio-Rad 200-400 mesh size) cation exchange column eluted with 0.01 M $Ca(OH)_2$ solution having a pH of approximately 3.5. The system is operated at room temperature at a flow rate of 9.9 ml per min (pressure ≃600 psi). The eluate is passed over a scintillation detector and the profile of radioactivity eluted is shown in Figure 3. When the glucose peak elutes (≃18 minutes) the eluate is collected for 6 minutes in a receiver containing a mixed bed resin. The eluate is then rotary evaporated to a volume of ≃3 ml buffered with a pH 7.5 buffer and millipore filtered prior to injection. An initial uptake of 100 mCi of carbon-11 carbon dioxide in the swiss chard leaves will lead to ≃3 mCi of glucose. After each preparation, a small aliquot of the solution is retained for analysis of the radiochemical purity of the radiopharmaceutical. This is carried out using a μ-Bondapak carbohydrate column (Waters Associates, Milford, MA) eluted with an acetonitrile (85%)/water (15%) mixture at a flow of 2 ml/min. A typical chromatogram is shown in Figure 4; the impurities are fructose and sugar phosphates. It should be noted that if a radiopharmaceutical is to be used for metabolic studies, radiochemical quality control procedures are very important, as impurities can greatly affect the measured metabolic rate.

The procedure used at Washington University for the production of 3-fluoro-3-deoxy-glucose production involves the reaction of the trifluoromethanesulphonate IV with cesium-fluoride (37).

IV

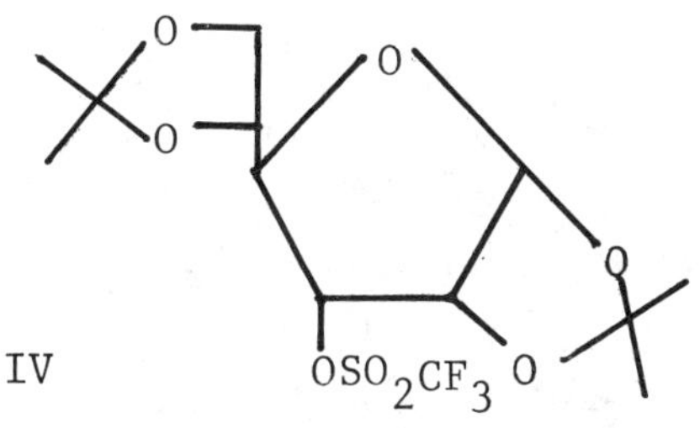

We have developed a procedure to produce carrier free cesium fluoride in which $H^{18}F$ produced by the deuteron bombardment of a neon hydrogen mixture is recirculated over a silver wool plug coated with cesium hydroxide. The silver wool plug with the activity adsorbed onto it is added to a solution of the trifluoromethanesulphonate (IV) in hexamethyl phosphoric triamide (HMPA) and heated to give 1,2-5,6-di-O-isopropylidene-3-deoxy-3-fluoro-α-D-glucofuranose. Treatment of this with boron trichloride and water gives fluorine-18 labeled 3-fluoro-3-deoxy-D-glucose which is purified on an ion exchange column. Yields on a typical preparation are 13 mCi of activity absorbed on the silver wool from a 10 μamp/hour bombardment with 5.8 MeV deuterons giving 4.5 mCi of carrier free purified product 90 minutes from the end of bombardment. Cerebral studies following an intracarotid injection in a rhesus monkey gave a cerebral extraction fraction of 31.4% compared to 31% obtained with ^{11}C-glucose. An example of initial human data is presented in Figure 5, which shows two tomographic slices obtained with the 3-fluoro-3-deoxy-glucose.

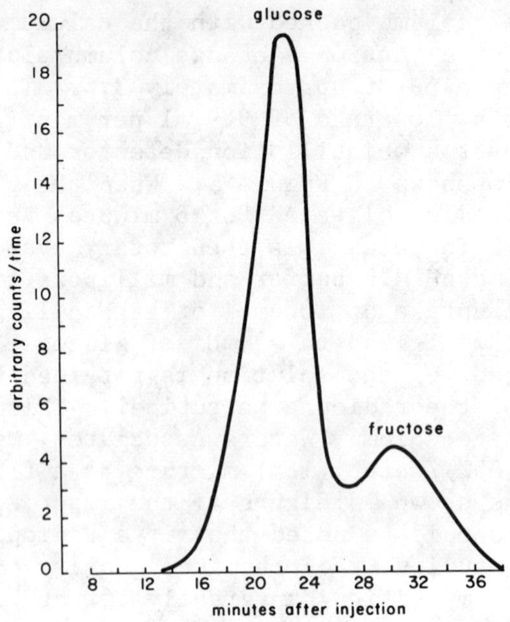

Figure 3.

Chromotogram obtained during the preparation of ^{11}C-glucose.

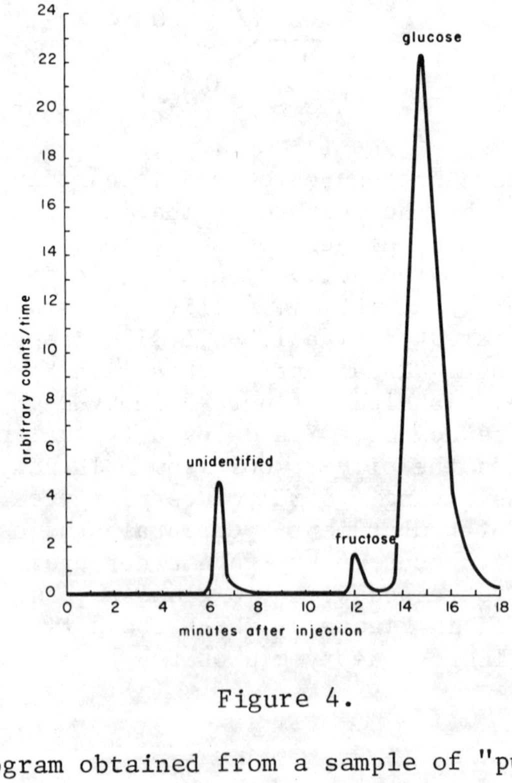

Figure 4.

Analytical chromotogram obtained from a sample of "pure" ^{11}C-glucose.

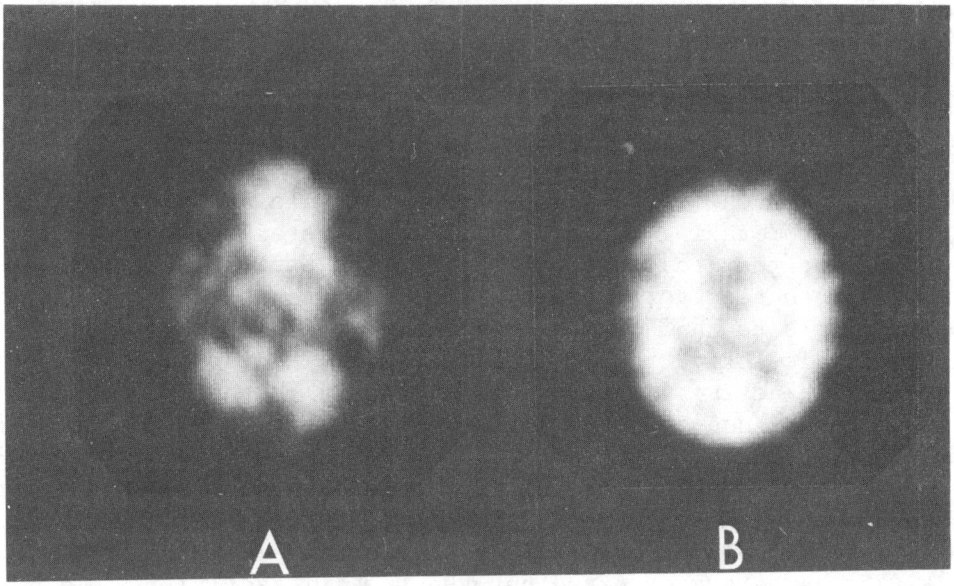

Figure 5.

Two tomographic slices obtained with PETT IV of a subject following administration of ^{18}F-3-fluoro-3-deoxy-D-glucose. Slice A is through the level of the nose with activity appearing in the sinus and the cerebellum while slice B is through the mid brain.

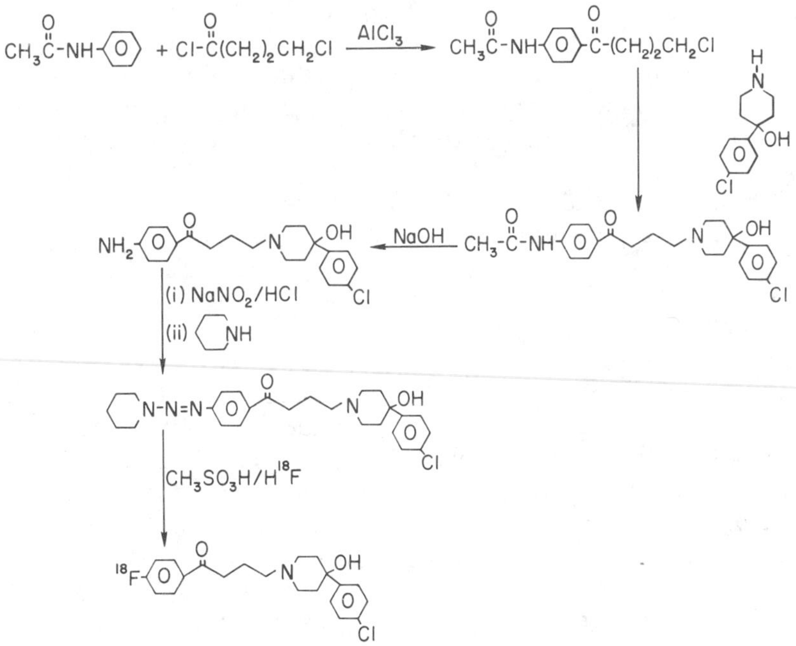

Figure 6.

Synthetic sequence for the preparation of carrier-free ^{18}F-haloperidol.

RECEPTOR STUDIES

One of the most exciting areas involving the use of radiopharmaceuticals for neurological studies is the mapping of specific receptor sites. Specific chemical compounds (agonists) which have a high affinity constant for the site, but a comparatively rapid metabolic turnover reversibly bind to the site and either directly cause the conformational change required or activate the site for a second compound which initiates the change. There are other compounds (antagonists) which also bind to the site with high affinity, but have much slower turnovers on the site and prevent the conformational change from taking place. There are several receptor systems in the brain however; the greatest interest from an isotopic approach is in the dopamine and opiate receptor systems. Autoradiographic techniques have been utilized in animals to map both the dopamine (42) and opiate (43) receptors using high specific activity ^{3}H-spiroperidol and ^{3}H-diprenorphine respectively. High specific activity material is necessary because of the presence of a large number of non-specific sites with low affinity constants for the antagonists that bind to the specific sites. The specific sites can be blocked by $\approx 10^{-9}$ moles of substrate (42) and excess substrate will simply bind to the non-specific sites and increase the background level. To date, fluorine-18 labeled haloperidol (44) and a fluorine-18 labeled dopamine derivative (45) have been prepared but the specific activities were too low to measure receptor concentrations. Carbon-11 labeled chlorpromozine, a dopamine antagonist, has been prepared (46), but little biological data has been reported. As discussed in another chapter (Wolf and Fowler) it is difficult to produce very high specific activity carbon-11 labeled compounds due to carbon-containing impurities in cyclotron targets and in the reagents used in the syntheses.

In the 19th century (47) fluoro compounds were prepared by pyrolysis of piperidyl triazine in concentrated hydrofluoric acid. We have utilized this approach to prepare the dopamine antagonist haloperidol (Figure 6).

Fluorine-18 as cesium fluoride is produced as previously described and is reacted with 2-N-piperidyl-p-diazo-4[4-(p-chlorophenyl)-4-hydroxypiperidyl] butyrophenone in refluxing 1,2-dimethoxyethane containing methane sulfonic acid. A 20-25% yield of carrier free haloperidol is obtained in a 4 minute reaction and the haloperidol can be separated from the reaction precursors by HPLC. This type of reaction should also be applicable to the labeling of spiroperidol.

PLATELET BLOOD VESSEL INTERACTIONS

There is substantial evidence that platelets play a major role in the formation of arterial thrombi, and that platelet emboli or platelet aggregates cause transient ischemic attacks (49,50). Platelets have been labeled with both indium-111 (51) and gallium-68 (12); platelets suspended in a mixture of 0.9% saline and Squibb Modified ACD solution give the most consistant labeling efficiency without platelet damage as assessed by in vitro ADP and collagen induced aggregation. Autologous platelets are labeled with indium-111 under aseptic conditions by a modification of the method discussed by Thakur et al. Platelets are isolated by differential centrifugation of 43 ml of whole blood anticoagulated with 7 ml of acid-citrate

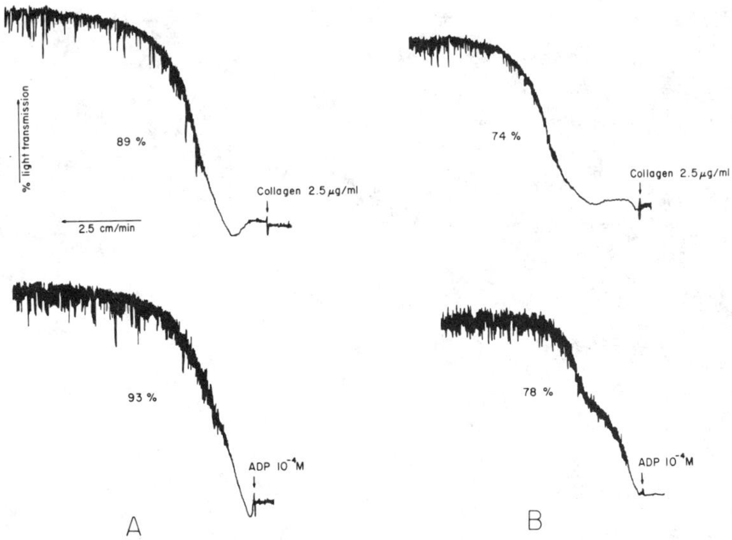

Figure 7.

Typical in vitro function studies obtained with platelet following stimulation by collagen and ADP. Unmanipulated platelets A, indium-111 labeled platelets B.

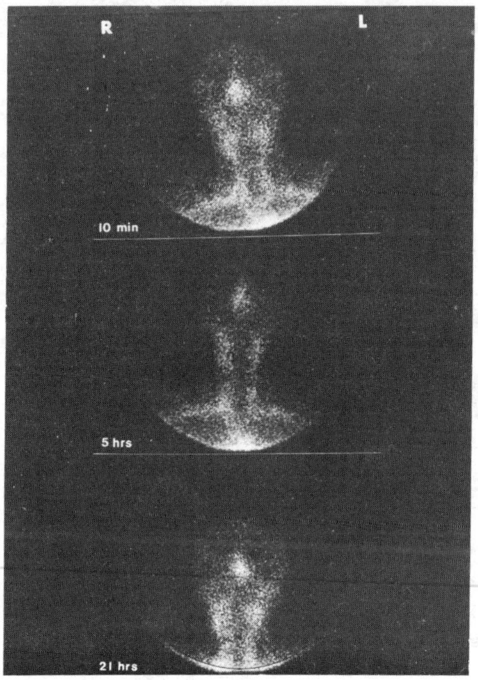

Figure 8.

Anterior images of the head and neck of a patient with angiographically documented fibromuscular hyperplasia done 10 minutes, 5 hours and 21 hours after injection of autologous, indium-111 labeled platelets. There is the normal central nasal pharyngeal activity due to naso-pharyngeal vascularity. Note the assymetric distribution of activity in the great vessels at the neck.

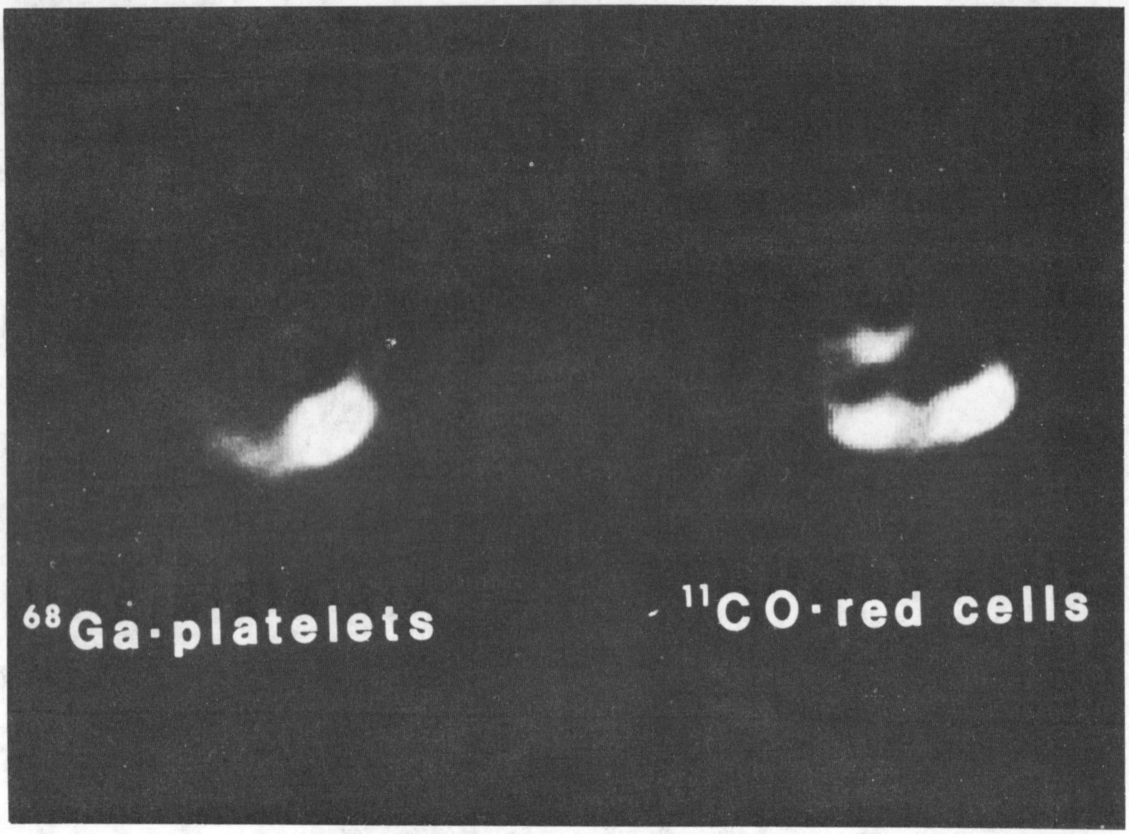

Figure 9.

Tomograms of the neck of dog viewed from below with induced endothelial damage of one carotid showing marked accumulation of ^{68}Ga-labeled platelets in abnormal carotid. Red cell image with ^{11}CO reveals only slight asymmetry of carotid which is related to residual ^{68}Ga activity in damaged carotid. ^{68}Ga-labeled platelets were administered intravenously; ^{11}CO by inhalation. (Reproduced from Welch, Thakur, Coleman, et al: J Nucl Med 18:558-562, 1977).

dextrose solution (Squibb). The platelets are washed once with ACD:normal saline (1:6 volume:volume at pH 6.5) and resuspended in 4 ml of this solution containing the indium-111 and 50 μg of oxine (8-hydroxyquinoline) at pH 6.5. After incubation at room temperature for 20 minutes, the platelet suspension is centrifuged, the supernatent decanted and the platelets washed with ACD/plasma to remove any unincorporated indium-111. The platelets are then resuspended in 4 ml of citrated plasma. Citrated plasma (obtained from 18 ml of whole blood anticoagulated with 2 ml of 3.8% sodium citrate) is used in order to assess in vitro platelet function. A typical in vitro aggregation study is shown in Figure 7; it can be seen that the unlabeled and labeled platelets behave similarly. To date we have evaluated the use of indium-111 labeled platelets in over 30 adult patients (52) with suspected cerebrovascular disease. The positive images obtained from one patient are shown in Figure 8. When compared to angiographic data, the sensitivity of the platelet imaging procedure was low; however, it should be noted that >80% of the patients studied were on either anticoagulant or antiplatelet drugs. Different pathophysiologic characteristics of arterial lesions may be demonstrated with indium-labeled platelets as compared to angiography. Gallium-68 labeled platelets can be utilized in conjunction with positron tomographs to demonstrate platelet deposition (12). A typical image obtained in a dog model is shown in Figure 9. Although the half-life of gallium-68 limits the length of a study, the sensitivity of tomography may give useful data with these labeled cells. It appears that gamma emitting labeled platelets have great potential for the study of platelet blood vessel interactions in patients with TIA or stroke.

CONCLUSIONS

Over the past few years, the number and type of radiopharmaceuticals utilized for neurological studies has increased dramatically. Some of these can be only utilized in centers possessing cyclotrons and positron imaging devices, while some (such as indium labeled platelets) are within the capabilities of most medical centers. It is to be anticipated that this review is only a progress report in that the next few years many other radiopharmaceuticals will be devised to assist in the study of neurological function.

ACKNOWLEDGEMENTS

This work was supported in part by U.S. PHS Grants 5 P01 HL13851 and 2 P50 NS06833 and DOE Contract EE-77-S-02-4318. The authors thank their many colleagues including M.E. Raichle, M.D., M.M. Ter-Pogossian, Ph.D., J.O. Eichling, Ph.D., B.A. Siegel, M.D., and H.H. Davis, III., M.D. who participated in the studies described, and Karen McElvany and Carla Mathias for assisting in editing an earlier draft of the manuscript.

REFERENCES

1. Holman B: The blood brain barrier. Anatomy and Physiology in Progress in Nuclear Medicine, Vol. 1, Potchen EJ, McCready VR, eds, Baltimore University Park Press, 1972, pp 236-247.

2. Homan BL, Davis DO: Radioisotopes assessment of cerebrospinal fluid pathways, in Progress in Nuclear Medicine, Vol. 1, Potchen EJ, McCready VR, eds, Baltimore University Park Press, 1972, pp 359-375.

3. Lassen NA, Ingvar DA: Radioisotopes assessment of regional cerebral blood flow, in Progress in Nuclear Medicine, Vol. L, Potchen EJ, McCready VR, eds, Baltimore University Park Press, 1972, pp 376-409.

4. Kuhl DE, Alavi A, Reivich M, et al: Computerized emission transaxial tomography and determination of local brain function in noninvasive brain imaging, in Computed Tomography and Radionuclides, DeBlanc HJ Jr, Sorenson JA, eds, The Society of Nuclear Medicine, New York, 1975, pp 67-79.

5. Ter-Pogossian MM, Phelps ME, Hoffman EJ, et al: A positron-emission transaxial tomograph for nuclear imaging (PETT). Radiology 114: 89-98, 1975.

6. Ter-Pogossian MM, Mullani NA, Hood JT, et al: Design considerations for a positron emission transverse tomograph (PETT V) for imaging of the brain. J Comput Assist Tomog 2:539-544, 1978.

7. Welch MJ, Eichling JO, Straatmann MG, et al: New short-lived radiopharmaceuticals for C.N.S. studies, in Non-Invasive Brain Imaging: Computed Tomography and Radionuclides, DeBlanc JH Jr, Sorenson JA, eds, Society of Nuclear Medicine, New York, 1975, pp 25-44.

8. Raichle ME: Quantitate in vivo autoradiography with positron emission tomography. Brain Res Rev, in press.

9. Greenburg JH, Alavi A, Reivich M, et al: Local cerebral blood volume response to carbon monoxide response in man. Circ Res 43:324-330, 1978.

10. Welch MJ, Ter-Pogossian MM: Preparation of short half-lived gases for medical studies. Rad Res 36:580-587, 1968.

11. Straatmann MG, Welch MJ: A general method for labeling proteins with ^{11}C. J Nucl Med 16:425-428, 1975.

12. Welch MJ, Thakur MJ, Coleman RE, et al: Gallium-68 labeled red cells and platelets: New agents for positron tomography. J Nucl Med 18: 558-562, 1977.

13. Raichle ME, Grubb RL Jr, Higgins CS: Measurement of brain tissue carbon dioxide content by emission tomography. Brain Res, in press.

14. Carter CC, Lifton JF, Welch MJ: Organ uptake and blood pH and concentration effects of ammonia in dogs determined with ammonia labeled with 10 minute half-lived nitrogen-13. Neurology 23:204-213, 1973.

15. Welch MJ, Lifton JF: The fate of nitrogen-13 produced by the ^{12}C

(d,n)^{13}N reaction in inorganic carbides. J Am Chem Soc 93:3385-3388, 1971.

16. Tilbury RS, Dahl JR, Monahan WG, et al: Production of nitrogen-13 labeled ammonia for medical use. Radiochem Radioanal Lett 8:317-323, 1971.

17. Straatmann MG, Welch MJ: Enzymatic synthesis of nitrogen-13 labeled amino acids. Radiat Res 56:48-56, 1973.

18. Lathrop KA, Harper PV, Rich BH, et al: Rapid incorporation of short-lived cyclotron-produced radiopharmaceutical, in Radiopharmaceuticals and Labelled Compounds, Vol. 1. I.A.E.A. Vienna, 1973, pp 471-481.

19. Phelps ME, Hoffman EJ, Raybaud C: Factors which affect cerebral uptake and retention of $^{13}NH_3$. Stroke 8:694-702, 1977.

20. Oldendorf WH: Molecular criteria for blood brain barrier penetration, in Non-invasive Brain Imaging: Computed Tomography and Radionuclides, DeBlanc JH Jr, Sorenson JA, eds, Society of Nuclear Medicine, New York, 1975, pp 17-23.

21. Eichling JO, Welch MJ, Straatmann MG, et al: unpublished data.

22. Robinson GD: Biologically active ^{18}F-fluoro-organic compounds: Rapid high specific activity synthesis of ^{18}F-fluorocarboxylates and derivates, in Radiopharmaceuticals and Labelled Compounds, I.A.E.A., Vienna, 1973, Vol 1, pp 423-430.

23. Saunders BC: Chemical characteristics of the carbon-fluorine bond, in Carbon Fluorine Compounds, A Ciba Foundation Symposium, 1972, pp 9-22.

24. Straatmann MG, Welch MJ: ^{18}F-diethylaminosulfur trifluoride (DAST), an F-for-OH fluorinating agent. J Nucl Med 18:151-158, 1977.

25. Tewson TJ, Welch MJ: unpublished data.

26. Lenzi GL, Jones T, McKenzie CG, et al: Study of regional cerebral metabolism and blood flow relationships in man utilizing the method of continually inhaling oxygen-15 and oxygen-15 labeled carbon dioxide. J Neurol Neurosurg Psych 41:1-10, 1978.

27. Subramanyam R, Alpert NM, Hoop BJ, et al: A model for regional cerebral oxygen distribution during continuous inhalation of $^{15}O_2$, $C^{15}O$, $C^{15}O_2$. J Nucl Med 19:48-53, 1978.

28. Yamamoto YL, Thompson CF, Meyer E, et al: Dynamic positron emission tomography for study of cerebral hemodynamics in a cross section of the head using positron-emitting ^{68}Ga-EDTA and ^{77}Kr. J Comput Assist Tomog 1: 43-56, 1977.

29. Diksic M, Yaffe L: Production of carrier-free ^{77}Kr and ^{123}Xe by proton irradiation. J Comput Assist Tomog 2:640, 1978.

30. Landau WM, Freygang WH Jr, Rowland LP, et al: The local circulation of the living brain: values in the anesthetized and unanesthetized cat. Trans Am Neurol Assoc 80:125-129, 1959.

31. Reivich M, Jehle J, Sokoloff L, et al: Measurement of regional cerebral blood flow with antipyrine-^{14}C in the awake cat. J Appl Physiol 27:296-300, 1969.

32. Sakurada O, Kennedy C, Jehle J, et al: Measurement of local cerebral blood flow with iodo[^{14}C] antipyrene. Am J Physiol 234:H59-H66, 1978.

33. Ter-Pogossian MM, Eichling JO, Davis DO, et al: The measure in vivo of regional cerebral oxygen utilization by means of oxyhemoglobin labeled with radioactive oxygen-15. J Clin Invest 49:381-391, 1970.

34. Jones T, Chesler DA, Ter-Pogossian MM: The continuous inhalation of oxygen-15 for assessing regional oxygen extraction in the brain of man. Brit J Radiol 49:339-343, 1976.

35. Raichle ME, Welch MJ, Grubb RL Jr, et al: Measurement of regional substrate utilization rates by emission tomography. Science 199:986-987, 1978.

36. Reivich M, Kuhl D, Wolf A, et al: The [^{18}F] fluoroglucose method for the measurement of local cerebral glucose metabolism in man. Circ Res 44:127-137, 1978.

37. Tewson TJ, Welch MJ, Raichle ME: ^{18}F-labeled 3-deoxy-3-fluoro-D-glucose: Synthesis and preliminary bio-distribution data. J Nucl Med 19:1339-1345, 1978.

38. Fowler JS, Lade RE, MacGregor RR, et al: Agents for the armamentarium of regional metabolic measurement in vivo via metabolic trapping: ^{11}C-2-deoxy-D-glucose and halogenated deoxyglucose derivative. J Labelled Cpds and Radiopharm 16:7-9, 1979.

39. Mestelan G, Aubert F, Beaucourt JP, et al: Rapid synthesis of 2-deoxy-D-[1-^{14}C]-glucose suitable for labelling with ^{11}C. J Labelled Cpds and Radiopharm 16:179-181, 1979.

40. Lifton JF, Welch MJ: The preparation of glucose labeled with 20 minute half-lived carbon-11. Radiat Res 45:35-40, 1971.

41. Straatmann MG, Welch MJ: The liquid chromatographi purification of carbon-11 labeled glucose. Int J Appl Radiat Isotopes 24:234-236, 1973.

42. Kuhar MJ, Murrin LC, Malouf AT, et al: Dopamine receptor binding in vivo: the feasibility of autoradiographic studies. Life Sci 22:203-210, 1978.

43. Pert CD, Kuhar MJ, Snyder SH: Opiate receptor: autoradiographic localization in the rat brain. Proc Natl Acad Sci (USA) 73:3729-3733, 1976.

44. Kook CS, Reed MF, Digenis GA: Preparation of (^{18}F) haloperidol. J Med Chem 18:533-535, 1975.

45. Garnett ES, Firnau G, Nahmios G, et al: [^{18}F] fluorodopa, an analog of dopa, and its use in direct external measurement of storage degradation and turnover of intracerebral dopamine. J Comput Assist Tomog 2:639-640, 1978.

46. Maziere M, Sainte-Laudy JL, Crouzel M, et al: Synthesis and distribution kinetics of ^{11}C-chlorpromezine in animals, in <u>Radiopharmaceuticals</u>, Subramanian G, et al, eds, Soc of Nucl Med, New York, 1975, pp 189-195.

47. Wallach O, Heusler Fr: Ueber organische fluoruerbindungen. Ann Chem 243:219-244, 1888.

48. Tewson TJ, Welch MJ: unpublished data.

49. Fieschi C, Battistini N, Volante F, et al: Animal model experimental study with intracarotid ADP infusion in rabbits. Stroke 6:617-621, 1975.

50. Wu K, Hoak JC: Increased platelet aggregates in patient with transient ischemic attacks. Stroke 6:521-524, 1975.

51. Thakur ML, Welch MJ, Joist JH, et al: Indium-111 labeled platelets: studies on preparation and evaluation of in vitro and in vivo functions. Thrombosis Res 9:345-357, 1976.

52. Davis HH III, Siegel BA, Welch MJ, et al: unpublished data.

RADIOASSAY METHODS

Roger Ekins

Department of Nuclear Medicine,
The Middlesex Hospital Medical School,
London, W.1.

The general principle of radioassay and other related labelled reagent techniques are discussed. Theoretical and practical considerations indicate that assay methods relying on 'excess' labelled antibody are likely to replace the current generation of 'competitive' radioassay methods within the foreseeable future. It is in relation to the 'excess' reagent techniques that labels other than radioisotopes are likely to provide increased sensitivity and speed and may replace isotopic techniques provided certain technological difficulties may be overcome.

INTRODUCTION

Perhaps the single most important contribution in medicine made by radioisotopic techniques in the last twenty-five years has been their application to the measurement of substances of biological interest - particularly those which are present in biological fluids at low concentration. Their most notable impact has been in endocrinology, and it is no exaggeration to claim that this area of medicine has been totally transformed - both in our basic understanding of the subject, and in the development of clinically valuable diagnostic techniques - in consequence of the radioanalytical methods for the assay of hormones which have emerged since the mid-1950's. The impact of radioassay methods in other fields has been later, and, as yet, less dramatic; nevertheless they are being adopted in such diverse areas as oncology, virology, pharmacology, immunology, haematology etc.: in short, in all areas in which the precise measurement of substances possessing biological significance or activity is essential to our understanding of the subject, or is helpful in clinical diagnosis.

In reviewing developments and trends in this wide domain, it would be both inappropriate and impossible for me to catalogue, other than in the broadest terms, the areas of application of radioassay methods in medicine, or to assess their clinical usefulness in each particular sector. In the short time available, I believe it to be more useful to discuss briefly the basic principles governing the radioassay techniques in order to identify both the particular reasons that have led to their overwhelming adoption, and the position they are likely to occupy in the future within the range of techniques available for use in medicine.

Before proceeding further, however, I feel I should confess to unease in isolating the 'radioassay' methods in a separate category; the techniques which are conventionally clustered under this umbrella form part of wider group which I would prefer to call 'labelled reagent' methods or something equivalent, since closely similar principles and methodologies pertain to this general group as a whole irrespective of the particular kind of label on which they depend. In particular, although this symposium is essentially concerned with radioisotopic techniques, I believe it would be valuable to discuss the particular advantages and disadvantages of the analogous, non--isotopic, methods in order better to identify the likely development of radioisotopically-based assays in the future.

In order to put the future development of this field into perspective, I think it is desirable first to examine some basic principles underlying this entire group of microanalytical methods. All analytical techniques depend upon interaction between the substance that we wish to measure - the 'analyte' - and a chemical or physical reagent. The specificity of an assay system as a whole - that is, its ability to single out and quantitate the single unique substance which represents the objective of the measurement - normally depends in part on a preliminary chemical or physicochemical isolation and purification procedure, and, in part, on the specificity of the reagent. Clearly, the more specific the reagent, the less rigorous are the demands placed upon the isolation procedure, and the simpler and more practicable the assay as a whole.

Following reaction between the analyte and the 'specific' reagent, the amount of the analyte is invariably inferred from an observation on the extent of the reaction since, generally speaking, the greater the amount of analyte in the system, the greater the amount of the reaction which takes place.

The particular advantages possessed by radioisotopes in this context is that they allow the observation of exceedingly small numbers (by conventional standards) of molecules in analytical reactions of this kind. This, in principle, permits the development of highly sensitive microanalytical methods. However this sensitivity potential could not be conveniently exploited were it not for the availability of highly specific reagents capable of distinguishing individual groups of molecules defined by a particular structure amidst the immense diversity of substances encountered in biological fluids.

The explosive development of the radioassay field since the late 1950's stems essentially from the combination of the sensitivity imparted to analytical methods by the use of radioisotopic techniques with the specificity characteristic of reagents of biological origin. In particular, antibodies constitute a class of proteins with a virtually infinite range of individual specificities for particular compounds. This implies that it is possible, in principle, to synthesise, by appropriate immunological techniques, antibodies which react with an individual substance and with no other. It must be emphasised, however, that there exist other classes of proteins in biological systems which are comparable to antibodies with regard to their specificity characteristics (for example, cell membrane receptors, transport proteins) which aside from the possible problems which may arise in isolating them in adequate quantity, offer the same fundamental basis as do antibodies for the development of highly sensitive, highly specific assay systems.

In order to understand the constraints on specificity and sensitivity inherent in radioassay methods, it is useful to subdivide them into two broad classes:

a) Reagent excess methods

Those in which the specific reagent is present in the assay system in excess (usually large) vis-a-vis the amount of the substance to be measured. In such assays the response variable comprises an observation reflecting the amount or proportion of the specific reagent which is converted (following reaction with the substance under test) to a second, readily identifiable form.

b) Limited reagent or 'saturation assay' methods

Those in which the specific reagent is present in a limited and carefully defined amount or concentration, usually less than the amount of the test substance in the system. In this type of assay, the analytical measurement rests on observation of the proportion of the test substance itself which, in consequence of interaction with the reagent undergoes change in its physico-chemical form.

GENERAL PRINCIPLES OF 'REAGENT EXCESS' METHODS

The majority of analytical techniques employed in biochemistry fall within this category. Their underlying principle can be expressed by the equation:

Test substance + excess reagent → 'converted' reagent + residual reagent

In the absence of limiting effects and other similar perturbations, the amount or fraction of the specific reagent converted (e.g. the amount of complex between reagent and test substance formed) will be proportional to the amount of the test substance in the system.

In the early 1950's, this general principle was linked to the sensitivity of detection of radioisotopic techniques for the measurement of a number of hormones, amino acids and proteins as exemplified in the development of assays for several steroids (1, 2, 3, 4) and thyroid hormones (5, 6). In their original form, such techniques relied on radioactive labelled reagents such as ^{3}H- ^{14}C-labelled acetic anhydride, ^{35}S-thiosemicarbazide, ^{14}C-fluorodinitrobenzene etc. which, following reaction with the test substances, yielded isotopically labelled derivatives whose radioactive content reflected the amount of the substance present. Because of the low intrinsic specificity of such reagents, and the stringency of the purification procedures required, these labelled excess reagent methods were displaced by the development, in the early 1960's, of the limited reagent or saturation assay techniques until interest in them was restored by the development, by Miles and Hales (7, 8), of the immunoradiometric assay of serum insulin. This form of assay relies on the use, as labelled reagents, of radioactively-labelled antibodies, specific in their reaction with the hormone, and isolated from antiserum by a specific immunoadsorbant. (See Fig. 1). This basic immunoradiometric technique can be both made technically more elegant, and increased in overall specificity, by preceding the reaction between labelled antibody and test substance by a preliminary immuno-extraction step - using a specific extracting antibody attached to a solid matrix - giving rise to what is popularly know as a 'sandwich' or 'two site' assay (Fig. 2). This technique which relies on two separate antibodies raised in different species against the test substance, clearly requires the presence, on the molecule of interest, of two separate binding sites, which imparts a very high degree of specificity to the overall procedure. The technique has been particularly useful in the measurement of intact hormone in the presence of hormone fragments, as exemplified in the measurement of parathormone, and has been commercially exploited in a widely employed assay for Australia antigen used in the screening of transfusion blood. The principle can also be used for the measurement of antibodies by attaching a specific antigen to the solid material. For example, the presence of syphilis antibodies may be detected by attachment of the specific antigenic material (a preparation of Treponema pallidium) to the walls of polystyrene tubes, incubation with the serum under test, and subsequent incubation with labelled sheep anti-human IgG (9).

Further methodological refinement can be attained by labelling, not the specific antibody directed against the test substance, but a third antibody as illustrated in Fig. 3. This device permits the development of 'universal' labels which can be used for the measurement of wide ranges of test substances. For example, if we so arrange that the second antibody comprising the 'sandwich' in such a system is invariably raised in sheep, then labelled anti-sheep γ-globulin can be employed for all such assays. Figure 4 shows results obtained in early trials in our

SIMPLE IMMUNORADIOMETRIC ASSAY

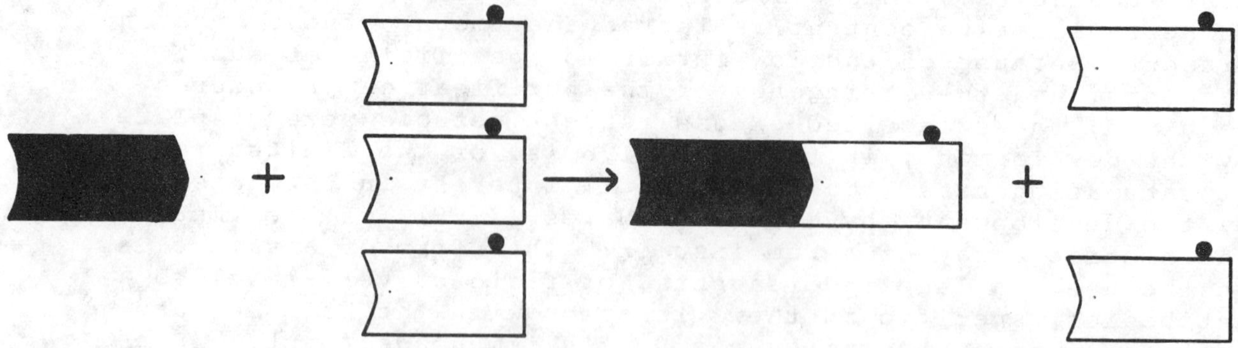

REMOVE RESIDUAL AB USING IMMUNO ADSORBENT: 'COUNT' ANTIGEN/AB COMPLEX IN SOLUTION.

Figure 1.

'SANDWICH' IMMUNORADIOMETRIC ASSAY

STAGE I

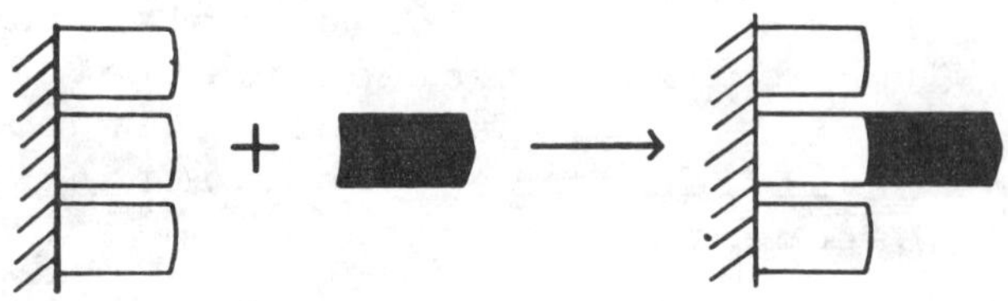

SOLID PHASE-LINKED AB

STAGE II

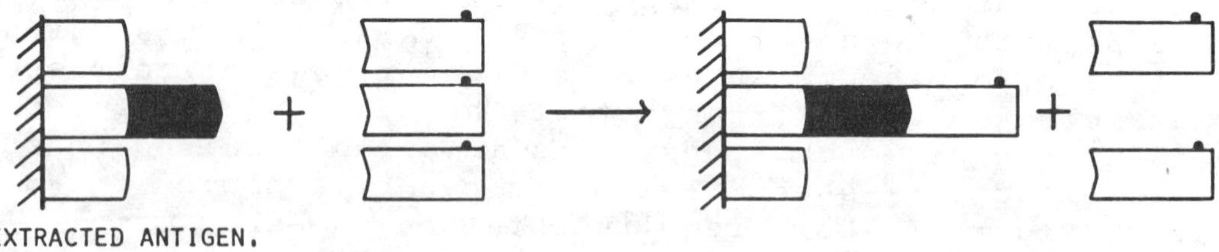

IMMUNO EXTRACTED ANTIGEN.

REMOVE RESIDUAL LABELLED AB BY WASHING.

'COUNT' SOLID PHASE LINKED AB/AG/AB+ COMPLEX.

Figure 2.

own laboratory using such a system for the measurement of TSH. In this system, the walls of the plastic tubes are coated with an excess of guinea pig TSH; following incubation with TSH, the tubes are exposed to rabbit anti-TSH, and finally to I-125 labelled donkey anti-rabbit γ-globulin. Aside from the particular advantage that the labelled anti γ-globulin can be employed in assays for a number of other substances, it is noteworthy that a typical preparation can be employed for many months without significant chemical deterioration, or need of replacement.

Labelled excess reagent systems thus allow us easily to combine highly specific immuno-extraction techniques with the fundamental analytical reaction; also, potentially, they permit us to prepare long lasting radioactive reagents which can be used in a wide variety of assays, which is obviously of great practical advantage for laboratories handling a range of assay techniques.

LIMITED REAGENT OR 'SATURATION ASSAY' SYSTEMS

The fundamental principle invoked in saturation assay can be expressed by the equation:

Test substance + limited reagent → Test substance/reagent complex
+ residual test substance

This principle is that which governs radioimmunoassay and a number of other methods which depend on other classes of specific reagent such as competitive protein-binding assay, radio-enzymatic assay, radioreceptor assay, radioligand assay etc. The two most important features of this general technique are:

(1) it depends upon observation of the behaviour, within the test system, of the test substance itself

(2) it necessitates the use of an amount of the specific reagent which is in the same order as the amount of the test substance (Fig. 5).

These two fundamental features which distinguish saturation assay, or 'limited-reagent' assay from the reagent excess system obviously carry a number of technical implications. First, to observe the reaction behaviour of the test substance it is in practice necessary (although the assay principle does not render it obligatory) to synthesise the analyte in relatively pure

INDIRECTLY LABELLED ANTIBODY 'SANDWICH' IMMUNORADIOMETRIC ASSAY

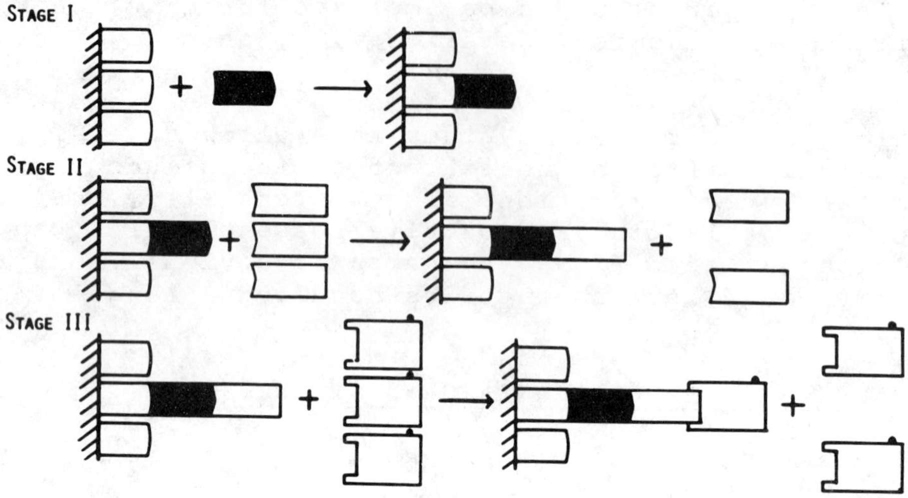

Figure 3.

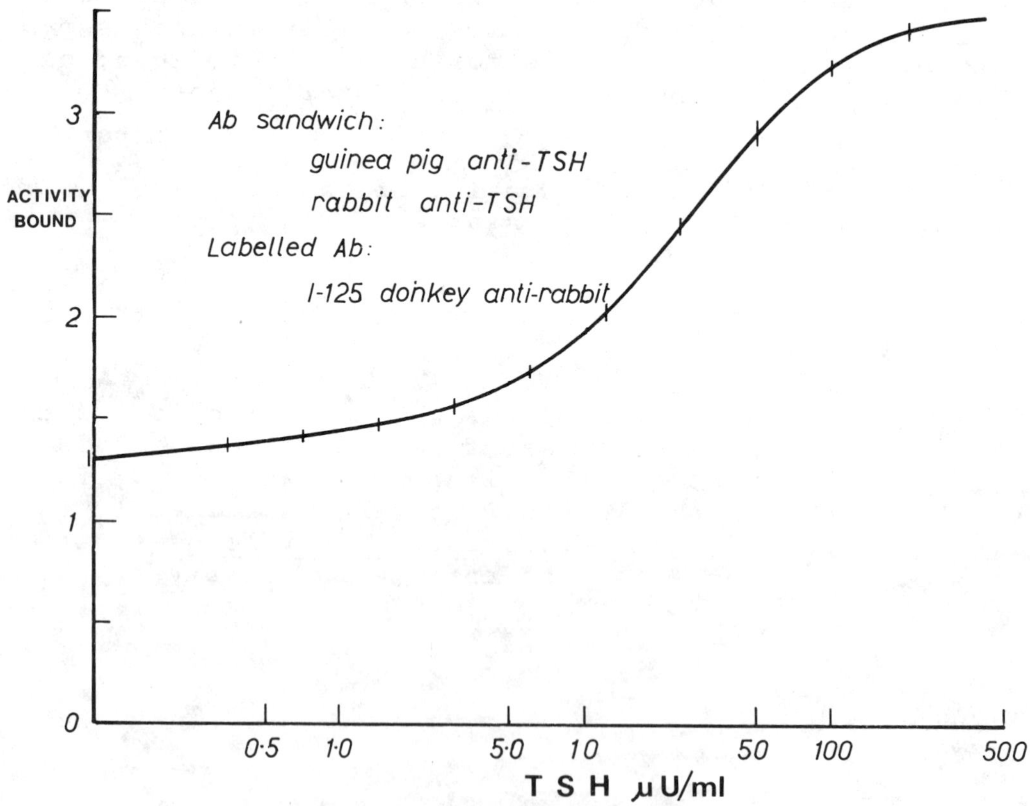

Figure 4.

labelled form. Secondly, the constraint on the amount of the specific reagent that can be used in such systems, restricts the sensitivities that are achievable. In general, under the most favourable conditions, a detection limit of the order of $\frac{0.01}{K}$ (where K is the equilibrium constant of the reaction) represents the highest sensitivity attainable. Moreover the use of a low concentration of specific reagent has other implications. It implies that saturation assay systems are particularly vulnerable to environmental influences which alter the equilibrium constant governing the basic reaction - especially as we approach the sensitivity limit of the particular system. It also implies that, since competing, or cross-reacting, substances present display relative potencies approximately equal to their relative energies of reaction, such a substance - assuming it displays a lower reaction energy than the test substance - will also exhibit a lower potency. Saturation assays, in short, are relatively specific in this particular sense.

RELATIVE MERITS OF EXCESS REAGENT AND LIMITED REAGENT SYSTEMS

As discussed above, the analytical principle underlying 'excess reagent' systems theoretically permits the use of an infinite amount of the specific labelled reagent. In contrast, the analytical principle governing limited reagent methods points in the direction of the use of zero amounts of the specific reagent used. In practice, neither of these extremes is optimal nevertheless amounts of specific reagent either considerably in excess of, or, conversely, somewhat less than, the amount of analyte in the system are conventionally used. This has important implications with respect to the sensitivity and specificity characteristics of these systems.

(i) Sensitivity

Because reagent excess systems permit the use of an infinitely large amount of the specific reagent, the reaction between analyte and reagent may be driven in a forward direction irrespective of the energy (or 'equilibrium' constant K) of the reaction. This implies, in practice, that the sensitivity limits of such techniques are defined by

a) the sensitivity of the system used to detect specific reagent 'bound' to analyte

b) the non-specific or 'blank' signal detected in the absence of analyte.

In contrast, because limited reagent systems necessarily depend on the use of an amount of reagent of the same order of magnitude, or less, than the amount of analyte, the effects of the Law of Mass Action become increasingly manifest as we attempt to penetrate to lower levels of analyte detection. The constraints defined under a and b above play some part in the limitation in sensitivity imposed on limited reagent assay systems; however they are overshadowed in practice by Mass Action effects, so that the sensitivities attained in these systems very frequently approximate to the theoretical limits ($\frac{0.01}{K}$) imposed by the Law of Mass Action.

The implications of these differences is that excess reagent assay system are potentially very much more sensitive than the limited reagent techniques. Since the highest equilibrium constants displayed by carefully selected antibodies and other binding reagents of biological origin are of the order of 10^{11}-10^{12} L/M, the corresponding sensitivities attainable using limited reagent systems is of the order of 10^{-13}-10^{-14} M/L. The ultimate sensitivities of reagent excess systems, however, are considerably greater than this (i.e. detection limits of the order of 10^{-18} M/L are potentially attainable). Another way of viewing this distinction is in terms of assay speed: because the presence of reagent in excess <u>drives</u> the underlying reaction in the forward direction, a given level of sensitivity (such as might be required for a diagnostic purpose) will generally be attained in a shorter time in a labelled reagent system.

ii) <u>Specificity</u>

Because excess reagent systems are less affected by Mass Action considerations, they are correspondingly more liable to respond to a variety of cross-reacting substances in biological fluids which react with the specific reagent, albeit possibly with low energies of reaction. In principle, therefore, they are less specific than the corresponding limited reagent systems. Nevertheless, because excess labelled reagents can be conveniently combined (as in a two-site, or 'sandwich' immunoassay technique), with a highly specific analyte purification step, and because the preparation of labelled reagents (e.g. radiolabelled antibody) for use in an excess reagent system often entails selection of a particular fraction of the reagent, the specificity of the excess reagent methods can, in practice, be made superior to the limited reagent techniques. Probably, therefore the highest specificity currently attainable for substances of large molecular size (such as proteins) is provided by the two-site excess labelled reagent techniques; for substances (such as hormones like thyroxine, drugs, etc.) which are of a molecular size insufficient to be simultaneously recognised by two reagent binding sites, then the limited reagent techniques are probably the more specific.

To summarise these concepts, I have endeavoured to demonstrate that, on the basis of theoretical considerations, the excess labelled reagent methods are preferable both with regard to sensitivity, speed and - for substances of large molecular size - specificity. What then is the reason for the far more widespread current use of the limited reagent assay techniques epitomised by radioimmunoassay? The answer to this question resides partially in history, partially in the technical problems respectively associated with these two approaches. At the time of the initial development, in the late 1950's, of the saturation assay methods (for insulin in the laboratory of Drs. Berson and Yalow in New York, and for thyroxine in my own laboratory) it was far easier to obtain labelled analyte than the corresponding labelled biological reagent, and it was inevitable that, in both laboratories, we should independantly have utilised the limited reagent approach in developing microassays for these two hormones. Subsequently, although methods for the isolation, purification and labelling of specific antibodies and other biological binding reagents have become available, the great popularity of, and technical expertise invested into, the limited reagent, labelled analyte systems have ensured a pre-eminent position for them in assay methodology. Moreover, labelled excess reagent systems are - inevitably - relatively expensive in terms of reagent usage (e.g. in respect of consumption of scarce antisera), so that practical constraints have continued to impel the majority of workers towards 'limited reagent' methodology. However the intrinsic merits of excess reagent methods, coupled with the advent of new methods of antibody production will ultimately, I believe, compel the abandonment of the limited reagent approach, particularly as applied to substances of large molecular size. Before discussing some of the technical developments which give support to this belief, it is relevant to examine the relative merits of the different types of label which are used in these techniques.

LABELS USED IN 'LABELLED REAGENT', AND 'LABELLED ANALYTE' ASSAY SYSTEMS

Radioisotopes have traditionally been employed in the present context primarily for the following reasons:

a) the sensitivity of detection of radiolabelled substances

b) their ruggedness in the face of environmental influences (chemical milieu, temperature, etc.)

c) the basic simplicity of current radioisotopic counting equipment, and the ease of counting techniques.

Nevertheless, they possess drawbacks, amongst which one may identify the following:

a) Radiation hazards: environmental and legal objections

b) restricted shelf life of radio-labelled reagents

c) cost of counting equipment

d) prolonged counting times sometimes required in circumstances requiring a high and rapid sample throughput

e) low detection efficiency

f) the need to separate physically the two components of the assay system (reacted and unreacted labelled material) following reaction

g) restriction to appropriately equipped laboratories: not readily translated to 'field use'

Before examining some 'alternative' labels it may be appropriate to examine briefly each of these disadvantages in turn.

(a) Environmental objections

In principle, a case may be made against any augmentation of the radiation burden in the environment; however, provided reasonable care is taken, the hazards to laboratory personnel arising from the use of radioassay methods are negligible, and far less than those arising from the presence in samples of virus or the use of carcinogenic chemicals which are typically encountered in clinical laboratories. Only in those laboratories synthesising their own labelled reagents is there a need for specialised facilities, and particular precautions of a kind not normally required in a conventional laboratory.

(b) Restricted shelf-life

Radioactively labelled compounds decay in consequence of

the decay of the radionuclide. ^{125}I is the isotope conventionally employed, because of its convenience, in the radioassay of the majority of proteins; ^{3}H is also used for substances such as steroids, cyclic nucleotides etc. where use of an iodine label is difficult though not impossible. However, the limited shelf-life of the 'labelled analytes' used in limited reagent methods is not entirely due to radioactive decay; it often arises largely in consequence of chemical degradation or instability of the analyte itself, either intrinsic or in consequence of damage sustained during the labelling process, or of the environment in which the material is maintained. There is, in short, no guarantee that the use of an alternative label of long intrinsic life will necessarily confer a long shelf-life upon the labelled compound.

Nevertheless, it must be recognised that variability of the chemical composition of the labelled material, both within a particular preparation and/or between preparations, probably constitutes the major source of variability and diagnostic imprecision in radioassay techniques, and any improvements in the long-term stability of the labelled materials employed is of critical importance. Insofar as alternative labels may contribute to this end, this constitutes probably the most important justification for the abandonment of radioisotpic labels.

(c) Cost of counting equipment

The basic cost of a radioisotope counter is extremely low in Europe, a simple manual counter for radioassay work is available at a cost of approx. $500. Costs escalate as soon as counting equipment is equipped with sample-changing and data-processing facilities: increase in cost for these reasons, however, would apply to any form of signal-detection method. Nevertheless, because a long counting time per sample may in certain circumstances be required, counting equipment may be occupied for overlong periods and render necessary the multiplication of instruments in the face of heavy sample loads. Under these circumstances, the use of labels such as fluorescent or enzyme markers which could increase instrument utilisation by a factor of ten or twenty fold or more would be of obvious advantage. The circumstances in which such demand arises, given 'intelligent' use of counting equipment (see below) are comparitively rare.

(d) Delays arising from long counting times

Related to the cost of counting equipment discussed above

are the problems that may arise because of the relatively long intervals of time, and consequent delays, that are associated with the measurement of radioactive samples. The problem is compounded to a significant extent by a misunderstanding of the statistical aspects of sample counting and the consequent misuse of counting equipment. With the advent of microprocessor controlled radioisotope counters (10) it is possible greatly to restrict sample counting time, and thereby to reduce delays, maximise the cost-effectiveness of the equipment used and minimise expenditure on instrumentation. A second approach to increasing sample throughput is by the construction of multiple sample counting equipment, and several manufacturers are now marketing counters capable of the simultaneous counting of batches of - for example - 16 samples. This development likewise eliminates delays due to sample counting and makes for a highly efficient and cost effective system capable of rapid sample throughput.

(e) Low detection efficiency

Although efficiencies of counting radioisotopically labelled substances is often presented as high, the actual detection efficiency of radionuclides is often extremely low. For example, using a conventional γ-counter, something of the order of 300,000 ^{125}I-labelled protein molecules (assuming mono-iodination with carrier free ^{125}I) are required to yield a count rate of 1 count per minute. In the presence of normal natural background levels it is clear that molecular numbers at least of the order of 100 times greater than this are required to yield count rates which are of practical use in a routine radioassay context.

In contrast, other labels are potentially capable of yielding much greater detection efficiencies. This may arise because of the possibility they offer of 'signal' amplification by virtue of an external energy input. For example, a single enzyme molecule in consequence of its catalytic effect on a chemical reaction, can lead to the generation of a very large number of product molecules from an appropriate substrate, the source of the energy required residing in the chemical energy stored in the substrate. Likewise, the external application of energy in the form of a light beam of appropriate integrity can ensure that, however low the fluorescence yield or conversion efficiency, an individual molecule of fluorescent label can (with near certainty) be induced to fluoresence in any specified time interval.

This implies that the efficiency of detection of labels other than conventional radioisotopic markers can, in principle, be rendered far higher, and the numbers of molecules so labelled which can be detected in practical time intervals far smaller. This concept can be expressed in terms of the 'specific activities' of different labels. Thus the specific activity of a ^{125}I labelled substances can be represented as approximately 1 detectable event/minute/150,000 molecules. The specific activity of other forms of label can however - given appropriate technology - be increased significantly above this figure.

What is the significance of these concepts with regard to the sensitivity and precision of radioassay and related analytical techniques? In the case of limited reagent assays, such as conventional radioimmunoassay, it is now well recognised that the specific activity of the labelled analyte does not - in most systems - limit assay sensitivity or precision and that no practical improvement stems from the synthesis of very high specific activity labels (other than a reduction in sample counting times). This follows from the constraint on sensitivity imposed by the equilibrium constant of the governing the analyte specific reagent reaction. In excess reagent systems, on the other hand, sensitivity is fundamentally limited by the detectability of the label and implicitly by the specific activity of the labelled reagent. It is in this assay of this type, therefore, that the high potential specific activities characterising non-isotopic labels will ultimately be exploited by the development of techniques of far higher sensitivity and speed than the current generation of radioassays.

(f) Separation of 'bound' and 'free' label

Radioisotopic labels exhibit their activity irrespective of their chemical environment: for this reason the distinction between reacted and unreacted, or 'bound' and 'free' forms of either analyte (as in the technique of radioimmunoassay) or of reagent (as in immunoradiometric assay) requires physical isolation of the two moieties. Non-isotopic labels can be selected, however, in which the bound and free moieties can be distinguished as a result of the effect of the microenvironment on the activity of the label. For example, the fluorescence of a substance tagged with a fluorescent marker can be quenched or enhanced, or otherwise altered, in consequence of the binding to molecules of the labelled substance of another molecule - e.g. antibody. In this way, the relative amounts of 'free' and 'bound' label can be assessed without physical separation. This possibility is of considerable practical advantage: nevertheless there are a number of other important practical reaons which frequently necessitate physical separation of bound and free

RADIOIMMUNOASSAY

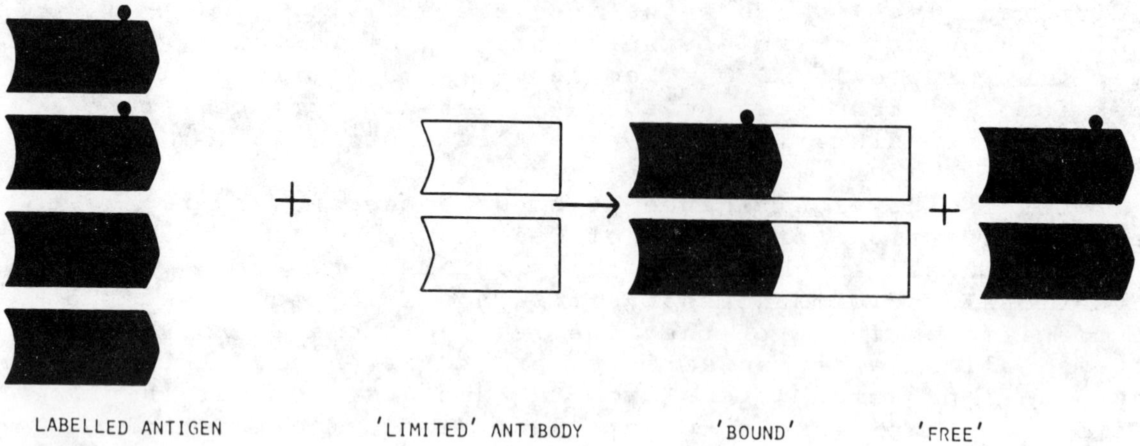

SEPARATE 'BOUND' AND 'FREE' LABELLED ANTIGEN.

DISTRIBUTION OF LABELLED ANTIGEN BETWEEN FREE AND BOUND FRACTIONS IS A FUNCTION OF TOTAL ANTIGEN PRESENT.

Figure 5.

ENZYME IMMUNOASSAY

ENZYME INHIBITION TECHNIQUE.

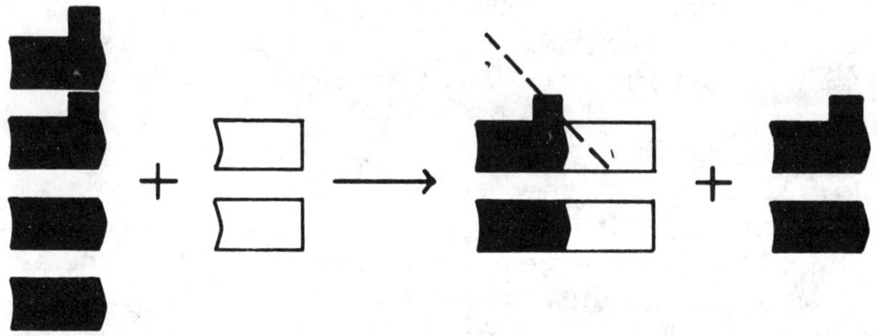

ADVANTAGE

DOES NOT REQUIRE PHYSICAL SEPARATION OF FREE AND BOUND FRACTIONS

DISADVANTAGES

1. LIMITED TO SMALL LIGANDS
2. POSSIBLE REDUCTION OF REACTION ENERGY BETWEEN ANTIBODY AND LABELLED LIGAND LEADING TO LOSS IN SENSITIVITY
3. 'EFFECTIVE' ANTIBODY IS THAT WHICH REACTS WITH LABELLED MOLECULES AND IMPEDES ENZYME SUBSTRATE INTERACTION. POSSIBILITY OF LARGE MISCLASSIFICATION ERRORS

Figure 6.

moieties in assay systems of this type (such as the avoidance of non-specific effects), so that the disadvantage associated with radioactive labels is of less practical importance than might at first appear.

(g) Limitation to a laboratory environment

Radioistope techniques inevitably require electronic equipment which implies that the radioanalytical methods are virtually confined to laboratory use. There are circumstances, however, in which labelled analyte or reagent methods might be of great use in circumstances in which even the simplest electronic equipment would be inadmissable. The WHO has recently launched projects intended to develop domestic assay techniques whereby women, by monitoring urinary steroid glucuronide levels, can assess the time of ovulation and the fertile period. Likewise the investigation under field conditions of the incidence of a variety of tropical diseases - for example schistosomiasis - may often only be carried out using non-isotopic labels. It is in these circumstances that labels which can give rise to a colour or some other readily perceived signal are clearly of particular value.

SOME NON-ISOTOPIC LABELS

(a) Enzyme labels

Enzymes can be used either as a 'conventional' label - i.e. in exactly the same way as a radioactive label in which the 'reacted' and unreacted' (or 'free' and 'bound' moieties) are separated prior to assay of the label content of either or both moieties - or, in saturation assay techniques, by the use of enzyme inhibition methods (see Fig. 6). The latter do not require physical separation of free and bound fractions, but rely on steric hindrance of the enzyme when the molecule to which it is attached is bound to antibody or to other specific reagent. In this situation, only enzyme existing in the free moiety is available for interaction with enzyme substrate. This strategem clearly represents a practical advantage; it also potentially offers an increase in sensitivity insofar as the misclassification errors incurred in conventional techniques can contribute to a loss of sensitivity and precision. Conversely a loss in sensitivity can result because of reduction in the equilibrium constant of the labelled ligand; moreover the method is not itself without sources of 'misclassification error' (e.g. some labelled molecules may be bound to non-enzyme-inhibiting antibody, and appear to be 'free') with a possible consequent loss in sensitivity. Moreover the method is essentially restricted to ligands of small molecular size, and in practice has not attained sensitivities comparable with those of the radioactive techniques.

Enzyme labels can also be attached to the specific reagents used in 'reagent excess' methods as in the technique referred to as ELISA (enzyme-linked, immunosorbant-assay) as originally developed by Engvall and Perlmann (11, 12). This technique which is of the sandwich type discussed above, has been successfully applied to the measurement of a number of substances, including antibodies, and has been claimed to be of such promise as to offer the prospect of replacing radioimmunoassay as the general method of the future. As originally described, the labelled antibody is prepared by conjugation with alkaline phosphatase using glutaraldehyde. Following reaction with the antigen (which in turn is bound by extracting antibody attached to the walls of polystyrene tubes), the enzyme is 'developed' by exposure to p-nitrophenyl phosphate to yield yellow p-nitrophenolate - a reaction which takes some 1 to 2 hours.

This particular technique has many attractive features, not the least of which is that the enzyme reaction takes place in a milieu from which the original biological sample has been removed, and thus is not subject to the chemical effects to which enzyme labels in other forms of assay may be exposed.

(b) Fluorescent labels

Fluorescent labels such as fluorescein can likewise be employed in both saturation assay and reagent excess systems. As with enzyme assay, they offer the theoretical possibility of techniques in which no separation of free and bound moieties is required by virtue of changes in the nature of the fluorescence when the labelled molecule is bound to specific reagent. Such modification of the fluorescence can take the form of fluorescence inhibition, fluorescence enhancement, or alteration in the polarisation characteristics of the fluorescent radiation when exposed to polarised light in consequence of alteration in rotational velocity of bound molecules.

Amongst the advantages that can be claimed for fluorescent label methods is the relative ease of labelling of many biological molecules, including antibodies; the stability of the labelled products and the simplicity of the measuremetts and instrumentation. As discussed above, they also offer the possibility of saturation assay techniques not requiring separation of free and reagent bound moieties; nevertheless this theoretical advantage is not realisable in practice - except in the case of relatively insensitive methods - because of the high fluorescence background contributed by the wide variety of naturally occurring fluorescent materials found in biological fluids. In practice, therefore, the successful use of fluorescent labels usually requires the separation of reacted and unreacted moieties as in conventional radioactive techniques.

(c) Other labels

A number of other labels have been proposed as substitutes for radioactive isotopes in radioassay techniques: for example free radicles, phages, chemiluminescent labels etc. It would be wrong to be complacent about such developments, and one or other may ultimately offer a viable alternative to isotopes as a general method of labelling. My personal belief, however, is that the majority of the other suggested alternative labelling methods offer little by way of notable practical or logistic advantages, or the wide applicability that is possessed by the radioisotopic methods, and are certainly not sufficiently technically advanced to merit extended discussion at this meeting. There exists perhaps one technique, however, which is worthy of mention primarily because it is technically elegant and is potentially usable in domestic situations for the kind of purpose discussed earlier. This depends upon the change in surface tension characteristics of a plastic surface consequent on the presence to the surface of an antibody - antigen - antibody sandwich (13). The presence and amount of antigen in a biological fluid can thus be deduced visually simply, for example, by exposing the surface to humid air - indeed, breathing on the surface is sufficient to yield a semi- quantitative measure of the amount of antigen in the system.

RELATIVE MERITS OF ISOTOPIC AND NON-ISOTOPIC LABELS

In summarising some of the points that have been discussed earlier regarding the relative advantages of different forms of label, it is clearly important to distinguish their use in the limited reagent labelled analyte techniques from that in the excess reagent labelled reagent methods.

With regard to the former, the particular advantages of non-isotopic techniques are the long-term stability they potentially offer, their unprovocative status vis-a-vis environmental and legal objections, and the possibility they offer for the development of non-separation techniques with the consequent simplification of laboratory protocols that this confers. On the other hand, non-separation methods are currently limited to substances of small molecular size such as drugs, thyroid hormones, etc.

Moreover, some of the non-radioisotopic labels are themselves of relatively large molecular size and this has implications with respect to their reactions with binding reagents such as antibodies and on the energies of these reactions. This fact, taken together with the increase in measurement errors frequently associated with 'signal detection' as compared with radioisotope counting, implies a loss of sensitivity in the non-isotopic methods. This prediction has generally proved true, albeit if the analyte is present in body fluids at high concentration (such as in the case of many drugs) this objection may be of minor importance.

In regard to the excess reagent methods, non-isotopic labels are potentially able to offer far higher sensitivities than are attainable with radioisotopically-tagged reagents. The realisation of this possibility is dependant on the development of labels and instrumentation which give a high signal/noise ratio. In pursuing this objective, our own approach is centred on the development of pulsed fluorescence techniques using fluorescent labels with prolonged fluorescence decay characteristics. By this means, it is possible to discriminate between the fluorescence of the tagged reagent from all other sources of fluorescence in the system, thereby largely eliminating the major sources of background which limit sensitivity using current fluorescent label techniques.

FUTURE DEVELOPMENT OF RADIOASSAY TECHNIQUES

I have endeavoured, in this review, to outline the fundamental principles governing limited reagent and excess reagent analytical methods. My objective in so doing is to identify concepts which are currently governing the strategy of future development of radioassay and related techniques in my own laboratory. I have attempted to show that, on grounds of sensitivity, specificity and ruggedness, in the face of minor differences in the chemical composition of the biological samples assayed, the labelled, excess reagent, two-site assay systems possess significant advantages as compared with the labelled analyte, limited reagent (saturation or 'competitive') assay systems - particularly for the measurement of substances of large molecular size - and are likely to replace them in the foreseeable future. This process is already taking place: the Ausria kit marketed by Abbott, and the TSH kit made by Corning, both represent examples of 'two-site' immunoradiometric assays. Many other examples of the same analytical concept are to be found under the designation ELISA - assays which, as I have shown, rely on enzyme labelled antibody as the top layer of the antibody-antigen-antibody sandwich on a two-site system.

In pursuing this objective, two major technical obstacles remain to be overcome.

(1) the reduction of 'background' signals, thus enhancing signal/noise ratios and thereby increasing assay sensitivity and reducing assay times.

(2) improvement of the synthetic and purification methods involved in the preparation of the labelled reagent i.e. - in general - labelled antibody.

Resolution of the first of these problems may involve, as I have indicated, the abandonment of radioactive labels when high sensitivity coupled with high sample throughput - perhaps by pulsed fluorescence techniques in the way in which my own group is investigating. Resolution of the second requirement is in sight in consequence of the exciting new developments pioneered by Milstein and others involving in-vitro fusion of mouse myeloma and antibody producing spleen cells. These techniques, now under intensive examination in my own laboratory holds out the promise of in-vitro monoclonal antibody production in unlimited amount, giving us for the first time the opportunity to tailor molecules possessing exactly the structural-recognition capacity which we require for use in two-site assay systems. Given the resolution of these two technical objectives, I have no doubt that we will witness, in the next decade, the development of a new generation of antibody-based analytical methods far surpassing in speed, sensitivity and precision the radioassay methods that are currently available.

REFERENCES

(1) Avivi, P., Simpson, S.A.S., Tait, J.F. and Whitehead, J.K. (1954). In: Proceedings 2nd Radioistope Conference, Vol. 1, p. 313. Edited by Butterworths, London.

(2) Peterson, R.E. (1959) Recent Progress in Hormone Research, 15, 237.

(3) Riondell, A., Tait, J.F., Gut, M., Tait, S.A.S., Joachiun, E. and Little, B. (1963) J. of Clin. Endocrinology & Metabolism, 23, 620.

(4) Riondel, A., Tait, J.F., Tait, S.A.S., Gut, M. and Little, B. (1965) J. of Clinical Endocrinology & Metabolism, 25, 229.

(5) Whitehead, J.K. and Beale, D. (1959) Clinica Chimica Acta, 4, 710.

(6) Whitehead, J.K. and Beale, D. (1960) Clinica Chimica Acta, 5, 150.

(7) Miles, L.E.H. and Hales, C.N. (1968) Biochemical Journal, 108, 611.

(8) Miles, L.E.H. and Hales, C.N. (1968) In: Protein and polypeptide Homrones, Part 1, p. 61. Edited by M. Margoulies Excerpta Medica Foundation, Amsterdam.

(9) Veldkamp, J. and Visser, A.M. (1975) Br. Jounral Ven. Dis., 51, 227.

(10) Ekins, R.P., Sufi, S. and Malan, P.G. (1978) In: Radioimmunoassay and Related Procedures in Medicine, Vol. I, p. 437. I.A.E.A. Vienna.

(11) Engvall, E. and Perlmann, P. (1971) Immunochemistry, 8, 871.

(12) Engvall, E. and Perlmann, P. (1972) J. Immunology, 109, 129.

RADIONUCLIDE LYMPHOSCINTIGRAPHY - AN UPDATE

Günes N. Ege, Ann Warbick-Cerone,
and Michael J. Bronskill

Department of Nuclear Medicine,
The Princess Margaret Hospital,
Toronto, Ontario, Canada M4X 1K9

ABSTRACT

Lymphoscintigraphy, utilizing the interstitial injection of radiocolloids of optimal physical and chemical properties, constitutes a safe, simple technique for visualizing lymph node groups inaccessible for study with radiocontrast lymphography. In a large group of patients with breast carcinoma, the technique has provided a means for (a) accurate localization of parasternal nodes for optimum radiation therapy; (b) evaluating the status of the internal mammary nodes prior to arriving at a management decision; and (c) following patients to assess the response to therapy or progression of disease. Preliminary results obtained show good correlation with clinical stage and prognosis.

PRINCIPLES

Radiocolloids of suitable particle size and properties introduced into appropriate tissue planes, will follow lymphatic pathways and provide a scintigraphic image of drainage lymph nodes. Since direct intralymphatic introduction is not required, there is less morbidity associated with the procedure and greater versatility in application to anatomic sites inaccessible for contrast lymphography. While the lymphogram demonstrates the distribution of passively introduced contrast material, the radionuclide lymphoscintigram reflects physiological processes such as lymphatic transport, filtration, phagocytosis and reticuloendothelial function.

HISTORIC PERSPECTIVE

Sherman and Ter-Pogossian (1) in 1953 introduced colloidal ^{198}Au into the rabbit abdominal wall as well as parametrial tissues to determine the feasibility of utilizing interstitially injected radiocolloids for therapeutic applications. Using radioautography, they were able to demonstrate accumulation of colloidal ^{198}Au in drainage lymphatics.

Seaman and Powers (2), studying the distribution of radioactive colloidal gold in regional lymph nodes containing metastases, injected normal breast tissue adjacent to tumor prior to mastectomy and axillary dissection, and examined 356 nodes from 12 patients. Even in normal lymph nodes there was a marked variation in the distribution of radiocolloid. Hultborn et al (3) confirmed these findings, demonstrating sporadic uptake of colloidal ^{198}Au in both normal and cancerous lymph nodes.

Thomas (4) concluded that lymphatic tissue is essential for deposition of radiogold in cancerous lymph nodes. Even with lymphoid elements present, most cancerous lymph nodes do not sequester radiocolloids. Lymph node metastases may alter lymphatic flow so that not only is there interference with the deposition of radiogold in carcinomatous nodes, but there is also less deposition in nodes uninvolved by tumor.

Where feasible, correlation was carried out with lymphography. In a study of patients with carcinoma of the cervix, Lang (5) demonstrated good correlation in patients with Stage III and IV disease with poor pelvic lymphatic uptake of radiocolloid injected into the paracervical tissues. However, in patients with Stage I and II disease, 25% of cases failed to show normal distribution of radiocolloid in lymph nodes which, on pathological examination, showed no evidence of tumor replacement.

Such data proved that the therapeutic outcome of intralymphatic radiocolloids was unpredictable. However, the diagnostic implications of visualizing lymphatics has led to the development and application of radionuclide lymphoscintigraphy in different anatomic sites.

Up until 1966 the agent most widely used was colloidal ^{198}Au. Sites injected included (6):

1. Dorsum of the foot in the first two interdigital spaces, to demonstrate inguinal, iliac, and paraaortic nodes.

2. Dorsal surface of the hand in the first two interdigital spaces and perinipple area for demonstration of axillary nodes.

3. Perinipple and parasternal region for sub- and supraclavicular nodes and internal mammary nodes.

4. Buccal mucosa and mastoid region for jugular nodes.

5. Hepatic capsule for esophageal, mediastinal, and internal thoracic nodes.

6. Splenic capsule for splenic hilar nodes.

7. Bronchial mucosa for hilar nodes, mediastinal and internal thoracic nodes.

8. Vaginal mucosa for external iliac and hypogastric nodes.

Clinical applicability of the injection of many of the sites listed above has been limited. The two areas studied most extensively with radiocolloid lymphoscintigraphy have been the external iliac and paraaortic nodes - visualized after the dorsal pedal subcutaneous injection first described by zum Winkel in 1963 (7), and the diaphragmatic and parasternal nodes - demonstrable with the subcostal injection into the posterior rectus sheath, described simultaneously by Schenck (8), Diethelm (9), and Rossi (10), in 1966.

AGENTS

Developments in radiopharmaceuticals and instrumentation have altered significantly the quality of lymphoscintigraphic images, and reservations concerning radionuclide lymphoscintigraphy on the basis of the less adequate studies of a decade ago are no longer justified.

Of the variety of radiopharmaceuticals utilized in interstitial lymphoscintigraphy, radiocolloids are the agent of choice (Table 1). The ideal agent is characterized as follows:

1. A sterile, pyrogen-free, homogeneous, mono-disperse, physiologically and chemically inert colloidal dispersion, with a small, narrow particle size distribution (less than 100 nm), and optimal peak at less than 24 nm.

2. Stability which precludes flocculation or particle aggregate formation by added electrolytes or in vivo tissue electrolytes.

3. Optimal mobilization of colloid from the injection site into drainage lymphatics within the first 6 hours of the study, and corresponding minimal tissue retention attributable to stabilizers or additives.

4. Physiological pH or ionic strength which is also unfavourable for particle growth.

5. Absence of biological toxicity of both the radiocolloid and the

TABLE 1. LYMPHOSCINTIGRAPHIC AGENTS AND THEIR PARTICLE SIZE CHARACTERISTICS

Agent	Particle Size	Reference
^{198}Au colloid	2-10 nm	11-14
^{51}Cr & ^{99m}Tc human serum albumin	mw 60,000	15-16
^{110}Ag oleate	-	17
^{99m}Tc sulphide colloid	> 100 nm	23,46-47
^{197}Hg sulphide	10-150 nm	19
^{99m}Tc antimony sulphide	2-15 nm	23
^{99m}Tc stannous phytate	ionic	25,48
^{125}I E. Coli 70S ribosomes	20 nm	49
^{99m}Tc liposomes	24-25 nm	50
^{99m}Tc HSA minimicroaggregates	< 90 nm	†

† New England Nuclear, Billerica, Massachussetts, U.S.A. - Personal communication.

stabilizer in the dosage to be administered, with eventual metabolism and removal from the lymph nodes.

6. Incorporation of a radioactive label with suitable half-life and energy characteristics, resulting in a low patient radiation dose and permitting quantitative or dynamic measurements with a computer interfaced scintillation camera.

7. Optimization of absorption of the lymphoscintigraphic agent by characterization of the effects of colloid particle size, shape, charge, chemical nature, and additives (including the stabilizer).

Since the early application of ^{198}Au colloid (11-14) as an interstitial lymphoscintigraphic agent, the search has continued for an agent with more desirable energy characteristics and suitable colloidal properties. The success of early ^{198}Au colloid studies was attributed to the small particle size of the agent. High molecular weight compounds, such as radio-labelled human serum albumin (15-16), delineate the lymphatic pathways and provide information on lymph flow, but high body background and poor retention in the lymph nodes limit the use of this agent to dynamic studies. In 1967 Anghileri (17) proposed the use of ^{110}Ag oleate (100-400 nm), which had a similar pattern of distribution and advantages of faster elimination as compared with ^{198}Au colloid (3-7 nm). It is noteworthy that these two chemically different colloids with different particle sizes show similar tissue distribution patterns.

The disappointing results obtained with ^{99m}Tc sulphide colloid (18) again identified colloid particle size as a major consideration in the selection of an interstitial lymphoscintigraphic agent. Cox (19) confirmed the importance of particle size in studying ^{197}Hg sulphide for lymphoscintigraphy and obtained satisfactory results with this preparation. The long half-life of ^{197}Hg and the lack of commercial availability of ^{197}Hg bichloride as starting material for the 'in house' radiocolloid prevented its widespread use.

From 1976 Ege (18, 20-21) utilized ^{99m}Tc antimony sulphide (Figure 1), developed by Garzon et al (22), as a small, uniform radiocolloidal dispersion, in internal mammary lymphoscintigraphy. The particle size distribution for this product was thoroughly investigated and the chemical nature resolved (23) (Table 2).

The application of a rabbit animal model by Strand et al (24) and Ege et al (25) has enabled them to evaluate and compare the various potential lymphoscintigraphic agents. Quantitative data on radiocolloid mobilization and absorption from interstitial injection sites and lymph node uptake has demonstrated that ^{99m}Tc antimony sulphide, compared with new agents such as ^{99m}Tc stannous phytate, is at present the agent of choice (25-26).

The mobilization and distribution of radiocolloids in the lymphatics are now known to be affected by a number of factors:

1. Uniform, pesudo-spherical particles appear to show greater mobilization than non-uniform aggregates or particles.

2. Negatively charged colloids are probably absorbed more efficiently

TABLE 2. CHARACTERISTICS OF TECHNETIUM-99m ANTIMONY SULPHIDE COLLOID

Distribution	log normal, +ve skew
Distribution limits	2 up to 40 nm, depending on variations in the manufacture of reagent antimony sulphide
Distribution peak	5-12 nm, depending on variations in the manufacture of reagent antimony sulphide
Particle shape	pseudo-spherical
Charge	negative
Chemical nature	Sb_2S_3
Stabilizer	polyvinylpyrrolidone mw 40,000
Storage of reagent	room temperature or refrigeration
Shelf-life	longer than 1 year
Labelling efficiency	greater than 95%
Label stability	8 hours

than positively charged colloids. Wilkins (27) demonstrated that intravenously injected positively charged colloids are probably converted to negatively charged colloids by binding with biological substances such as fibrin.

3. The chemical nature of the colloid may play an important role, with preferential phagocytosis of different colloids (28-29).

4. Proliferation of new macrophages may be the principal factor in recovery from R.E.S. blockade by non-metabolized particles (30).

5. Particle size not only is important for the mobilization of colloids from the injection site (31), but may determine how long the particle remains trapped in the lymph node (32).

6. Diminution in the volume of interstitially administered radiocolloid has been shown to enhance uptake (33-34).

TECHNIQUE

The optimum injection site for the delineation of a group of lymphatics must be unequivocally established to ensure technical adequacy of the procedure, and for the study to be of any value, it must depict consistently under normal circumstances a sequence of lymphatics, with allowances for normal distortion of the pattern.

Lymphatic visualization following interstitial injection of radiocolloid can be simplistically interpreted to suggest that tumor embolization pathways can be accurately assessed by injection of the tumor bed with radiocolloid, or that direct intraorgan injection can define with greater certainty the potential pathways of tumor spread. Our own experience indicates that direct intraorgan injection does not demonstrate all the relevant lymphatic drainage pathways, and by the time tumor is manifest lymphatic kinetics may be so altered and pathways sufficiently compromised that no useful information can be obtained.

Currently the agent of choice is ^{99m}Tc antimony sulphide colloid*. The injected activity may vary, depending upon available instrumentation and other factors. The volume of injection medium should be kept as small as possible, not exceeding 0.3 ml. The necessity of hyaluronidase has never been established. It is likely that such an agent may aid dispersion without necessarily enhancing absorption from the injection site. All lymphoscintigraphic procedures are well tolerated without the use of local anesthesia.

For visualization of femoral, inguinal, iliac, and paraaortic lymphatics, 500 μCi of radiocolloid should be introduced subcutaneously into the first and second web spaces on the dorsum of both feet. By alternating injections, one can demonstrate the presence or level of cross drainage, which may be of

---

* Philips-Duphar, Petten, Holland.
Union Carbide Corporation, Tuxedo, New York, U.S.A.

interpretative significance. A definitive image may be obtained within three hours, and delayed images are rarely contributory. The study should not be performed immediately following a contrast lymphogram because of blockage of sinusoids by contrast material. In view of the very small volume of radionuclide utilized, the lymphoscintigram will not interfere with a subsequent lymphogram.

Internal mammary lymphoscintigram (Figure 2) requires the introduction of 500 μCi of radiocolloid into the posterior rectus sheath bilaterally, approximately 3 cm lateral to the midline below the costal margin (Figure 3). Diaphragmatic and parasternal lymphatics originate at the insertion of the diaphragm into the rectus sheath. This network of lymphatics will ensure optimum distribution of radiocolloid and good visualization of both parasternal chains. A single subxiphoidal injection obscures xiphisternal and diaphragmatic lymph nodes, visualization of which is critical for the interpretation of the study.

The depth of injection depends upon the body habitus of the patient. For the average patient, a 22 gauge 1½" needle is satisfactory; for the obese patient, a 25 gauge 2½" needle may be required. With the patient supine and relaxed, slight tension should be exerted on the tissues overlying the costal margin and the needle guided till the resistance of the posterior rectus sheath is detected.

The visualization of at least one lymphatic aggregate ensures a technically adequate injection. In the absence of a clearly distinguished lymph node, the injection should be repeated before the study can be interpreted. To monitor the success of the injection, Kaplan (35) utilizes the following technique: immediately after injection the patient is placed in front of the detector in the lateral position with the empty syringe overlying the injection site as a marker. In the case of a superficial injection, the apposition between injection site and skin marker is apparent. In the case of a deep, intraperitoneal injection the freely distributed radiocolloid can be clearly recognized. The satisfactory injection is seen as an interstitial bolus clearly distinct and separated from the overlying skin marker.

In patients with breast carcinoma, we inject the subcostal site corresponding to the side of the tumor in the first instance. An image is obtained at 3 hours, the contralateral subcostal injection is carried out, and another image is obtained three hours later. The ipsilateral lymphatics are therefore studied with two images and the contralateral lymphatics with one image. At six hours a lateral image is obtained as well. If the initial three hour image does not demonstrate any lymph nodes, the ipsilateral injection is repeated at the time of the contralateral injection. By alternating injections cross drainage between parasternal lymphatic can be demonstrated, identifying situations in which contralateral lymphatics may be at risk through direct continuity with lymphatics of the tumor-bearing side (Figure 4). This technique also provides a means of examining in great detail lymphatic kinetics, and assists in identifying unusual anatomic variants such as the unilateral chain of lymphatics.

Another anatomic site currently being studied is the pararectal-internal iliac lymphatics demonstrable with the bilateral perianal injection of radiocolloid (Figure 5).

Figure 1.

An electron micrograph of technetium-99m antimony sulphide colloid taken with the Siemens Model 102 Transmission Electron Microscope.

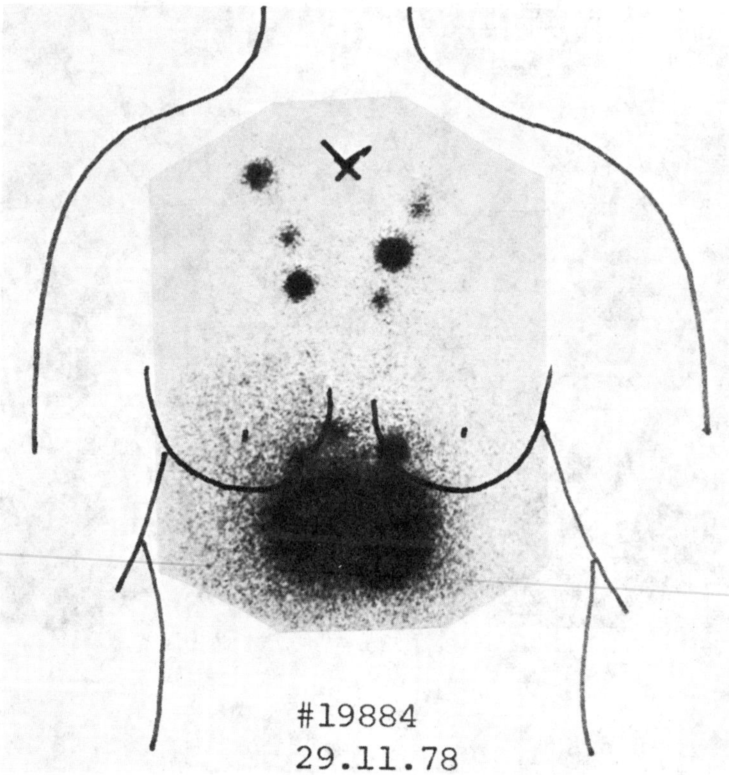

Figure 2.

Normal internal mammary lymphoscintigram 3 hours after the bilateral subcostal injection of ^{99m}Tc antimony sulphide colloid.

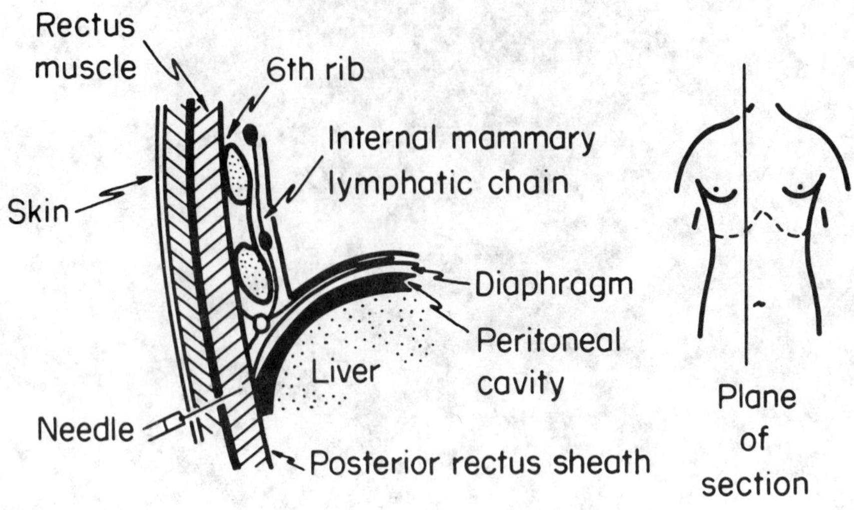

Figure 3.

Sagittal section of the anterior abdominal wall at the costal margin, showing site of injection of radiocolloid.

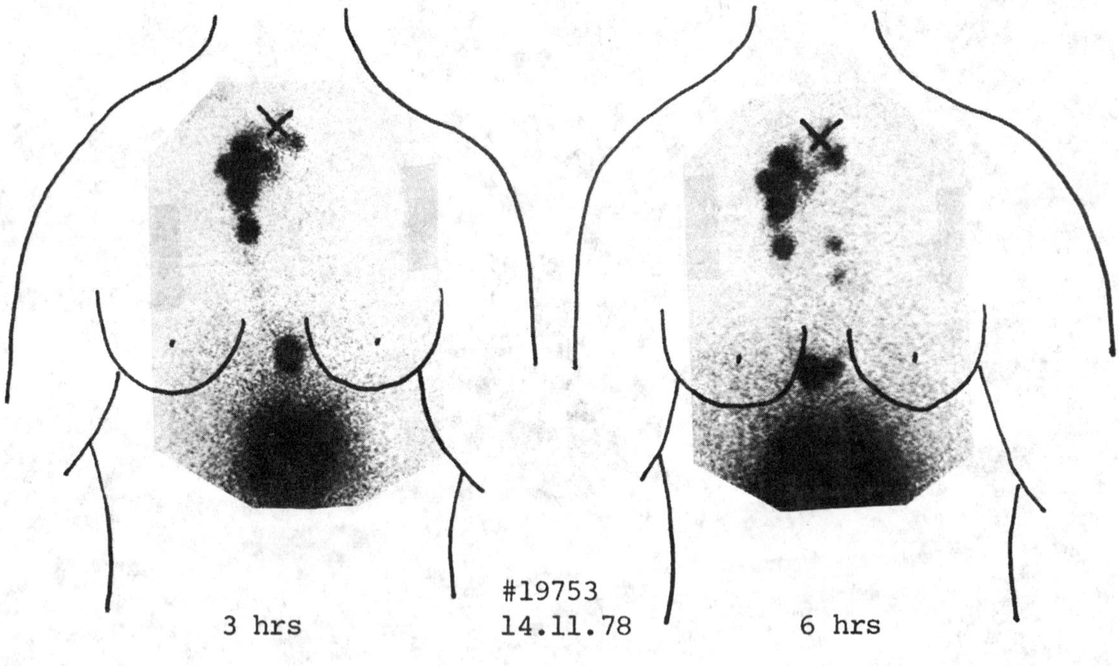

Figure 4.

Demonstration of sternomanubrial cross drainage following the unilateral right subcostal injection of radiocolloid. Proximal left parasternal nodes are visualized after the left subcostal injection.

RADIATION DOSIMETRY

The radiation dose to the injection sites and to individual lymph nodes has been estimated for internal mammary lymphoscintigraphy using the procedures detailed in the pamphlets published by the MIRD committee of the Society of Nuclear Medicine. We find considerable variation from patient to patient in the size of the volume around the injection site into which the injected activity diffuses. There is also considerable variation in the amount of activity which drains away from the injection site during 24 hours. For radiation dosimetry purposes we have calculated that the maximum dose to the injection site assuming all the injected activity remains at the injection site for 24 hours, contained in the smallest diffusion volume we have measured at 6 hours, could be as high as 100 rads. When the continuing expansion of the injection site volume is considered, together with the typical drainage rate observed, a more reasonable injection site dose estimate is 40 rads. The radiation dose to individual lymph nodes in the drainage chain is much lower, and only approaches the dose to the injection site if 4 or 5 percent of the injected activity is concentrated in a single node.

INSTRUMENTATION

At present the instrument of choice for internal mammary lymphoscintigraphy is a scintillation camera with a large field of view which can include the injection sites, xiphisternum and sternal notch markers, and axillary nodes in a single image. A general purpose low energy collimator is used which yields a scintillation camera efficiency of approximately 5 counts per second per microcurie for ^{99m}Tc, using a 20% energy window. Patiens are imaged in the supine position with the collimator parallel to the floor and as close to the sternum as possible.

Small spot markers (^{57}Co or ^{133}Ba) are used to indicate the positions of the sternal notch and xiphisternum, and an anatomical landmark device transfers these positions to the scintigram. We use a total imaging time of 450 seconds and see approximately 600,000 counts in the image obtained at 3 hours and approximately 1,000,000 counts in the image obtained at 6 hours. The oscilloscope intensity which we use is approximately the same as that which yields a satisfactory liver image with 1,000 counts per centimeter squared.

We have tested many types of film, and recommend the use of a low contrast type film such as Kodak PF. High contrast films tend to show individual lymph nodes with less intensity relative to the injection site. There is sufficient variation between different film types that clinical interpretation of the image can be influenced.

Because internal mammary lymphoscintigraphy has shown that lymph nodes often lie outside conventional radiation therapy portals (18), we have developed a system for locating individual lymph nodes accurately with respect to anatomical landmarks. The scintillation camera is interfaced to a remote PDP 11/40 computer and individual lymphoscintigrams are recorded as 64 x 64 matrices in the computer memory. The positions of the sternal notch and xiphisternum are obtained by imaging superimposed ^{133}Ba spot markers prior to the lymphoscintigram frame. These matrices are then processed with a computer program in which the individual lymph nodes are accurately located with respect to the sternal notch and xiphisternum (Figure 6).

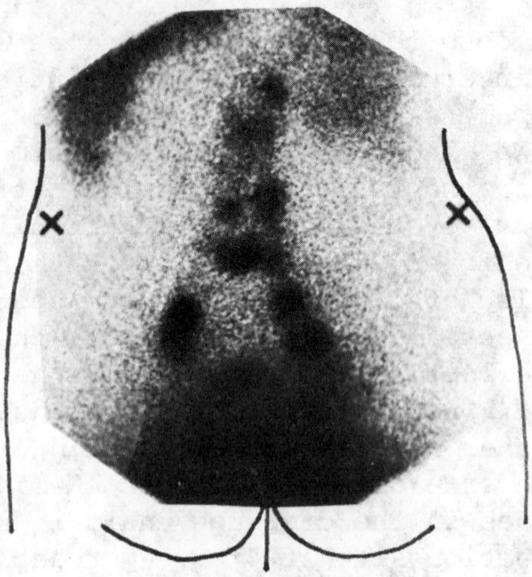

#16246
5.8.77

Figure 5.

Pararectal and internal iliac lymphatics visualized after bilateral perianal injection of radiocolloid.

#18313
25.4.78

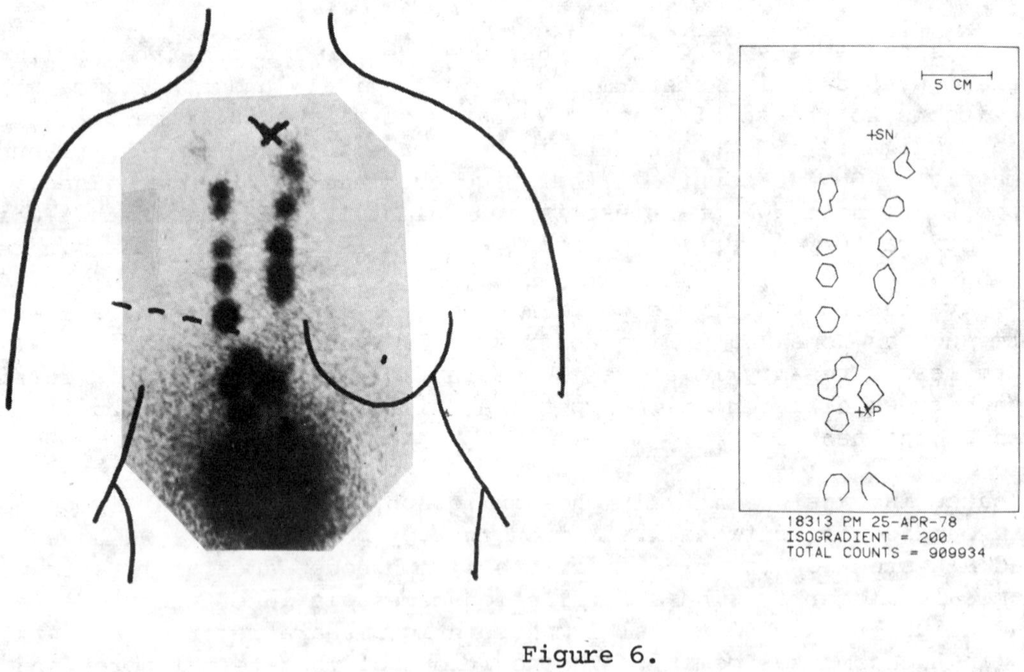

Figure 6.

Internal mammary lymphoscintigram with corresponding computerized isocontour. The isocontour, normally to full scale, provides accurate localization of individual lymph nodes.

PHYSIOPATHOLOGY

Transport of radiocolloid from any injection site and the subsequent sequestration within drainage lymph nodes reflects passive as well as active physiological processes and pathological disruption of these mechanisms. Immediately after injection there is egress of activity from the injection site, the speed of which depends upon the agent, the anatomic site under study, and the conditions of the study. Species differences should be borne in mind when data are extrapolated from an experimental model. By three hours after injection, activity within the lymph nodes has reached a plateau, and delayed images seldom contribute further anatomic or physiological information. ^{99m}Tc therefore serves as a suitable radionuclide both from the physical and physiological point of view.

Percentage of the radiocolloid retained in all visualized nodes averages less than 3% of the injected activity. Pathological processes affect lymph nodes through replacement and ultimate destruction of normal architecture, with decreasing radiocolloid capacity.

INTERPRETATION AND CLINICAL APPLICATION

Although the resolution capabilities of the radionuclide lymphoscintigram are limited, criteria have been established for distinguishing normal from abnormal lymphatic kinetics and lymph nodes (18).

The widest application of radiocolloid lymphoscintigraphy will be in the study of different lymphatic drainage areas for more accurate staging and management of patients with malignant disease (21). Visualization and localization of lymphatic structures ensures precision in planning radiation therapy in the dimensions relevant to the particular technique (36-37).

Radiocolloid introduced intracutaneously into the periphery of centrally located malignant melanoma nodules has provided an indication of the pathways of most likely spread, assisting the surgeon with the decision regarding the most relevant site of lymphatic extirpation (38).

The technique of interstitial radionuclide lymphoscintigraphy lends itself readily to the study of congenital lymphedema, especially in the pediatric age group where radiocontrast lymphography is difficult and inconvenient (39-41).

DISCUSSION

The ultimate objective of any diagnostic modality is to distinguish between a normal and abnormal state with reliability and reproducibility. Pearlman (42) compared paraaortic lymphoscintigraphy with contrast lymphography in 100 patients with lymphoma and found good correlation. Although there appear to be a small number of false positive lymphoscintigrams, false negative studies are unusual.

Duplicate internal mammary lymphoscintigrams carried out on 140 patients five days to three weeks apart have demonstrated 94% reproducibility, with both images essentially identical in the number of lymph nodes visualized and the intensity of radiocolloid uptake. In only 6% of the studies was there sufficient disparity between the two studies to give rise to differing

TABLE 3. CORRELATION OF ANATOMIC STUDIES AND LYMPHOSCINTIGRAPHY

Author	No. of Subjects	Methods	Mean per Subject	1st inter-space	2nd inter-space	3rd inter-space	4th inter-space	5th inter-space	Lower (Total)
Stibbe	60	Autopsy	8.5	202	201	203	18	25	280
Abrao	100	Autopsy	16.7	447	367	212	138	86	56
Ju	100	Autopsy	6.0	200	168	111	60	32	38
Soerensen	39	Autopsy	7.0	120	323	153	94	2	-
P.M.H.	100	Lympho-scintigraphy	7.3	202	146	107	57	103	34

(A) In the cases of Stibbe and Soerensen, the results have been calculated on the basis of 100 subjects for purposes of comparison.

(B) In the P.M.H. study all nodes below the 5th interspace were totaled together. This included diaphragmatic and xiphisternal nodes.

(C) References to support these figures may be found in reference 18 (Ege 1976).

interpretations. The anatomic validity of the internal mammary lymphoscintigram has also been established in comparison with series based on cadaver dissections by four authors (Table 3) (18).

In the absence of any other diagnostic modality applicable to the parasternal lymphatics, the sensitivity and specificity of the internal mammary lymphoscintigram must be evaluated in the context of histological examination. In a study by Matsuo (43), lymphoscintigrams utilizing ^{198}Au colloid were carried out preoperatively in 106 patients with breast carcinoma who subsequently underwent extended radical mastectomy, and correlated with histological examination of the parasternal nodes. 90% true positive and 100% true negative results are reported. In a recent study using ^{99m}Tc sulphide colloid, Aspegren et al (44) carried out preoperative lymphoscintigrams and histologically examined 19 parasternal nodes from 6 patients undergoing modified radical mastectomy for breast carcinoma. They report a high incidence of false positive results (7/16) and conclude that the study must be repeated with a small particle colloid before the diagnostic value of the procedure can be fully assessed.

In a large group of patients with breast carcinoma, studied with internal mammary lymphoscintigraphy and followed for up to three years, we found good correlation with clinical stage and prognosis, and results commensurate with those based on operative and postmortem examination of the parasternal lymph nodes (20). In approximately 5000 studies carried out in over 4000 patients with malignant disease, no complication has been encountered.

Currently available noninvasive ultrasonographic and CT techniques cannot distinguish normal from abnormal lymphatic structures (45). It is anticipated that with recent developments in radiopharmaceuticals and imaging techniques, there will be a resurgence in diagnostic radionuclide lymphoscintigraphy as a simple, safe, reliable procedure contributing to patient investigation and management.

ACKNOWLEDGEMENTS

Without the congenial collaboration of the entire staff of the Department of Nuclear Medicine, this subject could not have been developed. We are particularly grateful to Mrs. Elizabeth Capon for her timely and unique assistance in the preparation of the manuscript. We are also grateful to the editors of "Radiology" for permission to reprint Figure 3 and Table 3, which previously appeared in Radiology 118:101-107, January 1976.

REFERENCES

1. Sherman Al, Ter-Pogossian M: Lymph node concentration of radioactive colloidal gold following interstitial injection. Cancer 6:1238-1240, Nov 1953

2. Seaman W, Powers WE: Studies in the distribution of radioactive colloidal gold in regional lymph nodes containing cancer. Cancer 8:1044, 1955

3. Hultborn KA, Larsson LG, Ragnhult I: The lymph drainage from the breast to the axillary and parasternal lymph nodes, studied with the aid of colloidal ^{198}Au. Acta Radiol 43:52-64, 1955

4. Thomas CG Jr: Lymphatic dissemination of radiogold in the presence of lymph node metastases. Surg Gyn Obst 103:51, 1956

5. Lang EK: Demonstration of blockage and involvement of the pelvic lymphatic system by tumor, with lymphangiography and scintiscinogram. Radiol 74:71-73, 1960

6. Delaloye B, Magnenat P: La lymphographie indirecte et son intérêt. Medical Radioisotope Scanning, Vol II, IAEA, Vienna 1964, p 227

7. Winkel K zum: Zur technik der indirekten abdominellen lymphknoten-szintigraphie mit ^{198}Au colloidale. Nuclearmedizin 3:148-152, May 1963

8. Schenck P: Scintigraphische darstellung des parasternalen lymphsystems. Strahlentherapie 130:504-508, May-Aug 1966

9. Diethelm L, Buchwald W, Haas J-P et al: Erweiterung der diagnostik des mammakarzinoms mit helfe von isotopen. Strahlentherapie 131:69-78, Sep-Dec 1966

10. Rossi R, Ferri O: La visualizzazione della catena mammaria interna con ^{198}Au. Presentazione de una nuova metodica: la linfoscintigrafia. Minerva Med 57:1151-1155, Mar 1966

11. Kazem I, Nedwich A, Mortel R, Honda T: Comparative histological changes in the normal lymph node following ethiodol lymphography and colloidal gold-198 lymph scanning. Clin Radiol 22:382-388, 1971

12. Langhammer H, Hor G, Pabst HW, Hertel E: Scanning mediastinal lymph nodes. J Nucl Med 10: 749-751, 1969

13. Breit A, Czempiel H, Koernig-Kron K: Die lymphszintigraphie in der tumortherapie (Erfahrung an über 800 tumorpatienten). Strahlentherapie 138: 74-84, 1969

14. Kazem I, Brady LW, Croll MN, Faust DS, Wolferth CCW: The parasternal lymph node scan as a prognostic test in breast carcinoma. Radiol 92:617-619, 1969

15. Cox PH: Dynamic studies of lymph drainage using labelled human serum albumin. Proc 12th Ann Meeting Jugoslav Soc Nucl Med Ohrid, pp 76-82, 1972

16. Schenck P, Winkel K zum, Becker J: Die szintigraphie des parasternalen lymphsystems. Nucl Medizin 5:388-396, 1966

17. Anghileri LJ, Harbst H: A new radiocompound with possibilities in lymphographic studies: radiosilver oleate. J Nucl Biol Med 11:107-110, 1967

18. Ege GN: Internal mammary lymphoscintigraphy - the rationale, technique, interpretation and clinical application: a review based on 848 cases. Radiol 118:101-107, Jan 1976

19. Cox PH: The preparation of mercuric sulphide suspension for lymphatic investigations and the influence of particle size on its physiological properties. Radioisotopy 12:997-1010, 1971

20. Ege GN: Internal mammary lymphoscintigraphy in breast carcinoma: a study of 1072 patients. Int J Radiat Oncol Biol Phys 2:755-761, 1977

21. Ege GN: Internal mammary lymphoscintigraphy: a rational adjunct to the staging and management of breast carcinoma. Clin Radiol 29:453-456, 1978

22. Garzon OL, Palcos MC, Radicella R: A technetium-99m labelled colloid. Int J Appl Rad Isotop 16:613, 1965

23. Warbick A, Ege GN, Henkelman RM, Maier G, Lyster DM: An evaluation of radiocolloid particle sizing techniques. J Nucl Med 18:827-834, 1977

24. Strand SE, Persson RBR: Radiopharmacological and kinetic studies of various ^{99m}Tc colloids in the parasternal lymph nodes of rabbits. Presented at First International Symposium of Radiopharmacology, Innsbruck, Austria, May 1978

25. Ege GN, Warbick A: Lymphoscintigraphy - a comparison of ^{99m}Tc antimony sulphide colloid and ^{99m}Tc stannous phytate. Brit J Radiol - in press

26. Kaplan WD: Personal communication

27. Wilkins DJ: Interaction of charged colloids with the RES. Advances in Experimental Medicine and Biology, Vol 1, Editors: De Luzio and Paoletti, 1971, pp 25-31

28. Biozzi G: The comparative effect of certain colloids on the phagocytosis of other colloids by cells of the reticuloendothelial system and the phenomena of phagocytic preference. Reticuloendothelial Soc Bull 3:31, 1957

29. Chaudhuri TK, Evans TC, Chaudhuri TK: Autoradiographic studies of distribution in the liver of ^{198}Au and ^{99m}Tc-sulfur colloids. Radiol 109: 633-637, 1973

30. Wagner HN, Iio M: Studies of the reticuloendothelial system (RES) III. Blockade of the RES in man. J Clin Invest 43:1525-1532, 1964

31. Anghileri LJ: Lymph nodes distribution of several radiocolloids: migration ability through the tissues. J Biol Nucl Med 10:180-184, 1967

32. Ludwig J: Trapping of calibrated microspheres in rat lymph nodes. Lymphol 1:18-24, 1971

33. Walker LA: Localization of radiocolloids in lymph nodes. J Lab Clin Med 36:440-449, 1950

34. Chan ASK: Personal communication

35. Kaplan WD: Personal communication

36. Bronskill MJ, Harauz G, Ege GN: Computerized internal mammary lymphoscintigraphy in radiation treatment planning of patients with breast carcinoma. Int J Radiat Biol Oncol Phys - in press

37. Rose CM, Kaplan WD, Marck A: Lymphoscintigraphy of the internal mammary lymph nodes. Int J Radiat Oncol Biol Phys 2:102, 1977 (Abstract)

38. Robinson DS, Sample WF, Fee HJ, Holmes EC, Morton DL: Regional lymphatic drainage in primary malignant melanoma of the trunk determined by colloidal gold scanning. Surg Forum 28:147-148, 1977

39. Gates GF, Dore EF: Primary congenital lymphedema in infancy evaluated by isotope lymphangiography: case report. J Nucl Med 12:315, 1971

40. Vieras F, Boyd CM: Radionuclide lymphangiography in the evaluation of pediatric patients with lower-extremity edema. J Nucl Med 18:441, 1977

41. Jackson FI, Bowen P, Lentle BC: Scintilymphangiography with ^{99m}Tc antimony sulfide colloid in hereditary lymphedema (Nonne-Milroy disease). Clin Nucl Med 3:296-298, Jul 1978

42. Pearlman AW: Abdominal lymph node scintiscanning with radioactive gold (^{198}Au) for evaluation and treatment of patients with lymphoma. Am J Roentgen 109:780-792, 1970

43. Matsuo S: Studies on the metastasis of breast cancer to lymph nodes - II Diagnosis of metastasis to internal mammary nodes using radiocolloid. Acta Med Okayama 28:361-371, 1974

44. Aspegren K, Strand SE, Persson BRR: Quantitative lymphscintigraphy for detection of metastases to the internal mammary lymph nodes - biokinetics of ^{99m}Tc sulphur colloid uptake and correlation with microscopy. Acta Radiol Oncol 17:17-26, 1978

45. Munzenrider JE, Tchakarova I, Castro M, Carter B: Role of CT scan in radiation treatment planning for patients with breast cancer. Int J Radiat Oncol Biol Phys (Suppl 2) 2:148-149, 1977

46. Davis MA, Jones AG, Trindade H: A rapid and accurate method for sizing radiocolloids. J Nucl Med 15:923-928, 1974

47. Szymendera J, Zoltowski T, Radwan M, Kaminska J: Chemical and electron microscope observations of a safe PVP-stabilized colloid for liver and spleen scanning. J Nucl Med 12:212-215, 1971

48. Subramanian G, McAfee JG, Mehter A, Blair RJ, Thomas FD: ^{99m}Tc stannous phytate: a new in vivo colloid for imaging the reticuloendothelial system. J Nucl Med 14:459, 1973

49. Chan ASK, Warbick A, Ege GN, Ottensmeyer FP: Iodinated E. Coli 70S ribosomes as a radiocolloid of uniform particle size for lymph node and liver scanning. J Nucl Med 18:1034-1036, 1977

50. Ryman BE, Jewkes RF, Jeyasingh K, Osborne MP, Patel HM, Richardson VJ, Tattersall MHN, Tyrell DA: Lipid vesicles as carriers for introducing drugs and other biologically active materials into cells. II In vivo systems. Annals N. Y. Acad Sci 308:281-307, 1978

THE SYNTHESIS OF [^{18}F]-5-FLUOROURIDINE (F-18-5-FUR) AS A PROBE FOR MEASURING RNA SYNTHESIS AND TUMOR GROWTH RATES *IN VIVO*

Chyng-Yann Shiue, Joanna S. Fowler
Robert R. MacGregor, and Alfred P. Wolf

Chemistry Department
Brookhaven National Laboratory
Upton, New York 11973

ABSTRACT

A method for the rapid synthesis of high specific activity of [^{18}F]-5-fluorouridine is described. The $^{20}Ne(d,\alpha)^{18}F$ nuclear reaction is used to produce high specific activity, anhydrous [^{18}F]-F_2 at the Brookhaven National Laboratory 60" cyclotron. Fluorination of 2',3',5'-tri-*O*-acetyluridine with [^{18}F]-F_2 in glacial acetic acid at room temperature followed by hydrolysis with sodium methoxide in methanol gives [^{18}F]-5-fluorouridine with a radiochemical yield of 5-7% in a synthesis time of 90 minutes from EOB. The compound is required for the study of RNA synthesis and tumor growth rates *in vivo*.

INTRODUCTION

Labeling of organic compounds with positron emitting nuclides where the labels are in non or in minimally bioactive positrons so as not to alter the functionality of the compound, has become especially attractive with the advent of the new positron emission tomographs. The localization of these compounds or more importantly their involvement in specific metabolic pathways can serve as a probe for metabolism *in vivo*. This principle of "metabolic trapping" has been used to design and synthesize the needed radiopharmaceutical. Of particular interest in this research is the preparation of a nucleotide or nucleoside labeled with a positron emitting nuclide and its use to study ribonucleic acid synthesis *in vivo*. The uracil analog, 5-fluorouracil (5-FU) has been used as an antineoplastic agent for the treatment of a variety of disseminated solid tumors, especially colorectal, gastrointestinal, and breast, with about 30% of the patients showing objective tumor remission (1-3). The mechanism of anti-tumor action of 5-FU is well documented (4). It is principally due to a competitive inhibition of thymidylate synthetase by its metabolite 5-fluoro-2'-deoxyuridine-5'-monophosphate. This produces thymidylic acid deficiency and thus blocks the synthesis of DNA. In mammalian and bacterial systems, 5-FU is converted to its ribonucleoside (5-FUR) by uridine phosphorylase. The ribonucleoside is then phosphorylated to its nucleotide, 5-fluorouridine-5'-monophosphate (5-FUMP) by uridine kinase. Formation of 5-FUMP may also result from direct conversion of 5-FU to 5-FUMP by uracil phosphoribosyltransferase. 5-FUMP is then readily incorporated into RNA (5). Therefore, if we label 5-FUMP or its precursors with a positron emitting nuclide, such as ^{18}F, it might be feasible to measure RNA synthesis in certain organs which might be applicable to tumor RNA synthesis measurements *in vivo*.

[^{18}F]-5-Fluorouridylate (^{18}F-5-FUMP) was recently synthesized enzymatically (6) from [^{18}F]-5-fluorouracil (^{18}F-5-FU)(7) and the tissue uptakes of these two compounds were studied. It was shown that the nucleotide was incorporated into most of the tissues studied to a greater extent than the free base (6). This was particularly evident in tissues with a known high rate of RNA synthesis such as the spleen and small intestine and agreed with the findings of Heidelberger that 5-fluorouridine was incorporated into RNA more efficiently than 5-fluorouracil (8) and that 5-fluorouridylate was dephosphorylated *in vivo* to give 5-fluorouridine (9). Therefore, it was of interest to synthesize ^{18}F-5-fluorouridine by chemical methods for use as a probe for measuring RNA synthesis and tumor growth rates *in vivo*.

METHODS

The target, consisting of neon (Matheson Research Grade) containing 0.1% (60 μmol) of fluorine (F_2) carrier, was irradiated with deutrons at the Brookhaven National Laboratory 60" cyclotron (10). The ^{18}F-labeled fluorine was produced from the $^{20}Ne(d,\alpha)^{18}F$ nuclear reaction. The beam was degraded from 13.6 to 0 MeV in the target. Approximately 60-80% of the ^{18}F-F_2 produced was slowly purged through the target chamber (flow rate ∿ 45 ml/min, total time 25-30 min) into the solution of 5.88 mg (15.9 μmol) of 2',3',5'-tri-*O*-acetyluridine (**1**) (Sigma) in 6 ml of glacial acetic acid in a reaction vessel (11) at room temperature. The reaction mixture was then transferred to a round-bottom flask, allowed to stand at room temperature for 10 min and evaporated *in vacuo* to give [^{18}F]-2',3',5'-tri-*O*-acetyl-5,6-dihydro-5,6,-difluorouridine (**2**). Compound **2** was dissolved in 10 ml of sodium methoxide-methanol solution and evaporated to dryness. The residue was dissolved in water for chromatographic separation of [^{18}F]-5-FUR.

Chromatographic separations were obtained on cation exchange column (AG 50 X-8, H^+ form), eluted with water. The eluate was evaporated <u>in vacuo</u> and the residue was extracted with ethyl acetate:acetone:water (v/v 70/40/5) and then passed through a silica gel column, eluted with same solvent system, and the eluate was evaporated <u>in vacuo</u> to give 1.11 mg of [^{18}F]-5-FUR (**3**)(53.5% chemical and 6.6% radiochemical yield).

RESULTS AND DISCUSSION

5-Fluorouracil, 5-fluorouridine, and 5-fluoro-2'-deoxyuridine have been synthesized by direct fluorination of uracil, 2',3',5'-tri-O-acetyluridine and 3',5'-di-O-acetyl-2'-deoxyuridine with trifluoromethyl hypofluorite (12). However, this method is not suitable for the synthesis of high specific activity of [^{18}F]-labeled fluoropyrimidines. Recently, these fluoropyrimidines were synthesized by fluorination of the pyrimidines with molecular fluorine (13,14). Since we are able to produce very high specific activity [^{18}F]-F_2, we have synthesized [^{18}F]-5-fluorouridine by direct fluorination of 2',3',5'-tri-O-acetyluridine (**1**) with [^{18}F]-F_2.

Fluorination of (**1**) with [^{18}F]-F_2 in glacial acetic acid gave a clean reaction product (**2**) as shown in Fig. 1. There are several reagents used to hydrolyze the acetyl groups, such as triethylamine (12,14) or ammonia (13). We, however, find that sodium methoxide in methanol is more efficient than tertiaryamines. Hydrolysis of **2** with NaOMe/MeOH followed by passage of the mixture in water through a cation exchange resin and silica gel column results in the production of [^{18}F]-5-FUR (**3**) in a radiochemical yield of 5-7% in a synthesis time of 90 minutes from EOB (equation 1). A silica gel column was used to separate [^{18}F]-fluoride and [^{18}F]-5-FUR, as shown in Fig. 2.

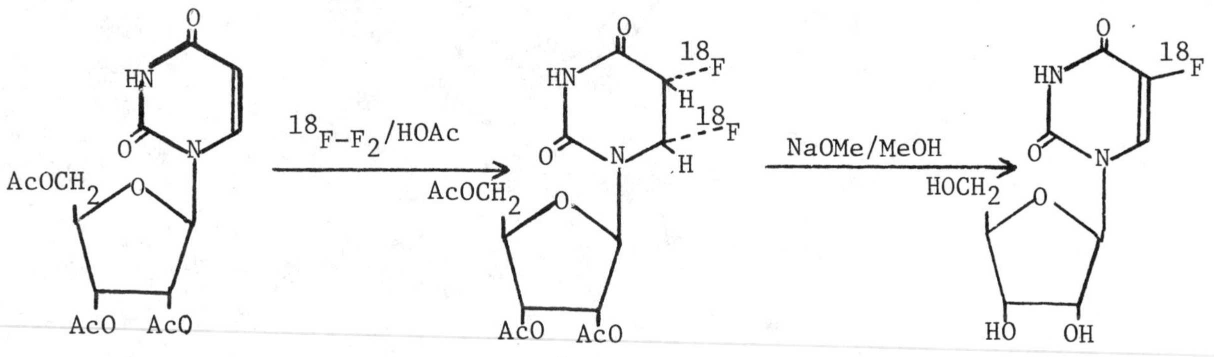

1 **2** **3**

In order to achieve the maximum radiochemical yield of [^{18}F]-5-FUR (**3**) and without the contamination of uridine, the molar ratio of [^{18}F]-F_2 with substrate **1** should keep between 2.5-3.0. The purity of compound **3** was checked by tlc and UV.

This convenient method thus simplifies the production of [^{18}F]-5-FUR (**3**) and eliminates the possibility of macromolecular contamination introduced in the enzyme catalyzed synthetic sequence. It also can readily be used to provide

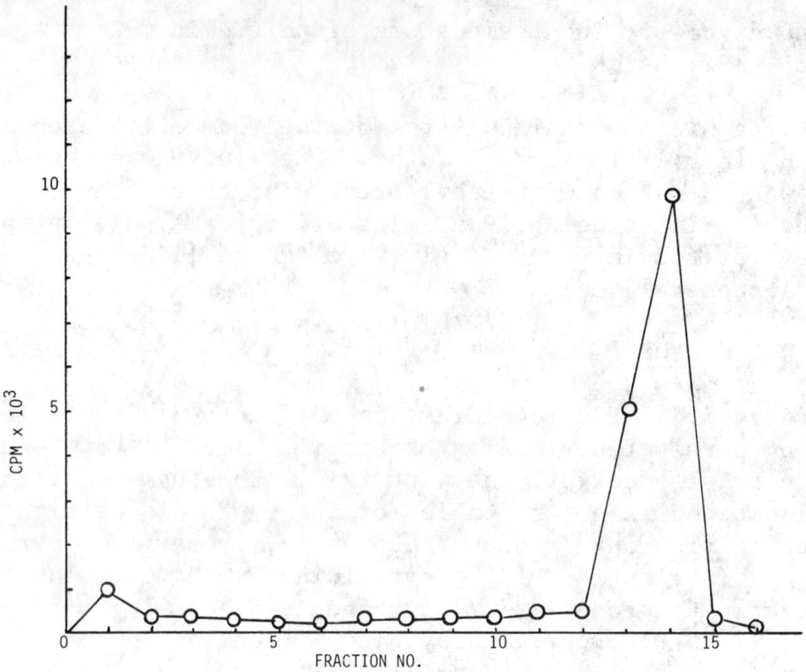

Fig. 1. Radiochromatograms of the reaction mixture of 2',3',5'-tri-O-acetyluridine with $^{18}F-F_2$.

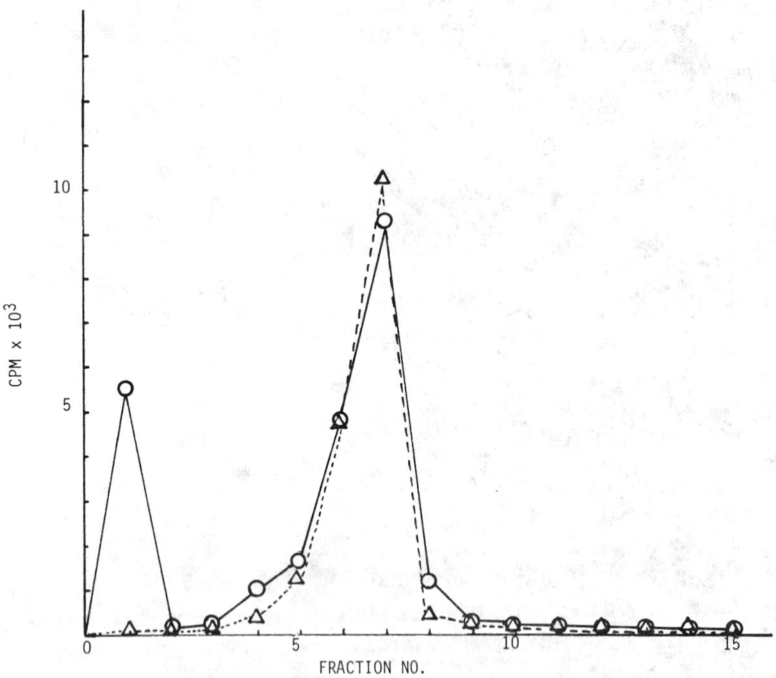

Fig. 2. Radiochromatograms of ^{18}F-5-FUR (-O-O-O-) before silica gel column; (--Δ-Δ-Δ-) after silica gel column. Separation were achieved by eluting a silica gel column with ethyl acetate:acetone:water (70:40:5).

compound 3 with very high specific activity. Indeed, the specific activity of 3 depends on total irradiation dose (Table 1). [^{18}F]-5-FUR is currently being used as a probe for measuring RNA synthesis and tumor growth rates *in vivo*.

ACKNOWLEDGEMENTS

Research carried out at Brookhaven National Laboratory under contract with the U.S. Department of Energy and supported by its Office of Basic Energy Sciences and the Office of Biomedical and Environmental Sciences.

REFERENCES

1. Davis, H. L., Ramirez, G., and Ansfield, F. J.: Adenocarcinomas of stomach, pancreas, liver, and biliary tracts. *Cancer* *33*: 193 (1974).

2. Regelson, W.: Chemotherapy of gastrointestinal cancer, in *Cancer Chemotherapy*, Brodsky, I., Kahn, S. B., and Moyer, J. H., eds., New York Grune and Stratton, 1967, 84.

3. Gailani, S., Holland, E. F., Falkson, G., et al.: Comparison of treatment of metastatic gastrointestinal cancer with 5-fluorouracil (5-FU) to a combination of 5-FU with cytosine arabinoside. *Cancer* *29*: 1308 (1972).

4. Heidelberger, C.: Cancer Chemotherapy with purine and pyrimidine analogs. *Ann Rev. Pharmacol*. *7*: 101 (1967).

5. Heidelberger, C.: The nucleotides of fluorinated pyrimidines and their biological activities, in *Carbon-Fluorine Compounds* (Ciba Foundation Symposium), New York, Associated Scientific Publishers, 1972, 125.

6. Crawford, E. J., Friedkin, M., Fowler, J., et al.: ^{18}F-5-fluorouridylate: A positron-emitting tracer of *in vivo* RNA metabolism. *J. Nucl. Med.*, in press.

7. Fowler, J. S., Finn, R. D., Lambrecht, R. M., et al: The synthesis of ^{18}F-5-fluorouracil. VII. *J. Nucl. Med*. *14*: 63 (1973).

8. Chaudhuri, N. K., Montag, B. J., and Heidelberger, C.: Studies on fluorinated pyrimidines. III. The metabolism of 5-fluorouracil-2-^{14}C and 5-fluoroorotic-2-^{14}C acid *in vivo*. *Cancer Res*. *18*: 318 (1958).

9. Heidelberger, C., Kaldor, G., Mukherjie, K. L., et al.: Studies on fluorinated pyrimidines. XI. *In vitro* studies on tumor resistance. *Cancer Res*. *20*: 903 (1960).

10. Casella, V., Ido, T., and Wolf, A. P.: Anhydrous ^{18}F-F_2 for radiopharmaceutical preparation. Manuscript in preparation.

11. Ido, T., Wan, C-N., Casella, V., et al.: Labeled 2-deoxy-D-glucose analogs. ^{18}F-labeled 2-deoxy-2-fluoro-D-glucose, 2-deoxy-2-fluoro-D-mannose, and ^{14}C-2-deoxy-2-fluoro-D-glucose. *J. Labelled Cmpds and Radiopharm*. *14*: 175 (1978).

12. Robins, M. J., MacCoss, M., Naik, S. R., et al.: Nucleic Acid Related Compounds. 21. Direct fluorination of uracil and cytosine bases and nucleosides using trifluoromethyl hypofluorite. Mechanism, stereochemistry, and synthetic applications. J. Am. Chem. Soc. 98: 7381 (1976).

13. Cech, D., Mernert, H., Etzold, G., et al.: Uber die reaktion von uracil und seinen nucleosiden mit elementarem fluor. J. Prakt. Chem. 315: 149 (1963).

14. Cech, D. and Holy, A.: Preparation of some 2'-deoxy-5-fluorouridine derivatives by a direct fluorination. Coll. Czech Chem. Commun. 41: 3335, (1976).

TABLE 1. IRRADIATION DOSE DEPENDENCE OF THE SPECIFIC ACTIVITY OF $[^{18}F]$-5-FUR

Run No.	Compound 1 (μM)	$[^{18}F]-F_2$ Activity (mCi)	Specific Activity of $[^{18}F]$-5-FUR (μCi/μM)
5	13.8	16.4	94.7
6	13.5	8.6	50.1
7	13.9	8.5	58.8
10	17.5	48.5	424
11	18.1	16.9	128
12	15.9	102.9	642

EVALUATION OF RADIOLABELED RUTHENIUM COMPOUNDS AS TUMOR-LOCALIZING AGENTS*

S. C. Srivastava, P. Richards,
G. E. Meinken, P. Som, H. L. Atkins
Brookhaven National Laboratory
Upton, New York 11973

S. M. Larson, Z. Grunbaum, J. S. Rasey
V. A. Hospital
Seattle, Washington 98108

M. Dowling, M. J. Clarke
Boston College
Chestnut Hill, Massachusetts 02167

ABSTRACT

This work introduces a new class of radiopharmaceuticals based on ruthenium-97. The excellent physical properties of Ru-97, the high chemical reactivity of Ru, the potential antitumor activity of several Ru coordination compounds, and BLIP production of Ru-97, provide a unique combination for the application of this isotope in nuclear oncology. A systematic study was undertaken on the synthesis, characterization, and evaluation of a number of ruthenium-labeled compounds. In a variety of animal tumor models, several compounds show considerable promise as tumor-localizing agents when compared to gallium-67 citrate. The compounds studied (with Ru in different oxidation states) include ionic Ru, a number of hydrophilic and lipophilic chelates, and various ammine derivatives.

*This work was supported by the following: Brookhaven National Laboratory - Contract No. EY-76-C-02-0016 with the UNITED STATES DEPARTMENT OF ENERGY; VAH, Seattle - Medical Research Service, Veterans Administration; Boston College - grant from the American Cancer Society, MA Div.

INTRODUCTION

Diagnosis of cancer at early stages using noninvasive nuclear medicine procedures is highly desirable but so far has shown limited potential because of the lack of suitable radiopharmaceuticals based on radionuclides with optimal physical and chemical properties. Even though gallium-67 citrate is presently the agent of choice, it does not concentrate selectively in tumors, the background levels remain high, and its imaging characteristics are far from ideal. As clinical trials have progressed, it has become evident that gallium-67 citrate concentrates in a variety of normal structures and in several diverse benign and malignant pathologic lesions. This lack of specificity has made the interpretation of gallium-67 scans difficult.

In general, slow kinetics of uptake of the radiopharmaceuticals by cancerous tissue necessitates the use of intermediate half-life nuclides, and for good quality scans, a radiolabel with optimum physical and biological properties is required. Ruthenium-97 is a potential radionuclide for applications in nuclear oncology because of its excellent physical properties -- an essentially monoenergetic gamma emission of 216 keV, and a half-life of 2.9 days -- and the high chemical reactivity of the element ruthenium (1). Because of the relatively high fission yield of ruthenium nuclides, various studies of biological interest have been reported in the literature. Usefulness of various ruthenium nuclides in nuclear medicine has been pointed out by some authors (2); possible application of ruthenium-97 was first suggested by Subramanian et al in 1970 (3). Therapeutic potential of ruthenium compounds has also been described (4,5).

A method was recently developed for an efficient production of clinically useful quantities of ruthenium-97 using the Brookhaven Linac Isotope Producer (BLIP) (6). In an ongoing program at Brookhaven, systematic studies were undertaken to exploit the chemistry of ruthenium for the design of new radiopharmaceuticals for a number of applications in nuclear medicine. A variety of representative agents (some developed and others under active development) is described in Table 1.

A major thrust has recently been on the development of tumor-localizing agents. Several classes of compounds (with ruthenium in various oxidation states) including ionic ruthenium, hydrophilic and lipophilic chelates and a variety of ruthenium-ammine derivatives were prepared using ruthenium tracer, and evaluated in normal and tumor-bearing animals. Several of these compounds have shown antitumor activity in recent NCI animals screens (5). Results with the more promising ruthenium-labeled tumor-localizing agents are reported.

MATERIALS AND METHODS

Most initial investigations were done using ruthenium-103 due to its convenient 39.6 day half-life. Ruthenium-103 chloride was obtained as a carrier-free solution in 3.5N hydrochloric acid from the Oak Ridge National Laboratory. Ruthenium-97 was prepared as needed at the BLIP from proton spallation of high purity (> 99.9%) rhodium foils (6). The target was bombarded with 200-MeV protons from the Linac. After the bombardment, the target was transferred

TABLE 1

RADIOLABELED RUTHENIUM COMPOUNDS AND THEIR POTENTIAL APPLICATIONS IN NUCLEAR MEDICINE

Compounds	Application
Simple salts (ionic)	
Ruthenium chloride	
Potassium *tris*-oxalatoruthenium(III)	Tumor localization;
Potassium pentachloroaquoruthenate(III)	myocardial agents
Hydrophilic chelates	
Tartrate	Kidney agents;
Citrate	tumor localization;
DTPA	cisternography
Glucoheptonate	
Lipophilic chelates	
8-hydroxyquinoline(oxine)	Hepatobiliary
Oxine 7-carboxylic acid	agents; labeling cells
Oxine 7-carboxylic acid acetate	and platelets, etc.; study
1,10-phenanthroline derivatives	of cellular transport
β-diketones	mechanisms; brain perfusion;
Lidocaine iminodiacetic acid(HIDA) and analogs	tumor localization
Dimethylsulfoxide	
Phosphate compounds	
Pyrophosphate	Bone agents (for normal
EHDP	uptake and for physiological
EDTMP	studies); agents for
MDP	myocardial-infarct localization
Ammine compounds	
Chloropentaammineruthenium(III)chloride	Tumor localization
cis-Dichlorotetraammineruthenium(III) chloride	
cis-Dichlorobis(ethylenediamine)-ruthenium(III) chloride	
Trichlorotriammineruthenium(III)	
Ammineruthenium(III) complexes	
Purines, pyrimidines, nucleosides, nucleotides	Tumor localization
Colloidal Preparations	
Ruthenium-sulfur colloid	Lymphoscintigraphy

to a processing hot cell and dissolved by a.c. electrolysis in 6N hydrochloric acid (0.3A/cm^2, ~15 hr). After evaporating the solution to near dryness, radiochemical separation of ruthenium-97 was achieved by distillation of RuO_4 from a sulfuric acid medium in the presence of an oxidizing agent (permanganate). The distillate was collected in ice-cold ethanol/hydrochloric acid (1:1). Recovery of ruthenium from the target was almost quantitative.

The complex-forming reactions were optimized with regard to different variables -- total ligand, time of reaction, temperature, pH, etc. In general, due to slow initial kinetics, a heating period of 10 to 30 min at 90° was included in most preparations as a last step. In-vitro analysis of all agents was performed using paper and thin-layer chromatography, gel filtration, cellulose acetate electrophoresis, and spectrophotometry and/or elemental analysis, when appropriate. A typical method of preparation was as follows. To a suitable quantity of the tracer (^{103}Ru or ^{97}Ru, both available in 3.5N HCl) neutralized to pH 2, an appropriate amount of the ligand was added, and following an incubation period of 5-20 min, the pH was raised to between 5 and 7.5. The resulting solution was heated on a boiling water bath for 10-30 min. In some preparations, 0.05-50 mg ruthenium was added to study the carrier effect on biological distribution. The complexes with 1,10-phenanthroline derivatives were prepared by reducing the ruthenium solution with hydroxylammonium chloride in the presence of the ligand, and subsequent heating for 30 min after pH adjustment. Oxine (8-hydroxyquinoline) 7-carboxylic acid acetate was crystallized from a solution of oxine 7-carboxylic acid in acetic acid. Radiosynthesis of potassium pentachloroaquoruthenate(III), potassium tris-oxalatoruthenium(III), chloropentaammineruthenium(III) chloride, cis-dichlorotetraammineruthenium(III)chloride, cis-dichloro bis(ethylenediamine)-ruthenium(III) chloride, and several other ammine complexes was carried out using suitably modified literature procedures (7).

Organ distribution data on the various compounds were obtained in normal mice, rats and hamsters. Biodistribution studies in tumor-bearing animals were carried out in the following tumor models: (i) EMT-6 sarcoma in Balb/c mice (8), (ii) squamous-cell carcinoma in Fisher CDF/female rats (9), and (iii) Greene melanoma in Golden Syrian hamsters (10).

RESULTS

For most ruthenium complexes prepared without added carrier, cellulose acetate electrophoresis in a borate/NaOH/KCl buffer of pH 8 was found to be the analytical method of choice. Paper and thin-layer chromatography also provided good resolution. Gel filtration generally did not give satisfactory separations. The ammine compounds were characterized spectrophotometrically and by elemental analysis.

Blood clearance data on various ruthenium compounds were obtained in normal dogs. Ruthenium chloride (pH 2, a mixture of 3 and 4 oxidation states) clears (Figure 1) from the blood slowly and at 30 minutes after i.v. injection, ca. 40% of the activity remains in circulation. At later time periods, most of the activity remains diffused over various parts of the body and the muscle uptake is quite high (~20% in mice). Upon injection of hydrophilic chelates

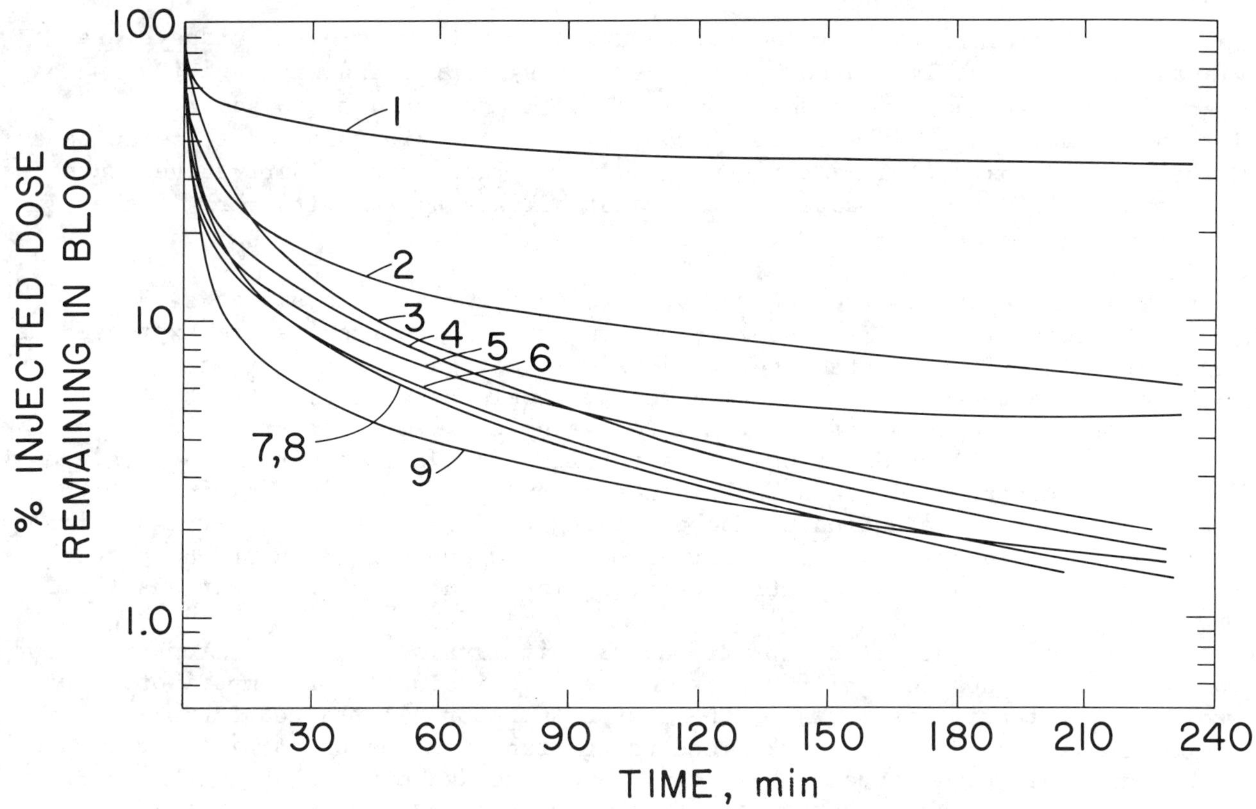

FIG.1. Blood clearance of several ruthenium complexes in dogs. Technetium-99m-DTPA and technetium-99m-MDP are included for comparison. (1) Ruthenium chloride, pH 2; (2) Ru-EHDP; (3) Ru(II)tris-(3,4,7,8-tetramethyl 1,10-phenanthroline)dichloride; (4) Technetium-99m-DTPA; (5) Ru-citrate; (6) Ru-EDTMP; (7) Ru-MDP; (8) Technetium-99m-MDP; (9) Ru-DTPA.

such as ruthenium citrate, ruthenium tartrate or ruthenium-DTPA, kidneys are visualized immediately. Blood clearance is very rapid, and within 5-10 minutes, most of the compound leaves the blood via the kidneys and the bladder. Ruthenium phosphate complexes clear rapidly from the blood and localize in bone. Complexes with oxine, phenanthroline, and iminodiacetic acid derivatives such as HIDA and PIPIDA, are extracted by the hepatobiliary system with very little excretion in the urine.

Table 2 describes representative data on the tissue uptake of several ruthenium compounds in EMT-6 sarcoma-bearing mice. The results are expressed as per cent injected dose uptake per g by the tissue; gallium-67 citrate is included for comparison. Data on ruthenium complexes with citrate, tartrate, diethylenetriaminepentaacetic acid (DTPA), ethylenediaminetetramethylene phosphonate (EDTMP), and various phenanthroline derivatives were also obtained but are not described in this table since tumor uptake was insignificant with these compounds. Evaluation of the promising compounds in the other two animal tumor models is presently being carried out. Compounds 2, 3, 5 and 6 in this table have shown chemotherapeutic potential in recent NCI animal screens (5).

Tumor-to-tissue ratios at various time intervals after injection are shown in Table 3. Values for gallium citrate are also included for comparison. The whole body retention is shown in the last column for all the compounds. Several compounds, as noted, were synthesized using stable ruthenium carrier and the ruthenium dose corresponded to between 2 and 6 mg/Kg body weight. Table 4 describes the tumor concentration index (TCI) of various compounds. This index which provides a useful correlation between the tumor uptake and the mean body concentration is defined as the ratio of per cent injected dose per g of the tumor to per cent injected dose per g remaining in the whole body at any given time period.

DISCUSSION

Ruthenium (Z=44) is a member of the second triad of the group VIII transition metals. It is a highly reactive element and its chemistry is quite complicated. It displays as many as ten oxidation states in its compounds, the common ones being Ru(II), Ru(III), Ru(IV) and Ru(VIII). The high reactiveness of ruthenium and the desirable physical characteristics of ruthenium-97, a pure gamma emitter, provide a valuable combination for application to nuclear medicine. The chemistry of ruthenium can be manipulated much as desired to obtain useful compounds for specific applications. The 2.9 day half-life of ruthenium-97 makes many studies and procedures possible, not otherwise practical with technetium-99m. Efficient and economical production (>100 mCi per day) of ruthenium-97 in the BLIP facility at Brookhaven has been found to be feasible using the rhodium-103(p,2p5n)ruthenium-97 reaction, and this should make widespread use of this nuclide possible subsequent to the development of suitable radiopharmaceuticals.

In the area of tumor-localizing agents, ruthenium-97 appears to show substantial promise. A number of ruthenium coordination compounds have shown antitumor activity in recent animal screens; several of these appear useful as tumor localizing agents when labeled with ruthenium-97. It is likely that dif-

TABLE 2. TISSUE UPTAKE OF VARIOUS PROMISING RUTHENIUM COMPOUNDS AND OF GALLIUM-67 CITRATE IN EMT-6 SARCOMA-BEARING MICE (% DOSE PER G)[a] (n=6)

Compound	Time (hr)	Tumor	Blood	Muscle	Liver	Kidney	Whole Body Retention (% Dose)
1. Ruthenium chloride	24	5.42	4.27	1.57	6.11	12.24	55.25
	48	6.10	3.20	1.15	9.17	10.48	51.67
2. $K_2[RuCl_5(H_2O)]$[b]	24	6.76	5.46	2.06	10.89	11.76	61.86
3. $K_3[Ru(Ox)_3]$ X H_2O[c]	24	5.97	3.81	2.03	5.22	6.32	57.77
4. $[Ru(NH_3)_5Cl]Cl_2$[d]	24	3.29	1.97	0.79	2.37	5.89	25.70
	96	2.30	0.67	0.80	1.63	3.81	17.40
5. cis-$[Ru(NH_3)_4Cl_2]Cl$[e]	24	5.38	5.18	1.54	5.12	9.23	49.61
	72	5.93	2.10	1.32	4.14	7.21	43.80
	96	4.92	1.15	1.20	3.23	5.35	37.06
6. cis-$[Ru(en)_2Cl_2]Cl$[f]	24	5.97	5.62	1.69	5.76	11.15	52.80
7. Ru-Oxine 7-CA acetate[g]	24	2.38	2.34	0.39	3.03	2.88	19.01
	72	1.68	0.72	0.24	1.49	1.46	9.61
	96	1.48	0.47	0.21	1.54	1.38	8.42
8. Gallium-67 citrate	24	7.09	1.81	0.53	9.30	9.21	59.58
	48	5.05	0.51	0.47	9.49	8.03	50.68
	96	3.54	0.36	0.37	8.19	7.14	38.29

[a]Compounds 2-6 were synthesized using stable ruthenium carrier; ruthenium dose 2-6 mg/Kg body weight

[b]Potassium pentachloroaquoruthenate(III)

[c]Potassium tris-(oxalato)ruthenium(III)

[d]Chloropentaammineruthenium(III) chloride

[e]cis-Dichlorotetraammineruthenium(III) chloride

[f]cis-Dichloro bis(ethylenediammine)ruthenium(III) chloride

[g]8-Hydroxyquinoline 7-carboxylic acid acetate

TABLE 3. TUMOR-TO-BLOOD AND TUMOR-TO-TISSUE RATIOS OF VARIOUS RUTHENIUM-103-LABELED COMPOUNDS IN EMT-6 SARCOMA BEARING MICE (n=6)

Compound[a]	Time Post Injection, Hr	Ratio, Tumor to Blood	Muscle	Liver	Kidney
Gallium-67 citrate	24	4.2	13.4	0.8	0.8
	48	10.7	11.0	0.6	0.6
	96	10.3	10.6	0.4	0.5
Ruthenium chloride	24	1.0	2.9	0.6	0.4
	48	1.9	5.3	0.7	0.6
$[Ru(NH_3)_5Cl]Cl_2$	24	1.7	4.2	1.4	0.6
	48	1.9	4.5	1.2	0.5
	96	3.5	4.1	1.4	0.6
<u>cis</u>-$[Ru(NH_3)_4Cl_2]Cl$	24	1.1	3.5	1.1	0.6
	48	1.7	4.0	1.3	0.7
	96	4.4	4.2	1.5	0.9
$K_2[RuCl_5(OH_2)]$	24	1.3	3.3	0.6	0.6
$K_3[Ru(Ox)_3]$	24	1.6	3.0	1.1	0.9
<u>cis</u>$[Ru(en)_2Cl_2]Cl$	24	1.1	3.6	1.1	0.5
Ru-oxine 7-CA Acetate	24	1.0	6.3	0.8	0.8
	48	1.7	7.2	1.0	1.0
	96	3.2	7.7	1.0	1.2

[a]For abbreviation and other explanations, see Table 2

TABLE 4. TUMOR CONCENTRATION INDEX[a] OF VARIOUS RUTHENIUM-103-LABELED COMPOUNDS IN EMT-6 SARCOMA (Balb/c mice)

Compound[b]	TCI 24 hr	48 hr	72 hr	96 hr
Ru-Oxine 7-CA acetate	1.90	2.39	2.68	2.50
Gallium-67 citrate	1.81	1.48	1.57	1.45
<u>cis</u>-$[Ru(en)_2Cl_2]Cl$	1.70	-	-	-
$[Ru(NH_3)_5Cl]Cl_2$	1.70	1.46	1.37	1.49
$K_2[RuCl_5(OH_2)]$	1.65	-	-	-
$K_3[Ru(Ox)_3]$	1.64	-	-	-
<u>cis</u>-$[Ru(NH_3)_4Cl_2]Cl$	1.56	1.74	2.03	2.00
Ruthenium chloride	1.33	1.86	-	-

[a]Tumor concentration index (TCI) is defined as the ratio of per cent injected dose per g of the tumor to per cent injected dose per g remaining in the whole body at any given time period. For inter-species comparisons, and for normalizing for different animal body weights, TCI could be expressed as per Kg body wt.

[b]For abbreviations and other explanations, see Table 2

ferent ruthenium agents follow different localization mechanisms and in order to develop successful tumor avid compounds, a whole series of ligands would have to be studied. Mechanistic studies on model compounds have recently been initiated which would undoubtedly aid in the design of more specific agents.

The preliminary results in the EMT-6 sarcoma in mice demonstrate the potential of a number of ruthenium compounds as tumor-scanning agents. The tumor uptake of ruthenium compounds described in Table 2, expressed as per cent dose per gram, ranges between 1.5 and 6.8, and compares favorably with the gallium uptake (3.5-7.1). The general body background of cis-dichlorotetraammine-ruthenium(III) chloride and Ru-oxine 7-carboxylic acid acetate is much lower than that of gallium citrate; the other compounds are comparable to gallium in this respect. The tumor-to-blood (T/B) and tumor-to-muscle (T/M) ratios with several compounds are high enough for delineating tumors by imaging. The uptake of gallium citrate by the tumor reaches a plateau at about 24 hr; thereafter, significant loss of the activity is seen without comparable excretion from the body. This results in a decrease in the value of the tumor concentration index (Table 4). This index reaches a maximum value at different time periods with different compounds. The TCI of seven most active compounds at 24-96 hr ranges between 1.33 and 2.68 and is comparable to or better than that of gallium citrate (1.45-1.81). Using various important criteria, ruthenium-oxine 7-carboxylic acid acetate (TCI max 72 hr 2.68; T/B 2.4; T/M 7.2) and cis-dichlorotetraammineruthenium(III) chloride (TCI max 96 hr 2.0; T/B 4.4; T/M 4.2) look particularly promising when compared to gallium citrate (TCI max 24 hr 1.81; T/B 4.2; T/M 13.4). The other compounds have favorable tumor uptake; however, the blood and body background remains quite high. By exploiting the chemistry of ruthenium, it may be possible to reduce nontarget uptake of these and other compounds and thus improve the tumor-to-background ratios. Chemical modification of the compounds is also expected to increase their tumor specificity thus making them as good or better than gallium citrate. Another approach that presently is being evaluated is the use of mixed-ligand chelating agents (11) or other competitive preparations (12) to selectively reduce the blood and normal tissue background without affecting the tumor uptake. A variety of other ruthenium compounds, in addition to those described here, are presently under investigation.

In summary, results so far on the tumor uptake of various ruthenium compounds have been very encouraging and the potential for developing highly specific tumor-localizing agents based on ruthenium-97 appears very great. Use of this nuclide could represent a distinct advantage over most presently used radioscintigraphic agents for tumor localization.

REFERENCES

1. Srivastava SC, Som P, Meinken G, Sewatkar A., et al: Ruthenium-97 labeled compounds -- A new class of radiopharmaceuticals. Read before the World Federation of Nuclear Medicine and Biology, 2nd International Congress, Washington, D.C., Sept. 18-21, 1978

2. Anghileri LJ: J Nucl Biol Med 18, 155 (1974); ibid., Strahlentherapie 149, 173 (1975); Wenzel M, Subramanian N, and Nipper E: Naturwiss 63,

341 (1976); Wenzel M, Nipper E, and Klose W: J Nucl Med 18, 367, 1977; Wenzel M, Klose W: Klin Wochenschrift 55, 559, 1977; Tanabe M, Yamamoto G: Acta Medica Okayama 29, 431, 1975; Tanabe M: Radioisotopes 25, 44, 1976

3. Subramanian G, McAfee JG, and Poggenburg JK: Ru-97: Preliminary evaluation of a new radionuclide for use in nuclear medicine. J Nucl Med 11, 365, 1970 (abst)

4. Kellman AD, Clarke MJ, Edmonds SD, et al: Proc. 3rd Intl. Symp. Pt Coord. Complexes in Cancer Chemotherapy, Dallas, Texas, Oct. 18-20, 1976; Dwyer FP, Mayhew E, Roe EMF, et al.: Brit J Cancer 19, 195, 1965; White DO, Harris AW, Dwyer FP, et al.: Austr J Exp Biol 41, 517, 1963; White DO, Harris AW, and Shulman A: Austr J Exp Biol 41, 527, 1963; Cleare MJ: Coord Chem Rev 12, 349, 1974; Durig JP, Danneman J, Behnke ND, et al.: Chem Biol Interactions 13, 287, 1976

5. Clarke MJ: Oncological implications of the chemistry of ruthenium. Metal Ions in Biological Systems, Vol. 11, H. Sigel, Marcel Dekker, Inc., New York (in press).

6. Ku TH, Richards P, Srivastava SC, et al: Production of ruthenium-97 for medical applications. Read before the World Federation of Nuclear Medicine and Biology, 2nd International Congress, Washington, D. C., Sept. 18-21, 1978

7. Mercer EE, Buckley RR: Inorg Chem 4, 1692, 1965; Raichart DW, Taube H: Inorg Chem 11, 999, 1972; Broomhead JA, Kane-MacGuire LAP: J Chem Soc (A), 546, 1967; Krentzien H: Ph.D. Dissertation, Stanford Univ., Stanford, CA, 1976; Lever FM, Powell AR: J Chem Soc (A) 1477, 1969; Ford PC, Rudd DFP, Gaunder R, et al.: J Amer Chem Soc 90, 1187, 1968

8. Larson SM: Mechanisms of localization of gallium-67 in tumors. Seminars Nucl Med 8, 193, 1978

9. Srivastava SC, Larson SM, Richards P, et al: Radiolabeled ruthenium compounds as tumor localizing agents. Read before the 4th Annual Scientific Meeting, Society of Nuclear Medicine, Greater NY Chapter, Philadelphia, PA, Nov. 10-12, 1978

10. Packer S, Lambrecht RM, Christman DR, et al: Amer J Ophthalmology 64, 80, 1977

11. Schubert J, Derr SK: Mixed ligand chelate therapy for plutonium and cadmium poisoning. Nature 275, 311, 1978

12. Larson SM, Rasey JS, Grunbaum Z, et al: Pharmacologic enhancement of gallium-67 tumor-to-blood ratios for EMT-6 sarcoma (Balb/c mice). Radiology 130, 241, 1979

In-111 BLEDTA: A CONJUGATE OF BLEOMYCIN WITH A BIFUNCTIONAL CHELATING AGENT FOR TUMOR LOCALIZATION

David A. Goodwin[*], Claude F. Meares[#], Leslie H. DeRiemer[#], Carol I. Diamanti[*] and Richard L. Goode[*]

[*]Department of Nuclear Medicine, V.A. Medical Center, and Stanford University School of Medicine, Palo Alto, CA

[#]Department of Chemistry, University of California at Davis, Davis, California 95616

ABSTRACT

BLEDTA, a bleomycin A_2 analog containing an EDTA group was labeled in the EDTA group to a specific activity of 70 mCi/mg and used for animal and human studies. KHJJ mouse tumor uptake was higher than the orange Co-57 bleomycin isomer and the tumor/organ ratios were > 1 for all organs except the kidney. In 29 biopsy proven cancer patients the scan done 24 hours post I.V. with 1-2 mCi of In-111 BLEDTA was positive in all clinically known sites in 18, and in some clinically known sites in 11. The In-111 BLEDTA clinical results are similar to reported Co-57 bleomycin clinical studies, and the method is proposed as an alternate way to produce a stable biologically active bleomycin labeled with In-111.

INTRODUCTION

Since the introduction of Co-57 labeled bleomycin for tumor imaging in 1972 (1) there has been a search for an alternate radiolabel providing similar tumor affinity but improved physical characteristics. The long physical half life of Co-57 (270 days) has posed serious contamination problems, and unfortunately none of the other isotopes of cobalt have the desired physical properties. Of the other metal chelates of bleomycin investigated (In-111, Cu-64, Cu-67, Tc-99m, Zn-62, Ga-67) (2) none have the stability of the cobalt chelate in vivo, and it appears this unique stability of Co-57 bleomycin is related to its favourable tumor imaging properties.

MATERIALS AND METHODS

Our approach to this problem was to add a much more powerful metal chelating group to bleomycin. We chose a para-substituted 1-phenyl-ethylene dinitrilotetraacetic acid group which is known to form metal chelates that are unusually stable both thermodynamically and kinetically (3,4). In order to enhance the tumor localization and to block the intrinsic metal-binding site of bleomycin, the stable cobalt (III) bleomycin A_2 was first prepared, then the EDTA-containing moiety was covalently attached to the sulfur atom in the "terminal amine" group of bleomycin by successive demethylation and alkylation, to yield the product shown in Fig. 1 (5). The EDTA group of the product may be labeled with any of a variety of useful metal ions; for example, mixing in solution with a simple salt of In^{3+} leads to quantitative chelation within a few minutes (4,5). The product was stored as a 0.55 x 10^{-3} solution in 25 μl portions (26 μg BLEDTA in each) at -80° C. When required, an aliquot was radiolabeled by adding 50 μl of 0.01 M citric acid containing 1-2 mCi of carrier-free In-111 chloride allowing the solution to stand 5 minutes at room temperature, and then diluting with normal saline and millipore filtering for injection. The specific activity was approximately 70 mCi/mg.

Whole body retention, organ distribution and tumor uptake was studied at 24 hours in BALB/c mice bearing KHJJ adenocarcinoma in the flank. Whole body scanning of rabbits bearing VX 2 adenocarcinomas implanted in the right hind leg, was done following I.V. injection. Control experiments included In-111 Cl_3, In-111 Bleo (directly chelated), In-111 phenyl-EDTA and Co-57 bleomycin.

For the human studies one to two mCi of In-111 BLEDTA was given I.V. and the patients scanned at 18 to 24 hours following injection. Blood samples and 24 hour urine collections were taken in several patients for calculation of blood disappearance and renal excretion. Samples of tumors and normal tissues were obtained in three patients from surgery 24 hours following injection and in one patient extensive tissue samples were obtained from autopsy 8 days following injection. In-111 BLEDTA identification by thin layer chromatography in 10% ammonium acetate: methanol (1:1) was done following radiolabeling, and on urine samples.

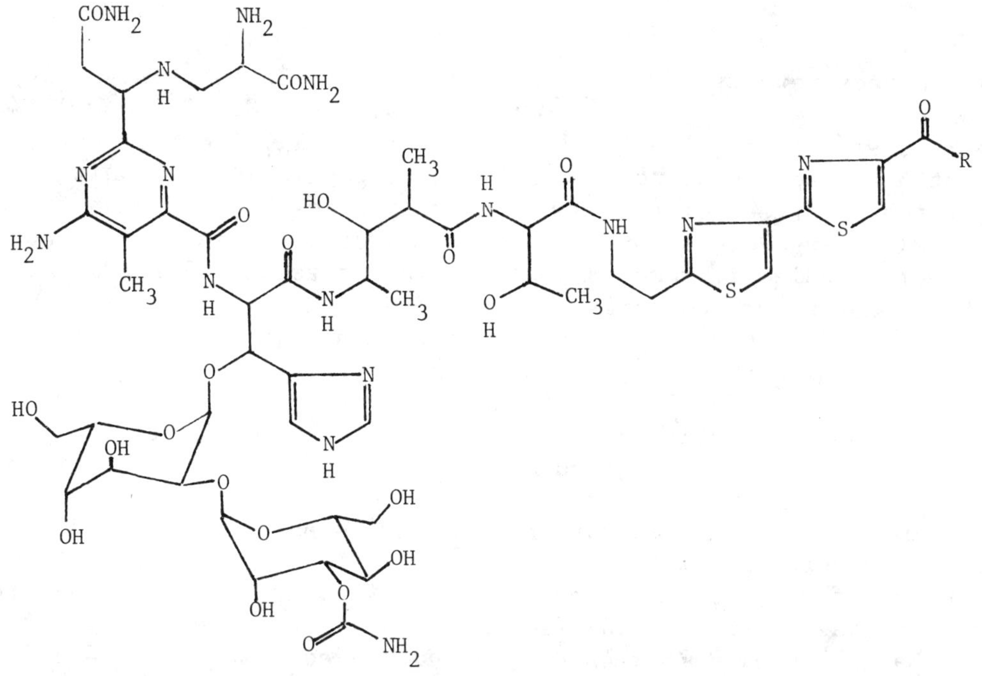

2, $R = NH(CH_2)_3\overset{+}{S}(CH_3)_2$

3, $R = HN(CH_2)_3SCH_3$

4, 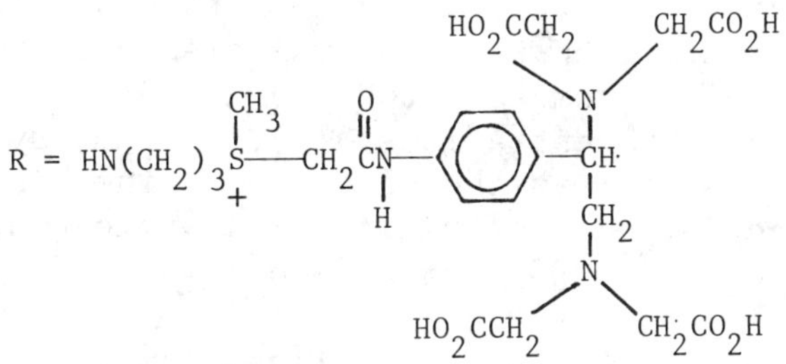

5 $R = HN(CH_2)_4\underset{H}{N}C(NH_2)=\overset{+}{N}H_2$

Figure 1.

Structure of the bleomycin nucleus, and the various bleomycin compounds studied: 2 = Co(111) bleomycin A_2, 3 = Co(111) bleomycin demethyl A_2, 4 = Co(111) BLEDTA, 5 = Co(111) bleomycin B_2.

RESULTS

Whole body retention in mice was inversely related to stability: In-111 Bleo (directly chelated); 20%, In-111 BLEDTA; 4.4%, Co-57 Bleo; 2.7%. Tumor uptake (% dose/gm) was; In-111 Cl_3; 5.0%, and In-111 Bleo (directly chelated); 2.0%, but high background reduced tumor to organ ratios. Cobalt bleomycin A_2 and B_2 formed a green colored isomer shortly after the addition of cobalt, which slowly changed to a more stable orange colored isomer: (5) both of these isomers were studied in tumor mice and compared to In-111 BLEDTA (Tables 1,2, 3). The green isomers had significantly higher tumor concentrations ($P < .01$) as well as blood, liver and spleen concentrations ($P < .05$) (Table 1,2). In-111 BLEDTA had higher organ and tumor concentrations than orange A_2, but not green A_2 (Table 3). In-111 BLEDTA had tumor/organ ratios > 1 for all organs except the kidney. Rabbit scans using 0.7 mCi In-111 BLEDTA showed significant tumor uptake at 2 and 24 hours over the control In-111 phenyl-EDTA scans.

The In-111 labeled preparations from two separate batches of BLEDTA were studied in humans. Chromatography of batch 1 (mean of 7 preparations) showed 71 ± 4.6% at Rf 0.5 (BLEDTA) with a 17.5 ± 3% impurity at Rf 0.4. The rest of the activity was at the origin, (3 ± 1%) and in the free chelate Rf > 8, (9.3%). Batch 2 showed 78 ± 4% Rf 0.5 and reduced impurity 7.4 ± 0.4% Rf 0.4. The patient kinetic studies were done with the first batch of BLEDTA. Chromatography of the 24 hour urine (mean of 6 patients) showed 87 ± 5% Rf 0.5 (BLEDTA) with no activity at Rf 0.4 (impurity). Ten percent of the urine activity was in the "free" chelate Rf > .8, and 3% at the origin. The 18-24 hour urine (mean of 4 patients) contained 75 ± 18% of the injected dose. The blood levels fell rapidly (mean of 4 patients); 28 ± 11% of the injected dose remaining in the total blood volume at 15 minutes, and only 3 ± 0.8% at 24 hours. Most of the blood activity remaining at 24 hours was in the cellular fraction (84%).

Patient tissue samples showed from 2.3×10^{-3} to 6.0×10^{-3} percent of injected dose per gram of tumor with tumor/normal tissue ratios of 3/1 to 11/1 (Table 4). Tissue samples from an autopsy of a patient injected 8 days previously had very low organ concentrations. (6). Dosimetry calculations based on these tissue data as well as the blood and urine kinetics showed the following results expressed in rads per millicurie: whole body .067, red marrow .566, testes .029, ovaries .088, and bladder wall .242.

The scan results of 29 patients are summarized in Table 5. In all patients with malignant tumors the scan was positive in at least one clinically proven site. In 11 patients (38%), however, only some of the clinically known sites were visualized. The tumor to background ratios computed from digitized images were high enough (1.7 - 2.7 to one) to provide good contrast on the scans and 50K counts per view were obtained in an average of 5-8 minutes. False positive scans occurred in areas of recent surgery undergoing healing, including tracheostomy wounds, (4 patients), and in osteomyelitis and pulmonary tuberculosis (2 patients). Typical scan results are shown in Fig. 2, 3, 4.

DISCUSSION AND SUMMARY

The results of this study show that a bifunctional chelating agent may be attached to the terminal amine group of bleomycin without greatly modifying

TABLE 1. COMPARISON OF ABSOLUTE ORGAN UPTAKE

% Dose/gm

Green Co-A_2 vs Orange Co-A_2

(Mean of 3 mice each ± S.D.)

Organ	Green ^{57}Co-A_2	Orange ^{57}Co-A_2	P Value <.05	P Value <.01
Blood	.013 ± .004	.003 ± .001	+	-
Lungs	.10 ± .04	.075 ± .008	-	-
Liver	.40 ± .11	.20 ± .03	+	-
Spleen	.15 ± .03	.093 ± .011	+	-
Kidneys	1.43 ± .46	1.69 ± .33	-	-
Tumor	.55 ± .14	.19 ± .03	+	+
Muscle	.020 ± .008	.015 ± .008	-	-
Bone	.089 ± .048	.035 ± .016	-	-
Skin	.059 ± .015	.064 ± .009	-	-
Stomach & Gut	.053 ± .019	.091 ± .092	-	-

TABLE 2. COMPARISON OF ABSOLUTE ORGAN UPTAKE

% Dose/gm

Green Co-B_2 vs Orange Co-B_2

(Mean of 3 mice each ± S.D.)

Organ	Green ^{57}Co-B_2	Orange ^{57}Co-B_2	P Value <.05	P Value <.01
Blood	.024 ± .002	.006 ± .005	+	+
Lungs	.18 ± .05	.10 ± .003	+	-
Liver	.65 ± .04	.23 ± .004	+	+
Spleen	.19 ± .03	.12 ± .02	+	-
Kidneys	2.12 ± .66	2.80 ± .82	-	-
Tumor	.54 ± .05	.26 ± .04	+	+
Muscle	.017 ± .003	.011 ± .003	+	-
Bone	.094 ± .013	.031 ± .009	+	+
Skin	.066 ± .026	.209 ± .228	-	-
Stomach & Gut	.073 ± .005	.152 ± .142	-	-

TABLE 3. ABSOLUTE ORGAN UPTAKE: COMPARISONS
% dose/gm

^{111}In-BLEDTA vs Orange Co-A_2 Mean of 3 mice ± S.D.

Organ	^{111}In-BLEDTA	Orange Co-A_2	P Value < .01
Blood	.55 ± .05	.003 ± .001	+
Lungs	.42 ± .05	.075 ± .008	+
Liver	.35 ± .03	.20 ± .03	+
Spleen	.27 ± .03	.093 ± .011	+
Kidneys	1.72 ± .26	1.69 ± .33	-
Tumor	.68 ± .16	.19 ± .03	+
Muscle	.063 ± .010	.015 ± .008	+
Bone	.077 ± .060	.035 ± .016	-
Skin	.21 ± .03	.064 ± .009	+

TABLE 4. In-111 BLEDTA ASSAYS ON SURGICAL/SAMPLES-24 HOUR POST INJECTION

Patient No.	Sample	%Dose/gm	Tumor/Normal Tissue
1.	"In Situ" Sq CA	0.0032	-
2,	Normal soft palate	0.0009	
	Sq CA soft palate	0.0023	2.6/1
	Sq CA tongue	0.006	6.7/1
3.	Normal tissue (mean of 3 biopsy sites)	0.00023	
	Squamous Ca (mean of 5 biopsy sites from the larynx)	0.00253	11/1
	Whole blood	0.000474	5.3/1

its transport properties or tumor concentrating ability in vivo. This suggests the terminal amine group, which is variable in nature, is not responsible for this biological property. Once concentrated in the tumor in the first few hours, In-111 BLEDTA remained bound to the tumor cells while background fell over the next 18-24 hours. Preliminary studies in vitro suggest that, unlike bleomycin, BLEDTA does not bind to DNA. Further work is planned to elucidate the mechanism of cellular uptake and retention of BLEDTA.

Of great interest is the finding that the two isomeric forms of both cobalt bleomycin A_2 and B_2 had significantly different tumor uptake, as well as organ concentrations in vivo. This suggests that the tertiary conformation of bleomycin as stabilized by cobalt is important for its transport into tumor cells, and may partly explain the enhanced uptake of the cobalt chelate of bleomycin in animal tumors previously reported by Kono et al. (7).

In-111 BLEDTA was stable in vivo as shown by the rapid drop in blood levels and large urinary excretion of unchanged In-111 BLEDTA in the first 24 hours in humans. The majority of the urine activity (87.3%) had an identical Rf (0.5) to In-111 BLEDTA on TLC.

Noteworthy is the observation that all 29 patients with known cancer had at least one site visualized with In-111 BLEDTA. A significant number of patients (38%) had only some of their disease shown on the scan. The significant incidence of positive scans in healing tissue and inflammation suggests the BLEDTA scan used alone will not be useful in the diagnosis of cancer. Used in combination with other studies, however, the specificity may be significantly increased (see Fig.'s 1 and 2). The high sensitivity of the BLEDTA suggests it may be very valuable in screening patients with known cancer for distant metastases, and in planning surgical node dissections in head and neck cancer.

The biological activity of In-111 BLEDTA observed in these studies suggests that the principle of labeling with bifunctional chelating agents may be generally useful with polypeptides and other relatively small molecules. In particular, this technique opens the possibility of labeling tracer amounts of a variety of hormones and drugs to high specific activity for use as radiopharmaceuticals.

ACKNOWLEDGMENTS

We thank Carol Whiteleather, Vincent Dentamaro, Gary Wayman, David Sherman and Robert Campbell for their technical help, Dr. William T. Bradner of Bristol Labs for the generous gift of blenoxane used in these studies, and Sheila Smith for performing the dosimetry calculations.

REFERENCES

1. Nouel JP, Renault M, Robert J, et al: La bléomycine Marquée au Co-57. Nouv Presse Med 2: 95, 1972

2. Goodwin DA, Meares CF: Radiolabeled anti tumor agents. Seminars in Nuclear Medicine VI: 3, 1976

3. Meares CF, Goodwin DA, S-H Leung C, et al: Covalent attachment of metal chelates to proteins: The stability in vivo and in vitro of the conjugate of albumin with chelate of 111-Indium. Proc Nat'l Acad Sci USA, 73:11, 3803-3806, 1976

4. Meares CF, DeRiemer LH, Goodwin DA: Conjugation of bifunctional chelating agents to bleomycin for use in nuclear medicine. In Bleomycin: chemical, biochemical, and biological aspects. S. M. Hecht, ed. (Springer-Verlag, 1979) in press

5. DeRiemer LH, Meares CF, Goodwin DA: BLEDTA: tumor localization by a bleomycin analog containing a metal-chelating group. J Med Chem (submitted)

6. Goodwin DA, Meares CF, DeRiemer LH, et al: Conjugates of bleomycin and bifunctional chelating agents (In-111 BLEDTA) for clinical tumor scanning. N Engl J Med (submitted)

7. Kono A, Matsushima Y, Kojima M, et al: Cobalt chelate of bleomycin. 1. Physiochemical properties and distribution in tumor bearing mice. Chem Pharm Bull 25: 1725, 1977

TABLE 5. In-111 BLEDTA SCAN RESULTS OF 29 PATIENTS

	Diagnosis	No. of Patients	Some Disease Shown	All Disease Shown
I	Squamous Ca, Head and neck	21	4	17
II	Lung Ca	3	2	1
III	Unknown 1°	2	2	0
IV	Other:			
	Adenocarcinoma (Metastatic)	1	1	0
	Thyroid Ca (anaplastic)	1	1	0
	Mucoepidermoid Ca	1	1	0
	TOTALS	29	11	18

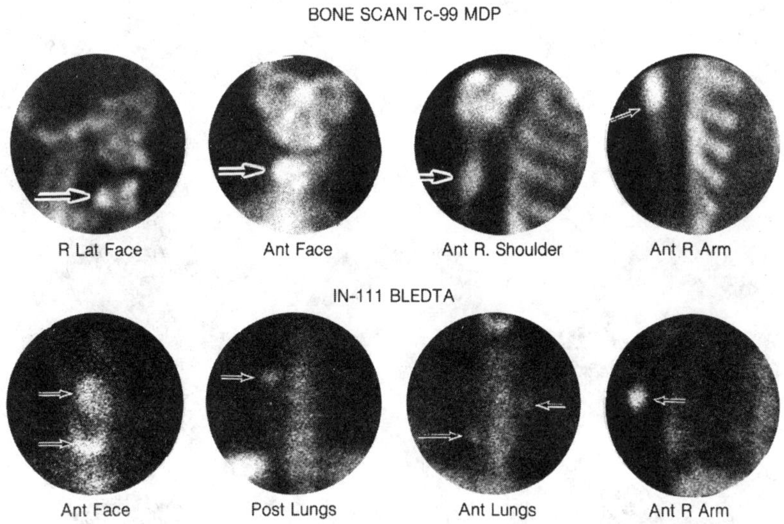

Figure 2.

Fifty-three year old white male, one year post resection of squamous ca base of tongue with recent weight loss, fatigue and trauma 2° to falling. Bone x-rays negative, bone scan positive in jaw, upper R humerus. Intense In-111 BLEDTA uptake in jaw and R humerus (arrows) and focal uptake in both lungs. Bone metastases confirmed by pathological fracture R humerus two months later. Chest x-ray: suspicious nodule L post lung; no nodules seen in anterior chest.

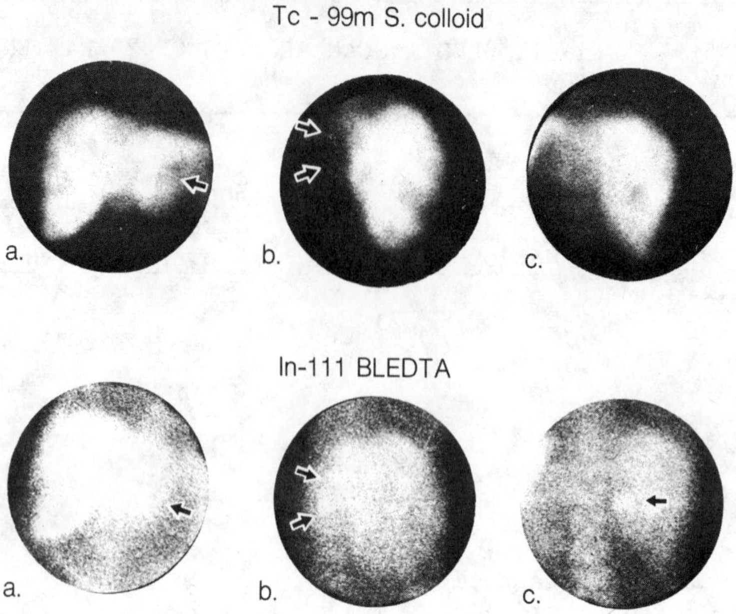

Figure 3.

Thirty-eight year old white male, progressive weight loss and ascites. Liver biopsy: metastatic adenocarcinoma. a) Ant, b) R Lat, c) Posterior. Tc-99m sulfur colloid liver views show filling defects (arrows a & b) which fill in on In-111 BLEDTA scan.

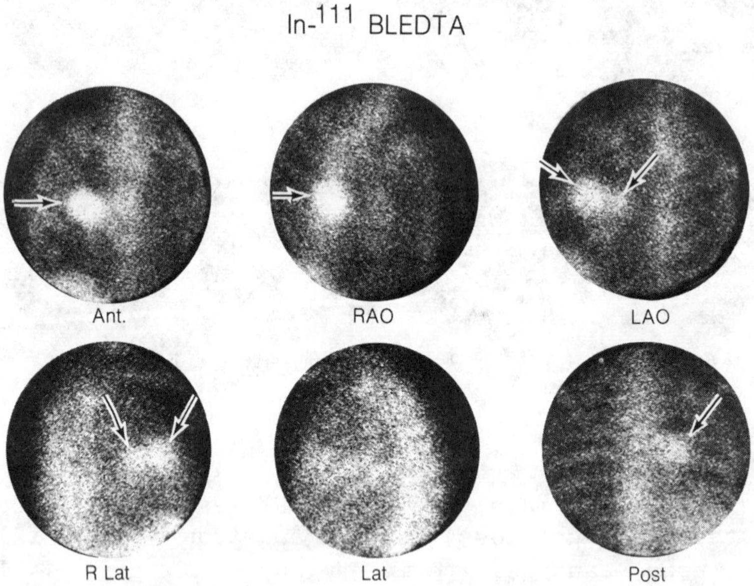

Figure 4.

Sixty-eight year old white male with a large R peri-hilar lung ca and malignant squamous cells in sputum. In-111 BLEDTA scan shows intense uptake in a large tumor, with a smaller central extension (arrows).

COMPARTMENTAL STUDIES OF A MOUSE BRAIN TUMOR

Howard J. Glenn, Muhammed R. Atallah, Thomas P. Haynie,
James H. Anderson, and Gilbert L. Raulston.

The University of Texas System Cancer Center
M. D. Anderson Hospital and Tumor Institute
Houston, Texas 77030

ABSTRACT

The purpose of this study was to determine the hematologic compartments, vascular, plasma, and extracellular fluid volumes of a brain tumor, normal brain, and other tissues in a well established mouse brain tumor model. Tumor vascular volume, determined using Cr-51 labeled red blood cells, was found to be 0.019 ml/g tumor over the first hour. Plasma volumes were estimated using I-131 labeled albumin. Only the 10-minute value (when corrected by the plasmacrit, 0.497) of 0.019 ml/g tumor agreed with the vascular volume. Extracellular fluid space, measured using ^{35}S sulfate, was found to be 0.51 ml/g tumor after an equilibration time of 6 hours. Relatively constant compartmental space values were obtained for other tissues except liver.

INTRODUCTION

The use of animal tumor models in cancer research, while not a perfect research tool, permits the investigator to develop experimental techniques and to generate information from the biosystems or models which may give a better understanding of the behavior of cancer. Malignant disease, whether in man or animal, is not a single entity, and each tumor model will have certain features that are unique. Neoplastic cells are part of a complex organized physiological structure; they exist in an environment consisting of at least two major compartments, the vascular and the interstitial spaces. The existence and survival of the tumor cells depends upon the characteristics of these compartments. Information relating to the physiology of these compartments in which tumor cells exist could be useful in the selection of substances for transport to the tumor for diagnostic or therapeutic purposes. This report relates to the study of the hematologic compartments of a mouse brain tumor model which has been used in many studies of the uptake of radionuclides by the tumor. Chromium-51 sodium chromate labeled red blood cells have been used to measure vascular volume, ^{131}I labeled human serum albumin to measure plasma volume, and ^{35}S sodium sulfate to measure extracellular fluid volume. Studied were tumor, blood, plasma, brain, and other selected tissues such as skin, muscle, kidney, and liver. The dynamics of these compartmental volumes with time are reported.

MATERIALS AND METHODS

Mouse red blood cells were labeled with ^{51}Cr sodium chromate with a specific activity of 100-400 mCi/mg Cr by a method closely related to that of Song and Levitt(1). Iodine-131 HSA containing 10 mg albumin/ml with a specific activity of 25 uCi/mg albumin was used. The percentage of free iodide was less than 1% as determined by paper chromatography. Sulfur-35 Sodium sulfate with a specific activity of 559 uCi/millimole sulfur and 99% radiochemical purity was also used.

The mice used were of the Yale-Swiss (Hale-Stoner) line. The procedure for the intracranial tumor transplant and a description of the tumor have been previously published (2). The techniques for quantitation of tumor and tissue radionuclide content have also been previously published (3,4).

For gamma counting, all blood, plasma, tumor, and tissue samples were dissolved in 1 ml of concentrated nitric acid in tightly closed gamma counting vials and counted in an automatic gamma counter. All samples for liquid scintillation counting were dissolved in commercially available solubilizing agents, decolorized, allowed to equilibrate to light and temperature for 24 hours, and counted in a liquid scintillation counting system. Blank vials containing all reagents except tissues were used for background counts, and correction for quenching was performed using the external standard method.

At least 6 mice were used for each point. The radioactivity content of all blood, plasma, tumor, and tissue samples is expressed as percent injected dose per gram of tissue. Vascular volumes and plasma volumes of tumor and tissues have been calculated using the formulas of Song and Levitt (1). Vascular volume has been expressed in terms of ml of blood/gram tissue, a value obtained by dividing the ^{51}Cr activity per gram of tissue by the activity per ml of blood. Plasma volume has been expressed in terms of ml of plasma per gram of tissue, a value obtained by dividing the ^{131}I human serum albumin activity per gram of tissue by the activity per ml of plasma. It was assumed that the red blood cells occupied uniformly the entire vascular volume. Since 1 ml of plasma occupies more space than the plasma in 1 ml of vascular volume, the plasma volume was converted to vascular volume by multiplying by the plasmacrit, 0.497. The extracellular fluid volumes of the tissues were calculated by the method of Manery and Bale (5) and are expressed in units of ml of water per gram of tissue.

RESULTS

In Figure 1 are depicted the blood or plasma concentration curves in tumor mice for ^{51}Cr labeled red blood cells (blood concentration), ^{131}I labeled serum albumin (plasma concentrations) and ^{35}S sodium sulfate (plasma concentrations) in which percent injected dose per ml of blood or plasma is plotted against time. Values for ^{35}S sodium sulfate are recorded for only 12 hours because of the small plasma concentration thereafter. The essentially constant concentration of ^{51}Cr activity in the blood after 2 hours indicates a good label of ^{51}Cr to the red blood cells.

In Table I are given the percent injected dose per gram of tissue and the calculated vascular volumes, ml of blood per gram of tissue, for tumor, brain, muscle, liver, and kidney in this mouse brain tumor model, from 10 minutes to 48 hours post-injection of ^{51}Cr labeled red blood cells. The tumor vascular volume averages 0.0194 ml/g tumor over the first hour, a constant and acceptable value, but shows progressive increase in "apparent volume" thereafter. The vascular volume of the brain shows little significant change averaging 0.0080 ml/g over the period studied. Vascular volumes for muscle and skin average 0.0171 ml/g for muscle and 0.0151 ml/g for skin over the duration of the study. The vascular volume for kidney is relatively constant, averaging 0.0632 ml/g tissue. The ^{51}Cr activity in liver peaks at about 4 and again at 48 hours. The vascular volume also is greatest at these time intervals. The 4-hour vascular volume for liver is 0.134 ml/g tissue and that at 48 hours is 0.1599 ml/g tissue. These values are significantly different (P=.005) from the average of the vascular volumes of the liver at other time periods, 0.0796 ml/g liver.

In Table II are given the percent injected dose per gram tissue of ^{131}I human serum albumin, and the calculated tissue plasma volumes for tumor, brain, muscle, skin, and liver of tumor-bearing mice. Apparent plasma volume for tumor increases very rapidly through 6 hours, the increase becoming slower and the rate more constant thereafter. The same is true to a lesser degree for the apparent plasma volume of brain. The apparent plasma volume of muscle and skin increases in an irregular manner over the first six hours, the rate of increase becoming less after 12 hours. The apparent plasma volume of liver is the only tissue volume that decreases throughout the first four hours and then shows a slow but regular increase thereafter. When calculated from the plasma volume, the vascular volume for the tumor at the 10 minute point is 0.0190 ml/g tumor, a value in excellent agreement with the vascular volume of 0.0194 ml/g tumor as determined by the use of ^{51}Cr labeled red blood cells.

In Table III are found the percent injected dose per gram tissue of ^{35}S sodium sulfate and the extracellular fluid volume values for tumor, brain,

TABLE I. TISSUE CONTENT AND VASCULAR VOLUMES FROM Cr-51 LABELED RBC.

Tissues		Time after injection								
		10 min	30 min	1 hr	2 hr	4 hr	6 hr	12 hr	24 hr	48 hr
TUMOR	a	1.23	1.72	3.19	3.96	4.80	8.07	9.01	12.25	14.47
	b	.0156	.0212	.0214	.0705	.0466	.1300	.1176	.1667	.2612
BRAIN	a	.555	.43	.45	.45	.51	.90	.73	.70	1.43
	b	.0066	.0057	.0064	.0063	.0072	.0096	.0067	.0095	.0139
MUSCLE	a	1.01	1.12	.97	1.03	1.06	.79	1.06	.92	1.31
	b	.0250	.0315	.0131	.0135	.0176	.0122	.0134	.0116	.0159
SKIN	a	1.16	1.36	1.13	1.36	1.22	.81	1.19	.98	1.57
	b	.0143	.0176	.0143	.0179	.0188	.0110	.0149	.0126	.0149
LIVER	a	4.84	6.56	5.38	5.96	8.15	5.91	6.15	8.21	11.10
	b	.0601	.0858	.0682	.0885	.1336	.0729	.0746	.1066	.1599
KIDNEY	a	3.89	3.83	4.45	6.08	5.06	4.58	4.04	4.85	4.14
	b	.0476	.0495	.0579	.0838	.0855	.0668	.0490	.0648	.0637
BLOOD	a	80.94	77.42	75.26	71.92	68.51	71.37	74.00	71.93	68.74

a. Percent of the injected dose per gram tissue.
b. Vascular Volume, (ml) blood per gram tissue.

TABLE 2. TISSUE CONTENT AND PLASMA VOLUMES FROM I-131 LABELED HSA

TISSUES		Time after injection								
		10 min	30 min	1 hr	2 hr	4 hr	6 hr	12 hr	24 hr	48 hr
TUMOR	a	4.32	7.24	9.59	11.72	11.36	12.28	12.19	7.08	4.14
	b	.0382	.0678	.1093	.1424	.1699	.1768	.2545	.2288	.2691
BRAIN	a	.83	1.11	1.11	1.76	2.08	1.99	3.77	1.87	1.07
	b	.0072	.0102	.0113	.0229	.0317	.0305	.0814	.0583	.0694
MUSCLE	a	1.77	1.55	2.27	1.39	1.85	1.40	1.67	1.55	0.98
	b	.0132	.0143	.0209	.0167	.0271	.0233	.0376	.0465	.0610
SKIN	a	2.88	3.25	2.63	2.51	4.73	3.70	4.65	5.22	4.67
	b	.0201	.0298	.0261	0299	.0725	.0676	.1104	.1453	.2706
LIVER	a	15.35	11.76	11.83	6.89	5.49	5.54	3.91	3.12	1.72
	b	.1222	.1061	.1006	.0874	.0791	.0866	.0841	.0912	.1074
PLASMA	a	127.8	108.91	105.69	86.21	68.57	71.08	46.64	33.98	16.22

a. percent of the injected dose per gram tissue.
b. Plasma volume, (ml) plasma per gram tissue.

muscle, liver, kidney, blood, and plasma, and in units of ml water/g tissue for varying time periods. There is a rather constant extracellular fluid volume in tumor for the first hour, averaging 0.33 ml/g tissue, and this volume increases to about 0.50 ml/gram tissue after 2 hours. Brain tissue seems to undergo a rather steady increase in extracellular fluid volume from 0.052 ml/g tissue to 0.34 ml/g tissue over the time period studied. The extracellular fluid volume of muscle remains constant, averaging 0.11 ml/g tissue over the 12-hour period studied. Extracellular fluid volume of kidney is constant over the life of the study, averaging 0.5 ml per gram of tissue. Five of the 7 liver values are essentially constant, averaging 0.40 ml/g tissue, while the 2 and 12-hour points are statistically different from the others, averaging 0.61 ml/g liver.

In Figure 2 are plotted the apparent vascular, plasma, and extracellular fluid volumes of tumor with time. Both the vascular volumes and the extracellular fluid volumes follow the same general pattern after the first hour which is quite different from the curve depicting the increase in the apparent plasma volume. If the apparent vascular volume is subtracted from the apparent extracellular fluid volume, the calculated value of the interstitial space is remarkably constant over 12 hours averaging 0.39 ml/g tumor if the 10-minute point is omitted. The increase in apparent tumor plasma volume with time appears to be the result of a far more regular physiologic process than are the increases in apparent vascular and extracellular fluid volumes.

DISCUSSION

In tracer kinetics, "compartment" has been defined as "an anatomical, physiological, chemical or physical subdivision of a system throughout which the ratio of concentration of labeled to unlabeled substance is uniform at any given time. Space has been defined as "an apparent volume obtained by dividing the amount contained in the compartment by the concentration of the tracer" (6). These definitions apply to the determination of compartmental volumes measured in this mouse brain tumor model. The vascular space of a tumor can thus be estimated from the concentration of a molecule, ^{51}Cr labeled red blood cells, which must be uniformly distributed through the vascular bed, but which should not have penetrated into the interstitial fluid. Each tumor has its own vascular pattern, that part of the internal milieu which is in direct connection with the blood stream. The vascular space increases proportionally with the size of the tumor. Most brain tumors contain more blood vessels per unit of tissue than normal brain because tumor growth requires more blood. As the tumor grows, the pre-existing arteries, veins, and capillaries in the area where the tumor is growing adapt for vascularization of that tumor. New vessels, usually thinwalled, fragile, and lacking in both muscular coat and elastic lamella, start forming in various directions surrounding the tumor. These abnormalities of the neoplastic blood vessels are responsible for the changes in the vascular permeability and are considered to play an important role in the accumulation of radionuclide tracers in the lesion. In this model the apparent increase in vascular volume after one hour is probably not a true increase since visual observation of the tumor shows an increasing number of cells packing into the tumor with time(sludging). Other contributing factors to this increased apparent vascular volume with increase in time could be tumor growth, an

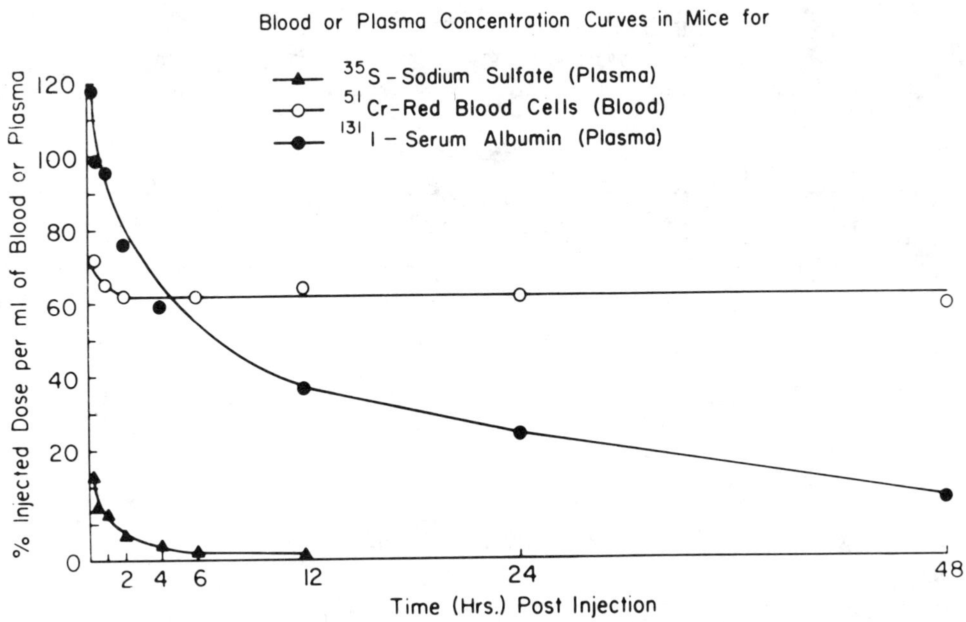

Figure 1

Blood or Plasma concentration curves in Tumor-bearing Mice for ^{35}S-Sodium sulfate (Plasma), ^{51}Cr-Red Blood Cells (Blood) and ^{131}I-Serum Albumin (Plasma) for 12 or 24 hours.

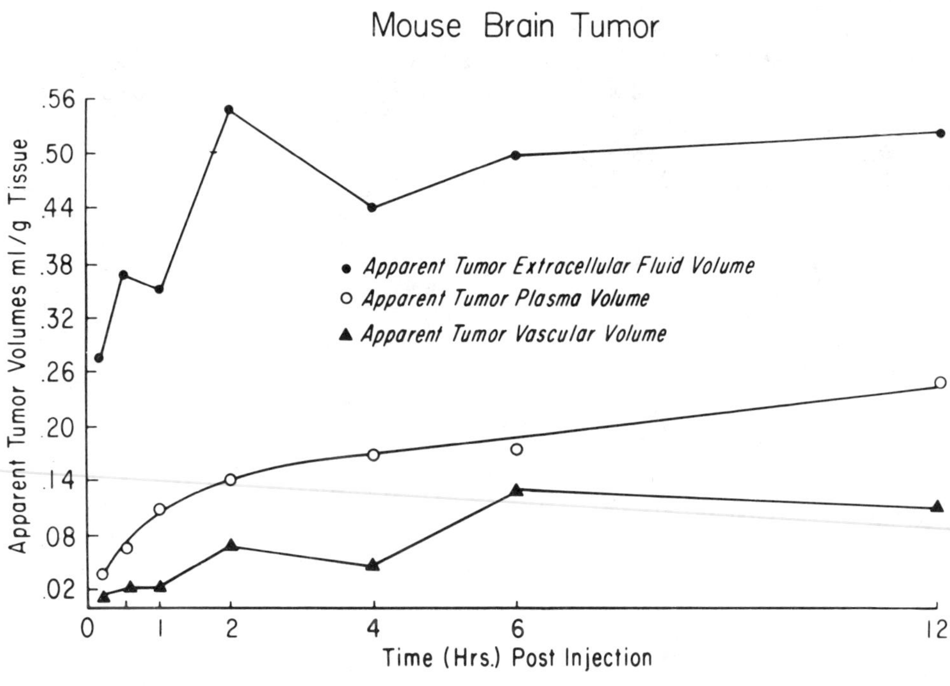

Figure 2

Apparent Tumor Extracellular Fluid, Plasma and Vascular Volumes in Brain Tumor-bearing Mice for 12 hours.

TABLE 3. TISSUE CONTENT AND ECF VOLUMES FROM S-35 SODIUM SULFATE

TISSUES		Time after injection						
		10 min	30 min	1 hr	2 hr	4 hr	6 hr	12 hr
TUMOR	a	5.90	5.03	4.09	3.55	1.60	1.23	0.74
	b	.2735	.3663	.3498	.5472	.4404	.4959	.5198
BRAIN	a	1.17	1.69	1.30	1.07	0.67	0.55	0.49
	b	.0524	.1216	.1095	.1618	.1865	.2232	.3438
MUSCLE	a	2.48	1.55	1.20	0.88	0.39	0.24	0.17
	b	.1135	.1112	.1019	.1369	.1094	.0955	.1192
LIVER	a	7.32	4.78	4.80	3.75	1.43	1.21	0.88
	b	.3359	.3441	.4122	.6081	.3946	.4911	.6111
KIDNEY	a	9.86	6.29	5.01	3.83	1.76	1.32	1.05
	b	.4529	.4446	.4146	.5916	.4851	.5342	.7355
PLASMA	a	22.27	14.01	12.54	7.10	3.71	2.62	1.50

a. Percent of the injected dose per gram tissue.
b. Extracellular Fluid Volume $(H_2O)E$, ml per gram tissue.

increase in tumor capillary permeability, or metabolic reaction between tumor cells and the red blood cells. Gullino (7) has shown that knowledge of the size of the vascular space can help in three areas of study: the measurement of the vascular pool of a given substance in the tumor; the evaluation of permeability of the vascular net of a tumor; and the compartmentalization of the environment of the neoplastic cell population.

The vascular volume of the brain is much less than tumor and remains low and essentially constant throughout the entire lifetime of the study. The vascular volume of the skin, muscle, and kidney also remains relatively constant. The apparent peak changes in vascular volume within the liver are consistent with the theory that the liver may act as a storage depot for blood, holding or releasing it on demand of organs and tissues, or when the blood volume is markedly changed (8).

Iodine-131 labeled human serum albumin was used for the determination of tissue plasma volume in this mouse brain tumor model. Within the tumor, labeled albumin accumulates largely in the extracellular space, but is also found intracellularly, probably due to pinocytosis. It is known that proteins cross capillary membranes through pores and that other factors such as vascularity, size of interstitial space, and metabolic activity of tumor cells all affect the tumor uptake of radioiodinated human serum albumin. When tumor plasma volume is converted to vascular volume using the plasmacrit, 0.497, the value obtained for the vascular volume for the 10 minute point, 0.019 ml/g tumor, is in excellent agreement with the vascular volume obtained using the labeled red blood cells. However, immediate "leakage" from the vascular system into the tumor, probably caused by capillary permeability and perfusion, commences and the apparent vascular volume grows larger with time. The vascular volume for brain and other tissue, with the exception of liver, share this same general pattern of increase in apparent plasma with time. Only the liver, which shows a decrease in apparent plasma volume for the first 4 hours, changes this general pattern.

The third hematological compartment measured in this study was extracellular fluid volume as measured by ^{35}S sulfate. Although there is no ideal agent for measuring extracellular fluid space, ^{35}S sulfate is probably the best radioactive substance available despite its limitation of being only a weak beta emitter. The results presented here indicate that the distribution of sulfate approximates that of extracellular fluid volume. The ^{35}S sulfate content of all tissues, blood, and plasma, except brain, decreases rapidly to a lower and more constant level during the 12 hour time period covered by this study. The initial rapid disappearance rate has been attributed to the mixing of the sulfate in the plasma and equilibration with the water outside the vascular bed, while the slower second part of the curve represents equilibrium between plasma and extracellular fluid. In humans, the rapid phase of blood and plasma disappearance curve has been reported to be about 20 minutes, and in dogs, about 30 minutes (9). In cats, the slow disappearance rate of the radiosulfate from plasma seems to start after 4 hours (10). In humans, it has been reported that sulfate equilibrium between plasma and tumor ascites fluid occurs in 5.5 hours (11). In our animal model, using tumor-bearing mice, this constant blood and plasma concentration occurs after 4 hours. Therefore the best data to calculate extracellular fluid volume will be the data from the later time periods.

Tumor extracellular fluid volume is 0.51 ml/g tissue. The best value for brain is a smaller value, 0.28 ml/g tissue. For muscle, extracellular fluid space is 0.11 ml/g tissue. The values for brain and muscle agree well with values reported for normal brain and muscle for mice, rats, rabbits, and dogs (12). No comparative data for tumor has been found. There is greater variation between the values for extracellular fluid space obtained in this study for kidney and liver from values reported in literature (12). This is probably because our animals were tumor-bearing animals and the tumor undoubtedly influences kidney and liver function. Our values for extracellular space in the tumor-bearing mouse is about 0.59 ml/g kidney. The values fround in literature for the normal mouse, rat, rabbit, and dog, vary from 0.36 to 0.50 ml/g. For liver, two statistically significant (P = 0.006) points of increased extracellular fluid volume at 2 and 12 hours were noted. Our average value for non-peak points, 0.40 ml/g liver, is in close agreement with the ^{35}S sulfate value of 0.36 ml/g liver (12) reported for the rat, but is somewhat greater than the values 0.18 - 0.24 ml/g, reported for normal mice, rats, rabbits, and dogs.

Data generated by this study make possible the determination of one other physiological parameter often of interest, that of interstitial space. Interstitial space is the difference between the extracellular fluid space and the vascular space. The interstitial space for this tumor can thus be calculated to give a best value of 0.49 ml/g tumor. In a like manner, the best value for interstitial space for brain can be estimated to be 0.28 ml/g tissue. With reference to Figure 2, the interstitial space remains surprisingly constant over the first 12 hours of the study, averaging 0.40 ml/g tumor if the 10 minute point is not included. This work emphasizes the close relationship between the vascular volume and the extracellular fluid space volume, even in tissue, and lends credence to the statement that the extracellular volume is the prime regulator of the vascular volume, while intracellular volume plays little or no important role in this regulation.

Compartmental studies of this nature, measuring vascular, plasma, and extracellular fluid volumes in tumor and tissues serve to clarify problems of the dynamics of circulation. They also add dimension to blood and plasma measurements and help describe the characteristics and distributions of various fluids between compartments in neoplasms and in normal tissues.

REFERENCES

1. Song CW and Levitt SH: Effect of X irradiation on vascularity of normal tissue and experimental tumor. Radiology 94:445-447, 1970

2. Farr LE and Konikowski T: The effect of regional thermal neutron exposure upon the growth and transplantability of a malignant tumor in the mouse. In Biological Effects of Neutron and Proton Irradiations 2, IAEA, Vienna, 1964, pp 159-160.

3. Haynie TP, Konikowski T and Glenn,HJ: The kinetics of ^{99m}Tc, ^{113m}In and ^{169}Yb DTPA compounds in brain sarcoma and kidneys of mice. J Nucl Med 13:205-210, 1972

4. Konikowski T, Glenn HJ and Haynie TP: Kinetics of ^{67}Ga compounds in brain sarcoma and kidneys of mice. J Nucl Med 14:164-171, 1973

5. Manery JF and Bale WF: The penetration of radioactive sodium and phosphorous into the extra and intra-cellular phase of tissues. Am J Physiol 132:215-231, 1941

6. Brownell GL, Berman M and Robertson JS: Nomenclature for tracer kinetics. Int J Appl Radiat Isot 19:249-262, 1968

7. Gullino PM and Grantham FH: The vascular space of growing tumors. Cancer Res 24:1727-1732, 1964

8. Albert SN: Blood Volume, Thomas, Springfield, Ill., 1963

9. Walser MW, Seldin DW and Grollman A: An evaluation of radiosulfate for the determination of the volume of extracellular fluid in man and dogs. J Clin Invest 32:299-311, 1953

10. Barlow, CF, Domek NS, Goldberg MA, et al: Extracellular brain space measured by S-35 sulfate. Arch Neurol 5:102-110, 1961

11. Madison LL: Evaluation of radiosulfate as a measure of extracellular volume in edematous patients. Am J Med 19:147, 1945

12 Locksley HB: Suitability of tumor bearing mice for predicting relative usefulness of isotopes in brain tumors. AMA Arch Neurol and Psychiatry 71:684, 1954

THE KINETICS OF UPTAKE AND MACROMOLECULAR BINDING OF GA-67 AND FE-59 BY THE EMT-6 SARCOMA-LIKE TUMOR OF BALB/C MICE

Steven M. Larson,* Janet S. Rasey†
Norma J. Nelson,† Zdenka Grunbaum*
David R. Allen,† George D. Harp*
and David L. Williams*

*Nuclear Medicine Service,
Veterans Administration Medical Center
and
†Division of Radiation Oncology and Nuclear Medicine,
Department of Radiology,
University of Washington Medical School
Seattle, Washington 98195

ABSTRACT

In order to compare the subcellular metabolism of ^{67}Ga and ^{59}Fe, we have studied the EMT-6 sarcoma of BALB/c mouse, a tumor with both in-vitro and in-vivo growth forms. In-vitro, ^{67}Ga and ^{59}Fe show similar temporal uptake profiles into exponentially growing EMT-6 cells, with a gradual increase over the first 24 hours of incubation and subsequent decline thereafter. In-vivo, 24 hours after intravenous administration of ^{67}Ga-citrate and ^{59}Fe-citrate to mice bearing subcutaneous EMT-6 tumors, most of the subcellular ^{67}Ga and ^{59}Fe is bound to discrete macromolecular components of approximate molecular weight, as follows: >250,000 MW, 80-90,000 MW, and 40-60,000 MW.

INTRODUCTION

The EMT-6 sarcoma-like mammary tumor of the BALB/c mouse grows both in-vivo, as a solid subcutaneous tumor, and in-vitro, as a tissue monolayer in long-term tissue culture (1). In previous communications, we reported an important role for transferrin in the cellular concentration of gallium-67 (^{67}Ga) by the EMT-6 sarcoma-like tumor of the BALB/c mouse (2,3). Competitive binding studies in-vitro, which employed iodine-125 labelled transferrin, ^{59}Fe-citrate and ^{67}Ga-citrate, implied that both iron and gallium are predominantly concentrated by the cell via a saturable transport process that involves a transferrin receptor (4). Because iron and gallium seem to be taken-up by a common intracellular transport system, we have sought further comparisons of the metabolism of iron and gallium by tumor. In the present study, we compared: 1) the detailed time-course of ^{59}Fe and ^{67}Ga uptake by EMT-6 sarcoma growing as a monolayer in long-term tissue culture, and 2) the macromolecular binding of the ^{59}Fe and ^{67}Ga which was concentrated by the EMT-6 sarcoma-like tumor growing subcutaneously in-vivo.

MATERIALS AND METHODS

In order to study the time-course of uptake of ^{125}I-transferrin, ^{67}Ga-citrate, and ^{59}Fe-citrate by EMT-6 sarcoma tumor cells growing in tissue culture, the following experiments were employed.

EMT-6 cells were grown in tissue culture in Waymouth's MB752/1 medium supplemented with 15% fetal bovine serum and 100 units of penicillin, and 100 μg of streptomycin per milliliter, as described by Rockwell and associated (1). 2.5 x 10^5 cells were innoculated in 5 milliliters of Waymouth's medium into a 25 cm^2 Falcon flask (10^4 cells/cm^2 of growth area). Cells were incubated at 37°C for 24 hours. At this time, the cells were growing exponentially. To each flask was added, in 0.5 ml Waymouth's, the following: 0.5 μCi of ^{67}Ga-citrate (Mediphysics), 0.25 μCi of ^{59}Fe-citrate (ferrous citrate, New England Nuclear), and 0.1 μCi of ^{125}I-transferrin (^{125}I-TF). (The ^{125}I-TF was prepared using a lactoperoxidase technique [5] and the specific activity was about 10 μCi ^{125}I per μg of transferrin; very high specific activity, iodine-125 [New England Nuclear], lactoperoxidase [Sigma], and human transferrin [Grade B, iron-free, >90%, Sigma] were used). Human transferrin (Grade B, greater than 90%, iron-free, Sigma) was added to the flasks in the following concentrations: 0 mg/ml (control); 0.0062 mg/ml, 0.025 mg/ml, 0.25 mg/ml, and 2.5 mg/ml. The flasks were then incubated for an additional 48 hours at 37°C. Duplicate flasks were harvested at the following intervals after beginning the second incubation: 30 minutes, 1 hour, 2 hours, 4 hours, 6 hours, 24 hours and 48 hours. At the time of harvest, cells were treated with a dilute solution of magnesium- and calcium-free HBSS, containing 0.05% trypsin EDTA (ethylene diamine tetra-acetic acid) in order to free them from the monolayer and create a cellular suspension for easier separation. The cells were then washed three times with nonradioactive Waymouth's medium, using centrifugation to remove surface absorbed radioactivity. The mixture was centrifuged for 10 minutes at 200 x G at 20°C. Final cell counts in each flask were determined. The harvested cells were assayed for radioactivity content in a Packard 5900 series automatic gamma counter and the fraction of the total dose present in the medium, which was incorporated into the cells, was determined by reference to the appropriate standard. Samples were counted to less than 1% relative error. The results were expressed as the percent of the total dose of ^{125}I-TF, ^{67}Ga, and ^{59}Fe, taken-up per 10^7 EMT-6 tumor cells as a function of transferrin concentration, at the various time intervals after addition of the radiotracers.

Female BALB/c mice (Simonsen) were obtained at 5 weeks of age, and were held for 2 weeks prior to use. Approximate weight of these animals was 20 g. Tumors were transplanted into the back by injecting 2×10^5 EMT-6 tumor cells in 0.1 ml of Hanks balanced saline solution. "Takes" were extremely reproducible, but spontaneous regression also occurred. After about 10 days, tumors of about 100 mg were obtained.

A comparison of the macromolecular binding of ^{59}Fe and ^{67}Ga was performed as follows: 20 µCi of ^{59}Fe (ferrous) citrate (New England Nuclear) and 50 µCi of ^{67}Ga-citrate (Mediphysics) were injected intravenously via the tail vein into each of 8 to 10 BALB/c mice with tumors of between 50-200 mg. Twenty-four hours later, the animals were sacrificed and about 8-10 tumors (total weight about 1 g) were removed from the animals. An outline of the separation of these tumors into various fractions is shown in Fig. 1. The tumors were washed with cold 0.25 M sucrose, and then the tumors were pressed through a fine mesh wire screen to obtain a uniform cellular suspension and remove gross fibrous elements. After washing the cells further with saline, the cells were homogenized with about 10 ml of distilled water, using about 40 strokes at 400 RPM of homogenizer. The resulting cellular suspension was then centrifuged for 10^6 g-minutes in a refrigerated centrifuge. About 50% of the ^{59}Fe and ^{67}Ga activities stayed in the supernatant fraction ("Lysate"), and the remaining 50% was found in the sediment ("Stroma"). The stroma was then incubated with 1% teric acid in water for 1 hour at 0°C. The stroma in teric acid was agitated and the suspension re-centrifuged for 10^6 g-minutes. About 25% of the original radioactivity in the tumor was made soluble by this treatment, and stayed in the supernatant ("teric extract"). After concentration of the sample to about one-half of the original volume, using an Amicon ultrafiltration cell and PM-30 membrane, both the lysate fraction and the teric extract fraction were then passed over a Sephacryl S-200 SF column in 0.1 M Tris-HCl at pH 7.4, and the eluate was collected in 2 milliliter fractions. The optical density at 280 nm and the radioactivity content for ^{59}Fe and ^{67}Ga was assayed in each fraction. An experiment designed to assess the rate of loss of radioactivity from the EMT-6 sarcoma cells growing in tissue culture was performed. In a separate experiment, after only 4 hours of incubation with medium containing ^{125}I-TF, ^{59}Fe, and ^{67}Ga, the radioactive medium was removed, and nonradioactive medium was added to the tissue culture flasks. Duplicate flasks were harvested at 0, 30, 60, 90, and 120 minutes of incubation in nonradioactive condition medium. The results are shown in Table 1.

RESULTS

A summary of the uptake data for ^{125}I-TF, ^{67}Ga, and ^{59}Fe by exponentially growing EMT-6 sarcoma is shown in Figs. 2A, B, and C, as a function

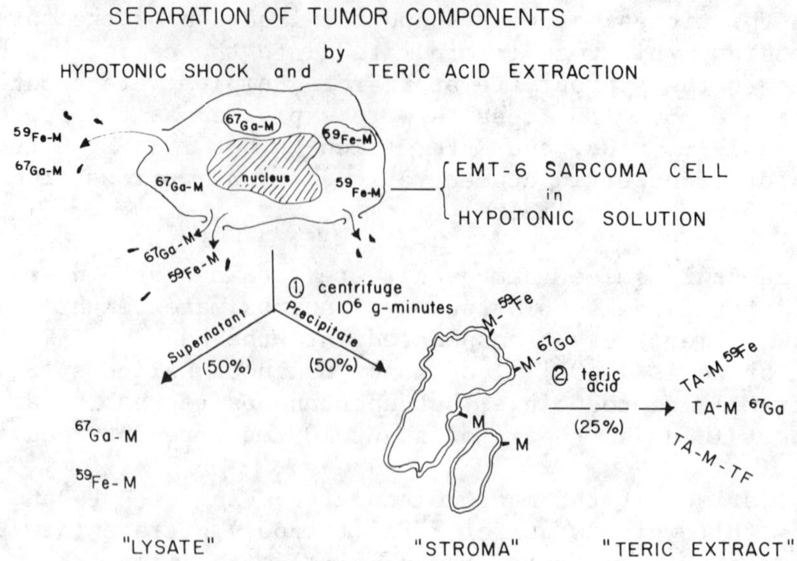

FIG. 1. Diagram of the hypotonic shock method (after Hayes and Carlton [12] for separating the tumor cells into various subfractions. Stromal elements and soluble (intracellular) macromolecules are separated by first lysing the cells and centrifugation. Teric acid treatment may solubilize some macromolecules which are membrane bound.

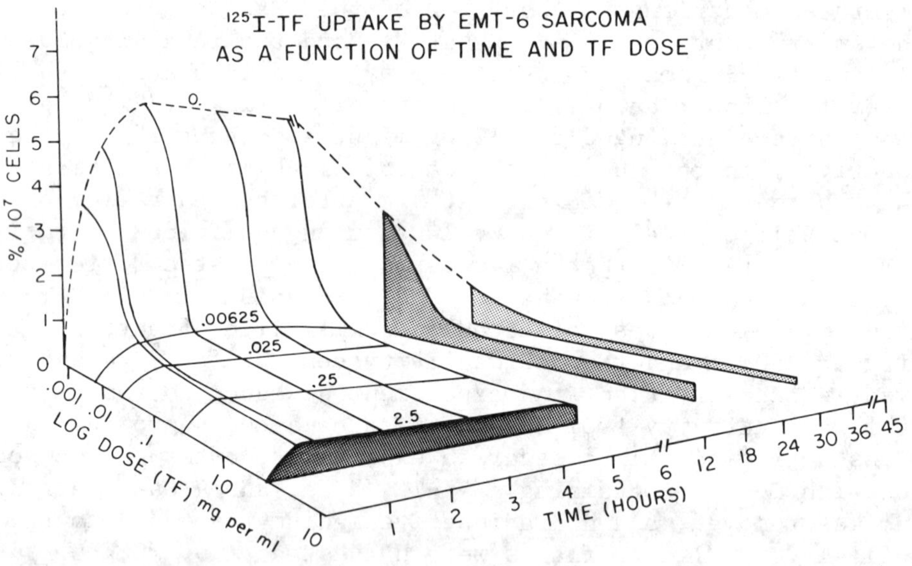

FIG. 2. Uptake of ^{125}I-labelled transferrin (^{125}I-TF), ^{67}Ga-citrate, and ^{59}Fe-citrate by exponentially growing EMT-6 tumor cells in long-term tissue culture as a function of time and human transferrin (TF) medium concentration,

FIG. 2A. ^{125}I-TF uptake reaches a plateau after 1-1/2 to 2 hours of incubation in these exponentially growing cells. Increasing TF medium concentration displaces ^{125}I-TF from the cells.

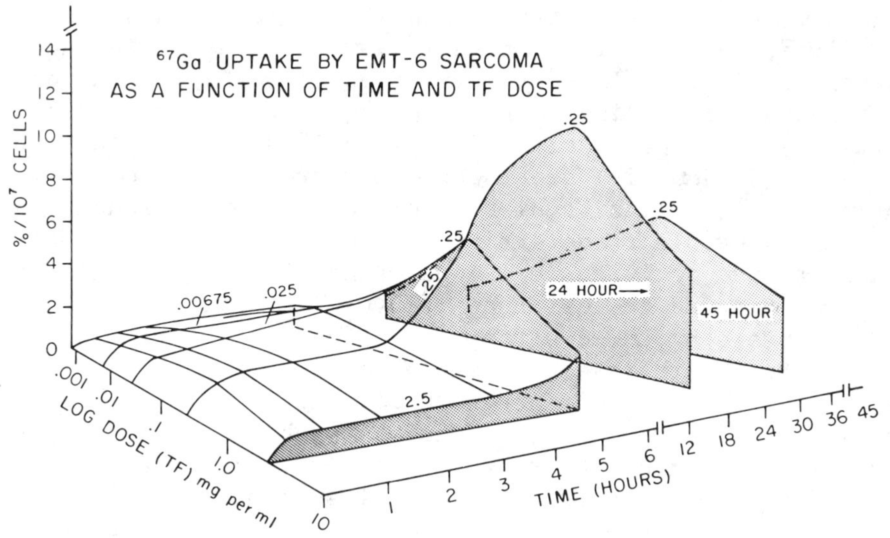

FIG. 2B. ^{67}Ga uptake increases at all times with increasing TF concentration up to 0.25 mg/ml. There is a decline with further increase in TF concentration. Cellular concentration of ^{67}Ga reaches a peak at 24 hours of incubation and falls thereafter.

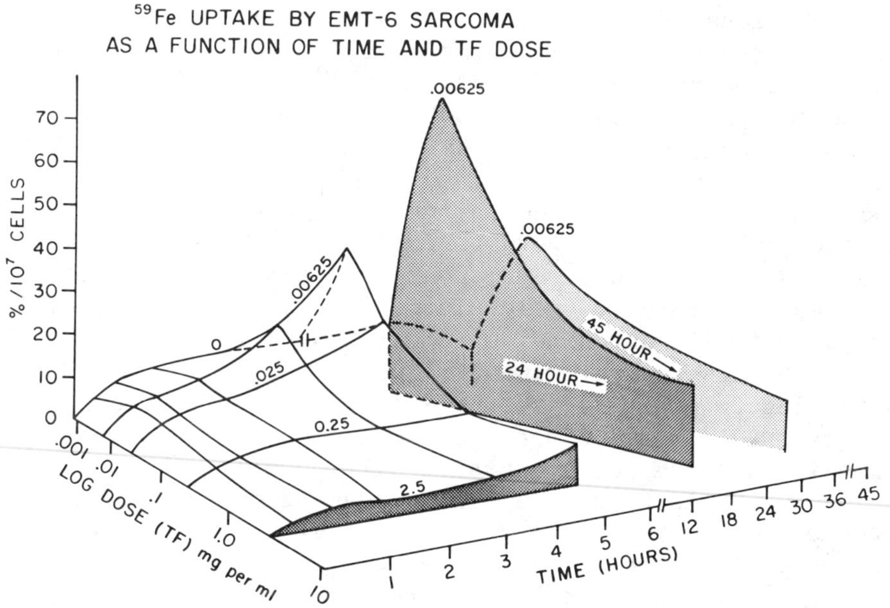

FIG. 2C. ^{59}Fe uptake increases at all times with increasing TF concentration up to only 0.00625 mg/ml.

There is a sharp decline as TF concentration increases further. Cellular concentration of ^{59}Fe reaches a peak at 24 hours of incubation, and falls thereafter.

of the log dose of human transferrin, and the time of incubation. For the labelled ^{125}I-TF, the cellular uptake is always greatest in the absence of any added unlabelled TF. The uptake reaches an early plateau by 2 hours, followed by a gradual decline so that uptake at 24 hours is about one-half the peak value. A further decline in uptake is seen at 45 hours, so that maximum uptakes (0 added TF) are about one-fourth of the maximum uptakes at 2 to 6 hours. At all times studies, there was a marked reduction of ^{125}I-TF uptake with added stable transferrin. For 0.00625 mg/ml, about one-third of the original maximum activity (0 TF) was taken-up by the cells; at 2.5 mg/ml, this was reduced to less than 1/10 of the uptake at 0 add TF. The data for ^{67}Ga uptake by EMT-6 sarcoma as a function of time and TF dose is shown in Fig. 2B. There is a gradual increase to a peak in uptake, of about 12%/10^7 cells as TF concentration is increased from 0 to 0.25 mg/ml, with a decline at higher concentrations. A peak in uptake is achieved at 24 hours after addition of the radioactivity. There is a decline in uptake thereafter, so that by 45 hours uptake is less than one-half of the uptake at 24 hours. Uptake of ^{59}Fe as a function of time and TF dose is shown in Fig. 2C. As TF concentration is increased from 0 to .00625 mg/ml, there is a marked increase in ^{59}Fe uptake, so that 60% of the total added radioactivity is concentrated by the cells. As TF concentration is increased, there is a marked reduction in the amount of radioactivity taken-up. Also, there is a definite fall in uptake after 24 hours, to about one-half at 45 hours.

Radioactivity that is taken-up into the cells is not totally retained. As shown in Table 1, there was a precipitous loss of ^{125}I-TF from the cells when the radioactive medium was replaced with nonradioactive medium, so that within 2 hours, less than 10% of the original activity was retained. Loss of ^{67}Ga and loss of ^{59}Fe was less rapid, with about one-third of the ^{67}Ga, and one-fourth of the ^{59}Fe being lost over the 2 hour interval. A comparison of the macromolecular binding of ^{59}Fe and ^{67}Ga in the "lysate" (Fig. 1) is shown in Fig. 3A and 3B. Fig. 3A shows the optical density at 280 nm as a function of fraction number for the eluate from a Sephacryl S-200 column. This column had a molecular weight range of about 10,000 to 250,000. There were 5 macromolecular protein peaks (#1 through #5). The approximate molecular weights were as follows: peak #1 (void volume), >250,000 MW; peak #2, 80-90,000 MW; peak #3, 40-60,000 MW; peak #4, <10,000 MW; peak #5, very low molecular weight (total bed volume). For ^{67}Ga, there was significant binding to all 5 of these protein peaks, but in particular, binding was prominant to peaks #1, 2, and 3. For ^{59}Fe, there was binding to peaks #1, 2, and 3 (Fig. 3B). No binding of ^{59}Fe was evident to peaks #4 and 5. Since the sample was concentrated by ultrafiltration using a large pore membrane, all low molecular weight components (less than 30,000 MW) should have been removed from the sample. Also, there is an increase in the proportion of radioactivity in peaks #4 and 5 if the same sample is compared before and after prolonged incubation. Thus, we believe that peaks #4 and 5 represent breakdown products of uncertain type.

DISCUSSION

Sephton and Harris (6,7) were the first to describe the importance of transferrin to the uptake of ^{67}Ga in tumors growing in long-term tissue

TABLE 1. LOSS OF PEAK RADIOACTIVITY* FROM EMT-6 CELLS

	^{125}I-TF[1] %/10^7 Cells (Fraction**)	^{67}Ga[2] %/10^7 Cells (Fraction**)	^{59}Fe[3] %/10^7 Cells (Fraction**)
0	3.15 (1.00)	2.29 (1.00)	12.55 (1.00)
30	1.01 (0.32)	1.62 (0.71)	11.82 (0.94)
60	0.58 (0.18)	1.45 (0.63)	10.37 (0.83)
90	0.41 (0.13)	1.32 (0.58)	10.61 (0.85)
120	0.27 (0.09)	1.49 (0.65)	8.81 (0.70)

*4 hours after addition of radioactivity to exponentially growing cells, radioactive medium was removed, nonradioactive medium was added, and cells were harvested at intervals thereafter. Average of duplicate measurements are shown.

**Fraction of activity remaining, with radioactivity at 0 time taken as 1.0

1 ^{125}I-TF uptake at 0 added TF

2 ^{67}Ga uptake at 0.25 mg/ml added TF

3 ^{59}Fe uptake at 0.00625 mg/ml added TF

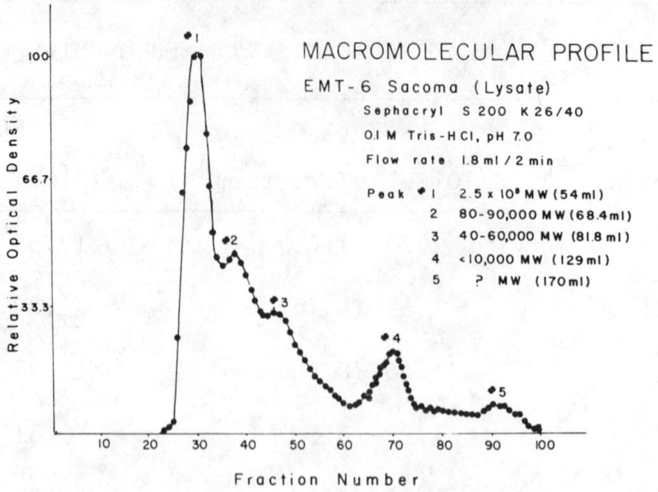

FIG. 3. Elution profile of EMT-6 sarcoma tumor lysate on Sephacryl S-200 SF.

FIG. 3A. Relative optical density at 280 nm, reflecting relative protein concentration, as a function of eluted fraction. Larger molecules are eluted first from the column. There were five molecular weight peaks. Since the sample was placed on the column only after concentration using an Amicon ultrafiltration cell with PM-30 membrane (passes all material less than 30,000 MW), the significance of peaks #4 and 5 is uncertain. These peaks may represent low molecular weight breakdown products, or proteins which interact with the Sephacryl column and thus have a filtration rate which is slower than expected.

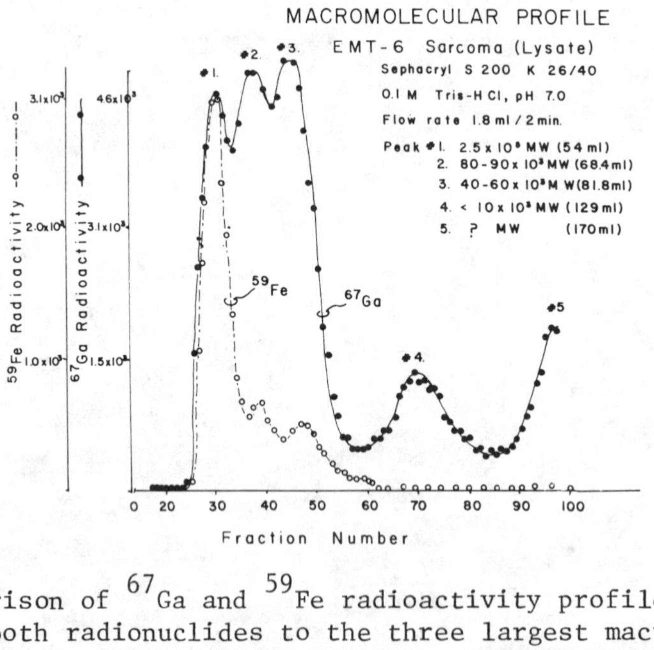

FIG. 3B. Comparison of ^{67}Ga and ^{59}Fe radioactivity profile. There is binding of both radionuclides to the three largest macromolecular fractions. ^{67}Ga is bound to peaks #4 and 5. The significance of this binding is uncertain.

culture. Our experiments have confirmed their findings that transferrin enhanced ^{67}Ga uptake in-vitro. We extended these observations to include the role of a tumor cell transferrin receptor in the uptake process (2,3) for both ^{59}Fe and ^{67}Ga (4). According to our view ("the transferrin receptor hypothesis"), at a given time of incubation, the shape of the uptake curve for both ^{67}Ga and ^{59}Fe as a function of TF medium concentration is related to two interacting processes: the formation of the metal transferrin complex (increases with increasing TF concentration), and the binding of the metal transferrin complex to the EMT-6 tumor cell (follows saturation kinetics and decreases with increasing TF medium concentration). We have shown that over a wide range of transferrin medium concentrations, the uptake curves for ^{67}Ga and ^{59}Fe are the result of these two independent processes (4). Differences in the shape of the curves for ^{59}Fe and ^{67}Ga uptake as a function of TF medium concentration, are principally related to the fact that ^{59}Fe has a much greater affinity for TF than does ^{67}Ga.

In discussion of the EMT-6 cellular "transferrin receptor," we understand that the term "receptor" is still principally an operational concept, and until we are in possession of a chemically purified, active cellular component with specific binding properties, we cannot be certain of the precise physical meaning of "transferrin receptor." We have used this term to mean a high affinity, saturable component of the intracellular transferrin transport system. Analogous "transferrin receptors" have been described for hematopoietic tissues (8-11).

When taken together with the "transferrin receptor hypothesis," the concept that emerges from the present experiments is that both ^{67}Ga and ^{59}Fe are actively metabolized by the EMT-6 sarcoma cell. This metabolism is the net result of several dynamic processes, including rapid intracellular transport of metal ion via a high turn-over transferrin transport system and the subsequent binding of ^{67}Ga and ^{59}Fe to several macromolecular components of the tumor cell. There is also some loss (back-diffusion?) of both ^{67}Ga and ^{59}Fe from the tumor cell as a function of time. However, the loss of ^{59}Fe and ^{67}Ga is low, relative to uptake, and there is a net retention and resulting progressive accumulation with time up to 24 hours. The loss of ^{125}I activity from the cell is very rapid (Table 1) and this, taken together with the plateau in ^{125}I-TF activity, must mean that uptake is equally rapid and between 2 to 6 hours of incubation (Fig. 2A) uptake and loss approximately balance. Thus, metal transferrin complex is rapidly taken-up into the cell; the metal ion is retained for the most part and the ^{125}I-TF is either regurgitated back into the medium or metabolized with resulting back diffusion of ^{125}I label into the medium.

In temporal terms, the uptake of ^{67}Ga and ^{59}Fe is maximal at 24 hours, and declines thereafter. Whether this phenomenon is related to changes in the biologic properties of the cells as they age in culture, or due to gradual loss of ^{59}Fe and ^{67}Ga from the cells, or a combination of both factors, is uncertain at present. In any event, the fact that ^{59}Fe and ^{67}Ga show very similar temporal uptake sequences, suggests major parallels in dynamic intracellular metabolism of the two metal ions.

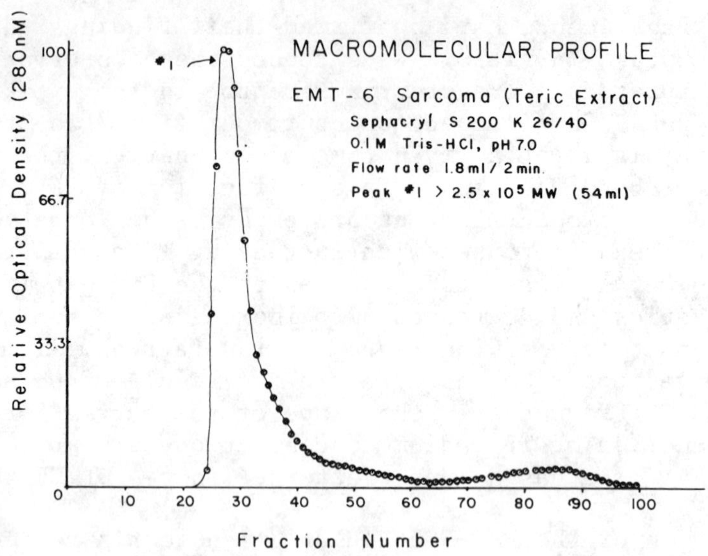

FIG. 4. Teric acid extract of EMT-6 sarcoma; elution profile on Sephacryl S-200 SF.

FIG. 4A. Relative optical density at 280 nm. Most of the optical density is in the largest peak >250,000 MW.

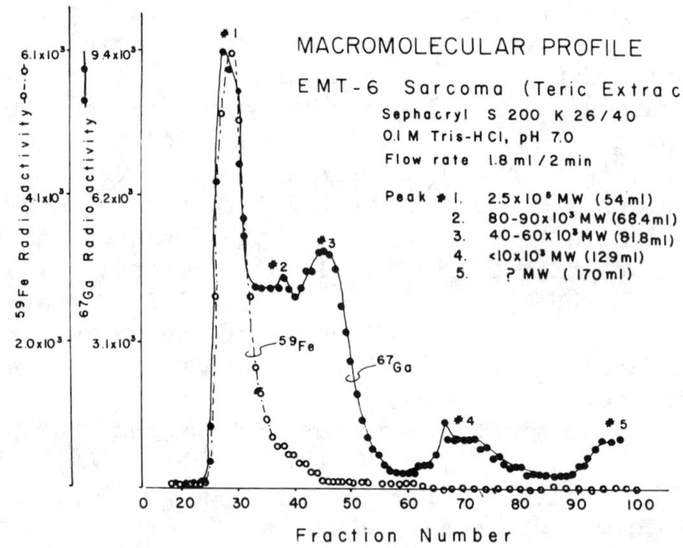

FIG. 4B. Comparison of ^{67}Ga and ^{59}Fe radioactivity profile. There is binding of both radionuclides to the largest macromolecular peak. ^{67}Ga is also evident in lower molecular weight peaks. Because of the relative abundance of ^{67}Ga binding to these peaks in the lysate fraction, this may reflect some contamination of the teric extract by the lysate fraction. The significance of binding in peak #4 and 5 is uncertain. See Figure Legend 3A.

Hayes and Carlton (12) demonstrated that ^{67}Ga in distilled water extracts of a variety of normal and malignant animal tumors was mainly present in macromolecular form. These authors have subsequently focused their attention on the 40-50,000 molecular weight macromolecule, because this component is present in greater concentration in 5123C minimal deviation hepatoma than in normal liver. However, these authors also observed five macromolecular peaks, corresponding to those found in the EMT-6. We modelled our experiments after theirs, because we were interested in comparing the behavior of ^{59}Fe to ^{67}Ga in terms of macromolecular components. ^{59}Fe elution profiles are qualitatively similar to ^{67}Ga elution profiles, in that ^{67}Ga and ^{59}Fe binds to three macromolecular fractions of similar molecular weight. Whether these binding substances are identical for the two metal ions will require further study.

In conclusion, ^{59}Fe and ^{67}Ga follow similar temporal patterns of uptake by exponentially growing EMT-6 sarcoma tumor cells growing in tissue culture. Also, at 24 hours after intravenous administration of ^{67}Ga-citrate and ^{59}Fe-citrate in-vivo, most of the ^{67}Ga and ^{59}Fe radioactivity is bound to macromolecules. ^{59}Fe and ^{67}Ga bind to three macromolecular fractions of similar molecular weights: >250,000; 80-90,000; and 40-50,000. In addition ^{67}Ga activity is found in association with lower molecular weight components (breakdown products). Thus, there are major similarities in the metabolism of ^{67}Ga and ^{59}Fe by EMT-6 sarcoma-like tumor of the BALB/c mouse.

REFERENCES

1. Rockwell SC, Kallman RF, and Fajardo LF: Characteristics of serially transplanted mouse mammary tumor and its tissue culture adapted derivative. J Natl Cancer Inst 49: 735, 1974

2. Larson SM, Rasey JS, Allen DR, Nelson NJ: A transferrin mediated uptake of Gallium-67 by EMT-6 sarcoma. I. Studies in-vitro (tissue culture). J Nucl Med (in press)

3. Larson SM, Rasey JS, Allen DR, Grunbaum Z: A transferrin mediated uptake of Gallium-67 by EMT-6 sarcoma. II. Studies in-vivo (BALB/c mice). J Nucl Med (in press)

4. Larson SM, Rasey JS, Allen DR, Nelson NJ, Grunbaum Z, Harp GD, and Williams DH: The transferrin receptor hypothesis: Mechanism of tumor uptake of carrier-free gallium-67. In Horst W. and Wagner HN Jr (ed) Frontiers in Nuclear Medicine. Springer-Verlag (in press)

5. Thorell JI, Johansson BG: Enzymatic iodination of polypeptides with ^{125}I to high specific activity. Biochim Biophys Acta 251: 363, 1971

6. Sephton RG, Harris AW: Gallium-67 uptake by cultured tumor cells, stimulated by serum transferrin. J Natl Cancer Inst 54: 1263, 1974

7. Harris AW, Sephton RG: Transferrin promotion of ^{67}Ga and ^{59}Fe uptake by cultured mouse myeloma cells. Cancer Res 37: 3634, 1977

8. Baker E, Morgan EH: The kinetics of the interaction between rabbit transferrin and reticulocytes. Biochemistry 8: 1133, 1968

9. Hemmaplardh D, Morgan EH: The role of endocytosis in transferrin uptake by reticulocytes and bone marrow cells. Br J Haematol 36: 85, 1977

10. Glass J, Nunez MT, Fischer S, Lavidor LM: Transferrin receptors and iron transport in normal murine erythroid precursors and friend leukemia cells. Fed Proc 36: 930, 1977

11. Sullivan AL, Weintraub LR: Identification of ^{125}I-labelled rat reticulocyte membrane proteins with affinity for transferrin. Blood 52: 436, 1978

12. Hayes RL, Carlton JE: A study of the macromolecular binding of ^{67}Ga in normal and malignant animal tissues. Cancer Res 33: 3265, 1973

DIFFERENTIAL UPTAKE OF ^{67}Ga IN

CANINE TUMORS AND ABSCESSES

U. K. Terner, H. Wong, A. Noujaim, B. C. Lentle
and J. R. Hill

Division of Bionucleonics and Radiopharmacy
University of Alberta, Cross Cancer Institute, Dept.
Pathology, University Hospital, Edmonton, Alberta, Canada

ABSTRACT

Differentiation between the radiogallium uptake in canine tumor and abscesses was effected by using a double-labeled ^{67}Ga-^{125}I-transferrin and by measurement of the ratio of isotopes at various time intervals. An alternative procedure consisted of the injection of a mixture of ^{67}Ga-transferrin and ^{68}Ga-citrate and the estimation of the ^{67}Ga/^{68}Ga ratio over the region of interest. Transferrin receptor sites which are only present on tumor cell membranes are responsible for the differential uptake. This technique could be of potential value in distinguishing tumors from abscesses upon clinical imaging. While the mechanism of gallium uptake in tumors was found to be specific and mediated solely through transferrin this was not the case in abscesses. Incorporation of radiogallium in inflammatory conditions appears to be non-specific and not due to the *in vivo* labeling of leucocytes.

Supported by the National Cancer Institute of Canada.

INTRODUCTION

The application of ^{67}Ga-citrate in oncological Nuclear Medicine is the result of the observations of Edward and Hayes(1) that the radionuclide localizes in various soft tissue tumors, and focal inflammatory lesions (2,3). The mechanism of localization of this nuclide in tumors is uncertain (9,10,11,12). Equally difficult has been the explanation of the radionuclide incorporation in inflammatory foci(2,3,4,5,6,7). In the present study an attempt is made to elucidate the differences in uptake of gallium in these two different pathological conditions.

METHODS

Tumor induction

The model used for our experiments is the canine Transmissible Venereal Tumor(TVT). Tumor cells were obtained from the Veterans Administration Westside Hospital in Chicago, Ill. A colony of TVT-bearing dogs was established and maintained in our laboratories by injecting simple cell suspensions which were prepared by trypsinization of solid tumor tissues at 37°C and their subsequent cryopreservation. Tumor growth regularly occurs following experimental inoculation of random dogs with tumor cells(18). The neoplasm follows a variable clinical course under experimental conditions where logarithmic growth is usually followed by spontaneous regression after six months(13). Healthy mongrel dogs of both sexes, weighing approximately 20 kg were chosen as tumor hosts. TVT tumor induction was accomplished by the subcutaneous injection of cell suspensions in the lateral abdominal region. The dogs were inspected weekly to ensure tumor development at the injection sites. The size of tumor reached about 6 cm in diameter 20 days after transplantation and remained relatively constant for three to six months. A total of six dogs were used and experiments were repeated up to 4 times on some dogs.

Abscess Formation

Dogs were first anesthetized with Na Pentobarbital (30 mg/kg body wt), I.V. and sterile inflammatory lesions were produced on the hind limbs by intramuscular injection of 0.5 ml turpentine about 24 hours prior to the administration of radiogallium. An inflammatory lesion was defined as an area of erythema and induration. During each experiment animals were maintained by I.V. drip of 3 1/3% dextrose - 0.3% NaCl injection. Body temperature was also monitored and maintained at 38°C. A total of six experiments were conducted.

Materials

Human apotransferrin was iodinated by the iodine monochloride method as described by Boullanger(14). ^{67}Ga-^{125}I-double labeled transferrin was prepared by a method described previously(27).

Gallium-68 was obtained in the form of ^{68}Ga-EDTA from a ^{68}Ge-^{68}Ga generator(New England Nuclear). The eluted material was mixed with an equal volume of 8N HCl and passed through a Dowex 1 x 4 anion exchange column. The column bed was then washed with 10 ml of 8N HCl. Radiogallium retained by the resin was then eluted by the addition of 0.1N HCl. ^{68}Ga-citrate was subsequently prepared by the addition of 15 - 20 mg of Na-citrate and adjusting the pH of the solution to 7.4.

^{99m}Tc-labeled canine leucocytes were prepared by the method described by Lentle _et al_(_15_). Approximately 300-500 μCi of ^{99}Tc^{m}-labeled cells were injected in each dog.

Radioactivity was measured by three (3/4" x 3/4") collimated NaI(Tl) crystals precalibrated for the radionuclides of interest. The detectors were positioned over the heart, the site of inflammation or tumor, and the thigh to provide dynamic observations. The detector outputs were fed into separate amplifier/analyzers and then into a 4-input time scaling multichannel analyzer. Data were recorded continuously and later corrected for channel spillover and radioactive decay. Spillover of ^{67}Ga and ^{68}Ga counts into ^{125}I channels were about 2.5% and 7% respectively. Spillover of ^{67}Ga counts into ^{99m}Tc channels was 25%. Blood clearance of a radionuclide was measured by the changes in radioactivity levels in the region of the heart and was compared to its accumulation in the abscess or tumor.

In each experiment, the ^{67}Ga radioactivity was first monitored for about 2.5 hours after the intravenous injection of ^{67}Ga-^{125}I-transferrin (~200 μCi and ~100 μCi). This was followed by the intravenous administration of about 50-100 μCi of ^{68}Ga-citrate. By performing ^{68}Ga-citrate control runs on the same day as the administration of double-labeled transferrin the possibility of day-to-day changes in the biologic and physiologic conditions of the tumor or abscess is eliminated. In the case of labeled leucocytes, both cells and labeled transferrin were injected and monitored simultaneously in abscess-bearing dogs.

RESULTS

In vitro studies by a number of investigators(_12_,_16_-_19_) have shown that in the presence of transferrin there is a significant increase in the uptake of ^{67}Ga by cultured cells. The importance of gallium-transferrin complexation in the rate of tumor uptake is illustrated in Figs. 1 and 2. The results in Fig. 1 indicate that when radiogallium, which is complexed to transferrin,is administered intravenously into dogs with TVT tumors there is an almost immediate, as well as rapid, uptake of ^{67}Ga by the tumor. This differs from the tumor uptake of gallium following the injection of ^{68}Ga-citrate(Figure 2). The time lag observed in the ^{68}Ga tumor uptake study may well reflect the time required for an amount of gallium to complex to plasma transferrin before uptake occurs. Such observations give strong credence to the concept that gallium must first be complexed to transferrin for uptake of gallium to occur by tumors as has been suggested by a number of workers(_12_,_16_-_19_). The importance of the

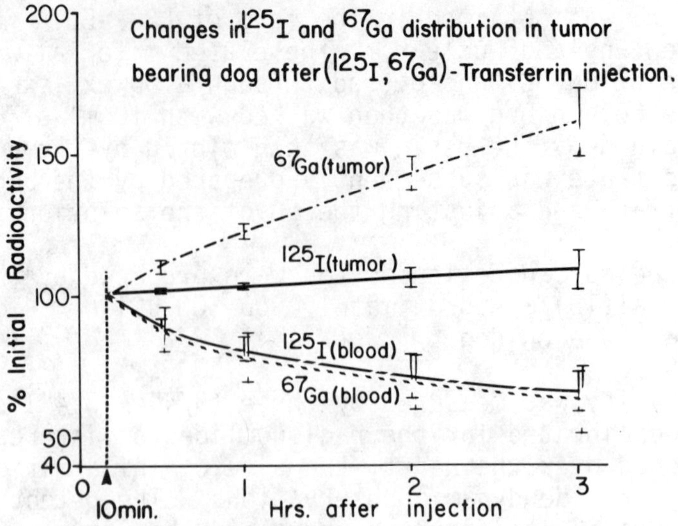

Figure 1.

Changes in ^{125}I and ^{67}Ga distribution in tumor bearing dog after ^{67}Ga-^{125}I-transferrin injection. The ^{67}Ga activity increased rapidly in the tumor while the ^{125}I activity remained relatively constant over three hours after injection. The blood levels of both ^{125}I and ^{67}Ga decreased in a similar manner.

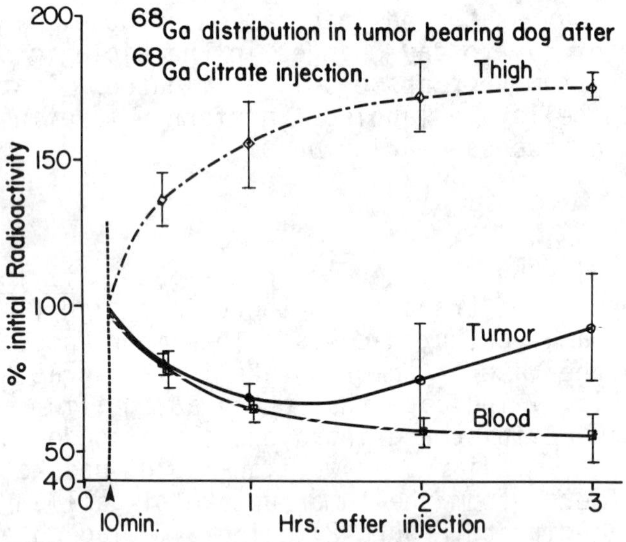

Figure 2.

^{68}Ga distribution in tumor bearing dog after ^{68}Ga citrate injection. Initially after injection, the activity in the tumor followed that of the blood while after about 1 1/2 hours the tumor demonstrated an increased uptake of ^{68}Ga(o). The blood level of ^{68}Ga declined steadily(□), while the activity in the thigh rose over the measured time period(◇).

TABLE 1. RELATIVE RADIOACTIVITY RATIOS OF ^{67}Ga, ^{68}Ga, AND ^{125}I AT VARIOUS TIMES FOLLOWING INJECTIONS OF [^{125}I, ^{67}Ga]-TRANSFERRIN AND ^{68}Ga-CITRATE IN A CANINE TVT MODEL

Post Injection Time	^{68}Ga tumor / ^{68}Ga blood	^{67}Ga tumor / ^{67}Ga blood	^{67}Ga tumor / ^{68}Ga tumor	^{67}Ga tumor / ^{125}I tumor	^{125}I tumor / ^{125}I blood
10 min.	1.00	1.00	1.00	1.00	1.00
30 min.	.99 ± .05	1.25 ± .04	1.41 ± .03	1.08 ± .01	1.16 ± .01
1 hr.	1.06 ± .06	1.60 ± .05	1.81 ± .05	1.19 ± .01	1.27 ± .01
2 hrs.	1.31 ± .03	2.07 ± .08	1.91 ± .17	1.33 ± .05	1.50 ± .06
3 hrs.	1.69 ± .06	2.51 ± .12	1.72 ± .15	1.45 ± .06	1.65 ± .07
7 hrs.	1.75*	3.50*			

Results in mean ±S.D. from 6 experiments   * Data from one experiment

TABLE 2. RELATIVE RADIOACTIVITY RATIOS OF ^{67}Ga,^{68}Ga and ^{125}I AT VARIOUS TIMES FOLLOWING INJECTION OF [^{125}I,^{67}Ga]-TRANSFERRIN AND ^{68}Ga-CITRATE IN A CANINE MODEL

Post Injection Time (Minutes)	^{67}Ga Abscess / ^{67}Ga Blood	^{125}I Abscess / ^{125}I Blood	^{67}Ga Abscess / ^{68}Ga Abscess	^{67}Ga Abscess / ^{125}I Abscess
10	1.00	1.00	1.00	1.00
30	1.28 ± .08	1.20 ± .06	.98 ± .07	1.01 ± .08
60	1.59 ± .09	1.45 ± .09	.96 ± .08	1.02 ± .12
90	1.88 ± .13	1.69 ± .10	.94 ± .10	1.01 ± .13
120	2.18 ± .15	1.91 ± .09	.90 ± .16	1.00 ± .14
150	2.38 ± .18	2.08 ± .10	.85 ± .15	.99 ± .14

Results in mean ± S.D. from 6 experiments

gallium-transferrin complexion is shown in Table 1. The fact that the $^{67}Ga/^{125}I$ tumor ratio shows a constant increase over the period of the experiment is indicative of the carrier-role that transferrin plays in the transport of radiogallium to the tumor. Furthermore, the ^{125}I tumor/blood ratio which equally shows significant increase during the same period reflects the presence of specific transferrin receptors on the tumor cell membranes. On the other hand, the uptake pattern of either gallium-citrate or gallium-transferrin appears to be identical in canine abscesses (Figure 3). Indeed, no dissociation of radiogallium from the transferrin complex is evident at the abscess site. This is demonstrated by the fact that the $^{67}Ga/^{125}I$ ratio in the abscess was approximately 1 throughout our experiments(Table 2) thus indicating the passive role contributed by transferrin. Given that the Abscess/Blood ratio of ^{125}I is almost doubled(Table2) it could then be hypothesized that inflammatory foci either contain transferrin cell receptors or that the radioiodine is cleaved off by proteolytic enzymes once the transferrin molecule reaches the abscess formation. Because of the possibility that the gallium may have labeled the canine leucocytes *in vivo* we conducted an experiment in which the canine cells were labeled with ^{99m}Tc sulfur colloid and injected simultaneously with ^{67}Ga-transferrin. Figure 4 shows that the gallium uptake is completely independent of white cell uptake in the abscess.

DISCUSSION

The existence of cell surface transferrin receptors which bind transferrin to cell membranes, thus allowing various metal ions to be transported with the cell has long been recognized(20). Furthermore, a number of workers have recently characterized the receptor(21,22,23) The results shown in Table 1 which summarize the distribution of ^{67}Ga, ^{68}Ga and ^{125}I in tumor and blood suggest the existence of a finite number of binding sites on the cells of the tumor. This is indicated by the relative constant ^{125}I activity associated with the tumor after 10 minutes post-injection. This does not appear to be the case where abscesses are concerned where radiogallium, regardless of its mode of complexation, seems to diffuse into the inflammatory foci in a non-specific fashion. Advantage could be taken of these findings by plotting either the $^{67}Ga/^{125}I$ ratios (Figure 5) or the $^{67}Ga/^{68}Ga$ ratios(Figure 6) observed over the regions of interest. A ratio greater than 1 after 2-3 hours is indicative of a tumor. It has been suggested that endocytosis of the gallium-transferrin complex by the cell and subsequent release of gallium from the protein is the mechanism of uptake(24,25) while others have suggested that the metal ions are released from the transferrin-receptor complex and then enter the cell(22, 23,26). Our data tend to support the latter theory as confirmed in a malignant melanoma tissue culture system(27).

It has been reported that the injection of a small dose of turpentine can cause a rapid increase in vascular permeability and elicit a rapid accumulation of polymorphonuclear leucocytes(28). Although one reported investigation found alterations in vascular permeability not to be a major mechanism of radio-gallium localization(29), it has also been suggested that accumulation of gallium-transferrin complex inside cells in inflammatory lesions is due to altered cell membrane permeability(30).

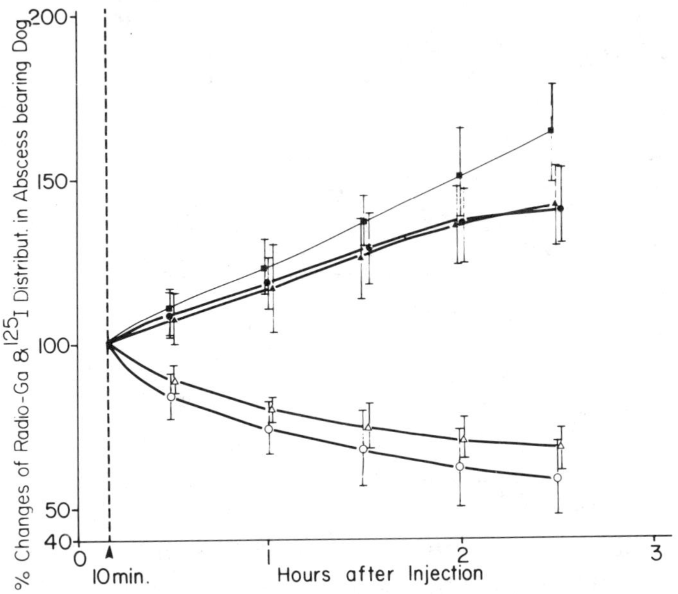

Figure 3.

Relative distribution of ^{67}Ga, ^{125}I and ^{68}Ga after injection of ^{67}Ga-^{125}I-transferrin or ^{68}Ga citrate, respectively, in abscess-bearing dogs. The uptake of ^{68}Ga in abscess(■), ^{67}Ga in abscess(●), ^{125} I in abscess(▲), ^{125}I in blood(△) and ^{67}Ga in blood(o) as recorded in 6 experiments. The behavior of ^{68}Ga is similar to that of ^{67}Ga in blood.

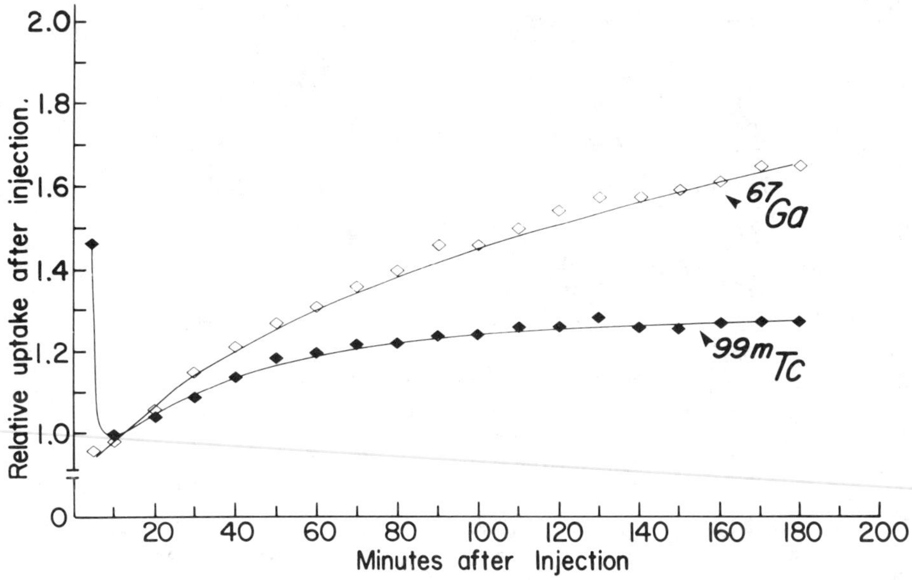

Figure 4.

Differential uptake of ^{99m}Tc and ^{67}Ga in canine abscesses after simultaneous injection of ^{99m}Tc-labeled leukocytes(◆) and gallium-transferrin(◇). Different rates of incorporation indicates that radiogallium does not necessarily bind to leukocytes for incorporation in abscesses.

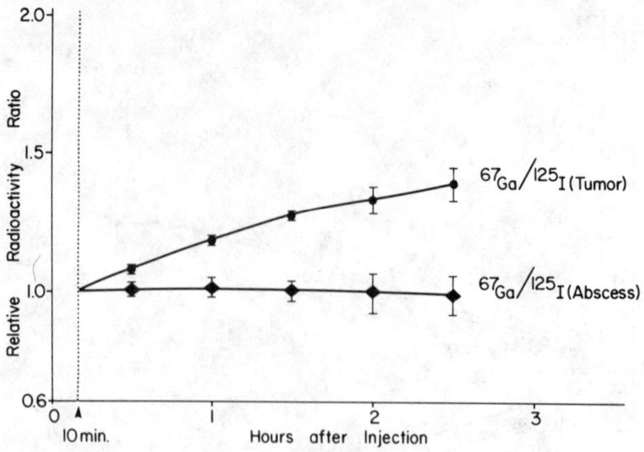

Figure 5.

The relative radioactivity uptake ratios of tumors and abscesses in dogs injected with ^{67}Ga-^{125}I-transferrin. The ratio of ^{67}Ga to ^{125}I in tumors increased while the ratio of ^{67}Ga to ^{125}I in abscesses remained constant. This suggests that, in tumors, selective accumulation of ^{67}Ga occurs, while in abscesses, there is a non-specific binding of the entire double-labeled transferrin molecule. The change in ratio also affords a possible differentiation between internal abscess and tumors.

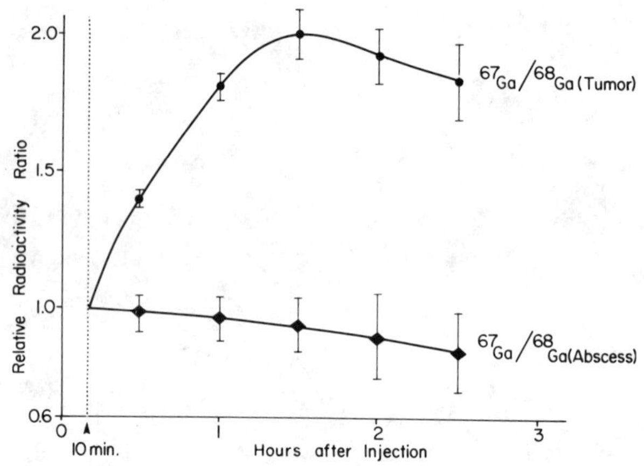

Figure 6.

The relative radioactivity uptake ratios of tumors and abscesses in dogs injected with ^{67}Ga,^{125}I transferrin and ^{68}Ga-citrate. There was a sharp increase in the ^{67}Ga to ^{68}Ga ratio in tumors during the first 1 1/2 hours, indicating that for tumor uptake of gallium to occur, complexation to transferrin is essential. The decrease in the ^{67}Ga to ^{68}Ga ratio after 1 1/2 hours suggested that the ^{68}Ga-citrate first complexes to transferrin before uptake occurs in the tumor. On the other hand, the uptake of gallium in abscesses does not require prior complexation to transferrin.

Very recent reports have shown that gallium can still accumulate in abscesses in agranulocytic patients(31) or in experimental animals made agranulocytic by pretreatment with myleran(32). Since we have observed a very rapid uptake of ^{68}Ga, complexed either to citrate or possibly to canine plasma proteins such as transferrin, as well as a rapid uptake of ^{67}Ga-^{125}I-transferrin by the abscess, with the ^{125}I to ^{67}Ga ratio remaining constant, it would suggest that both the administered ^{68}Ga-citrate and ^{67}Ga-^{125}I-transferrin bind to abscess components other than granulocytes. The ^{68}Ga-citrate can localize at the site of inflammation by binding, non-specifically, to proteins which have accumulated due to altered vascular permeability. Ga-transferrin can also bind to some intracellular components of fragmented cells. We have observed in our laboratories that fragmented dead cells from tissue culture, do bind a significant amount of both Ga-citrate and Ga-transferrin(34). It was also found that ^{67}Ga was associated in the non-cellular fraction in abscesses induced by either E. coli or turpentine(33). Thus, from our observations as well as those of others, it may be concluded that while the uptake in abscesses is non-specific, that in tumors is specific and differentiation can be established between these two pathologic conditions.

REFERENCES

1. Edward CL, Hayes RL: Scanning malignant neoplasms with gallium-67. JAMA 212:1182-1190, 1970.

2. Littenberg RL, Taketa RM, Alazraki NP, et al: Gallium-67 for localization of septic lesions. Ann Intern Med 79:403-406, 1973.

3. Blair DC, Carroll M, Silva J, et al: Localization of infectious processes with gallium citrate Ga-67. JAMA 230:82-87, 1974.

4. Harvey WC, Rodoloff DA, Kapp DT: ^{67}Ga scanning in 68 consecutive infection searches. J Nucl Med 16:204, 1975.

5. Kaplan LR, Griep RJ, Schaffler MD et al: Gallium-67 scanning at 6 hour in active inflammatory bowel disease: case report. J Nucl Med 18:448-449, 1977.

6. Burleson RL, Johnson MC, Head H: In vitro and in vivo labeling of rabbit blood leukocytes with ^{67}Ga-citrate. J Nucl Med 15:98-101, 1974.

7. Hoffer BH, Huberty J, Khayan-Bashi H: The association of Ga-67 and lactoferrin. J Nucl Med 18: 713-717, 1977.

8. Reif AE: A simple procedure for high-efficiency radioiodination of proteins. J Nucl Med 9:148-152, 1968.

9. Aulbert E, Gebhardt A, Schulz E, Haubold U: Mechanism of ^{67}Ga accumulation in normal rat liver liposomes. Nucl Med 15:185, 1976.

10. Bichel P, Hansen HA: The incorporation of ^{67}Ga in normal and malignant cells and its dependence on growth rate. Br J Radio 45:182, 1972.

11. Winchell HS: Mechanism for localization of radiopharmaceuticals in neoplasims. Nucl. Med. 6:371, 1976.

12. Larson SM: Mechanism of localization of gallium-67 in tumors. Sem Nucl Med 8:193-203, 1978.

13. Epstein RB, Bennett BT: Histocompatibility typing and course of canine venereal tumors transplanted into unmodified random dogs. Cancer Res 34:788-793, 1974.

14. Boullanger PH, Nagpurkar A, Noujaim AA et al: Application of ^{125}I-Radioimmunoassay to measure inhibition of precipitin reactions using carbohydrate-specific antibodies. Can J Biochem 56:1102-8, 1978.

15. Lentle BC, McPherson TA, Scott JR, et al: Localization of acute inflammatory foci in man with ^{99}Tc^{m} sulfur colloid labeled granulocytes. Clin Nucl Med 1:118-21, 1976.

16. Septon RG, Kraft N: ^{67}Ga and ^{59}Fe uptakes by cultured human lymphoblasts and lymphocytes. Cancer Res 38:1213-1216, 1978.

17. Noujaim AA, Lentle BC, Hill JR, et al: On the role of transferrin in the uptake of gallium by tumor cells. Submitted for publication.

18. Sephton RG, Harris AW: ^{67}Ga-citrate uptake by cultured tumor cell, stimulated by serum transferrin. J Natl Cancer Inst 54:1263-1266, 1975.

19. Harris AW, Sephton RG: Transferrin promotion of ^{67}Ga and ^{59}Fe uptake by cultured mouse myeloma cells. Cancer Res 37:3634-3638, 1977.

20. Jandl JH, Katz JH: The plasma-to-cell cycle of transferrin. J Clin Invest 42:314-326, 1963.

21. van Bockxmeer FM, Morgan EH: Identification of transferrin receptors in reticulocytes. Bioch Biophy Acta 468:437-450, 1977.

22. Leibman A, Aisen P: Transferrin receptor of the rabbit reticulocyte. Biochem 16:1268-1272, 1977.

23. Verhoef NF, Noordeloos PF: Binding of transferrin and uptake of iron by rat erythroid cells in vitro. Clin Sci Molec Med 52:87-96, 1977.

24. Morgan EH, Appleton TC: Autoradiographic localization of ^{125}I-labeled transferrin in rabbit reticulocytes. Nature 223:1371-1372, 1969.

25. Hemmaplardh D, Morgan EH: The role of endocytosis in transferrin uptake by reticulocytes and bone marrow cells. Br J Haemat 36: 85-96, 1977.

26. Loh TT, Yeung YG, Yeung D: Transferrin and iron uptake by rabbit reticulocytes. Biochem Biophys Acta 471:118-124, 1977.

27. Terner UK, Wong H, Noujaim AA, et al: ^{67}Ga and ^{59}Fe uptake in human melanoma cells. Int J Nucl Med Biol in press, 1979.

28. Harris H: Role of chemotaxis in inflammation. Physiol Rev 34:528-562, 1974.

29. Gelrud LG, Arseneau JC, Milder MS, et al: The kinetics of ^{67}Gallium incorporation into inflammatory lesions, experimental and clinical studies. J Lab Clin Med 83:489-495, 1974.

30. Hayes RL, Brown BH: Biokinetics of radiogallium in Nuklearmedizin: Fortschritte der Nuklearmedizin in klinischer und technologisher Sicht, 12th Annual Meeting, Gesellschaft fur Nuklearmedizin, Munchen, Sept. 11-14, 1974 Pabst HW, Hor G, Schmidt HAE, eds. Stuttgart-New York Schattaner, 1975 837-847.

31. Dhawan VM, Sziklas JJ and Spencer RP: Localization of Ga-67 in inflammation in the absence of circulating polymorphonuclear leukocytes. J Nucl Med 19:292-294, 1978.

32. Tsan MF: Studies on gallium accumulation in inflammatory lesions: III. Roles of polymorphonuclear leukocytes and bacteria. J Nucl Med 19:492-495, 1978.

33. Tsan MF, Chen WY, Scheffel U, et al: Studies on the gallium accumulation in inflammatory lesions: l. Gallium uptake by human polymorphonuclear leukocytes. J Nucl Med 19: 36-43, 1978.

34. Terner UK, Noujaim AA, Lentle BC. et al: unpublished observations.

DEVELOPMENT OF A LYMPHSCINTIGRAPHY AGENT: Tc-99m-Sn-PO_4

David L. Gilliland, Garo P. Basmadjian, Alan S. Kirschner
Joe C. Leonard*, Rodney D. Ice

College of Pharmacy
The University of Oklahoma HSC, Childrens Memorial Hospital*,
Oklahoma City, OK. 73190

Stannous chloride dissolved in a solution of Na_2HPO_4 pH 9.0 gives a stable solution of Sn(II) ions. When pertechnetate-99m is added to such a solution, a stable, soluble complex of technetium-99m-Sn-PO_4 results. This is a mixed metal complex in which the Tc is present in the $(III)^+$ oxidation state and the Sn is an integral part of the complex.

This complex, when injected I.V., localizes in the RES due to the formation of an in-vivo colloid with Ca(II) ions in the blood. When injected subcutaneously (SC), the complex travels through the lymphatic channels and localizes in the lymph nodes.

In rabbits, axillary and popliteal nodes were visualized after injecting SC into the dorsum of the hind and fore paws. In addition, uptake by the liver, kidneys and bone marrow was noted. After intratesticular injection in dogs, the para-aortic lymph nodes were imaged at 3-30 minutes.

INTRODUCTION

Contrast media are currently used for lymphography (1). Despite the high resolution and precision of the standard ethiodol (an iodinated fatty acid ester), lymphography suffers from several known complications and shortcomings. Studies using ^{131}I labeled ethiodol show lung, hepatic, and cerebral oil emboli occuring in patients. Although ethiodol is rarely symptomatic, reports of fatal reactions have been noted (2,3). Hypersensitivity reactions occur in approximately 0.1 percent of the procedures (4). In addition, direct lymphography is applicable to only a limited number of anatomic sites (5).

In attempts to overcome the various complications and shortcomings of the standard lymphography procedures, lymph node scintigraphy using various radiopharmaceuticals has been developed. While lymph node scintigraphy lacks the resolution of contrast lymphography, improved scintigraphic equipment and increased specificity of radiopharmaceuticals for the lymphatics may provide acceptable and clinically useful images with the advantage that it is: (a) relatively innocuous,(b) non-invasive, (c) requires little time on the part of personnel and physician, (d) requires only a few hours to complete (6).

Colloidal gold Au-198, was the first of many radiopharmaceuticals employed to visualize the lymphatic system using scintigraphic techniques (7-9). This tracer has two major disadvantages: (a) high radiation dose to the site of injection and (b) unfavorable high gamma energy for collimation.

Technetium-99m labeled sulfur colloid has been used by several workers in attempts to overcome the problems of radiation dosage and collimation (10-12), and appears to be as effective as Au-198 colloid for lymphscintigraphy. However, due to the large colloidal size of Tc-99m sulfur colloid, i.e. 0.3 μ (11), much of the colloid is trapped at the injection site, thus limiting its usefulness. To overcome the problem of colloid size, Ege evaluated Tc-99m antimony colloid for use as a lymphscintigraphy agent (12). According to Ege, Technetium-99m antimony sulfide colloid has a particle size of 4-12 nm, and is thus more suitable for lymph node visualization due to increased amounts of uptake from injection sites.

Alavi has evaluted Tc-99m stannous phytate as a lymphscintigraphy agnet (9). This agent differs from the previously mentioned colloids in one major aspect: whereas gold colloid, Tc-99m sulfur colloid, and Tc-99m antimony sulfide colloid are all preformed *in-vitro* colloids, Tc-99m stannous phytate forms an *in-vivo* colloid, presumably by reacting with interstitial calcium ions. The insoluble Tc-99m calcium phytate particles are then presumably engulfed by phagocytic cells in the lymph nodes much the same as the *in-vitro* colloids. Comparable images of the lymph nodes have been obtained with Tc-99m stannous phytate and Tc-99m antimony sulfide colloid (13-14).

The objective of this research was to evaluate Tc-99m-Sn-PO_4 as a radiopharmaceutical for lymphscintigraphy. This radiopharmaceutical has been shown to behave very similar to Tc-99m stannous phytate upon I.V. injection. This evaluation was performed in rats, rabbits, and dogs injected subcutaneously.

MATERIALS & METHODS

The stannous phosphate for complexation with technetium (99&99m) was prepared as follows:

(a) Nitrogen was bubbled into 85 ml of sterile water for injection for 30 minutes to purge all oxygen;

(b) After dissolving 8.1 gm of Na_2HPO_4 into oxygen free water, 190.1 mg of $SnCl_2$ (anhydrous) was added with continuous stirring and N_2 purging (a slight drop in pH was noted);

(c) The pH was adjusted to 9.0 with 10N NaOH;

(d) Sufficient oxygen purged sterile water was added to make the final volume 100 ml. The final stock solution concentration was 54.2 mg/ml HPO_4^{-2} ions, and 1.18 mg/ml Sn(II) ions;

(e) Following filtration through 0.22 μ filter, the solution was stored in 30 ml evacuated vials under nitrogen.

The technetium-99m-Sn-PO_4 complex was prepared by the addition of Tc-99m pertechnetate (10-100 mCi) to 1 ml of the prepared $SnPO_4$ solution. This solution was diluted with sodium carbonate-bicarbonate buffer, pH 9.0, to yield the desirable specific concentration of Tc-99m-Sn-PO_4 complex for injection.

The following quality control procedures were performed.

(a) Reductive capacity of the stock solution: To 1 ml of $SnPO_4$, was added 1 ml of technetium-99 NH_4TcO_4 (1.44 X 10^{-2} M). A dark reddish-brown color appeared, indicating the formation of technetium-99m-Sn-PO_4.

(b) Radiochemical purity: The Tc-99m-Sn-PO_4 complex was chromatographed on thin layer cellulose chromatogram strips with acetone as the solvent. Free pertechnetate had an R_f of 0.6-0.7, Tc-99m-Sn-PO_4 had an R_f of 0.

(c) Charge: Electrophoresis of Tc-99m-Sn-PO_4 was performed on Sepraphore III in tris-barbital-sodium barbital pH 8.8 buffer, (30 min. at 275 volts) for evaluation of charge and radiochemical purity. Technetium-99m-Sn-PO_4 is uncharged and has an $R_{TcO_4^-}$ value of 0, in relation to $R_{TcO_4^-}$ of 1.0.

(d) Stability: The shelf life of the stannous phosphate stock solution has been shown to be greater than 120 days at room temperature under air or nitrogen atmospheres (Table 1).

ANIMAL BIODISTRIBUTION

Intravenous injection of Tc-99m-Sn-PO_4 in rabbits shows uptake primarily in the RES system: liver, spleen, and bone marrow (Table 2).

Biodistribution of Tc-99m-Sn-PO_4 in rats injected S.C. was also studied as follows:

Five mCi of $^{99m}TcO_4^-$ were added to 0.6 ml of stannous phosphate stock solution and sufficient sodium carbonate-bicarbonate buffer added to yield 5 ml final volume Twelve rats were injected with 100 μCi per 0.1 ml of Tc-99m-Sn-PO_4 S.C. into the dorsum of a hind paw. Two rats were sacrificed at 5, 15, 30, 60, 120, and 360 minute intervals. Blood, liver, kidneys, and paws were removed and counted using a Searle 1135 gamma counter. The extent of product localization was recorded as % Kg dose/gm (Table 3).

TABLE I - CHARACTERISTICS OF Tc (-99 & 99m)-Sn-PO_4

STABILITY:	Very stable under air or N_2 atmospheres. Once formed, inert to substitution.
VISIBLE SPECTRUM:*	Reddish-brown; E_{max} 523 mμ.
CHARGE:**	Uncharged - By electrophoresis and chromatography on anion and cation exchange resins.
OXIDATION STATE OF Tc:†	$Tc(III)^+$ - Iodometric titration.
PRESENCE OF Sn IN COMPLEX:††	Yes - Sn-113 shown to be involved in the complex.
IN-VIVO DISTRIBUTION:	_In-vivo_ colloid - Uptake by liver, spleen and bone marrow.

*Solution containing 1 ml of stannous phosphate (1.18 mg/ml Sn^{++} ions) and 1 ml of a 1.44 X 10^{-2} M NH_4TcO_4 solution pH 9.0 using an Acta III spectrophotometer.

**Electrophoresis on Sepraphore III in Tris-barbital-sodium barbital pH 8.8 buffer for 30 minutes at 275 volts. Greater than 95% activity eluted off columns of anion exchange resin Dowex 1-X8 and cation exchange resin Cellex-CM.

†Procedure of Steigman et. al. (16).

†† (15,16).

TABLE 2 - BIODISTRIBUTION OF Tc-99m-Sn-PO_4 IN RABBITS*

FORMULATION**: 1 ml of Na_2HPO_4 containing 1.18 mg of Sn^{++} ions + 1 ml of 99mTcO_4^- (10-100 mCi). Inject 0.2 ml (2-10 mCi) into rabbit ear vein. Sacrifice 30 minutes post injection.

		% REMAINING ACTIVITY						
Days post prep	Storage cond.	Blood	Lung	Liver	Spleen	GI	Carcass	Kids/Bladder
3	50°C	0.45	0.98	79.35	1.88	0.62	15.96	0.69
3	RT	1.05	2.67	79.65	1.25	0.90	13.36	0.69
5	50°C Inverted	0.88	0.38	77.64	4.45	0.75	15.14	0.63
11	50°C Inverted	1.75	0.14	71.19	2.12	0.97	22.89	0.89
21	50°C Inverted	0.35	0.13	80.19	1.65	0.45	16.67	0.51
21	50°C Inverted	0.96	0.91	74.79	2.20	0.44	11.13	0.58
28	50°C Inverted	3.37	0.26	73.95	6.58	0.69	13.96	1.14
28	RT	1.28	0.75	69.49	5.02	1.01	21.41	0.98
74	RT	1.01	0.13	72.37	4.68	0.72	20.52	0.58
105	RT	0.71	0.15	69.76	8.16	0.46	19.91	0.57
125	RT	1.86	0.28	71.66	2.43	1.06	20.12	1.99
	$\overline{X}$	1.24	0.62	74.57	3.67	0.73	17.36	0.84

*Two rabbits sacrificed at each time interval
**Formulation stable for 10-12 hours post reconstitution.

TABLE 3 - DISTRIBUTION OF Tc-99m-Sn-PO_4 IN NORMAL MALE RATS*

	% Kg dose/gm Tissue**					
	5 min.	15 min.	30 min.	60 min.	120 min.	360 min.
Blood (x)	.044	.044	.075	.104	.059	.025
(s)	.002	.009	.041	.077	.031	.003
Liver (x)	.043	.047	.024	.282	.249	.026
(s)	.022	.022	.244	.271	.252	.003
Kidney (x)	.086	.123	.331	.510	.511	.442
(s)	.004	.070	.144	.204	.147	.026
	% dose absorbed from injection site+					
(x)	14	20	16	30	51	50

*0.1 ml injected S.C. into a hind paw of each animal
**Mean value (x) and, standard deviation (s) for two rats. Corrected for actual activity absorbed from injection site.
+Mean value for two rats. Obtained by comparing activity remaining in injected paw to known administered dose.

To understand the mechanism and rate of uptake of Tc-99m-Sn-PO_4 in the lymphatics after S.C. injection the following γ-camera studies were performed:

To 0.2 ml of $SnPO_4$ stock solution was added sufficient buffer to make 1:5, 1:10, 1:50, and 1:100 dilutions prior to the addition of pertechnetate. The pH was varied between 4.0-9.0 by using different buffers. After performing quality control, 500 µCi was injected S.C. into the dorsum of the paws of rabbits. Using a Searle LFOV camera, images were obtained following the injection of various specific concentrations of Tc-99m-Sn-PO_4. A 1:100 dilution of $SnPO_4$ with pH 9.0 sodium carbonate-bicarbonate buffer appeared to provide best results. One minute post injection, radioactivity localization was sufficient to image lymph vessels. Within 3 minutes there was sufficient radioactivity in the axillary and popliteal nodes to allow imaging (fig. 1-2). The Tc-99m-Sn-PO_4 continued to concentrate in the lymph vessels and nodes, reaching peak concentration in 30 minutes. As the Tc-99m-Sn-PO_4 enters the vascular system through the thoracic duct, kidney activity begins to appear around 5 minutes. Images at 15 and 30 minutes post injection show a cluster of axillary nodes and lymph vessels draining injection sites along with kidney and faint liver uptake (fig.3-4). Some bone marrow radioactivity can be seen 3 hours post injection.

Studies in dogs, injected (with a 1:100 dilution of a pH 9.0 preparation of Tc-99m-Sn-PO_4) intratesticular to follow the drainage of the lymphatic channels into the para-aortic lymph nodes were also performed using a Searle LFOV camera and images obtained at various intervals (fig.5-6).

DISCUSSION

Inorganic phosphates chelate technetium in the presence of reducing agents such as Sn(II) salts. The phosphate group has a stabilizing effect on the reductive capacity of the stannous ion at physiological pH. The orthophosphate anion chelates Sn(II) ions at pH 1-12 without affecting its hydrolysis or oxidation (15).

Stannous chloride when dissolved in a solution of Na_2HPO_4, pH 9.0, gives a stable solution of Sn(II) ions. Pertechnetate (-99&99m) when added to such a solution is reduced and a stable soluble complex of Tc-99m-Sn-PO_4 results. This complex is a mixed metal complex, in which the technetium is present in the $(III)^+$ oxidation state and the tin is an integral part of the complex (15). Steigman, concurs with this observation and concludes "- - - that the reaction between $Na^{99}TcO_4$ and excess $SnCl_2$ in 1 M phosphate at pH 7 produces a mixed metal phosphate complex which contains Sn(II), $Tc(III)^+$ and $Tc(IV)^+$, in unknown proportions" (16).

The in-vivo distribution of this complex depends very much on the pH of the injected solution. At a pH less than 7, the complex, upon I.V. injection, distributed through out the extravascular fluid and was slowly excreted through the kidneys. At a pH greater than 7, upon I.V. injection, an in-vivo colloid was formed with Ca(II) ions in the blood and was taken up by the RES organs: liver, spleen and bone marrow (15).

When Tc-99m-Sn-PO_4 was injected S.C., the in-vivo colloid was formed with interstitial Ca(II) ions, and uptake occured via the lymphatics.

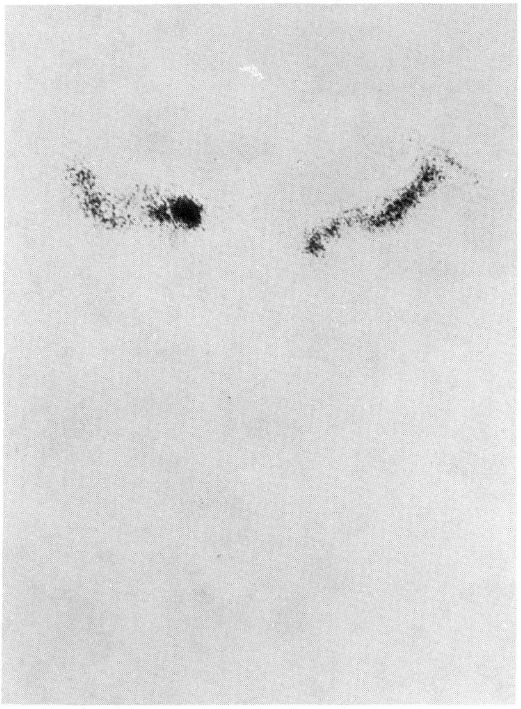

Fig. 1 Axillary nodes 3 min. post injection. Anterior view. 0.5 mCi Tc-99m-Sn-PO_4, S.C. front paws, rabbit.

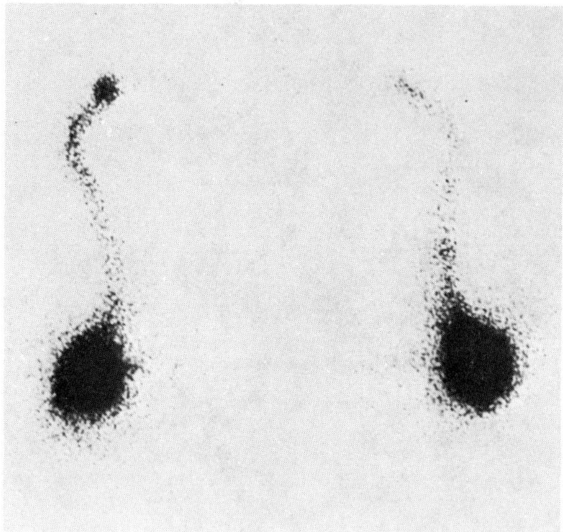

Fig. 2 Popliteal nodes 3 min. post inj. Anterior view. 0.5 mCi Tc-99m-Sn-PO_4, S.C. hind paws, rabbit.

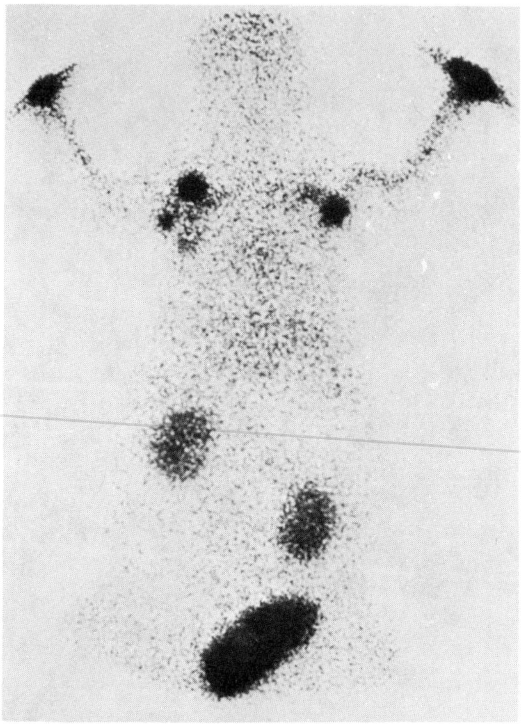

Fig. 3 Anterior view 30 min. post inj. 0.5 mCi Tc-99m-Sn-PO_4 S.C. front paws, rabbit.

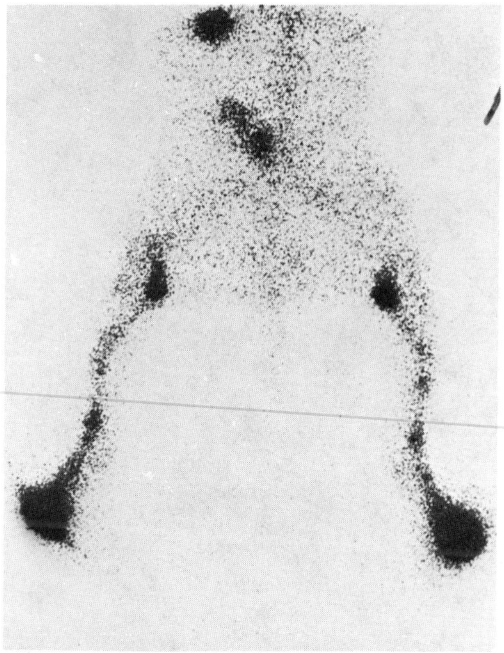

Fig. 4 Lower anterior view 15 min. post inj. 0.5 mCi Tc-99m-Sn-PO_4 hind paws, rabbit.

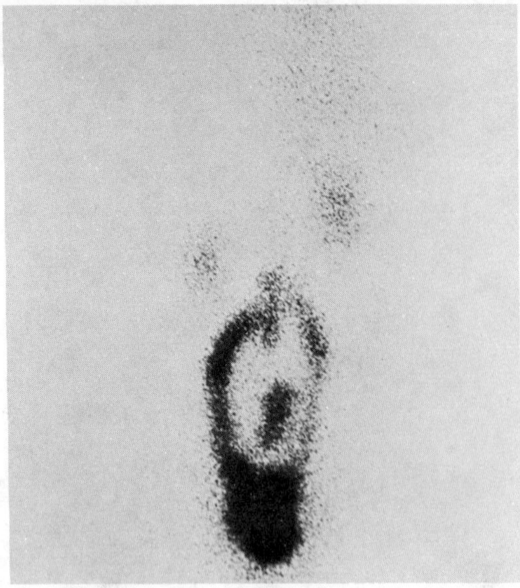

Fig. 5 Posterior view. Dog inj. intratesticular, 0.5 mCi Tc-99m-Sn-PO_4. Para-aortic nodes visualized 15 min. post inj.

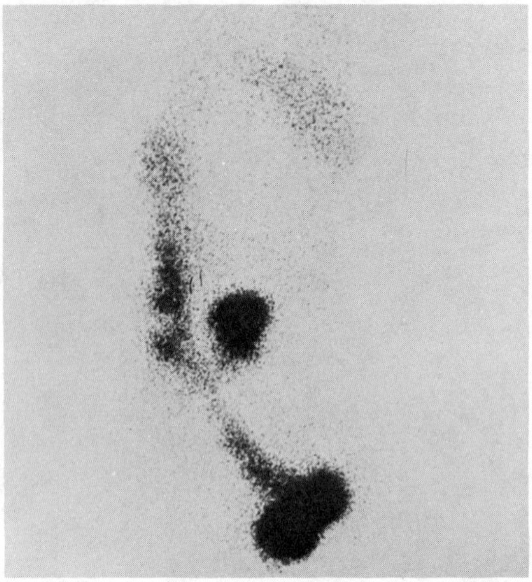

Fig. 6 Lateral view. Dog inj. intratesticular, 0.5 mCi Tc-99m-Sn-PO_4. Para-aortic nodes visualized 30 min. post inj.

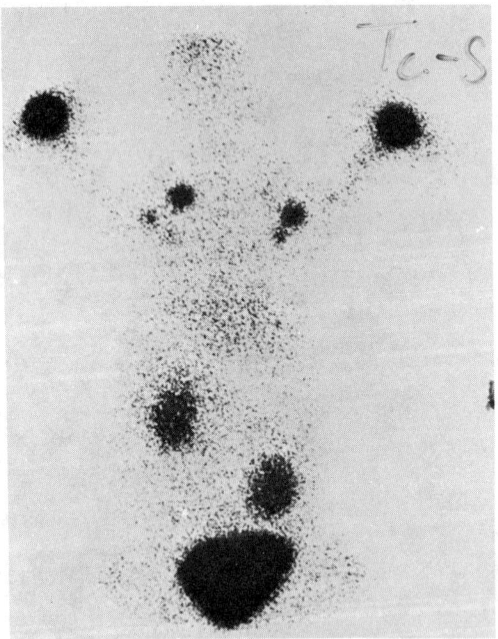

Fig. 7 Anterior view Tc-99m-Sn-PO4 30 min. post inj., S.C. front paws, rabbit.

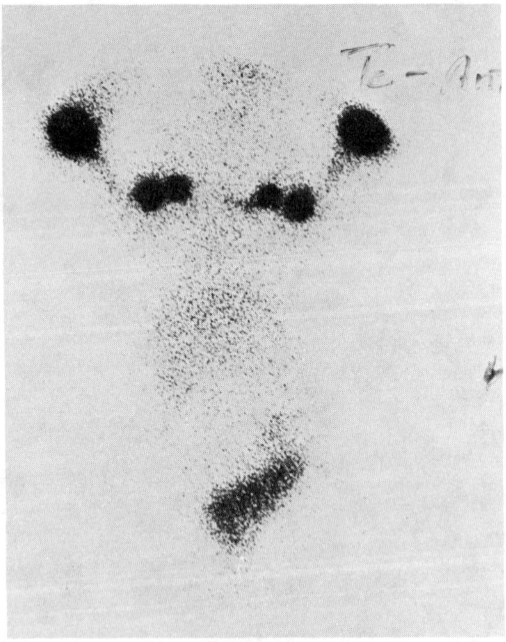

Fig. 8 Anterior view Tc-99m-antimony sulfide colloid 30 min. post inj. S.C. front paws, rabbit.

The Tc-99m-Sn-PO_4 appears to image rabbit and dog lymphatics comparably with other lymphscintigraphy agents. Comparison studies in rabbits between Tc-99m-Sn-PO_4 and Tc-99m antimony sulfide colloid appeared to visualize axillary and popliteal nodes equally will. Technetium-99m-Sn-PO_4 and Tc-99m antimony sulfide colloid both showed kidney and bladder uptake 5 minutes post injection. However, Tc-99m antimony sulfide colloid showed significant renal clearence 1 hour post injection, Tc-99m-Sn-PO_4 did not (fig.7-8). This suggests smaller particle size of Tc-99m-Sn-PO_4, which might be trapped in the kidneys.

The small colloidal particle size of Tc-99m-Sn-PO_4 facilitates uptake from the injection sites, and increases mobility within the lymphatics. This advantage combined with the ease and minimal time required for preparation of Tc-99m-Sn-PO_4 provides a distinct improvement over *in-vitro* colloidal agents.

In our animal studies to date, Tc-99m-Sn-PO_4 has provided quality and consistant images and warrants research in humans to validate its usefulness as an alternate lymphscintigraphy agent. Evaluation of parasternal lymphatics in breast carcinoma (12), para-aortic node involvement in carcinoma of the prostate and testicles, hyperplasia and loss of parasternal chain uniformity in lymphoma, obstruction of diaphragmatic and mediastinal lymphatics corresponding to clinically intractable ascites, should be seen using technetium-99m-Sn-PO_4.

ACKNOWLEDGEMENTS

The authors of this paper are greatly indebted to Mr. Ronald G. Raftery and Mr. George C. Hoebing, Nuclear Medicine Dept., Childrens Memorial Hospital. Without their technical assistance this paper would not have been possible. Their time and patience are greatly appreciated. The authors also appreciate the expert help and advise of Hrair T. Kurkjian, M.D.

REFERENCES

1. Fuchs WA: Complications in lymphography with oily contrast media. Acta Rad 57: 427, 1962.

2. Steckel FJ, Furmanski S, Dunham R, Collins JD, Ross N, Haorld SD, Poe N: Radionuclide perfusion lymphangiography. Amer J of Roentgenology 124: 600-609, 1957.

3. Koehler PR: Typical fatal reactions after lymphography. Cancer Chemother Rep 52: 113-115, 1968.

4. Fuchs WA: Technique and complications of lymphangiography. In Angiography, 2nd ed., Abrams HL: Boston, Little Brown and Co., vol. 2: 1325-1135, 1971.

5. Kinmonth JB: The Lymphatics; Diseases, Lymphography and Surgery. Baltimore, The Williams & Wilkins Co., 1972, pp 1-20.

6. Fairbanks VF, Tauxe WN, Kiely JM, Miller WE: Scintigraphic visualization of abdominal lymph nodes with Tc99m sulfur colloid. J Nucl Med 13: 185-190, 1972.

7. Sherman AJ: Lymph node concentration of radioactive colloidal gold following interstitial injection. Cancer 6: 1238-1240, 1953.

8. Kazem I, Antioiades J, Brady LW: Clinical evaluation of lymph node scanning utilizing colloidal gold 198: Radiology 90: 905-911, 1968.

9. Alavi A, Staum MM, Sheosl FB, Bloch PH: Technetium-99m stannous phytate as an imaging agent for lymph nodes. J Nucl Med 19: 422-426, 1978.

10. Hauser W: Lymph node scanning with technetium-99m sulfur colloid. Radiology 92: 1369-1371, 1969.

11. Goranson LR: Scintigraphy of the parasternal lymphatics in rabbits using technetium-99m sulfur colloid. Acta Rad Diag 14: 169-178, 1974.

12. Ege GN: Internal mammary lymphscintigraphy. Radiology 118: 101-107, 1976.

13. Ege GN: Tc-99m phytate as an imaging agent for lymph nodes. Letters to the Editor, J Nucl Med 19: 1363, 1978.

14. Alavi A: Tc-99m phytate as an imaging agent for lymph nodes. Reply, J Nucl Med 19: 1364, 1978.

15. Basmadjian GP, Mills SL, Hart DL, Kirschner AS: Chemistry of technetium phosphate and phosphonate complexes: Applications to radiopharmaceuticals: In proceedings of the 15th International meeting of the Society of Nuclear Medicine, Gronigen, the Netherlands, 13-16, 1977 (in press)

16. Steigman J, Meinken G, Richards P: The reduction of pertechnetate -99 by stannous chloride-II. The stoichiometry of the reaction in aqueous solution of several phosphorus (V) compounds. Int J of Appl Rad Isotopes 29: 653, 1978.

RELATIVE STABILITY OF In-111 AND Ga-67 DESFERRIOXAMINE AND HUMAN TRANSFERRIN COMPLEXES

Ronald E. Weiner, Mathew L. Thakur, Mark Goodman, and Paul B. Hoffer

Yale University School of Medicine
Department of Nuclear Medicine
New Haven, Connecticut 06510

ABSTRACT

The relative stabilities of Ga-67 and In-111 desferrioxamine B (DF) complexes in relation to human transferrin have been investigated by chromatographic and equilibrium dialysis techniques. The influence of pH, temperature and chelate concentrations have been studied on the formation of DF metal chelates. The Ga:DF complex is much stronger than the Ga:TF or the In:DF chelate. The results support the use of DF to reduce the Ga-67:TF radioactivity in vivo and suggest the possibility that siderophores (a class of DF-like compounds) may be responsible for the accumulation of Ga-67 in abscess.

INTRODUCTION

Desferrioxamine B (DF) is one of a general class of iron transporting chelators (siderophores) which are synthesized and secreted by many prokaryotes in response to an iron-deprived environment (1). These compounds have a marked and selective avidity for ferric ion and are used by aerobic microorganisms to circumvent the low solubility of ferric ion at physiological pH (10^{-18}M) (2). The two most common classes of siderophores contain either hydroxamate [-N(OH)-CO-] or catecholate [1,2 dihydroxybenzene] functional groups. Ferrioxamine B (DF) isolated from Streptomyces pilosus (3) and ferrichrome, isolated from the smut fungus, Ustilago sphaerogena (4) are typical of the hydroxamate class of siderophores. Each compound chelates a single ferric ion via the bidentate hydroxamate group forming a hexacoordinate complex (1) (Fig. 1).

STRUCTURE OF FERRIOXAMINE B

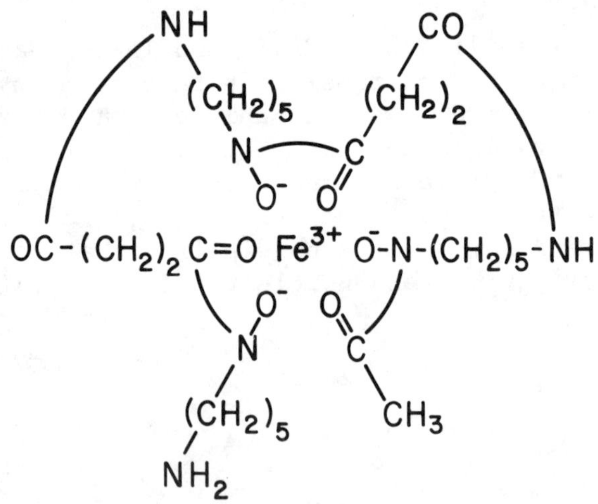

Figure 1

DF is currently used in the treatment of acute and chronic iron overload precisely because of its selective affinity for ferric ions (5-8).

Our current interest in siderophores stems from the work of Emery (9) who demonstrated that desferriferrichrome(DFC) facilitates fungal uptake of not only ferric ions but also the group III elements Al^{3+}, Ga^{3+} and, to much lesser extent, Tl^{3+}. These data support the concept that Ga-67 behaves physiologically, in part, as non-reducible iron analogue (10). Preliminary investigations in our laboratory demonstrate the relative stability of DF complex was: GaDF> FeDF>InDF>CuDF at 22°C and pH 7.1 (11). The high affinity of DF for Ga-67 has prompted its use as an agent to reduce Ga-67blood pool background in clinical imaging. We (12) and others (13-14) have shown that DF injected I.M. binds to Ga-67 in vivo, reduces blood concentration of Ga-67 and enhances lesion to background ratio for both tumor and abscess.

Emery (9) observed that C-14 DFC mediated uptake of ferric ion by Ustilago sphaerogena resulted in the trapping and accumulation of iron, but not the chelate, within the organism. However, when another group III metal, Al^{3+}, was substitued for ferric iron, the entire aliuminum-siderophore complex was trapped. Subsequently, it has been shown that Ga^{3+} DFC is also trapped by the U. sphaerogena (15). Menon et al., (16) have shown that bacteria selectively take up Ga-67 in vitro. Thus, one possible mode of Ga-67 accumulation in abscesses could be chelation of Ga-67 by bacterial siderophores and entrapment of the siderophore Ga-67 complex in the bacteria.

We have undertaken to examine some of the factors which influence the formation of In-111 and Ga-67 complexes of DF. Moreover, in order to investigate the possible role of direct siderophore binding in abscess localization of gallium, we have investigated the relative stability of the Gallium-DF complex compared to the Gallium-human transferrin (TF) complex (TF is the major plasma binding protein for Gallium(17,18). In parallel experiments, the relative affinity of Indium for DF and TF was also studied.

MATERIALS AND METHODS

Gallium-67 and Indium-111 DF complex formation: To examine the influence of incubation time and reaction temperature, a 40 mM solution of DF (Ciba-Geigy) in 1 ml of .1 M acetate buffer pH 6.1 was incubated at the requisite temperature with 8-11 μCi of either carrier free Ga-67 citrate (New England Nuclear) of In-111 chloride (Diagnostic Isotopes). At predetermined times a 5 μl aliquot of the mixture was spotted on Whatman No. 1 chromatographic paper strips and allowed to air dry. The strips were developed using n-butanol:n-propanol: H_2O 9:6:5 as a solvent system (19). The solvent was allowed to ascend for 10 cm, paper was removed from the solvent, air dried, cut into 1 or ½ cm sections and counted in an automatic well-type scintillation counter. The R_f's were DF^*M = 0.5, and M = 0.0. The percentage of radioactivity in each peak was determined by the sum of the radioactivity in the particular sections divided by the total activity. To check the affect of pH on DF radionuclide formation, a 40 mM DF solution in .1 M acetate buffer, adjusted to the desired pH, was used. Each solution was incubated at 22°C for 1 hour after the addition of Ga-67 citrate on In-111 chloride and then subjected to chromatographic analysis as before.

Ga-67 and In-111 TF formation: With the solutions containing TF the experimental conditions approaching those found physiologically were used. The final TF concentration was maintained at 13μM and the incubation buffer was .05 M N-2-hydroxyethylpiperazine-N′-2-ethanesulfonic acid (HEPES, Calbiochem Inc.) buffer .1 M NaCl pH 7.4. To promote In-111 and Ga-67 TF complex formation the solution always contained a final concentration of 10μM nitrilotriacetic acid (NTA) and 1.2 mM bicarbonate ion (21-23). To 500μl of a TF solution of 8-11μCi of Ga-67 citrate was added and incubated for 3 hours at 22°C. In similar experiments with In-111 chloride, the incubation temperature was raised to 37 C and incubation time prolonged to 18 hours. To check the formation of the TF-Ga-67 complex under the conditions used in the paper chromatography experiments, aliquots of a TF-Ga-67 solution incubated in the same manner were applied to a column containing the anion exchanger AG 1X2 100-200 mesh (BioRad Laboratories Inc.).This column was used to separate TF bound from unbound Ga-67 (24).

Competition experiments: To obtain a better estimated of the relative stabilities of the TF and DF radionuclide complexes the competition was performed at different DF:TF metal site ratios. The site ratio was used since TF contains two metal binding sites (21) while DF has only one(1). To challenge the In-111 or Ga-67 TF complex, 500μl of a HEPES buffer solution containing various DF concentrations was added to 500μl of a solution containing the preformed TF complex. At specific times, 5μl of the mix was withdrawn and chromatographed on paper as described. The TF *M R_f is 0.0. To attempt to dissociate the DF preformed complex, 500μl of a TF solution was added to 500μl of a Ga-67 or In-111 DF solution in buffer. To form the DF complex, Ga-67 and In-111 were incubated for 25 minutes at 22°C and 1 hour at 37°C respectively.

We utilized equilibrium dialysis (ED) to examine the formation and dissolution of the In-111 or Ga-67 TF complex. This was done because free and TF complexed radionuclide had identical R_f's on paper chromatography. The ED experiments were performed using Spectrum Medical Industries equilibrium dialysis unit consisting of 5 teflon dialysis cells stacked together which rotate at a constant speed. The apparatus is placed in a waterbath to maintain a constant temperature. The Ga-67 and In-111 experiments were carried on at 22°C and 37°C, respectively. The final concentrations in a typical experiment were: 13μM TF, 10μM NTA, 1.2 mM bicarbonate ion, .05 M HEPES buffer .1M NaCl pH 7.4 and 1-4μCi Ga-67 or 1-4μCi In-111/cell. The samples were run at least in duplicate and averaged. 30μl were removed from each chamber at periodic intervals and the radioactivity was counted in a automatic well counter. TF has a high affinity for ferric and other metal ions therefore care was taken to minimize spurious metal ion contamination. The water used was distilled then deionized. The membranes and plastic items were soaked overnight in a 50 mM solution of 8-hydroxyquinoline. All glassware was acid washed and all buffer solutions were passed through a column containing chelex 100, a divalent cation exchange, before use. The pH was checked before use.

RESULTS

Ga-67 and In-111 DF formation: Figure 2 shows the influence of time and temperature on the formation of Ga-67 and In-111 DF complexes as examined by paper chromatography. It can be seen the Ga-67 DF is rapidly and almost completely formed at pH 6.1 and 22°C using 40 mM DF. The formation of Ga-67 DF is so rapid that 61% of the activity was bound to DF immediately after additions of Ga-67 citrate. In contrast, under identical conditions the In-111-DF complex is only slowly and poorly formed. Upon raising the incubation temperature to 60°C, the rate of In-111 DF complex formation and the % yield increased markedly.

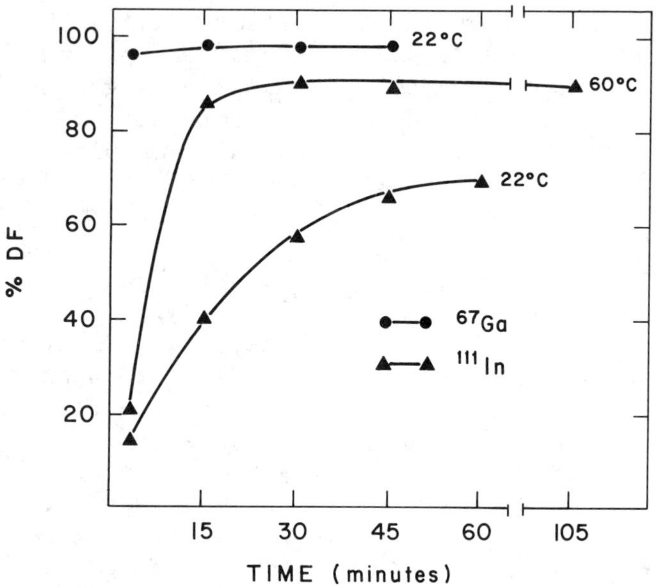

Figure 2

The influence of time and temperature on the formation of Ga-67 and In-111-DF. 40 mM DF in 1ml .1M acetate buffer pH 6.1 was incubated with Ga-67 or In-111 and chromatographed using butanol: n-propanol: H_2O, 9:6:5, as the solvent. The % DF represents the radioactivity present in the DF chromatographic peak.

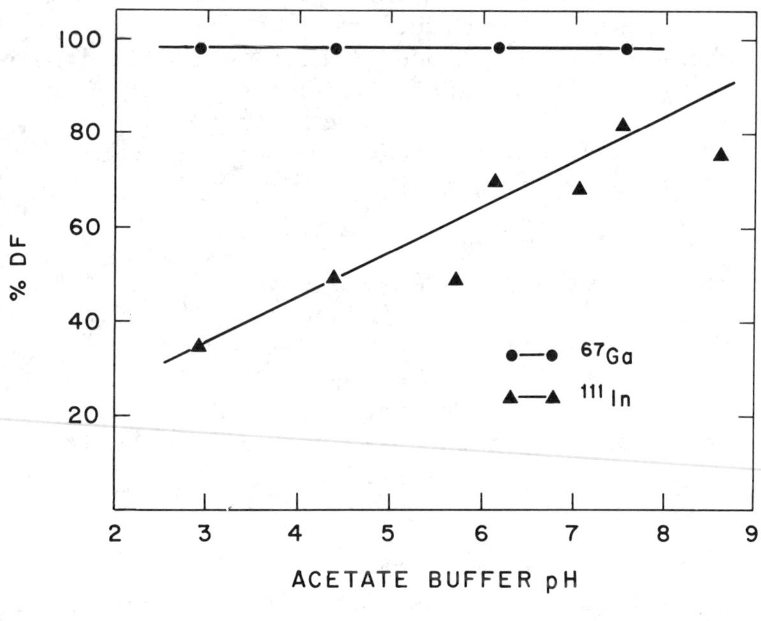

Figure 3

The influence of pH on the formation for the Ga-67 and In-111 DF complex. A solution of 40mM DF in .1M acetate buffer was incubated at 22°C for 1 hr with Ga-67 (●) and In-111 (▲) and subjected to paper chromatographic analysis.

Figure 3 indicates Ga-67 DF formation is affected little by changes in pH from 2.9 to 7.4. In contrast, the In-111-DF formation appears to be largely pH dependent. However, at physiological pH, the percentage of complex formed with both radionuclides is comparable. It was also observed that at physiological pH Ga-67 DF complex formation was essentially complete at 26μM DF, where as, 130μM DF and 37°C incubation temperature were needed to achieve 70% In-111-DF (Table 1).

Competitions experiments - paper chromatography: Table 1 shows the percentages of In-111 and Ga-67 found in association with DF under a variety of conditions. It appears that at a 1:1 DF:TF ratio or higher Ga-67 DF once formed can not be dissociated by TF even after 22 hrs of incubation. At the low DF:TF ratio only 50% of the DF-Ga-67 is formed before challenge by TF. This is because the DF concentration at this ratio is very low, .26μM. However, even in competition with a 100-fold TF sites excess, the percentage of the Ga-67 DF complex continues to increase over time. Moreover, when the Ga-67 is incubated with the TF prior to the addition of DF (40-75% TF -Ga-67) within 2 hours of DF addition almost all the activity binds to DF.

The preformed In-111-DF was relatively stable when TF was added at a 5:1 DF:TF ratio and the incubation continued at 37°C. There was about a 10-12% loss of activity to TF within 1-4 hours with little loss, thereafter. When TF was incubated with In-111 DF added and the incubation continued at 37°C, only about 25% In-111-DF was formed. This probably represented In-111 not bound to TF when the DF was added. There was only a slight increase in radioactivity bound to DF over 22 hours, suggesting that DF could not remove the TF bound In-111.

Competition experiment - Equilibrium Dialysis: Figure 4 shows the addition of DF at the arrow, rapidly dissociated the Ga-67 TF complex (▲). In a seperate experiment, where DF was not added, the Ga-67 remained bound to the TF during the same time period (■). When DF was added initially no Ga-67 binding to TF occurred (●). Therefore it appears that Ga-67 once bound to TF can be easily removed by a relatively small concentration of DF. In the presence of the both chelating agents in the mixture,Ga-67 binds to DF in preference.

The identical experiments were performed with In-111 except that the incubation temperature was raised to 37°C (Figure 5). The DF at a 1:1 DF:TF ratio can not remove the In-111 from the TF irrespective to the DF added at the start of the experiment or at the arrow (Fig. 5). The TF molecule just continues the slow uptake of In-111 under these conditions. Thus, these data confirm the results obtained by paper chromatography that the In-111 DF is a relatively weaker complex than the In-111 TF.

DISCUSSION

We have shown that under varying conditions of pH, incubation time, reaction temperature and DF concentration the Ga-67-DF complex molecule is more stable and more kinetically favored than the In-111-DF complex. The data also show that DF at a relatively low concentration can remove TF bound Ga-67. Therefore DF in plasma will probably complex both the TF bound and non-TF bound Ga-67, causing the chelated Ga-67 to be rapidly excreted (25).

TABLE 1. THE PERCENTAGE OF In-111 OR Ga-67 DESFERRIOXAMINE B(DF) WHEN In-111 OR GA-67-DF IS PREFORMED AND CHALLENGED WITH HUMAN TRANSFERRIN(TF) OR VICE VERSA AS A FUNCTION OF DF/TF RATIO AND INCUBATION[a]

Preformed Ga-67-DF[b]	PERCENT ACTIVITY BOUND TO DF AS A FUNCTION OF INCUBATION TIME[d]		
DF:TF[c] ratio	0 hrs	3 hrs	22 hrs
1:100	49.8	65.5	81.6
1:1	92.4	90.7	87.3
4.5:1	91.4	94.8	95.2

Preformed Ga-67-TF[e]	PERCENT ACTIVITY BOUND TO DF AS A FUNCTION OF INCUBATION TIME		
DF:TF ratio	0 hrs	2 hrs	23 hrs
1:1	1.4	88.0	92.6

Preformed In-111-DF[f]	PERCENT ACTIVITY BOUND TO DF AS A FUNCTION OF INCUBATION TIME				
DF:TF ratio	0 hrs	1 hrs	4 hrs	23 hrs	35 hrs
5:1	70.5	61.0	57.6	57.5	53.5

Preformed In-111-TF[g]	PERCENT ACTIVITY BOUND TO DF AS A FUNCTION OF INCUBATION TIME				
DF:TF ratio	0 hrs	1 hrs	4 hrs	23 hrs	35 hrs
5:1	23.5	-	27.9	29.6	-

a Incubation buffer was 50mM HEPES 100 mM NaCl pH 7.4 and incubation temperatures 22 °C for Ga-67 and 37 °C for In-111 complexes.
b Ga-67-DF incubated 25 min at 22 °C before TF addition.
c Ratio of metal binding sites, TF concentration was always 13μM.
d Time after addition of competing chelator.
e Ga-67-TF incubated 3 hours at 22 °C before DF addition.
f In-111-DF incubated 1 hour at 37 °C before DF addition.
g In-111-TF incubated 18 hours at 37 °C before DF addition.

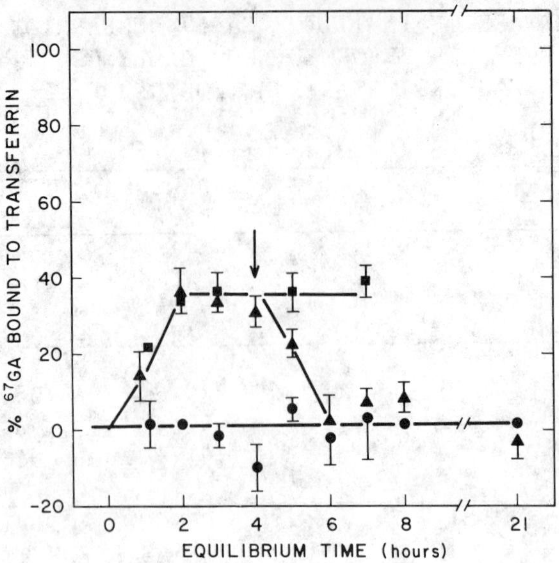

Figure 4

Equilibrium dialysis of Ga-67 TF complex challenged by DF. The DF was either added at the beginning of the experiment (●) or at the point indicated by the arrow after the TF-Ga-67 complex had come to equilibrium (▲). In a separate experiment used as a control the Ga-67-TF equilibrium was allowed to continue without DF addition (■). The incubation temperature was 22 C. The error bars represent standard deviation. Some error bars were eliminiated for clarity. The final concentration of constituents was: 10μM NTA, 13μM TF, 1.2 mM bicarbonate ion, 26μM DF and 1-4μCi Ga-67 citrate per dialysis cell in 1 ml.

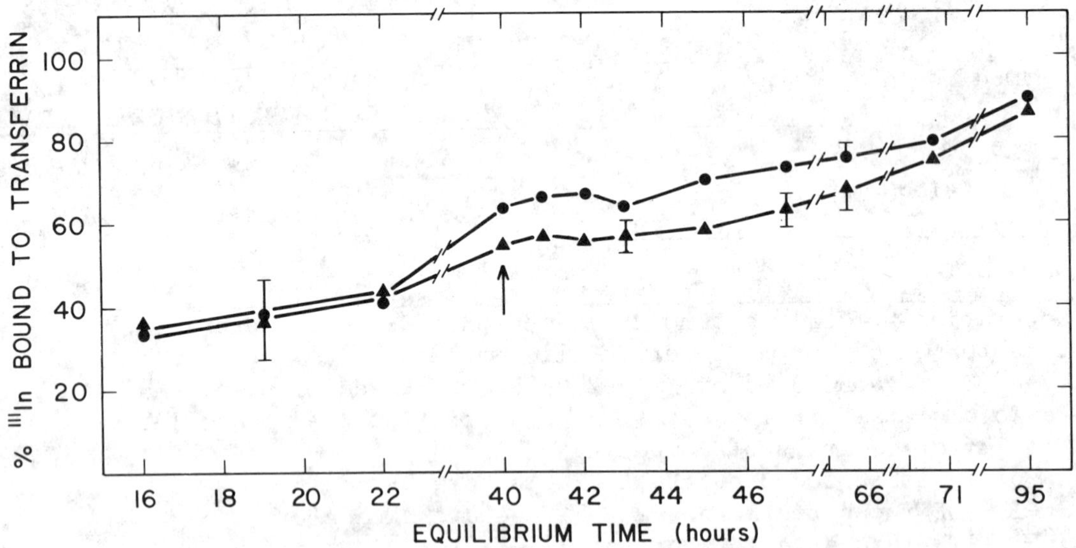

Figure 5

Equilibrium dialysis of In-111-TF complex challenged by DF. The DF was either added at the experiments out set(●) or at the point indicated by the arrow (▲). The incubation temperature was 37°C and all other conditions are as described in Figure 4.

DF thus shows promise as an effective agent for reducing the Ga-67 blood pool background (12-14). In contrast, we anticipate DF would make a very poor agent to reduce In-111 activity in the blood. The results suggest DF could not dissociate In-111 bound to TF in the blood.

The fact that DF can remove Ga-67 associated with TF is also consistent with our contention, (10) that siderophores play a role in the mechanism of Ga-67 sequestration in abscesses. We theorize Ga-67 transported to the inflammatory site bound to TF (17,18) would be removed from TF by bacterial siderphores. The siderophore could then facilitate the transfer of the Ga-67 into the invading organism where the radionuclide would eventually be trapped. The weakness of the In-111 DF complex compared to the strength of the radio-indium TF complex implies bacterial sideriphores would not be as efficient in removing In-111 from TF as Ga-67. Thus, the In-111 would not concentrate in bacteria and partially explain why In-111 is less effective than Ga-67 as an abscess scanning agent(26). Moreover, there are a number of pieces of data which lend support to this hypothesis. Each could influence the rate of Ga-67 uptake at the pus site. For example, the rate of DFC mediated cation transfer into the U. sphaerogena is more rapid for Ga-67 than for Fe^{3+} (15). Another important factor is the dissociation rate of Ga-67-TF. The transfer of Ga-67 to a siderophore would be enhanced if the pH of the abscess tissue was lower than normal blood pH. The pH of abscesses is most likely acidic.Experiments have indicated, as the pH is lowered TF has a lower affinity for Ga-67 (27). Another kinetic factor suggested by our data is a low siderophore concentration present at the site would preclude rapid Ga-67 siderophore formation. However, over a 24 hour time span, we would anticipate the formation would continue (see table 1). The increase in body temperature which usually accompanies an inflammation would increase the rates for all biochemical reactions. Hence, this would stimulate Ga-67 deposition by the siderphores and may overcome the problem of concentration. Therefore in conclusion, our data and that of others points to an important role of bacterial siderophores as one of the components in the mechanism of Ga-67 deposition in abscesses.

ACKNOWLEDGEMENT

The work was supported by the DOE contract No's. Ey-S-02-4078 and EP-78-S-02-4625.

REFERENCES

1. NIELANDS JB: Microbial iron transport compounds (siderochromes). In Inorganic Biochemistry, Eichhorn G, (ed), Amsterdam, Elsiver 1973, 167-202
2. COOPER SR,McARDLE JV, RAYMOND KN: Siderophore electrochemistry: Relation to intracellular iron release mechanism. Proc Natl Acad Sci USA 75: 3551-3554, 1978
3. BICKEL H, HALL GE, KELLER-SCHIERLEIN W, et al: Stoffwechselprodukte von Actinmycin 27 Mitteilung. Über die Konstitution von Ferrioxamine B.Helv Chem Acta 43: 2129-2138, 1960
4. NEILANDS JB: A crystalline organo-iron pigment from a rust fungus (Ustilago sphaerogena). J Am Chem Soc 74: 4846-4847, 1952
5. BARRY M, FLYNN DM, LETSKY EA, et al: Long-term chelation therapy in thalassemia major: Effect on liver iron concentration liver histology and clinical progress. Br Med J 2:16-20, 1974

6. MODELL CB, BECK J: Long-term desferrioxamine therapy in thalassemia. Ann NY Acad Sci 232: 201-210, 1974
7. PROPPER RD, SHURIN SB, NATHAN DG: Reassessment of the use of desferrioxamine B in iron overload. N Engl J Med 294: 1421-1423, 1976
8. SESHADVI R, COLEBUTCH JH, GORDON P, et al: Long-term administration of desferrioxamine in thalassemia major. Arch Dis Child 49: 621-626, 1974
9. EMERY T: Role of ferrichrome as a ferric ionophore in Ustilago sphaerogena. Biochemistry 10: 1483-1488, 1971
10. HOFFER PB, WEINER R, MILLER R, et al: The Role of Lactoferrin and other iron binding proteins in the Localization of Ga-67. Second International Congress of the World Federation of Nuclear Medicine and Biology, p 48, September, 1978, Washington D.C. (abst)
11. GOODMAN MM, THAKUR ML, HOFFER PB, et al: Indium-111 Desferrioxamine Complexes: Preparation and Stability studies. J. Label Comp Radiopharm 16: 1 (abst) 1979
12. HOFFER PB, SAMUEL A, BUSHBERG JT, et al: Desferrioxamine Mesylate (Desferal): A Contrast Enhancing Agent for Ga-67 imaging. J Nucl Med, in press
13. LARSON SM, RASEY JS, GRUNBAUM Z, et al: Pharmacologic enhancement of Ga-67 tumor-to-blood ratios for EMT-6 sarcoma (BALB/c) mice. Second International Congress of the World Federation of Nuclear Medicine and Biology, p. 48, September, 1978, Washington, D.C. (abst)
14. OSTER ZH: The effect of desferrioxamine mesylate (desferal) on the biodistribution of gallium-67 citrate. J Nucl Med 19: 732 (abst), 1978
15. EMERY T, HOFFER PB: Gallium can mimic iron in microbial transport systems. Society of Nuclear Medicine 26th Annual Meeting, Submitted 1979
16. MENON S, WAGNER HN Jr., TSAN M-F: Studies on gallium accumulation in inflammatory lesions: 11 Uptake by staphylococcus aureus: Concise Communication J Nucl Med 19: 44-47, 1978
17. GUNASEKERA SW, KING LJ, LAVENDER PJ: The behavior of tracer gallium-67 toward serum proteins. Clin Chim Acta 39: 401-406, 1972
18. LARSON SM, ALLEN DR, RASEY JS, et al: Kinetics of binding of carrier free Ga-67 to human transferrin. J Nucl Med 19: 1245-1249, 1978
19. SNOW AG: Metal complexes of mycobactin P and desferrisideramines. Bacter Rev 34: 99-120, 1970
20. DIEM K, LENTNER C Editors: Ciba-Geigy Scientific Tables. 7th ed. Geigy Pharmaceuticals, Macclesfield England, 975, p. 583.
21. HARRIS DC, GRAY GA, ASIEN P: ^{13}C Nuclear Magnetic Resonance Study of the Spatial Relation of metal-and anion binding sites of human transferrin. J Biol Chem 249: 5261-5264, 1974
22. BATES GW, BILLUPS C, SALTMAN P: The kinetics and mechanism of iron (III) exchange between chelates and transferrin I. The complexes of citrate and nitrilotriacetic acid. J Biol Chem 242: 2810-2815, 1967
23. BATES GW, Schlabach MR: The synergistic binding of metal ions and anions by transferrin. In proteins of iron storage and transport in biochemistry and medicine. R.R. Crighton (ed) Amsterdam, North Holland 1975, 51-58.
24. LEHMANN HP, KAPLAN A: Use of an anion-exchange column to determine serum iron-binding capacity. Clin Chem 17: 941-947, 1971
25. PETERS G, KEBERLE H, Schmid K, et al: Distribution and renal excretion of desferrioxamine and ferrioxamine in the dog and the rat. Biochem Pharmacol 15: 93-109, 1966
26. THAKUR ML: Gallium-67 and Indium-111 radiopharmaceuticals. Int J Appl Radiat 28: 183-201, 1977
27. HOFFER PB, HUBERTY J, KHAYAM-BASHI H; The Association of Ga-67 and Lactoferrin. J Nucl Med 18: 713-717, 1977

ROLE OF RADIOPHARMACEUTICALS IN NUCLEAR HEMATOLOGY

Mathew L. Thakur, Ph.D.
Alexander Gottschalk, M.D.

Yale University School of Medicine
Department of Diagnostic Radiology
Section of Nuclear Medicine
New Haven, Connecticut 06510

ABSTRACT

The continued reassessment and refinement of techniques during the past few decades have generated a series of radioactive agents that thave been evaluated as tracers for cellular blood elements. A few of the radiopharmaceuticals which have contributed a great deal to nuclear hematology have been concluded in detail. The mode of their actions, merits and disadvantages have been discussed. None of the compounds currently available appear to meet all the ideal requirements in that it is still impossible to selectively label and follow a cohort of cells for their entire life span with external monitoring by currently available instrumentation. Nevertheless, the newer agents show great promise as potential radiopharmaceuticals designed to investigate a variety of hematologic disease in humans.

Szirmai has designated the use of radioactive compounds as tracers in the study of specific aspects of hematology as "Nuclear Hematology" (1). This constitutes a broad field of endeavor including study of blood forming organs, evaluation of radiation effect on hematopoietic tissues, estimation of blood cell survival, and study of kinetics and pathophysiology of other cellular blood elements. During the past two decades, numerous in vivo and in vitro studies have been conducted in nuclear hematology, and a voluminous literature has been published. In clinical nuclear medicine, procedures for imaging liver, spleen and bone marrow are performed routinely. Much of this can be attributed to the availability of Tc-99m sulfur colloid, which reduced the radiation dose to the liver and bone marrow from 4-10 rads and 0.3-0.7 rads/scanning dose of Au-198 colloid to 1.8 and 0.026-0.34 rads/mCi, respectively (2,3). The status of bone marrow imaging and bone blood flow has been explicitly evaluated by a number of authors (2-5). Since Tc-99m colloid is rapidly phagocytosed by the reticuloendothelial system in the liver and spleen, it is nearly an ideal agent for imaging these organs. The use of Tc-99m colloids as lymphangiographic agents has been discussed by Dr. Ege and does not warrant a further discussion.

Recent methods for labeling red cells in vivo, with Tc-99m have added a great deal to nuclear cardiology (6-8). This technique has become widely used for imaging the cardiac blood pool at short intervals during the cardiac cycle, thus providing the means for either calculating the left ventricle ejection fraction, or analyzing regional myocardial wall motion (9,10). RBC labeling is performed by intravenous injection of "cold" commercially available bone scanning agents to impregnate red cells with Sn^{2+} ions followed by administration of Tc-99mO_4-, which then binds circulating red cells selectively. Because it is simple, the technique is widely employed.

Using the same principle, red cells also can be labeled with Tc-99m in vitro. These red cells can then be deformed by a controlled exposure to heat and administered to a patient for imaging the spleen selectively. The spleen, which functions as an on line filter in the portal circulation with a flow rate of about 300 ml/min (approximately 5% cardiac output) (11), sequesters the denatured labeled red blood cells and permits excellent splenic images. An elegant kit for labeling red blood cells in vitro has been designed by Smith and Richards (12), but it is not available commercially at the time of this writing. Armas, Thakur, and Gottschalk (13) have developed a modification of the Smith Richard concept. Approximately 15 mg of commercially available cold stannous pyrophosphage is intravenously injected into an adult patient and 30 minutes later 6-8 ml venous blood is obtained in a heparinized syringe. To this blood the desired dose of Tc-99mO_4 - is added, followed by a 35 minute incubation in a water bath at 49°C. The incubation at elevated temperatures promotes erythrocyte damage while Tc-99m activity binds to the cells. The labeled and deformed cells are then reinjected into the patient. Excellent spleen images may be obtained one to two hours later.

IMPORTANCE OF OTHER CELLULAR BLOOD ELEMENTS IN NUCLEAR HEMATOLOGY

Because of the clinical success of reticuloendothelial imaging, techniques to label the rest of the cellular blood elements (14, 15) have been investigated only recently. Many pathophysiological processes result in an excessive accumulation of one or more types of cellular blood elements. Polymorphonuclear leukocytes, for example, localize in abscesses and are the major cellular constitutents in the lesions of rheumatoid arthritis and in myocardial infarction (16). Patients with myocardial infarction also have decreased platelet survival and increased platelet aggregates (17). Furthermore, platelet aggregates have been found trapped in the myocardial microcirculation in patients suffering from coronary arterial thrombosis. Arterial emboli rich in platelets (18) are a primary source of morbidity and mortality in patients with severe mitral valve disease. These conditions are particularly serious when an enlarged left atrium and decreased cardiac blood flow are present. An increased risk of embolism is also associated with mitral and aortic valve prosthesis, porcine heterografts, cardiac catheterization, and pacemakers implantation.

It has long been recognized that the lymphocytic immune system plays an important defensive role in neoplastic disease and certain types of tumors frequently invoke a mononuclear cell infiltrate (20). Estimation of peripheral blood cell counts is a routine screening test to evaluate a patient's health and disease. To capitalize fully on changes in production and migration of blood cell elements, non-invasive techniques are needed to identify areas of cell localization. Labeling blood cell elements with radionuclide provides such a technique. This chapter evaluates the role of some of the radiopharmaceuticals employed for labeling leukocytes, platelets, and lymphocytes.

PROPERTIES OF A SUITABLE RADIOACTIVE TRACER FOR CELLULAR BLOOD ELEMENTS

The ideal radioactive tracer for cellular blood elements would be one that could be used to label the desired cell type within the marrow prior to release as a cohort into the circulating blood. This radioactive label should emit appropriate energy photons and have a physical half-life long enough to permit observations of in vivo events during the life of the cohort cells by external monitoring of the radioactivity. It would then be possible to study the in vivo cell distribution, the rate and quantification of cell appearance and disappearance in healthy and diseased tissue. Additionally, the tracer should neither elute from the labeled cells nor be reutilized after the biological destruction of the cells. Ideally, the radioactive tracer should have a half-life long enough to follow the life span of the labeled cohort, thereby minimizing irradiation to the subject.

The first attempt to label blood cells, in this case circulating erythrocytes, was made by Chapin and Ross (21) in 1942. They employed Fe-59 (($t_{1/2}$ - 45.6 d, β^-, γ -1.095 MeV (56%) 1.292 MeV (44%)) which is known to bind to heme. Although a discussion on the advantages and disadvantages of this specific erythrocyte label is beyond the scope of this article, this pioneering concept stimulated sufficient interest in labeling the other blood cell elements such as polymorphonuclear leukocytes, (PMN), platelets and lymphocytes.

RADIOPHARMACEUTICALS FOR CELL LABELING

Most of the early work in this field consisted of methods to determine the life span or intravascular survival time of the cellular blood components. Several radioactive tracers were employed. These can be classified in two major groups: (1) DNA or cohort labels, and (2) Non-DNA or random labels.

DNA OR COHORT LABELS

These radioactive tracers are made available to the bone marrow stem cells and are incorporated during maturation of the cell cycle when DNA is being synthesized. The radioactivity thus binds exclusively to cells before they are released from bone marrow and not to mature cells already released in the circulating blood. After a delay required for maturation, the radioactivity appears in circulation as a cohort, or pulse, of labeled cells of uniform age. DNA is considered metabolically stable, and the radioactivity is thus bound within the nucleus until the cell dies. The life span of cells can be determined by monitoring the appearance and loss of radioactivity in the blood.

Phosphorous - 32 ($t\frac{1}{2}$ - 14.5 d, β^-) as sodium phosphate, and carbon-14 ($t\frac{1}{2}$ - 5500 y, β^-) labeled adenine, guanine and orotic acid have been reported (22) as cohort labels. However, these nucleic acid precursors are highly metabolically reactive and are not suitable pulse labels.

H-3 THYMIDINE

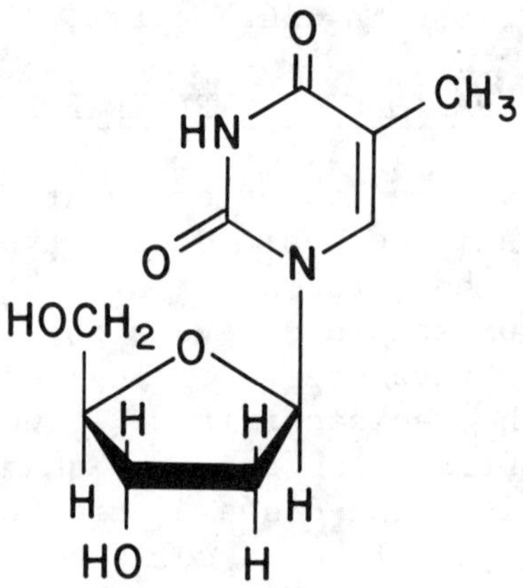

Tritiated (H-3, $t\frac{1}{2}$ - 12.34 yr, β^-) thymidine was first synthesized by Taylor (23) et. al. in 1957. In their studies in humans, Rubine (24) et. al. reported that when H-3 thymidine was administered intravenously, approximately 90% of the radioactivity cleared from plasma within one minute. The rapid plasma clearance of the radioactivity was associated with incorporation of thymidine into the newly-formed DNA of proliferating cells. Labeling of the bone marrow was nearly complete within ten minutes after injection, and the label appeared to remain in these cells. Residual H-3 thymidine was either excreted in the urine or catabolized to ^3H_2O. The question of the long-term reutilization of the tracer was not clarified in this work, but in the initial phase, tritiated thymidine served as a true cohort or flash label.

Attempts (24) to label isolated canine bone marrow and thoracic duct lymphocytes as well as human leukemic blood cells were unsuccessful, since 90% of the added activity remained in the cell free supernatant. This was attributed to the rapid degradation of most of the H-3 thymidine to H-3 thymine in the presence of the intracellular thymidine phosphorylase.

The efficient in vivo incorporation of thymidine into proliferating cells suggests the possibility of labeling thymidine with a suitable gamma emitting radionuclide to allow monitoring of a labeled cohort of granulocytes. There are, however, several serious problems. Since thymidine incorporates directly into the DNA of the dividing cells, serious potential genetic consequences are raised which weight strongly against its use in humans. Reutilization of the label by lymphocytes, regenerating liver cells, and skin after the cell death is also possible (22). The cytocidal effects in human leukemia after 10 injections of 0.25 μCi/g body weight, over a period of 5 days (24) have almost prevented its further application in humans.

SE-75-SELENOMETHOININE:

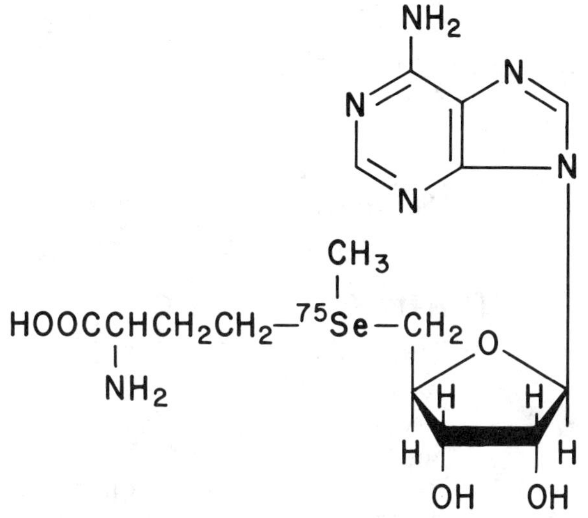

Methoinine labeled with gamma emitting Se-75 ($t_{1/2}$ - 120 d, γ -265 KeV, 60%) provided the first biologically active amino acid labeled with a gamma emitting radionuclide (26). When injected intravenously, the tracer labeled cohort leukocytes, erythrocytes and platelets (27, 28). This work suggested Se-75 -selenomethoinine would be a suitable agent for cell labeling. Following this report, the compound was injected (i.v.) into 10 healthy human volunteers (29) and daily blood samples were obtained for 21 days. Sufficient quantities of platelets, leukocytes and erythrocytes were isolated and concomitant radioactivity measured. The results indicated that there was a considerable reutilization of the label by blood cells and the compound did not provide a true cohort label. The inability of the compound to label isolated leukocytes in vitro was confirmed later (14).

NON-DNA OR RANDOM LABELS

Several radioactive compounds have been investigated as random labels for leukocytes and platelets. Many of these have not had extensive applications, and few reports in which they have been employed are found in the literature. The most prominent of these agents are discussed below.

S-35 SODIUM SULPHATE

When carrier free S-35 ($t\frac{1}{2}$ 87.9 d, β^-) sodium sulphate is injected intravenously, most of the radioactivity "promptly" binds to uronic acid, which is excreted in the urine (30). A very small amount of the inorganic sulphur binds to amino acids and proteins. The major proportion of the remainder of the radioactivity binds to sulphated mucopolysaccharides which incorporate into megakaryocytes. In man, this process is completed within 24 hours after each dose of S-35 and thus provides a selective label for platelets. Therefore, although it is a non-DNA label, it can be classified as a cohort label for platelets. Vodopick and Knisley (30) employed this agent for platelet survival studies in six patients. Due to the need for beta counting, however, the agent has never achieved a wide popularity.

P-32-DIISOPROPYLEFLUOROPHOSPHATE (DFP-32)

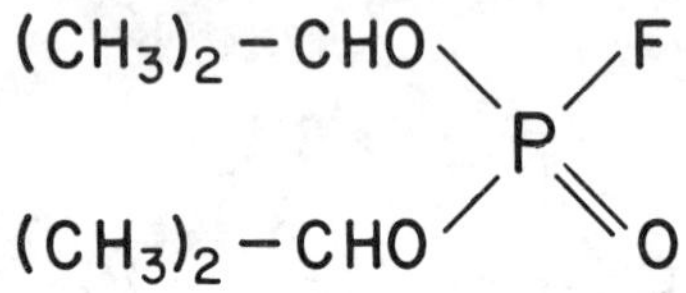

The well known toxicity of the alkylesters of fluorophosphate was demonstrated to be due to the irreversible inhibition of cholinesterase (Che) enzymes. One member of this family of compounds is DFP. During an extensive study of this compound Grob, et. al. (31) suggested that DFP binds to red cells. Seven years later, Cohen and Warringa (32) reported the in vivo use of the radioactive DFP. When DFP-32 was injected intramuscularly, the radioactivity appeared in peripheral blood within 90 minutes. Thirty-three percent of the radioactivity was excreted through the urine within 24 hours and a total of 60-70% within 10 days. The remainder of the radioactivity was associated with plasma proteins and erythrocytes. Leeksman and Cohen later reported intramuscularly administered DFP-32 also binds to leukocytes and platelets (33). Following this report, Athens and co-workers used the tracer to make fundamental contributions to the knowledge of granulocyte kinetics (34-37). When DFP-32 is used for labeling granulocytes (PMN) in isolated whole blood, the labeled granulocytes maintain their functional ability. The compound is neither eluted from the cells nor reutilized after cell destruction. Using DFP-32 Athens and co-workers have studied leukocyte physiology extensively. However, neither human nor canine lymphocytes incorporate the compound (36). The reasons for this latter phenomenon are not clear.

The sites of binding of DFP-32 to PMN or to platelets have not been investigated. The probable mechanism of cell binding, however, is believed to be enzymatically mediated. The DFP-32 are known to form irreversible bonds with a number of enzymes and other aklyesters of fluorophosphoric acid. In this reaction the phosphorous atom binds to the hydroxy group of an amino acid, serine, as follows;

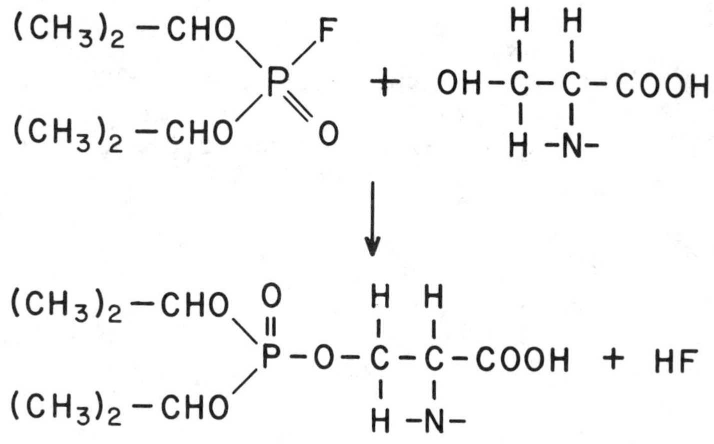

and liberates fluoride ion as hydrogen fluoride (32, 38). The diisoprophyl-phosphoserine then must be enzymatically incorporated into blood cells. Contrary to earlier observations, reutilization may be a problem, since when either the enzymes or the labeled cells are biodegraded, the released P-32 may bind to certain lipids which, in turn, bind to platelets (39).

When DFP is administered daily for up to five days in quantities of 1.5 mg or more, toxic effects in man have been reported (31). All symptoms were attributed to irreversible inhibition of cholinesterase. In spite of these problems, if the compound could be labeled with high specific activities and a suitable gamma emitting isotope, it could be a useful tracer for cell labeling. Unfortunately, phosphorous has no other suitable radioisotopes and no work has been carried out with phosphorous analogues such as arsenic.

Cr-51-chromate

Cr-51 has been the most popular gamma emitting non-DNA cell label. Gray and Sterling (40) hypothesized in 1950 that radioactive chromium would be taken up by erythrocytes and plasma proteins since stable chromium is found in both human tissue and normal blood. Since then, numerous investigators have employed the racter, in oxidized form, to label erythrocytes, granulocytes, lymphocytes, platelets and eosinophils (41-43). Today, it still remains an agent of choice for clinical survival studies of platelets and erythrocytes (44). If high specific activity sodium chromate is used, no reduction in cell viability is reported. However, poor labeling efficiency and low cell recovery (due to loss of radioactivity in organs rich with reticuloendothelial cells) are common regardless of the cell type labeled. Diffusion through the cell membrane of anionic hexavalent chromium (CrO_4-) followed by enzymatic recution to Cr^{3+} has been suggested as the probable labeling mechanism (40). Unfortunately, a spontaneous elution of radioactivity at a rate greater than 1% per day generally has been observed. Since the released radioactivity is in the Cr^{3+} form, no reutilization of tracer by other type of cells occurs. However, the released activity is known to bind to plasma proteins which must be removed from the cell sample to ensure an accurate measurement of cell activity.

Despite its availability and long shelf life ($t_{1/2}$ 27 days) Cr-51 is far from an ideal label because of its lack of specificity and unsuitable photon flux for external detection (only 9% of the total decay produces 320 KeV gammas).

TC-99M PERTECHNETATE

More recently, radioactive agents with better physical characteristics have been introduced for cell labeling. The use of sodium pertechnetate as a tracer for lymphocytes (45) and platelets (46) has been explored. Addition of a reducing agent, $SnCl_2$ was essential in platelet labeling, but toxic effects of $SnCl_2$ on lymphocytes have been reported (47). The availability and the ideal photon energy of Tc-99m render it a very attractive radionuclide, but its relatively short half-life of six hours may limit its use in certain applications involving platelets and lymphocytes where physiologic responses lasting for days may need to be followed.

TC-99M S-COLLOID

Tc-99m-sulfur colloid has been investigated as a label for leukocytes (48, 48a). This provided the first physiologic and seelctive label for neutrophils, since phagocytosis of radioactive colloid was the most likely mechanism by which the radioactivity was incorporated. Further investigations of this compound however, revealed that approximately 60% of the colloid particles were not engulfed (15). A pure labeled PMN preparation could be obtained only by also allowing the leukocytes to engulf iron carbonyl particles, thereby permitting isolation with a strong magnetic attraction. This procedure cannot be performed if leukocytes are to be used for in vivo applications. The authors also reported that a significant proportion of the radioactive particles in association with the cells could be exchanged with addition of "cold" sulphur colloid to the cell suspension. This indicates that many of the colloid particles are simply adsorbed on the cell surface and not truly phagocytosed. It is also unclear whether or not leukocytes once stimulated for phagocytosis will maintain their normal survival and functional parameters in vivo.

Ga-67-CITRATE

In 1973 Burleson and co-workers (49) reported the use of Ga-67-citrate as a label for leukocytes obtaining labeling efficiencies of 2.8% (50). However, since Ga-67-citrate localizes in inflammatory lesions very effectively anyway (51), no further investigations have been made with this compound as agent for labeling leukocytes.

OTHER RADIOACTIVE COMPOUNDS

McAfee and Thakur (14, 15) assessed several soluble radioactive compounds and particles for incorporation into leukocytes by either pinocytosis or phagocytosis. This work indicated that soluble agents have certain practical advantages over radioactive particles. Among the soluble agents In-111 or Tc-99m chelated with 8-hydroxyquinoline (oxine), In-111-tetraphenylporphyrine, and Br-77-DNA conjugated with adriamycin appeared to be promising agents for labeling isolated neutrophils in vitro.

Indium-111-oxine is easy to prepare. It has a longer shelf life than Tc-99m oxine and has a more suitable gamma photon emission for external detection than the gamma photons of Br-77 ($t_{1/2}$ - 56 hr. γ -240 KeV, 30%, 520 KeV, 24%). Although In-111 tetraphenyl porphyrine appeared potentially useful, it could not be readily synthesized.

Following the introduction of In-111-oxine, Ga-68-oxine ($t_{1/2}$ - 68 min) has also been prepared and bound to erythrocytes and platelets (52). Gallium-68 is a generator produced radionuclide (parent Ge-68, $t_{1/2}$ - 275 days), and a readily available source of β^+ for emission tomographic studies. The cellular blood components labeled with this compound will have certain applications where the pathophysiological cell response to a lesion is rapid and tomographic examinations are desirable. The use of Tc-99m-oxine, as a tracer for platelets has been evaluated by Wistow, et. al. (62). However, when compared to In-111-oxine "considerably lower survival times" (0.87 days) and "lower initial recovery" (47%) has been noted. At the present time, how, ever, more work has been done with In-111-oxine. This will be briefly reviewed.

INDIUM-111-OXINE

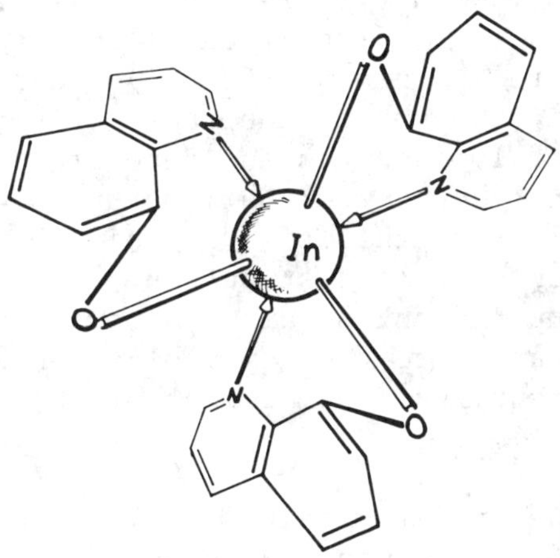

Indium-111 was chosen as a tracer because of its physical characteristics. It is produced in a carrier free form with charged particle accelerators. When in ionic or weakly chelated form, it is known to bind to transferrin (53). 8-hydroxyquinoline is a bidentate ligand and binds many metal ions. The bactericidal properties of oxine have been known for many years (54), but it has been suggested that oxine itself is not toxic to staphylococci unless traces of iron or copper are present (55). When oxine dissolved in ethanol is added to a carrier free In-111 solution of a certain pH, a neutral complex extractable in lipophilic solvents is formed (56). An excess of oxine is usually necessary to scavenge other metal ions that may be present in the solution.

STABILITY OF IN-111-OXINE

The thermal stability of a compound is an important parameter which determines its shelf life and its suitability to a specific use. The stability constants of In-111-oxine have been estimated to be approximately 10^{11}. This is high enough for the compound to be stored until the radioactivity is decayed. However, if the complex is added to a solution containing a relatively large quantity of certain ionic metals, e.g., Fe^{3+}, the oxine will bind to these metal ions in preference and release indium-111 atoms free in solution. Similarly, in the presence of strong chelating agents such as EDTA or plasma transferrin, indium-111 no longer remains In-111-oxine and requires erythrocytes (53), leukocytes (56), platelets (57), and lymphocytes (47) suspended in plasma free medium for high efficiency labeling. Consequently, labeling methods using solutions containing plasma or serum lead to a much lower yield of radioactive cells (58-62).

LOCATION, ELUTION AND REUTILIZATION OF THE TRACER

In the case of a gamma emitting radionuclide such as indium-111 which permit efficient external detection, the integrity of the tracer is an important parameter if meaningful conclusions are to be drawn regarding in vivo cell districution. The label should not only remain affixed to labeled cells in vivo, but also must not be reutilized after cell death. Various groups agree that only a small proportion of the radioactivity is eluted from labeled cells administered to humans (63, 64) and animals (47, 56, 58). Furthermore, it has been shown that the radioactivity, administered to humans as labeled leukocytes or lymphocytes and deposited in the liver, spleen and marrow remains fixed in location for periods up to 48 hours post injection (63, 65). It is not clear at this time if this phenomenon occurs due to natural cell margination or cell destruction. It has been observed that the radioactivity in the supernatant obtained after homogenization of labeled cells does not incorporate into freshly isolated blood cells. These data indicate that in vivo reutilization of most of the cell bound radioactivity is unlikely.

The location of the tracer within a cell has been explored in the case of human neutrophils (66). It has been concluded that the lipid soluble complex indium-111-oxine passively penetrates the cell membrane where the vast majority of indium-111 binds to three cytoplasmic components of apparent m.w. 540,000, 80,000 and 3,600. Only a negligible proportion of the radioactivity binds to DNA and a major proportion of oxine is released from the cells. At this writing, the intracellular location of indium-111 in the other types of blood cells remains unknown. A similar mechanism and tranfer of indium-111 radioactivity to subcellular platelet components other than the storage granule constituents has been proposed (59).

If indium-111-oxine were a much stronger complex, it might penetrate the cell membrane in the presence of other competing chelating agents. In that event, the indium-111 probably would be unavailable for binding intracellularly and would be released from cells spontaneously.

Cell Toxicity

Cells may show toxic effects following labeling with indium-111-oxine. At least four factors in the labeling process may affect cells adversely. The isolation procedures which involve repeated centrifugation and resuspension unavoidably cause some damage to the cell membrane; the degree of damage may vary from preparation to preparation. The other three sources of toxicity arise from the exposure of cells to oxine, ethanol and radioactivity. Oxine alone is not toxic to the cells (54), but when chelated with metal ions it may cause toxic effects either by introducing undesirable metal ions (67) or by depleting essential metal ions in the cells.

No adverse effects of either oxine, ethanol or radiation on platelet aggregability have been reported (59). The functional viability of labeled platelets is also supported by in vivo results obtained by several workers (57, 58, 62, 64, 68-72). Lymphocytes are known to be more susceptible to radiation damage (73). It has been shown in vivo and in vitro tests that their function is impaired as a direct result of exposure to radiation (47, 60). These effects were more dramatic when the lymphocytes were tested 24-48 hours after labeling. When 10^6 lymphocytes are labeled with approximately 2 μCi indium-111-oxine and left in suspension in a test tube, each cell will receive a radiation dose of 200 rads/day when decay of indium-111 is taken into account (60). Yet when approximately 10^8 lymphocytes were labeled with up to 500 μCi of indium-111-oxine and administered soon after labeling minimal, if any, direct radiation effects on the lymphocyte survival can be detected (47, 74). Since only a neglible proportion of indium-111-oxine binds to DNA, no mutagenic consequences should be expected. More systematic data are required however, to elucidate the optimal dose of radioactivity that can be incorporated without adversely effecting characteristic lymphocyte function.

Polymorphonuclear leukocytes are resistant to radiation (75). We (76) have shown that when 10^8 human leukocytes, suspended in 5 ml 0.9% NaCl are exposed to up to 2.9 mCi indium-111, (integrated radiation dose/cell = 50 k rad (77)), 650 μg oxine,and 500 μl ethanol for up to 15 minutes, their chemotaxis, random imgration bactericidical capacity and ultrastructure remain unaffected. These examinations were performed within 2-3 hours after the cells were exposed. Contrary to these results, Segal, et. al. (60) have reported that as little as 75 μg oxine per hundred million PMN reduces their chemotaxis by nearly 50% at 24-48 hours following labeling. (The half clearance period of PMN from the blood is approximately 8 hours). The difference in time following cell labeling and a different source of indium-111 may account for these functional descrepancies. However, even 75 μg of oxine is higher than the 50 μg of oxine routinely employed in our laboratory for labeling 10^8 leukocytes for in vivo use. Leukocytes exposed to up to 50 μg oxine appear to maintain their viability (63, 78, 79) and respond to chemotactic stimuli in vivo (80).

CURRENT STATUS

IN VIVO APPLICATIONS

Table I summarizes the current status of the applications of cellular blood elements labeled with indium-111-oxine, and indicates their potential in nuclear hematology.

Effective labeling also permits studies of in vivo cell kinetics. The kinetics of granulocytes in man have been studied primarily by measuring the fate of DFP-32 radioactivity in the blood and have been reviewed by Cartwright and colleagues (84). The granulocytes are the non-dividing end cells, and mean value of their turnover rate as estimated in 56 normal males has been reported to be 163 x 10^7 granulocytes per kg body weight per day. The granulocytes leave the total body granulocyte pool in an exponential fashion with a mean half time disappearance of 6-7 hours. From the circulating blood they enter many tissues, mainly, liver, spleen, and lungs. Many of those that leave circulation enter gastrointestinal tract and are lost from the body. There is no appreciable pool of granulocytes in the extramedullary tissue of normal subjects. Unlike lymphocytes, they do not return from the tissues to a blood pool.

Similarly, the half survival time of indium-111 labeled granulocytes in dogs was 8-9 hours (56). When the labeled cells were injected intravenously radioactivity immediately appeared in the liver, spleen and lungs. The activity in the lungs cleared with 3 to 4 hours. During this time, increased radioactivity was noted in circulating blood. At 24 hours after injection, there was 17.1 ± 7.22 in the liver and 32.32 ± 6.4 times more activity in the spleen than in the blood.

Similar observations were made in man (63). Following i.v. injection of labeled cells, indium-111 immediately was detected in the lungs and was 50% cleared within 15 minutes. Four hours after injection, no detectable radioactivity was noted in normal lungs. In contrast, there was a rapid accumulation of radioactivity in the spleen and in the liver. This radioactivity was quantified using whole body scans and a computation techique described by Williams, et. al. (85). The data indicated that the radioactivity in the liver varied between 12% and 40% of the administered dose, depending upon the method used for the isolation of leukocytes. When a method more likely to cause cell damage was used, a greater proportion of the radioactivity accumulated in the liver. When pure PMN suspensions were injected, the radioactivity in the spleen was minimal (6%). The percentage increased to 22% when PMNs contaminated with red cells and lymphocytes were administered. Data obtained from sequential scans of each subject indicated that the quantity of tracer accumulated in the spleen and liver within the first 30 minutes after injection remained essentially unaltered for up to 48 hours. There was less variation of indium-111 lymphocyte radioactivity in the liver and spleen of human recipients compared to PMNs. This may be due to the lesser degree of contaminating cells normally found in isolated lymphocytes (65). These data need further clarification before it is safe to assume they represent true cell margination kinetics of leukocytes and lymphocytes.

FUTURE POSSIBILITIES

Despite the several advantages of indium-111-oxine as a tracer in nuclear hematology, its disadvantages cannot be disregarded. Indium-111-oxine is a non-specific label. Selective cell labeling require pure cell type suspension preferably deprived of natural nutrient medium during labeling. Such preparation invariably causes a certain degree of damage to cells even if the greatest care is exercised in handling them. Cells labeled in this manner have shown promising results in localization studies, but their use to study cell kinetics and the normality of their ultimate fate in extravascular and vascular compartments must be questioned.

Much rewarding research remains to be carried out. It might be directed toward simplification of cell separation techniques so that the required cell types could be obtained "pure and healthy". Potentially, it could be directed toward replacing indium-111 by another radionuclide which, when chelated with oxine or other moiety will label cells as effectively in the presence of natural nutrient medium. Oxine, of course, is not a unique molecule. Could it be replaced with any chelating agent that may form a saturated complex of an adequate stability? Ultimately, we should aim for several high photon yield radioactive agents, each of which will label only one type of cohort cell in vivo. To succeed in this quest will demand much effort and coordination by radiochemists, radiologists, cell biologists, hematologists and immunologists.

REFERENCES

1. Szirmai E and Pinkerton JB: Introduction to nuclear hematology. Nuclear Hematology, Edited by Szirmai E. Academic Press, New York and London, 1965

2. Kniseley RM: Marrow studies with radiocolloids. Seminars in Nuclear Medicine 2:71-85, 1972

3. McIntyre PA: Newer developments in nuclear medicine applicable to hematology, 377. Progress in Hematology Vol. X, Brown EB Grune and Stratton, New York, London 1977

4. Van Dyke D, Dobson EL, Harker H, et al: Radiopharmaceutical techniques for the study of bone and bone marrow blood flow. Radiopharmeuticals. Subramanian G., Rhodes BA, Cooper JF, Sodd VJ, Society of Nuclear Medicine 1975

5. McIntyre PA: Agents for bone marrow imaging: An evaluation of radiopharmaceuticals. Subramanian G, Rhodes BA, Cooper JF, Sodd VJ, Society of Nuclear Medicine 1975

6. Harwig JF, Alderson PO, Primeau JL, et al: Development and evaluation of a rapid and efficient electrolytic preparation of Tc-99m labeled red blood cells. Int J Appl Radiat and Isotopes 28:113-121, 1977

7. Ducasson D, Arnaud D, Bardy A, et al: A new stannous agent kit for labeling red blood cells with Tc-99m and its clinical applications. Br J Radiol 49;344-347, 1976

8. Malmud H: Dosimetry of Tc-99m labeled blood pool scanning agents. Clinical Nuclear Medicine 3:420-421, 1978

9. Wackers FJTH, Berger HJ, Johnstone DE: Multiple gated cardiac blood pool imaging for left-ventricular ejection fraction: validation of the technique and assessment of variability. Am J Cardiol, In press

10. Ashburn WL, Schelberg HR, Verba JW: Left ventricular ejection fraction: A review of several radionuclide angiographic approaches using scintillation camera. Progress in Cardiovascular Disease 20:267-284, 1978

11. Crosby WH: Structure and function of the spleen. Hematology - 2nd Edition, edited by Williams WJ, Beutter E, Ersler AJ, et al. McGraw Hill Book Co. 1977

12. Smith TD and Richards P: A simple kit for the preparation of Tc-99m labeled red blood cells. J Nucl Med. 17:126-132, 1976

13. Armas R, Thakur ML and Gottschalk A: A simplified method for spleen scanning with Tc-99m labeled erythrocytes. Radiology. In press

14. McAfee JG and Thakur ML: Survey of radioactive agents for in vitro labeling of phagocytic leukocytes - I, Soluble Agents. J Nucl Med 17: 480-487, 1976

15. McAfee JG and Thakur ML: Survey of radioactive agents for in vitro labeling of phagocytic leukocytes - II, Particles. J Nucl Med, 17:488-492, 1976

16. Mallory GK, White PD, Salcedo-Salgar J: The speed of healing of myocardial infarction. American Heart Journal 18:647-671, 1939

17. Vik-Mo H: Effect of acute myocardial ischemia on platelet aggregation in the coronary sinus and aorta in dogs. Scand J Haematol 19:68-74, 1977

18. Gjesdal K: Platelet function and plasma free fatty acids during myocardial infarction and severe angina pectoris. Scand J Hematol. 17:205-212, 1976

19. Edmiston WA, Harrison EC, Duick GF, et al: Thromboembolism in mitral procine valve recipients. Am J Cardiol 41:508-511, 1978

20. Mechanism of Tumor Immunity. Green I, Cohen S, McCluskey RT. John Wiley and Sons - New York, London, Sydney, Toronto, 1977

21. Chappin MA, Ross JF: The determinations of the true cell volume by dye dilution by protein dilution and with radioactive iron. The error of the centrifuge hematocrit. Am J Physiol 137:447-455, 1942

22. Perry S, Godwin HA, Zimmerman TS: Physiology of granulocyte - Part I JAMA 203:937-944, (1968)

23. Taylor JH, Woods PS, Hughes WL: The organization and duplication of chromosome as revealed by autoradiographic studies using tritium labeled thymidine. Proceedings National Academy Sci. USA. 43:122-218, 1957

24. Rubini JR, West-cott E, Keller S: In vitro DNA labeling of bone marrow and leukemic blood leukocytes with tritiated thymidine - II ^{3}H thymidine biochemistry in vitro. J Lab Clin Med 68:566-576, 1966

25. Greenberg ML, Chanana AD, Cronkite ED, et al: Tritiated thymidine as cytocydal agent humen leukemia. Blood 28:851-859, 1966

26. Blau M: Biosynthesis of (^{75}Se) selenomethoinine and (^{75}Se) selenocystine. Biochem Biophys Acta 219:389-390, 1961

27. Cooley H, Gardner FH: The use of selenomethoinine (^{75}Se) as a label for canine and human platelets. Am Soc of Clin Inv 44:1036-1037, 1965 (Abstract)

28. Penner JA and Meyers MC: Methonine - ^{75}Se uptake in circulating blood cells and plasma proteins. J Lab Clin Med 68:1005, 1966 (Abstract)

29. McIntyre PA, Evatt B, Hodkinson BA, et al: Selenium-75, Selenomethoinine as a label for erythrocytes, leukocytes and platelets in man. J Lab Clin Med 75:472-480, 1970

30. Vodopick HA, Kniseley RM: Sulfur-35 studies in man: Platelet and plasma and urinary radioactivity assayed by beta liquid scintillation spectrometry. J Lab Clin Med 62:109-120, 1963

31. Grobb D, Lilienthal JL Jr, Harvey AM, et al: The administration of di-isopropylfluorophosphate (DFPO)to man. Bull. Johns Hopkins Hospital 81:217-243, 1947

32. Cohen JA, Warringa MGPJ: The fate of P^{32} labeled diisopropyfluorophosphate in the human body and its use as labeling agent in the study of the turnover of blood plasma and red cells. J Clin Inv 33:459-467, 1954

33. Leeksman CHW, Cohen JA: Determination of the life span of human platelets using labeled DFP. J Clin Inv 35:964-969, 1956

34. Athens JW, Mauer AM, Ashenbrucker H, et al: Leukokinetic studies I. A method for labeling leukocytes with DF^{32}P. Blood 14:303-333, 1959

35. Mauer AM, Athens JW, Ashenbrucker H, et al: Leukocytes kinetics II. A method for labeling granulocytes in vitro with radioactive DFP^{32}. J Clin Inv 39:1481-1486, 1966

36. Athens JW, Haab OP, Raab SO, et al: Leukokinetic Studies IV. The total blood, circulating and marginal granulocyte pool and the granulocyte turnover rate in normal subjects. J Clin Inv 40:989-995, 1961

37. Raab SO, Athens JW, Haab OP, et al: Granulokinetics in normal dogs. Am J Pathology 206:83-88, 1964

38. Schaffer NK, May SC Jr, Summerson WH: Serine phosphoric acid from diisopropyl phosphonyl chymotrypsin. J Biol Chem 202:67-76, 1953

39. Mizuno NS, Perman V, Bates FW, et al: Life span of thrombocytes and erythrocytes in normal and thrombopenic calves. Blood 14:708-719, 1959

40. Gray SJ, Sterling K: The tagging of red cells and plasma proteins with radioactive chromium. J Clin Invest 29:1604-1613, 1950

41. McMilan R, Scoll JL: Leukocyte labeling with ^{51}Cr. Technique and results in normal subjects. Blood 32:739-754, 1968

42. Bjö̈rnson J: The effect of ^{51}Cr - labeling procedure on platelet aggregation. Scand J Haemat 13:252-259, 1974

43. Dale DC, Hubert RT, Fauci A. Eosinophil kinetics in hypereosinophilic syndrome. J Lab Clin Med 87:487-495, 1976

44. Belcher EH, Berlin NI, Eerhisse JG, et al: Recommended methods for radioisotope platelet survival studies. Blood 50:1137-1144, 1977

45. Barth RF, Singla OM, Gillespes GY: Use of Tc-99m as a radioisotopic label to study the migrating patterns of normal and neoplastic cells. J Nucl Med 15:656-661, 1974

46. Uchida T, Tasunaga K, Kariyone S, et al: Survival and sequestration of ^{51}Cr and TcO_4-99m labeled platelets. J Nucl Med 15:801-807, 1974

47. Rannie GH, Thakur ML, Ford WL: An experimental comparison of radioactive labels with potential applications to lymphocyte migration studies in patients. Clin Exp Immunol 29:509-514, 1977

48. English D, Anderson BR: Labeling of phagocytes from human blood with Tc-99m - sulphur colloid. J Nucl Med 16:5-10, 1975

48a. Anderson BR, English D, Akalin HE, et al: Inflammatory lesion localized with Tc-99m labeled leukocytes. Arch Int Med 135:1067-1071, 1975

49. Burleson RL, Johnson MD, Head H: Scintographic demonstration of Experimental abscesses with intravenous ^{67}Ga citrate and ^{67}Ga labeled blood leukocytes. Annals of Surgery 178:446-452, 1973

50. Burleson RL, Johnson MC, Head H: In vitro and in vivo labeling of rabbit blood leukocytes with ^{67}Ga - citrate. J Nucl Med 15:98-101, 1973

51. Ga-67 Imaging. Edited by: Hoffer PB, Bakerman C, Henkin R. Wiley, 1978

52. Welch MJ, Thakur ML, Coleman RE, et al: Gallium - 68 labeled red cells and platelets. New agents for position tomography. J Nucl Med 18:558-562, 1977

53. Thakur ML: Gallium-67 and indium-111 radiopharmaceuticals. Int J Appl Radiat Isotopes 28:183-201, 1977

54. Rubbo SD, Albert A, Gibson MI: The influence of chemical constitution on antibacteridal activity Part V. The antibacterial action of 8-hydroxyquinoline oxine. Brit J Path 31:425-441, 1950

55. Albert A, Hampton A, Selbi FR, et al: The influence of chemical constituent on antibacterial activity. Part VII. The site of action of 8-hydroxyqerinoline (oxine). Brit J Pathol 35:75-84, 1954

56. Thakur ML, Coleman RE, Welch JJ: Indium-111 labeled leukocytes for the localization of abscesses. Preparation analysis, tissue distribution and comparison with gallium 67 citrate in dogs. J Lab Clin Med 89:217-228, 1977

57. Thakur ML, Welch MJ, Joist JH, et al: Indium-111 labeled platelets. Studies on preparation and evaluation of in vitro and in vivo functions. Thrombosis Research 9:345-357, 1976

58. Scheffel U, McIntyre PA, Evatt B, et al: Evaluations of In-111 as a new high photon yield gamma - emitting "physiological" platelet label. The Johns Hopkins Med J 140:285-293, 1977

59. Joist HJ, Baker RK, Thakur ML, et al: Indium-111 labeled human platelets; uptake and loss of label and in vitro functions of labeled platelets. J Lab Clin Med 92:829-836, 1978

60. Segal AW, Deteix P, Garcia R, et al: Indium-111 labeling of leukocytes: A detrimental effect on neutrophil and lymphocytic functions and an improved method of cell labeling. J Nucl Med 19:1238-1244, 1978

61. Ferluga J, Allison AC, Thakur ML: The use of indium-111 for studies of cytotoxicity mediated by lymphocytes or by antibodes and complement. J Clin Lab Immunol. In press

62. Wisto BA, Grossman ZD, McAfee JG: Labeling platelets with oxine complexes of Tc-99m and In-111. In vitro studies and survival in the rabbits. J Nucl Med 19:483-487, 1978

63. Thakur ML, Lavender JP, Arnot RN, et al: In-111 labeled leukocytes in man. J Nucl Med 18:1014-1021, 1977

64. Goodwin DA, Bushberg JT, Doherty PW, et al: Indium-111 labeled antigens platelets for localization of vascular thrombi in humans. J Nucl Med 19:626-634, 1978

65. Lavender JP, Goldman JM, Arnot RN, et al: Kinetics of indium-111 labeled lymphocytes in normal subjects and patients with hodgkins disease. Brit J Med 2:797-799, 1977

66. Thakur ML, Segal AW, Welch MJ, et al: Indium-111 labeled cellular blood components. Mechanism of labeling and intracellular localization in human neutrophils. J Nucl Med 18:1022-1026, 1977

67. Phillips JP: The reaction of 8-quinolinol. Chem Res 56:271-297, 1956

68. Riba AL, Thakur ML, Gottschalk A, et al: Imaging experimental infective endocarditis with indium-111 labeled blood cellular components. Circulation 59:336-363, 1979

69. Riba, AL, Thakur ML, Gottschalk A, et al: Imaging acute coronary thrombosis with indium-111 labeled platelets. Circulation. In press

70. Davis HH, Heaton AW, Segal BA, et al: Scintigraphic detection of atherosderotic lesions and venous thrombi in man by indium-111 labeled autologous platelets. Lancet 1185-1187, 1978

71. McIlmoyle G, Davis HH, Welch MJ, et al: Scintigraphic diagnosis of experimental pulmonary embolism with In-111 labeldd platelets. J Nucl Med 18:910-914, 1977

72. Dewanjee MK, Fuster V, Kaye MP, et al: Imaging platelet deposition with In-111 labeled platelets in coronary artery bypass graft in dogs. Mayo Clin Proc 53:327-331, 1978

73. Anderson RE, Warner NL: Ionizing radiation and the immune response. Adv Immunol 24:215-235, 1976

74. Frost P, Smith J, Frost H: The radiolabeling of lymphocytes and tumor cells with indium-111. Proc of Soc for Expt Biol and Med 157:61-65, 1978

75. Holley TR, Van Epps DE, Harvey RL, et al: Effects of high doses of radiation on human neutrophil chemotaxis, phagoerythosis and morphology. Am J Pathol 75:61-72, 1974

76. Zakhireh B, Thakur ML, Malech HL, et al: Viability random migration chemotaxis, bactericidal capacity and ultra structure of indium-111 labeled human polymorphonuclear leukocytes. J Nucl Med. In press

77. Silvester DJ: Consequences of indium-111 decay in vivo: Calculated absorbed radiation dose to cell labeled with indium-111-oxine. J Labeled Compounds and Radiopharm 16:1, 1979, In press

78. Segal AW, Thakur ML, Arnot RN, et al: Indium-111 labeled leukocytes for localization of abscesses. Lancet 1056-1058, 1976

79. Thakur ML: Indium-111: A new tracer for leukocytes. Exp Hemat Suppl 5:145-150, 1977

80. Thakur ML, Gottschalk A, Zaret BL: Imaging experimental myocardial infarction with indium-111 autologous leukocytes: Effects of infarct age and residual regional myocardial blood flow. Circulation. In press

81. Doherty P, Goodwin DA: Indium-111-oxine labeled autologous leukocytes in the diagnosis of inflammatory disease. J Nucl Med 19:742, 1978 (Abstract)

82. Alavi JB, Alavi A, Staum M: Scintigraphic evaluation of indium-111 labeled filtration granulocytes. J Nucl Med 19:672, 1978 (Abstract)

83. Forstrom L, Gomex L, Weibler B, et al: Clinical use of indium-111 oxine labeled granulocytes in the detection of inflammation or abscess. J Nucl Med 19:672, 1978 (Abstract)

84. Cartwright GE, Athens JW, Wintrobe MM: Analytical review: The kinetics of granulopoiesis in normal man. Blood 24:780-802, 1964

85. Williams ED, Merrick VM, Lavender JP: The distributions and dosimetry of 111 In-bleomycin in man. Brit J Radiol 48:275-278, 1975

86. Grossman ZD, Wisto BW, McAfee JG, et al: Platelets labeled with oxine complexes of Tc-99m and Indium-111. Part 2. Localization of experimental induced vascular lesions. J Nucl Med 19:488-491, 1978

87. Coleman RE: Private communication

ACKNOWLEDGEMENTS

We would like to thank Ms. Gael Maher for carefully typing the manuscript and to Ms. Joan Wallace for editorial comments.

The work was supported by DOE contract No. EY-76-S-02-4078 and the American Heart Association award no. 78-957.

TABLE I. IN VIVO APPLICATIONS OF IN-111-OXINE LABELED LEUKOCYTES, LYMPHOCYTES AND PLATELETS

Cell Type	Application	In Animals Ref.	In Humans Ref.
Leukocytes	Location of abscess	56	78, 79, 81, 83, 87
	Myocardial infarction	80	
	Bacterial endocarditis	68	
Lymphocytes	Visualization of lymph nodes	47.74	
	Hodgkin's disease		65
Platelets	Detection of venous thrombosis	57.86	64
	Atherosclerosis		70
	Pulmonary emboli	71	
	Bacterial endocarditis	68	
	Coronary thrombosis	69	
	Coronary artery bypass graft	72	

NONINVASIVE RADIOISOTOPIC TECHNIQUE FOR DETECTION OF PLATELET DEPOSITION IN CORONARY ARTERY BYPASS GRAFTS IN DOGS AND ITS REDUCTION WITH PLATELET-INHIBITORS

M. K. Dewanjee, V. Fuster, M. P. Kaye,
and M. Josa

Mayo Clinic and Mayo Foundation
Diagnostic Nuclear Medicine
Cardiovascular Diseases, Internal Medicine,
and Cardiovascular Surgical Research
Rochester, MN 55901

ABSTRACT

At 8 and 32 hours after saphenous vein bypass graft surgery in dogs, in vivo images (174 and 247 keV) were obtained with a gamma camera after I.V. injection, 2 hours postoperatively, of autologous Indium-111-labeled platelets (0.25-0.50 mCi) and Iodine-125-labeled fibrinogen (0.08-0.1 mCi). Under identical conditions, in dogs treated with dipyridamole (2.5 mg/kg/day) plus Aspirin (15 mg/kg/day), (D + ASA), the grafts and hemorrhagic cardiac muscle had significantly less platelet deposition as estimated by imaging. The radioactivity ratio of In-111 in isolated graft sections to nontarget tissues of blood, lung, and cardiac muscle in control and treated animals, sacrificed at 32 hours after administration, demonstrated a threefold reduction in platelet deposition in the bypass graft, and graft vs blood, lung, and myocardium ratios of 16.3 ± 10.8, 29.96 ± 25.6, 37.1 ± 29.4 were obtained in the control group of seven dogs. The corresponding values in the treated group of 12 dogs were 4.4 ± 3.2, 6.7 ± 5.5, 30.8 ± 28.1 respectively. On the other hand, fibrinogen uptake in the graft was not reduced by platelet-inhibitors and graft-to-blood, -lung, and -myocardium ratios of 10.5 ± 3.4, 21.9 ± 9.3, and 85.6 ± 34.9 respectively were obtained in both control and treated animals. This technique may be a promising tool for a better understanding of the role played by platelets and fibrinogen in the process of occlusion of saphenous vein bypass grafts and arteriosclerosis and its reduction with platelet-inhibitors.

INTRODUCTION

Aortocoronary bypass surgery with autologous saphenous vein grafts (1) is at present used widely in patients with coronary artery disease. A major limitation of this procedure is the high frequency of occlusion of the vein graft (2) secondary to thrombosis or intimal proliferation or both (3-5). There is now suggestive clinical and experimental evidence that platelets may play an important role in the pathogenesis of the occlusion (6, 7), and it has been considered that the use of platelet inhibitory drugs might be useful for interrupting this process. However, no in vivo technique has been available for direct demonstration of this process of platelet deposition in the vein grafts and its possible reduction with platelet-inhibitory drugs.

Labeling of platelets with Indium-111 (^{111}In) and its application in thrombosis research have been described by Thakur et al and other investigators (8-13) and Scheffel and associates (14). Recently, by means of ^{111}In-labeled platelets and imaging with a gamma camera, we have been able to detect in vivo platelet deposition in saphenous vein bypass grafts in dogs (15). Since our initial report, we have expanded our study, and this report presents a comparison between the in vivo platelet deposition and in vitro quantification in coronary artery bypass grafts in a larger group of normal control dogs and in another group treated with platelet-inhibitor drugs (16).

Since fibrinogen also participates along with platelets in the denuded endothelium in the stabilization of platelet fibrin thrombus, we have used double labeling techniques (12, 17) for studying the effect of platelet inhibitors on platelet and fibrinogen deposition in the aortocoronary bypass graft.

METHODS

Aortocoronary Bypass Grafting. With the use of cardiopulmonary bypass, 14 mongrel dogs weighing 18 to 26 kg underwent aorto-left anterior descending coronary artery bypass grafting carried out through a thoracotomy in the fourth intercostal space (Fig. 1). The unoperated, undistended, reversed right saphenous leg vein was used for the bypass graft. In all animals, the distal anastomosis was carried out with a running suture of 2-0 Prolene, and the use of systemic hypothermia at 25° C. After completion of this distal anastomosis, the dog was rewarmed and the proximal anastomosis of vein to aorta was carried out with a running 6-0 Prolene stitch. Air was vented from the vein graft by aspiration through a 25-gauge needle, and the animals were defibrillated with a single shock; when suitable cardiac action had been reestablished, bypass was discontinued. The effect of heparin was reversed with protamine, a chest tube was placed, and the chest was closed. When adequate spontaneous respiration occurred in the postoperative period, the chest tube and the monitoring catheters were removed, and the animals were returned to their cages.

Medication. The dogs were distributed into two different groups. Six animals underwent bypass grafting and received no medication other than systemic antibiotics during surgery. Eight animals were given oral dipyridamole (55 mg/day) for two days preoperatively and then dipyridamole (55 mg) plus Aspirin (323 mg) postoperatively after extubation and again the day after surgery.

Angiographic Studies. About 24 hours after surgery, five untreated and three treated animals, under light sodium pentobarbital anesthesia, were subjected to catheterization for the purpose of visualizing the patency of the bypass graft. An automated delivery system injected a constant volume of Renografin with a 7-F catheter in the bypass graft, and a spot film was taken to demonstrate the anatomic location and the patency of the bypass graft.

Preparation of In-111-Platelets, Imaging, and Tissue Distribution. In-111-labeled 8-hydroxyquinoline (oxine), prepared according to the method of Thakur and associates (8), was supplied by Diagnostic Isotopes (Upper Saddle River, New Jersey). The labeling of the dog platelets was performed during surgery according to the procedure of Scheffel and associates (14). I-125-labeled fibrinogen was obtained from Amersham Corporation (Arlington Heights, Illinois). About 2 hours after the operation, the dogs were injected with 300 to 500 μCi of autologous In-111-labeled platelets in 4 to 8 ml of acid citrate dextrose-plasma solution followed by an injection of I-125-labeled fibrinogen (90-100 μCi). At 6 and 30 hours after intravenous administration of platelets, dogs anesthetized with sodium pentobarbital were imaged with a gamma camera (Searle, large field of view) fitted with a medium-energy parallel-hole collimator. The camera spectrometer was adjusted to cover both the 174-keV and 247-keV peak of the In-111-radioisotope. The relative distribution of In-111-labeled platelets in the thorax and abdomen was then obtained. One hundred thousand counts were accumulated at the anterior, left and right lateral, and left and right anterior oblique positions. Pinhole views of the graft with a Searle camera were also obtained simultaneously.

After the in vivo imaging, the dogs were killed with an overdose of sodium pentobarbital and the heart, lung, liver, spleen, kidneys, blood, and tissue samples from skeletal muscle, bone, and marrow were obtained and weighed; from this, the distribution of radioactivity in the organ and whole body was determined with a dose calibrator (Capintec-CRC-5) and an automatic gamma well counter (Beckman Gamma-8000). Finally, chambers of the heart were washed free of blood and postmortem clots, the interventricular septum was removed, and the right and left ventricles were spread flat and imaged in the same orientation as used for the in vivo imaging.

It is important to perfuse the graft through the aorta with isotonic saline to remove residual blood and postmortem clots immediately after isolating the heart. The adherent tissue and thrombus surrounding the graft was carefully removed. The graft, along with the two sites of anastomosis, was removed from the aorta and myocardium; the proximal and distal sites of anastomosis 2-3 mm away from each end of the graft were cut and saved for measurement of radioactivity. The graft was cut open by a midline incision, rinsed with saline, and divided into three equal (≃ 1 cm) sections: proximal, middle, and distal. Each section, along with a similar section from the contralateral saphenous vein, was weighed with a microbalance, and the radioactivity per unit weight of the graft sections and surrounding right ventricle, left ventricle, right atrium, and interventricular septum was determined. Then the radioactivity in the graft was compared with that in blood, lung, cardiac muscle, and skeletal muscle. A crossover correction (≃ 30%) of gamma rays of In-111 in the Iodine-125 channel was made for all tissues.

Scanning Electron Microscopy of Bypass Graft. The perfusion fixation of the graft was performed by the modified method of Davies et al (18). The graft and coronary arteries were perfused successively with isotonic glucose solution buffered with hyproxyethylpiperazine 5-ethane sulfonic acid, isotonic silver nitrate (1 min) and 3% glutaraldehyde (60 min) at a pressure of 120 mm of Hg. The graft was dehydrated by drying over silica gel and coated with gold, palladium and carbon and examined by a scanning electron microscope at low and high magnifications and photographed.

RESULTS

Imaging with In-111-Platelet and Angiography. In the untreated dogs, at 8 and 32 hours after surgery and 6 and 30 hours after intravenous administration of radioactive platelets, scintiphotos of the distribution of In-111-labeled platelets clearly delineated the bypass grafts (Fig.2). The intensity of radioactivity in the bypass graft as visualized by imaging was not as great as that found in the liver and spleen but was greater than that in the nontarget tissues, namely, blood, lung, cardiac muscle, and skeletal muscle. The values for the ratio of radioactivity in the graft to that in blood, lung, cardiac muscle, or skeletal muscle increased considerably after 30 hours compared with those at 6 hours as a result of clearance from the blood and platelet deposition. The coronary angiograms of the bypass grafts showed patency of the graft in all studied dogs. For confirmation of the results of external imaging, the isolated heart without the interventricular septum was imaged in the orientation of the in vivo imaging, and this again demonstrated the graft as the major site of platelet deposition in the heart. Small other "hot" spots indicated platelet deposition at the sites of cannulation for the surgical procedure. Figure 3 shows the scintiscan and the photograph of the isolated saphenous vein graft specimen indicating the deposition of labeled platelets in the damaged graft wall.

In the treated as compared with the untreated dog, scintiphotos of the distribution of In-111-labeled platelets showed less intense or no evidence of increased radioactivity in the bypass graft (Fig. 4). The coronary angiograms of the bypass grafts showed patency of the grafts in all dogs studied.

Biodistribution of In-111-Platelets and I-125-Fibrinogen. The biodistribution of In-111-labeled platelets 30 hours after intravenous administration in the control (6) and treated animals (8) is shown in Table 1. It is interesting to note that there is no statistically significant difference in the biodistribution of labeled platelets between the control and treated animals except graft. The graft in the control group has two or three times higher value than the treated group. In general, liver, spleen, skeletal muscle, and blood retain (35 ± 8)%, (25 ± 10)%, (6 ± 5)%, (29 ± 10)% of the administered dose. The lung tissue in the treated group retains slightly less platelets than the treated group and slightly more platelets are in the circulatory system in the treated group. On a weight basis, for In-111-platelet distribution, splenic tissue has 9-12 times more In-111 radioactivity than hepatic tissue and radioactivity is almost the same in lung and kidney. Hemorrhagic lung has two to three times higher activity than normal lung tissue. When platelets are aggregated during labeling, (8-10)% of platelets was observed in the lung. Usually, damaged platelets are sequestered by the reticuloendothelial system, i.e., liver and spleen. Surgically damaged skeletal muscle has three to six times more activity than normal muscle.

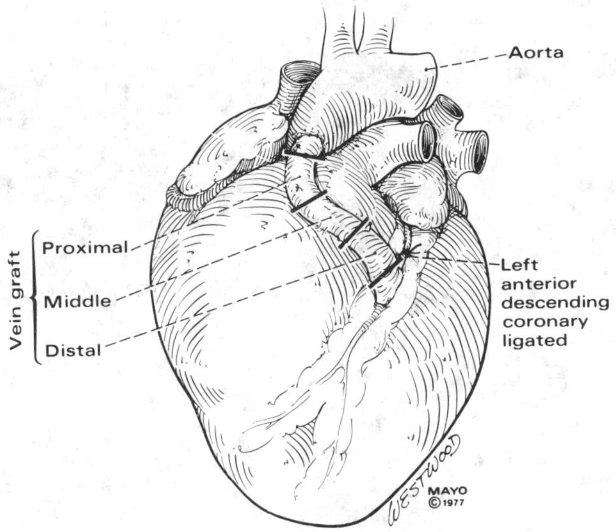

Figure 1.

Anastomosis of saphenous vein bypass graft with aorta in proximal end and left anterior descending coronary artery in distal end. After sacrificing animal, graft was isolated and cut into three sections: proximal, middle, and distal, and platelet deposition was quantitated.

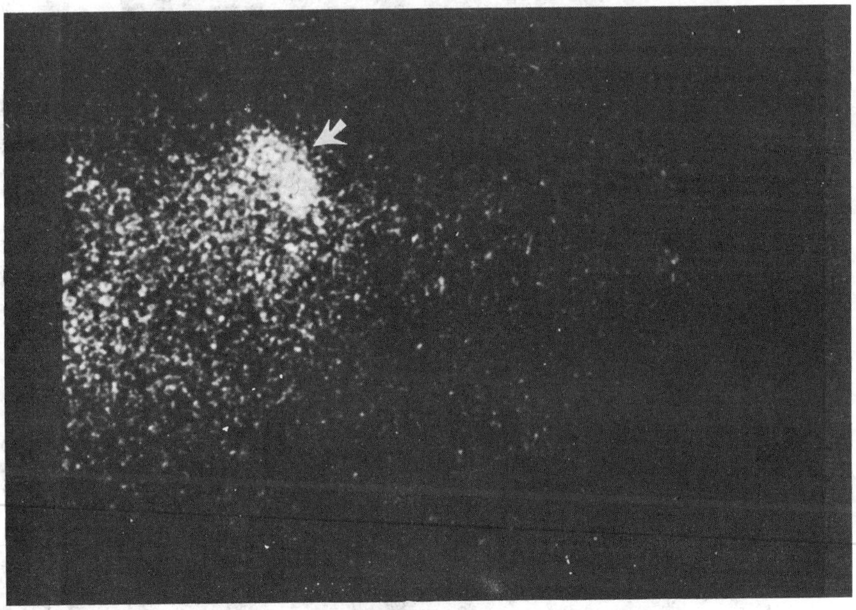

Figure 2.

Control dog. Scintiphoto of thorax of intact dog in anterior position at 30 hours after intravenous injection of In-111-labeled platelets (32 hours after surgery). Uptake shown by arrow corresponds to graft. Radioactive platelets in upper part of liver also evident.

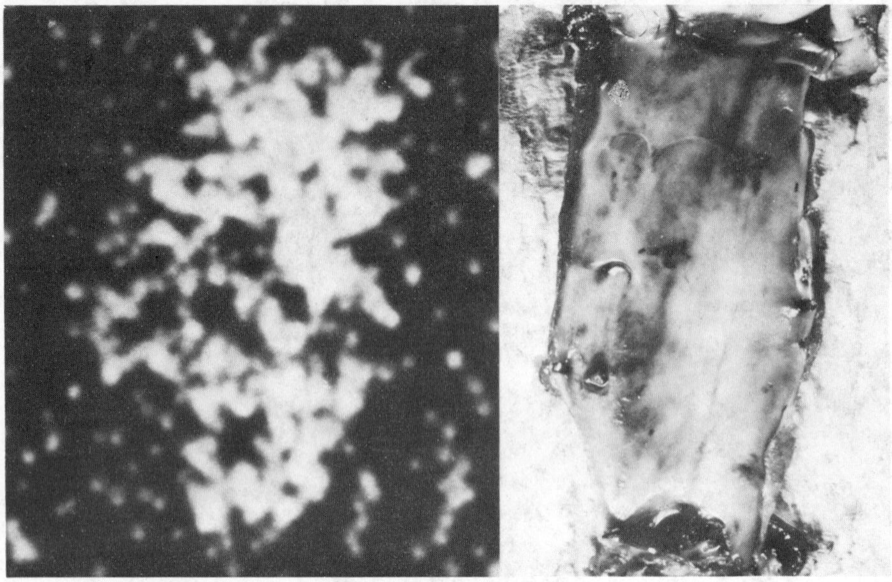

Figure 3.

Control dog. Scintiscan of isolated graft spread on flat surface (left). Isolated graft specimen (right).

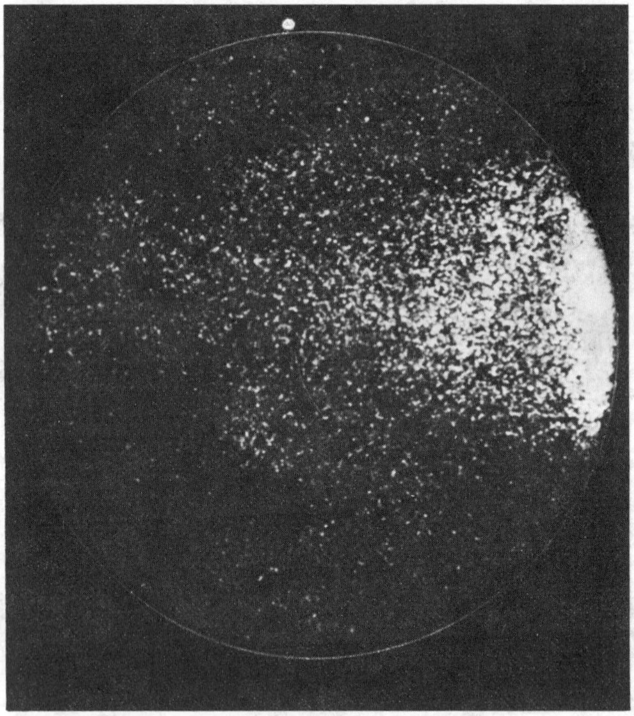

Figure 4.

Dog treated with dipyridamole and Aspirin. Scintiphoto of thorax of intact dog in anterior position at 30 hours after intravenous injection of In-111-labeled platelets (32 hours after surgery). Observe diffuse uptake in cardiac region; however, radioactive platelets in upper part of liver are also evident (lower part of photograph).

The infarcted myocardium has only two to three times more activity than the normally perfused myocardium. In general, myocardial muscle has four to seven times more platelets than skeletal muscle.

Both the proximal and distal sites of anastomosis lead to thrombus formation and have higher In-111 radioactivity than the nontarget tissues of lung and blood. This may be due to the traumatic injury at the terminus of the graft during graft harvesting, suturing, and leakage of blood. In the control group, the graft has 0.2-0.4% of the administered acvivity. Because platelets retain the Indium-111 inside the membrane during the lifetime of the labeled platelets, no significant radioactivity was found in bone marrow, urine (< 1%), and feces; (2-4)% localized in the kidneys. The mean values and standard deviations of the relative distribution of In-111-labeled platelets in the proximal, middle, and distal sections of the grafts in comparison with the nontarget tissues (blood and lungs) at 30 hours after injection are shown in Table 2. Radioactivity determination averaged 16 ± 11 times greater activity in the graft than blood, 29 ± 26 times greater activity than the lung, and 37 ± 29 times greater activity in the graft than the cardiac muscle. The graft-to-blood and graft-to-lung ratios of radioactivity are the two determinant factors for external imaging. The favorable ratios account for the imaging of the grafts. However, in treated dogs the radioactivity ratios of In-111 platelets in the proximal, middle, and distal sections of the grafts in comparison with the nontarget tissues were significantly lower ($p < 0.01$) (Table 3). Radioactivity counting averaged 5 ± 3 times greater activity in the overall graft than the blood and 8 ± 6 times greater activity than the lung.

Tables 4 and 5 demonstrate that the platelet inhibitors do not affect the deposition of fibrinogen in the denuded endothelium. In all the three graft sections, the graft-to-blood, -lung, and -myocardium ratios remain the same in both the control and treated groups. Both the platelet and fibrinogen depositions at the site of anastomosis were higher than the graft area.

Scanning Electron Micrographs of Bypass Graft in the Control and Treated Dogs. High resolution scanning electron micrographs of bypass graft clearly demonstrate the areas of injury to endothelial lining as craters; this injury occurs during harvesting and implantation. The craters are full of platelets and a few red blood cells (Figs. 5 and 6). The injured endothelium is healed after 2 months of treatment with dipyridamole and Aspirin, and continuous layers of endothelial cell junctions stained with silver nitrate stain are clearly demonstrated in Fig. 7.

DISCUSSION

The method of platelet labeling with In-111 radioisotope provides a new tool for the study of external imaging of the process of platelet deposition. Thus, the half-life of 2.8 days along with essentially 186% yield of gamma emissions of two gamma ray energies of 147 and 247 keV makes it an excellent platelet marker for in vivo imaging and platelet survival.

In our study in dogs after saphenous vein aortocoronary bypass surgery, we were able to image the deposition of 0.2-0.4% of In-111-labeled platelets

TABLE 1. MEAN VALUES (± SD) OF BIODISTRIBUTION* OF In-111-LABELED PLATELETS IN CONTROL AND TREATED DOGS†

Tissue	Control (n = 7)	Treated (D + ASA) (n = 12)
Liver	35.63 ± 8.42	35.04 ± 5.44
Spleen	25.28 ± 8.36	23.29 ± 11.37
Lungs	4.09 ± 3.00	2.19 ± 1.28
Kidneys	1.47 ± 0.41	1.28 ± 0.32
Skeletal muscle	8.30 ± 5.85	5.63 ± 3.75
Blood	29.56 ± 10.97	30.87 ± 8.47
ACBG‡	0.22 ± 0.08	0.07 ± 0.5

*Percentage of injected dose
†30 hours after administration and 32 hours after ACBG surgery
‡ACBG = Aortocoronary Bypass Graft

TABLE 2. MEAN VALUES* (± SD) OF RADIOACTIVITY RATIOS OF In-111-LABELED PLATELETS IN SECTIONS OF GRAFT WITH RESPECT TO NONTARGET TISSUES OF BLOOD, LUNG, AND CARDIAC MUSCLE IN 7 DOGS

Graft	Graft/Blood	Graft/Lung	Graft/Cardiac Muscle
Proximal	14.01 ± 8.77	28.40 ± 25.14	37.78 ± 30.06
Middle	14.17 ± 10.27	27.79 ± 25.91	37.64 ± 29.50
Distal	18.49 ± 13.56	31.49 ± 25.67	36.41 ± 28.69

*30 hours after administration and 32 hours after ACBG surgery

TABLE 3. MEAN VALUES* (± SD) OF RELATIVE DISTRIBUTION OF In-111-LABELED PLATELETS IN SECTIONS OF GRAFT WITH RESPECT TO NONTARGET TISSUES OF BLOOD, LUNG, AND CARDIAC MUSCLE IN 12 DOGS TREATED WITH DIPYRIDAMOLE AND ASPIRIN

Graft	Graft/Blood	Graft/Lung	Graft/Cardiac Muscle
Proximal	4.28 ± 2.40	6.53 ± 5.35	29.83 ± 27.52
Middle	5.82 ± 2.13	9.02 ± 5.59	38.54 ± 18.21
Distal	4.46 ± 4.03	7.12 ± 6.16	30.74 ± 28.54

*30 hours after administration and 32 hours after ACBG surgery

TABLE 4. MEAN VALUES* (± SD) OF RADIOACTIVITY RATIOS OF I-125-LABELED FIBRINOGEN IN SECTIONS OF GRAFT WITH RESPECT TO NONTARGET TISSUES OF BLOOD, LUNG, AND CARDIAC MUSCLE IN 5 DOGS

Graft	Graft/Blood	Graft/Lung	Graft/Cardiac Muscle
Proximal	10.59 ± 4.29	21.49 ± 8.15	71.00 ± 36.86
Middle	12.74 ± 7.58	26.80 ± 20.80	102.24 ± 86.42
Distal	10.21 ± 4.69	20.19 ± 7.48	71.34 ± 44.14

*30 hours after administration and 32 hours after ACBG surgery

TABLE 5. MEAN VALUES* (± SD) OF RELATIVE DISTRIBUTION OF I-125-LABELED FIBRINOGEN IN PROXIMAL, MIDDLE, AND DISTAL GRAFT SECTIONS WITH RESPECT TO NONTARGET TISSUES OF BLOOD, LUNG, AND CARDIAC MUSCLE IN 5 DOGS TREATED WITH DIPYRIDAMOLE AND ASPIRIN

Graft	Graft/Blood	Graft/Lung	Graft/Cardiac Muscle
Proximal	11.42 ± 4.63	22.16 ± 14.81	87.50 ± 44.97
Middle	14.11 ± 4.60	26.09 ± 14.48	107.02 ± 32.02
Distal	9.07 ± 4.57	18.72 ± 15.54	72.79 ± 25.35

*30 hours after administration and 32 hours after ACBG surgery

within the grafts. Platelets do not adhere to normal endothelium. There is good experimental evidence that platelets tend to adhere to damaged endothelial vascular surfaces (19-22), especially the deeper structure of the vessel wall, particularly the basement membrane and the collagenous and possibly other tissue components. Since, during the surgical procedure the saphenous vein graft becomes diffusely de-endothelialized, most likely this explains the deposition of platelets as visualized by imaging and confirmed morphologically by scanning electron micrographs (Figs. 5 and 6).

Knight et al (12) demonstrated that In-111-labeled platelets produce higher thrombus-to-blood ratios than radioiodinated fibrinogen in thrombi less than 24 hours old. They also described that injured vessel wall near the clot retained 24 times more In-111-labeled platelets than the normal contralateral vessel. The age of thrombus is the predominant factor for the incorporation of platelets and fibrinogen; neither agent is taken up well by thrombi older than 24 hours. Highest uptake of both the tracers was obtained by thrombi that are 6 hours old or less. Ritchie et al (23) and Steele et al (6) both report that platelet survival times are frequently shortened in patients with venous or arterial thrombosis, although fibrinogen consumption is not drastically affected. They also demonstrated platelet-inhibitors increase platelet survival time.

The detection by in vivo imaging of an extensive platelet deposition in the coronary artery bypass graft in dogs is possibly of significant importance since there is now clinical (6) and experimental (7) evidence that platelets may play an important role in the pathogenesis of the highly frequent graft occlusion in man. However, although the usefulness of the external imaging of platelet deposition in the bypass graft is evident in dogs, its suitability in studying the same phenomenon in man needs to be cautiously evaluated for three reasons:

First, we do not know yet for certain whether this early deposition of platelets with angiographic evidence of graft patency is only a transient response to graft endothelial injury or is a continuous process that may lead to the occlusion of the graft, although the preliminary studies suggest less platelet deposition one week after surgery. Second, we do not know whether the small "hot" spots detected in dogs indicating platelet deposition at the sites of cannulation might lead, in future studies in man, to a false positive interpretation of platelet deposition in the coronary graft. Third, since in occluded bypass grafts blood flow is impaired, we do not know whether this will prevent the deposition of In-111-labeled platelets to the site of occlusion and therefore be a limitation for the diagnosis of graft occlusion by this imaging technique.

In our study in dogs pretreated with the regimen of dipyridamole plus Aspirin, at comparable dosages to the ones used in humans (17), the radioactivity ratio in the graft was significantly reduced to approximately one-third as compared with the untreated dogs. There is good morphologic experimental indication that these agents may interrupt the process of platelet deposition to damaged arterial endothelial surface (20) and to coronary artery saphenous vein bypass grafts (3). The fact that the deposition of platelets, as assessed by our technique, was only reduced at the graft and cannulation sites, but not in the other nontarget tissues, is an intriguing observation.

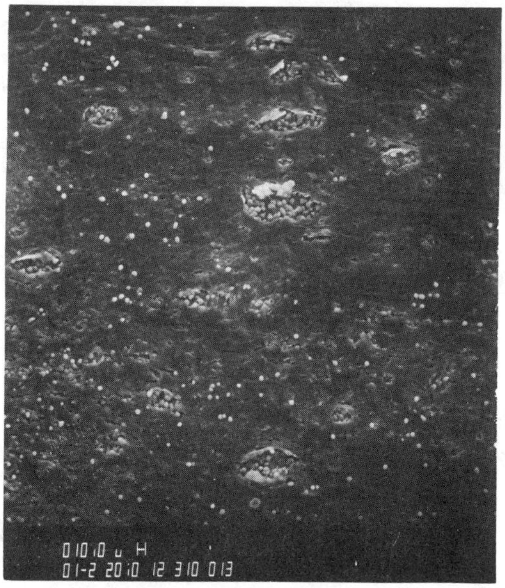

Figure 5.

Scanning electron micrograph of endothelial lining of saphenous vein of dog. Craters represent sites of denuded endothelial cells (x 400). Platelets in craters have extended pseudopods to adhere to basement membrane.

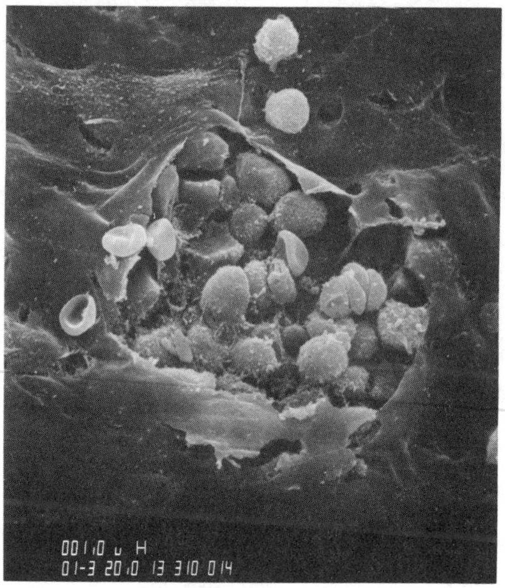

Figure 6.

Scanning electron micrograph of magnified view of crater (x 1000) full of platelets and few red blood cells (discoid cells).

This indicated that these agents only inhibited the deposition of platelets to the damaged vascular sites and did not affect the circulation and accumulation of platelets in the other organs. The most current information suggests that low dose of Aspirin, as used in our experimental model, affects the platelet by preventing its generation of the intraplatelet proaggregating thromboxane A_2 (24), and the dose does not appear to be sufficient to have an anti-inflammatory action (25).

The finding, with this new technique, that platelet deposition in the grafts was reduced by treatment with dipyridamole and Aspirin, is of importance. Thus, this radioisotopic technique may be a promising tool for better and rapid assessment of potential platelet inhibitory drugs that can be used in future preventive studies in man. We realize, however, that a serious limitation of the present external imaging method is that it only provides qualitative information. Quantitative data were obtained with the postmortem determination of radioisotope in the different tissues.

ACKNOWLEDGMENTS

We greatly appreciate the encouragement of Dr. H. W. Wahner and the technical assistance of Mr. James M. Byrne, Mrs. LaVonne Lund, Mr. O. Arlan Hildestad, and Mr. Gregory S. Anderson. We would also like to express our thanks for the excellent secretarial assistance of Mrs. Noreen Johnson. The Persantine (dipyridamole) was generously supplied by Boehringer Ingelheim Ltd.

REFERENCES

1. Favolaro RG: Saphenous vein autograft replacement of severe segmental coronary artery occlusion: Operative technique. _Ann Thorac Surg_ 5:334-339, 1968

2. Assad-Morall JL, Frye RL, Connolly DC, et al: Aortocoronary artery saphenous vein bypass surgery, clinical and angiographic results. _Mayo Clin Proc_ 50: 379-386, 1975

3. Unni KK, Kottke BA, Titus JL, et al: Pathologic changes in aortocoronary saphenous vein grafts. _Am J Cardiol_ 34: 526-532, 1974

4. Vlodaver Z, Edwards JE: Pathologic changes in aortic-coronary arterial saphenous vein grafts. _Circulation_ 44: 719-728, 1971

5. Spray TL, Roberts WC: Status of the grafts and the native coronary arteries proximal and distal to coronary anastomotic sites of aortocoronary bypass grafts. _Circulation_ 55: 741-749, 1977

6. Steele P, Battock D, Pappas G, et al: Correlation of platelet survival time with occlusion of saphenous vein aorto-coronary bypass grafts. _Circulation_ 53: 685-687, 1976

7. Metke MP, Lie JT, Fuster V, et al: Reduction of intimal thickening in canine coronary artery bypass vein grafts with dipyridamole and aspirin. _Am J Cardiol_ 41: 411, 1978 (abst)

8. Thakur ML, Welch MJ, Joist JH, et el: Indium-111 labeled platelets: Studies on preparation and evaluation of in vitro and in vivo functions. Thromb Res 9: 345-357, 1976

9. McIlmoyle G, Davis HH, Welch MJ, et al: Scintigraphic diagnosis of experimental pulmonary embolism with In-111 labeled platelets. J Nucl Med 18: 910, 1977

10. Goodwin DA, Bushberg JT, Doherty P, et al: The use of In-111 oxine labeled platelets for localization of vascular thrombi in humans. J Nucl Med 18: 626-634, 1978

11. Riba AL, Thakur ML, Gottschalk A, et al: Imaging of acute coronary artery thrombosis with Indium-111 platelets. 58: II-4, 1978 (abst 1)

12. Knight LC, Primeau JL, Siegel BA, et al: Comparison of In-111-labeled platelets and iodinated fibrinogen for the detection of deep vein thrombosis. J Nucl Med 19: 891-896, 1978

13. Davis HH, Siegel BA, Joist JH, et al: Scintigraphic detection of atherosclerotic lesions and venous thrombi in man by indium-111-labeled autologous platelets. Lancet, 1185-1187, 1978

14. Scheffel U, McIntyre PA, Evatt B, et al: Evaluation of indium-111 as a new high photon yield gamma-emitting "physiological" platelet label. Johns Hopkins Med J 140: 285-293, 1977

15. Dewanjee MK, Fuster V, Kaye MP, et al: Imaging platelet deposition with ^{111}In-labeled platelets in coronary artery bypass grafts in dogs. Mayo Clin Proc 53: 327-331, 1978

16. Dewanjee MK, Fuster V, Kaye MP, et al: Noninvasive isotopic technique for detection of platelet deposition in coronary artery bypass grafts in dogs and its prevention with platelet inhibitors. Circulation 58, 59: Suppl No II, 1978 (abst 11)

17. Brinkhous KM, Johnson JH: Thrombus. Kalamazoo, Michigan, The Upjohn Co, 1974, pp 9-25

18. Davies PF, Reidy MA, Goode TB, et al: Scanning electron microscopy in the evaluation of endothelial integrity of the fatty lesion in atherosclerosis. Atherosclerosis 25: 125, 1976

19. Baumgartner HR, Haudenschild C: Adhesion of platelets to subendothelium. Ann NY Acad Sci 201: 22-36, 1972

20. Stemerman MB, Ross R: Experimental arteriosclerosis. I. Fibrous plaque formation in primates, an electron microscope study. J Exp Med 136: 796-798, 1972

21. Harker LA, Ross R, Slichter SJ, et al: Homocystine-induced arteriosclerosis. The role of endothelial cell injury and platelet response in its genesis. J Clin Invest 58: 731-741, 1976

22. Fuster V, Bowie EJW, Lewis JC, et al: Resistance to arteriosclerosis in pigs with von Willebrand's disease. Spontaneous and high cholesterol diet-induced arteriosclerosis. _J Clin Invest_ 61: 722-730, 1978

23. Ritchie JL, Harker LA: Platelet and fibrinogen survival in coronary atherosclerosis. Response to medical and surgical therapy. _Am J Cardiol_ 39: 595-598, 1977

24. Burch JM, Stanford N, Majerus PW: Inhibition of platelet prostaglandin synthesis by oral aspirin. _J Clin Invest_ 61: 314-319, 1978

25. Moncada S, Korbut R: Dipyridamole and other phosphodiesterase inhibitors act as antithrombotic agents by potentiating endogenous prostacyclin. _Lancet_ 1: 1286-1289, 1978

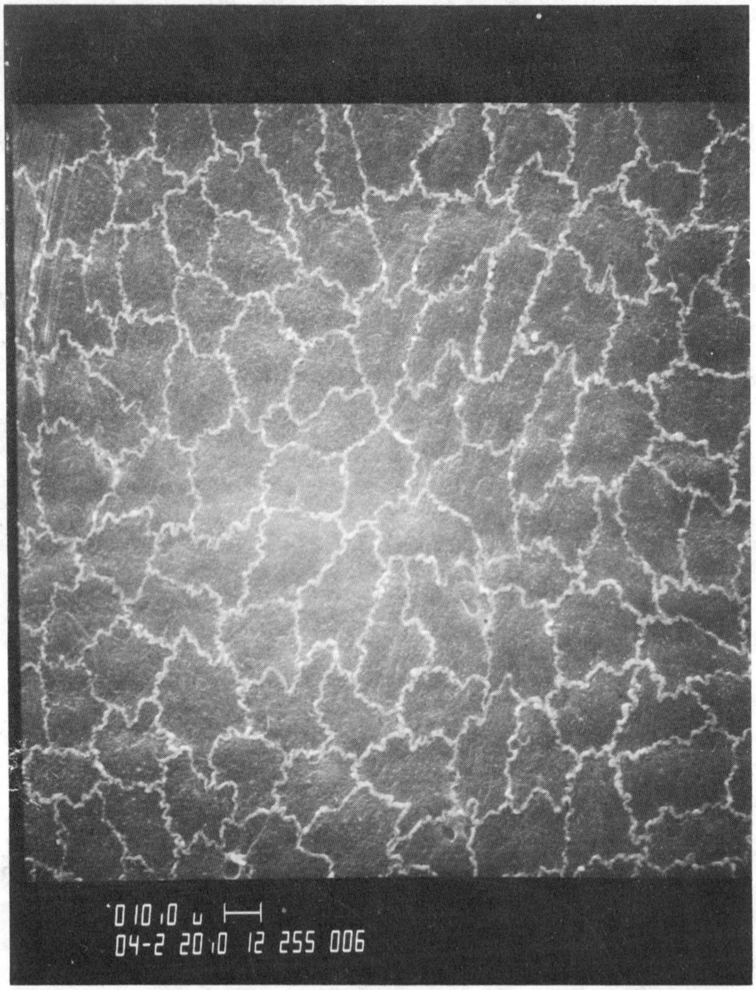

Figure 7.

Scanning electron micrograph of treated endothelium of saphenous vein of dog. Injured areas of endothelial lining completely recuperated and re-endothelialized 2 months after surgery (x 400).

AN IN VITRO STUDY OF THE MECHANISM OF THE IN VIVO RED BLOOD CELL LABELING TECHNIQUE

M. W. Billinghurst
E. A. Somers

Health Sciences Centre
Winnipeg, Manitoba - Canada

ABSTRACT

This study shows that the labeling of red blood cells is primarily dependent on the stannous ion concentration with efficient labeling occurring when the stannous ion concentration is between 0.08 and 0.04 ugm per ml of blood. If the stannous ion concentration was below 0.04 ugm per ml the Tc-99m was found to be present in the plasma as free pertechnetate while at concentration over 0.08 ugm per ml the Tc-99m was increasingly present in the plasma as Tc-99m labeled compounds. Both pyrophosphate and stannous ions were shown not to associate with the red blood cells.

INTRODUCTION

The recent interest in cardiovascular nuclear medicine has resulted in considerable interest in ways of obtaining a good radioactive tracer which stays in the blood stream throughout the course of the study. While technetium-99m labeled albumin enjoyed some early interest[1,2] it was soon superseded by labeled red blood cells which were shown to remain in the blood stream for a longer period of time. The convenience and the good labeling efficiency of the in vivo labeling technique[3] quickly led to the in vivo labeling of red blood cells with technetium-99m becoming the predominant radiopharmaceutical used in cardiovascular nuclear medicine. Although there are several investigations of this technique reported in the literature[3,4] relatively little is known of the exact mechanism of the labeling of the red blood cells. This study was undertaken to try to establish the mechanism by which a preinjection of stannous pyrophosphate results in the tagging of Tc-99m to the red cells upon injection of Tc-99m pertechnetate.

METHOD AND MATERIALS

The standard stannous pyrophosphate was prepared by mixing 0.4 ml of a 2.5% w/v solution of sodium pyrophosphate decahydrate, 3.0 ml of physiological saline and 0.5 ml of 0.25 N acetic acid in a multidose vial. Two sterile tin electrodes were then aseptically inserted through the rubber septum with the aid of a 16 gauge needle. These electrodes were connected to a power supply, the vial placed in an ultrasonic bath and a 4 mA current passed through the electrodes for 2 minutes, at the end of which time the vial was removed from the ultrasonic bath and the electrodes immediately removed. Such a preparation contains 2.25×10^{-5} moles of pyrophosphate and 2.5×10^{-6} moles of stannous ion in 3.9 ml. In those parts of the study where the pyrophosphate concentration was varied this was achieved by varying the volume of sodium pyrophosphate used and making the corresponding adjustment in the volume of saline so that the total volume remained the same. Alterations in the stannous ion concentration were achieved by varying the time for which the 4 mA current was passed between the two tin electrodes.

In those studies in which tin-113 was used as a tracer for the stannous ion approximately 250 uCi of tin-113 was added in the form of stannous chloride in hydrochloric acid prior to the insertion of the tin electrodes. The specific activity of the tin-113 used was approximately 15 mCi/mgm, ie the amount of tin added this way was approximately 17 ugm or less than 6% of the normal amount of tin used in a preparation. In those studies in which P-32 pyrophosphate was used to follow the pyrophosphate approximately 100 uCi of P-32 sodium pyrophosphate with a specific activity of approximately 6 Ci/m Mole was added with the sodium pyrophosphate solution. This corresponds to approximately 0.1% of the quantity of pyrophosphate in the total normal preparation.

The procedure used for the in vitro studies, was to draw the blood fresh from a single human donor and add heparin as an anticoagulant, then to 2 ml portions of this heparinised blood, 0.2 ml of various dilutions of the stannous pyrophosphate were added, the mixture allowed to incubate at room temperature for one hour prior to the addition of approximately 1 mCi of Tc-99m pertechnetate. The samples were then incubated for a further hour before being centrifuged to separate the cells and plasma. The cells were washed with 1 ml of ACD solution, centrifuged and the wash removed. Equal volumes of the plasma and the packed red blood cells were then counted. Samples of the plasma were spotted on silica gel thin layer chromatograms and the chromatograms developed in methyl ethyl ketone.

Gel chromatography on Bio-Gel P-10 was carried out on selected plasma samples to demonstrate the binding of the Tc-99m to plasma components. The gel columns used were 60 x 0.9 cm columns from which 40 drops (1.3 ml) samples were collected.

In one series the plasma and cells were separated prior to the addition of Tc-99m pertechnetate, and the Tc-99m pertechnetate added to both fractions. These fractions were then allowed to incubate for one hour. The plasma fraction was then sampled and spotted on a thin layer chromatogram which was developed in methyl ethyl ketone. The cell fraction was centrifuged, the supernant removed and the cells washed with a solution of ACD, centrifuged and the wash removed. The cells and the combined supernant and wash were then counted.

The counting of the Tc-99m and the In-113m daughter of the tin-113 was done on a 2" sodium iodide detector coupled to a single channel analyser. When the tin-113 was used as a tin tracer the samples to be counted were stored several days prior to counting for the In-113m, this allowed for the Tc-99m to decay and for any In-113m which was present at the time of separation to decay so that the observed In-113m count was a true indication of the tin-113 present in the sample. The P-32 samples were counted by taking 1 ml of the sample and mixing with 10 ml of PCS liquid scintillation solvent (Amersham) loaded into a Nuclear Chicago liquid scintillation counter, allowed to dark adapt for one hour and then counted for one minute.

RESULTS

The percentage of the Tc-99m associated with the red blood cells in the initial study in which the standard pyrophosphate preparation was used and various dilutions of it added to 2 ml blood samples are shown in Table 1. This table also shows the amount of Tc-99m in a form other than free pertechnetate in the plasma samples which were separated from the cells prior to the addition of Tc-99m pertechnetate. When pertechnetate was added to the red blood cells which had been treated with stannous pyrophosphate and the plasma removed it was found that the pertechnetate was always bound to the red blood cells provided that the initial quantity of stannous pyrophosphate corresponded to more than 0.2 ml of a 1/100 dilution of the standard stannous pyrophosphate preparation.

The effect of varying the concentration of the tin in the original stannous pyrophosphate preparation is shown in Figure 1, while Figure 2 shows that the effect of similarly varying the pyrophosphate concentration. The distribution of tin-113 and P-32 in tracer experiments using the standard concentrations of tin and pyrophosphate are shown in Table 2. Further studies in which the tin concentration was increased, in the case of the tin-113 tracer studies, and the pyrophosphate concentration increased, in the case of the P-32 tracer studies confirmed that neither the tin nor the pyrophosphate are significantly associated with the red blood cells. Gel chromatography studies were carried out on the plasma fractions of selected samples together with a variety of starting materials to serve as standards. The results of these studies are shown in Figures 3 and 4.

Thin layer chromatograms of the plasma fractions on silica gel using methyl ethyl ketone as the solvent showed that at stannous pyrophosphate concentrations above those which gave maximum binding of the Tc-99m to the red blood cells the Tc-99m remaining in the plasma was not in the form of free pertechnetate,

TABLE 1. DISTRIBUTION OF TECHNETIUM-99M

Dilution of Sn-Pyro	% Tagged to R.B.C.	% Not in Free TcO_4^- Form when TcO_4^- Added to Separated Plasma	Sum of % Bound to R.B.C. and % Bound to Separated Plasma
1/1000	36.9		
2/1000	47.1		
3/1000	71.1		
4/1000	83.2		
5/1000	89.6		
1/100	89.5	13.1	102.6
1.2/100	86.5	10.8	97.3
1.4/100	86.3	16.1	102.4
2/100	56.2	48.2	104.4
3/100	32.7	77.3	100.0
5/100	11.9	95.5	107.4
8/100	7.7	99.1	106.8
1/10	6.3	99.1	105.4

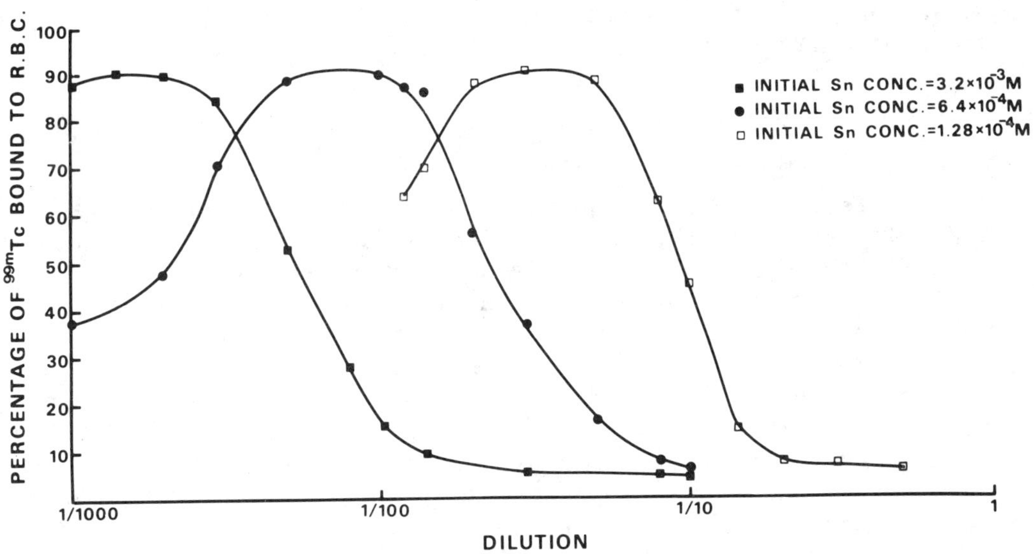

Figure 1

The effect of the tin concentration on the labeling of red blood cells. Five fold variations in the concentration of the stannous ion in the stannous pyrophosphate preparation showing corresponding five fold variations in dilutions necessary to obtain efficient red blood cell labeling.

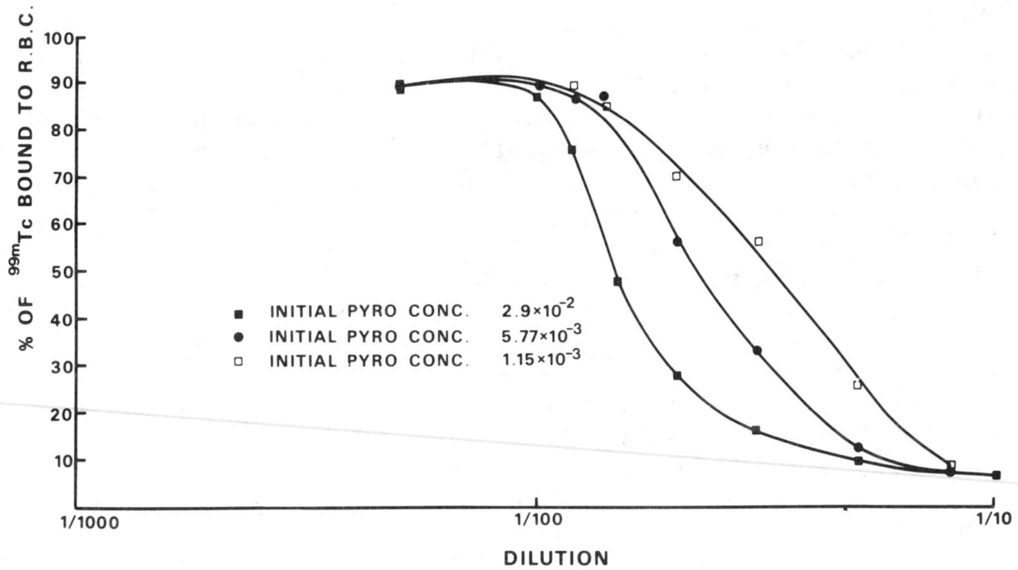

Figure 2

The effect of the pyrophosphate concentration on the labeling of red blood cells. Five fold variations in the concentration of the pyrophosphate in the stannous pyrophosphate preparation show relatively little variations in the dilutions which yield efficient labeling of the red blood cells.

while at lower stannous pyrophosphate concentrations the Tc-99m remaining in the plasma was free pertechnetate.

DISCUSSION

It was observed that when the quantity of stannous pyrophosphate added to the blood sample was less than that required to give optimum labeling of the red blood cells the Tc-99m in the plasma fraction was in the form of free pertechnetate indicating that there was insufficient stannous ion to cause the reduction of the pertechnetate and the resultant binding to the red blood cells. However at concentrations higher than that required to give optimum labeling of the red blood cells the Tc-99m remaining in the plasma was no longer in the form of pertechnetate but rather some labeled species indicating that at high concentrations of stannous pyrophosphate the red blood cells do not become labeled because there is preferential labeling of the components in the plasma. Gel chromatography of these plasma samples on Bio Gel P-10, Figure 3, suggests that the portion bound in the plasma consists of two fractions, one of which is eluted at void volume and is probably a labeled protein and one which is eluted after the technetium pyrophosphate peak but before the pertechnetate peak indicating a labeled species of molecular size somewhere between technetium pyrophosphate and pertechnetate. It will also be noted in Figure 3 that there is a shoulder at the technetium pyrophosphate position indicating that a small amount of the pyrophosphate present is also labeled. The contention that the reason that labeling of the red blood cells does not take place at higher concentrations of stannous pyrophosphate is that plasma species are preferentially labeled is further supported by the fact that if the plasma and cells are separated prior to the addition of the pertechnetate and then pertechnetate added to each fraction essentually complete labeling of the red blood cells is obtained while in the plasma portion the fraction of the pertechnetate which did not remain as free pertechnetate but became bound to various species in the plasma is virtually the exact compliment of the fraction which become bound to the red blood cells in the identical samples where the plasma and cells were not separated prior to addition of the pertechnetate, ie only that pertechnetate which does not become bound to the plasma components is available for the labeling of the red blood cells. Presumably this is principally due to the fact that labeling in the plasma fraction occurs relatively instantaneously while red blood cell labeling requires a certain amount of time for the technetium to cross the cell membrane.

The variation of the quantity of stannous ion in the pyrophosphate, Figure 1, clearly shows that both the completion of the binding of the technetium to the red blood cells, ie the removal of all the free pertechnetate, and the onset of competition for the technetium by constituents of the plasma are directly dependent on the concentration of the stannous ion in the preparation. Approximately 0.04 ugms of stannous ion per ml of blood being adequate to provide complete binding of pertechnetate to the red blood cells while a concentration in excess of 0.08 ugms per ml of blood causes the pertechnetate to start to become bound to constituents of the plasma. A similar variation in the concentration of the pyrophosphate, Figure 2, failed to show a similar effect. The effect observed is relatively secondary with the slope of the curve representing the increase of binding to the plasma constituents being somewhat

TABLE 2. DISTRIBUTION OF ^{113}SN AND ^{32}P PYROPHOSPHATE

Dilution of Sn-Pyro	% Tagged To The Red Blood Cells		
	^{99m}Tc	^{113}Sn	^{32}P
3/1000	71.1	6.6	2
4/1000	83.2	6.2	4
5/1000	89.6	7.6	4
1/100	89.5	5.5	0
1.4/100	86.3	8.5	0
2.0/100	56.2	5.6	5
3.0/100	32.7	6.2	0
5/100	11.9	9.2	7

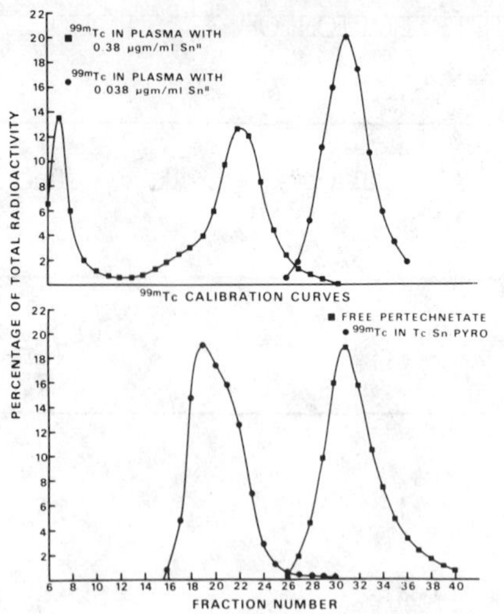

Figure 3

Bio Gel chromatography of plasma samples. Plasma samples chromatogramed on 60 x 0.9 cm Bio Gel P-10 columns. The upper curves show typical plasma samples, one above the optimum stannous ion concentration for red blood cell labeling, and one below. The lower curves show the chromatograms of pertechnetate and technetium stannous pyrophosphate.

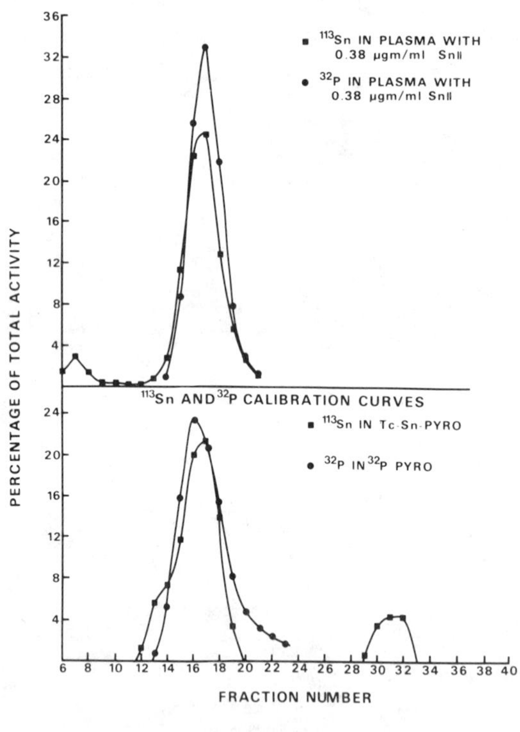

Figure 4

^{113}Sn and ^{32}P chromatograms on Bio Gel P-10. The upper curves show the chromatograms of plasma samples which contained ^{113}Sn and ^{32}P added to the whole blood in the form of stannous pyrophosphate. The lower curves show the chromatogram of ^{113}Sn in the form of stannous pyrophosphate and the chromatogram of ^{32}P in the form of sodium pyrophosphate.

steeper for higher pyrophosphate concentrations, possibly due to the pyrophosphate itself becoming increasingly active as a receptor for the reduced technetium. As shown in Table 2, neither tin nor pyrophosphate were significantly absorbed onto the red cells. The low levels that were observed probably reflect the incomplete washing out of the plasma from the packed red blood cells. In the experiment where the plasma and red blood cells were separated prior to addition of pertechnetate the red blood cells were not washed prior to the addition of the pertechnetate and it was probably the tin in the plasma which was trapped in the packed red blood cells which led to the binding of the pertechnetate to the red blood cells. The gel chromatography on Bio Gel P-10 of plasma samples containing tin-113 and P-32, Figure 4, clearly show that the tin and the pyrophosphate in the plasma remain in the form of stannous pyrophosphate and are not associated with any of the components of the plasma.

Since we have shown that the stannous ion does not become associated with the red blood cells the question as to how the technetium becomes bound in the cell remains to be answered. Pertechnetate alone when added to blood appears to be able to cross the cell wall but is not subject to binding to the cells unless stannous ion is present in the blood, suggesting that it is probably a lower oxidation state which crosses the cell wall to become bound to the cell. However the amount of reducing agent necessary is clearly less than that necessary to cause binding of the technetium to constituents in the plasma under the conditions present in the blood. From such an observation we should expect that the presence of any drug in the blood which very readily becomes labeled with technetium might well significantly interfere with the technique of in vivo labeling of red blood cells.

CONCLUSIONS

It has been shown that neither the pyrophosphate nor the stannous ion are taken up by the red blood cells, but rather that it is only the technetium that becomes associated with the red blood cells. The efficient binding of the technetium to the red blood cells is shown to be dependent on the concentration of stannous ion in the blood, insufficient stannous ion leading to the presence of free pertechnetate in the plasma while an excess leads to the binding of the technetium to constituents of the plasma thus decreasing the labeling of the red blood cells.

REFERENCES

1. Meinken G. Srivastava S.C. Smith T.D. et al. Is there a "good" Tc-99m albumin? J. Nucl. Med. 17:537, 1976.
2. Rhodes B.A. Considerations in the radiolabeling of albumins. Semin. Nucl. Med. 4:281-293, 1974.
3. Pavel D.G. Zimmer A.M. Patterson V.N. In vivo labeling of red blood cells with ^{99m}Tc - A new approach to blood pool visualization. J. Nucl. Med. 18:305-308, 1977.
4. Hamilton R.G. Alderson P.O. A comparative evaluation of techniques for the rapid and efficient in vivo labeling of red blood cells with [^{99m}Tc] pertechnetate. J. Nucl. Med. 18:1010-1013, 1977.

STUDIES ON THE ACCUMULATION OF RADIOPHARMACEUTICALS IN INJURED TISSUES

E.B. Robbins, L.R. Chervu, and M.D. Blaufox

Albert Einstein College of Medicine
Bronx, NY

ABSTRACT

Localization of ^{67}Ga and ^{99m}Tc-radiopharmaceuticals in various rat tissues was studied following diverse insults. The lack of specificity of radiopharmaceutical uptake in these injured tissues lends support to the probable importance of tissue injury and/or changes in blood vessel permeability in the localization of these agents. Studies of ^{67}Ga uptake in cultured cell monolayers as a function of temperature and in the presence of cytochalasin B indicate that the mechanism may involve both diffusion and pinocytosis.

INTRODUCTION

Since the observations by Edwards and Hayes (1) showing ^{67}Ga uptake by certain tumors, several theories have been proposed to explain the mechanism of this localization (2). Suffice it to say that none has been generally applicable. The problem is complicated by preferential accumulation of ^{67}Ga in many types of inflammatory processes as well as its inconsistent uptake by specific tumor types, even within the same patient at different sites. These vagaries have limited its usefulness if not its use. On the basis of experiments presented here we have concluded that the pattern of ^{67}Ga localization in pathological tissue is at the very least the resultant contribution of two separate dynamic processes. The first involves the efflux of the radionuclide from a partially compromised vascular system and the second involves its transfer, probably both actively and passively, across the cell membrane. The relative contribution of these two processes is difficult to ascertain and may well be different from case to case.

EXPERIMENTAL

^{67}Ga localization was studied both in vivo using Sprague-Dawley rats and in vitro using malignant (HeLa) cells and a strain of fibroblasts that display contact inhibition (Balb C 3T3).

IN VIVO EXPERIMENTS

Localized injury of rat skin, liver, kidney and muscle was inflicted by diverse means. These include crush injury, abscess induction, thermal injury and infarct production. ^{67}Ga citrate (10 μCi) was injected via tail vein. At appropriate time intervals, samples of injured tissue as well as comparable normal tissue samples were excised, weighed and activities measured using a NaI (Tl) well detector. Since rat skin lends itself to multiple samplings with minimal inconvenience, this tissue was used for many of the in vivo studies. However, abscess induction was more readily effected in muscle and infarction was easily produced in the kidney. In some studies ^{67}Ga uptake in injured liver was also evaluated. Injuries were inflicted as follows:

Crush; The flank was denuded with a commercial BaS preparation and approximately 1 sq. cm. area was crushed with a square tipped surgical hemostat. Degree of crush was crudely controlled by notches on the hemostat handle.

Cryogenic: Approximately 1 cc of dry ice was applied to a denuded area of flank skin or to surgically exposed liver for timed intervals at constant pressure.

Thermal: Denuded flank skin was exposed to a 1 sq.cm. heated metal branding iron at red heat for 1 second.

Abscess induction: 0.1 cc of a water/rat feces suspension (3/1 v/v) was injected into each thigh muscle and a purulent abscess was allowed to develop over a 6 day interval.

Infarction: Either the main or a segmental renal artery was exposed through a small flank incision and clamped for 2 hours. An interval of 30 min. was allowed to elapse following release of the clamp and ^{67}Ga was then injected.

In order to test the specificity of our data, all of the above experiments also were performed using ^{99m}Tc-glucoheptonate, ^{99m}Tc-pyrophosphate or ^{99m}Tc-sulfur colloid, in place of ^{67}Ga.

IN VITRO EXPERIMENTS:

HeLa cell and Balb C 3T3 cell monolayer cultures in Eagle's Minimum Essential Medium (MEM) plus 5% fetal calf serum (Grand Island Biologicals*) were plated on 60 mm plastic petri dishes. Cells were grown to confluence for most experiments and while there were slight differences in cell numbers from one experiment to another, the numbers within the confines of an individual experiment were the same from dish to dish. 0.1 to 10 μCi/ml of ^{67}Ga was added with or without cytochalasin B†(10^{-4}M) and human transferrin (0.3 mg/ml). The effects of time and temperature were studied.

RESULTS

Crush injury caused a significant accumulation of ^{67}Ga in the injured tissue compared to normal counter parts. Fig. 1 shows that the ratio of ^{67}Ga activity in injured vs normal skin increases with increasing time of crush up to 5 minutes at which point it plateaus. Table 1 shows that the ratios also increase with increasing degree of crush. Since injury must compromise vascular integrity these data are consistent with the hypothesis that increased vascular permeability following injury leads to increased ^{67}Ga activity.

If this assumption is correct, we would expect to find a lack of specificity regarding accumulation of radionuclide in injured tissues. Table 2 shows that this is in fact the case. Irrespective of the mode of injury, ^{99m}Tc glucoheptonate as well as ^{67}Ga show a significant increase in injured tissue compared to normal controls. Traumatized skin, liver, kidney and muscle all display the same basic pattern with the ratios of radionuclide falling in the range of 3.6 to 5.2.

Traumatized rat skin usually weighed 1.5 to 2 times as much per sq.cm. of surface area as the normal skin and the thickness of the tissue was significantly greater than normal skin. This is due to edema and again reflects increased vascular permeability secondary to injury. Since the ratios are expressed on a per gram of tissue basis, the ^{67}Ga concentration in injured tissue on a cellular basis is actually higher than indicated, i.e. the increased weight is due to fluid while the number of cells remains the same.

While ^{67}Ga is clinically useful for abscess detection, its non-specific nature in this regard is emphasized by the results shown in Figs. 2 and 3. If injected Tc-99m glucoheptonate activity is high enough to compensate for the rapid loss of radionuclide via the kidneys, it too can effectively demonstrate purulent abscesses by scintillation camera imaging. As anticipated we have also found ^{99m}Tc compounds to accumulate in partially necrotic tumors. Figure 4 shows significant above background activity in a transplanted Walker sarcoma following injection of ^{99m}Tc-glucoheptonate.

In confirmation of others (3,4) we have noted that human transferrin markedly stimulates cellular ^{67}Ga uptake. This does not appear to be a

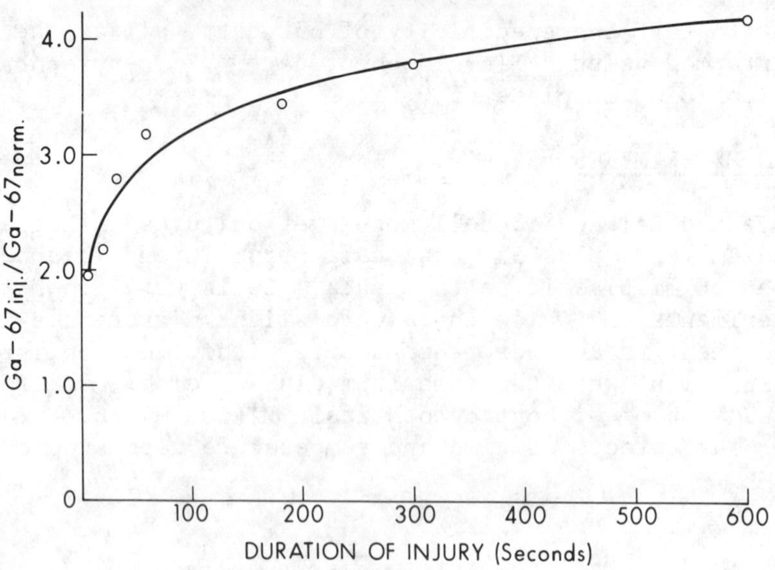

Figure 1.

Ratios of ^{67}Ga activity in injured skin vs normal skin as a function of duration of injury. Increasing periods of crush leads to increased ratios with gradual attainment of a plateau.

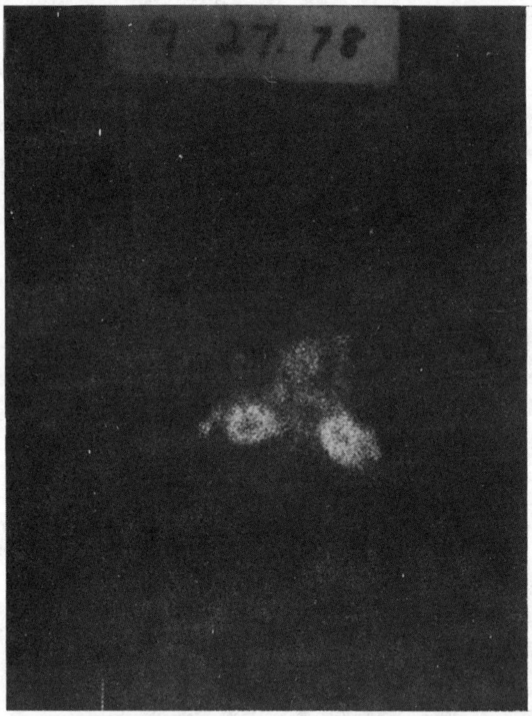

Figure 2.

Following induction of purulent abscesses in both thighs, 100 μCi of Ga-67 was injected intravenously into rats. Imaging by gamma camera was performed after 24 hours.

TABLE 1. EFFECT OF DEGREE OF INJURY ON ^{67}Ga ACCUMULATION

Relative Degree of Crush	Activity in injured tissue / Activity in normal tissue
1+	2.7
2+	3.7
3+	4.2

TABLE 2. EFFECT OF INJURY ON RADIOPHARMACEUTICAL ACCUMULATION

Injury	Activity in Injured Tissue/ Activity in Normal Tissue	
	^{67}Ga	^{99m}Tc-Glucoheptonate*
Crush (Skin)	4.2	3.6
Cryogenic (Skin)	3.6	2.9
Thermal (Skin)	3.8	3.6
Abscess (Muscle)	5.2	4.6
Infarction (Kidney)	4.4	4.7
Cryogenic (Liver)	4.6	4.0

* Similar results were obtained with ^{99m}Tc-pyrophosphate and ^{99m}Tc-S-colloid.

process that is specific for tumor cells since both 3T3 and HeLa cells exposed to MEM plus transferrin (0.3 mgm/ml) show a marked stimulation of uptake after 5 hours (Table 3).

Observations on the uptake of ^{67}Ga at 37°C and 4°C indicate that the process is probably composed of two components (Table 4). In the absence of transferrin, there is an appreciable intracellular accumulation of ^{67}Ga in both HeLa and 3T3 cells, and this is only slightly decreased by lowering the ambient temperature from 37° to 4°C. In contrast, the presence of transferrin causes a sharp drop in the accumulation of ^{67}Ga at 4°C. This suggests that lacking transferrin, ^{67}Ga enters the cell by simple diffusion while the much enhanced accumulation that takes place in the presence of transferrin is mediated by metabolic processes. Since Larson has observed that transferrin as well as ^{67}Ga become intracellular (2) we consider the possibility that passage of the complex into the cytoplasm was effected by pinocytosis. In the presence of 10^{-4} cytochalasin B (an inhibitor of pinocytosis) the uptake of transferrin mediated ^{67}Ga was reduced by approximately 50% (Table 5). While it is known that cytochalasin B affects other cellular activities including glucose transport, these results may well be related to the disruption of pinocytotic activity.

DISCUSSION

The data presented here suggest that the in vivo localization of ^{67}Ga in many tumors and inflammatory processes is due to the increased vascular permeability that accompanies these pathological processes and not to any special tumor or abscess seeking quality inherent in the radionuclide. Alternatively stated abnormal vascular permeability due to any cause including trauma, inflammation of tumor leads to non-specific radionuclide localization at the site of pathology. Once the ^{67}Ga (or ^{67}Ga-transferrin complex) reaches the extracellular fluid it probably does not have to be transported intracellularly to provide useful information. This is indicated by the fact that tissue sample excised within 30 minutes of intravenous administration of radiopharmaceutical reflect the ultimate distribution while we and others have found that intracellular accumulation in vitro - even with maintenance of high medium concentration is a relatively gradual process. If the radionuclide does enter the cells from the extracellular pool, it appears that both passive diffusion and metabolically mediated transport involving pinocytosis is involved. That pinocytosis plays a role is indicated not only by our data but by the electron microscopic studies of Swartzendruber et al (5). They showed intracellular ^{67}Ga to be concentrated in cytoplasmic dense bodies. Since dense bodies are thought to originate from fusion of primary lysosomes with pinocytotic vacuoles, the association between pinocytosis and ^{67}Ga uptake is clear.

FOOTNOTES

* Grand Island Biological Co., Grand Island, NY 14072.
† Sigma Chemical Company, St. Louis, Mo.

ACKNOWLEDGEMENTS

This work has been supported in part through funding from American Cancer Society Grant No. VC-172.

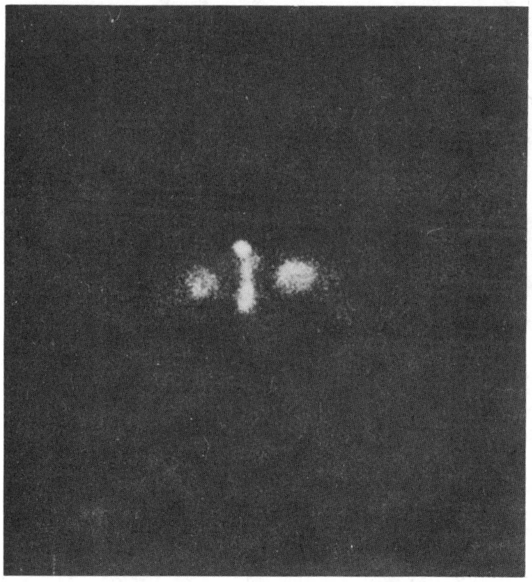

Figure 3.

Following induction of purulent abscesses in both thighs, 200 μCi of Tc-99m Glucoheptonate was injected intravenously into rats. Imaging by gamma camera was performed 2 hours post injection. The kidneys were masked with lead strips.

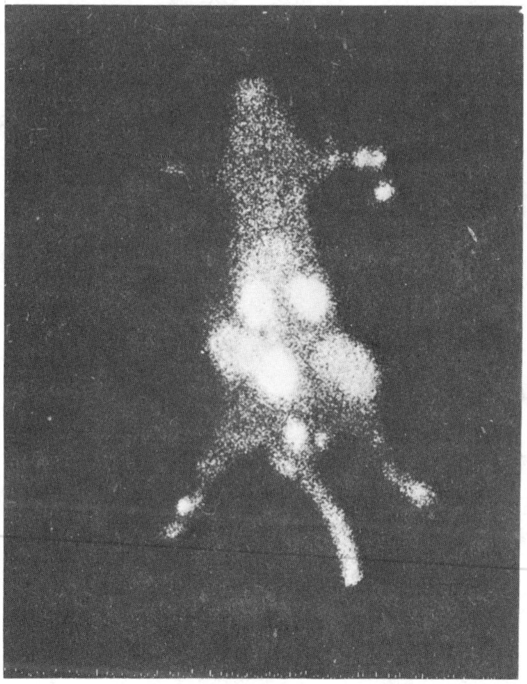

Figure 4.

Walker sarcoma was implanted into both thigh muscles. After 1 week, 250 μCi of Tc-99m Glucoheptonate was injected i.v. Imaging was performed 1 hr post injection. The excellent visualization of the tumors again illustrates the non-specific nature of radionuclide localization.

TABLE 3. EFFECT OF HUMAN TRANSFERRIN ON ^{67}Ga UPTAKE AT 37°C.

Medium	3T3 cells cpm	HeLa cells cpm
Eagle's MEM	8935	7500
Eagle's MEM + Transferrin	16,843	18,547

TABLE 4. EFFECTS OF TEMPERATURE ON ^{67}Ga UPTAKE IN HeLa CELLS

Medium	37°C cpm	4°C cpm
Eagle's MEM	13,265	8800
Eagle's MEM + transferrin	33,800	9800

TABLE 5. EFFECTS OF CYTOCHALASIN B ON ^{67}Ga UPTAKE BY HeLa CELLS AT 37°C

Medium	cpm
Eagle's MEM + Transferrin	51969
Eagle's MEM + Cytochalasin B + Transferrin	25796

REFERENCES

(1) Edwards CL, Hayes RL: Tumor Scanning with ^{67}Ga citrate. J Nucl Med 10: 103-105, 1967

(2) Larson S: Mechanisms of Localization of Gallium-67 in Tumors. Seminars in Nucl Med 8: 193-203, 1978

(3) Sephton RG, Harris AW: Gallium-67 citrate uptake by cultured tumor cells, stimulated by serum transferrin. J Natl Canc Inst 54: 1263-1266,1974

(4) Harris AW, Sephton RG: Transferrin promotion of ^{67}Ga and ^{59}Fe uptake by cultured mouse myeloma cells. Canc Res 37: 3634-3688, 1977

(5) Swartzendruber DC, Nelson B, Hayes RL: Gallium-67 localization in lysosomal like granules of leukemic and non-leukemic murine tissue. J Natl Canc Inst 46:941-952, 1971

PHARMACOKINETICS: CURIOSITY OR CURE?

Robert E. Notari

College of Pharmacy
The Ohio State University
Columbus, Ohio 43210

What is the fate of a drug from the time of its introduction into the body to the end of its duration? Pharmacokinetic studies are often designed to provide an answer to this question. But this question may be asked of any drug and research that is limited to answering it will remain empirical. Pharmacokinetic studies can provide answers to many other drug-related questions. In doing so pharmacokinetic research has the potential of improving drug therapy as well as the design and evaluation of drugs. While significant contributions can be cited, the future of pharmacokinetics depends upon its increased impact on _clinical practice_ and _drug design_. How can a molecule be tailored for site specificity? Can chemical modification selectively alter absorption, distribution, metabolism, binding or excretion? In what new ways can pharmacokinetic information increase the predictability of drug therapy? Such questions, to which pharmacokinetics should provide answers, are numerous and easily identified. But the definitive studies are difficult both to create and conduct. Whether or not pharmacokinetics can achieve its full potential will depend upon the extent to which it can provide answers to these currently unanswered questions.

ORIGINS

In an editorial, Wagner (1) has briefly traced the origins of pharmacokinetics back to 1924 when Widmark and Tandberg (Sweden) published two theoretical papers on the one compartment open model. The Michaelis-Menten equation, used today in connection with elimination kinetics, was published to describe enzyme kinetics in 1913. In 1937, Teorell in Sweden, published two theoretical papers which are regarded today as classical papers and perhaps the beginnings of what we call pharmacokinetics. This work was largely disregarded for many years due perhaps to the complexity of the equations and the lack of analytical tools. The advent of computers and the proliferation of sophisticated analytical methodology has removed these obstacles. Teorell was probably too innovative for the technological development of the time. Today the limitation in pharmacokinetics is perhaps innovation rather than technology.

One example of early experimental success is that of Marshall, Emerson and Cutting (2) who, in 1937, determined sulfanilamide blood levels as a function of time. This work was unique at that period of time but 40 years later the literature still contains such studies under the auspices of pharmacokinetics.

EMPIRICAL PHARMACOKINETICS

Many pharmacokinetic studies past and present have the simplistic goal of quantitatively accounting for drug in the body as a function of time. Drug and metabolites are measured in accessible fluids, generally blood and urine, and kinetic analysis of the data is employed to make deductions regarding inaccessible regions. Pharmacokinetics is the study of the kinetics of drug absorption, distribution and elimination (often termed ADE) as illustrated in Scheme I. Biopharmaceutics is the study of those factors which influence the rate and amount of drug which passes from the site of administration into the blood. When a pharmacokinetic study begins with a rapid intravenous injection of drug (thus avoiding the absorption process) the results are often described as the kinetics of drug disposition (DE kinetics). Whereas the biopharmaceutical phase in Scheme I is limited to absorption, all of the ADE rate processes may be influenced by the chemistry of the drug itself and the condition of the patient (i.e. age, sex, weight, disease state).

DRUG BLOOD LEVELS WITHOUT DATA REDUCTION

Physical science involves observation, description and prediction. After observing the behavior of the system a method for describing this to other scientists is postulated and its adequacy is tested by its ability to predict the outcome of situations not yet studied. The description should be limited to the smallest statement required to describe the area. For example, linear systems may be described by one simple linear equation together with equations for predicting the slope and intercept. An entire field might thereby be described by three equations. Many examples of published studies bearing the word pharmacokinetics in the title represent the antithesis to this approach. Reports that are limited to a display of the drug time course in blood are little more than the sharing of a laboratory notebook. The ability to analyze a drug in the blood as a function of time was no doubt a major breakthrough in 1937 when Marshall *et al.* (2) described sulfanilamide. Today it represents

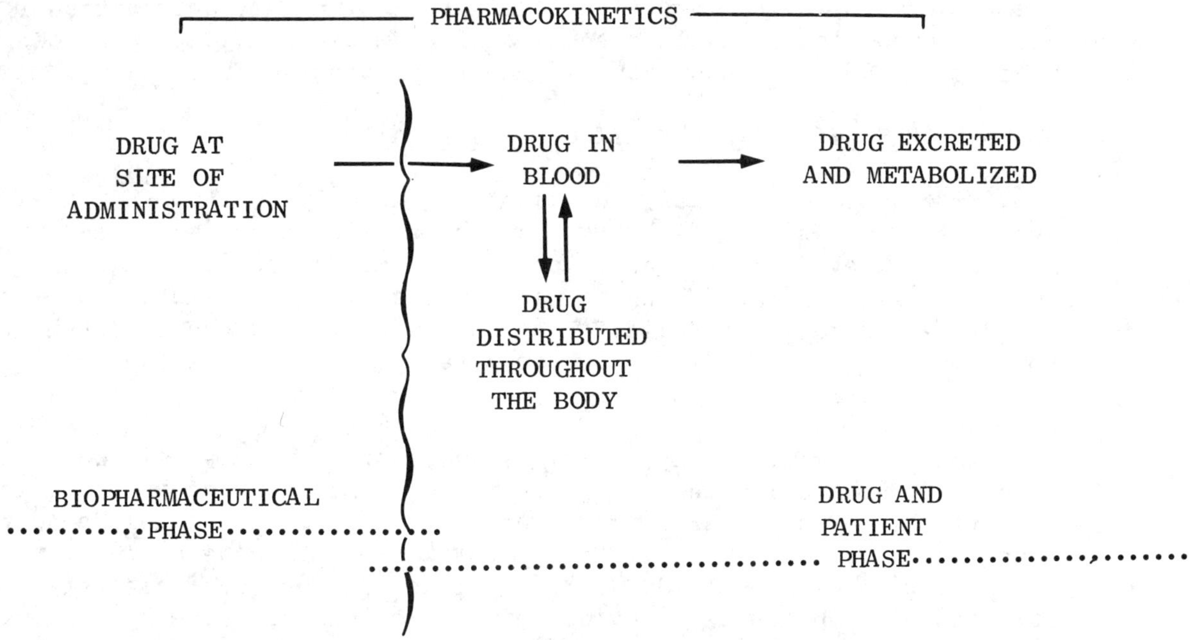
PHARMACOKINETICS
DRUG AT SITE OF ADMINISTRATION
DRUG IN BLOOD
DRUG EXCRETED AND METABOLIZED
DRUG DISTRIBUTED THROUGHOUT THE BODY
BIOPHARMACEUTICAL PHASE
DRUG AND PATIENT PHASE

SCHEME I

only the first step in the investigational process, namely observation. A research area cannot count itself a science if communication between those in the field is limited to the sharing of empirical observations.

DRUG BLOOD LEVELS WITH DATA REDUCTION: THE EQUATION

The first approach to data reduction is generally an attempt to describe many examples of experimental observations by a minimum number of equations. This has been achieved in pharmacokinetics. In most cases, the equations which now comprise the field, can be used to describe the time course for drug in the body as a function of time. Results can be reduced to a statement of the appropriate equation and the numerical values for the constants together with the statistical limits.

The process is most simply described by considering disposition studies using Scheme I. Commonly, the decrease of drug concentration in blood is often describable by an equation representing monoexponential, biexponential or triexponential decay. Specific drug behavior may then be defined by stating the appropriate equation and the values for the coefficients and apparent rate constants in the exponential terms. For example, the biexponential equation is usually written

$$P_t = Ae^{-\alpha t} + Be^{-\beta t} \qquad (1)$$

where the concentration in blood as function of time, P_t, can be predicted from the experimentally determined values for A, α, B, β.

Such systems are referred to as linear since they behave like first-order processes thus providing blood levels that are linear functions of the size of the dose. Drugs which cannot be adequately described by these simple exponential functions are often accommodated by the mathematics associated with Michaelis-Menten kinetics (3).

$$V = V_{max}[P_t]/(K_m + [P_t]) \qquad (2)$$

A rate process whose velocity, V, is described in this way is said to behave according to non-linear kinetics. Values for K_m and V_{max} are generally reported to describe the drug but have no simple theoretical significance.

The adequate description of drug plasma levels by an appropriate equation has become commonplace. The ability of a research worker to describe a drug for the first time using one of the now standard approaches is routine. It does not, in itself, advance the state of the art any more than a Beer's Law Plot for a previously unstudied drug advances either drug or analytical research.

THE ADDITION OF MODELS: FACT OR FANCY?

In empirical chemical kinetics the use of models is a well developed approach (4,5). Any student of chemical kinetics learns to recognize the concentration-time profiles associated with $A \rightarrow B \rightarrow C$, $A \rightleftharpoons B$, $B \leftarrow A \rightarrow C$, $A \rightleftharpoons B \rightarrow C$ etc. In chemical kinetics, unlike pharmacokinetics, we are dealing with a closed system of known volume which places some useful restrictions on the data. The models can be tested by mass balance, isolation, initiating the reaction with an isolated product etc. The fact that the volume is both known and constant provides a basis for mass balance as a function of time.

Pharmacokinetics has employed similar modeling without the benefits of those restrictions which make the models meaningful in chemical kinetics. In terms of the previous discussion regarding drug disposition, the number of exponentials required has been used to define the minimum number of compartments required. Thus, monoexponential decay is a one-compartment, biexponential is a two-compartment, and triexponential is a three-compartment open model. For the case wherein drug loss occurs from the central compartment (blood and easily perfused tissues) models are normally represented as shown in Scheme II. These compartmental schemes can be modified to include the absorption process, Michaelis-Menten behavior, first-pass metabolism and other experimental variations. There is no doubt that such pictures have been useful in visualizing and teaching the observed kinetic behavior. They are often the source of misunderstanding and misinterpretation of results. The models are only mathematical descriptions and should not be confused with anatomical interpretations. Typically, the drug is assayed in the blood (compartment 1). Other compartments therefore represent the entire remainder of the unassayed body which is no doubt heterogeneous with respect to drug levels. The mathematical pooling of the non-blood drug into the additional compartments (sometimes called tissues) is not meant to imply anything beyond a mathematical description. The model is compatible with the equation and nothing more.

A problem arises when conclusions rely on the predicted levels in a compartment where $n > 1$. Drug response, for example, may be due to the drug concentration in a particular organ within the compartment $n > 1$. The time course in that organ may not resemble the predicted time course for the total compartment. This is due to the fact that the drug may not be homogeneously distributed and the target organ may have levels that are too low to affect the pharmacokinetic analysis. This has been observed in mice wherein the maximum amount in the brain (target organ) for the highest analog was only 5% of the I.V. dose (6,7).

Further doubt regarding predicted time profiles in compartment $n > 1$ arises from the uncertainty of the model when only blood level data are used. Benet (8) has shown that the number of solvable rate constants, R, in this case is

$$R = 2n - 1 \tag{3}$$

where n is the total number of compartments at the donor end of one or more arrows. For example, in the three-compartment model (Scheme II) $n = 3$ so that $R = 5$. The values for each of the 5 rate constants in this model can be calculated from blood level data. However, Wagner (9) has pointed out that this model is only one of 13 kinetically equivalent linear three-compartment open models which adequately describe the data for triexponential decay. This multiplicity can be demonstrated by adding the possibility shown in Scheme III to that in Scheme II and counting all the potential models that result from removing the restriction that elimination occurs only from the central compartment. In Scheme II elimination may occur from: (1) all 3, (2) 1 and 3, (3) 2 and 3, (4) 1 only, (5) 2 only, (6) 3 only. In Scheme III we can have: (1) all 3, (2) 1 and 3, (3) 1 and 2, (4) 2 and 3, (5) 1 only, (6) 2 only, and (7) 3 only. Since all 13 of these models may be described by

$$P_t = Ae^{-\alpha t} + Be^{-\beta t} + Ce^{-\gamma t} \quad (4)$$

the prediction of drug behavior in the sampling compartment (blood) using equation 4 is reliable. Predictions for $n > 1$ compartments are questionable. Wagner emphasized that most of the current applications of pharmacokinetics can be accomplished using blood level equations.

In addition to the fact that all of the above possibilities having more than one elimination route have more rate constants than can be legitimately solved from blood level data (equation 3) the predictions for time profiles in compartments ($n > 1$) are model dependent. Since both the rate constant values and the time course in compartment ($n > 1$) is highly model dependent then conclusions based on those calculated profiles are also model dependent. If the model is arbitrary or speculative then any additional information derived from it may be incorrect and the apparent increase in the sophistication of data analysis may be a step in the wrong direction.

PHARMACOKINETIC RESEARCH: CURIOSITY OR CURE?

When a new research technique is introduced there is generally a period of time devoted to cataloging the observed results by their publication in research journals. When spectroscopy was young one could publish the UV spectra of two or three compounds and discuss the potential reasons for differences. The analytical journals are full of examples of techniques that never went beyond this initial flurry of activity. They became mere curiosities. Those which remained and became our armamentarium for today's research are those which can be used to solve problems other than those associated with the development of the technique itself. Today UV spectroscopy is as common as a pH meter. No longer is the publication of spectra regarded as an achievement unto itself. But it may be a significant part of the solution of a larger research problem.

While analogies are always inadequate, pharmacokinetics may be regarded in much the same way. We can obtain the spectra (P_t vs time), we can mathematically describe it and we can make fairly reliable predictions. While the system is not thoroughly "debugged" in that no research method is free from improvement, it is at the point wherein attention should be focused primarily outside of the technique itself. It must move into the fields of application and not become a curiosity. This movement has already begun but is far from reaching its full potential since the experimental obstacles are severe. It is much easier to move on to another drug and get a new spectrum. Pharmacokinetics holds significant potential to improve three areas of drug therapy. These are shown in Scheme IV wherein the areas are arranged in rank order (most to least) of current research activity and feasibility.

Product evaluation and design first felt the impact of primarily biopharmaceutic research (Scheme I). Dramatic problems such as the high risk accompanying the lack of bioequivalency of digoxin and the recall of 200 million nonequivalent chloramphenicol capsules by various manufacturers (other than the innovator) served to convince medical science of the significance of bioavailability requirements. Today bioavailability in drug product design is commonplace. Reliable assessment of bioavailability and of the performance of specialized dosage forms (such as long acting formulations) requires the technique or

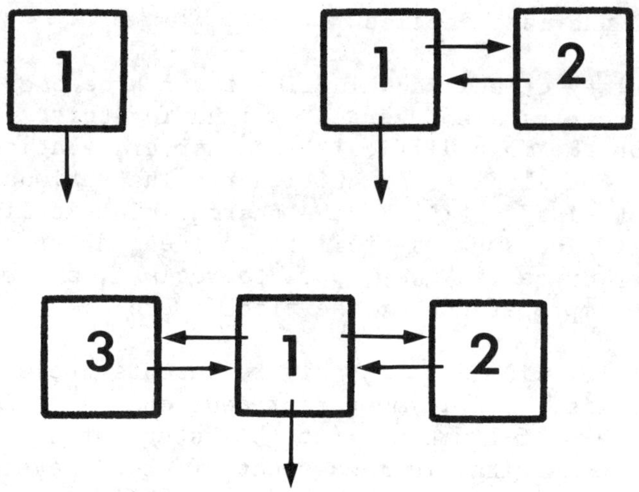

SCHEME II

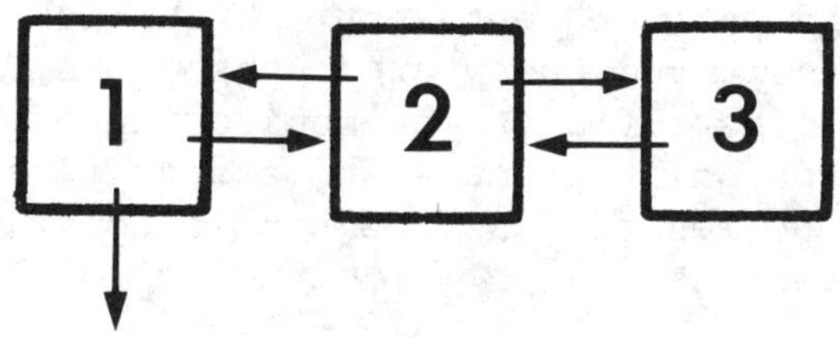

SCHEME III

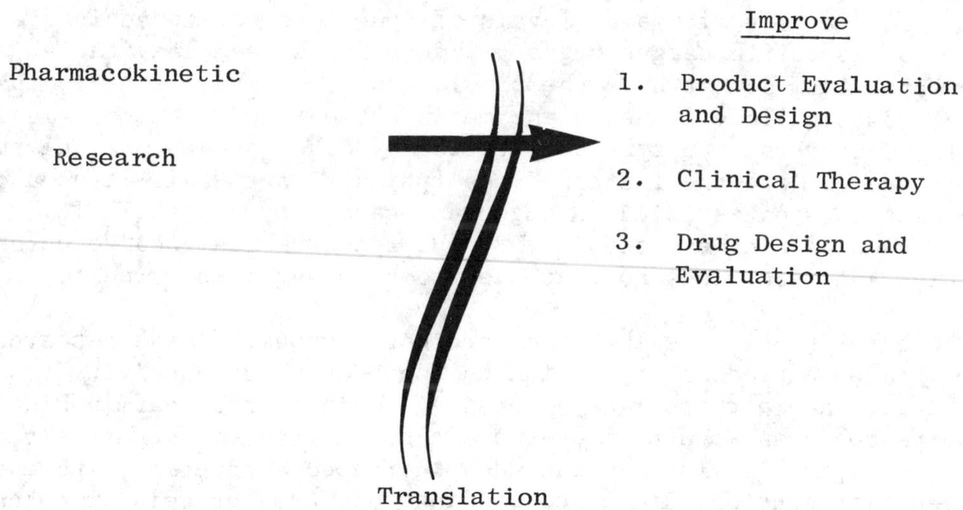

SCHEME IV

tool that we call pharmacokinetics.

The wide variety of pharmacokinetic research becomes an academic curiosity if the results are reversed by ignoring this knowledge in the clinic. This may represent the most difficult area for translation since it interfaces the scientist and the clinical practitioner. The rational selection and use of a drug may prove ineffective if the pharmacokinetic limitations are not known by the user. The drug may fail to be ideal due to: drug interactions with food or other drugs, an inappropriate regimen, an improper route of administration, the improper use of the product etc.

Clinical pharmacokinetics probably represents the area of greatest progress along this axis. But it must represent only a beginning since the activities of today are self-limiting. At this stage of development, clinical pharmacokinetics deals with the management of blood levels of those drugs which are essential but have a narrow safety margin. Dosage regimens are individualized to the patient and blood levels are monitored to ensure success (10). Creatinine clearance values, used as a measure of glomerular filtration rate, have been correlated with the elimination rate constants of many drugs that undergo renal excretion. This technique has provided a means for adjusting the dosage regimen in renal failure thus improving therapy with many critically needed drugs. However the effect of disease, age, sex, habits etc. on the pharmacokinetics in an individual is just beginning to surface (11). Regimens for the average patient represent a high risk for many drugs. Typical of drugs that are monitored today are: phenytoin, gentamicin, theophylline, digoxin, lidocaine and others (12). The monitoring of plasma levels and the adjustment of dose can become routine. While significant in improving therapy this activity must not be regarded as the single contribution that pharmacokinetics can make to medicine.

Drug design and evaluation offers the biggest challenge. While some progress has been made (12,13) the predictability of structural effects is low. The problems are easily identified but solutions are not. Drug specificity, resistance to metabolism, high predictable absorption and controllable time profiles, are but a few of the often considered problems in drug design. Most doses flood the body with high levels of drug in order to achieve an effective level in one specific target organ. Side effects seem inevitable. Drugs which resist metabolism do not have the problems of first-pass effect, toxic metabolites, low drug levels due to enzyme induction, and high drug levels due to metabolic disorders or enzyme competition by other agents. Among the notable areas of success are the increased absorption, longer half-lives and increased tissue distribution achieved through analogs and prodrugs (12,13). Other successful examples can be cited (12,13) but research designed to establish basic predictive knowledge does not yet seem to have begun in earnest.

The experimental possibilities are not obvious. These research goals have probably taken a secondary position to those structure-activity studies which appear easier and perhaps less expensive. Much of the pharmacokinetic progress has been made in an effort to overcome a known problem with an already successful drug. Typical goals have included increased absorption, increased duration, increased distribution, decreased binding to plasma protein or potential gastrointestinal contents, decreased metabolism and/or first-pass effects, control of the time course including lower peaks and finally the achievement of optimum drug levels in a target organ. Many successful examples can be cited. The poor absorption of ampicillin ($F \simeq 0.45$) has been overcome by prodrugs (pivampicillin, bacampiccillin, talampicillin) and analogs (amoxicillin). Oto- and nephro-toxicity of aminoglycosides has been studied relative to their pharmacokinetics and their tissue binding to the organs. Newer tetracyclines are better absorbed, more widely distributed throughout tissues, less bound in the presence of food or milk and longer in duration with half-lives in excess of 12 hours. Semi-synthetic penicillins are better absorbed and more stable to stomach acidity and penicillinase. Recent cephalosporins show good oral absorption, increased stability in the stomach and longer biological half-lives. Molecular modifications have successfully increased renal clearance thus providing high urinary levels of anti-infective agents. Known cases of first-pass metabolism have been reduced or eliminated. Prodrugs have increased the absorption of many drugs and, in other instances, have increased duration. Some repository intramuscular prodrugs have prolonged drug blood levels in excess of 30 days. But all of these studies are typical of the introduction of pharmacokinetic research _a posteriori_.

Pharmacokinetics, if it is to be a cure and not a curiosity must move into roles that impact on drug design and evaluation in the same way that pharmacology and medicinal chemistry have become integral steps in the process. Future directions must address more basic questions. We can currently obtain the spectra (P_t _vs_ time) but can we design the molecules to provide an ideal spectra? Such questions do not yet seem to have been asked.

REFERENCES

1.. Wagner, J.G., Editorial: The history of pharmacokinetics, _Drug Intel. Clin. Pharm._ _11_:747-748, 1977.

2. Marshall, E.K., Jr., Emerson, K. and Cutting, W.C., The distribution of sulfanilamide in the organism, _J. Pharmacol. Exptl. Therap._ _61_:196-204, 1937.

3. Wagner, J.G., Properties of the Michaelis-Menten equation and its integrated form which are useful in pharmacokinetics, _J. Pharmacokin. Biopharm._ _1_:103-121, 1973.

4. Rodiguin, N.M. and Rodiguina, E.N., _Consecutive Chemical Reactions_, D. Van Nostrand Co., Inc., Princeton, N.J., 1964, 136 pages.

5. Capellos, C. and Bielski, B.H.J., _Kinetic Systems_, Wiley Interscience, N.Y., 1972, 138 pages.

6. Notari, R.E., Burkman, A.M. and Van Tyle, W.K., Structural effects in drug distribution: whole animal pharmacokinetics, _J. Pharm. Pharmac._ _26_:481-492, 1974.

7. Burkman, A.M., Notari, R.E. and Van Tyle, W.K., Structural effects in drug distribution: comparative pharmacokinetics of apomorphine analogues, _J. Pharm. Pharmac._ _26_:493-507, 1974.

8. Benet, L.Z., General treatment of linear mammillary models with elimination from any compartment as used in pharmacokinetics, J. Pharm. Sci. 61: 536-541, 1972.

9. Wagner, J.G., Do you need a pharmacokinetic model and if so, which one? (Scientific commentary), J. Pharmacokin. Biopharm. 3:457-478, 1975.

10. Levy, G., ed., Clinical Pharmacokinetics, A.Ph.A., Washington D.C., 1974, 180 pages.

11. Benet, L.Z., ed., The Effect of Disease States on Drug Pharmacokinetics, A.Ph.A., Washington, D.C., 1976, 252 pages.

12. Notari, R.E., Biopharmaceutics and Pharmacokinetics, An Introduction to Theory and Clinical Practice, 3rd ed., Marcel Dekker Inc., N.Y., 1979.

13. R.E. Notari, in Design of Biopharmaceutical Properties through Prodrugs and Analogs, Roche, E.B., ed., A.Ph.A., Washington, D.C., 1977, pp. 68-97.

NUCLEAR PHARMACOLOGY/RADIOPHARMACOKINETICS,
A NEW DIMENSION OF RADIOPHARMACY/NUCLEAR MEDICINE

Walter Wolf, Randall C. Manaka, David Young,
Alan Schumitzky, Jan K. Siemsen and Jashovam Shani

University of Southern California
Radiopharmacy Program,
Department of Mathematics and Division of Nuclear Medicine
Los Angeles, California 90033

ABSTRACT

Radiopharmacokinetics is defined as the mathematical and experimental method pertinent to the use of gamma labeled drugs, whereby chemotherapy can be optimized for an individual patient through the use of the non-invasive techniques of nuclear medicine, by providing input specifically useful to clinical pharmacology. The present paper lays the foundation and presents the basic methodology necessary to utilize information gleaned from this novel use of gamma labeled chemotherapeutic agents. The kinetic "area of interest" results are analyzed with the aid of identification theory methods, so as to help identify uniquely the amount of a gamma labeled drug present in a given organ or tissue. The problem of controllability of drug administration is then to be determined with the use of these techniques. We propose that this new subspecialty of nuclear medicine be known as "Nuclear Pharmacology."

INTRODUCTION

Some years ago, we targeted as one of our major research objectives the study of the potential of labeled chemotherapeutic agents in clinical pharmacology. To accomplish such a goal, it was necessary to define the conditions under which non-invasive kinetics and product identifiability could be determined.

The first systematic study dealing with drug levels in the body was published in 1937 by T. Teorell (1,2). Since then, pharmacokinetics or the study of the kinetics of absorption, distribution, metabolism, and excretion of drugs (KADME) has developed into a field which ahs established its value in the understanding of drug action and drug utilization. The definition of pharmacokinetics as presented above, however, may be somewhat limited; the ultimate goal is to characterize and understand the interaction of a drug within a biological system so as to predict the system response to the drug and to design optimum strategies for the control of the system. In reality, however, the components of a biological system are so numerous and the interactions so complex that a full analysis is difficult, if not impossible. Accordingly, the real system is reduced to a low dimensional model system which is simple enough to describe in detail yet retains essential features of the real system. Thus, a mathematical model is homomorphic to the biological system.

One approach to the modeling of biological systems has been to group the biological components into a finite number of homogeneous components or compartments. Numerous texts and publications have dealt with "classical pharmacokinetics" (3-8) where heterogeneous biological components have been compartmentalized into homogeneous mathematical components. The simplest case is the representation of the system as a one compartment open model. A drug is introduced into the compartment and elimination is assumed to be governed by the Law of Mass Action. Additionally, all components of the system are considered to be in excess to the drug and the amount of drug is considered rate limiting. Elimination is thereby first order and the system can re represented as an ordinary differential equation. This model has been successful in representing a large number of drugs (9-11).

In many instances the system cannot be groups into a single homogenous compartment and two compartment (12-14) and multicompartment (15-17) models have been used where the system is represented as a set of ordinary linear differential equations. Furthermore, when the assumption of the Law of Mass Action is no longer valid, saturation or high order models can be used representing the system as a set of non-linear differential equations.

A second approach has been to formulate a completely different homomorphic model in which the biological components are abstracted into homogeneous organ compartments. Transfer of drug between organs is generally assumed to be blood flow limited, with protein binding taken into account. These "Bischoff-Dedrick" models have been successfully applied to chemotherapeutics, notably the pharmacokinetics of methotrexate (18). Again, the system is represented mathematically as a set of non-linear differential equations.

It should be mentioned that the solution of a set of differential equations past a three dimensional system is an extremely rigorous mathematical procedure. Furthermore, an explicit analytical solution is not possible for even the simple one dimensional system with Michaelis-Menton (saturable) elimination. Thus, parameter identification for the so-call "inverse problem" is essentially a problem for digital computers. A variety of programs using different numerical methods are available for this task (19-24).

A third approach which has only recently been applied to pharmacokinetics is Systems Control/Realization Theory. Systems theory is a relatively new mathematical area formalized by Kalman in 1963 (25) and has been used to describe highly complex systems in engineering and physics. Its application to pharmacokinetics is based on the argument that if one wishes to formulate a model which is useful for prediction, fits data well, and can be used to design dosage regimens, why not use a model which relies only upon the input-output relationships and not on an internal description of the system? Systems theory represents the biological system as a homomorphic model reduced to a single heterogeneous entity, i.e., a "black box." Techniques are available to optimize this description of the system so that it has properties of minimal dimension, minimal number of parameters, and maximal linear parameters. This representation can be termed a "compartment free" model.

We have used radioactive isotopes to study the pharmacokinetics of selected chemotherapeutic agents, namely, ^{195m}Pt-labeled *cis* platin and ^{18}F-labeled 5FU. In the past, there has been a reluctance to use isotopes for pharmacokinetic studies in that isotope technique yields information proportional to the number of radioactive atoms present and not the chemical state in which they exist (3,26). Indeed, there is the potential to misinterpret the data if one is not aware of the limitations of the technique. In addition, the use of carbon-14 and tritium, which are soft beta emitters, subjects the patient to a significant radiation burden. On the other hand, the use of radionuclides allows one to readily quantitate small amounts of substances and, if the proper gamma emitting radionuclide is chosen, monitor levels externally by non-invasive techniques, possibly in a continuous mode. Thus, it may be possible to use the relatively inexpensive and rapid techniques of nuclear medicine to characterize the system in terms of its controllability invariants (27). Then, with as few as possible invasive measurements, one can define a unique solution and, thereby, establish compartment free control for the duration of the study. The advantages of radioisotopic studies, with care in the selection and incorporation of the isotope into the drug molecule, should allow us to formulate a new mathematical treatment of kinetic data. This treatise, when applied to drug systems shall be termed "radiopharmacokinetics."

THEORY

In modeling a biological system, *s*, mathematically, one can characterize it in terms of:

1) System dynamics, as a set of differential equations

$$\frac{d}{dt} X(t) = \dot{X}(t) = f(X,U,\kappa,t) \quad t_o \leq t \leq \tau$$

2) Initial conditions

$$X(t_o) = X_o(\kappa)$$

and 3) Output

$$Y(t) = g(X,\kappa,t)$$

To simplify the representation we have used vector notation where X(t) is an n-dimensional vector representing the "state" of the system at time t (in compartmental models n usually reflects the number of compartments). $X_o(\kappa)$ is the initial state of the system at time $t = t_o$ and is n-dimensional. κ is a p-dimensional vector representing the parameters of the system. In compartmental models, κ is in terms of rate constants or other pharmacokinetic parameters such as clearance, volume of distribution, etc. U(t) is an r-dimensional vector representing the input of drug to the system at time t. It can incorporate delta functions, discontinuous functions and multiple dose inputs. Y(t) is an m-dimensional vector representing the output of the system.

Knowledge of U(t) and κ for all $t \leq \tau$ uniquely defines a solution for the system over the interval $[t,\tau]$. Most drug studies deal with the "inverse problem." That is, given a set of experimental measurements, find the solution (i.e., parameters and functions) to which this set belongs.

That is, q-observations are made of the system:

$$4) \quad Z_i = g(x(t_i),\kappa) + n_i \qquad i = 1,2,3, \ldots, q$$

The n_i's are the additive noise included in the observation (measurement procedure). The parameter vector, κ, is then estimated by fitting the observations to the output, equation 3, while incorporating (of course) the differential equations and initial conditions, equations 1 and 2. Either the method of least squares or maximum likelihood is usually used for this purpose [example (28)].

In actuality, there is a more fundamental problem associated with the problem of parameter estimation than that described above. This is the problem of structural identifiability (29,30,31,32). Namely, given exact observations (i.e., no noise) and as many observations as you want (i.e., $q = \infty$), just what can be uniquely identified? (Precise definitions appear in 31,32).

An example will illustrate this concept. Consider the first order absorption, first order elimination compartment model of drug kinetics given schematically by:

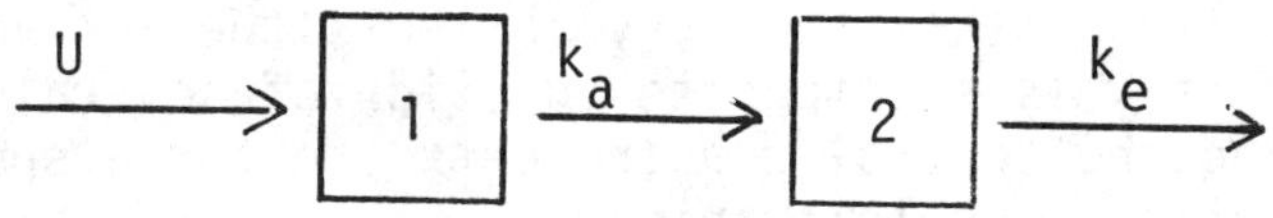

and mathematically by:

$$5)\quad \frac{d}{dt} x_1(t) = -k_a x_1(t) + u(t) \qquad x_1(0) = 0$$

$$\frac{d}{dt} x_2(t) = k_a x_1(t) - k_e x_2(t) \qquad x_2(0) = 0$$

$$6)\quad y(t) = x_2(t)/V$$

where $x_1(t), x_2(t)$ are the amounts of drug in compartments 1 and 2 at time t; $y(t)$ is the concentration of drug in compartment 2 at time t; $U(t)$ is the rate of IV input administered to compartment 1 at time t; k_a, k_e are the absorption and elimination rate constants; and V is the volume of distribution of compartment 2. In this case, it is assumed that $U(t)$ is known and that the parameters k_a, k_e, and V are unknown.

It is not hard to verify that if $\kappa = (k_a, k_e, V)$ is one set of parameters, then the same output $y(t)$ will be obtained for the parameter set $\tilde{\kappa} = (k_e, k_a, k_e V/k_a)$. Thus no experiment involving an input into compartment 1 and an output concentration in compartment 2 will be able to distinguish between the parameter vectors κ and $\tilde{\kappa}$.

We say, in this case, that the model is not structurally identifiable. Of course, if additional measurements are taken in compartment 1; or if additional input is administered to compartment 2; or if we know in advance that $k_a > k_e$, then, this lack of identifiability can be resolved.

The system S may be characterized by certain invariants. These invariants may not define κ uniquely and thus may only define families of solutions to the state equation. In order to resolve these families to obtain a unique solution one may resort to additional constraints on κ, additional outputs from S, or a reparameterization of the system.

As mentioned previously, using systems control theory it is possible to assume a "black box" relationship between the input function U(t) and the output function $Y(x,\kappa,t)$ [hereafter represented as u and y, respectively].

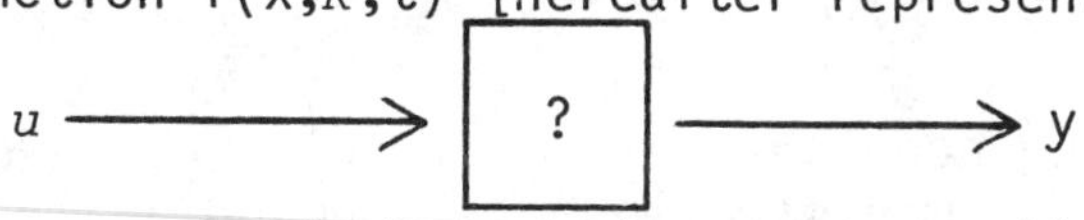

Various analytical techniques are available to describe the relationship between u and y, depending upon assumptions, and compartment free control can be established. Additionally, prediction of y based on any input u can be made.

This approach works fine for pharmacokinetic studies if the output y is the function one wishes to control (e.g., y reflects the active drug in the central compartment). If y is a composite function of several states of the system, however, and the control requirements are for a specific element of the output function, problems arise.

Consider an n-dimensional system where the output is a summand of S states, i.e., $y = x_1 + x_2 = . . . x 1$ Once it has been ascertained that control of y for the system S can be established, the goal is to relate cahnges in y with the ith state x_i. An example for S = 2 would be a measurement of radioactivity in the plasma where the drug exists in two forms: free and plasma protein bound. As it has been postulated that drug effectiveness is dependent upon the levels of free, active drug in the plasma (33), one might wish to control this component. (In its general form, this could be extended to unchanged drug and its metabolites in the blood).

STUDIES AND DISCUSSION

Active chemotherapeutic agents that could be studied by these techniques have to be labeled with radionuclides in such a manner as not to modify their chemical nature. Among the convenient agents we have chosen to develop these concepts with is *cis* dichlorodiammine platinim(II), a potent new anti-tumor agent, which can be labeled with platinum radionuclides. The 195m nuclide has been selected because:

a) Its γ energies, while on the low side, are in a range that is easily detectible with current nuclear medicine instrumentation.

b) Its photon yield is acceptable.

c) Its half life of 4 days makes it practical for transportation, synthesis, biochemical studies, etc.

d) The absence of β-radiation should keep its dosimetry down.

In our early studies on the radiopharmacokinetics of *cis* platin in blood using serial sampling techniques we reported (34), as did several others (35,36), that in rats and humans, the clearance of platinum appeared to follow bi-exponential, i.e., two compartment clearance, with rapid distribution half life on the order of minutes and a slow excretion half life on the order of hours or days.

In a clinical study utilizing non-radioactive *cis* platin, we obtained samples of blood during the first administration of drug and analyzed the plasma platinum levels by AA spectroscopy (37). When we used the classical pharmacokinetic parameters obtained from a two compartment analysis of this first study to predict drug levels in subsequent administrations, we were unable to successfully predict the peak and trough levels, suggesting that the model was not appropriate.

Experiments using an extracorporeal blood loop (38) in rats, where the platinum levels in whole blood were continuously monitored, suggested that a three compartment analysis might be more appropriate, with a very rapidly clearing component. However, these formulations lacked input as to the chemical nature of *cis* platin present in blood.

In a paper in 1978, we reported a significant association of *cis* platin with both the plasma proteins and the erythrocytes (39). These results led to the formulation of a five compartment model. We have now determined that the uptake of *cis* platin by both the erythrocytes and plasma proteins is essentially irreversible, with bound *cis* platin being metabolized and apparently unavailable to return to the free drug pool.

Hence, it is clear that a higher dimensional system is necessary to represent the pharmacokinetics of *cis* platin. Inasmuch as higher dimensional systems become increasingly complex to analyze, we have initiated a stepwise study of this system, starting with a 3 dimensional model, that was a reasonable reduction of the more complex system. We will now describe, as our example of how to assess structural identifiability and to illustrate the analysis of a radiopharmacokinetic model, the system shown below. This system has been formulated from the experimental data, i.e., the administered drug is bound by a first order process, with a rate constant k_{12} to plasma proteins which are then eliminated from the system by a first order process, rate constant k_{20}; simultaneously, the free drug is transferred to tissues by a process with rate constant k_{13} and subsequent elimination by process rate constant k_{30}.

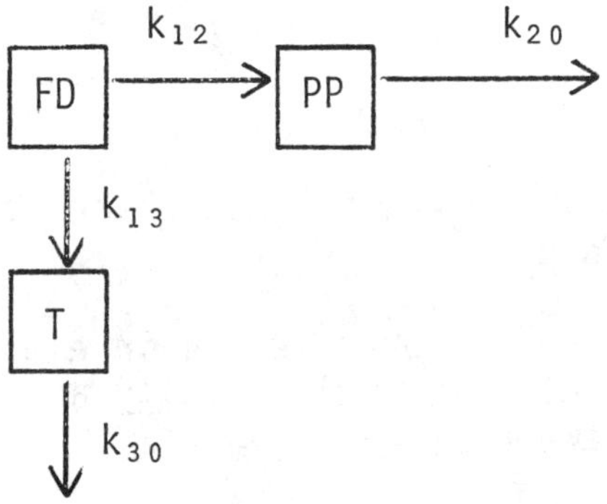

Let us analyze this system's structural identifiability for three different procedures:

CASE I

The total amount of free drug and plasma protein bound drug in the system is measured non-invasively as a function of time.

CASE II

The total amount of drug in tissues is measured non-invasively as a function of time.

CASE III

The combination of Case I and Case II, where both measurements are made as a function of time.

The rigorous mathematical treatment of these three cases is given in the appendix. From these results it can be shown that in Case I the system is not minimal and can in fact be represented by the alternate system:

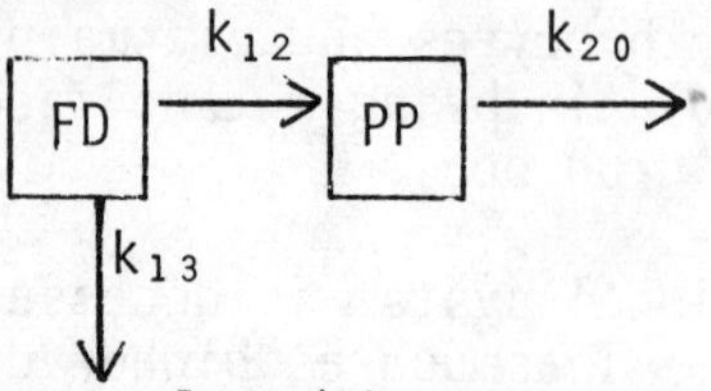

The tissue pool and k_{30} are non-observable and a twofold multiplicity exists for k_{20} and k_{12}. A graphical representation is given in Figure 1. A single invasive measurement of the amount of free drug or the amount of plasma protein bound drug or "a priori" knowledge that $k_{12} > k_{20}$ will resolve this ambiguity. In no instance, however, will information be gained on either k_{30} or the amount of drug in the tissue pool.

In Case II, the system again is non-minimal and can be represented by the system:

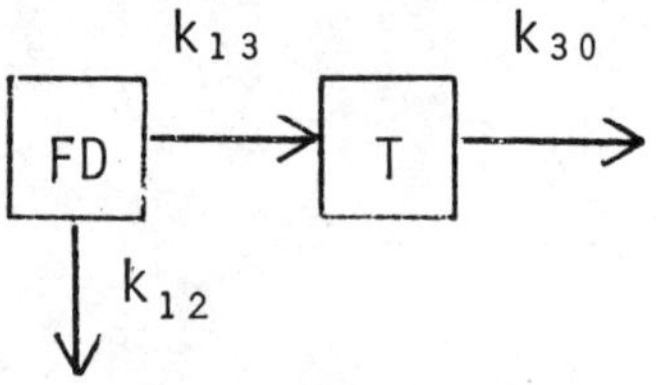

Similar to Case I, the plasma protein pool and k_{20} are non-observable, and a twofold multiplicity exists for k_{30} and k_{12}.

Likewise, a single invasive measurement of drug levels in either the free drug pool or the tissue pool can resolve the multiplicity as can "a priori" knowledge of the relationship between k_{12} and k_{30}.

For Case III, if measurements are made as a function of time for both cases, theoretically by ROI analysis, the system can be shown to be completely identifiable, as shown in the appendix. Thus, through two separate non-invasive measurements as a function of time, this full system can be completely analyzed and drug levels in each pool can be predicted at any time.

The approaches presented above can now be extended to higher dimensional systems. Inasmuch as the analysis of systems without an a priori determination of structural identifiability can lead to erroneous results, we feel that such an analysis is essential before accepting a "model" as valid representation of a pharmacological system.

In summary, through the proper utilization of information obtained from non-invasive nuclear medicine techniques one may characterize and predict the behavior of drugs, in a manner that cannot be obtained from conventional pharmacokinetics.

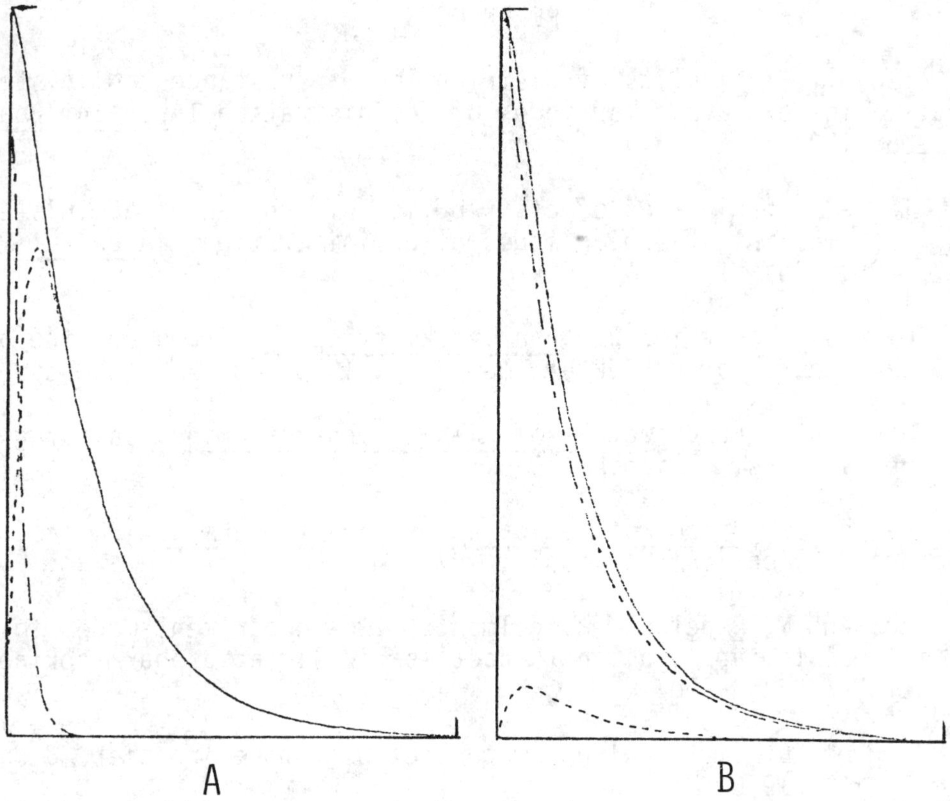

FIGURE 1.

A graphical illustration of the twofold multiplicity is shown above where the drug levels in each individual pool is simulated for each set of equivalent parameters.

The dashed line represents X_1, the amount of free drug, and the dotted line represents X_2, the amount of plasma protein bound drug. Both systems have identical output, $Y = X_1 + X_2$ (solid line); however, graph A is simulated with parameters $k_{13} = 10$, $k_{12} = 190$ and $k_{20} = 30$. Graph B is simulated with parameters $k_{13} = 10$, $k_{12} = 20$ and $k_{20} = 200$.

REFERENCES

1. Teorell T: Kinetics of distribution of substances administered to the body, I. The extravascular modes of administration. Arch Int Pharmacodyn 57: 205, 1937

2. Teorell T: Kinetics of distribution of substances administered to the body, II. The intravascular modes of administration. Arch Int Pharmacodyn 57: 226, 1937

3. Gibaldi M, Perrier D: Pharmacokinetics, I. Drugs and the pharmaceutical sciences. Marcel Dekker, New York, 1975

4. Wagner JG: Fundamentals of clinical pharmacokinetics. Drug Intelligence Publications, Illinois, 1975

5. Wagner JG: Biopharmaceutics and relevant pharmacokinetics. Drug Intelligence Publications, Illinois, 1971

6. Rowland M, Benet LZ, Riegelman S: Two-compartment model for a drug and its metabolite: application to acetylsalicylic acid pharmacokinetics. J Pharm Sci 59: 364, 1970

7. Garret ER: Classical pharmacokinetics to the frontier. J Pharm Biopharm 1: 341, 1973

8. Vaughan DP, Trainor A: Derivation of general equations for linear mammillary models when drug is administered by different routes. J Pharm Biopharm 3: 203, 1975

9. Jusko WJ: Pharmacokinetic principles in pediatric pharmacology. Pediatric Clinics of North America 19: 81, 1972

10. Garrett ER, Lambert HJ: Pharmacokinetics of trichloro-ethanol and metabolites and interconversions among variously referenced pharmacokinetic parameters. J Pharm Sci 62: 550, 1973

11. Taylor JA: Pharmacokinetics and biotransformation of chloropropamide in man. Clin Pharmacol Ther 13: 710, 1972

12. Sidell FR, Groff WA: Intramuscular and intravenous administration of small doses of 2-pyridiniun aldoxime methochloride to man. J Pharm Sci 60: 1224, 1971

13. Jusko WJ, Lewis GP, Schmitt GW: Ampicillin and hetacillin pharmacokinetics in normal and anephric subjects. Clin Pharmacol Ther 14: 90, 1973

14. Mitenko PA, Oglivie RI: Rapidly achieved plasma concentration plateaus, with observations on theophylline kinetics. Clin Pharmacol Ther 13: 329, 1972

15. Loo TL, Tanner BB, Householder GE, et al: Some pharmacokinetic aspects of 5-(dimethyltriazeno)-imidazole-4-carboxamide in the dog. J Pharm Sci 57: 2126, 1968

16. Doherty JE, Perkins WH, Flanigan WJ: The distribution and concentration of tritiated digoxin in human tissues. Ann Int Med 66: 116, 1967

17. Nagashima R, Levy G, O'Reilly RA: Comparative pharmacokinetics of coumarin anticoagulants IV. Application of a three-compartment model to the analysis of the dose-dependent kinetics of bishydroxycoumarin elimination. J Pharm Sci 57: 1338, 1968

18. Bischoff KB, Dedrick RL, Zaharko DS: Preliminary model for methotrexate pharmacokinetics. J Pharm Sci 59: 149, 1970

19. Pfeffer M: A time-sharing program for nonlinear regression analysis of compartment models of drug distribution. J Pharm Biopharm 2: 137-163, 1973

20. Buell J, Kalaba R, Yakush A, et al: A program for identification of linear systems. Comput Programs Biomed 2: 8-15, 1971

21. Feldman H: A short parameter-fitting routine for compartment models. Comput Programs Biomed 7: 135-144, 1977

22. Marquardt D: Least squares estimation of nonlinear parameters. IBM Share Library Program, No. 309401, 1966

23. Berman M, Weiss M: A users manual for SAAM. NIH, Bethesda, Maryland, 1968

24. Metzler C: NONLIN, a computer program for parameter estimation in situations. Technical Report 7292/69/9292/0005. The Upjohn Company, 1969.

25. Kalman RE: Mathematical description of linear dynamical systems. SIAM J Control 1: 152, 1963

26. Riegelman S, Rowland M, Benet LZ: Use of isotopes in bioavailability testing. J Pharm Biopharm 1: 83, 1973

27. Kalman P, Falb P, Arbib M: Topics in mathematical system theory. McGraw Hill, New York, 1969

28. Bard Y: Nonlinear parameter estimation. Academic Press, New York, 1974

29. Bellman R, Astrom KJ: On structural identifiability. Mathematical Biosciences 7: 329-339, 1970

30. Cobelli C, Romanin-Jacur G: On the structural identifiability of biological compartment systems in a general input-output configuration. Mathematical Biosciences 30: 139-151, 1976

31. Glover K, Willems JC: Parametrizations of linear dynamical systems: canonical forms and identifiability. IEEE Trans Automatic Control 19: 640-646, 1974

32. Herman R, Krener AJ: Nonlinear controllability and observability. IEEE Trans Automatic Control 22: 728-740, 1977

33. Scatchard G, Hughes WL Jr, Gurd FRN, et al: The interaction of proteins with small molecules and ions. In Chemical Specificity in Biological Interactions. FRN Gurd, ed., pp. 193-219, Academic Press, New York, 1954

34. Wolf W, Manaka RC, Ingalls RB: Radiopharmaceuticals in clinical pharmacology: $^{195}Pt^{m}$ -*cis*-diammine dichloroplatinum. In New Developments in Radiopharmaceuticals and Labelled Compounds, Vol. II, 205-221, IAEA, Vienna, 1973

35. Lange RC, Spencer RP, Harder HC: Synthesis and distribution of a radiolabeled antitumor agent *cis*-diamminedichloro platinum(II). J Nucl Med 14: 191, 1973

36. Litterst CL, Gram TE, Dedrick RL, et al: Distribution and disposition of platinum following intravenous administration of cis-daimminedichloroplatinum(II) (NSC 119875) to dogs. Cancer Res 36: 240, 1975

37. Wilcox S, Leeb J, Simon T, et al: The microdetermination of serum platinum by flameless atomic absorption. (Manuscript submitted for publication.)

38. Wolf W, Manaka RC: Synthesis and distribution of ^{195m}Pt-*cis*-dichlorodiammine platinum(II). J Clin Hematol Oncol 7: 79-85, 1977

39. Manaka RC, Wolf W: Distribution of *cis*-platin in blood. Chem-Biol Interact 22: 353-358, 1978

APPENDIX I - STRUCTURAL IDENTIFIABILITY OF RADIOPHARMACOKINETIC SYSTEMS

In a biological system s, assuming constant volume, linearity and time invariance the system dynamics can be described as a set of differential equations:

$$\dot{x}_j = \sum_{\substack{j \neq i \\ j \neq o}}^{n} k_{ji} x_j + u_i - \sum_{\substack{j \neq i \\ j \neq o}} k_{ij} x_i - k_{io} x_i$$

where x_i is the state variable associated with the ith compartment, k_{ji} is the rate constant for the transfer of material from the jth compartment to the ith compartment, k_{io} is the rate constant for transfer of material from the ith compartment to the external environment and u_i is the input to the ith compartment.

The system can be written as an input-output relation of a "noiseless" dynamical linear time invariant system by the equations:

$$\dot{X} = A(\kappa)X(t) + B(\kappa)U(t)$$

$$Y(t,\kappa) = C(\kappa)X(t)$$

X, U, Y are the state, input and output vectors and $A(\kappa)$ is the system matrix whose elements are given by:

$$a_{ii} = -\sum_{j=1}^{n} k_{ij} - k_{io}$$

$$a_{ij} = k_{ji} \; : i \neq j$$

Matrix A is n x n where n is the dimension of the system (i.e., number of compartments), B is of $n \times m_B$ dimension, $m_B \leq n$ corresponding to inputs into the $i_B{}^{th}$ compartment, and C is of $m_C \times n$ dimensions, $m_C \leq n$.

Systems theory states that given two systems, $s(A,B,C)$ and $\tilde{s}(\tilde{A},\tilde{B},\tilde{C})$, the systems are identical if given the same inputs U, they both give the same response Y. If the system s is both completely observable and completely controllable (a system of minimal dimension) the problem of structural identification of the κ parameters in s is to find an equivalent system $\tilde{s}$ whose parameters θ are known to be identifiable and then to map κ on θ. If a one to one mapping is found, then s is said to be structurally identifiable.

$$\text{If } s(A,B,C) \simeq \tilde{s}(\tilde{A},\tilde{B},\tilde{C}) \; ; \; A = A(\kappa),\ B = B(\kappa),\ C = C(\kappa)$$
$$\tilde{A} = \tilde{A}(\theta),\ \tilde{B} = \tilde{B}(\theta),\ \tilde{C} = \tilde{C}(\theta)$$

And s is minimal.

Then, $\theta = f(\kappa)$.

CASE I

The full system of dimension 3, although controllable, can be shown to be not observable, i.e.:

$$\det c \neq 0; \quad c = [B | AB | A^2B]$$

$$\det o = 0; \quad o = [C^T | A^TC^T | A^{2T}C^T]$$

Thus the system s was reduced to a lower dimension s_B, a subsystem of dimension 2, and was parametrically mapped on an equivalent identifiable cannonical system $\tilde{s}$.

$$\theta = (\alpha_1, \alpha_2, \beta_1, \beta_2), \ \kappa = (k_{12}, k_{13}, k_{20}); \ C = (1,1)$$

One set of invariants (identifiable) for this subsystem is:

$$\alpha_1 = k_{12} + k_{13} + k_{20}$$

$$\alpha_2 = (k_{12} + k_{13})\, k_{20}$$

$$\beta_1 = 1$$

$$\beta_2 = k_{13}$$

and a map of κ on θ gave:

$$k_{13} = \beta_2$$

$$k_{12} = \alpha_1 - \beta_2 - k_{20}$$

$$k_{20} = \frac{\alpha_1 \pm \sqrt{\alpha_1^2 - 4\alpha_2}}{2}$$

A twofold multiplicity exists for k_{12} and k_{20} even within the pharmacokinetic constraint $\kappa \in \mathbb{R}_+^p$.

CASE II

Again the full system was not observable and was reduced to a subsystem of dimension 2.

$$\theta = (\alpha_1, \alpha_2, \beta_1, \beta_2); \ \kappa = (k_{12}, k_{13}, k_{20}); \ C = (0,1)$$

$$\alpha_1 = k_{12} + k_{13} + k_{30}$$

$$\alpha_2 = k_{30}(k_{12} + k_{13})$$

$$\beta_1 = 0$$

$$\beta_2 = k_{13}$$

In mapping, again note that a twofold multiplicity exists for k_{12} and k_{30}.

$$k_{13} = \beta_2$$

$$k_{12} = \alpha_1 - \beta_2 - k_{30}$$

$$k_{30} = \frac{\alpha_1 \pm \sqrt{\alpha_1^2 - 4\alpha_2}}{2}$$

CASE III

The full system was controllable and observable as determined by:

det $[cc^T] \neq 0$; det $[oo^T] \neq 0$.

and $\theta = (\alpha_1, \alpha_2, \alpha_3, \beta_1, \beta_2, \beta_3, \beta_4, \beta_5, \beta_6)$; $\kappa = (k_{12}, k_{13}, k_{20}, k_{30})$

$$C = \begin{pmatrix} 110 \\ 001 \end{pmatrix}$$

A set of identifiable invariants for this system is (Reference 27; section 10.11):

$$\alpha_1 = k_{12} + k_{13} + k_{20} + k_{30}$$

$$\alpha_2 = (k_{12} + k_{13})\, k_{20} + (k_{12} + k_{13})\, k_{30} + k_{20}k_{30}$$

$$\alpha_3 = k_{20}k_{30}\,(k_{12} + k_{13})$$

$$\beta_1 = 1$$

$$\beta_2 = 0$$

$$\beta_3 = -k_{13}$$

$$\beta_4 = k_{13}$$

$$\beta_5 = k_{12}k_{13} - k_{12}k_{20} + k_{13}^2$$

$$\beta_6 = -k_{13}(k_{12} + k_{13} + k_{30})$$

and a mapping of κ on θ gives:

$$k_{13} = \beta_4$$

$$k_{20} = \frac{\alpha_1\beta_4 + \beta_6}{\beta_4}$$

$$k_{12} = \frac{\beta_4\beta_5 - \beta_4^3}{\beta_4^2 - \alpha_1\beta_4 + \beta_6}$$

$$k_{30} = \alpha_1 - \beta_4 - k_{20} - k_{12}$$

One can see that a one to one mapping occurs indicating that k_{12}, k_{13}, k_{20} and k_{30} are identifiable.

PHARMACODYNAMICS OF STANNOUS-CHELATES ADMINISTERED WITH TECHNETIUM-99m-CHELATES FOR RADIONUCLIDE IMAGING

Mrinal K. Dewanjee, Gregory S. Anderson, and Heinz W. Wahner

Mayo Clinic and Mayo Foundation
Section of Nuclear Medicine
Rochester, MN 55901

ABSTRACT

The pharmacodynamics of tin compounds, given as tin-human serum albumin (Sn-HSA), tin-glucoheptonate (Sn-GH), tin-dimercaptosuccinate (Sn-DMSA), tin-pyrophosphate (Sn-PP), tin-diethylenetriaminepentaacetate (Sn-DTPA), and tin-hydroxyethylene diphosphonate (Sn-HEDP), all labeled with ^{113}Sn, were studied in rabbits. Biodistribution, equilibrium dialysis with human serum albumin and red cell hemolysate, and autoradiography were used. The results show that (1) Sn-chelates of PP, HEDP, DTPA, HIDA, and GH are unstable in vivo and give rise to free stannous ion; (2) free stannous ion localizes mainly in bone, and most of the rest is excreted in the urine; (3) Sn-DMSA is a stable chelate, and its biodistribution pattern is more like that of the corresponding Tc-99m-DMSA chelate; (4) binding of stannous chelates with serum protein, specifically human serum albumin, is weak when compared with that of the corresponding Tc-99m chelates; (5) binding of stannous ion with hemoglobin is higher than with human serum albumin; (6) skeletal uptake, intrarenal distribution and renal excretion of Sn-chelates are similar to other divalent cations. Tin-chelates, administered by necessity together with Tc-99m, show pharmacodynamics different from those chelates of the corresponding Tc-99m chelates.

INTRODUCTION

Stannous compounds are being administered intravenously to patients undergoing nuclear medicine imaging procedures with certain Tc-99m-labeled radiopharmaceuticals. The stoichiometric ratio of stannous ion to chelating agents in the most frequently used Sn(II)-chelates is 1:1, 1:2, or 2:1. The elemental tin content in different commercial tin-chelate kits is listed in Table 1. The chemistry of these tin-complexes has been studied by several investigators (1-5) and reviewed by Donaldson (6). In contrast to these relatively large quantities of tin, total body content of tin is only 17 mg. In man, 4 mg of tin are ingested daily but only 20 µg are absorbed. The concentrations of tin in liver and kidneys are 0.3 and 0.2 µg/g respectively (7). Thus, pharmacologic doses of tin, 2 to 100 times higher than the normal daily absorption, are administered intravenously (8-16) to these patients.

Toxic effects, however, have not been observed. Subtle effects have been reported. Alterations in the distribution of Tc-99m-pertechnetate in tissue due to the prior administration of Sn(II)-complexes were observed by several investigators (16, 17), and Kappas and Maines (18) showed the formation of heme in liver and kidneys resulting from a toxic effect of stannous chloride on heme oxygenase enzyme.

The exact chemical structures of tin-chelates (the chelating agents being pyrophosphate (PP), glucoheptonate (GH), diethylenetriaminepentaacetate (DTPA), 2,3-dimercaptosuccinate (DMSA), or hydroxyethylene diphosphonate (HEDP)) are still not known. Using Tc-99m and Sn-119-labeled tetracycline, Dewanjee et al (10) first demonstrated that the metabolism of its technetium-chelate and tin-chelate is different. Tofe and Francis (12), using three labels (Tc-99m, Sn-113, C-14-HEDP), suggested that a Tc-99m-Sn-chelate, multinuclear complex, is formed. However, other observations provided evidence that Sn-chelates and Tc-99m-chelates are formed side by side and that they are different chemical entities with different metabolic pathways.

The present study was designed to investigate the radiopharmacology of such water-soluble Sn(II)-chelates in different organs of the rabbit and to study the binding of stannous-chelates with human serum albumin (HSA) and red blood cell (RBC) hemolysate.

MATERIALS AND METHODS

Chemically pure stannous chloride, DTPA, PP, DMSA, and GH from Sigma Chemical Company, HEDP from Montsanto Chemical Company, and Sn-PP kits from Mallinckrodt Nuclear Company were used for studying the effect of increasing carrier. Human Serum albumin from Armour Laboratories, and $^{113}SnCl_2$ in 4N HCl (specific activity 12 to 16 mCi/mg) from New England Nuclear Company were utilized.

Sn(II)-chelates were prepared by titrating $^{113}SnCl_2$ in 4N HCl solution with 20 to 25 mg of each chelating agent to a pH of 6. Removal of colloid and sterilization were performed by membrane filtration (0.22 µm, Millipore).

Biodistribution Studies. A dose of 500 to 600 µCi of freshly prepared ^{113}Sn(II)-chelate was administered intravenously to New Zealand albino rabbits

TABLE 1. ELEMENTAL TIN CONTENT IN COMMERCIAL Sn(II)-CHELATES

Sn(II)-Chelate	Content* (mg/vial)
Sn-pyrophosphate	0.63-2.1
Sn-diphosphonate	0.1 -0.63
Sn-dimercaptosuccinate	0.26
Sn-human serum albumin	0.05-0.07
Sn-diethylenetriaminepentaacetate	0.32
Sn-glucoheptonate	0.05-0.07
Sn-dimethylphenylcarbamoylmethyliminodiacetate	0.06-0.63
Sn-microsphere	0.1 -0.25

*Values obtained from data supplied by the manufacturers.

(2.5 and 3 kg). Rabbits were sacrificed with sodium pentobarbital 0.5, 1, 3, 6, and 24 hours in studies of Sn-PP and 3 and 24 hours in studies of the other Sn-chelates. Tissue samples from bone, bone marrow, blood, and muscle and the entire kidneys, liver, spleen, and lung were collected and weighed. The radioactivity was determined 24 hours after death using a gamma well counter and ionization chamber. This length of time was necessary for the buildup of ^{113m}In from the parent isotope of ^{113}Sn. Organ distribution of radioactivity was expressed as percent of the dose for entire organs. Blood volume and the weight of total mass for muscle, bone, and bone marrow were calculated using normal value tables, as previously described (10). Bile and urine were collected. The same experiments were repeated after ^{113}Sn(II)-PP was mixed with varying amounts of commercially obtained Sn-PP, keeping the maximal amount of tin within 80 to 90 μg/kg of body weight. Again, the concentration of chelating agent was always maintained at 10 to 20 times that of stannous ion on a weight basis.

Contact Autoradiography of ^{113}Sn-PP in Canine Femur and Rabbit Kidneys. Since rabbit bones are too brittle for thin-sectioning, we have used dogs for skeletal autoradiography. Small dogs weighing 2 to 3 kg were injected intravenously with 10 mCi of freshly prepared ^{113}Sn-PP and sacrificed 4 hours later. Femur and tibia were removed, scraped free of adherent tissue, dehydrated, and defatted with absolute alcohol and acetone-ether (1:1). The bone was then embedded in polymethylmethacrylate, sectioned with a rotary milling machine to a thickness of 150 μm, and polished to a thickness of 100 μm with a mirror grinding kit. The sections of dry bone were kept in contact with Kodak Type A autoradiography emulsion slides for three days by means of spring-loaded clamps and the exposed film developed. The predominant K conversion electron (36 keV) from the ^{113m}In radioisotope was well suited for contact autoradiography.

Intrarenal distribution of ^{113}Sn-PP was studied by autoradiography of thin transverse sections of rabbit kidneys three hours after I. V. injection of 1 mCi of ^{113}Sn-PP. The sections were exposed for 48 hours to single emulsion Nuclear Medicine Film (Kodak).

Protein-Binding Studies of ^{113}Sn-Chelates with HSA and RBC-Hemolysate. Protein binding to HSA and red cell hemolysate was studied by equilibrium dialysis. Sections of 15-cm cellulose tubing immersed in distilled water were heated at 75° C for 30 minutes in an incubator and rinsed in distilled water three times to remove impurities before dialysis. Human red blood cells were washed free of plasma with isotonic saline. 10 ml of red cell mass was lysed with 100 ml of distilled water. Red cell membranes were lysed with 2-3 ml of chloroform and removed by centrifugation. For dialysis, HSA or hemolysate (100 mg) and ^{113}Sn-PP (10-15 μCi) were added to cellulose tubing and the final volume adjusted to 6 ml with 0.9% saline. This was dialyzed against 5 liters of isotonic saline solution at 4° C in an Oxford Multiple Dialyzer. Control dialysis experiments without HSA or hemolysate were performed to evaluate nonspecific binding by dialysis tubing. Samples were dialyzed for 24 hours. Radioactivity in the retentate, cellulose tubing, and dialysate was determined.

RESULTS

Biodistribution. The distribution of radioactivity in different organs was similar for all Sn-chelates, including Sn-HSA (the stannous ion reference tracer) with the exception of Sn-DMSA (Table 2). For the former Sn-chelates, at 3 and 24 hours after injection, 35 to 50% of the radioactivity localized in bone and most of the residual radioactivity was excreted in the urine. For Sn-DMSA, only 5% of the radioactivity localized in bone, but 15% was found in the liver. Blood clearance of Sn-DMSA was significantly slower. A very small fraction of the injected dose was recovered from the bile, and about 50% was excreted in the urine. Addition of carrier Sn-chelate (up to 3.0 mg of stannous ion) did not significantly change skeletal uptake and urinary excretion, but hepatic uptake increased from 3 to 20% and bone marrow from 1.0 to 6.5% due to colloid formation (Table 3).

Uptake in the normal heart was 0.02 to 0.04% of the injected dose and again was similar for ^{113}Sn-HSA and the Sn-chelates tested. Myocardial uptake of ^{113}Sn(II) was slightly higher than that of skeletal muscle, most likely a result of higher blood flow to myocardium.

The plasma-to-red-cell ratio of ^{113}Sn-PP was studied 30 minutes and 1, 3, and 24 hours after intravenous administration (Table 4). Only a small amount of stannous ion entered the red cells, and at 30 minutes, only about 0.5% of the dose was found in the red cell mass.

Equilibrium Dialysis. The results of equilibrium dialysis of ^{113}Sn(II)-chelates in presence of HSA and RBC hemolysate are shown in Table 5. The percentage of radioactivity in the retentate, dialysate, and adsorption unto the cellulose tubing itself was determined, using a known standard. The results showed that the hemolysate of red blood cells binds unstable Sn-chelates more strongly than HSA. Binding of Sn-DMSA with hemoglobin was slightly lower than other Sn-chelates. Among the ^{113}Sn-chelates, Sn-DMSA demonstrated the strongest binding to HSA.

Autoradiography. Three hours after intravenous administration, the intrarenal distribution of ^{113}Sn(II)-PP in rabbits showed a cortex/medulla ratio of 8:1. Activity localized preferentially in renal cortex and collecting system (Fig. 1).

Autoradiographs and microradiographs from transverse sections of canine femurs are shown in Fig. 2. Deposition of stannous ion is seen in the areas of immature bone formation similar to that with ^{85}Sr-ion and ^{99m}Tc-PP.

DISCUSSION

The metabolism of tin administered to humans with ^{99m}Tc-labeled radiopharmaceuticals is poorly understood and the chemical structures of tin-chelates in these radiopharmaceuticals is not known exactly. These radiopharmaceuticals are probably mixtures of Tc-99m-chelates and Sn-chelates. The controversy regarding the structure of these chelates has been reviewed by us in the following paper in these proceedings.

The stability constants of stannous complexes with simple inorganic anions (OH^-, Cl^-) do not exceed 10^3. Some of the more common organic ligands, such as acetate, oxalate, or citrate, form more stable Sn(II)-complexes, with K-values ranging from 10^3-10^6 (1-6). Most of the Sn(II)-chelates used in nuclear medicine, with the exception of Sn(II)-DMSA, show very weak binding and the Sn(II)-complexes are not very stable in vivo. The bonding appears to be through the

TABLE 2 .--BIODISTRIBUTION IN RABBITS OF STANNOUS ION, STANNOUS HUMAN SERUM ALBUMIN (^{113}Sn-HSA), STANNOUS PYROPHOSPHATE (^{113}Sn-PP), AND STANNOUS 2,3-DIMERCAPTOSUCCINATE (^{113}Sn-DMSA), ALL WITHOUT ADDED CARRIER

Tissue	3 Hours*			24 Hours*		
	^{113}Sn-HSA (n = 15)	^{113}Sn-PP (n = 15)	^{113}Sn-DMSA (n = 6)	^{113}Sn-HSA (n = 15)	^{113}Sn-PP (n = 15)	^{113}Sn-DMSA (n = 6)
Liver	2.7 ± 1.6	2.6 ± 0.4	12.9 ± 2.4	1.7 ± 1.2	1.6 ± 0.5	19.4 ± 5.5
Kidneys	4.2 ± 1.8	2.1 ± 0.5	3.2 ± 0.5	4.0 ± 1.7	3.3 ± 0.6	1.7 ± 1.2
Skeletal muscle	3.3 ± 1.0	2.3 ± 3.5	3.6 ± 0.7	2.3 ± 1.0	2.0 ± 0.6	3.1 ± 0.7
Bone	43.2 ± 8.3	45.1 ± 8.0	4.1 ± 1.5	33.7 ± 5.6	36.9 ± 6.2	5.6 ± 1.7
Bone marrow	0.3 ± 0.2	---	0.1 ± 0.1	0.2 ± 0.1	---	0.1 ± 0.1
Blood	1.6 ± 1.0	7.2 ± 1.2	2.0 ± 0.6	0.7 ± 0.4	0.7 ± 0.1	0.9 ± 0.3
Urine	---	---	51.1 ± 5.8	---	---	68.7 ±15.2

*Organ uptake, percent of injected dose (M ± SE).

TABLE 3.--CARRIER EFFECT OF Sn-PYROPHOSPHATE ON BONE, BONE MARROW, PLASMA, AND RED BLOOD CELL DISTRIBUTION

	With Added Carrier (0.04 mCi/mg)		No Added Carrier (10 mCi/mg)	
Ratio*	3 Hours	24 Hours	3 Hours	24 Hours
Bone/blood	9.9 ± 2.3	25.9 ± 4.3	11.6 ± 2.7	34.3 ± 8.6
Bone/marrow	0.7 ± 0.2	0.8 ± 0.3	14.9 ± 3.8	20.6 ± 4.2
Bone/skeletal muscle	78.8 ±10.8	166.7 ± 9.7	37.3 ± 6.7	75.2 ±12.7
Plasma/red cell mass	4.6 ± 1.2	14.6 ± 3.4	6.1 ± 1.3	5.5 ± 1.2

*Ratios comparing equal weights.

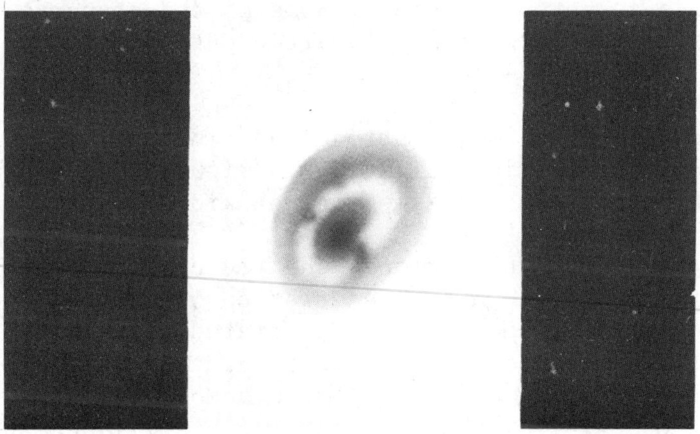

Figure 1.

Autoradiograph of transverse section of rabbit kidney three hours after I.V. injection of 1 mCi of ^{113}Sn-pyrophosphate. Concentration in renal cortex is higher than in medulla.

oxygen atom. In contrast, Sn(II)-DMSA complex is stable in vitro and in vivo.

We have systematically studied the metabolism of these water-soluble tin-chelates. Our observations suggest that the tin-chelates, with the exception of Sn-DMSA, are unstable in vivo and that tin is distributed as free metal ion and localized mostly into bone. The rest is almost entirely excreted into urine. Regardless of whether PP, HEDP, GH, HIDA, or DTPA were used as a chelating agent, their organ distribution was similar and identical to that of ^{113}Sn-HSA. Thus, the metabolism of ^{113}Sn-chelates was found to be different from that of the corresponding ^{99m}Tc-labeled complexes. Because the organ distribution of free stannous ion (^{113}Sn-HSA) is similar to that of ^{99m}Tc-HEDP and ^{14}C-HEDP, tissue distribution studies with triple-labeled diphosphonate as performed by Tofe et al could not reveal the true nature of tin-chelates and ^{99m}Tc-chelates (12, 19, 20).

With increasing amounts of carrier tin-chelates, there was an increase in uptake by liver and bone marrow resulting from the formation of a colloid with preferential localization in the reticuloendothelial system. The ^{113}Sn-DMSA chelate showed a different organ distribution than ^{113}Sn-HSA and the other chelates tested. Tin-DMSA was more stable in vivo because of the presence of two sulfhydryl groups, which participate in stable tin-chelate formation. Studies by us have shown that the association constant of ^{113}Sn-DMSA with HSA is two times higher than that of ^{113}Sn-PP.

The studies on subcellular distribution show that stannous ion is present in a protein-bound and a free form in homogenates of both normal and infarcted myocardial tissues. Uptake in infarcted tissue was higher than in normal tissue for all Sn-chelates and the free stannous ion. Activity in the infarcted tissue was always higher than in blood, indicative of fixation of the stannous ion at the necrotic tissue site. The ratio of ischemic-to-normal myocardium was about 4 and the ratio of infarcted-to-normal myocardium was about 8. Considering the similarity of calcium and stannous compounds in stability and in pharmacologic action, the localization of stannous compounds in ischemic and infarcted myocardia is not unexpected. Some of the stannous ion in the mitochondria of infarcted myocardium might participate in hydroxyapatite formation (16).

The effect of stannous ion, a strong reducing agent, on oxygen in ischemic myocardium may need further investigation in view of the known effect of stannous ion on heme oxygenase enzymes in rat kidney (18).

As found with other metal cations, the red cell membrane acted as a semipermeable barrier, slowing down the penetration of stannous ion into red cells (21). Maximally, 0.5% of the dose was associated with the total red cell mass at 30 minutes after intravenous administration. Dialysis experiments showed a 2-3 times stronger binding of ^{113}Sn-ion to hemoglobin than albumin. This would explain the observation that red cells can still be labeled with $^{99m}TcO_4^-$ ion when the plasma levels of stannous ion are already low. Though Sn(II)-PP is generally used, the similarity in behavior suggests that all the described tin-chelates with the exception of Sn(II)-DMSA could be used for red cell labeling.

Gel filtration results indicate the major tin-binding protein is hemoglobin.

TABLE 4.--TIME COURSE OF ORGAN DISTRIBUTION* OF ^{113}Sn-PYROPHOSPHATE WITHOUT AND WITH ADDED Sn-PYROPHOSPHATE CARRIER

Tissue	0.5 Hour	1 Hour	3 Hours	6 Hours	24 Hours
Liver	2.1 ± 0.5	1.2 ± 0.2	2.6 ± 0.4	1.3 ± 0.3	1.6 ± 0.5
	16.7 ± 4.8	18.3 ± 4.0	21.2 ± 6.3	27.5 ± 3.9	14.6 ± 4.3
Kidneys	2.3 ± 0.2	2.6 ± 0.3	2.1 ± 0.5	3.4 ± 0.4	3.3 ± 0.6
	3.6 ± 1.1	3.0 ± 0.7	2.0 ± 0.4	1.4 ± 0.2	1.8 ± 0.1
Muscle†	13.0 ± 1.7	7.7 ± 1.1	2.2 ± 3.5	9.7 ± 1.1	1.9 ± 0.6
	6.2 ± 0.8	4.6 ± 1.1	2.6 ± 0.4	1.5 ± 0.2	0.7 ± 0.1
Bone	46.8 ± 4.9	38.8 ± 9.8	45.1 ± 7.9	55.3 ± 9.6	36.9 ± 6.2
	32.0± 6.1	36.3 ± 5.9	37.5 ± 4.2	42.9 ± 5.5	41.0 ± 5.2
Blood	7.4 ± 1.7	5.8 ± 0.7	7.1 ± 1.2	2.4 ± 0.7	0.7 ± 0.1
	5.1 ± 1.1	1.9 ± 0.8	0.8 ± 0.2	0.6 ± 0.4	1.1 ± 0.2
Skin	6.0 ± 0.8	6.3 ± 1.1	6.2 ± 1.2	5.0 ± 0.8	1.9 ± 0.5
	4.2 ± 0.6	5.8 ± 1.2	4.8 ± 1.9	1.2 ± 0.1	1.0 ± 0.1
Urine	20.5 ± 6.0	22.8 ±10.3	16.9 ±15.8	26.3 ± 9.6	35.3 ±10.9
	19.2 ± 7.3	24.6 ±10.6	26.5 ± 9.3	17.8 ±12.3	26.6 ±11.3

*Percent of injected dose (M ± SE(in 15 rabbits; top values represent without carrier; bottom values represent with carrier.

†Skeletal muscle.

TABLE 5. (MEAN ± S.E.)* OF PERCENTILE RADIOACTIVITY IN RETENTATE AND CELLULOSE TUBING FROM EQUILIBRIUM DIALYSIS OF $^{113}Sn(II)$-CHELATES WITH AND WITHOUT HUMAN SERUM ALBUMIN AND RBC-HEMOLYSATE†

Dialysis	^{113}Sn-DMSA	^{113}Sn-PP	^{113}Sn-GH	^{113}Sn-DTPA
Control				
Retentate	0.72 ± 0.53	0.82 ± 0.42	2.40 ± 1.16	0.64 ± 0.41
Cellulose tubing	1.06 ± 0.92	5.93 ± 2.11	10.48 ± 2.26	2.61 ± 1.91
Albumin				
Retentate	35.21 ± 5.23	8.38 ± 2.26	3.71 ± 1.23	3.15 ± 1.43
Cellulose tubing	3.51 ± 1.21	2.14 ± 1.02	1.65 ± 0.76	2.73 ± 1.16
Hemolysate				
Retentate	38.69 ± 6.73	53.87 ± 9.51	45.20±10.87	41.36 ± 8.63
Cellulose tubing	4.73 ± 2.21	18.69 ± 5.26	15.23± 4.76	17.85 ± 7.94

*4 sets of experiments.
†24 hours at 4° C.

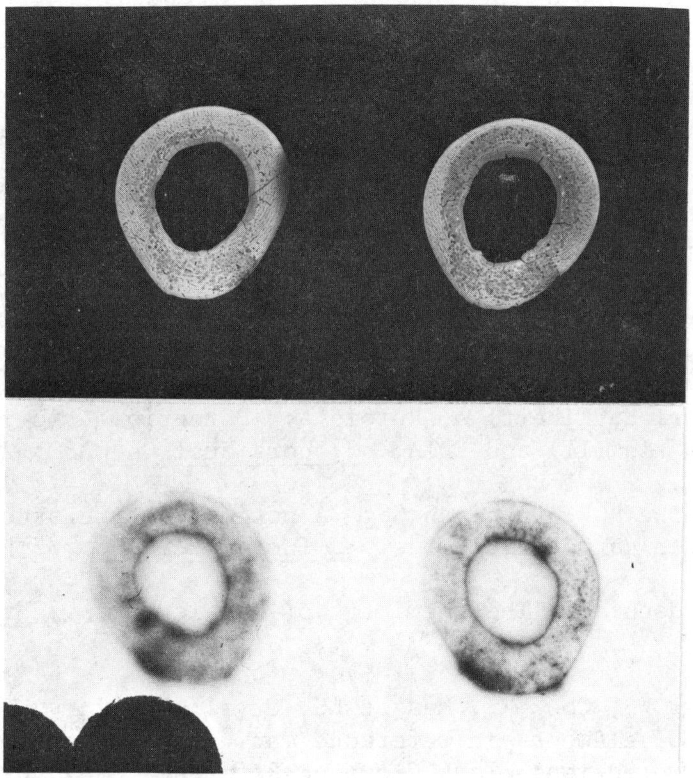

Figure 2.

Midshaft of transverse section of canine femur four hours after intravenous injection of ^{113}Sn-pyrophosphate. Top, microradiograph; Bottom, autoradiograph. Uptake is observed in areas of higher blood flow and immature bone formation.

Dewanjee et al (22) demonstrated that the β-chain of hemoglobin binds reduced Tc-99m ion. Since the iron is tightly bound in the porphyrin ring in heme, the possibility of replacing iron with cation of larger ionic radius, such as Tc-ion or Sn-ion, is small; hence, the globin fraction is the most likely site of binding of stannous ion. Thus, the study of pharmacodynamics of water-soluble tin-chelates suggests that they behave like other alkaline earth metal compounds.

ACKNOWLEDGMENTS

The authors appreciate the technical assistance of Mrs. Min-Chu Chen and Mrs. Sunanda Sengupta and the excellent typing of Mrs. Noreen Johnson. Radioactive ^{113}Sn as $^{113}SnCl_2$ was kindly supplied by New England Nuclear Corporation. The study of skeletal autoradiography was performed in collaboration with Dr. Jennifer Jowsey in the Section of Orthopedic Research.

REFERENCES

1. Mesmer RE, Irani RR: Metal complexing by phosphorus compounds-VIII, complexing of Tin(II) by polyphosphates. J Inorg Nucl Chem 28: 493-502, 1966

2. Langer HG: The chemistry of tin(II) chelates: I. Solid chelates of tin(II) with ethylenediamine tetra-acetic acid (EDTA). J Inorg Nucl Chem 26: 767-772, 1966

3. Langer HG, Bogucki RF: The chemistry of tin (II) chelates. II. The tin(II)-EDTA system in aqueous solution. J Inorg Nucl Chem 29: 495-502, 1967

4. Bottari E, Liberti A, Rufolo A: Potentiometric investigation of complexes between tin (II) and EDTA. J Inorg Nucl Chem 30: 2173-2179, 1968

5. Elbourne RGP, Buchanan GS: A potentiometric study of complexes of tin(II) and with carboxylic acids. J Inorg Nucl Chem 32: 3559-3567, 1970

6. Donaldson JD: The chemistry of bivalent tin. Prog Inorg Chem 8: 287-356, 1967

7. Snyder WS, Cook MJ, Nasset ES, et al: Summary of model values for daily balance of elements in reference man. In International Commission on Radiological Protection: Task Group on Reference Man. Oxford, Pergamon Press, Publication No. 23, 1975, p. 318

8. Charamaza O, Budikova M: Die ^{99m}Tc-markierung des Serumalbumins für die Plazenta-, Lungen-, und Leberzintigraphie. Isotopenpraxis 4: 322-324, 1968

9. Eckelman W, Richard P: Instant ^{99m}Tc-DTPA. J Nucl Med 11: 761, 1970

10. Dewanjee MK, Fliegel CF, Treves S, et al: ^{99m}Tc-tetracyclines: Preparation and biological evaluation. J Nucl Med 15: 176-182, 1974

11. Hiles RA: Absorption, distribution and excretion of inorganic tin in rats. Toxicol Appl Pharmacol 27: 366-379, 1974

12. Tofe AF, Francis MD: Optimization of the ratio of stannous tin: Ethane-1-hydroxy-1, 1-diphosphonate for bone scanning with ^{99m}Tc-pertechnetate. J Nucl Med 15: 69-76, 1976

13. Furchner JE, Drake GA: Comparative metabolism of radionuclides in mammals--XI. Retention of ^{113}Sn in the mouse, rat, monkey and dog. Health Phys 31: 219-224, 1976.

14. Loberg MD, Fields AT: Chemical structure of technetium-99m-labeled N-(2,6-dimethylphenylcarbamoylmethyl)-iminodiacetic acid (Tc-HIDA). Int J Appl Radiat Isot 29: 167-173, 1978

15. Ikeda I, Inoue O, Kurata K: Chemical and biological studies on ^{99m}Tc-DMS: II. Effect of Sn(II) on the formation of various Tc-DMS complexes. Int J Appl Radiat Isot 27: 681-688, 1976

16. Dewanjee MK, Wahner HW: Localization of stannous ion from technetium-99m labeled agents in infarcted myocardium. J Nucl Med 19: 717,1978 (abst)

17. McRae J, Sugar RM, Shipley B, et al: Alterations in tissue distribution of ^{99m}Tc-pertechnetate in rats given stannous tin. J Nucl Med 15: 151-155, 1974

18. Kappas A, Maines MD: Tin: A potent inducer of heme oxygenase in kidney. Science 192: 60-62, 1976

19. Dewanjee MK, Brueggemann P, Wahner HW: Affinity constants of technetium-99 pertechnetate and technetium-chelates with human serum albumin. Proc Second Int Symposium on Radiopharmaceuticals, Seattle, Washington, 1979

20. Deutsch E, Elder RC, Lange BA, et al: Structural characterization of a bridged ^{99}Tc-Sn-dimethylglyoxime complex: Implications for the chemistry of ^{99m}Tc-radiopharmaceuticals prepared by the Sn(II) reduction of pertechnetate. Proc Natl Acad Sci USA 73: 4287-4289, 1976

21. Sachs JR, Knauf PA, Dunham PB: Transport through red cell membranes. In The Red Blood Cell, Surgenor DM, ed, Vol. II, New York, Academic Press, 1975, pp. 644-653

22. Dewanjee MK: Binding of ^{99m}Tc ion to hemoglobin. J Nucl Med 18: 703-706, 1974

AFFINITY CONSTANTS OF TECHNETIUM-99-PERTECHNETATE AND Tc-CHELATES WITH HUMAN SERUM ALBUMIN

Mrinal K. Dewanjee, Paul Brueggemann,
and Heinz W. Wahner

Mayo Clinic and Mayo Foundation
Section of Nuclear Medicine
Rochester, MN 55901

ABSTRACT

The affinity constants (AC) characterizing the binding of pertechnetate (TcO_4-) and Tc-chelates with human serum albumin (HSA) have been determined using equilibrium dialysis and Tc-99m tracer. Tc-99-Hydroxyethylene diphosphonate (Tc-HEDP), Tc-99-pyrophosphate, Tc-99-glucoheptonate, Tc-99-dimercaptosuccinate, and Tc-99-N (dimethylphenylcarbamoylmethyl)-iminodiacetate were prepared by stannous chloride reduction in nitrogen-purged saline. Scatchard plot from TcO_4^- showed a single slope; AC was ≈ 1.96×10^3 and molar ratio of TcO_4^- and HSA was 1:1. Also, for Tc-HEDP, a single slope was obtained (AC = 7.8×10^4). For the other Tc-chelates tested, two slopes were found. The ranges of low and high AC values were $(1.9\text{-}22.9) \times 10^3$ and $(0.6\text{-}1.4) \times 10^5$ respectively. These AC values reflect the relative strength of interaction between HSA and Tc-99m-chelates and explain differences among Tc-99m-chelates in initial volume of distribution, renal clearance, and mechanism of localization in certain tumor and necrotic tissue.

INTRODUCTION

Although technetium-99m-labeled radiopharmaceuticals have been used for nearly 15 years, no systematic study has been made to investigate the biochemical interactions of the water-soluble Tc-99m-chelates with serum proteins and other macromolecules. Since proteins are present in the vascular, interstitial, and intracellular space, knowledge of such protein-binding interactions will enhance our understanding of the mechanism governing localization of these radiopharmaceuticals in health and disease. Insufficient knowledge of the chemical structure of these complex Tc-labeled compounds has contributed to the delay in such studies. Several investigators have studied the chelation of reduced Tc-99m ion with albumin and hemoglobin and of the formation of weak complexes of Tc-99m pertechnetate ion and Tc-polyphosphate with serum proteins (1-10). The mechanism of localization of Tc-labeled tetracycline and pyrophosphate has been studied using an in vitro cellular necrosis and experimental myocardial infarct model (6). These studies demonstrated that chelation of pyrophosphate with reduced Tc-ion increases serum protein binding of pyrophosphate.

Scatchard (11, 12) reported a method for evaluation of electrostatic attractions of proteins for small molecules and ions. With this technique, it is possible to calculate the stoichiometry and binding affinities of ligands and macromolecules. The side chains of amino acids on the protein molecules possess functional groups for formation of association complexes with a variety of drugs and metal chelates (13-18). Based on the law of mass action, a mathematical formulation was developed that answered the question of strength of binding in terms of an affinity constant. Berson and Yalow (13) applied Scatchard analysis to quantitate the binding between antigen and antibody in the radioimmunoassay procedure. Klotz et al (14, 15) quantitatively studied this interaction between organic anions and proteins by spectrophotometric and dialysis techniques. Equilibrium dialysis techniques were used for the determination of the affinity constant of Ga-67 and Co^{2+} with transferrin and other proteins (19-21). In an attempt to further quantitate these interactions, we have determined the affinity constants of pertechnetate and water-soluble Tc-chelates with human serum albumin, using Tc-99m, Tc-99-carrier and dialysis techniques.

MATERIALS AND METHODS

Crystalline powder of technetium-99-labeled ammonium pertechnetate, obtained from Oak Ridge National Laboratory, was dissolved in distilled water containing a drop of hydrogen peroxide. Hydrogen peroxide was then removed by careful boiling and the final volume was adjusted with nitrogen-purged saline. Anhydrous stannous chloride was obtained from M & T Chemicals Inc. and salt-poor human serum albumin (25% protein) from Armour Pharmaceutical Company. The following chelating agents were tested: hydroxyethylene diphosphonate (HEDP), pyrophosphate (PP), glucoheptonate (GH), dimercaptosuccinate (DMSA), and N-dimethylphenylcarbamoylmethyl-iminodiacetate (HIDA), all obtained from Sigma Chemical Company.

A. Preparation of Technetium-99m and Technetium-99-Chelate. Tc-99m and Tc-99-chelates were prepared separately. Anhydrous stannous chloride (6.3 mg in 0.1 ml of conc. HCl) and 20 to 125 mg of each of the chelating agents were dissolved in 1.5 ml of nitrogen-purged saline in vacutainer tubes and pH was adjusted to 6.1. Of this stock solution, 0.1 ml was mixed with 8-10 mCi of Tc-99m pertechnetate ion in 0.2-0.5 ml of N_2-purged saline. The volume was adjusted to 3 ml, and the solution was membrane filtered (0.22 μM Millipore filter paper). Five hundred micrograms of technetium-99-pertechnetate was added to the residual stannous chloride solution containing chelating agent, and pH was adjusted to 6.5 in a total volume of 5 ml. This was membrane filtered as before. The labeling efficiency of Tc-99m-chelates was checked with ITLC-silica gel and acetone. Tc-99m-pertechnetate was obtained from a Mo-99-Tc-99m Generator (Squibb Inc).

B. Sequential Dialysis. Cellulose dialysis tubing (25.5 cm x 1 cm) was boiled in distilled water at 75° C in an incubator for 30 minutes and rinsed three times with distilled water to remove sulfur containing impurities in the membrane. To determine optimum time of equilibrium in these dialysis experiments, 16 samples of each type of freshly prepared Tc-99m-chelate (14 containing 100 mg HSA and two without HSA as controls) were dialyzed in an Oxford multiple dialyzer containing 5 liters of isotonic saline solution. Each tube contained 100 μCi of Tc-99m-chelate. The final volume was adjusted to 6 ml with N_2-purged saline, and duplicate samples were withdrawn at 0.5, 1, 2, 3, 6, 18, and 24 hours after dialysis. Radioactivity in 2 ml aliquots of retentate, dialysate, and cellulose tubing was determined (22).

C. Equilibrium Dialysis and Determination of Affinity Constants. Equilibrium dialysis was performed in cellulose tubing with a constant amount of Tc-99m-chelate (100 microcuries) and variable amounts of Tc-99-chelates (5, 10, 15, 20, 40, 60, 100, 120 micrograms) in presence of 100 mg of human serum albumin. The volume of solutions containing Tc-99m-chelate, Tc-99-chelate, and HSA (0.4ml) was adjusted to 4 ml and immersed in 13 ml of nitrogen-purged isotonic saline in a screw-top plastic vial. Dialysis was performed at room temperature in a semicircular rocking motion. After 24 hours of dialysis, radioactivity in 2-ml aliquots of retentate, dialysate, and total tubing was determined with a gamma-well NaI(Tl) detector (Beckman Gamma 8000).

D. Theoretical Considerations. From the law of mass action, Equation (1), the affinity constant K_2, for the binding of a Tc-chelate with human serum albumin (HSA) can be described by the following equations:

$$\text{Tc-chelate}_{\text{free}} + \text{HSA}_{\text{free}} \underset{k_2}{\overset{k_1}{\rightleftharpoons}} \text{Tc-chelate-HSA} \qquad (1)$$

$$K = \frac{k_1}{k_2} = \frac{[\text{Tc-chelate-HSA}]}{[\text{Tc-chelate}_{\text{free}}] \times [\text{HSA}_{\text{free}}]} \qquad (2)$$

Where $[\text{Tc-chelate}_{\text{free}}]$ is the molar concentration of free Tc-chelate; [Tc-chelate-HSA] is the molar concentration of bound Tc-chelate; k_1 and k_2 are the forward and reverse rate constant of the association reaction respectively. At equilibrium, the concentration of free HSA could be written as follows:

$$[\text{HSA}_{\text{free}}] = \underset{(\text{Total})}{[\text{HSA}]} - \underset{(\text{bound})}{[\text{Tc-chelate-HSA}]} \qquad (3)$$

Rearranging (2) above, we get

$$K([HSA_{free}]) = \frac{[Tc\text{-chelate-HSA}]}{[Tc\text{-chelate}_{free}]} \tag{4}$$

Substituting for HSA_{free} in equation (4), we obtain

$$K([HSA] - [Tc\text{-chelate-HSA}]) = \frac{[Tc\text{-chelate-HSA}]}{[Tc\text{-chelate}_{free}]} \tag{5}$$

$$\text{Hence, } \frac{[Tc\text{-chelate}_{bound}]}{[Tc\text{-chelate}_{free}]} = \frac{[B]}{[F]} = K([HSA] - [Tc\text{-chelate-HSA}] \tag{6}$$

Therefore, at equilibrium, we have

$$[B]/[F] = K.\ [HSA]\ -K.\ [Tc\text{-chelate}_{bound}] \tag{7}$$

$$= \text{constant } -K.\ [Tc\text{-chelate}_{bound}] \tag{8}$$

This is the simplest form of Scatchard relationship which states that the bound-to-free ratio of a Tc-chelate is a function of the product of the affinity constant K, and the bound molar concentration of Tc-chelate at equilibrium. Knowing the amount of bound Tc-chelate, one can calculate the amount of free Tc-chelate from the following relationship:

$$Tc\text{-chelate}_{free} = \text{Total counts} - \text{bound counts}$$

$[B]/[F]$ is then obtained by dividing the bound counts by the free counts. $[B]/[F]$ is plotted on the ordinate as the response variable versus $[B]$ on the abscissa which is the dose variable.

RESULTS

Competitive Nature of Binding of Tc-Chelates with HSA and Cellulose. Figure 1 graphically demonstrates observations on competitive binding between Tc-99m-PP, HSA, and cellulose tubing. As the amount of HSA was increased from 25 mg to 250 mg, binding of Tc-99m-PP with cellulose decreased from 45% to 17%, and binding with HSA increased from 45% to 78%. Total amount of Tc-99m-PP bound to HSA and cellulose together increased from 45% to 76% and maintained a plateau level around 80%. Hence, in most of the experiments on sequential dialysis and equilibrium dialysis for the determination of AC, the amount of HSA was maintained at 100 mg.

Sequential and Equilibrium Dialysis of Tc-99m-Chelates. Sequential dialysis of Tc-99m-chelates in presence of human serum albumin showed equilibrium for all chelates at 24 hours and at room temperature (Fig. 2). The retentate value in the control experiment without HSA was smaller than with HSA and the cellulose tubing retained most of the radioactivity. The range varies from

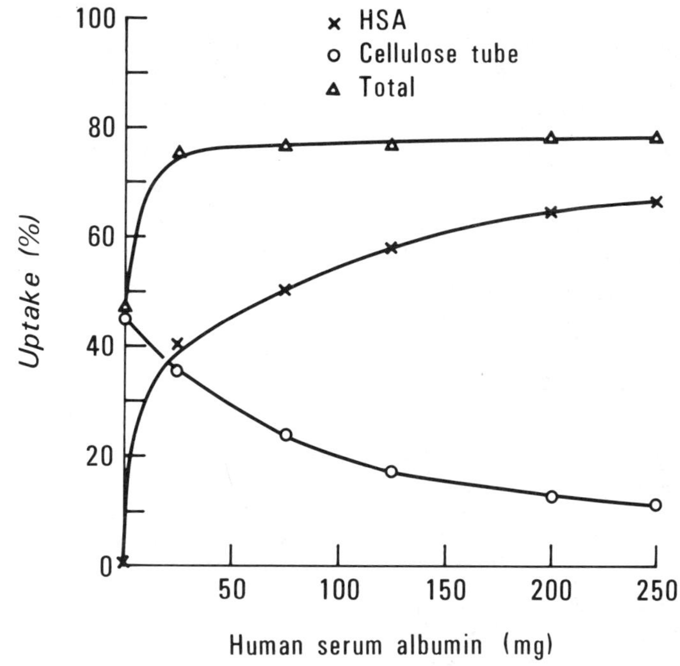

Figure 1.

Competitive binding of human serum albumin and cellulose tubing with Tc-99m-pyrophosphate.

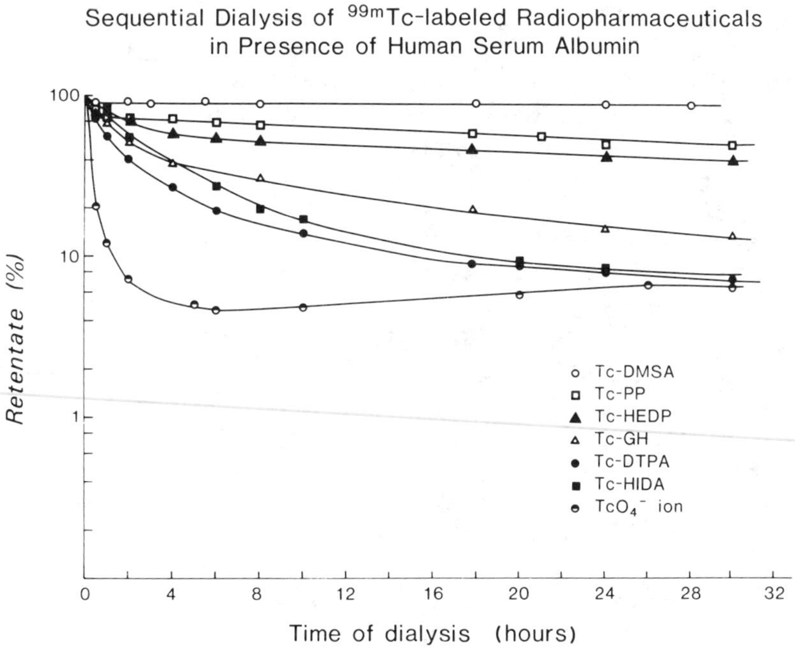

Figure 2.

Sequential dialysis of Tc-99m-chelates in presence of human serum albumin.

45.63 to 6.41%. Binding to cellulose tubing was in the following order: Tc-DMSA > Tc-PP > Tc-GH > Tc-MDP > Tc-HIDA > Tc-DTPA > TcO_4^-. The retentate in presence of HSA varied from 82.59% to 8.76%. The order of HSA binding was as follows: Tc-DMSA > Tc-PP > Tc-MDP > Tc-DTPA > Tc-HIDA > TcO_4^- (Table 1).

Affinity Constants. A Scatchard plot of the binding of Tc-99-pertechnetate with HSA showed a linear relationship (Fig. 3). At the intercept on the x-axis where [B]/[F] is zero, one obtains the maximum concentration of Tc-99-pertechnetate bound to 100 mg of HSA. A molar ratio of 1:1 was found indicating that one ion of pertechnetate binds to one molecule of HSA. Similarly, a straight line was obtained on the Scatchard plot for Tc-99-HEDP and HSA (Fig. 4). Hyperbolic curves were obtained for all other Tc-99-chelates and HSA. These curves were fitted with two slopes. The higher slope represents binding of Tc-chelates with higher avidity. Such a Scatchard plot is shown in Fig. 5. The affinity constant values for the binding of TcO_4^- ion and Tc-99-chelates with HSA are summarized in Table 2. The cellulose tube retained molecules with molecular weights of 12,000 daltons and above. The molecular weights of most of the Tc-99-chelates we have studied are less than 1000. Without the formation of association complexes with HSA (molecular weight: ≃ 67,000), most of these Tc-chelates would be dialyzable.

DISCUSSION

Tc-99m-labeled radiopharmaceuticals obtained by stannous ion reduction are mixtures of Tc-99m-chelates and Sn-chelates. In the interpretation of the results, it should be recognized that other compounds such as Sn-chelates and excess chelating agents are present and may compete with Tc-99m-chelates for binding with HSA. Tc-99m-DMSA contains several chemical species (23), and Tc-99m-PP consists of a mixture of Tc(III) and Tc(IV) pyrophosphates with distinct biological properties (24). Deutsch et al (25) described a bridged Tc-99-Sn-dimethylglyoxime complex obtained from organic solvent. Similar bridged complex formations from tracer amounts of Tc-99m-pertechnetate in aqueous solutions are unlikely. The formation of separate Sn-chelates and Tc-chelates with distinct physical, chemical, and biological properties has been demonstrated (26, 27). It is very unlikely for reduced oxo-Tc-99m-cation to bind to Sn-complex with reduced electron-density in their orbitals when free chelating agents with higher electron-density in the coordinating groups are already available in excess.

Also from biodistribution studies, speculations regarding the status of Tc and Sn complexes have been made. Tofe and Francis (28), using three labels (^{14}C-HEDP, ^{113}Sn, ^{99m}Tc), found that all three labels were deposited in about equal portions into bone and therefore concluded that a 99m Tc-Sn-HEDP complex was formed which was stable in vivo and deposited intact at the site of bone. However, ^{14}C-HEDP, ^{99m}Tc-HEDP, ^{113}Sn-HEDP, and ^{113}Sn(11)-ion, all individually localize into bone and no multinuclear complex formation has to be postulated. In support of this, Eckelman et al (29) showed that technetium-labeled DTPA could be prepared with several nonstannous reducing agents. Tin-free electolyte preparations have biological properties similar to the tin-labeled pyrophosphates (30). Loberg and Fields (27) have synthesized Tc-HIDA using $SnCl_2$ as reducing agent and shown that separate Tc-HIDA and Sn-HIDA complexes

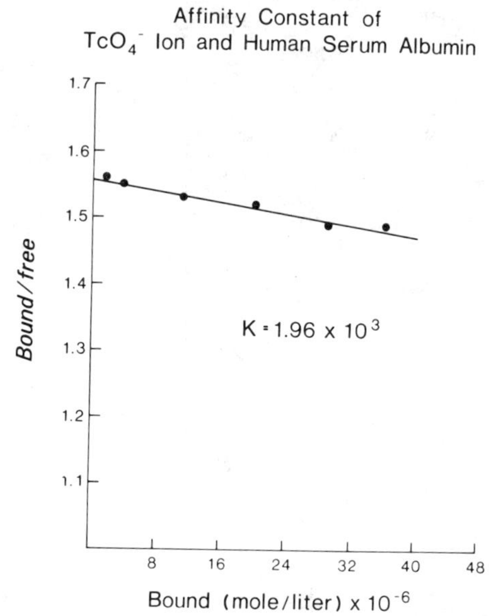

Figure 3.

Scatchard plot of the binding of human serum albumin with Tc-99-pertechnetate ion. Straight line indicates one true affinity constant value of one homogeneous binding substance and one ligand.

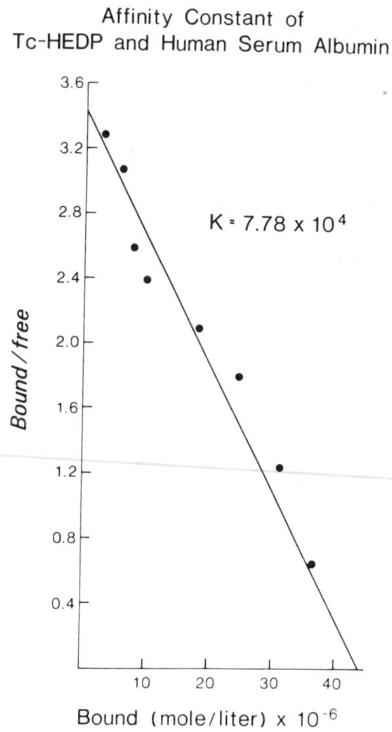

Figure 4.

Scatchard plot of the binding of Tc-99-HEDP with human serum albumin.

TABLE 1. (MEAN ± S.E.)* OF PERCENTILE RADIOACTIVITY IN RETENTATE AND CELLULOSE TUBING FROM EQUILIBRIUM DIALYSIS OF Tc-99m-PERTECHNETATE AND Tc-99m-CHELATES WITH AND WITHOUT HUMAN SERUM ALBUMIN†

Dialysis	$^{99m}TcO_4^-$	^{99m}Tc-DMSA	^{99m}Tc-PP	^{99m}Tc-MDP	^{99m}Tc-GH	^{99m}Tc-DTPA	^{99m}Tc-HIDA
Control							
Retentate	1.13 ± 0.34	3.94 ± 1.91	2.15 ± 0.76	2.63 ± 1.14	1.42 ± 0.56	1.67 ± 0.71	1.46 ± 0.61
Cellulose tubing	1.82 ± 0.42	45.63 ± 15.61	35.76 ± 9.68	21.24 ± 8.63	29.42 ± 10.46	6.41 ± 2.16	14.23 ± 4.18
Albumin							
Retentate	4.59 ± 1.89	82.59 ± 7.63	55.71 ± 6.73	41.38 ± 8.45	22.75 ± 7.25	10.86 ± 3.42	8.76 ± 2.63
Cellulose	1.02 ± 0.64	12.54 ± 1.63	11.67 ± 2.64	3.32 ± 1.19	6.29 ± 2.31	5.46 ± 1.94	5.73 ± 2.14

* 4 sets of experiments
† 24 hours at 4° C

TABLE 2. AFFINITY CONSTANTS OF TcO_4^- ION, Tc-CHELATES AND HUMAN SERUM ALBUMIN

AC	Tc-HIDA	Tc-DMSA	Tc-PP	Tc-HEDP	Tc-GH	Tc-DTPA	TcO_4^-
LAC x 10^3	4.07 ± 1.21	22.86 ± 5.42	5.50 ± 1.24	-----	6.36 ± 1.21	1.88 ± 0.93	1.96 ± 0.47
HAC x 10^5	1.38 ± 0.35	1.24 ± 0.25	0.91 ± 0.17	0.78 ± 0.25	1.13 ± 0.43	0.56 ± 0.34	-----

were formed. Dewanjee et al (26) demonstrated that the biodistribution of Tc-99m -tetracycline is different from that of Sn-119-tetracycline. Thus, evidence has accumulated that the fate of the stannous-complexes is different from the technetium-complexes and would have to be evaluated separately.

Some information of the nature of the complexes is available. Owunwanne et al (31) demonstrated the formation of oxo-technetium ion after reduction of pertechnetate with stannous ion. Although each oxo-technetium or technetium cation obtained by stannous ion reduction may carry a positive charge of 3, 4, or 5, the presence of multiple negatively charged coordinating groups of -OH, -COOH, $-PO_3H$, and -SH in the chelating molecules makes the whole Tc-complex negatively charged. Tc-complexes during electrophoresis migrate to the anode (9). The determination of partition coefficients of Tc-99m-compounds by McAfee and Subramanian (32) indicates that most of Tc-compounds are hydrophilic. At a pH of 7.4, the HSA molecule is above the isoelectric point of 5.5, and the net charge is negative. At a pH of 7.4, each molecule of human serum albumin contains approximately 100 positively charged and 100 negatively charged sites. Each individual site or a combination of sites could bind positively and negatively charged ions, molecules, and metal-chelates (14-20).

The binding of cations by albumins is less than its binding of organic and inorganic anions. Only Cu(II) and Ni(II) appear to bind at a specific site on the albumin molecule giving rise to purple Cu(II) or yellow Ni(II) compounds at neutral pH. The association constant, K, is 6×10^5 for Ni^{2+} and 9×10^6 M^{-1} for Cu^{2+}. Other serum and cellular proteins bind cations and anions; since albumin is both the most acidic and the most abundant protein in plasma, it has the greatest capacity for binding.

Most of the radiopharmaceuticals after intravenous injection bind non-specifically with serum protein in a reversible way and thus stabilize the metal complex. Both hydrolysis and oxidation of Tc-chelates are inhibited over a period of 1-2 days, once the radiopharmaceuticals are mixed even with only small amounts of serum protein. We have observed that even in the absence of ascorbic acid, Tc-MDP is stable over a period of 24 hours (Tc-99m-pertechnetate = 2.3% and $TcO(OH)_2$, $Sn(OH)_2$ colloid $\simeq$ 1-2%) in the presence of rabbit or human serum. Biodistribution experiments showed no evidence of transchelation after 24-hour incubation with HSA. This type of weak bond formation may involve H-bond, induced dipole and other electrostatic interactions, and the bond energies may be of the order of a few kilocalories per mole. Other types of bond formations are possible. The hydrophobic methylene group of HEDP reduces the protein binding of Tc-labeled diphosphonate with respect to that of Tc-labeled pyrophosphate.

Among the physico-chemical techniques for studying the protein-binding of Tc-chelates, equilibrium dialysis appears most suitable (4-8). Although the tubing material invariably binds most of the Tc-compounds, sufficient amount of protein ($\sim$100 mg) will reduce this membrane binding. Gel-filtration provides a large surface for interaction with tracer molecules, leading to nonspecific adsorption and some possibilities of transchelation (2). Electrophoresis disrupts the weak bonds formed between Tc-compounds and proteins resulting in a falsely low estimate of protein-binding. The precipitation of proteins by trichloracetic acid, ammonium sulfate solution or organic solvents was also used (16); but coprecipitation and hydrolysis of Tc-compounds at low pH may lead to misleading results. It is important to wash the dialysis tubing, check the stability of the radiopharmaceuticals during dialysis, and determine the time of equilibrium for consistent results.

In the present study, the AC values were found to be highest for Tc-DMSA and lowest for Tc-DTPA. Similar binding was also observed with bovine serum albumin and bovine gamma globulin and does not seem to be unique to HSA. There was good agreement between AC values (Table 2) of Tc-99-chelates and retentate values (Table 1) of the corresponding Tc-99m-chelates. This suggests that the results performed with carrier Tc-99 are applicable to trace experiments. An exception was Tc-HIDA. The higher AC here may be due to the presence of a benzene ring with substituents, which might bind to some hydrophobic center in the albumin molecule.

Binding of Tc-99m-pertechnetate ion with HSA is very weak (4). The negatively charged ion can only bind at one positively charged site of HSA. In contrast, the Tc-99m-chelates form multiple bonds with neighboring sites on the HSA molecule. This results in a higher binding affinity as seen from our data. When compared with Yb-169-DTPA, a metal chelate with one oxidation state and therefore only one metal complex, there are significant differences with Tc-DTPA. With Yb-DTPA less than 1-2% is retained bound to HSA. This was 7-8% for Tc-DTPA and suggests the presence of Tc ion in multiple oxidation states forming at least two Tc-chelates, one dialyzing like Yb-DTPA and the other being retained with higher affinity.

Among the Tc-compounds, the binding of Tc-99-pertechnetate with cellulose and HSA is the weakest. Similar results of the strength of binding between Tc-compounds and HSA were observed with the modified Hummel-Dreyer technique (8). Poor protein binders like Tc-99m-pertechnetate ion and Tc-99m-DTPA distribute fast into the extracellular space and are excreted faster by glomerular filtration. Strong protein binders like Tc-99m-DMSA clear slowly from the blood pool. Protein binding slows down distribution into ECF space and glomerular filtration and thus gives the bone seekers enough time to be absorbed by calcium phosphate in bone tissue. Results from the dialysis of Tc-99m-HEDP and Tc-99m-PP in presence of HSA and calcium phosphate indicate that the adsorption of Tc-99m-labeled phosphate and phosphonates by calcium phosphate is very high and works as a sink for the removal of Tc-chelates from the Tc-chelate-serum protein complex. Binding of Tc-chelates by bone matrix is higher than that of serum protein. The clinical implications of these are obvious. A poor protein binder like TcO_4^- ion is retained in the extracellular space in the tumor bed; a fraction will diffuse into tumor cells. Then, as the radioactivity from blood is reduced with time, the level of activity in the lesion will decrease concomitantly. Stronger protein binders will be retained for a longer time in the tumor bed with a higher tumor-to-normal tissue ratio. In the presence of calcification with tumor, the uptake of bone-seeking Tc-chelates will even be higher. In infarcts, other factors prevail. The extracellular space increases by a factor of 3 in the infarcted tissue and leaky (ischemic and dead) cells retain Tc-chelates by protein binding and amorphous calcium phosphate (7).

ACKNOWLEDGMENTS

The authors highly appreciate the critique of Franklyn G. Prendergast,

M.D., Ph.D., in revising the manuscript, the technical assistance of Mrs. Sunanda Sengupta and Mr. Gregory S. Anderson, and the excellent typing of Mrs. Noreen Johnson.

REFERENCES

1. Lin MS, Winchell HS, Shipley BA: Use of Fe(II) or Sn(II) along for technetium labeling of albumin. J Nucl Med 12: 204-211, 1971

2. Richards P, Steigman J: Chemistry of technetium as applied to radiopharmaceuticals. In Radiopharmaceuticals, Subramanian G, Rhodes BA, Cooper JF, Sodd VJ, eds, Society of Nuclear Medicine, pp. 23-25, 1975

3. Dewanjee MK: Binding of ^{99m}Tc ion to hemoglobin. J Nucl Med 18: 703-706, 1974

4. Hays MT, Green FA: In vitro studies of Tc-99m-pertechnetate binding by human serum and tissues. J Nucl Med 14: 149-158, 1973

5. Dewanjee MK, Prince EW: Cellular necrosis model in tissue culture: Uptake of ^{99m}Tc-tetracycline and pertechnetate ion. J Nucl Med 15: 577-581, 1974

6. Dewanjee MK, Kahn P: Mechanism of ^{99m}Tc-labeled pyrophosphate and tetracycline in infarcted myocardium. J Nucl Med 17: 639-646, 1976

7. Dewanjee MK: Function of muscle cells and myocardial imaging. First Intl. Congress on Radiopharmacology, Colombetti LG, ed, Innsbruck, Austria, CRC Press, 1978

8. Dewanjee MK, Brueggemann P: Dissociation constants of Tc-99m chelates with human serum albumin. J Nucl Med 18: 625, 1977 (abst)

9. Krishnamurthy GT, Thomas PB, Tubis, M, et al: Comparison of ^{99m}Tc-polyphosphate and ^{18}F. 1. kinetics. J Nucl Med 15: 832-836, 1974

10. Saha GB, Boyd CM: Plasma protein binding of ^{99m}Tc-pyrophosphate. Intl J Nucl Med Biol 5: 236-239, 1978

11. Scatchard G: The attractions of proteins for small molecules and ions. Ann NY Acad Sci 51: 660-672, 1949

12. Scatchard G, Coleman JS, Shen HL: The binding of small anions to serum albumin. J Am Chem Soc 79: 12-20, 1957

13. Berson SA, Yalow RS: Quantitative aspects of the reaction between insulin and insulin-binding antibody. J Clin Invest 38: 1996-2016, 1959

14. Klotz IM: Spectrophotometric investigation of the interactions of proteins with organic anions. J Am Chem Soc 68: 2299-2304, 1946

15. Klotz IM: The binding of organic ions by proteins. J Am Chem Soc 70: 2935-2941, 1948

16. Peters T: Serum albumin. In *The Plasma Proteins*, Putnam FW, ed, vol. 1, New York, Academic Press, 1975, pp. 154-159

17. Goldstein A: The interaction of drugs and plasma proteins. *Pharmacol Rev* 1: 102-165, 1949

18. Koch-weser J, Sellers EM: Binding of drugs to serum albumin, *New Engl J Med* 294: 311-316, 526-531, 1976

19. Clausen J, Edeling C-J, Fogh J: ^{67}Ga binding to human serum proteins and tumor components. *Cancer Res* 34: 1931-1937, 1974

20. Larson SM, Allen DR, Rasey JS, et al: Kinetics of binding of carrier-free Ga-67 to human transferrin. *J Nucl Med* 19: 1245-1249, 1978

21. Nandekar AK, Basu PK, Friedberg F: Co^{++} binding by plasma proteins. *Bioinorganic Chem* 2: 149-157, 1972

22. Stewart KK: Thin film dialysis. In *Advances in Protein Chemistry*, vol. 31, Anfinsen CB, Edsall JR, Richards FM, eds, New York, Academic Press, 1964, p. 135

23. Ikeda I, Inoue O, Kurata K: Preparation of various Tc-99m dimercaptosuccinate complexes and their evaluation as radiotracers. *J Nucl Med* 18: 1222-1229, 1977

24. Russell CD, Majerik JE, Cash AG, et al: Technetium pyrophosphate: A mixture?--Preparation of Tc(III) and Tc(IV) pyrophosphates and their comparative biologic properties. *Intl J Nucl Med Biol* 5: 190-195, 1978

25. Deutsch E, Elder RC, Lange BA, et al: Structural characterization of a bridged ^{99}Tc-Sn-dimethylglyoxime complex: Implications for the chemistry of ^{99m}Tc-radiopharmaceuticals prepared by Sn(II) reduction of pertechnetate. *Proc Natl Acad Sci USA* 73: 4287-4289, 1976

26. Dewanjee MK, Fliegel CF, Treves S, et al: ^{99m}Tc-tetracyclines: Preparation and biological evaluation. *J Nucl Med* 15: 176-182, 1974

27. Loberg MD, Fields AT: Chemical structure of technetium-99m-labeled N-(2,6-dimethylphenylcarbamoylmethyl)-iminodiacetic acid (Tc-HIDA). *Intl J Appl Radiat Isotope* 29: 167-173, 1978

28. Tofe AJ, Francis MD: Optimization of the ratio of stannous tin: Ethane-1-hydroxy-1,1-diphosphonate for bone scanning with ^{99m}Tc-pertechnetate. *J Nucl Med* 15: 69-74, 1974

29. Eckelman WC, Meinken G, Richards P: The chemical state of ^{99m}Tc ion in biomedical products. II. The chelation of reduced technetium with DTPA. *J Nucl Med* 13: 577-581, 1972

30. Hosain P: Technetium-99m labeled pyrophosphate: A simple and reproducible bone scanning agent. *Br J Radiol* 46: 724-728, 1973

31. Owunwanne A, Marinsky J, Blau M: Charge and nature of technetium species produced in the reduction of pertechnetate by stannous ion. J Nucl Med 18: 1099-1105, 1977

32. McAfee JG, Subramanian G: Radioactive agents for imaging. In Clinical Scintillation Imaging, Freeman LM, Johnson PM, eds, New York, Grune and Stratton, 1975, p. 36

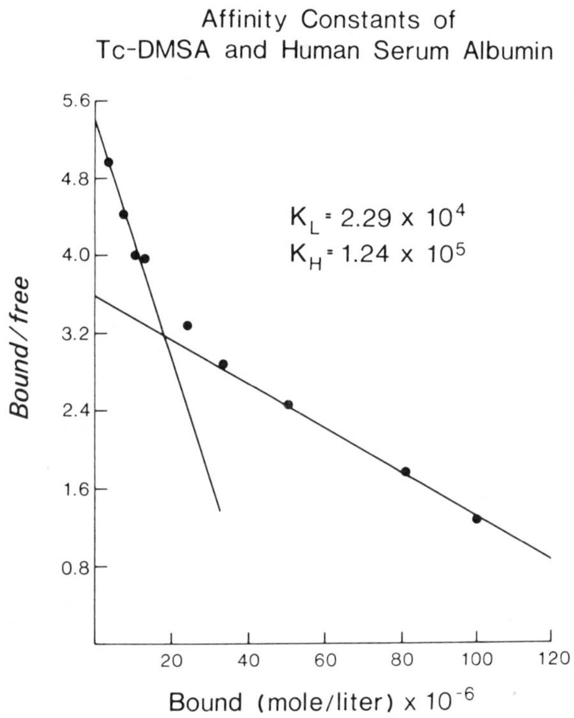

Figure 5.

Scatchard plot of the binding of human serum albumin and Tc-99-DMSA. Biphasic curve indicates presence of two slopes representing high and low affinity constants.

TRANSPORT OF ^{99m}Tc COMPLEXES THROUGH THE BLOOD-BRAIN BARRIER

Michael D. Loberg, Elizabeth H. Corder,
Anna T. Fields, and Patrick S. Callery

University of Maryland Hospital
Departments of Medicine and Medicinal Chemistry
Baltimore, Maryland 21201

ABSTRACT

Thirteen ^{99m}Tc complexes have been synthesized and used to determine the relationships between protein binding, lipophilicity and membrane transport. The lipophilicity of the ^{99m}Tc complexes was altered by adding substituents to either IDA, EDTA, DTPA or oxine; membrane transport was estimated using the brain uptake index (BUI) method. The BUI of the ^{99m}Tc complexes was found to vary directly with lipophilicity and inversely with protein binding. These results demonstrated that ^{99m}Tc-oxine derivatives are better suited for use in the development of intracellular tracers than are the ^{99m}Tc derivatives of aminopolycarboxylates.

INTRODUCTION

For a compound to reach intracellular sites of biological action, lipoidal membrane barriers must be penetrated. One of the least studied barriers to the free biological distribution of ^{99m}Tc radiopharmaceuticals is the blood-brain barrier (BBB); no existing ^{99m}Tc radiopharmaceutical is known to cross the intact BBB. The characteristics of this barrier have been the topic of numerous reviews (1-3). Only compounds with specific physicochemical properties are able to gain access to the extracellular fluid of the central nervous system (CNS). Some of the least restricted compounds for CNS penetration are those with an ideal lipid/water partition coefficient (4). When a substrate's olive oil/water partition coefficient is greater than about 0.03, measurable penetration of the BBB is observed. In addition to passive transport into the CNS based on lipid solubility, eight carrier-mediated transport systems have been described for the transport of hydrophilic metabolic substrates through the BBB (5). This study reports the synthesis of thirteen ^{99m}Tc complexes with varying degrees of lipophilicity. The membrane transport properties of each complex were assessed relative to its ability to penetrate the intact BBB.

METHODS

Complexing agents containing N-substituted carbamoylmethyliminodiacetate, substituted oxine, N-substituted DTPA, and N,N'-diesters of EDTA were synthesized (Table 1). All commercially available chemicals were used as received except for 8-hydroxyquinoline (IIA) which was recrystallized from medium boiling petroleum ether. 5-(1-Hydroxy-2-isopropylaminoethyl)-8-hydroxyquinoline (IId, quinterenol) was kindly provided by Pfeizer Inc. Derivatives IIIb and IVa were obtained from Eckelman. 2,6-Dimethylphenylcarbamoylmethyliminodiacetate (Ia) was prepared as previously reported (6). Compounds Ib, Ic and Id were prepared according to the method of Burns and co-workers (7) from the anhydride of nitrilotriacetic acid and the appropriate amine. Compound IIe was synthesized using the method of Matsumara et al (8). All complexes were prepared using the standard stannous reduction method in aqueous solution except for ^{99m}Tc-$(Id)_2$, ^{99m}Tc-IIa and ^{99m}Tc-IIb which were prepared in absolute methanol. ^{99m}Tc-IIIb and ^{99m}Tc-IVa were prepared according to the method of Eckelman (9). The mixed ligand complex, ^{99m}Tc-IbIc was synthesized in 50% yield utilizing a 7:1 molar ratio of Ib:Ic and was purified using high pressure liquid chromatography (HPLC) on a μBondapak C_{18} column using solvent programming from 16% to 70% CH_3CN/phosphate buffer (pH 6.8, 0.05M) at a flow rate of 1 ml/min.

The extent of protein binding was determined for the ^{99m}Tc complexes, ^{14}C-nicotine (NEN) and ^{14}C-iodoantipyrine (NEN), using equilibrium dialysis under a nitrogen atmosphere. All values

TABLE 1. BIFUNCTIONAL CHELATING AGENTS

I $R\text{-}NHC(=O)CH_2N(CH_2COOH)_2$

Ia -2,6-DIMETHYLPHENYL-

Ib $-(CH_2)_{10}COOH$

Ic $-(CH_2)_{10}CH_3$

Id $-(CH_2)_{17}CH_3$

II

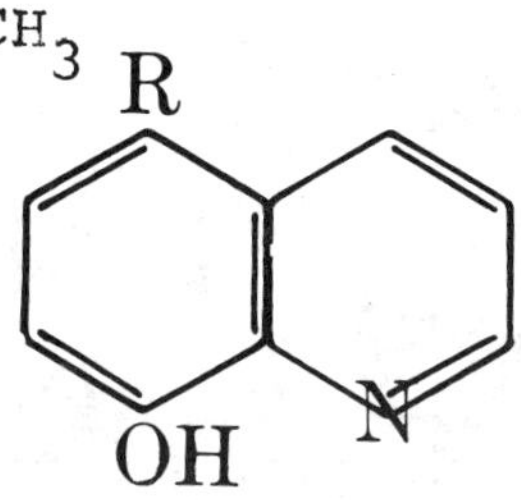

IIa -H

IIb -I*

IIc $-SO_3H$

IId $-CHOHCH_2NHCH(CH_3)_2$

IIe $-CH_2CHNH_2COOH$

III $(HOOCCH_2)_2NCH_2CH_2N(CH_2COOH)CH_2CH_2N(CH_2COOH)R$

IIIa $-CH_2COOH$

IIIb $-(CH_2)_6CH{=}CH(CH_2)_7COOH$

IV $(ROOCCH_2)(HOOCCH_2)NCH_2CH_2N(CH_2COOR)(CH_2COOH)$

IVa $-(CH_2)_{14}COOH$

*Substituted in both 5 and 7 positions

reported are for a dialysis time of 16 hours. The lipophilicity of each ^{99m}Tc complex was determined relative to its partition coefficient between octanol and 0.05M phosphate-buffered saline, pH 6.8. These partition coefficients were normalized to a constant weight of solvent. The ability of each radiolabeled test substance to penetrate the BBB was assessed in terms of its brain uptake index (BUI) according to the method of Oldendorf (10). Briefly, it consisted of injecting into the right carotid artery 5 µCi of tritium-labeled water and 10 µCi of either the ^{14}C or ^{99m}Tc-labeled compound whose BUI was to be determined. Fifteen seconds postinjection the animal was sacrificed by decapitation, the brain excised, and the relative radioactivity in 0.1 to 0.2 gram brain samples compared to that of a standard of the injectate as shown in Equation 1.

$$BUI(\%) = \frac{^{99m}Tc\text{-}R(Br)/^{99m}Tc\text{-}R(Std)}{^{3}H_2O(Br)/^{3}H_2O(Std)} \times 100 \qquad (1)$$

In this manner, the ability of the test substrate to cross the BBB is compared to that of the highly diffusible internal standard, ^3H_2O. The radiochemical purity of each of the ^{99m}Tc complexes was determined using HPLC (µBondapak C_{18}, silica, and µCN columns) and by paper chromatography in saline, chloroform and methanol.

RESULTS

Paper chromatography indicated that $^{99m}TcO_4^-$ and ^{99m}Tc-colloid were less than 2% of the total radioactivity in all cases. However, using HPLC, the presence of more than one ^{99m}Tc complex was indicated for ^{99m}Tc-IIIb, ^{99m}Tc-IVa and ^{99m}Tc-IIe. ^{99m}Tc complexes of IIIb and IVa were determined to each yield three radiochemicals with 60% and 80% of the total radioactivity attributed to one complex, respectively. Figure 1a,b shows an example of the multiple ^{99m}Tc complexes obtained from ^{99m}Tc-IIe. The HPLC chromatograms were obtained 5 minutes and 20 hours after preparation. The relative abundance of each complex was seen to change with time. The radiochemical purity of each of the other ^{99m}Tc complexes was invariant out to 3 hours postpreparation. The mixed ligand, ^{99m}Tc-IbIc (Figure 2) which was prepared in pure form by collection of the eluent fraction from the HPLC, showed maximum stability when the HPLC eluent was collected in an evacuated vial containing the lyophilized stannous complexes of Ib and Ic in the same 7:1 ratio used in preparing the initial radiochemicals. After 5 hours of storage, the initially pure mixed ligand was found to contain 85% ^{99m}Tc-IbIc, 6% ^{99m}Tc-$(Ib)_2$, and 9% ^{99m}Tc-$(Ic)_2$.

Figure 3 is a graph of BUI for each ^{99m}Tc complex, ^{14}C-nicotine, and ^{14}C-iodoantipyrine plotted against its partition coefficient. In general, as the lipophilicity of the compounds

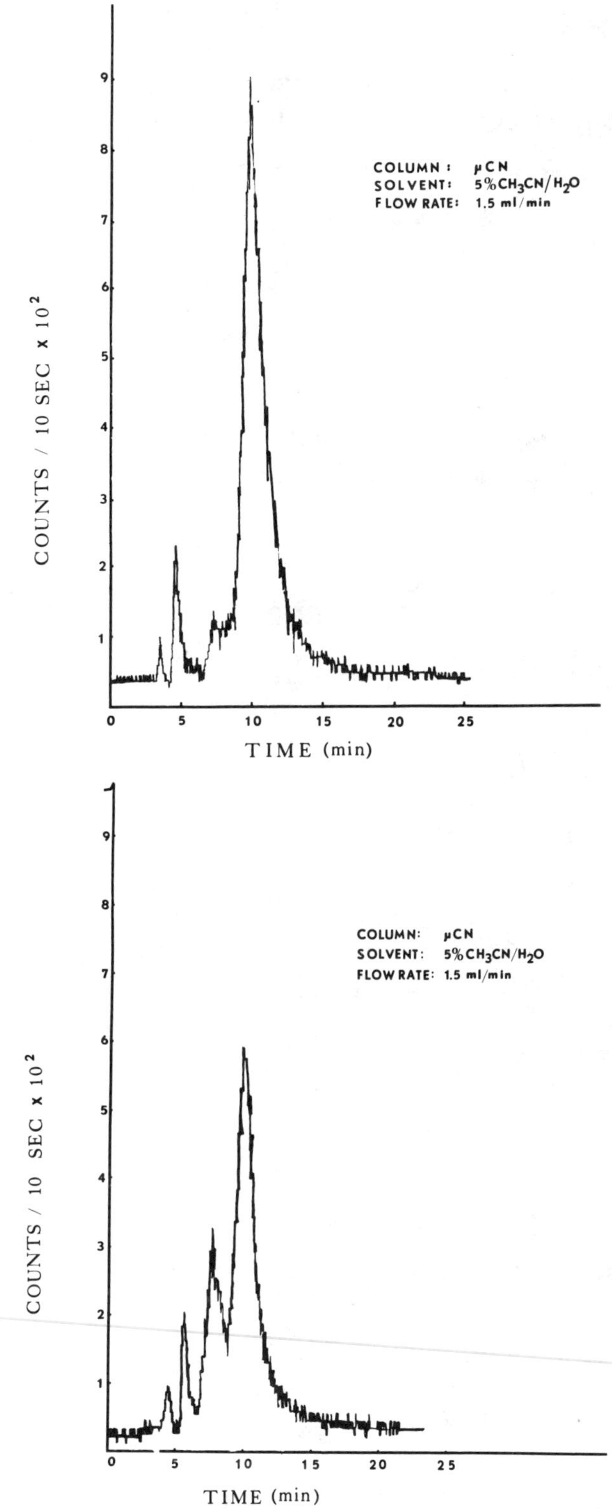

Figure 1.

High pressure liquid radiochromatograms resulting from the reaction of $^{99m}TcO_4^-$ with the stannous complex of IIe. Chromatograms were obtained at (A) 5 min. and (B) 20 hrs. postpreparation.

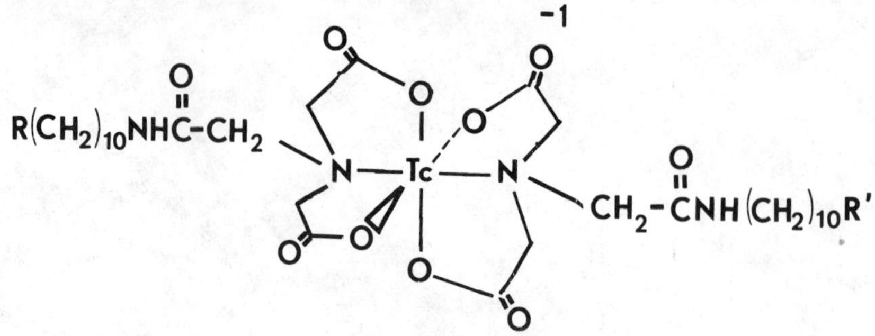

Tc-(Ib)$_2$: R = R' = COOH
Tc-(Ic)$_2$: R = R' = H
Tc-(IbIc): R = H; R' = COOH

Figure 2.

Presumed radiochemical structures of ^{99m}Tc-(Ib)$_2$, ^{99m}Tc-(Ic)$_2$, and ^{99m}Tc-IbIc.

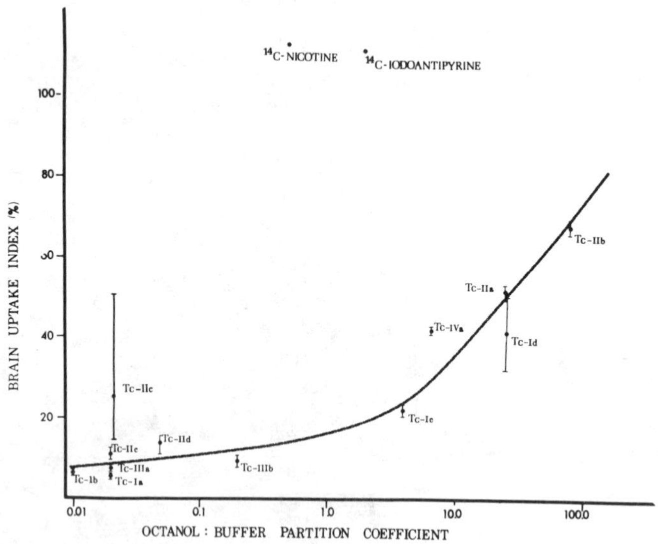

Figure 3.

Variation in brain uptake index (BUI) as a function of substrate lipophilicity for a series of ^{99m}Tc complexes.

increased, as evidenced by increasing partition coefficients, so too did their protein binding and BUI. All of the ^{99m}Tc complexes, irrespective of which complexing agent was used, lie on the same curve, whereas ^{14}C-iodoantipyrine and ^{14}C-nicotine lie on the same curve. ^{99m}Tc-IIe was found to also lie substantially off the curve. The extremely large standard deviation associated with this compound is attributed to its radiochemical impurities and instability. It would appear that only one of the ^{99m}Tc complexes shown in Figure 1 is capable of crossing the BBB and that the relative percentage of this compound in the reaction mixture is a function of the method of synthesis and time postpreparation. Hydrophilic compounds exhibited BUI of approximately 6%, a value that can be assumed to reflect the vascular space within the brain.

DISCUSSION

The BBB is a complex physiological system which limits the exchange between the cerebral capillary fluid and the brain extracellular fluid. Transport across the BBB is accomplished by passive diffusion through the membrane's phospholipid bilayer or by facilitated diffusion via binding to proteins within the membrane. This work reports the synthesis of lipophilic ^{99m}Tc-labeled complexes with octanol/saline partition coefficients as high as 78.5. Further, it has shown that these radiochemicals can cross the intact BBB in proportion to their degree of lipophilicity. The graph of BUI versus partition coefficient, shown in Figure 3, has the same general shape as that determined by Oldendorf for ^{14}C-labeled drugs and nutrients. Since all the ^{99m}Tc complexes except ^{99m}Tc-IIe fall on the same curve irrespective of which of the four general chelating agents was used in preparing the radiolabeled compound, it appears that membrane transport by passive diffusion is dependent on lipophilicity and is independent of each complex's individual stoichiometry and charge distribution.

The deviation between the ^{99m}Tc complexes and both ^{14}C-nicotine and ^{14}C-iodoantipyrine may be due to the increased protein binding of the ^{99m}Tc complexes. Preincubation of ^{99m}Tc-IIa with albumin prior to carotid injection decreased its BUI(%) from 58.3% to 5%. This is in contrast to what was observed for ^{14}C-iodoantipyrine in which no change in BUI was observed when it was preincubated with protein prior to injection. The development of a totally diffusible tracer labeled with ^{99m}Tc therefore requires not only sufficient lipophilicity for membrane transport but also reduced protein binding to allow for adequate concentration of the free drug in solution.

Table 2 contains a summary of the data for the ^{99m}Tc-labeled derivatives of I, carbamoylmethyliminodiacetate. In general, it was found that these derivatives required a long N-substituted alkyl chain in order to offset the negative charge associated with chelate bond itself. Since it is necessary to add large, highly lipophilic substituents to derivatives of I in order to obtain cellular transport, it would appear that such derivatives have limited utility in serving as ^{99m}Tc bifunctional radiopharmaceuticals. This statement can be generalized to the other aminopolycarboxylates shown in Table 3 in which, once again, long alkyl substituents are required for transport across the BBB. While the radiochemical structure of the ^{99m}Tc complexes of III and IV have yet to be reported, the significance of such information can be appreciated by a comparison of the partition coefficients and BUI for ^{99m}Tc-IIIb and ^{99m}Tc-IVa. ^{99m}Tc-IVa has a much higher partition coefficient and corresponding brain uptake than ^{99m}Tc-IIIb, a discrepancy that would have been difficult to predict upon an examination of the chelating agents alone.

The ^{99m}Tc-labeled derivatives of II (Table 4) offer promise for the development of new intracellular tracers. ^{99m}Tc-IIa and ^{99m}Tc-IIb were highly lipophilic and extracted by the brain, whereas the substitution of either a negative or a positive charge in the 5 position served to greatly decrease both lipophilicity and brain uptake. The observed lipophilicity of ^{99m}Tc-IIa is similar to that found for the cell-labeling agent ^{111}In-IIa. However, the ^{99m}Tc complexes with II have been observed to be inert with respect to ligand exchange reactions, unlike the labile ^{111}In-II complexes (11). The ^{99m}Tc complex of IIe is particularly noteworthy since this compound appears to cross the BBB despite its low lipophilicity. Similar brain uptake properties have been observed for Oldendorf for substances undergoing carrier-mediated uptake into the CNS. Studies involving ^{99m}Tc complexes with IIe are complicated by the presence of multiple radiochemicals formed between ^{99m}Tc and IIe and by the fact that the relative abundance of these compounds changes with time.

SUMMARY

This work has demonstrated that ^{99m}Tc complexes can cross the BBB in proportion to their lipophilicity. It has been shown that many of the ^{99m}Tc-labeled derivatives of fatty acids developed for imaging the myocardium (12,13) are incapable of doing so because of restrictions in their intracellular transport. The necessity of substituting long lipophilic alkyl chains onto the aminocarboxylates in order to obtain intracellular transport will diminish their usefulness as bifunctional radiopharmaceuticals for the examination of nonexcretory organs. The feasibility of developing a nonvolatile, diffusible ^{99m}Tc-labeled tracer has been demonstrated and the problem reduced to a physicochemical one of retaining lipophilicity while minimizing protein binding. The significance of lipophilicity in the membrane transport of ^{99m}Tc compounds has been demonstrated and new mechanisms are proposed for developing radiopharmaceuticals specific for nonexcretory organs.

TABLE 2. SUMMARY OF ^{99m}Tc COMPLEXES WITH SUBSTITUTED CARBAMOYLMETHYLIMINODIACETATES

COMPLEXING AGENT	OCTANOL / BUFFER	PROTEIN BINDING (%)	BUI (%)
$^{99m}Tc-(Ia)_2$	0.02	51.3	4.8
$^{99m}Tc-(Ib)_2$	0.01	80.9	6.2
$^{99m}Tc-(Ic)_2$	4.0	87.6	22.2
$^{99m}Tc-(Id)_2$	25.7	84.1	41.6

TABLE 3. SUMMARY OF ^{99m}Tc COMPLEXES WITH AMINOPOLYCARBOXYLATES

COMPLEXING AGENT	OCTANOL / BUFFER	PROTEIN BINDING (%)	BUI (%)
$^{99m}Tc-(Id)_2$	25.7	84.1	41.6
^{99m}Tc-IVa	6.7	93.9	42.4
^{99m}Tc-IIIb	0.2	93.5	7.0

TABLE 4. SUMMARY OF ^{99m}Tc SUBSTITUTED OXINES

COMPLEXING AGENT	OCTANOL / BUFFER	PROTEIN BINDING (%)	BUI (%)
^{99m}Tc-IIa	23.6	96.8	53.6
^{99m}Tc-IId	0.05	64.2	13.9
^{99m}Tc-IIb	78.5	98.6	68.6
^{99m}Tc-IIe	0.02	80.8	24.3

REFERENCES

1. Rapoport SI: Blood Brain Barrier in Physiology and Medicine. New York, Raven Press, 1976.

2. Rall DP, Zubrod CG: Mechanisms of drug absorption and excretion: Passage of drugs in and out of the central nervous system. Ann. Rev. Pharmacol. 2:109-128, 1962.

3. Barlow CF, Lorenzo AU: Studies on membrane permeability, hematoencephalic permeability. In Radionuclides in Pharmacology, Cohen Y (ed.) Pergamon, New York, 1971, pp 539-563.

4. Oldendorf WH: Lipid solubility and drug penetration of the blood-brain barrier. Proc. Soc. Exp. Biol. Med. 147:813-816, 1974.

5. Pardridge WM, Oldendorf WH: Transport of metabolic substrates through the blood-brain barrier. J. Neurochem. 28:5-12, 1977.

6. Callery PS, Faith WC, Loberg MD, et al: Tissue distribution of technetium-99m and carbon-14 labeled N-(2,6-dimethylphenylcarbamoylmethyl)iminodiacetic acid. J. Med. Chem. 19:962-964, 1976.

7. Burns HD, Sowa DT, Marzilli LG: Improved synthesis of N-(2,6-dimethylphenylcarbamoylmethyl)iminodiacetic acid and analogs. J. Pharm. Sci. 67:1434, 1978.

8. Matsamura K, Kasai T, Tashiro H: 3-[8'-Quinolinol-(5')-yl] alanine and related compounds. Bull. Chem. Soc. Jap. 42:1741-1743, 1969.

9. Eckelman NC, Karesh SM, Reba RC: New compounds: Fatty acid and long-chain hydrocarbon derivatives containing a strong chelating agent. J. Pharm. Sci. 64:704-706, 1975.

10. Oldendorf WH: Measurement of brain uptake of radiolabeled substances using a tritiated water internal standard. Brain Res. 24:372-376, 1970.

11. Loberg MD, Fields AT: Unpublished results.

12. Karesh SM, Eckelman WC, Reba RC: The biological distribution of chemical analogs of fatty acid and long chain hydrocarbons. J. Pharm. Sci. 66:225-228, 1977.

13. Loberg MD, Fields AT, Porter DW: Radiochemically pure Tc-99m labeled fatty acid analogs. J. Nucl. Med. 19:717, 1978.

RADIOPHARMACEUTICALS IN EVALUATION OF KIDNEYS

H. S. Winchell, M. D., Ph. D.

Medi-Physics, Inc.
Emeryville, California 94608

ABSTRACT

The kidney may be thought of as a system which first filters low and medium molecular weight water-soluble materials from the blood and then either pumps or allows diffusion of such materials from the filtrate back into the blood after their passage through a "chromatographic-like column" contained within the renal tubules. Additionally, water-soluble materials in the blood may diffuse or be pumped in the opposite direction through the "chromatographic column" in the tubules to reach the filtrate. Materials whose renal handling is limited to filtration processes can be employed to measure glomerular filtration (e. g., radioinulin and possibly radiocations chelated by EDTA or DTPA). Materials whose renal handling is by a combination of filtration and quantitative pumping or diffusion from the blood to the filtrate, and for which retention in the chromatographic-like system of the tubules is negligible, can be employed for measuring renal plasma flow (e. g., radioiodohippurate). Materials which have prolonged retention times in the renal tubular "chromatographic system" can be used to evaluate renal anatomy and to infer aspects of renal physiology (e. g., Tc-99m DMSA).

OUTLINE

I. Introduction

II. Pertinent Physiology of Renal Excretory Function

III. Chemical Nature of Materials Excreted or Retained by the Kidney

IV. Radiotracer Materials Excreted or Retained by the Kidneys

   A. Materials Excreted Without "Retention"

   B. Materials Significantly "Retained" by the Kidneys

V. Clinical Information Concerning Renal Anatomy or Physiology Which Can Be Obtained Through the Use of Radiopharmaceuticals

INTRODUCTION

The kidney serves both as an endocrine and an excretory organ. While its endocrine functions (e.g., renin and erythropoietin production) have been studied in vitro employing radioimmunoassay techniques, thus far radiopharmaceuticals have not been developed for in vivo assessment of such endocrine function. Consequently, the present discussion of the use of radiopharmaceuticals in evaluation of the kidneys is limited to radiopharmaceuticals whose handling by the kidneys is related to that organ's excretory function.

The kidney is the principal organ normally involved with excretion of water and water-soluble materials from the body. As a rule, all parenterally administered highly water-soluble materials, or materials whose metabolites are highly water soluble, will be excreted by the kidneys and found in the urine. The major exception to this rule occurs where renal reabsorption mechanisms exist for an agent such that renal reabsorption approximates renal excretion, and net urinary excretion is negligible (e.g., glucose, amino acids). To a degree, all parenterally administered radiopharmaceuticals will be excreted by the kidneys, and, to a greater or lesser extent, the pattern and amount of such renal excretion can be used to evaluate renal function. The present discussion briefly outlines the physiology and biochemistry involved in renal handling of radiopharmaceuticals and relates the information gathered through observation of radiopharmaceutical passage through, or retention in, the kidney to questions arising in clinical practice.

PERTINENT PHYSIOLOGY OF RENAL EXCRETORY FUNCTION

The nephron, consisting of the glomerulus and its associated renal tubular system, is the primary anatomical element of renal excretory function. Arterial blood entering the kidney normally traverses the glomerulus where it is subjected to ultrafiltration processes which result in passage of water and water-soluble low and medium molecular weight materials from the blood to the lumen of the nephron. Cellular elements and high molecular weight elements are normally retained in the blood. Blood leaving the glomerulus perfuses the renal tubule where diffusion or cellular transport may move molecules from the blood to the lumen of the nephron. Conversely, materials in the lumen of the nephron may diffuse or be transported across the renal tubule from the lumen of the nephron to the blood. Certain molecules such as inulin are cleared from the blood solely by glomerular filtration and are not subjected to significant net tubular secretion or reabsorption. Quantitative evaluation of the excretion of such materials can be quantitatively related to glomerular filtration processes. Other molecules are cleared from the blood by a combination of glomerular filtration and tubular secretion. Where clearance of a material from the blood is essentially complete during a single passage through the kidney, quantitative evaluation of its clearance rate can be employed to quantitate renal plasma flow (e.g., hippuric acid and its analogs). Most molecules are cleared from the blood by a variable combination of glomerular filtration and trans-tubular transport and diffusion, and it is difficult to quantitatively relate the clearance of such materials to isolated aspects of kidney function. Additionally, some materials in this latter group may be retained on binding sites in their passage across renal tubular cells (e.g., cationic mercury and certain organomercurials). The pattern, degree, and duration of such retention may provide some information concerning the nature and distribution of such binding sites.

CHEMICAL NATURE OF MATERIALS EXCRETED OR RETAINED BY THE KIDNEY

Those materials which appear to be cleared by the kidneys through glomerular filtration alone are generally of low or medium molecular weight, do not bind strongly to blood elements or materials in tissues, are biochemically inert, do not interact with any renal tubular transport mechanisms, and are not bound by any of the "chromatographic-like elements" of the renal tubules. With these materials, to the degree that diffusion across the renal tubules occurs, the diffusion from blood to nephron lumen approximates diffusion from nephron lumen to blood, and net diffusion in either direction is small. Inulin, a biochemically inert polysaccharide of molecular weight of approximately 5000, is the classical example of a material in this category. Other, less clear, examples include heavy metal cations chelated to EDTA and DTPA. EDTA and DTPA themselves can be classified as organic acids and one would anticipate that they would be subjected to the organic acid renal tubular transport system. That this apparently is not the case for heavy-metal-cation EDTA and DTPA moieties is attested to by the observation that their renal clearance values approximate those for glomerular filtration alone. Consequently, any renal tubular transport or trans-tubular diffusion which may be occurring does not appear to effect net excretion. This may be due, in part, to occupation of the coordination shell of the heavy metal cation by the electronegative carboxylic acid groups of the chelate, thus limiting their availability to interact with the organic acid transport systems of the renal tubules.

Those materials which are cleared "quantitatively" in a single passage through the kidney are characterized by a high affinity for efficient transport across the renal tubule from blood to nephron lumen by renal tubular transport mechanisms operating in that direction. Exemplary of this category is hippuric acid and its analogs which efficiently interact with the organic acid excretory mechanism of the renal tubule.

Those materials which are cleared from the blood by glomerular filtration but do not normally appear in the urine are characterized by a high affinity for efficient transport across the renal tubule from nephron lumen to blood by renal tubular transport mechanisms operating in that direction. Exemplary of this category is glucose.

Those materials which show prolonged retention in the kidneys have a high affinity for binding sites in the renal tubules. These materials gain access to the renal tubular binding sites by diffusion or transport from the nephron lumen following clearance from the blood during glomerular filtration or by diffusion or transport from the blood perfusing the renal tubules. The nature of binding sites ("chromatographic-like elements") in the renal tubules may be diverse, but include thiol and disulfide moieties having high affinity for cationic forms of heavy metals. In the case of most heavy metals, the renal tubular binding sites for such cations is such that marked concentration and prolonged retention in the kidneys characteristically occur unless the cations are administered as strong chelates (i.e., very low dissociation constant of the metal chelate). Where one wishes to purposely maximize localization of a heavy metal cation in the kidney, the heavy metal can be administered in a form which limits its binding in non-renal tissue but allows for intrarenal retention, with or without dissociation, by binding to renal tubular sites. Examples of this are chlormerodrin and Tc-99m DMSA.

Most materials are handled by the kidney through a complex combination of glomerular filtration and trans-tubular diffusion and transport. Consequently, quantitative interpretation of the handling of such materials by the kidney is difficult.

RADIOTRACER MATERIALS EXCRETED OR RETAINED BY THE KIDNEY

Materials Excreted without "Retention"

As can be inferred from that stated previously, this group includes metabolically inert materials such as inulin labeled with suitable radiotracers (e.g. ^{3}H, ^{14}C, radioiodine). It also includes suitable radiometals chelated by EDTA or DTPA (e.g., Tc-99m, In-113m, or In-111). This group exemplifies those agents whose renal excretion reflects glomerular filtration processes.

Radiolabeled hippuric acid analogs (e.g., radioiodine-labeled orthoiodohippuric acid) exemplify agents whose renal excretion reflects total renal plasma flow.

Materials Significantly "Retained" by the Kidney

This group includes all filterable polyvalent radiocations which are not strongly chelated or whose chelate form interacts with renal tubular binding sites. This group is very large and includes most free polyvalent radiocations or those chelated by weak chelating agents (e.g., acetate, amino acids, etc.) as well as those chelated by somewhat stronger chelates (e.g., citrate, ascorbate, glucoheptonate, dimercaptosuccinate). It excludes certain radiocations bound by exogenously administered strong chelating agents (e.g., EDTA or DTPA) or endogenous strong chelating agents (e.g., transferrin). In certain pathological states, the kidney may be so altered as to allow large quantities of chelated radiocations to gain greater access to tubular binding sites (e.g., increased vascular permeability to transferrin-bound radiocations, allowing a larger amount of radiocation to accumulate in the kidney in the area of altered vascular permeability), or tubular binding sites may be so altered as to increase their affinity for radiocations or certain of their chelated forms (e.g., Tc-99m bone localizing agent accumulation in areas of kidney suffering from radiation damage).

In addition to chelated and nonchelated radiocations, other materials may be retained during their passage across the renal tubular cell (e.g., certain peptides and peptide fragments). However, none of the non-radiocation, non-organometallic materials showing renal retention have proven useful in nuclear medicine studies thus far.

CLINICAL INFORMATION CONCERNING RENAL ANATOMY OR PHYSIOLOGY WHICH CAN BE OBTAINED THROUGH THE USE OF RADIOPHARMACEUTICALS

The size, shape, and location of functioning renal tissue can be obtained from imaging the in vivo distribution of practically any radionuclidic material which is concentrated and excreted by the kidneys. The efficacy of the agent employed for this purpose will depend upon the degree of concentration of the agent in the kidney at the time of the study and the compatibility of the radiation emitted by the radionuclide employed with the imaging device being used.

Size, shape, and location of functioning renal tissue is best assessed employing agents which show a high degree of concentration and retention in the renal parenchyma.

Abnormalities in the form and distribution of functioning renal tissue are often present in neoplastic, cystic, and inflammatory disease affecting the kidneys.

The anatomy and function of the renal drainage system can be inferred from serial imaging of the passage of radionuclidic materials excreted by the kidney into the urine. This is best assessed employing agents which are quantitatively cleared from the blood and excreted in the urine.

Measurement of specific aspects of renal function can be best obtained employing radiotracers whose handling by the kidney can be related to a single process (e.g., glomerular filtration, renal plasma flow). However, many aspects of renal function can be inferred from analysis of the temporal distribution of agents which are retained in the kidney but whose retention is a function of multiple parameters. For example, differential renal function can, indirectly, be inferred from the differential uptake of radio-organomercurials and certain Tc-99m chelates in the two kidneys.

PHYSIOLOGIC BASIS AND UTILITY OF MYOCARDIAL PERFUSION IMAGING

Gerald M. Pohost,
Kenneth A. McKusick, H. William Strauss

Massachusetts General Hospital
Depts. of Cardiology and Nuclear Medicine
Boston, Massachusetts

Myocardial imaging with $^{201}Tl^{+}$ allows improved detection of coronary artery disease. The distribution of $^{201}Tl^{+}$ in heart muscle is related primarily to blood flow and cellular integrity. Serial imaging of $^{201}Tl^{+}$ heart muscle distribution demonstrates gradual fill in of initial $^{201}Tl^{+}$ defects which are due to ischemia (reduced blood flow and/or increased metabolic demand), and persistence of defects which are due to acute infarct or scar. Improvement in radionuclide detection of coronary artery disease will likely occur over the next decade through: 1) computer approaches to quantification of serial changes in $^{201}Tl^{+}$ distribution; 2) gamma camera tomographic approaches using multiaperture collimators; 3) addition of the multigated acquisition blood pool imaging approaches; 4) application of positron transverse section emission tomography; and 5) development of other radiopharmaceuticals.

INTRODUCTION

Approximately one third of all deaths in the United States are related to disease of the coronary arteries. Coronary artery disease is almost always related to the disease, atherosclerosis, which is due to deposition of cholesterol or plaque in the lining of the coronary arteries, the vessels which supply the heart muscle with its life sustaining blood flow. When enough plaque is deposited to produce blood flow reduction, myocardial ischemia can occur. Ischemia is a state in which the blood supply to a segment of myocardium cannot keep up with metabolic demands. This can occur in two ways: 1) blood flow reduction and; 2) increased metabolic demands through increased heart work such as that associated with exercise. Infarction or death of a segment of myocardium occurs when the blood flow is inadequate to sustain life.

MONOVALENT CATIONIC MYOCARDIAL IMAGING RADIOPHARMACEUTICALS

The major use of the myocardial imaging radiopharmaceuticals is to aid in the detection of coronary artery disease. Thallium-201 is the most widely applied myocardial imaging agent and the following discussion will be restricted to the pharmacophysiology of this agent. Although $^{201}Tl^+$ is a transitional metallic element in Group 111A of the periodic table and has atomic weight and number higher than that of potassium, it has properties that closely resemble potassium in biological systems. Like potassium, Tl^+ ion can activate the sodium-potassium ATPase system (1).

Other monovalent cationic radionuclides which have been employed for myocardial imaging with the gamma camera ($^{43}K^+$, $^{81}Rb^+$, and $^{129}Cs^+$) emit relatively high energy spectra which require specialized collimators. Accordingly, image quality suffers due to scatter leading to sacrifice of resolution and contrast. ^{201}Tl has a relatively low energy spectrum with 95% abundance of mercury K X-rays (67-82 KeV) and two higher energy peaks with 12% or less abundance of 135 KeV and 167 KeV.

PHYSIOLOGIC BASIS FOR THE UTILITY OF $^{201}Tl^+$ FOR MYOCARDIAL IMAGING

Serial imaging after intravenous injection of $^{201}Tl^+$ appears to be capable of detecting ischemia and infarction, and differentiating between them. Initial distribution of Tl^+ is related to: a) regional perfusion; and b) cellular extraction efficiency of the various organs. In the resting state there is obvious Tl^+ uptake in the splanchnic viscera, but very little in the skeletal muscle. During exercise, blood flow increases in the exercising muscle and to the heart in proportion to the amount of work performed, while splanchnic blood flow remains relatively low. Tl^+ uptake appears to shift from the splanchnic organs to the skeletal muscle. After intravenous injection, blood levels of Tl^+ reach a peak rapidly and then drop within approximately 10 to 15 minutes to a low level. Subsequently, blood Tl^+ decreases very slowly. In a canine model in which a coronary artery was ligated, initial $^{201}Tl^+$ distribution has been shown to be closely related to regional blood flow as defined by microspheres (2). Extraction of Tl^+ from the blood by the myocardium may be reduced by acidosis, hypercapnea, digitalis and propranolol and may be increased by insulin, glucose, and isoprotenenol (3). Myocardial extraction efficiency for $^{201}Tl^+$ is approximately 85% under basal conditions. It is this very efficient extraction of Tl^+ by the myocardium which causes the initial distribution of the agent to be so closely related to blood flow. Work from our laboratory demonstrates that myocardial $^{201}Tl^+$ distribution changes over time (4). In anesthetized dogs subjected to 20 minutes of coronary occlusion and 100 minutes of reperfusion, $^{201}Tl^+$ was given during the occlusion phase. Comparison of $^{201}Tl^+$ activity with activity of microspheres administered during occlusion revealed that $^{201}Tl^+$ activity in the inner half of the myocardium in the central ischemic zone was 56% of normal, while regional blood flow at the time of $^{201}Tl^+$ administration was reduced to 8% of normal blood flow. These data suggest that redistribution of $^{201}Tl^+$ occurs into the ischemic zone. A control group of animals with persistent occlusion revealed that the distribution of $^{201}Tl^+$ two hours after occlusion was not significantly different from that of microspheres. Serial gamma camera images obtained after $^{201}Tl^+$ administration demonstrated that early defects present on images made during occlusion filled in completely after two to four hours of reperfusion.

Disappearance of defects is not only related to active uptake of $^{201}Tl^+$ by the ischemic zone but also to washout of $^{201}Tl^+$ from normal myocardial zones. Studies undertaken in dogs employing a myocardial punch biopsy technique to obtain serial myocardial samples from ischemic and normal zones demonstrate a three- to four-fold increase in myocardial ^{201}Tl activity in ischemic regions after 20 minutes of reperfusion; whereas, washout of $^{201}Tl^+$ from normally perfused areas was minimal (5). By 90 minutes there was significant loss of $^{201}Tl^+$ from the normal zones with continued redistribution into the ischemic zone and by four to six hours of reperfusion, $^{201}Tl^+$ activity in both ischemic and normal zones was nearly equal. Schelbert, et al (6) also concluded from canine studies that filling in of defects due to transient ischemia was due primarily to active redistribution within the first 30 minutes after $^{201}Tl^+$ administration, but due to washout from adjacent normal regions later. This filling in of defects occurs after transient ischemia and is related to equilibration of residual $^{201}Tl^+$ contained in the blood with the ischemic zone myocardium and to washout of $^{201}Tl^+$ from normal myocardium. Regions of myocardial infarction or scar would not be expected to accumulate $^{201}Tl^+$ since dead cells or scar tissue cannot extract $^{201}Tl^+$. Thus, infarcted myocardium would be expected to demonstrate persistent defects and this postulate has been confirmed by clinical studies (7).

IMAGING TECHNIQUES WITH $^{201}Tl^+$

After intravenous administration of 1.5 - 2 mCi of $^{201}Tl^+$, imaging should begin as soon as possible. Background activity may preclude imaging for five to ten minutes when $^{201}Tl^+$ is administered in the resting state, while imaging can usually begin within one minute if administered during exercise. We have observed occasional exercise induced ischemic defects which disappeared within 15 minutes of administration of $^{201}Tl^+$ due to redistribution. Since the sensitivity of the technique depends on demonstrating ischemic defects, as the time between $^{201}Tl^+$ administration and imaging increases, the sensitivity decreases. The gamma camera is used with a 30% window set around the mercury K X-ray emission peak (67-82 keV). Images should be collected using a computer display system for a preset time (approximately five minutes) so that at least 300,000 counts are collected in the full field or at least 50,000 counts in the cardiac region of interest. Multiple projections are obtained to include an anterior and multiple left anterior obliques (e.g. 30°, 50°, and 70°).

SPATIAL RESOLUTION OF $^{201}Tl^+$ DISTRIBUTION

Resolution of regions of abnormally reduced myocardial $^{201}Tl^+$ activity due to diminished perfusion or infarct is related to several factors (8-10).

1) Intrinsic resolution of the gamma camera: Recent generation 37 photomultiplier tube gamma cameras result in substantially improved resolution of $^{201}Tl^+$ distribution compared to older 19 photomultiplier tube models.

2) Imaging efficiency: The further the myocardium is from the gamma camera crystal, the less $^{201}Tl^+$ activity will be imaged due to tissue attenuation. Thus, normal basal myocardium may appear to have less activity than the more proximal anterior or lateral myocardium.

3) Cardiac geometry: a) Since each region on an image is related to all activity in the solid angle of view, normal $^{201}Tl^+$ activity in over- or underlying myocardium may obscure defects. b) The left ventricular apex is normally thin and frequently appears to have reduced activity.

4) Inhomogeneity of $^{201}Tl^+$ in normal myocardium: A variation of $^{201}Tl^+$ activity of up to 20% can be observed in images after administration at rest (11). Exercise images demonstrate more homogeneous distribution.

5) Uptake of $^{201}Tl^+$ in subdiaphragmatic viscera: Uptake of $^{201}Tl^+$ in the liver and other splanchnic viscera adjacent to the heart can obscure myocardial deficits in regional $^{201}Tl^+$ uptake (11). Splanchnic activity is minimized in resting studies by having the patient fast before the study and by injecting $^{201}Tl^+$ with the patient in the upright posture. During exercise, redistribution of blood flow leads to diminished subdiaphragmatic activity relative to myocardial activity.

6) Uptake of $^{201}Tl^+$ in the lungs: Usually background $^{201}Tl^+$ activity located in the lungs is insignificant. However, a number of circumstances may lead to increased lung activity. These include severe chronic lung disease and pulmonary congestion. The latter factor can lead to dramatic increases in lung Tl^+ activity which is related to heart failure induced by stress occurring during exercise.

7) Cardiac motion: Cardiac motion leads to decreased resolution. However, myocardial activity of $^{201}Tl^+$ using 1.5 - 2mCi may make standard gating impractical due to necessary increase in imaging time.

The distribution of $^{201}Tl^+$ activity observed in a single myocardial image must be interpreted in the light of the many factors described above.

CLINICAL APPLICATIONS OF $^{201}Tl^+$

In practice, myocardial imaging is used to visualize the normal myocardium and delineate zones of myocardial ischemia and infarction, and to detect stress-induced regional perfusion abnormalities resulting from coronary artery disease. Regions of reduced $^{201}Tl^+$ activity are related to diminished perfusion or myocardial infarction (12).

I. Exercise studies (Table I)

When $^{201}Tl^+$ is injected intravenously during exercise stress, images obtained a short time later demonstrate the regional perfusion pattern at the time of exercise.

Myocardial imaging with $^{201}Tl^+$ has been shown to improve specificity and sensitivity of stress electrocardiography in the detection of coronary artery disease (13-15). A recent multicenter study comparing exercise and rest images demonstrated a sensitivity in the detection of coronary disease (one or more vessels with greater than 50% stenosis) of 76% when an abnormality was present on either exercise image alone (transient ischemia) or both exercise and rest images (myocardial scar) (14). In patients with less than 50% coronary stenosis, there were exercise defects not present at rest in only 2%, but rest defects in an additional 10%. In our early experience in 56 patients using the serial imaging approach, defects were observed on early $^{201}Tl^{+}$ images in all 45 patients with coronary disease (50% or greater stenosis) but in none of nine patients with no significant coronary artery disease (less than 40% stenosis) (16). Defects were also noted in two patients with 40 - 50% coronary stenosis. Of 65 defects which filled in totally by four hours, 53 (82%) occurred in segments with appropriate significant coronary artery disease and normal wall motion. Of 53 defects which filled in partially, 41 (77%) were observed in segments with appropriate significant coronary disease but with reduced motion.

Optimally, images should be collected on a computer so that contrast and maximum myocardial intensity can be matched from image to image. Also, the computer provides a means of quantifying activity in myocardial regions of decreased as well as normal activity. Since small shifts in obliquity may have important effects on apparent $^{201}Tl^{+}$ distribution, it is crucial that serial images be obtained in comparable projections. If activity is observed to increase over the imaging period, active redistribution and thus, ischemia is suggested. Normal washout indicates normal myocardium, and persistently decreased activity is observed in the presence of a myocardial scar. The preceding comments provide a potential method for objective assessment of $^{201}Tl^{+}$ imaging. Further work is needed to develop and apply the ideal quantitative approach. Recently, Vogel and his colleagues have reported a means of tomographic imaging of $^{201}Tl^{+}$ using a wide field gamma camera and a multi-aperture collimator (with seven pinholes) (17). These investigators reported enhanced sensitivity and specificity over conventional planar imaging when the multiaperture collimator was employed.

When myocardial ischemia occurs, the ability of the left ventricle to contract becomes compromised. Borer and his colleagues have demonstrated the potential utility of multigated blood pool scanning as a means of diagnosis of coronary artery disease (18). A fall in the overall contractile function of the left ventricle or in contractile function of a wall with exercise suggests coronary artery disease. No change or a fall in overall function of the ventricle with exercise, however, is not specific for coronary artery disease and has been shown to occur with valvular heart disease (19) or disease of the heart muscle (20). We have recently compared both the thallium and the multigated blood pool approaches in a series of 62 patients. In the 50 patients with coronary artery disease, the sensitivity of the Tl^{+} approach was 86%, while that of the multigated blood pool scan approach was 79% (p=nonsignificant). In the group of 12 patients without significant coronary artery disease, one was falsely positive by the Tl^{+} approach, while four demonstrated a fall or no change in ejection fraction with exercise.

We now utilize both techniques in clinical practice, but apply the Tl^{+} approach as a diagnostic test and the multigated blood pool imaging approach to help prognosticate. Once the diagnosis of coronary artery disease has been made with the exercise electrocardiogram, and if necessary, the Tl scan, the multigated scan approach is applied. We are more concerned about patients with coronary artery disease who do not improve their ejection fraction with stress than those who do. The exact prognostic significance of exercise induced ejection fraction change needs further study. Some laboratories use the multigated blood pool imaging approach as the primary radionuclide diagnostic method in part because of the lower cost and easier access to the radiopharmaceutical.

II. Studies at rest (Table II)

When $^{201}Tl^+$ is injected intravenously in the resting state, subdiaphragmatic visceral activity is frequently observed and sometimes obscures the activity in the inferior wall of the left ventricle. Thus, when possible, the patient should be fasting and the injection made in the upright posture in order to minimize splanchnic activity (12).

MYOCARDIAL INFARCTION

Since $^{201}Tl^+$ localization initially depends upon cellular extraction ability and perfusion, regions of myocardial infarction, both acute and old, demonstrate reduced or absent activity. Wackers, et al (21) reported observing defects in 82% of 200 patients with acute myocardial infarction. Images obtained earlier in the course of myocardial infarction (before 24 hours) were more frequently abnormal (90 of 96) than patients studied after 24 hours (75 of 104). Patients studied both early and late tended to show reduction in defect size on follow-up images consistent with the observation of Buja and coworkers (22) that these defects represent an admixture of ischemic and infarcted myocardium. Post-mortem evaluation in 23 patients demonstrated 91% agreement with imaging localization of acute myocardial infarction but only 70% agreement with electrocardiographic localization (23).

There are several situations in which the resting $^{201}Tl^+$ scan could appear abnormal in the absence of myocardial infarction. These include cardiomyopathy both congestive, in which segments of muscle may be replaced by scar (8) and infiltrative (e.g. sarcoidosis), in which infiltrating tissue (e.g. granulomata) displaces myocardium and leads to defects (24). As mentioned earlier, certain regions of the myocardium appear to have reduced activity under normal conditions. Thinning of the left ventricular apex frequently leads to an apparent defect in the absence of organic disease. Basal portions of the myocardium which are furthest from the gamma camera may appear to have reduced activity due to attenuation of gamma photon flux.

MYOCARDIAL ISCHEMIA

As mentioned earlier, defects on resting $^{201}Tl^+$ scans do not necessarily indicate myocardial scar in patients with stable coronary artery disease since transient defects after rest injection have been observed in association with severe coronary artery stenosis. Transient defects have also been reported in patients with variant angina (25) as well as in patients with unstable angina pectoris (26). Serial $^{201}Tl^+$ imaging may be useful in differentiating between normal, ischemic and infarcted myocardium in patients admitted with acute chest pain. If $^{201}Tl^+$ distribution is normal on initial images, important ischemia and/or infarction are unlikely causes. A significant defect demonstrated on an initial $^{201}Tl^+$ scan may represent ischemia and/or infarction. A significant defect which persists on late images most often represents infarction, while a transient defect most often represents ischemia (27,28).

CONCLUDING COMMENTS

$^{201}Tl^+$ is an agent which allows good quality myocardial imaging with the gamma camera. Its distribution in the myocardium is related primarily to perfusion and myocardial cellular integrity. Redistribution of $^{201}Tl^+$ into regions of decreased activity identifies ischemic myocardium whereas persistent reduction of regional activity occurs in areas of infarction. Over the next decade, myocardial imaging methods with $^{201}Tl^+$ should improve probably through quantitative computer analysis and tomographic approaches. Other radiopharmaceuticals may be developed for gamma camera imaging. Instrumentation such as the positron emission tomographic camera shows promise and allows the application of a new set of radiotracers for myocardial imaging (29,30).

TABLE I: POTENTIAL INDICATIONS FOR ^{201}Tl IMAGING WITH EXERCISE

**1) Identification of the true positive ischemic ECG response in patients with ST segment depression and no chest pain on previous stress test.

**2) Detection of other regions of jeopardized myocardium in patients after myocardial infarction.

3) Determination of extent of coronary artery disease in patients with stable angina.

   *a) By observing the myocardial distribution of ^{201}Tl during exercise and rest.

   b) By quantitative analyses of regional myocardial Tl distribution and redistribution.

   c) By evaluating relative pulmonary activity which appears to increase due to stress-induced congestive failure.

*4) Improved screening of persons in occupations in which the presence of coronary disease and possibility of sudden deaths might be hazardous to others (e.g. airline pilots).

*5) Detection of non-coronary heart disease.

*6) To assess the significance of angiographically documented coronary stenosis of borderline severity.

*7) Evaluation of the effect of coronary artery bypass surgery and detection of the occluded graft.

TABLE II: POTENTIAL INDICATIONS FOR ^{201}Tl IMAGING AT REST

**1) Detection of myocardial ischemia and/or infarction in patients with acute chest pain syndromes.

*2) Differentiation of normal, ischemic and infarcted myocardium by serial imaging in patients with acute chest pain syndromes.

*3) Detection of high grade coronary stenoses supplying non-infarcted myocardium in patients without significant symptoms of acute myocardial ischemia.

*4) Determination of extent of myocardial ischemia and of infarction in patients with acute ischemic syndromes (variant angina, unstable angina and myocardial infarction).

**5) Detection of pulmonary hypertension by visualization of the right ventricle.

**6) Detection of asymmetric septal hypertrophy.

*Potential

**Established

REFERENCES

1) Britten, J.S., and Blank, M.: Thallium activation of the (Na^+-K^+) - activated ATPase of rabbit kidney. Biochem. Biophys. Acta 159: 160, 1968 .
2) Strauss, H.W., Harrison, K., Langan, J.K., et al: Thallium-201 for myocardial imaging. Relation of thallium-201 to regional myocardial perfusion. Circulation 51: 641, 1975.
3) Schelbert, H., Ingwall, J., Watson, R., Ashburn, W.: Factors influencing the myocardial uptake of thallium-201. J. Nucl. Med. 18: 598, 1977 (Abstract).
4) Pohost, G.M., Zir, L.M., Moore, R.H., McKusick, K.A., Guiney, T.E., and Beller, G.A.: Differentiation of transiently ischemic from infarcted myocardium by serial imaging after a single dose of thallium-201. Circulation 55: 1977.
5) Beller, G.A. and Pohost, G.M.: Mechanism for thallium-201 redistribution after transient myocardial ischemia. Circulation 55 and 56 (Suppl III): 140, 1977 (Abstract).
6) Schelbert, H., Schuler, G., Ashburn, N., and Dovell, J.: Time course of redistribution of thallium-201 after transiently ischemia. J. Nucl. Med. 18: 598, 1977 (Abstract).
7) Blood, D.K., McCarthy, D.M., Sciacca, R.R., and Cannon, P.J.: Comparison of single-dose and double-dose thallium-201 myocardial perfusion scintigraphy for the detection of coronary artery disease and prior myocardial infarction. Circulation 58: 777, 1978.
8) Holman, B.L.: Radionuclide methods in the evaluation of myocardial ischemia and infarction. Circulation 53 (Suppl I): 1-112, 1976.
9) Mueller, T.M., Marcus, M.L., Ehrhardt, J.C., Chauduri, T., and Abboud, F.M.: Limitations of thallium-201 myocardial scintigrams. Circulation 54: 640, 1976.
10) Strauss, H.W.: Cardiovascular nuclear medicine. A new look at an old problem. Radiology 121: 257, 1976.
11) Cook, D.J., Bailey, I., and Strauss, H.W.: Thallium-201 for myocardial imaging: Appearance of the normal heart. J. Nucl. Med. 17: 583, 1976.
12) Zaret, B.L.: Myocardial imaging with radioactive potassium and its analogs. Progr. Cardiovasc. Dis. 20: 81, 1977.
13) Bailey, I., Griffith, I.S.C., Rouleau, J., Strauss, H.W. and Pitt, B.: Thallium-201 myocardial perfusion imaging at rest and during exercise: Comparative sensitivity to electrocardiography in coronary artery disease. Circulation 55: 1977.
14) Ritchie, J., Zaret, B., Strauss, H.W., Pitt, B., Berman, D.S., Schelbert, H., Ashburn, W., Berger, H., and Hamilton, G.: Myocardial imaging with thallium-201 at rest and exercise - a multicenter study. Circulation 55 and 56 (Suppl III): 230, 1977.
15) Ritchie, J.L., Trobaugh, G.B., Hamilton, G.W., Gould, K.L., Narahara, K.A., Murray, J.A., and Williams, D.L.: Myocardial imaging with thallium-201 at rest and during exercise: Comparison with coronary arteriography and resting and stress electrocardiography. Circulation 56: 67, 1977.
16) Pohost, G.M., McKusick, K.A., Dinsmore, R.E., Zir, L.M., Guiney, T.E., and Beller, G.A.: Redistribution of thallium-201 during sequential myocardial imaging after exercise: Relationship to coronary artery disease and wall motion abnormalities. Clin. Res. 25: 244A, 1977 (Abstract).
17) Vogel, R.A., Kirch, D.L., LeFree, M.T., Rainwater, J.O., and Steele, P.P.: Quantitative analysis of standard and tomographic myocardial perfusion scintigraphy. Circulation 58: II-9, 1978 (Abstract).
18) Borer, J.S., Bacharach, S.L., Green, M.V., Kent, K.M., Johnston, G.S., and Epstein, S.E.: Effect of nitroglycerin on exercise-induced abnormalities of left ventricular regional function and ejection fraction in coronary artery disease. Circulation 57: 314, 1978.

19) Borer, J.S., Backarach, S.L., Green, M.V., Kent, K.M., Rosing, D.R., Seides, S.F., McIntosh, C.L., Conkle, D., Morrow, A.G., and Epstein, S.E.: Left ventricular function aortic stenosis: Response to exercise and effects of operation. Am. J. Cardiol. 41: 382, 1978 (Abstract).

20) Borer, J.S., Bacharach, S.L., Green. M.V., Kent, K.M., Maron, B.J., Rosing, D.R., Seides, S.F., and Epstein, S.E.: Obstructive vs. nonobstructive asymmetric septal hypertrophy: Differences in left ventricular function with exercise. Am. J. Cardiol. 41: 379, 1978 (Abstract).

21) Wackers, F.J., Sokole, E.B., Samson, G., van der Schoot, J.B., Lie, K.I., Liem, K.L., and Wellens, H.J.: Value and limitations of thallium-201 scintigraphy in acute phase of myocardial infarction. New Engl. J. Med. 295: 1, 1976.

22) Buja, L.M., Parkey, R.W., Stokely, E.M., Boute, F.J., and Willerson, J.J.: Pathophysiology of technetium-99m stannous pyrophosphate and thallium-201 scintigraphy of acute anterior myocardial infarcts in dogs. J. Clin. Invest. 57: 1508, 1977.

23) Wackers, F.J., Becker, A.E., Samson, G., Sokole, E.B., van der Schoot, J.B., Vet, A.J., Lie, K.I., Durrer, D., and Wellens, H.J.: Location and size of acute transmural myocardial infarction estimated from thallium-201 scintiscans: A cliniopathologic study. Amer. J. Cardiol 56: 72, 1977.

24) Bulkley, B.H., Rouleau, J., Strauss, H.W., Pitt, B.: Sarcoid heart disease: Diagnosis by thallium-201 myocardial perfusion imaging. Amer. J. Cardiol. 37: 125, 1976 (Abstract).

25) Maseri, A., Parodi, O., Severi, S., and Pesola, A.: Transient transmural reduction of myocardial blood flow, demonstrated by thallium-201 scintigraphy, as a cause of variant angina. Circulation 54: 280, 1976.

26) Wackers, R.J., Lie, K.I., Lien, K.I., Sokole, E.B., Samson, G., van der Schoot, H., Wellens, H.J., and Durrer, D.: Thallium-201 scintigraphy in unstable angina. J. Nucl. Med. 18: 594, 1977.

27) Osborn, R.C. and Smitherman, T.C.: Differentiation of transiently ischemic from infarcted myocardium following acute myocardial ischemia in man by serial myocardial scintigraphy following a single dose of thallium-201. Clin. Res. 1977, (Abstract).

28) Pond, M., Rehn, T., Burow, R., and Pitt, B.: Early detection of myocardial infarction by serial thallium-201 imaging. Circulation 55 and 56 (Suppl III): 230, 1977 (Abstract).

29) Weiss, E.S., Siegal, B.A., Sobel, B.E., Welch, M.J., and Ter-Pogossian, M.M.: Evaluation of myocardial metabolism and perfusion with positron-emitting radionuclides. Princ. of Cardiovasc. Nucl. Med., Editor B.L. Holman, E.H. Sommenblock, M. Lesch, Grune & Stratton, New York.

30) Beller, G.A., Hoop, B., Parker, J.A., et al: Sequential myocardial imaging with rubidium-82, and ultra-short lived radionuclide. Circulation 51-52 (Suppl II): 110, 1975 (Abstract).

UPTAKE OF RADIOPHARMACEUTICALS IN DEVELOPING MYOCARDIAL LESIONS

Hank F. Kung, Monte Blau and Robert Ackerhalt

Department of Nuclear Medicine
State University of New York at Buffalo
Buffalo, New York 14215

ABSTRACT

Isoproterenol induced myocardial damage in rats is a convenient model for the study of the varying uptake of radiopharmaceuticals during the course of infarct development. Using several known agents we have shown that the uptake in damaged rat hearts parallels what is known about the localization of these agents in human infarcts. The lesion to normal heart ratio is highest for MDP and PPi, with MDP having a higher ratio but a lower absolute uptake. GH and GA have much lower lesion to normal heart ratios but they are able to concentrate in infarcts at a much earlier stage of development. This model should be useful as a guide to the development of radiopharmaceuticals which localize during the early hours of human infarct development.

Myocardial infarct localization for bone agents generally parallels bone localization. However, there are some subtle differences, for example; Tc-99m MDP is superior to Tc-99m PPi as a bone localizing agent, but the latter exhibits a higher infarct uptake (14). It is obvious that in addition to concentration in calcium deposits in the injured cells other factors will also affect the infarct localization.

There are several other bone localizing agents which also show infarct localizing properties, including Tc-99m labeled EHDP, polyphosphate, MDP and IDP and F-18 fluoride ion (15). Using a double-labeling technique (Tc-99m and Tc-95m), EHDP and pyrophosphate were compared in experimental infarcts in dogs (16). The myocardial infarct uptake data showed PPi to be a superior agent.

The accumulation of the radiopharmaceutical in the damaged myocardium is not only related to the severity of cell injury, but also to the degree of remaining blood flow. Parkey et al (7, 10) reported that a marked concentration of Tc-99m PPi can be demonstrated only in the outer peripheral regions of transmural myocardial infarcts. This explains the doughnut pattern observed on the heart images in dogs with proximal LAD coronary artery occlusion and in some patients with large anterior infarcts. Similarly, concentration of Tc-99m PPi throughout the region of extensive necrosis in dogs with acute endocardial infarcts correlates with the homogeneous patterns found in patients with acute subendocardial infarcts (17).

Accumulation of Tc-99m glucoheptonate (GH) and gluconate (GA) in the infarcted myocardium has been observed in a number of animal models and in humans (18-21). Ahlberg et al (19) reported that in the coronary artery occlusion dog model Tc-99m GA localized in the damaged area within one hour following the occlusion and uptake reached a maximum at 14 hours after the infarct initiation. Using the same dog model, Jacobstein et al (18) reported serial myocardial imaging studies with Tc-99m GH. One hour after the occlusion, the infarct-to-normal heart ratios are comparable to those obtained for 6-9 hour old infarcts. Good quality infarct images were obtained at any time within the first day of the infarction.

Since Tc-99m GH and GA are not bone localizing agents, the mechanism of localization is not associated with calcium deposits in the damaged myocardium. A separate mechanism probably involving altered cell permeability and tissue protein binding must be responsible for the uptake of Tc-99m GH and GA. The time of maximum uptake in the lesion (early for Tc-99m GA and GH; late for bone agents) is a reflection of the differing mechanisms of lesion uptake.

In order to develop and test radiopharmaceuticals useful for clinical infarct imaging, it is necessary to have a suitable animal infarct model. Several animal models have been used for this purpose (Table 1). However, these models are either impractical for screening large numbers of agents or not suitable for studying localization at different time intervals after injury. Coronary artery ligation in dogs is the commonly used model for studying myocardial infarct localization (9, 15, 16, 18). The infarct resembles the clinical situation in humans. However, the model is suitable for studying a few potentially useful agents, but it is impractical for screening a large number of radiopharmaceuticals or for detailed time course studies.

INTRODUCTION

The diagnosis of acute myocardial infarction is usually based on clinical symptoms and laboratory data including electrocardiograph changes, an altered pattern of serum enzymes, etc. These findings provide indirect evidence of myocardial injury and often the location and extent of an infarct cannot be obtained easily. Myocardial infarct imaging with gamma emitting radiopharmaceuticals is a simple and non-invasive technique which can add significantly to the clinical and laboratory findings.

Infarct imaging is an increasingly useful method for determining the location and size of infarcts. However, the imaging agent in current use, Tc-99m pyrophosphate, is unable to detect infarcts during the first 24 hours after the onset of symptoms. This early period is often of critical importance in patient management and a simple non-invasive test which provides direct evidence of infarction would be useful especially in situations where other diagnostic tests are equivocal.

It was first observed by Bonte et al (1, 2) that Tc-99m pyrophosphate (PPi), a bone localizing agent, accumulated in the acutely infarcted myocardium. The uptake in the infarcted area appears to follow the known calcium influx and deposition in damaged myocardium (3-6).

This agent has been used in various clinical situations (7) and is particularly valuable in detecting:

1) small transient infarcts,
2) new acute transmural infarcts (24-72 hours old),
3) acute subendocardial infarcts,
4) acute infarct with left bundle branch block, and
5) perioperative myocardial infarction.

One of the major disadvantages of Tc-99m PPi is that in the majority of patients with acute myocardial infarcts, localization is maximum only after 48-72 hours (8). Early infarcts (0-24 hours old) accumulate less Tc-99m PPi than older infarcts.

Several studies on the mechanism of localization of Tc-99m PPi in infarcts have been reported (9-12). The concentration of Tc-99m PPi is related to the abnormal influx of calcium into infarcted myocardium. This calcium is in the form of amorphous calcium phosphate, crystalline hydroxy apatite and calcium complexed with myofibrils and other organic macromolecules.

The calcium concentration in relaxed normal myocardium is close to 10^{-7}M, while the extracellular fluid and blood have a much higher concentration, close to 10^{-3}M (13). When the injured myocardial cells can no longer maintain cellular integrity, the extracellular calcium ions will follow the concentration gradient and move into the cytoplasm. Intracellular phosphate and various phosphate containing molecules (ATP, ADP, TTP, GTP and UTP) as well as the calcium binding proteins (myosin, actin and troponin) contribute towards the calcium accumulation.

TABLE 1. ANIMAL MODELS REPORTED FOR MYOCARDIAL INFARCT LOCALIZATION STUDIES

	Methods (Species)	Comments
1.	Coronary Artery Ligation (rats, rabbits, dogs) (9, 15, 16, 18)	Close to clinical situation; involves surgery; difficult to produce; extent of lesion not reproducible.
2.	Heat-Induced Local Necrosis (rats) (22-24)	Easy to produce but far from real clinical situation.
3.	Vassopressin-Induced Local Necrosis (rabbits) (25)	Relatively difficult to produce.
4.	Subcutaneous Epinephrine Injection (rats) (26)	Not always reproducible.
5.	Random Electric Foot-Shock Stress (rats) (26)	Relatively difficult to produce.
6.	Cultured Fetal Heart (mouse) (27)	Far from real clinical situation.

A heat-induced local necrosis model in rats (22-24) has been widely employed for screening various agents. The necrosis is easily produced by touching the exposed heart with a soldering-iron. However, this model is so far removed from clinical myocardial infarction that it is not reliable for studies of the time course of uptake during lesion development.

A vassopressin-induced local necrosis model has also been used for screening radiopharmaceuticals (25). This model is relatively difficult to produce. The direct local delivery of vassopressin to the heart is not always successful (∿50% success rate).

Subcutaneous epinephrine injection into rats (26) will induce myocardial lesion, however, due to the rapid metabolism of this agent the effect is not reproducible. Random electric foot-shock stress infarcts may be useful, but they are not very convenient to produce (26). The cultured fetal heart (27), on the other hand, is a convenient model, but is far from the clinical situation.

The search for radioactive agents which localize in myocardial infarcts has been hampered by the lack of a suitable convenient animal model. We have developed a new rat model suitable for radiopharmaceutical investigations (28). It is an adaptation of the well known isoproterenol model. Isoproterenol-induced myocardial necrosis in animals was first reported by Rona (29, 30). The biochemical and histologic changes occurring after administration of this agent in rats have been well documented (31-37). The pharmacologic effect of isoproterenol is believed to be associated with its β-adrenergic effect which increases heart rate while decreasing blood pressure thus diminishing the oxygen supply to the myocardium. As early as 6 minutes after intra-peritoneal injection histologic changes occur in the myocardium. These include myofilament fragmentation, contraction-band formation, and hyalinization associated with dilatation of the sarcoplasmic reticulum (36). Focal necrosis, extensive inflammation and infiltration by polymorphonuclear leucocytes are found in the damaged heart at 24 hours after the injection. These changes resemble those observed in human myocardial infarction. Comparable biochemical alterations such as shifting electrolytes, serum enzyme elevation, etc., have also been reported (31, 33, 34, 37). All of these studies suggest that this experimental lesion is comparable to the infarct found in humans.

This paper describes our studies of the pattern of uptake of various known myocardial infarct localizing agents (Tc99m labeled PPi, MDP, GH, GA and F-18 fluoride ion) in developing lesions. In particular we have used the isoproterenol model to investigate the effect of time after lesion initiation on the uptake.

MATERIALS AND METHODS

Animal model and biodistribution studies. Myocardial damage was induced by drug overdose as reported previously (28). Briefly, isoproterenol hydrochloride (dissolved in 0.2-0.5 ml of saline) was injected subcutaneously into Sprague Dawley rats (220-280 g) at a dose of 30 mg/kg of body weight. After various time intervals, radiopharmaceutical was injected intravenously and 1 hour after this injection the rats were sacrificed. Control rats received the radiopharmaceutical injection only. The whole heart was removed, squeezed and blotted. Other organs of interest (blood, liver and femur) were also dissected and counted in a well counter. The % dose per organ was calculated by

TABLE 2. NORMAL HEART UPTAKE AND DH/N RATIOS* FOR VARIOUS RADIOPHARMACEUTICALS (VALUE ± S.E.M.)

	% Dose Normal Heart (n)	1 hour DH/N Ratio (n)	6 hour DH/N Ratio (n)
Tc-99m PPi	0.05 ± 0.01(20)	7.01 ± 0.64(6)	16.37 ± 3.90(13)
Tc-99m MDP	0.03 ± 0.01(7)	14.81 ± 5.20(5)	22.96 ± 7.92(6)
F-18 NaF	0.03 ± 0.01(9)	1.94 ± 0.52(3)	6.25 ± 0.41(4)
Tc-99m GH	0.05 ± 0.01(9)	3.64 ± 1.09(6)	2.64 ± 0.33(4)
Tc-99m GA	0.06 ± 0.02(7)	4.06 ± 0.69(3)	3.28 ± 0.09(3)

*See text for explanation of DH/N ratios.

comparing the radioactivity in the organ to a standard prepared from the initial dose.

DH/N heart ratios were calculated by comparing the % dose in the whole damaged heart to that in the normal heart. This index (DH/N ratio) of radiopharmaceutical uptake gives "lesion" to normal heart ratios that are lower than those obtained in other models (15, 22-27). Our "lesion" consists of the whole heart of the drug treated animal which, of course, contains a large amount of normal myocardial tissue. Since the damaged heart contains many small regions of infarcted and necrotic tissue, it is not possible to dissect normal and abnormal areas. However, our experience has shown that the DH/N index is quite high (up to 23) and reproducible.

Radiopharmaceuticals. Tc-99m labeled compounds were prepared by adding the appropriate amount of TcO_4^- to commercial freeze-dried kits. The radiochemical purity of each preparation was determined with two ascending paper chromatography systems: 85% methanol on Whatman #1 paper and saline on TLC-SG. Labeling efficiency for Tc-99m preparations was always over 95%.

F-18 fluoride was obtained from Brookhaven National Laboratory. The radionuclidic purity (gamma emitters only) was over 98%.

RESULTS AND DISCUSSION

The changes in DH/N ratio during the development of the heart lesion were studied by varying the waiting period after isoproterenol injection (from 0 to 24 hours) while keeping constant the time between radiopharmaceutical injection and excision (1 hour).

Normal heart uptake and DH/N heart ratio at 1 hour and 6 hours after the isoproterenol injection for the various radiopharmaceuticals are shown in Table 2. At 6 hours after the lesion initiation, DH/N heart ratio for MDP (DH/N = 23) is higher than that for PPi (DH/N = 16); however, the total heart uptake for PPi (0.9%) is higher than that for MDP (0.6%). This is consistant with the findings in human infarcts (14). In a comparison time study (Fig. 1), Tc-99m MDP exhibits a higher DH/N ratio than that for Tc-99m PPi at all times during the development of the lesion. Both time curves exhibit a more or less biphasic pattern and DH/N heart ratios are higher in the late myocardial necrosis (6-8 hours) than those in the early necrosis (1 hour).

The peak at 6-8 hours for these bone agents is no doubt associated with the well-established mechanism of binding to calcium deposits in damaged myocardium. Bloom and Davis (33) studied calcium flux in isoproterenol lesions in rats and found a maximum calcium concentration at 4-8 hours. As expected the bone imaging agents show a much more pronounced uptake at 6-8 hours than at 1 hour.

Figure 2 shows the time curve for F-18 fluoride ion. There is a hint of a similar biphasic pattern, however, the late peak at 6-8 hours predominates. The DH/N heart ratios are generally lower than those for the Tc-99m labeled bone agents (PPi and MDP). Since fluoride ion has a high affinity for calcium deposits this suggests that in addition to concentration in calcium deposits in the injured cells other factors also affect infarct localization.

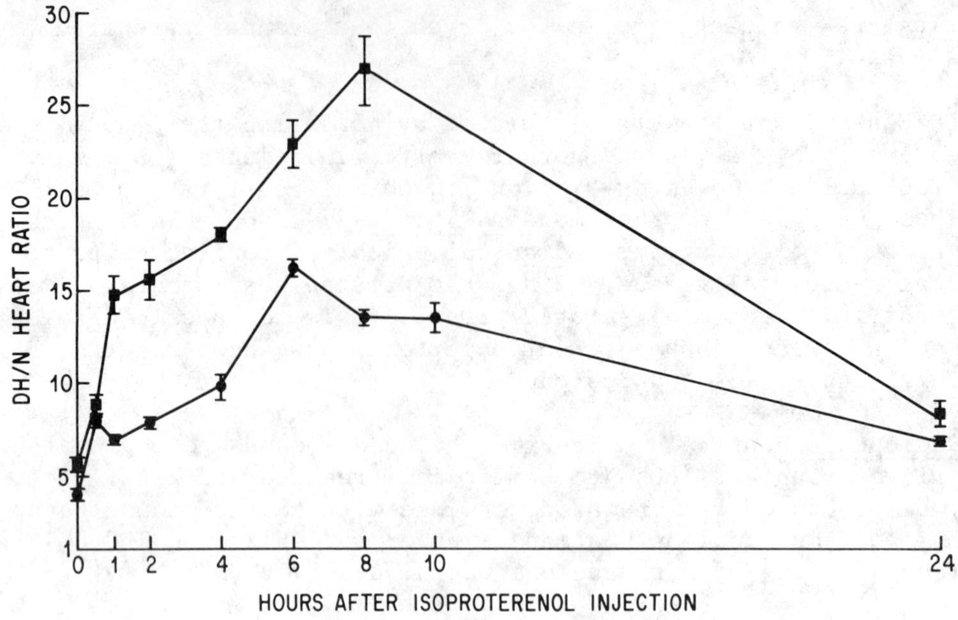

Figure 1.

DH/N ratios for Tc-99m PPi (●-●) and Tc-99m MDP (■-■) at different times after myocardial lesion initiation. Each point represents the mean from 3 to 13 rats ± S.E.M.

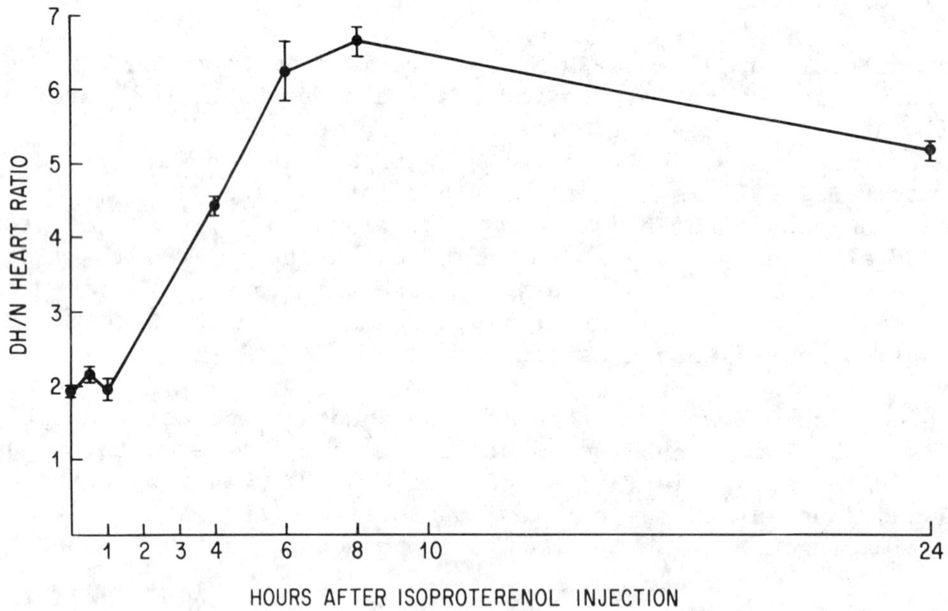

Figure 2.

DH/N ratios for F-18 NaF at different times after myocardial lesion initiation. Each point represents the mean from 3 to 6 rats ± S.E.M.

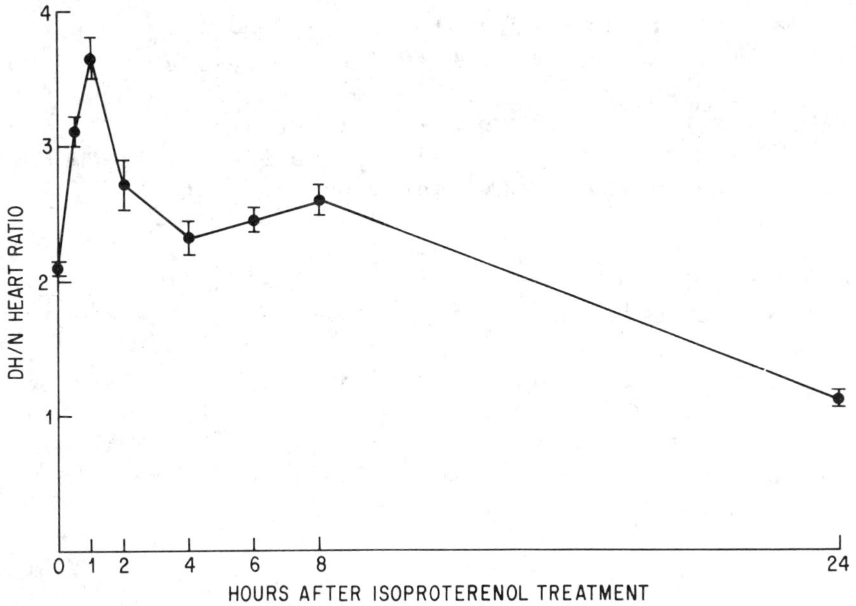

Figure 3.

DH/N ratios for Tc-99m glucoheptonate (GH) at different times after myocardial lesion initiation. Each point represents the mean from 3 to 6 rats ± S.E.M.

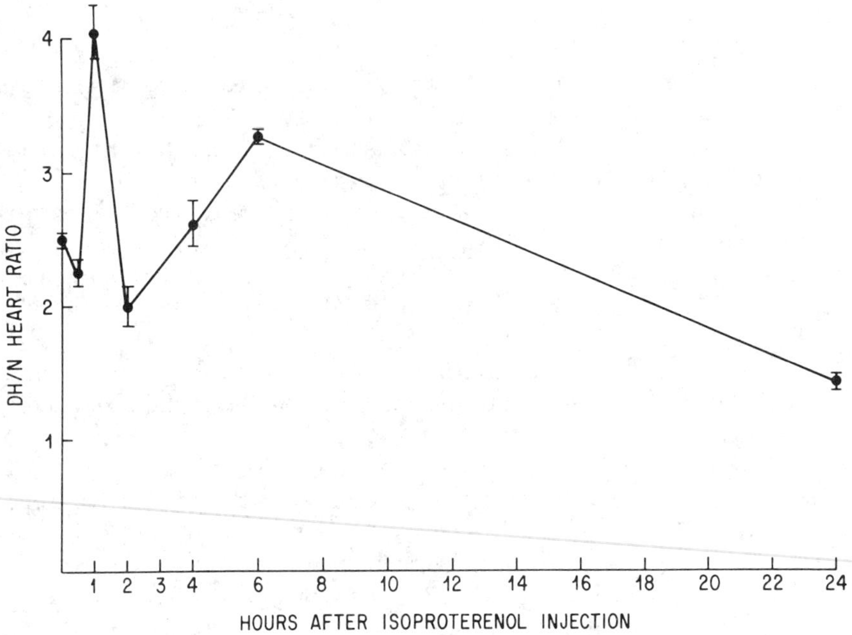

Figure 4.

DH/N ratios for Tc-99m gluconate (GA) at different times after myocardial lesion initiation. Each point represents the mean from 3 to 6 rats ± S.E.M.

The DH/N ratios for glucoheptonate and gluconate are substantially lower than for PPi and MDP (Table 2). In the time course studies Tc-99m glucoheptonate (Fig. 3) and gluconate (Fig. 4), in contrast to the bone agents, show the reverse pattern with the highest uptake at 1 hour and a small increase at 6-8 hours. The Tc-99m GH and GA data parallels the findings from coronary occlusion in dogs and from human scans showing early but low uptake (18-21) with these agents.

Thus, our preliminary experience with this convenient animal model using known agents leads us to conclude that it is suitable for developing new myocardial infarct localizing agents. In particular, it is applicable to measuring the uptake changes during lesion development. This is necessary for the study of new agents which are taken up in early lesions. The time course in this model is shorter than in human infarcts (PPi peak at 6-8 hours in rats, but 48-72 hours in humans). This compression of time scale is often the case when comparing data in small animals to humans.

REFERENCES

1. Bonte FJ, Parkey RW, Graham KD, et al: A new method for radionuclide imaging of myocardial infarcts. Radiology 110:473, 1974
2. Parkey RW, Bonte FJ, Meyer SL, et al: A new method for radionuclide imaging of acute myocardial infarction in humans. Circulation 50:540, 1974
3. D'Agostino AW: An electron microscopic study of cardiac necrosis produced by a 9-a-fluorocortisol and sodium phosphate. Am J Pathol 45:633, 1964
4. D'Agostino AW, Chiga M: Mitochondrial mineralization in human myocardium. Am J Clin Pathol 53:820, 1970
5. Shen AC, Jennings RB: Myocardial calcium and magnesium in acute ischemic injury. Am J Pathol 67:417, 1972
6. Shen AC, Jennings RB: Kinetics of calcium accumulation in acute myocardial ischemic injury. Am J Pathol 67:441, 1972
7. Parkey RW, Bonte FJ, Buja LM, et al: Myocardial infarct imaging with technetium-99m phosphates. Seminars in Nucl Med 7:15, 1977
8. Willerson JT, Parkey RW, Bonte FJ, et al: Technetium stannous pyrophosphate myocardial scintigrams in patients with chest pain of varying etiology. Circulation 51:1046, 1975
9. Buja LM, Tofe AJ, Kulkarni PV, et al: Sites and mechanism of localization of Tc-99m phosphorous radiopharmaceuticals in acute myocardial infarcts and other tissues. J Clin Invest 60:724, 1977
10. Buja LM, Parkey RW, Dees JH, et al: Morphological correlates of ^{99m}technetium stannous pyrophosphate of acute myocardial infarcts in dogs. Circulation 52:596, 1975
11. Dewanjee MK, Kahn PC: Mechanism of localization of ^{99m}Tc-labeled pyrophosphate and tetracycline in infarcted myocardium. J Nucl Med 17:639, 1976
12. Dewanjee MK: Localization of skeletal-imaging ^{99m}Tc chelates in dead cells in tissue culture. J Nucl Med 17:993, 1976
13. Nayler WG, Merrillees NC: Cellular exchange of calcium. In Calcium and the Heart by Harris P, Opie L, Academic Press, New York, 1971, pp25
14. Singh A, Usher M: Comparison of Tc-99m methylene diphosphonate with Tc-99m pyrophosphate in the detection of acute myocardial infarction. J Nucl Med 18:790, 1977
15. Bonte FJ, Parkey RW, Graham KD, et al: Distribution of several agents useful in imaging myocardial infarcts. J Nucl Med 16:132, 1975

16. Williams CC, Nishiyama H, Aldolph RJ, et al: A comparison of technetium-etidronate and pyrophosphate for acute myocardial infarct imaging. J Nucl Med 18:905, 1977

17. Willerson JT, Parkey RW, Bonte FJ, et al: Acute subendocardial myocardial infarction in patients: Its detection by technetium-99m stannous pyrophosphate. Circulation 51:436, 1975

18. Jacobstein JG, Alonso DR, Roberts AJ, et al: Early diagnosis of myocardial infarct in dog with ^{99m}Tc-glucoheptonate. J Nucl Med 18:413, 1977

19. Ahlberg N-E, Bois-Svensson I, Calissendorff B, et al: Uptake of ^{99m}Tc-gluconate and early histologic findings in induced myocardial infarcts in dogs. Acta Radiol Diagnosis 19:65, 1978

20. Rossman DJ, Strauss HW, Siegel ME, et al: Accumulation of ^{99m}Tc-glucoheptonate in acutely infarcted myocardium. J Nucl Med 16:875, 1975

21. Fink/Bennett D, Dworkin HJ, Lee Y-H: Myocardial imaging of the acute infarct. Radiology 113:449, 1974

22. Adler N, Camin LL, Shulkin P: Rat model for acute myocardial infarct: application to technetium-labeled glucoheptonate, tetracycline and polyphosphate. J Nucl Med 17:203, 1976

23. Davis MA, Holman BL, Carmel AN: Evaluation of radiopharmaceuticals sequestered by acutely damaged myocardium. J Nucl Med 17:911, 1976

24. Jones AG, Davis MA: The affinity of radiolabeled bone-seeking compounds for injured myocardium in the rat. Intl J Appld Rad Isotopes 28:203, 1977

25. Grossman ZD, Foster AB, McAfee JG, et al: Myocardial uptake (rabbit) of six ^{99m}Tc-tagged pharmaceuticals and ^{85}Sr after vasopressin-induced necrosis. J Nucl Med 18:51, 1977

26. Miller DG, Gilmour RF, Grossman ZD, et al: Myocardial uptake of Tc-99m skeletal agents in the rat after experimental induction of microscopic foci of injury. J Nucl Med 18:1005, 1977

27. Schelbert HR, Ingwall JS, Sybers HD, et al: Uptake of infarct-imaging agents in reversibly and irreversibly injured myocardium in cultured fetal mouse heart. Circulation Res 39:860, 1976

28. Kung HF, Blau M: Subcutaneous isoproterenol: A convenient rat model for early detection of myocardium necrosis. J Nucl Med 19:948, 1978

29. Rona G, Chappel CI, Balazs T, et al: An infarct-like myocardial lesion and other toxic manifestations produced by isoproterenol in the rat. Arch Path 67:553, 1959

30. Kahn DS, Rona G, Chappel CI: Isoproterenol-induced cardiac necrosis. Ann NY Acad Sci 156:285, 1969

31. Lehr D: Tissue electrolyte alteration in disseminated myocardial necrosis. Ann NY Acad Sci 156:344, 1969

32. Zbinden G, Moe RA: Pharmacological studies on heart muscle lesions induced by isoproterenol. Ann NY Acad Sci 156:294, 1969

33. Bloom S, Davis DL: Calcium as mediator of isoproterenol-induced myocardial necrosis. Am J Pathol 69:459, 1972

34. Yarom R, Ben-Ishay D, Zinder O: Myocardial cationic shifts induced by isoproterenol: Electron microscopic and electron probe studies. J Mol Cell Card 4:559, 1972

35. Handforth CP: Isoproterenol-induced myocardial infarction in animals. Arch Path 73:161, 1962

36. Csapo Z, Dusek J, Rona G: Early alterations of the cardiac muscle cells in isoproterenol-induced necrosis. Arch Path 93:356, 1972

37. Wexler BC, Kittinger GW: Myocardial necrosis in rats: serum enzymes, adrenal steroid and histopathological alterations. Circulation Res 13:159, 1963

SYNTHESIS AND BIOLOGIC DISTRIBUTION OF Tc-99m LABELED PALMITIC ACID DERIVATIVES

Eli Livni, Michael A. Davis and Victor D. Warner

Department of Radiology, Joint Program in Nuclear Medicine
Harvard Medical School and Children's Hospital Medical Center
and
Department of Medicinal Chemistry and Pharmacology
College of Pharmacy and Allied Health Professions
Northeastern University, Boston, Massachusetts 02115

ABSTRACT

Based on the hypothesis that Tc-99m derivatives would behave biologically like palmitic acid and thus serve as imaging agents for normal functioning myocardium, three palmitic acid derivatives have been prepared. The compounds synthesized, 2-mercaptopalmitic acid, 16-mercaptopalmitic acid, and 2-mercapto-1,16-hexadecanedioic acid were radiolabeled with Tc-99m. Biologic distribution studies of [16-^{14}C] palmitic acid and the Tc-99m labeled fatty acids were performed in rats with a heat inflicted myocardial infarction. Whereas [16-^{14}C] palmitic acid exhibited a normal myocardium to infarcted myocardium uptake ratio of 4.95, all the Tc-99m labeled mercapto derivatives had ratios ranging from 0.33 to 0.54.

INTRODUCTION

Fatty acids substituted with gamma emitting radionuclides such as I-131, I-123, C-11, F-18, and Te-123m have been suggested as potential imaging agents (1-6) for normal functioning myocardium (cold spot or negative imaging agents). At present [1-^{11}c] palmitic acid is being used clinically in the detection of myocardial ischemia. Due to the isotope's 20 min half life and its mode of decay (positron emission, resulting in penetrating gamma radiation of 511 keV) the use of this c-11-labeled compound is limited to medical centers having cyclotrons and positron cameras or positron emission tomographic instrumentation.

Technetium-99m is the isotope preferred for almost all imaging procedures. Therefore, substances combining fatty acid specificities for localization in the myocardium and chelating groups for technetium should be ideal for visualization of the myocardium. Eckelman et al. (7) have synthesized derivatives of palmitic acid containing polyaminocarboxylic acids as chelating groups for Tc-99m. The concentration in the myocardium of these derivatives was lower than that of [^{3}H]palmitic acid. This dissimilarity in myocardial localization was attributed to the difference in polarity between [^{3}H]palmitic acid and the Tc-99m labeled fatty acid derivatives. Recently Loberg and coworkers (8) labeled decanoic and undecanoic acids with Tc-99m using iminodiacetate or carbamoylmethyl iminodiaetate as the chelating group. The variation in biologic distribution of these compounds compared with that of [^{3}H]palmitic acid was thought to be due to the lower lipohilicity of the Tc-99m compounds.

It was observed (9) that when substituting the sulfhydryl function on palmitic acid, the specificity of the fatty acid analog for the myocardium is retained. Three mercapto derivatives of palmitic acid have been labeled with Tc-99m: 2-mercaptopalmitic acid I, 16-mercaptopalmitic acid II, and 2-mercapto-1,16-hexadecanedioic acid III. The biologic distribution of the Tc-99m labeled derivatives was compared with that of [16-^{14}C] palmitic acid ([16-^{14}C]PA) in rats with heat induced myocardial lesions.

$CH_3(CH_2)_{13}CH(SH)COOH$	$HSCH_2(CH_2)_{14}COOH$	$HOOC(CH_2)_{13}CH(SH)COOH$
I	II	III

MATERIALS AND METHODS

Chemical Syntheses: The synthesis of the mercapto derivatives of palmitic acid was achieved by reaction of the corresponding bromo compounds with thiourea (9). To prepare Tc-99m labeled I and III a solution of $^{99m}TcO_4^-$ (0.05ml, 2-5mCi) was added to 2 ml of the corresponding mercapto derivative (1mg/ml) in glacial acetic acid and the solution incubated 30 min at room temperature. An aliquot (0.5 ml) of the resultant solution was evaporated to dryness under nitrogen at 50°C and the residue dissolved in 5 ml propylene glycol. The solution was filtered through a 0.22μ membrane filter (Millipore Corporation) before injection. To prepare Tc-99m labeled II solution of $^{99m}TcO_4^-$ (0.05 ml, 2-5 mCi) was added to 2 ml (1mg/ml) of II in glacial acetic acid. The solution was incubated 1 hr at room temperature. An aliquot (0.5 ml) was evaporated to dryness under nitrogen at 50°C and the residue dissolved in 5 ml chloroform. The chloro-

TABLE 1. NORMAL AND INFARCTED MYOCARDIAL UPTAKE IN RATS FOLLOWING INTRAVENOUS ADMINISTRATION OF RADIOLABELED FATTY ACIDS

	5 min			15 min			60 min		
	N[a]	I[a]	N/I	N	I	N/I	N	I	N/I
$[16\text{-}^{14}C]$PA[b]	2.05±0.75	1.29±0.23	2.25±0.69	2.04±0.79	0.59±0.38	4.95±3.07	0.89±0.22	0.59±0.24	1.67±0.01
Tc-99m-I[b,d]	0.54±0.09	1.08±0.19	0.54±0.08	0.42±0.04	1.06±0.28	0.46±0.04	0.26±0.03	1.04±0.53	0.33±0.16
Tc-99m-II[b,d]	0.84±0.11	1.77±0.12	0.47±0.11	0.48±0.02	1.93±0.01	0.25±0.01	0.59±0.05	1.18±0.09	0.51±0.08
Tc-99m-III[b,d]	0.96±0.10	2.98±0.59	0.33±0.04	0.99±0.16	2.65±0.01	0.37±0.06	0.68±0.06	1.38±0.53	0.60±0.28

a. N=normal, I=infarcted.
b. PA=palmitic acid; I=2-mercaptopalmitic acid; II=16-mercaptopalmitic acid; III=2-mercapto-1,16-hexadecanedioic acid.
c. Data from reference 9.
d. Mean and standard deviation of 3 values expressed as percent injected dose per gram of fresh tissue.

form solution was transferred to another vial and evaporated to dryness. The residue was dissolved in a nitrogen purged solution of propylene glycol in ethanol (50% V/V). The solution was filtered through a 0.22μ membrane filter into a 6 ml evacuated vial.

Radiochemical purity: An aliquot (1-3μl) of the Tc-99m labeled compounds was spotted on ITLC/SG (Gelman Instrument Company) and the paper dried for 10 min at 1 mm pressure. The radiochemical purity of Tc-99m labeled I, II, and III was ascertained in two solvent systems: (a) isotonic saline, (b) methyl ethyl ketone; glacial acetic acid, 9:1 (V/V). The Tc-99m mercapto derivatives have Rf values of 0.0 in system (a) and 1.0 in system (b). $^{99m}TcO_4^-$ has an Rf value of 0.82 in system (a) and 1.0 in system (b). Colloidal Tc-99m has an Rf value of 0.0 in both systems.

Paper Electrophoresis: The Tc-99m labeled compounds in a 0.05M phophate buffer were subjected to paper electrophoresis on 3MM paper (Whatman Incorporated) for 1.2 hr at a constant voltage of 500V in water: n-propanol, 2:3 (V/V), pH 7.4. It was necessary to apply the tested solution in the middle of the paper, otherwise the compound migrated, as in chromatography, even without applying voltage.

Damaged heart model: The model used has been described previously (10). Sprague-Dawley rats (180-250g) were anesthetized with ether. A midline thoractomy was performed and an incision made between the seventh and eighth ribs to allow access to the ventricle. The ventricle was touched for 1 sec with a hot soldering iron (Wahl Iso-tip, regular size). This was repeated four or five times in adjacent areas to produce a circular pattern. The chest was then closed.

Tissue distribution studies: Two and a half to 3 hr after the infarct was made, 0.2 ml (5 to 10 μci) of the radiolabeled compound was injected through the saphenous vein. The rats were sacrificed by ether asphyxiation at 5, 15, and 60 min after dosing. Before myocardial activity ceased, blood samples were obtained by cardiac puncture. The appropriate organs were then excised, weighed, and the radioactivity was measured in a NaI (Tl) well scintillation counter.

RESULTS

Labeling efficiency: Chromatographic analyses showed that the injected solutions of Tc-99m-I and Tc-99m-II contained less than 3% radiochemical impurities. For Tc-99m-II less than 13% was present as either $^{99m}TcO_4^-$ or colloidal Tc-99m. Tc-99m-I and Tc-99m-III were found to be stable for at least 24 hr after preparation, while the radiochemical impurity in Tc-99m-II increased to 18% by 6 hr.

Electrophoretic mobilities: Tc-99m-I and Tc-99m-III migrated 2 cm towards the anode. Tc-99m-II streaked in the first 2 cm towards the anode. Under the conditions applied for the Tc-99m derivatives, $^{99m}TcO_4^-$ migrated 7.2 cm towards the anode.

Tissue distribution: The concentration of the radiolabeled fatty acids in normal and damaged rat myocardium as a function of time is shown in Table 1. In contrast to [16-^{14}C]PA, the technetium labeled compounds have greater uptake in infarcted myocardium than in normal at all times. The highest ratio of normal to damaged myocardium, among the Tc-99m mercapto derivatives is found with Tc-99m-III at 60 min. Tables 2 and 3 compare the biodistribution of the radiolabeled fatty acids at various time intervals. The blood levels of the Tc-99m compounds are much higher than those of [16-^{14}C]PA. The muscle uptake of [16-^{14}C]PA is 4 to 5 times greater than that of the technetium mercapto derivatives. The hepatic concentration is high at all times for all four compounds. With time liver uptake increases for Tc-99m-I, II, and III, while it decreases for [16-^{14}C]PA. activity of the Tc-99m labeled fatty acids in the intestines increases over the time period sampled.

DISCUSSION

An ideal fatty acid derivative for myocardial imaging should be extracted by the normal cardiac cells without being rapidly metabolized. In designing such a radiopharmaceutical, an understanding of the mechanism of fatty acid uptake into cardiac cells would be useful. Samuel at al. (11) have suggested a specific transport system for fatty acids in cardiac cells from chick embryos. The details of this transport system are still unknown.

The myocardial specificity of the Tc-99m mercapto derivatives is lower than that of [16-^{14}C]PA. This difference may be related to the structure of the Tc-99m chelates. Recently the structures of several Tc-99 compounds having technetium-sulphur coordination have been determined (12,13). Extrapolation of this information and the fact that the Tc-99m mercapto derivatives migrate toward the anode suggest that the structure of Tc-99m-I and Tc-99m-III may be

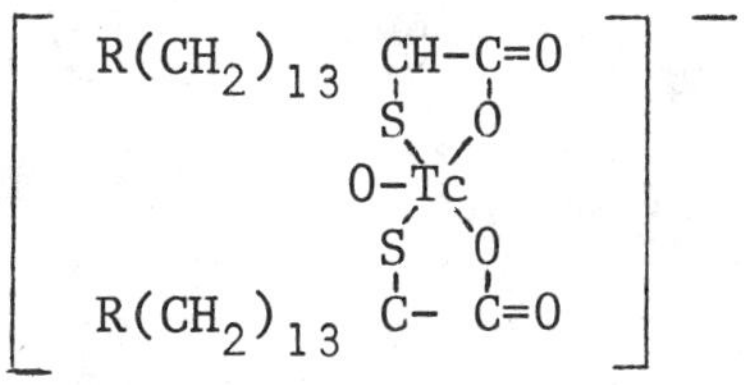

where R=CH_3 for Tc-99m-I and R=COOH for Tc-99m-III. If the hydroxyl oxygen is involved in binding to technetium, Tc-99m-I does not have a free carboxylate and the compound does not resemble fatty acid. In Tc-99m-III there is free carboxylate, but the molecular weight of the chelate is much higher than that of palmitic acid.

The chromatography and electrophoresis results demonstrate that the α-mercaptocarboxy group strongly binds technetium, while a single mercapto group does so more weakly.

Bilheimer at al. (14) have demonstrated that 24 hr after injection of [^{14}C] oleic acid into dogs with acute myocardial infarction the greatest concentration of radioactivity is in the peripheral and border zones of the lesion. The pharmacokinetics of uptake in the border zone indicate that optimal imaging of the infarcted region is 24 hr post injection. Their findings

TABLE 2. BIODISTRIBUTION OF PALMITIC ACID AND THREE Tc-99m MERCAPTO DERIVATIVES IN RATS

Compound	Time (min)	Blood	Liver	Lung	Spleen	Kidney	Muscle
[16-^{14}C]PA[a,b]	5	0.09±0.03[c]	4.84±1.13	2.78±1.56	0.43±0.16	1.75±0.83	0.34±0.09
	15	0.10±0.04	3.33±1.44	2.26±0.32	0.42±0.14	1.17±0.54	0.26±0.12
	60	0.09±0.01	2.01±1.02	1.80±0.12	0.39±0.18	0.73±0.33	0.20±0.11
Tc-99m-I[a]	5	2.96±0.60	4.29±0.85	1.11±0.25	1.08±0.30	1.16±0.05	0.07±0.02
	15	2.54±0.53	4.52±0.37	1.01±0.09	1.29±0.22	1.09±0.06	0.06±0.01
	60	1.49±0.17	4.89±0.18	0.74±0.20	0.75±0.14	1.40±0.15	0.05±0.01
Tc-99m-II[a]	5	1.44±0.04	2.60±0.33	1.69±0.12	1.15±0.33	0.91±0.16	0.05±0.01
	15	1.05±0.07	2.96±0.39	2.31±0.73	1.68±0.73	0.84±0.50	0.07±0.02
	60	0.76±0.17	3.28±0.04	2.54±0.14	2.09±0.39	1.42±0.15	0.04±0.01
Tc-99m-III[a]	5	4.47±0.91	3.18±1.00	1.32±0.07	0.73±0.17	1.93±0.21	0.10±0.01
	15	4.21±0.52	3.23±0.25	1.59±0.21	0.92±0.06	2.20±0.04	0.10±0.01
	60	4.04±0.39	3.48±0.53	1.42±0.27	0.59±0.27	1.89±0.10	0.08±0.01

a. PA=palmitic acid; I=2-mercaptopalmitic acid; II=16-mercaptopalmitic acid; III=2-mercapto-1,16-hexadecanedioic acid.
b. Data from reference 9.
c. Mean and standard deviation of 3-5 values expressed as percent injected dose per gram of fresh tissue.

TABLE 3. BIODISTRIBUTION OF PALMITIC ACID AND THREE Tc-99m MERCAPTO DERIVATIVES IN RATS

Compound	Time (min)	Blood[a]	Liver	Lung	Spleen	Kidney	Muscle[b]	Stomach	Intestines
[16-^{14}C]PA[c,d]	5	1.19±0.56	48.09±4.46	2.39±0.41	0.24±0.02	2.71±0.63	26.06±3.02		
	15	1.66±0.65	36.20±7.07	2.44±0.69	0.24±0.06	2.26±0.70	23.99±7.47		
	60	1.45±0.19	23.85±9.09	2.33±0.15	0.23±0.10	1.53±0.55	19.63±7.43		
Tc-99m-I[c,e]	5	33.35±5.26	41.17±3.68	1.64±0.22	0.72±0.18	2.14±0.11	5.68±1.88	0.18±0.01	1.77±0.07
	15	29.93±6.88	46.40±1.91	1.25±0.13	0.71±0.07	1.82±0.03	4.87±0.63	0.26±0.02	2.25±0.33
	60	18.54±2.17	51.43±0.34	0.89±0.26	0.48±0.14	2.74±0.20	4.32±0.49	0.31±0.07	5.66±0.54
Tc-99m-II[c,e]	5	26.72±1.90	34.05±4.28	2.78±0.18	0.86±0.24	2.18±0.33	5.76±0.23	0.35±0.29	3.09±0.32
	15	17.58±2.10	40.80±4.07	3.76±0.81	1.22±0.37	2.08±1.25	8.11±2.26	0.37±0.09	4.11±1.09
	60	11.97±4.66	41.20±1.14	3.53±0.64	2.09±0.39	3.15±0.28	5.39±0.49	0.44±0.01	8.92±1.19
Tc-99m-III[c,e]	5	43.08±7.75	26.47±4.45	1.66±0.32	0.47±0.12	3.03±0.23	6.48±0.66	0.33±0.07	2.81±0.34
	15	39.89±6.31	31.07±0.69	1.81±0.15	0.49±0.02	3.07±0.23	5.38±0.40	0.36±0.02	2.83±0.06
	60	38.41±2.42	33.71±1.05	1.86±0.32	0.36±0.11	3.37±0.21	5.43±0.68	0.29±0.11	3.51±0.25

a. Estimated as 6% of body weight.
b. Estimated as 40% of body weight.
c. PA=palmitic acid, I=2-mercaptopalmitic acid, II=16-mercaptopalmitic acid, III=2-mercapto-1,16-hexadecanedioic acid.
d. Data from reference 9.
e. Mean and standard deviation of 3 values expressed as percent injected dose per organ.

suggest that radiolabeled fatty acids may have application in delineating areas of infarction in addition to their generally accepted use as agents for imaging normal myocardium.

ACKNOWLEDGMENTS

We wish to express out thanks to Mrs. Alice Carmel for her expert technical assistance and to Mrs. Rebekah Taube for her editorial work on the manuscript.

Adapted from a dissertation to be submitted by E. Livni to the Graduate School of Pharmacy and Allied Health Professions, Northeastern University, in partial fulfillment of the Doctor of Philosophy degree requirements.

This work was supported in part by USPHS Grant GM23270 and Department of Energy Contract EY-76-S-4115.

REFERENCES

1. Evans JR, Gunton RW, Baker RG, et al: Use of Radioiodinated fatty acids for photoscans of the heart. Circ Res 16:1-10, 1965

2. Robinson GD, Lee AW: Radioiodinated fatty acids for heart imaging: Iodine monochloride addition compared with iodide replacement labeling. J Nucl Med 16:17-21, 1975

3. Robinson GD, MacDonald NS: Biological dynamics of ^{18}F and ^{11}C-labeled carboxylates and alcohols. J Nucl Med 14:446-447, 1973

4. Weiss ES, Hoffman EJ, Phelps ME, et al: External detection and visualization of myocardial ischemia with ^{11}C-substrates in vitro and in vivo. Circ Res 39:24-32, 1976

5. Basmadjian GP, Mills SL, Parker GR, et al: 16-Methyl Telluro-Te-123m-9-hexadecenoic acid: A potential myocardial agent. J Nucl Med 19:718, 1978

6. Knapp FF Jr, Ambrose KR: Tellurium - 123m. Nuclear Medicine Technology Progress Report for quarter ending June 30, 1978, ORNL/TM-6638

7. Karesh SM, Eckelman WC, Reba RC: Biological distribution of chemical analogs of fatty acids and long chain hydrocarbons containing a strong chelating agent: J Pharm Sci 66:225-228, 1977

8. Loberg MD: Recent advances in radiopharmaceuticals derived from bifunctional chelating groups. Presented at the Eighth Northeast Regional Meeting of the American Chemical Society, Boston, MA, June 25, 1978, published abstract

9. Livni E, Davis MA, Warner VD: Synthesis and biologic distribution of mercapto derivatives of palmitic acid. J Med Chem, in press

10. Davis MA, Holman BL, Carmel AN: Evaluation of radiopharmaceuticals se-

questered by acutely damaged myocardium. J Nucl Med 17:911-917, 1976

11. Samuel D, Paris S, Ailhaud G: Uptake and metabolism of fatty acids and analogues by cultured cardiac cells from chick embryos. Eur J Biochem 64: 583-595, 1976

12. DePamphilis BV, Jones AG, Davis MA, et al: Preparation and crystal structure of Oxotechnetium bis (thiomercaptoacetate) and its relationship to radiopharmaceuticals labeled with Tc-99m. J Am Chem Soc 100:5570-5571, 1978

13. Smith JE, Byrne EF, Cotton AF, et al: A thiol complex of technetium pertinent to radiopharmaceutical use of Tc-99m. J Am Chem Soc 100:5571-5572, 1978

14. Bilheimer DW, Buja LM, Parkey RW, et al: Fatty acid accumulation and abnormal lipid deposition in peripheral and border zones of experimental myocardial infarcts. J Nucl Med 19:270-283, 1978

IMPROVED PREPARATION AND APPLICATION OF Tc-99m-HEPARIN FOR SCINTIGRAPHIC DETECTION OF EXPERIMENTAL MYOCARDIAL ISCHEMIC INJURY

Padmakar V. Kulkarni,* Robert W. Parkey,* L. Maximilian Buja,**
Frederick J. Bonte,* James E. Wilson III***, and James T. Willerson***

The Departments of Radiology,* Pathology,** and Internal Medicine***
The University of Texas Health Science Center at Dallas, Texas

ABSTRACT

Our previous studies demonstrated the localization of technetium-99m-labeled heparin in experimental myocardial infarction. In this report we describe a one-step method for preparation of the labeled agent and further evaluation of the agent in dogs undergoing temporary left anterior descending (LAD) coronary artery ligation for 45 minutes, 3 hours, and 6 hours. Moderate myocardial damage and uptake of technetium-99m-heparin (Tc-99m-heparin) occurred in one of five dogs after 45 minutes of temporary LAD occlusion. Maximum uptake of Tc-99m-heparin occurred after 3 hours of temporary coronary artery occlusion, and exceeded 20 times the activity in the normal myocardium in 4 of 6 dogs with hemorrhagic infarcts. Uptake of Tc-99m-heparin in damaged myocardium was somewhat lower after 6 hours of temporary LAD occlusion in the four dogs studied, but was still in the 15 to 1 range in three with hemorrhagic infarcts. In the animals with prominent histologic damage, there was sufficient accumulation of the agent in the damaged tissue to permit scintigraphic detection of the damaged myocardium in vivo. These new data confirm that Tc-99m-labeled heparin is suitable for imaging damaged myocardium during early stages of ischemic injury if significant blood flow persists to the damaged myocardial area.

INTRODUCTION

A number of radiopharmaceuticals have been shown to be sequestered by acutely damaged myocardium (1), particularly, radiolabeled tetracycline and its analogs (2), Tc-99m-labeled glucoheptonate (3), and Tc-99m-labeled phosphates (4). Technetium-99m-phosphates have been proved to be clinically useful imaging agents in identification of myocardial infarcts (5-7). Recently, we showed that a new agent, heparin labeled with Tc-99m, also localized in damaged areas of myocardium in dogs with temporary ligation of the left anterior descending (LAD) coronary artery for 60 to 75 minutes and reperfusion (8). The method of labeling heparin was somewhat cumbersome. Thus, we have modified the tagging procedure to permit the preparation of the labeled agent by one step addition of $^{99m}TcO_4^-$ to the tin reagent. We evaluated this preparation in experimental animals (mongrel dogs) undergoing temporary LAD coronary artery ligation for various periods of time (45 minutes to 6 hours). Recently, a commercial manufacturer† has introduced a heparin kit for experimental purposes. This was also evaluated in the canine myocardial infarct models and found to be suitable for imaging experimental myocaridal injury.

MATERIALS AND METHODS

Preparation of Tc-99m-Heparin

The method for labeling heparin with Tc-99m was modified from the original procedure (8). It involves the preparation of "instant kits" whereby the labeled product is obtained by addition of pertechnetate to a stannous heparin preparation. To obtain a pH of approximately 2 to 3, 0.6 ml of hog mucosa heparin (20,000 U/ml, Abbott) is acidified with 0.1 N HCl, to which 1.5 mg of $SnCl_2.2H_2O$ in 1 ml of 0.1 N HCl is added. The pH of the solution is raised to 5.5 to 6.0 with .125 NaOH. The solution is diluted with saline to 6 ml and passed through a 0.2 μm membrane filter. The filtered solution is distributed (0.6 ml aliquots) into individual vials. All the above operations are carried out under nitrogen atmosphere and all solutions exhaustively bubled with nitrogen gas. The labeled material is obtained by addition of 10 to 30 mCi of $^{99m}TcO_4^-$ from a generator to a vial containing 1200 Units of heparin and 150 μg of $SnCl_2.2H_2O$. The reaction mixture is allowed to stand for 10 to 15 minutes. The amount of free pertechnetate ($^{99m}TcO_4^-$) is routinely monitored using instant thin layer chromatography (ITLC-SG) with acetone or methylethylketone (MEK) as the solvent system. Cellulose acetate electrophoresis is performed on some preparations with 0.1 M acetate buffer, pH 5.5 for 15 minutes at 300 volts. Heparin is stained with .5% toluidine blue. Activated partial thromboplastin times are also determined in some preparations as described in our earlier communication (9-10).

Myocardial Ischemia Model

Mongrel dogs of either sex weighing approximately 30 kg were anesthetized with sodium pentobarbital (30 mg/kg). An endotracheal tube was inserted and

†Solco Nuclear, Basle, Switzerland

the animal placed on a Harvard respirator. The chest was opened through a left lateral thoracotomy and the proximal left anterior descending coronary artery (LAD) visualized and gently dissected free. The LAD was then occluded with a purse-string suture passed through a narrow plastic tube so that the artery was pulled against the tube with the string tightly held in place. After occlusion for a period of time, the artery was gradually released to allow reflow of blood through it. The dog's chest was closed after releasing the LAD occlusion. A total of 15 dogs were divided into three groups: 1) five dogs had temporary LAD occlusion for 45 minutes; 2) six dogs for 3 to 3.5 hours; and 3) four dogs for 6 hours. Technetium-99m-heparin (3 to 6 mCi) was injected intravenously 5 to 10 minutes post-release of the LAD occlusion. Serial images in the anterior, left anterior oblique (LAO), and left lateral projections were obrained as described earlier (8). The scintigraphic images were graded (0 to 4+) according to the method of Parkey et al. (5).

Activated whole blood clotting time (AWBCT) was determined in some dogs prior to administration of the labeled agent and at 5 minutes and one hour after injection. The test utilizes Platelin[R*] plus activator which is a platelet Factor-3 reagent (partial thromboplastin) with celite for determination of activated partial thromboplastin times (APTT). For each test, 0.1 ml of Platelin[R] plus activator was added to glass tubes and brought to 37°C. Blood (0.5 cc) was collected in a tuberculin syringe and 0.2 cc was added to each of the tubes. The tubes were tilted every 15 seconds until a solid clot was formed, and the time to form the clot was recorded.

The animals were sacrificed 3 hours and 30 minutes post-injection of Tc-99m-heparin by administration of a large dose of sodium pentobarbital (5 gm/101bs). The hearts were removed, washed in saline, and imaged in vitro by placing them in a tray under the scintillation camera and 100,000 counts were collected.

Tissue Assay for Radioactivity, Histology

Each heart was divided into transverse slices, and multiple transmural blocks were obtained from grossly normal posterior left ventricular myocardium and anterior myocardium containing foci of gross damage. Sections for histologic examination were removed from selected blocks and fixed in phosphate buffered 10% formalin. The extent of light microscopic damage was graded from 0 to 4+ based on the extent of congestion, hemorrhage, neutrophilic infiltrate, muscle cell alteration including nuclear damage and contraction gand formation, and degree of extension of damage from subendocardium into subepicardium.

---

*Obtained from General Diagnostics, Division of Warner-Lambert & Co., Morris Plains, New Jersey

Tissue from each block was divided into epicardial and endocardial halves, weighed and counted for Tc-99m activity in a sodium iodide well counter. A standard was prepared by diluting a known amount of Tc-99m activity and 1.0 ml of the solution was counted along with the tissue and blood samples. Counts per minute per gram of the samples were calculated. Percent of administered dose per gram of the sample was evaluated from the known amount of administered activity. The data were tabulated as the ratio of cpm/gm of sample to cpm/gm of the normal myocardium.

RESULTS

Test for Radiochemical Purity

Radiochromatographic analysis indicated that Tc-99m labeled preparations had less than 5 percent of free pertechnetate. In some preparations examined, most of the radioactivity was associated with the same zone as that of unlabeled heparin identified by the blue stain of toluidine blue, which migrated 5 to 7 cm from the point of application in the cellulose acetate electrophoresis. Most (99%) of the radioactivity remained at the point of application in those preparations which did not contain heparin and had equivalent amounts of stannous chloride only, indicating reduced but unbound (hydrolyzed) technetium stayed at the point of application in this test system.

Activated whole blood clotting time was slightly elevated at 5 minutes postinjection. However, it returned to normal values at 1 hr after injection of the labeled agent (Table 1).

The rate of clearance of radioactivity from blood and the soft tissue uptake depended upon the source of heparin (Table 2). Radiolabeled heparin obtained from Abbott Labs (Panheparin^R) cleared much faster than that obtained from Upjohn Company. Activities remaining in blood (percent injected dose/gm blood) at 3 hours post-administration of the labeled agent (mean ± S.D.) were 1) 6.90 ± 2.10 (n = 16) for Upjohn heparin and 2) 2.65 ± 1.62 for Abbott Panheparin^R. The Abbott heparin cleared more rapidly by soft tissue uptake; liver uptake (percent I.D./g) for Abbott heparin was 86.49 ± 27.60 (n = 13) compared to 36.1 ± 12.4 (n = 16) for Upjohn heparin. Kidney uptake (percent I.D./g) was 70.3 ± 21.53 (n = 16) with Upjohn heparin compared to 25.17 ± 12.0 (n = 13) for Abbott heparin (Table 2).

Scintigraphic Infarct Localization Studies

As reported in our previous studies, the images were easily visible at one hour post-injection. Visibility improved with continued blood clearance, and remained visible throughout the entire 3 1/2 hours of the study. Figure 1 demonstrates the typical localization of the radioactivity in the ischemic zone of a dog with LAD occlusion for 3 hours and reperfusion.

In five dogs subjected to temporary LAD occlusion for 40 to 45 minutes, 4 had no gross visible damage and minimal histological damage (1 -2+), and 1 dog had moderately severe damage (3+) located primarily in the subendocardium. Only the latter dog showed modest accumulation of

TABLE 1

Tc-99m-Heparin

Activated Whole Blood Clotting Time (AWBCT)

Injection: 200-400 Units (Tc-99m) Heparin/dog

Dog #	Before Inj.	5 minutes postinjection	1 Hour postinjection
7929	2 min 15 sec	3 min 15 sec	1 min 30 sec
7931	1 min 15 sec	2 min 30 sec	1 min 45 sec
7849	2 min 30 sec	3 min 30 sec	2 min 8 sec
7789	1 min 58 sec	3 min	1 min 45 sec

Source of heparin: Hog mucosa, Abbott Laboratories, North Chicago, Illinois

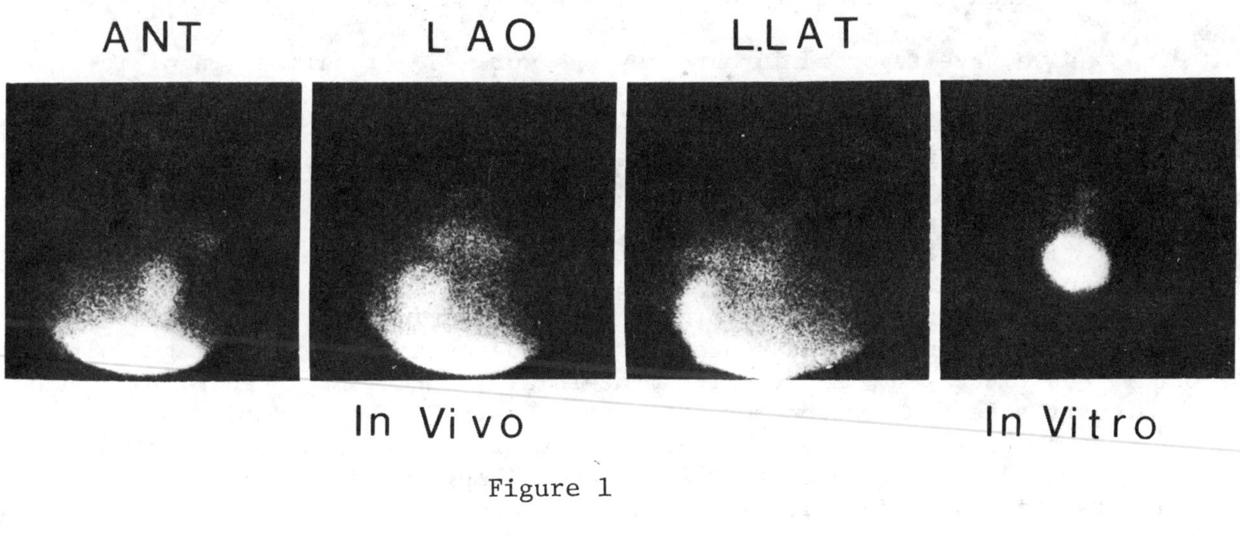

Figure 1

Anterior (ANT), left anterior oblique (LAO), and left lateral scintigrams showing increased concentration of Tc-99m labeled heparin in the freshly infarcted myocardial tissue in vivo. Notice activity in the liver. Far right panel demonstrates the heart imaged in vitro.

radioactivity in the ischemic region with damage to normal ratios 3.3 in epicardium and 9.1 in endocardium (Table 3).

In six dogs subjected to 3 to 3.5 hours of temporary LAD occlusion, 2 had minimal damage (1-2+ histologic damage) and the ratios of damaged-to-normal myocardium ranged 1.8 to 2.5 in the ischemic zones. Four dogs which showed hemorrhagic infarcts (4+ histologic damage) had marked localization of radioactivity in the damaged regions with correspondingly increased Tc-99m-heparin uptake, i.e., maximal damaged-to-normal tissue ratios of 16.6 to 50.4 (Table 3).

In four dogs with temporary LAD occlusion for 6 hours and reperfusion, 3 had hemorrhagic infarcts (4+ histologic damage), whereas one had patchy areas of damage (3+ histologic damage). The latter dog showed modest accumulation of radioactivity in the ischemic region with damaged-to-normal ratios of 4.6 to 8.0. The three dogs with the hemorrhagic infarcts showed higher damaged-to-normal ratios with epicardium of 6.4 to 9.9 and endocardium of 14.2 to 15.1 (Table 3).

DISCUSSION

Our data suggest that scintigraphic imaging of the myocardium and cardiovascular system with Tc-99m-heparin can be carried out as early as one hour post-injection and continued for several hours. Successful visualization of ischemically damaged myocardium in vivo can be achieved with the Tc-99m-heparin canine models of temporary LAD occlusion and reperfusion. The data obtained in the present study also suggest that maximal uptake of Tc-99m-heparin occurs after 3 hours of temporary LAD occlusion not only at a point when vessel and muscle injury were extensive, but also when the level of reperfusion was high (Table 3).

As we have reported in our previous work (8), localization of Tc-99m-heparin in dogs undergoing permanent LAD ligation was not sufficiently high to permit scintigraphic imaging of the damaged myocardium in vivo. Different results observed in the temporary and permanent coronary occlusion models may be related to the relatively large size of heparin molecules (compared to phosphate for example), reduced delivery of the agent via collaterals and/or to differences in the affinity of Tc-99m-heparin for damaged myocardium at later stages of infarction. Our very recent work with a modified form of Tc-99m-heparin prepared in phosphate buffer has shown improved concentration and positive in vivo images with this modified agent in dogs with myocardial infarcts produced by permanent coronary artery occlusion for 24 to 48 hours. This preparation uses a different heparin species that is obtained from sheep lung. It is not known whether differences in the origin of heparin make any difference in its ability to localize in damaged myocardium.

It is known that heparin has affinity for thrombin and antithrombin (11-13). The mechanism responsible for localization of labeled heparin in damaged myocardium needs further elucidation. Technetium-99m-heparin could have affinity for specific constituents of damaged myocardium, including amorphous and crystalline calcium deposits (14), denatured macromolecules or plasma proteins or constituents which accumulate in

TABLE 2

Organ Distribution of Tc-99m-Heparin in

Dogs 3-3.5 Hr Post I.V. Injection

% I.D./g of Organ x (10^3)

Source of Heparin	Upjohn	Abbott
No. of Dogs	n = 16	n = 13
Organ	Mean ± S. D.	Mean ± S.D.
Blood	6.9 ± 2.10	2.65 ± 1.62
Bone	4.8 ± 2.10	3.01 ± 1.93
Liver	36.1 ± 12.4	86.49 ± 27.60
Lungs	6.2 ± 2.70	5.08 ± 2.01
Kidneys	70.3 ± 21.53	25.17 ± 12.51
Normal Myocardium	2.7 ± 0.91	1.12 ± 0.54
	Ratio* to Normal Myocardium	
Blood	2.56	2.37
Bone	1.78	2.69
Liver	13.37	77.22
Lungs	2.30	4.54
Kidneys	26.04	22.47
Normal Myocardium	1.00	1.00

*(cpm/g tissue)/(cpm/g normal myocardium)

Table 3

RATIO (% Injected Dose per Gram) OF INJURED MYOCARDIAL TISSUE TO NORMAN MYOCARDIUM, Mean (Range)

MODEL: Temporary LAD Occlusion and Reperfusion

Period of Occl.	45 Minutes		3 Hours		6 Hours	
No. of Dogs	4	1	2	4	3	1
Gross Pathol.	No gross inf.	pallor	No gross inf.	Hemorrhagic inf.	Hemorrhagic inf.	pallor
Histologic Score* (0-4+)		3+	1+	4+	4+	3+
Scintigraphic Score (0-4+)						
In vitro	0 - 2+	3+	0 - 1+	4+	4+	N.A.
In vivo	0 - 2+	2+	0	3+ - 4+	3+ - 4+	N.A.
Perip. epi.	0.95 (0.7-1.2)	1.5	2.00 (1.8-2.2)	21.15 (5.2-39.9)	8.11 (6.8- 9.2)	2.97
Perip. endo.	1.53 (0.8-3.0)	6.1	2.35 (2.2-2.5)	25.13 (9.8-38.9)	14.90 (13.3-18.0)	5.90
Central epi.	1.38 (1.1-1.8)	3.3	1.05 (1.0-1.1)	23.55 (13.5-32.7)	9.03 (6.4- 9.9)	4.64
Central endo.	1.88 (1.2-3.4)	9.1	1.60 (1.2-2.0)	35.50 (16.6-50.4)	14.47 (14.2-15.1)	7.98
Normal myocardium % ID/gx(10^3)	1.47 (0.33-2.24)	0.44	0.86 (0.61-1.11)	1.13 (0.66-1.65)	1.11 (0.60-1.52)	1.81

*Histologic score (0-4+) is based on the extent of congestion, hemorrhage, neutrophilic infiltrate, muscle cell cell alteration including nuclear change and contraction band formation and degree of extension of damage from the subendocardium into the subepicardium.

N.A. = not available

the ischemically injured myocardial cells (15). Conformational change in the membranes of injured cells may result in binding sites for highly reactive molecules including polyanions such as heparin. Binding to the damaged microvasculature is yet another possible mechanism (16,17).

The modified procedure of labeling permitted us to inject less than 400 Units of heparin with 4 to 6 mCi of reduced Tc-99m per dog. The activated whole blood clotting times of blood drawn from the animals within 5 minutes post-injection of the agent were slightly elevated; however, they returned to normal (values prior to injection) within one hour of the injection (Table 1). There were no noticeable changes in bleeding from surgical wounds in the animals after administration of the small amount of labeled heparin. If required, further modifications in the prepration could be made to permit administration of less than 100 units of heparin per unit dose.

Heparin is a heterogeneous polymer. It has been observed that the rate of clearance of the labeled heparin depended upon the type and source of heparin. This might be due to variations in the composition of the different heparin preparations (18). Further studies are needed to evaluate the significance of this aspect as well as the relationship between structure, composition, and myocardial localization of the various heparin-type molecules. Some of the potential applications of these agents that need further exploration include scintigraphic localization of myocardial damage associated with trauma (contusion), or in the setting of open-heart surgery, as well as in the identification of damaged peripheral vessels associated with ischemia, trauma, or thromboemobolism.

ACKNOWLEDGMENTS

The authors thank Dorothy Gutekunst, Katie Wolf, Rose Sullivan, and Ruth Thornton for expert technical assistance, and Gayle Blust and Belinda Lambert for secretarial assistance.

This work was supported in part by the Ischemic Heart Disease Specialized Center of Research (SCOR) Grant HL-17669, NIH Grant HL-17777, the Southwestern Medical Foundation, Harry S. Moss Heart Fund, and Pulmonary Specialized Center of Research (SCOR) NIH Grant HL-1487.

This paper was presented in part at the World Federation of Nuclear Medicine and Biology, Second International Congress, 1978, Washington, D.C.

REFERENCES

1. Davis MA, Holman BL, Carmel AN: Evaluation of radiopharmaceuticals sequestered by acutely damaged myocardium. J Nucl Med 17:911, 1976.

2. Holman BL, Dewanjee MK, Iodine J, et al.: Detection and localization of experimental myocardial infarction with ^{99m}Tc-tetracycline. J Nucl Med 14:595, 1973.

3. Adler N, Camin L and Shulkin P: Rat model for acute myocardial infarction: Application to technetium-labeled glucoheptonate, tetracycline, and polyphosphate. J Nucl Med 17:203, 1976.

4. Bonte FJ, Parkey RW, Graham KD et al.: Distribution of several agents useful in imaging myocardial infarcts. J Nucl Med 16:132, 1975.

5. Parkey RW, Bonte FJ, Meyer SL et al.: A new method for radionuclide imaging of acute myocardial infarction in humans. Circulation 50:540, 1974.

6. Willerson JT, Parkey RW, Bonte FJ, et al.: Technetium stannous pyrophosphate myocardial scintigrams in patients with chest pains of varying etiology. Circulation 51:1046, 1975.

7. Zweiman FG, Holman BL, O'Keefe A, et al: Selective uptake of ^{99m}Tc complexes and ^{67}Ga in acutely infarcted myocardium. J Nucl Med 16:975, 1975.

8. Kulkarni PV, Parkey RW, Buja LM, Wilson JE III, Bonte FJ, Willerson JT: Technetium-labeled heparin: Preliminary report of a new radiopharmaceutical with potential for imaging damaged coronary arteries and myocardium. J Nucl Med 19:810, 1978.

9. Shapiro GA, Huntzinger SW, Wilson JE: Variation among commercial activated partial thromboplastin time reagents in response to heparin. Am J Clin Path 67:477, 1977.

10. Automated APTT: A platelet factor 3 reagent (partial thromboplastin) plus particulate activator for determination of activated partial thromboplastin times, in vitro diagnostic kit. General Diagnostics, Division of Warner-Lambert Company, Morris Plains, N.J.

11. Hook M, Bjork I, Hopwood J et al.: Anticoagulant activity of heparin: separation of high-activity and low activity heparin species by affinity chromatography on immobilized antithrombin. FEBS Letters 66:90, 1976.

12. Lam LH, Silbert JE, Rosenberg RD: The separation of active and inactive form of heparin. Biochem Biophys Res Commun 69:570, 1976.

13. Smith GF: The heparin-thrombin complex in the mechanism of thrombin inactivation by heparin. Biochem Biophys Res Comm 77:111, 1977.

14. Buja LM, Tofe AJ, Kulkarni PV, et al.: Sites and mechanisms of localization of technetium-99m phosphorus radiopharmaceuticals in acute myocardial infarcts and other tissues. J Clin Invest 60:724, 1977.

15. Dewanjee MK, Kahn PC: Mechanism of localization of ^{99m}Tc-labeled pyrophosphate and tetracycline in infarcted myocardium. J Nucl Med 17:639, 1976.

16. Sawyer PN, Pate JW: Electrical potential differences across the normal aorta and aortic grafts of dogs. Amer J Physiol 175:113, 1953.

17. Sawyer PN, Reardon JH, Ogoniak JC: Irreversible electrochemical precipitation of mammalian platelets and intravascular thrombosis. Proc Natl Acad Sci 53:200, 1965.

18. McDuffie MM, Dietrich CP, Nader HB: Electrofocussing of heparin: Fractionation of heparin into 21 components distinguishable from other mucopolysaccharides. Biopolymers 14:1473, 1975.

POTASSIUM-38, A 7.6 MINUTE HALF-LIVED RADIONUCLIDE FOR ASSESSMENT OF MYOCARDIAL FUNCTION

R. Chandra*, J.M. McDonald, R.E. Reiman and R.S. Tilbury

Memorial Sloan-Kettering Cancer Center
and *New York University Medical Center, New York, N.Y.

ABSTRACT

Potassium-38 in isotonic saline solution has been used to study the effect of cardio-active drugs, dipyridamole, propanolol and digoxin on the potassium uptake in the myocardium of dogs. The ^{38}KCl without added carrier was injected intravenously at about 0, 1, 2 and 3 hours after administration of the drug and the dog was scanned with a rectilinear scanner from 10-25 minutes after injection. The counts/sec /mCi corrected for decay were computed for a fixed number of scan elements and compared with controls. Dipyridamole produced a 50-80% increase in potassium uptake at 1 hour which returned to normal at 3 hours, Digoxin produced a 10-30% increase, and propanolol produced a 25% decrease at 1 to 2 hours. Results are compared with studies of Hamilton using Tl-201. Our results demonstrate that measurements of K-38 uptake can be made at hourly intervals to study the effects of cardio-active drugs.

INTRODUCTION

Advances in the instrumentation for conventional or tomographic imaging of positron emitting radionuclides (1,2) have given impetus to the development and production of these radionuclides. Recently, Lambrecht, et al. (3) and this laboratory (4) have reported the production of the positron emitter ^{38}K ($T_{1/2}$=7.6 Min.) in multimillicurie amounts for medical applications. Potassium-38 is a promising radiopharmaceutical for use in the evaluation of myocardial function, as the potassium cation is an essential element in the functional integrity of the myocardium. In the normal myocardium, uptake of potassium is proportional to the blood flow to the myocardium, whereas in the diseased state, it is a reflection of blood flow as well as the changes in the extraction efficiency of potassium by the myocardial cells. In addition to the relatively low radiation dose associated with K-38 (4), its 7.6 minute half life is an advantage over the other potassium isotopes or analogs, such as ^{42}K, ^{43}K, ^{82}Rb, or ^{201}Tl, because this makes it possible to study the effects of various cardio-active drugs or therapeutic surgical interventions on the myocardial uptake of potassium in vivo at intervals as short as an hour. The present study was undertaken to determine the feasibility of such an endeavor. Three cardioactive drugs-dipyridamole, digoxin and propranolol-were selected because of their wide clinical use and known cardio-effects. Another reason for their choice was that Hamilton et al. (5) have studied their effects on the myocardial uptake of ^{201}Tl, an analog of potassium and the current agent of choice for myocardial imaging. Therefore, it was felt that a comparison with these studies may yield useful information regarding similarities and dissimilarities in the biological behavior of potassium and thallium in the myocardium of the dog.

MATERIALS AND METHODS

Potassium-38 was produced in the Sloan-Kettering Institute cyclotron; the details of the method of production and processing will be reported elsewhere (4). Briefly, about 100 mg of NaCl was irradiated with a 40-50 μA beam of 14.7 MeV alpha particles for about 20 minutes. This typically generated 20-25 mCi of "no carrier added" ^{38}K at the end of irradiation, of which 10-12 mCi were available for injection after allowing for the processing time. This procedure could be repeated easily at hourly intervals.

Three normal dogs (one female and two males, approximately 20-25 kg) were used in this study, no dog being used more than once per week. The experimental protocol consisted of the following: the dog to be studied was anesthetized with sodium pentabarbitol and scalp vein sets inserted in superficial leg veins for the administration of ^{38}K or one of the above mentioned drugs and, in some cases, for sampling blood. Ten to 12 mCi of ^{38}K, accurately measured in a calibrated ionization chamber, were administered as an intravenous bolus three to four times at intervals of approximately one hour. The scanning of the dog's chest in lateral position was started ten minutes after the injection of ^{38}K and required approximately 15 minutes to complete. In some cases, a second scan was performed immediately after the completion of the first scan. The drug to be studied was injected slowly over a period of 2 to 3 minutes on the completion of the scans after the first ^{38}K injection, which served as a baseline for that dog. The amounts of the drugs used were 0.5 mg/kg for dipyridamole, 0.2 mg/kg for propranolol and 1.50 mg total for digoxin. To determine the variability in the quantitation of our baseline measurements, additional studies at different times were also obtained in each dog under anesthesia but without the administration of any additional drug. In order to determine the clearance of the ^{38}K, blood was sampled at selected intervals of time after the administration of ^{38}K in one anesthetized animal.

All quantitative ^{38}K scans were performed on the Sloan-Kettering Institute High Energy Gamma (HEG) rectilinear scanner. The HEG is equipped with two heavily shielded 4" x 4" NaI (Tl) detectors, constant response collimators which have a resolution of 2.0 cm (FWHM) at 511 keV, and a maximum of eight digital data channels with magnetic tape storage. Two energy windows were used for each detector in these studies: one set on the 511 keV photopeak due to the positron annihilation gamma rays, and a second window of the same width just above the 511 keV peak in order to correct for the scatter events in the 511 keV window due to the 2.17 MeV photons also emitted by ^{38}K.

Scan data were retrieved from the magnetic tape for analysis and display on a DEC PDP 11/70 computer. Counts from the two detectors were summed for each scan image element and corrected for the scatter contribution and the decay of ^{38}K during the scan time. The resulting image of the dog's heart was corrected for tissue background. The ^{38}K heart uptake was then quantitated by using a computerized algorithm for determining the organ boundary (6).

For one study, the TOKIM gamma camera system (7) was used to determine the time dependent uptake of ^{38}K by the myocardium *in vivo*.

RESULTS AND DISCUSSION

The counts per second per ml of venous blood, as determined at various times after the intravenous administration of ^{38}K, are given in Figure 1. As expected on the basis of known behavior of potassium, the clearance of ^{38}K from blood was observed to be rapid. Within five minutes, most of the ^{38}K radioactivity had been removed from the blood. A graphical analysis of the blood clearance data reveals the presence of at least two exponential components, one with a half time of 2.2 min and the other with a half time of about 37 min. Since blood samples earlier than 2.5 min were not obtained, it is possible that there may be another exponential component with a half life much shorter than 2.2 minutes.

The heart uptake of ^{38}K as a function of time as determined on the gamma camera is shown in Figure 2. These data show that there is no significant change in the ^{38}K activity in the heart area from about 2 minutes onwards after the ^{38}K injection. This, taken together with the blood clearance data, indicates that there was no appreciable biological excretion from the heart area during the time period in which scanning of the heart was performed with the rectilinear scanner for determining the myocardial uptake. This was further substantiated by the second rectilinear scans; these scans represented the ^{38}K myocardial uptake 25 minutes after ^{38}K injection (MAI). The decrease in ^{38}K heart uptake for scans started 25 MAI versus those performed at 10 MAI was found to be $2\% \pm 7\%$ (S.D., N=4) for a control dog with no drug other than the anesthetic.

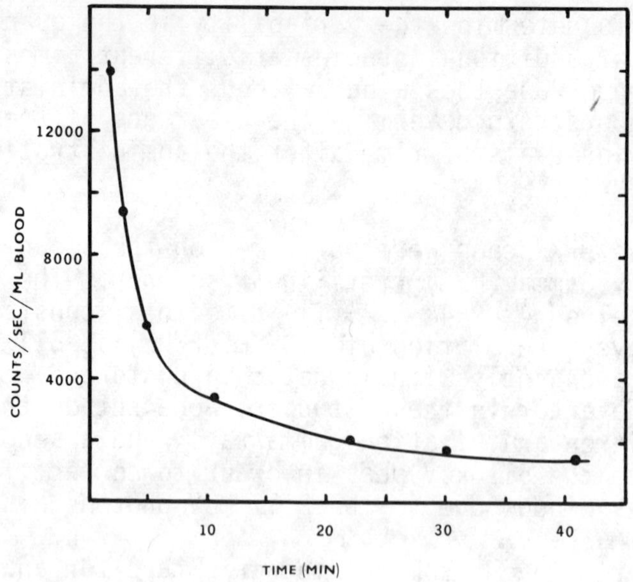

Figure 1.

Radioactivity in whole blood as a function of time following intravenous bolus administration of potassium-38 chloride in an anesthetized dog.

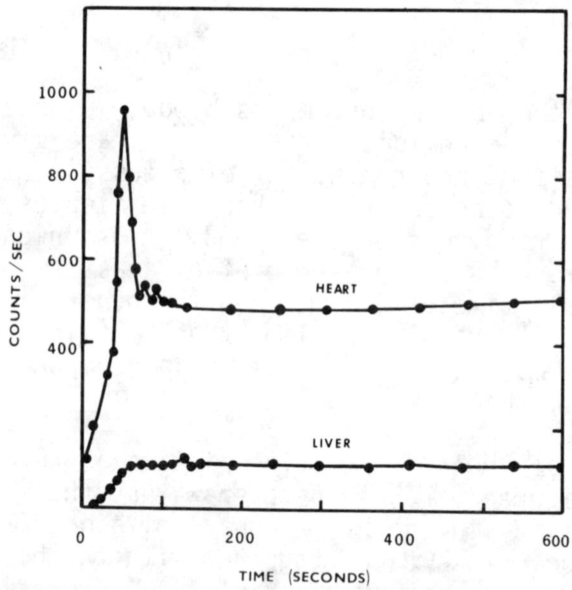

Figure 2.

Radioactivity in heart and liver regions as a function of time following K-38 administration in an anesthetized dog. Dynamic data were recorded and displayed using the TOKIM gamma camera system; regions of interest circumscribing heart and liver tissue were defined and the integrated count-rate, corrected for physical decay, was computed for appropriate time intervals.

Table 1 summarizes the results of the control scans for each of the three dogs. The standard deviations of the mean myocardial uptakes averaged 8%, indicating the reproducibility of the measurements and the computerized quantitation methods used. Our estimate of errors in the myocardial uptake measurement is about $\pm$ 7%, which is similar to the standard deviation of the mean counts obtained at different times in the anesthetized dogs. Within our experimental error, no effect of prolonged anesthesia was observed. To determine the fraction of injected activity which localized in the heart, a phantom containing a known amount of ^{38}K activity was scanned. Using this phantom calibration, it was found that approximately 5.1% of the injected activity localized in the heart in the control studies.

When dipyridamole (Figure 3) was administered to the anesthetized dog, there was an immediate and large increase (50-80%) in the myocardial uptake of ^{38}K. The effect of the drug persisted for about three hours. The difference in the myocardial uptake between the two studies may be due to some infiltration of the dipyridamole in one instance. Qualitatively, however, both cases are similar. The increase in myocardial uptake of ^{38}K in the case of dipyridamole is expected because it is a known coronary vasodilator and as such is thought to increase the perfusion of the myocardium.

In the case of digoxin (Figure 4), the effect of the drug on ^{38}K uptake by the myocardium is not as pronounced as in the case of dipyridamole, though it is noticably increased. In the case of propranolol (Figure 5), the overall effect seems to be a small decrease in the ^{38}K uptake by the myocardium. Both of these drugs exert their pharmacological effect on the myocardium in a complex fashion; therefore, in these cases it can not be said with certainty whether these effects are due to changes in blood flow to the myocardium or due to the changes in extraction efficiency of potassium by the myocardial cells. However, our findings are consistent with those of Hamilton (5) for the ^{201}Tl uptake by the myocardium of the dog, where they found an average 60% increase in the thallium uptake by the left ventricle of a dog after the administration of dipyridamole. In the case of propranolol, they found a decrease of 11% in the thallium myocardial uptake. From these results then it is evident that potassium and thallium uptakes by the myocardium are similarly effected by these two drugs. The study of Hamilton et al. consisted of two parts. In part one, four dogs were used with the four drugs-dipyridamole, propranolol, digoxin and furosemide. In the first week gamma camera images were obtained after the administration of ^{201}Tl from which the ratios of counts per unit area of myocardium to that of counts per unit of lungs were computed. These served as a base line. In the following week the gamma camera images were obtained in the same dogs after the administration of one drug to each dog and the same ratios were computed again. This ratio increased only in the case of dipyridamole which was due to the increased myocardial uptake of ^{201}Tl. In the second part, the uptake of ^{201}Tl in the myocardium and several other tissues was determined by sacrificing a number of dogs (12; 4 each for control, dipyridamole and propranolol) at 5 minutes post injection of the drug. Thus only one time point after the administration of each drug was determined. In our studies, no dog was sacrificed and it was possible to obtain three time points post drug injection which demonstrates one advantage of ^{38}K.

TABLE 1. BASELINE VALUES OF MYOCARDIAL UPTAKE IN DOGS A, B, AND C WITH PENTABARBITOL ANESTHESIA ONLY

Dogs (Sex)	No. Scans	Mean Myocardial Uptake (CPS/mCi) ±S.D.
A (M)	5	12,890 ± 1,059 (8%)
B (M)	3	12,576 ± 784 (6%)
C (F)	6	13,236 ± 1,184 (9%)

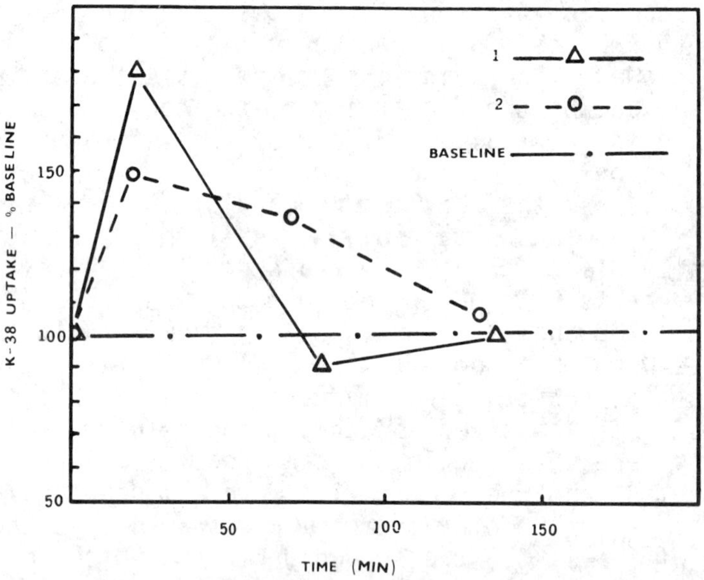

Figure 3.

K-38 uptake in the heart at various times following the administration of 0.5 mg/kg dipyridamole. Curves 1 and 2 represent two separate experiments with anesthetized dog A. The uptake values are expressed as percent of the average baseline value for this dog (see Table 1).

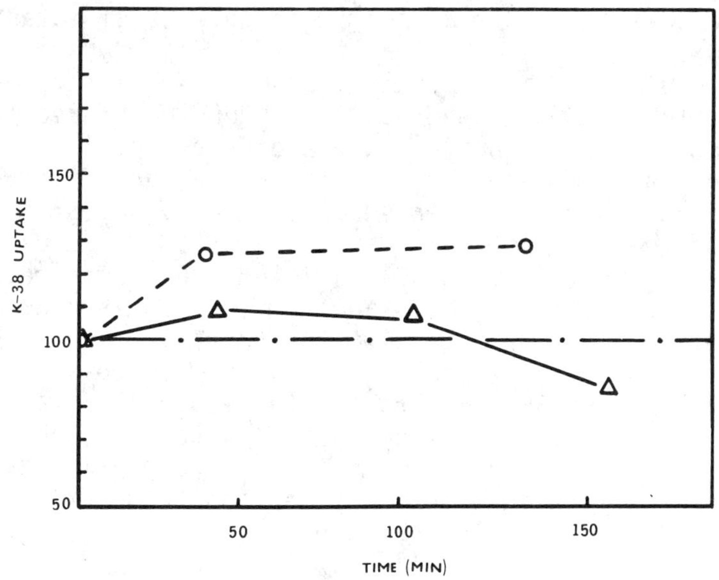

Figure 4.

K-38 uptake in the heart at various times following the administration of 1.5 mg digoxin. The two curves represent separate experiments in animal B.

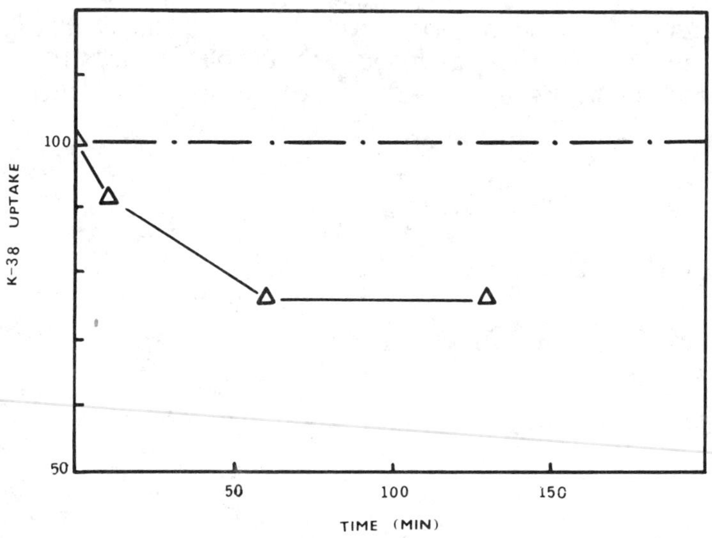

Figure 5.

K-38 uptake in the heart at various times following the administration of 0.2 mg/kg propanolol in dog C.

Sapirstein (8) had suggested that the organ uptakes of potassium and rubidium can be used to determine the fractional cardiac output to various organs except the brain. However, recently, other investigators (9) comparing the uptake of labeled microspheres with $^{86}RbCl$ concluded that the $^{86}Rb^+$ uptake does not give an accurate measure of the fractional distribution of cardiac output to many organs or tissues in the rat. They suggested that the fraction of injected $^{86}Rb^+$ that a tissue or organ extracts from the blood, depends not only on the fraction of cardiac output that the tissue received, but also on several other factors: the rate of blood flow through the tissue, the resistance of the capillary walls and cell membranes to transport of $^{86}Rb^+$ and the extraction ratio for $^{86}Rb^+$ in other tissues. Some or perhaps all of these factors vary with the physiological status of the animal. The uptake of $^{86}Rb^+$ by the myocardium was only 0.5 that of microspheres, though a change in blood flow to the myocardium was accurately reflected by changes in the $^{86}Rb^+$ uptake to the myocardium. Thus it seems that the uptake of rubidium can not be used to determine the absolute blood flow to the myocardium but can be used to determine the changes in blood flow to the myocardium.

CONCLUSIONS

The 7.6 minute half life of ^{38}K allows repeated measurements of potassium uptake by the myocardium in vivo at hourly intervals. This can be used to study the effect of a variety of cardio-active drugs on the potassium uptake non-invasively. Presently ^{201}Tl is widely used for heart scanning where studies are performed in patients with and without exercise. Because of the 72 hour half life of this radionuclide, these studies must be separated generally by about a week. With ^{38}K, both these studies can be performed in less than two hours and in one sitting. The use of positron tomography may increase the usefulness of these studies further as then it will be possible to determine the effect of various drugs or therapeutic surgical interventions on normal and diseased portions of the myocardium on a regional basis (e.g., roughly 1 cm^3).

REFERENCES

1. Phelps, M.E., Hoffman, E.J., Haung, S.C., et al. ECAT: A new computerized tomographic imaging system for positron emitting radiopharmaceuticals. J. Nucl. Med. 19:635-647 (1978).

2. Phelps, M.E., Emission Computed Tomography. Sem. in Nucl. Med 7:337-365, (1977).

3. Lambrecht, R.M., Hara, T., Gallagher, B.M., et al. Cyclotron isotopes and radiopharmaceuticals XXVII. Production of Potassium-38 for myocardial perfusion studies. Int. J. Appd. Radiat. & Isotopes 29:667-671 (1978).

4. Tilbury, R.S., Dahl, R.J., Chandra, R., et al. Production of 7.6 minute Potassium-38 for medical use. (submitted for publication - J. Nucl. Med.)

5. Hamilton, G.W. Narahara, K.A., Yee, Ho, et al. Myocardial imaging with ^{201}Tl: Effect of cardiac drugs on myocardial images and absolute tissue distribution. J. Nucl. Med. 19:10-16 (1978).

6. Reiman, R.E. - Unpublished Work

7. Monahan, W.G., Beattie, J.W., Powell, M.D., et al. Total organ kinetic imaging monitor; System design and applications. In Medical Radioisotope Scintigraphy, Vol 1, Proceedings of IAEA Symposium Monte Carlo, 1972, pp 285-297, IAEA, Vienna, 1973.

8. Sapirstein, L.A. Regional blood flow by fractional distribution of indicators. Am. J. Physiol. 193:161-168 (1958).

9. Foster, D.O. and Frydman, M.L. Comparison of microspheres and $^{86}Rb^{+}$ as tracers of the distribution of cardiac output in rats indicates invalidity of $^{86}Rb^{+}$ Based Measurements. Can. J. Physiol. Pharmacol. 56:97-109 (1978).

ACKNOLWEDGEMENTS

This work supported by DOE contract No. EE-77-5-02-4268. We thank R. Lee and J.R. Dahl for technical assistance.

REVIEW AND CURRENT STATUS OF HEPATOBILIARY IMAGING AGENTS

Michael D. Loberg and David W. Porter

University of Maryland Hospital
Departments of Medicine and Medicinal Chemistry
Baltimore, Maryland 21201

James W. Ryan

University of Chicago
Department of Radiology
Chicago, Illinois 60637

ABSTRACT

This review will outline aspects of the normal hepatobiliary anatomy and physiology relevant to an understanding of cholescintigraphy and discuss factors of hepatobiliary diseases which impact on the design of new hepatobiliary radiopharmaceuticals. The synthesis, pharmacokinetics, and structure-distribution relationship of the existing ^{99m}Tc hepatobiliary agents will be compared, and the clinical status of present cholescintigraphic agents will be summarized. Future trends in the design and use of hepatobiliary tracers will be presented.

INTRODUCTION

This chapter will review briefly the current status of radiopharmaceuticals used in scintigraphic studies of the hepatobiliary (HB) system. The initial ^{99m}Tc-labeled cholescintigraphic agent, ^{99m}Tc-penicillamine, developed by Tubis and co-workers was described at the First International Symposium on Radiopharmaceuticals in 1974 (1). Since that time, more than twenty new ^{99m}Tc cholescintigraphic agents have been proposed (Table 1) and several of these tracers have been investigated in clinical studies which have demonstrated their superior imaging features compared to the standard HB agent, ^{131}I-rose bengal (RB). This review will outline aspects of the normal HB anatomy and physiology relevant to an understanding of cholescintigraphy and discuss factors of HB diseases which impact on the design of new HB radiopharmaceuticals. The synthesis, pharmacokinetics, and structure-distribution relationship of the existing ^{99m}Tc HB agents will be compared, and the clinical status of present cholescintigraphic agents will be summarized. Future trends in the design and use of HB tracers will be presented.

ANATOMY AND PHYSIOLOGY

The liver is the largest organ in the body, averaging 2% of the total body weight in adults, and performs a complex variety of metabolic functions. It is highly vascular, receiving about 20% of the resting cardiac output, and its afferent flow is routed through both the portal vein (75%) and hepatic artery (25%). The metabolic functions of the liver occur primarily in the parenchymal cells or hepatocytes which contain a well-developed system of organelles and are richly perfused by the vast network of hepatic sinusoids (Figure 1). The products of hepatocyte metabolism may be secreted into either the plasma or the bile or stored within the cells for future use. The liver also contains active phagocytic cells, termed Kupffer cells, which are an integral part of the body's reticuloendothelial system. These cells are situated at intervals along the lining of the sinusoids and remove foreign particulates from the blood.

An active transport process exists at the hepatocytes which selectively removes compounds from the blood and concentrates them in the bile. Several detailed reviews of this complex system have appeared (2-4) and only a brief summary is presented here. As shown in Figure 1, four carrier-mediated transport pathways have been identified in the hepatocyte, each with separate sites of uptake at the cell membrane. These pathways accommodate either organic anions, organic cations, neutral compounds, or conjugated bile salts. In the transport processes there may be intermediate storage sites such as ligandin (Y protein) and Z protein which bind bilirubin and other organic anions, and metabolic processing may be required prior to secretion. Many of the compounds utilizing these uptake and metabolic pathways are secreted into

TABLE 1. Tc-99m HEPATOBILIARY AGENTS GROUPED ACCORDING TO TYPE OF COMPLEX

		Ref.
1.	THIOL COMPLEXES	
	Tc-99m-penicillamine	(17)
	Tc-99m-βmercaptoisobutyric acid (MIBA)	(18)
	Tc-99m-dihydrothioctic acid (DHT)	(19)
	Tc-99m-6-mercaptopurine	(20)
	Tc-99m-kethoxal-bis(thiosemicarbazone)	(21)
2.	AMINE COMPLEXES	
	Tc-99m-sulfonyl ureas	(22)
	Tc-99m-hydralazine	(23)
3.	8-HYDROXYQUINOLINE COMPLEXES	
	Tc-99m-Bioquin-7CA	(24)
4.	AMINO ACID SCHIFF'S BASE COMPLEXES	
	Tc-99m-pyridoxylidene glutamate	(25)
	Tc-99m other pyridoxylideneaminates	(26-28)
	Tc-99m-Sn-pyridoxylidene isoleucine	(29)
	Tc-99m-3-hydroxy-4-formyl-pyridine glutamate acid	(30)
5.	SUBSTITUTED IMINODIACETATE (IDA) COMPLEXES	
	Tc-99m-HIDA	(31)
	Tc-99m-diethylacetanilido IDA	(32)
	Tc-99m other acetanilido IDA derivatives	(33-35)
	Tc-99m-iminodicarboxymethyl-2-napthyl ketone	(23)
	Tc-99m-phthalein complexone	(36)
	Tc-99m-N-(5,pregnene-3βol-20yl-carbamoylmethyl)IDA	(37)
	Tc-99m-3α:7α:12αtrihydroxy-24norcholanyl-23 IDA	(37)
6.	PROTEIN COMPLEXES	
	Tc-99m-protamine	(38)
	Tc-99m-asialo-orosomucoid	(39)
7.	MISC. COMPLEXES	
	Tc-99m-tetracycline	(40)

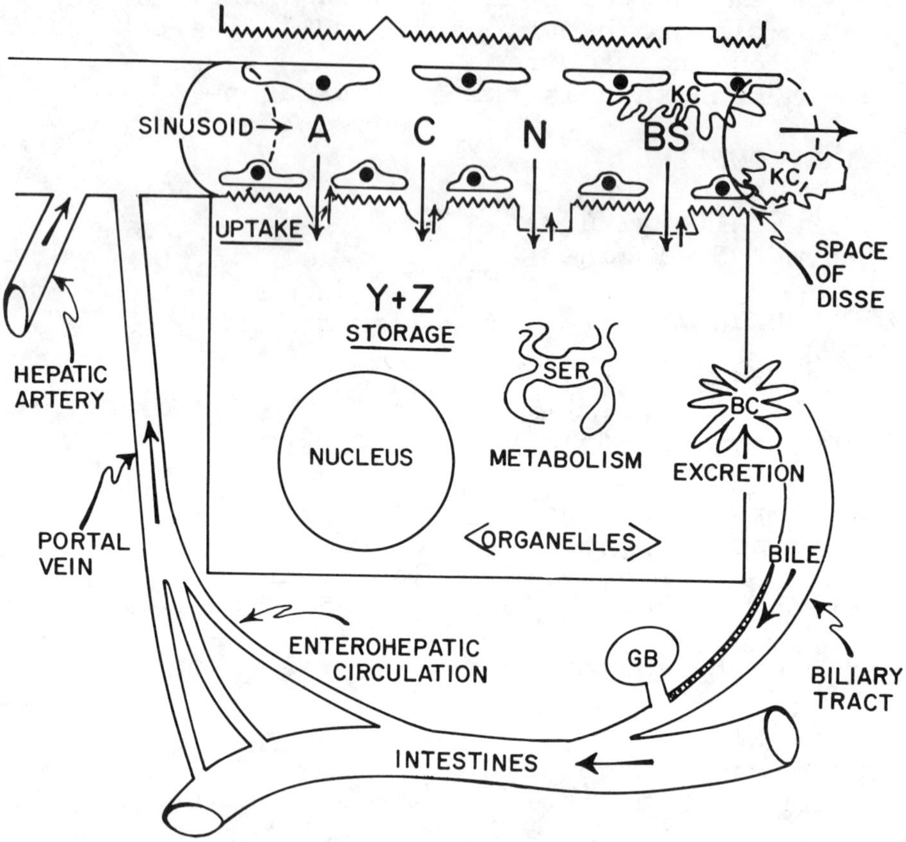

Figure 1

Schematic representation of an hepatocyte. Substrates in the blood diffuse through pores in the endothelial lining of the sinusoids and bind to the hepatocyte at one of four membrane-bound carriers: anionic (A), cationic (C), non-ionic (N), bile salt (BS). Inside the hepatocyte the substrate may be stored at specific binding sites such as Y and Z proteins, and it also may undergo metabolic conversion at other sites including the smooth endoplasmic reticulum (SER). Biliary excretion occurs at a biliary canaliculus (BC). Subsequently the substrate in the bile may be stored and concentrated in the gallbladder (GB) or excreted into the intestines. Some biliary components are reabsorbed from the intestines into the portal vein and reextracted by the hepatocyte ("enterohepatic circulation"). The sinusoids are lined by Kupffer cells (KC) which are a part of the reticuloendothelial system.

the bile at a bile canaliculus. While the exact mechanisms of transport and bile secretion are not well understood, it is agreed that the rate limiting step for the overall transport of compounds from the blood into the bile is the excretory process at the bile canaliculus. Bile flow is a function of osmotic forces, and two processes which actively transport solutes into the bile canaliculus have been reasonably well identified. They are the system for bile salt transport ("bile acid dependent flow") and the sodium pump ("bile acid independent flow"). Augmentation of bile flow (choleresis) may be caused by infusion of bile salts and some other agents which are secreted at the canaliculus, and it may also be produced by hormones that act on the secretory epithelium of the biliary ductules. After leaving the liver, bile may enter the duodenum directly or it may flow into the gallbladder where it is concentrated, stored, and subsequently passed into the intestines during gallbladder contractions. The primary determinant of gallbladder emptying is circulating cholecystokinin (CCK) that is secreted post-prandially into the blood by cells lining the upper intestinal tract. Bile is important for the efficient digestion and absorption of ingested fats. Some bile components may be resorbed from the intestines and returned to the liver via the portal circulation, a cyclic process that has been termed the enterohepatic circulation.

CLINICAL GOALS

The role of cholescintigraphic agents in the assessment of the HB system can be understood by a reexamination of Figure 1. In simplest terms, the clinical usefulness of hepatobiliary agents is based upon their ability to produce liver images reflecting the distribution of hepatocytes, to outline the biliary tract, and to trace the pathway of bile flow. To accomplish these goals a radiopharmaceutical must be capable of binding to receptors in membranes, concentrating within hepatocytes, and undergoing biliary excretion. There are several hepatocyte cell membrane receptors and a variety of intrahepatic transport pathways available for radiopharmaceutical uptake and transit to the excretory sites at the bile canaliculus. Both the nature of the physiological function to be measured and the underlying mechanisms of the HB diseases to be evaluated by these agents are important factors which must be considered in the design of clinically useful cholescintigraphic agents.

If images showing only hepatocyte distribution are needed, tracers with prolonged intracellular residence should be chosen, but procedures designed to outline both distribution of hepatocytes and biliary anatomy should use tracers with faster transit times. Both types of tracers may be useful in the evaluation of focal defects noted on radiocolloid liver images. Quantitation of overall liver function or regional intrahepatic function requires tracers which have only an hepatic clearance mechanism. Hepatic blood flow and bile salt kinetics are examples of functions that could be measured with appropriate tracers. For studies of biliary tract physiology, tracers which rapidly enter the biliary tract in sufficient quantity to outline the major hepatic ducts and the gallbladder are needed; and the evaluation of diseases with obstructions or deviations in the path of bile flow require tracers with the same features.

In the jaundiced patient in whom the sites of bilirubin uptake, storage and excretion in the liver are saturated, it will be necessary to develop tracers which either compete effectively with bilirubin for these sites or use alternate transport pathways from the blood into the bile.

For patients who have had liver transplants, it is important to assess the integrity of the surgical anastomoses, determine the current functional status of the transplanted liver, and detect episodes of rejection. The first two requirements may be fulfilled by one of the tracers previously described, but the third may require significant new tracer features.

In view of the diverse clinical requirements, it is unlikely that any single tracer using one hepatic uptake and transport process will be best suited for the evaluation of all HB diseases. It is more likely that the kinetics of individual hepatocyte agents will provide optimal data for only a few specific clinical disease entities or syndromes. These considerations have stimulated the development of multiple new cholescintigraphic agents.

HEPATOBILIARY AGENTS

The development of HB imaging agents began in 1955 when the known specific hepatic uptake and subsequent biliary clearance of the dye, rose bengal, led Taplin (5) to synthesize ^{131}I-rose bengal (RB). This agent achieved widespread clinical usage and prompted the development of ^{131}I-bromosulphalein (6), ^{131}I-ioglycamic acid (7), ^{131}I-toluidine blue (8), ^{131}I-thyroxine glucuronide (9), ^{131}I-indotricarbocyanine (10), ^{131}I-indocyanine green (11), and ^{131}I-asialoproteins (12,13) as well as other agents. The reduction in absorbed dose and the improvement in image quality associated with ^{123}I, prompted the subsequent synthesis of ^{123}I-bromosulphalein (14), ^{123}I-rose bengal (15), and ^{123}I-indocyanine green (^{123}I-ICG) (16). Recently, many ^{99m}Tc complexes have been found to concentrate in the polygonal cells of the liver and to be excreted into the bile. A partial listing of these ^{99m}Tc-labeled hepatobiliary agents is given in Table 1. A search of the literature reveals that the development of ^{99m}Tc-labeled hepatobiliary agents has been one of the most active areas in radiopharmaceutical development during the last few years. This sudden increase in the number and diversity of ^{99m}Tc hepatobiliary agents is not unexpected since ^{99m}Tc, for which there is no known metabolic requirement, should be most readily handled by one of the body's detoxification mechanisms. Perhaps the only surprise is that these ^{99m}Tc-labeled hepatobiliary agents were developed a number of years after the introduction of ^{99m}Tc agents for examining the elimination mechanisms of the lungs, kidneys, and RES cells within the liver.

Research toward the development of the optimal hepatobiliary agent is currently being conducted in three general areas. One approach involves a comparison of the existing ^{99m}Tc-labeled hepatobiliary agents with the previous standard, ^{131}I-rose bengal, in order to determine the optimum kinetic profile for use in each of the patient populations. Other research is directed toward the development of new, improved approaches to measuring hepatobiliary function. Lastly, research is being undertaken to determine the exact mechanism by which these radiopharmaceuticals undergo hepatobiliary clearance. As in all instances of radiopharmaceutical usage, a knowledge of the rate limiting step in a radiopharmaceutical's distribution is necessary before the scintiphotos can be correctly interpreted.

COMPARISON OF CURRENT HEPATOBILIARY AGENTS

The speed and extent of hepatobiliary clearance are important parameters in the development of new hepatobiliary agents. Scintigraphic representation of hepatobiliary function is best achieved with radiopharmaceuticals that are rapidly excreted into the bile in large quantities. This allows the radiopharmaceutical to enter the bile as a bolus and guarantees maximum contrast in the image. Once optimum kinetics have been developed, other significant criteria to be evaluated include the ease with which the radionuclide is incorporated into the drug and the stability of the bond between the radionuclide and drug. The latter is important since hepatobiliary studies often extend over 24 hours.

When comparing relative in vivo kinetics of various radiopharmaceuticals, it is important to use the same animal species. Hirom and co-workers (41) examined biliary excretion of sixteen organic ions in female rats, guinea pigs and rabbits. They reported the three hour cumulative biliary excretion of phenolphthalein glucuronide for the rat, guinea pig and rabbit as 82%, 44% and 22%, respectively; while the corresponding three hour cumulative urinary excretions were 0%, 20% and 53%. In general, it has been demonstrated that the rat, dog and hen are good biliary excreters, the rabbit and guinea pig are poor biliary excreters, and the cat and sheep are intermediate biliary excreters. Primates and humans appear to fall into the second category. Jansholt and co-workers (42) studied the hepatobiliary clearance of ^{99m}Tc-PYG in four animal species and observed similar variations in the degree of urinary clearance compared to biliary clearance with significantly greater urinary excretion in dogs than in monkeys and rabbits.

Table 2 compares the tissue distribution of four selected hepatobiliary agents in mice. The four agents selected, while not necessarily representing the optimal agents available, do

TABLE 2. ORGAN DISTRIBUTION OF HEPATOBILIARY AGENTS IN MICE

AGENT	BLOOD		LIVER		INTESTINES		Ref.
	5min	60min	5min	60min	5min	60min	
^{131}I-RB	4.48	1.99	71.0	15.8	6.5	67.2	(43)
^{123}I-ICG	5.54	3.87	14.9	5.79	13.2	36.7	(16)
^{99m}Tc-PYG	9.34	0.76	11.1	5.62	7.55	20.9	(43)
^{99m}Tc-HIDA	3.3	0.8	28.2	8.0	38.3	68.3	(44)

All values represent percent dose per organ.

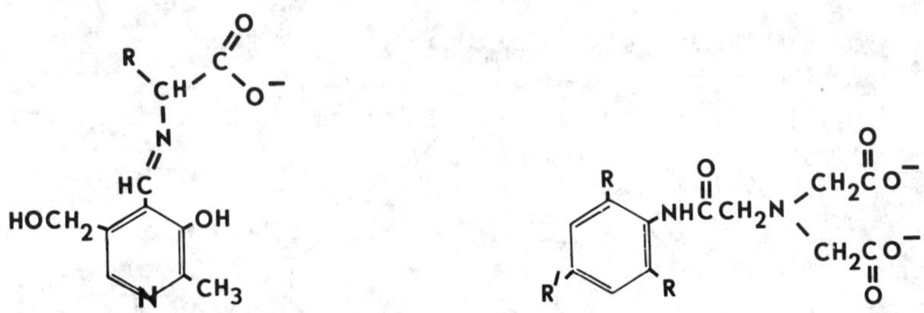

Figure 2

General structure of pyridoxylideneaminates and acetanilidoiminodiacetates with various substituents at R and R'.

indicate the historical trend in radiopharmaceutical development. ^{123}I-labeled agents were developed to replace ^{131}I agents, but both are being challenged currently by new ^{99m}Tc-labeled agents. The ^{99m}Tc-pyridoxylideneglutamate (^{99m}Tc-PYG) reported in Table 2 was prepared by autoclaving, and in this method, pyridoxal is presumed to reduce pertechnetate while forming a technetium complex with the Schiff's base generated from the reaction of pyridoxal and glutamic acid. ^{99m}Tc-PYG was determined to have kinetics superior to those of its technetium predecessor, ^{99m}Tc-penicillamine, but the extent and speed of its hepatobiliary clearance were still less than those of ^{131}I-rose bengal. Considerable variability between workers exists regarding the distribution of ^{99m}Tc-PYG in mice with Baker et al (45) reporting the sixty-minute intestinal activity to be 39.3%. ^{99m}Tc-HIDA was found to clear the blood more rapidly compared to RB, and to possess nearly identical intestinal activity at sixty minutes. Neither ^{99m}Tc-PYG nor ^{99m}Tc-HIDA are metabolized in vivo or extensively protein-bound in plasma.

Various analogs of ^{99m}Tc-PYG and ^{99m}Tc-HIDA have been developed in an effort to maximize their speed and extent of hepatobiliary clearance (Figure 2). A variety of ^{99m}Tc-pyridoxal amino acid complexes similar to ^{99m}Tc-PYG have been synthesized, but none was found to have better hepatobiliary clearance than ^{99m}Tc-PYG. However, Kato and Hazue (29) reported much greater hepatobiliary specificity and totally different radiochromatographic results if the ^{99m}Tc-labeled Schiff bases were prepared at room temperature with stannous ion as a reducing agent as opposed to the previously reported method involving a thirty-minute autoclaving. They also reported that ^{99m}Tc-(Sn)pyridoxylidene isoleucine (PYisoL) had a hepatobiliary specificity similar to that of ^{99m}Tc-HIDA and much greater than that of ^{99m}Tc-PYG. It would thus appear that the radiochemical structure of the complex formed between ^{99m}Tc and Schiff's bases is affected by either the nature of the reducing agent, the reaction temperature, the presence of ascorbic acid added in preparing ^{99m}Tc(Sn)-PYisoL, or a combination of these factors.

Various derivatives of ^{99m}Tc-HIDA have been prepared by altering the substituents on HIDA's aromatic ring. In 1976, Wistow et al (46) reported that the 2,6-diethyl derivative of HIDA had increased hepatobiliary specificity relative to that of ^{99m}Tc-HIDA. They subsequently reported that the 2,6-diisopropyl, 4-butoxy, 4-isopropyl and 4-butyl derivatives all possessed excellent hepatobiliary specificity, differing primarily in their speed of blood clearance and extent of urinary output (35). Burns and co-workers (34), studying a series of ^{99m}Tc-HIDA analogs reported that the 4-iodo derivative had the greatest hepatobiliary specificity. In general, the ^{99m}Tc-HIDA derivatives possess greater hepatobiliary specificity than do the ^{99m}Tc-PYG derivatives prepared by the pyridoxal reduction method. No comparisons have been reported between ^{99m}Tc-HIDA and ^{99m}Tc-(Sn)-PYisoL.

Other important factors to be evaluated when comparing hepatobiliary imaging agents include radiochemical purity and in vivo stability. Pertechnetate, the most common radiochemical impurity in technetium complexes, is particularly deleterious in ^{99m}Tc hepatobiliary studies because its intestinal clearance complicates image interpretations regarding overall biliary tract patency. It is also important that the ^{99m}Tc hepatobiliary agents should be stable in vitro and should be inert with respect to any in vivo ligand exchange reactions that might diminish the radiopharmaceutical's radiochemical purity with time. The factor of in vivo stability is particularly important with hepatobiliary agents since imaging studies may be performed up to 24 hours postinjection.

The percent of $^{99m}TcO_4^-$ in ^{99m}Tc-PYG was studied as a function of the reaction pH by Baker et al (45). They reported a minimum of 2% $^{99m}TcO_4^-$ at a pH of 8.0. Exchange reactions with ^{99m}Tc-PYG have yet to be reported. ^{99m}Tc-HIDA has been shown to be inert with respect to ligand exchange reactions at physiological pH and to exist as a single radiochemical with greater than 98% purity.

Finally, significant quantities of radiochemical impurities have been found in the chelate fraction of various ^{99m}Tc-labeled hepatobiliary agents. Yokoyama et al (47), studying ^{99m}Tc-penicillamine, were able to resolve the chelate fraction of the reaction mixture into five different radiochemicals. The relative abundances of these radiochemicals were found to vary with reaction conditions, but only one of the complexes had significant hepatobiliary specificity. Similar results have been observed by Fields (48) and co-workers when labeling substituted phenylethyliminodiacetates with reduced ^{99m}Tc. Similarly, Fritzberg et al (23) reported that ^{99m}Tc-iminodicarboxymethyl-2-napthyl ketone had greater biliary excretion than did ^{99m}Tc-2-(dicarboxymethylimino)methylnapthene, an observation that is perhaps best explained by the difference in their radiochemical structures. These results demonstrate the necessity of a good analytical system for determining the radiochemical purity of reaction mixtures. The chemical purity of the reactants should be considered a necessary but insufficient criterion for assuring the radiochemical purity of a radiopharmaceutical. High pressure liquid chromatography and high voltage paper electrophoresis have been found to be the most reliable analytical systems for determining radiochemical purity.

STRUCTURE-DISTRIBUTION RELATIONSHIPS

In 1978, Firnau (49) examined the existing ^{99m}Tc chelates to determine why they were effective as cholescintigraphic agents. He found that all of the ^{99m}Tc hepatobiliary agents had molecular weights between 300 and 1,000, existed as organic anions, contained at least two aromatic ring systems in the molecule, and were all bound to serum albumin. While this initial work was helpful in focusing attention on the general mechanisms of hepatobiliary clearance, it served also to place emphasis on the importance of radiochemical structure in determining in vivo distribution.

The first insight into the importance of radiochemical structure in determining in vivo distribution was provided by a comparative in vivo distribution study using ^{99m}Tc-HIDA ^{14}C-HIDA and ^{113}Sn-HIDA (50). ^{99m}Tc-HIDA had a 90 minute cumulative biliary excretion of 71% in dogs as opposed to 0.153% and 0.015% for ^{14}C-HIDA and ^{113}Sn-HIDA, respectively. It was apparent, therefore, that the ^{99m}Tc complex and not the complexing agent was the main compound undergoing hepatobiliary clearance. This finding prompted two different research groups to begin work in determining the structure of ^{99m}Tc-HIDA, resulting in their independent arrival at the structure in Figure 3. ^{99m}Tc-HIDA was found to exist as a bis structure with an overall charge of -1. The ^{99m}Tc was in the +3 oxidation state. Clearly, ^{99m}Tc-HIDA had a higher molecular weight than HIDA, 782 versus 294, and was substantially more lipophilic than HIDA itself. This enhanced lipophilicity has been confirmed by HPLC on reverse phase columns. Fields and co-workers (48) provided additional structure distribution information by comparing the in vivo distribution of ^{99m}Tc-HIDA with that of ^{99m}Tc-phenethyliminodiacetate. Unlike ^{99m}Tc, ^{99m}Tc-phenethyliminodiacetate was found to exist as three radiochemicals, only one (22%) of which was found to have the lipophilicity, HPLC retention time and hepatobiliary clearance characteristic of ^{99m}Tc-HIDA. Extending this work to other compounds, these workers were able to demonstrate that the carbamoylmethyl moiety in HIDA was necessary in order to maintain the bis structure shown in Figure 3. They speculated that iminodiacetates should be synthetically incorporated into drug analogs in such a fashion as to reduce the electron density on the imino nitrogen from a pKa value of 8 to about 6.

Wistow and co-workers (46) were the first to demonstrate the profound effect of aromatic substitution on the in vivo distribution of ^{99m}Tc-HIDA. Burns and co-workers (51) subsequently undertook a systematic study of nine different aromatically substituted derivatives of ^{99m}Tc-HIDA in mice. Their data are shown in Figure 4 in which the percent biliary excretion at 60 minutes is plotted as a function of the natural log of the molecular weight divided by the charge. The observed linear relationship between the percent biliary excretion and the natural log of the molecular weight divided by the charge is one of the most significant structure distribution relationships yet to be determined for ^{99m}Tc-labeled hepatobiliary agents. These workers provided an additional piece of information when they substituted $^{51}Cr^{+3}$ for $^{99m}Tc^{+3}$ in ^{99m}Tc-HIDA, which is known to form bis complexes with iminodiacetate. The percent biliary excretion of ^{51}Cr-HIDA is point J on Figure 4. Their work indicated that the addition of high molecular weight substituents on the aromatic ring increase hepatobiliary specificity and that the role of ^{99m}Tc in ^{99m}Tc-HIDA is simply that of a metal ion which shapes the overall structure of the complex. Once again, these data stress the importance of the radiochemical structure

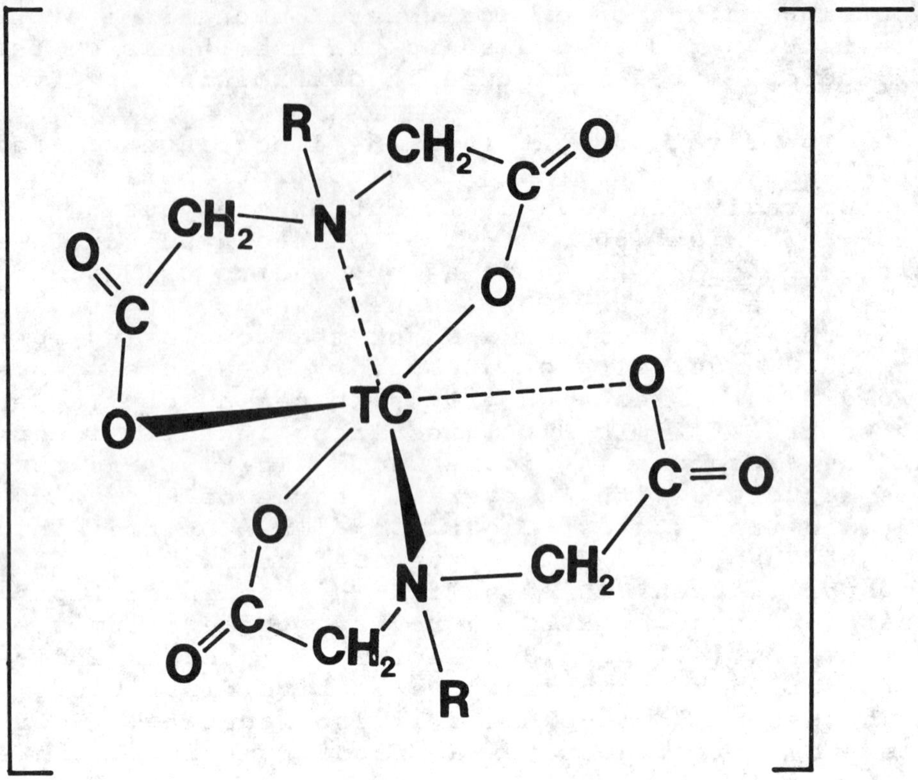

Figure 3

Bis comples of N-substituted ^{99m}Tc-iminodiacetate.

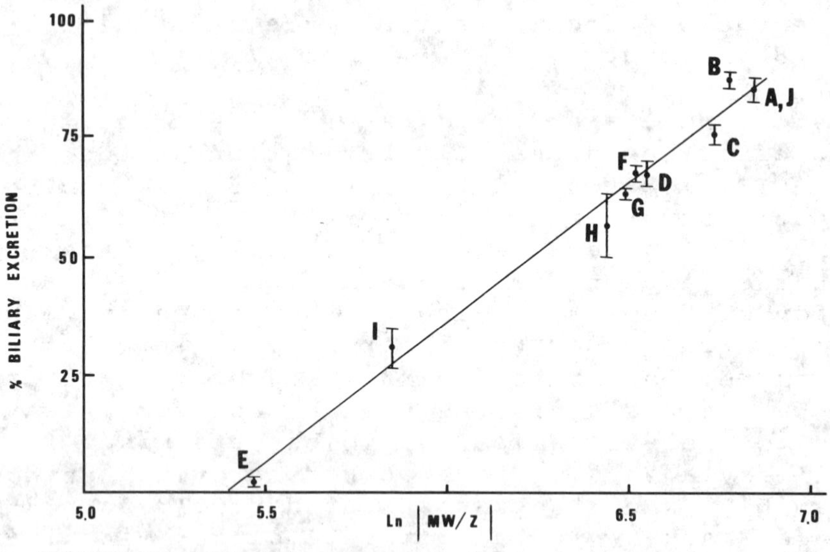

BURNS, MARZILLI, WAGNER, 1977

Figure 4

Biliary excretion of ^{99m}Tc-HIDA analogs (A-I) and ^{51}Cr-p-iodoHIDA (J) in mice at 60 minutes. Data are plotted as a function of the natural log of molecular weight of the complex divided by its charge.

of these agents. The development of these structure-distribution relationships would have been impossible without information on the radiochemical structure and charge of the complexes.

The mechanisms by which ^{99m}Tc complexes are concentrated by the hepatocytes and cleared into the bile have also been investigated. Most, if not all, of the existing hepatobiliary radiopharmaceuticals share the same general anionic structure and would probably be transported via the anionic pathway. (See Figure 1.) It has been shown by inhibition studies in dogs that ^{131}I-rose bengal and ^{99m}Tc-HIDA are eliminated through the liver's anionic clearance mechanism (52). The anionic compound, bromosulphthalein, reduced the cumulative ^{131}I-rose bengal hepatobiliary excretion from 20% to 6% and ^{99m}Tc-HIDA excretion from 55% to 1.5%. In contrast, saturation levels of the cationic compound, oxyphenonium, did not affect the hepatobiliary clearance of these agents.

Bilirubin is cleared from the serum by the same general anionic clearance mechanism and has been shown to competitively inhibit the hepatobiliary clearance of ^{99m}Tc-HIDA. Bilirubin and the various anionic hepatobiliary agents can compete for 1) carrier proteins located at the plasma-hepatocyte interface, 2) Y and Z storage proteins located intthe cytoplasm, and 3) carrier proteins at the hepatocyte-bile canalicular interface. All these competitive binding processes can be saturated and can be described by nonlinear kinetics. In some disease states, the concentration of serum bilirubin may rise to a level that it displaces radiopharmaceuticals from one or all of the essential binding sites. Under these circumstances, bilirubin will become the determinant of radiopharmaceutical distribution and elimination kinetics, independent of other clinical parameters. Such an occurrence not only decreases the radiopharmaceutical's usefulness as an imaging agent, but also complicates pharmacokinetic approaches to the quantitation of liver function. Equation 1 contains a simple pharmacokinetic model of the same general type used by several investigators to describe the in vivo distribution of radiolabeled hepatobiliary agents. Ideally, time-activity curves for the individual compartments would generate model-dependent rate constants that quantitate both hepatocyte function and biliary clearance. However, the variation of these rate constants with serum bilirubin invalidates the use of linear pharmacokinetic models in the jaundiced patient. The model in Equation 1 depicts the competition between hepatobiliary clearance and renal clearance in terms of $k_{2,4}$ and $k_{1,5}$, respectively. While attempts have been made to maximize hepatobiliary clearance and to minimize renal clearance, some degree of renal clearance appears essential for the scintigraphic representation of hepatobiliary function. Therefore, ^{131}I-rose bengal, which enjoys almost complete hepatobiliary specificity, may not possess the optimum imaging radiopharmaceutical kinetics. In general, hepatobiliary specificity can be increased by either increasing $k_{2,4}$ or decreasing $k_{1,5}$. A decrease in the rate constant for renal clearance has the

$$\begin{array}{ccccc} \text{tissue} & \underset{k_{3,1}}{\overset{k_{1,3}}{\rightleftharpoons}} & \text{blood} & \underset{k_{2,1}}{\overset{k_{1,2}}{\rightleftharpoons}} & \text{liver} \\ & & \downarrow k_{2,4} & & \downarrow k_{2,4} \\ & & \text{urine} & & \text{bile} \end{array} \qquad (1)$$

effect of elevating the radioactivity in the blood and thus decreasing the total body clearance, defined as administered dose divided by the area under the serum concentration curve from $t=0$ to $t=\infty$.

In the nonjaundiced patient, in whom at least 80% of the administered dose is cleared through the hepatobiliary system, there appears to be little advantage in increasing hepatobiliary specificity by decreasing the rate of renal clearance since the decreased rate of total body clearance ultimately delays the time at which the image could be obtained. Ideally, hepatobiliary agents pertinent to the nonjaundiced should maintain the rapid rate of renal clearance found in ^{99m}Tc-HIDA and should seek to improve hepatobiliary specificity by increasing the rate constant for hepatobiliary elimination. At this point, it is uncertain whether the increased hepatobiliary specificity of the various derivatives of ^{99m}Tc-HIDA is attributed to either increased hepatobiliary clearance or decreased renal clearance.

Different radiopharmacetuicals are required for studying the jaundiced and nonjaundiced patient. In studying the jaundiced patient, a radiopharmaceutical should be chosen whose rate constant for renal elimination is less than the corresponding rate constant for use in the nonjaundiced patient ($k_{1,5}$ - jaundiced < $k_{1,5}$ - nonjaundiced). The resultant slower renal clearance in the jaundiced patient would afford the liver's saturated anionic clearance mechanism a better opportunity to concentrate and secrete the radiopharmaceutical. Even in the jaundiced patient, however, some degree of renal clearance still appears essential since it is the main mechanism for reducing blood background and allowing scintigraphic visualization of intestinal radioactivity. Since secretion of ^{99m}Tc HB agents into the bile is kinetically irreversible, renal elimination should make possible the detection of even a few percent of an administered radiopharmaceutical dose in the colon at 18 hours postinjection. In such instances, the presence of even a few percent of $^{99m}TcO_4^-$ as a radiochemical impurity would have a most deleterious effect on overall diagnostic accuracy.

CLINICAL SIGNIFICANCE

The rapid hepatic uptake and biliary excretion and the high photon flux available with the newer ^{99m}Tc-labeled hepatobiliary agents provide excellent images of the main hepatic biliary radicles, the common duct, the gallbladder and the portion of the intestinal tract which receives bile flow. In nonjaundiced patients, hepatobiliary images of diagnostic quality may be obtained with all of the newer ^{99m}Tc hepatobiliary tracers, despite some variation in the rates of hepatic uptake and biliary excretion among the tracers.

Multiple reports evaluating the clinical usefulness of these tracers have appeared but space limitation precludes a detailed listing or analysis of all studies, and only a brief overview of the representative data will be presented. The newer tracers reliably enter the gallbladder in normal subjects within 20-30 minutes with only occasional exceptions noted which are related primarily to recent food intake. Cholescintigraphy has been used to clarify the nature of some defects on liver images obtained with radiolabeled colloids, and has been utilized as a diagnostic modality in a variety of disease entities. These include patients with cholecystitis, symptomatic patients who have undergone previous upper abdominal surgery, patients in whom abnormal biliary ductal anatomy or abnormal gallbladder physiology is suspected, and jaundiced patients. A special category is the use of these agents in patients who have undergone liver transplants. As yet it has not been possible to measure hepatic blood flow or bile salt kinetics reliably, and quantitative studies of liver function are difficult because of the renal clearance of tracer.

In patients with acute abdominal pain, cholescintigraphy has been useful in assessing cystic duct patency and establishing the diagnosis of acute cholecystitis if the gallbladder does not visualize on serial images (53-56). Some investigators have used an injection of CCK or its synthetic analogs prior to the injection of tracer to exclude causes of nonvisualizing gallbladder other than cystic duct occlusion. Very high accuracy rates in the diagnosis of acute cholecystitis have been reported with these techniques.

The pattern of scintigraphic images obtained in patients with chronic cholecystitis is less uniform. The gallbladder may visualize in the normal manner or there may be a delay in the transit of tracer to the gallbladder. Gallstones in the gallbladder are not detected by these techniques. In occasional patients with chronic cholecystitis and a patent cystic duct, the gallbladder may not visualize despite pretreatment with CCK.

Visualization of the biliary flow patterns may be useful in the assessment of symptomatic patients who have undergone biliary or upper abdominal surgery (57). Obstructions at the sites of biliary-enteric anastomoses have been detected and postoperative bile leakage has been demonstrated. Following gastroenteric bypass surgery, sites of obstruction have been identified by abnormal intestinal drainage patterns, even in the presence of jaundice. In some patients with the "post-cholecystectomy" syndrome abnormalities in bile flow have been identified (58).

Biliary ductal abnormalities have been recognized in the scintigrams of patients who have choledochal cysts and those with Caroli's disease.

In patients with elevated serum bilirubin levels, the quality of hepatobiliary imaging obtainable with these tracers is diminished; however, the various causes of jaundice may be distinguished with some success (54-56,59,60). The reported studies of jaundiced patients differ about the degree to which the elevated bilirubin levels affect the ability to visualize the biliary ducts, gallbladder, or intestines; but all agree that in severe jaundice of any cause, cholescintigraphy is markedly impaired. Some of the variation may be attributed to the intrinsic qualities of the different tracers, but the dosage of tracer used and the imaging techniques are equally important factors. We have used higher doses of tracer (up to 15 mCi ^{99m}Tc-HIDA) in moderately to severely jaundiced patients because of the decreased hepatic clearance of tracer at elevated serum bilirubin levels. All of these newer tracers have an alternate renal excretory pathway so that with time the background due to radioactivity in the blood will decrease to the point that it may be possible to detect even small amounts of tracer in the intestines with appropriate imaging techniques. In studies utilizing ^{99m}Tc-HIDA we have successfully detected intestinal radioactivity in patients who had serum bilirubin levels as high as 50 mg/dl (61). Some of the new ^{99m}Tc-HIDA analogs, however, may prove to be more feasible in the evaluation of jaundiced patients.

In adult patients with jaundice of recent onset, the major clinical decision focuses on whether the underlying disease process requires medical treatment or surgical intervention. Surgically treatable causes of jaundice are primarily focal extrahepatic obstructions which produce one or more of the following patterns in cholescintigraphic images: dilated biliary ducts proximal to the obstruction, delayed biliary flow patterns with localized increases in ductal activity, nonvisualization of the biliary tree or intestines, or a negative image of the gallbladder. The most accurate results in differentiating the types of jaundice are obtained when there are only minor elevations of serum bilirubin, and at higher serum bilirubin levels, it may not be possible to distinguish incomplete focal biliary obstruction from advanced hepatocellular disease. In severe jaundice, some investigators report markedly reduced hepatic concentration of the tracer and absence of the tracer in the intestine, regardless of the etiology of jaundice.

Neonatal jaundice was evaluated in one series of eleven infants who were imaged with ^{99m}Tc-PYG (100 μCi/kg) (62). The imaging studies arrived at the correct diagnosis in eight out of eleven infants, but they failed to show intestinal radioactivity at 18 hours in two cases which were not due to biliary atresia and provided "inconclusive" evidence of intestinal radioactivity in one baby with biliary atresia. In this study, the results were comparable to those obtained with either ^{131}I-rose bengal fecal excretion tests or liver biopsies. Iio studied patients with the Dubin-Johnson variety of familial jaundice. In these patients, there was markedly delayed hepatic excretion of both

^{99m}Tc-HIDA and ^{131}I-BSP, but almost normal hepatic excretion of ^{131}I-rose bengal and ^{99m}Tc-Sn-pyridoxylidene-isoleucine (63).

In one study of patients with liver transplants (64), cholescintigraphy was useful in assessing the patency of the surgical anastomoses and overall hepatic function but was not as accurate as the serum bilirubin level in evaluation of rejection episodes. Cholescintigraphy with ^{99m}Tc agents has been performed in combination with gastric emptying studies employing ^{111}In-DTPA to evaluate patients with abdominal complaints (65).

Combined modality imaging studies involving hepatobiliary agents have been reported infrequently so that the role of cholescintigraphy in clinical diagnosis is not yet well defined. It seems probable, however, that the hepatobiliary imaging studies will complement rather than totally replace any of the well-established radiological procedures in the diagnosis of hepatobiliary disease. Since cholescintigraphy does not visualize gallstones, other studies including cholecystography with radioopaque dyes and ultrasound examinations are preferred screening tests for detection of cholelithiasis or chronic cholecystitis. In acute cholecystitis, however, cholescintigraphy has several advantages over dye studies. The studies can be completed more rapidly, are more reliable in visualizing the gallbladder. The radiopharmaceuticals are nontoxic and are particularly useful in patients with elevated serum bilirubin levels because the nanogram quantities of ^{99m}Tc-HIDA present in the injectate do not stress the liver's uptake and elimination mechanisms for bilirubin. The administration of gram quantities of radioopaque contrast dyes to a system already strained because of elevated bilirubin levels increases the overall load of anions presented to the liver and may saturate anionic uptake and elimination pathways. Ultrasound studies have been reported to be less accurate than cholescintigraphy in the diagnosis of acute cholecystitis (66). In patients with acute cholecystitis, therefore, cholescintigraphy may become the preferred imaging modality.

The anatomic resolution of ultrasound and CT studies is better than that of cholescintigraphy, and these modalities are also reported to provide a high diagnostic accuracy in evaluation of jaundiced patients (67,68). Their results are based primarily on the presence or absence of detectable dilatations in the biliary ducts, but at times other useful regional anatomic information also may be present in the studies. In some cases, however, analysis of the anatomic data may not yield the correct diagnosis. Thus, in the neonates with jaundice, the biliary ducts characteristically are not dilated with biliary atresia, and in patients in whom upper abdominal surgical procedures have been performed, dilated ducts may be present as a residual of the operation and may not be a reliable indicator of biliary tract disease. In these cases, accurate assessment of biliary flow patterns provides a more useful diagnostic evaluation. Other techniques, such as endoscopic retrograde cholangiopancreatography and transhepatic cholangiography have been used with excellent results in diagnosis of jaundice. Each of the diagnostic modalities has a special contribution in the study of pathophysiology of jaundice, and the clinically useful role of the hepatobiliary tracers in conjunction with these other modalities remains to be defined in the jaundiced patient.

FUTURE TRENDS AND CONCLUSIONS

The development of ^{99m}Tc hepatobiliary agents has given new impetus to cholescintigraphy. These new agents provide excellent images in studies of biliary physiology and bile flow patterns in nonjaundiced patients. Further improvement in HB specificity in these patients would lower the patient's absorbed radiation dose but would probably yield only minor improvement in clinical information. However, improvement in HB specificity should be attempted only by increasing HB clearance and not by decreasing renal clearance.

The ideal radiopharmaceutical for studying biliary flow patterns in jaundiced patients has yet to be developed. In these patients, competition between bilirubin and the currently available ^{99m}Tc agents significantly degrades image quality. New agents should be developed that can either displace bilirubin from binding sites or utilize separate and distinct carrier-mediated processes. An in vitro isolated hepatocyte system (Figures 5 and 6) has been developed in our laboratory for the evaluation of competitive effects at two of the binding sites (uptake and storage). Figure 5 demonstrates the percent radiopharmaceutical uptake in the hepatocytes as a function of time for various labeled compounds. These agents were found to concentrate in the hepatocyte to a much larger extent than ^{99m}Tc-DTPA or $^{3}H_2O$. Figure 6 illustrates the percent radiopharmaceutical uptake in the hepatocytes at 30 minutes as a function if bilirubin concentration. In this system, ^{99m}Tc-labeled p-isopropyl derivative of ^{99m}Tc-HIDA is more effective than either ^{99m}Tc-HIDA or ^{131}I-rose bengal in competing with bilirubin for cellular binding sites.

Two additional factors should be considered in the design of new radiopharmaceuticals for studying patients with reduced HB clearance. The ideal agent for such studies should exhibit a compensatory reduction in renal clearance and thus allow the hepatocytes a continuing opportunity to bind the agent and excrete it in the bile. With delay of both hepatic and renal clearances, however, it will be necessary to image the patient at later times postinjection. Radiopharmaceuticals with a significantly longer physical half-life than 6 hours would be preferable for delayed cholescintigraphy, and tracers incorporating ^{111}In (69) may be suitable for this role. Even with significant improvements in the design of the HB agents, however, it is probable that the severity of the underlying disease processes affecting the HB system in some patients will impair clinical scintigraphy performed with any agent.

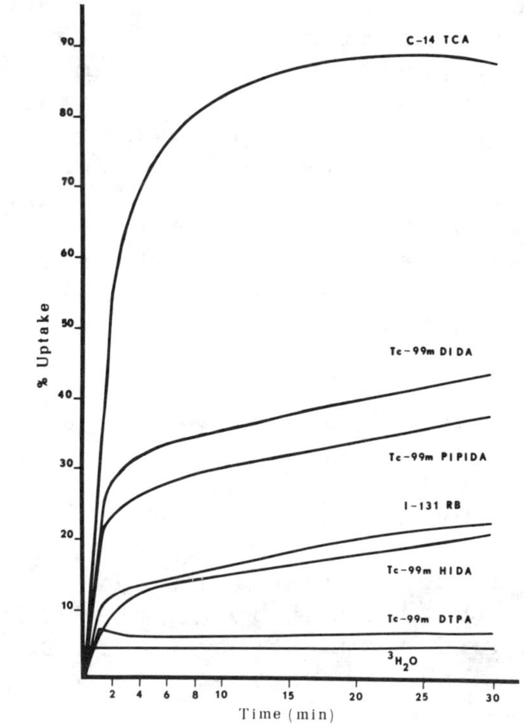

Figure 5

Timecourse of uptake of labeled compounds by isolated hepatocytes. Taurocholate (TCA), diethylIDA (DIDA), p-isopropyl-IDA (PIPIDA), rose bengal (RB), dimethylIDA (HIDA), diethylene triaminepentaacetic acid (DTPA).

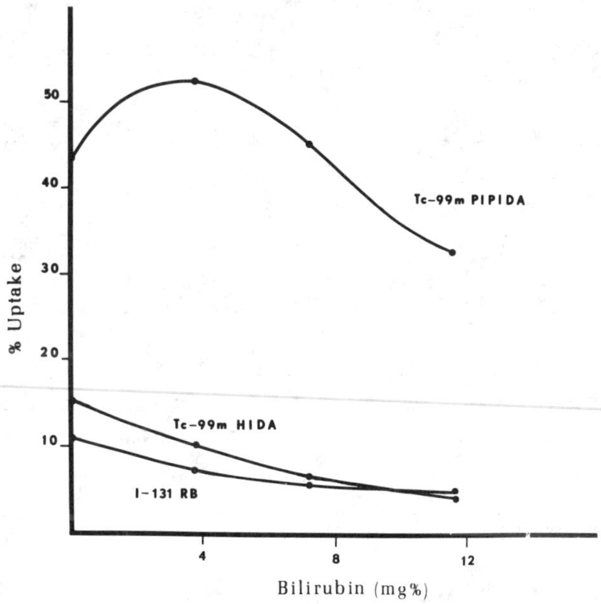

Figure 6

Uptake of labeled compounds by isolated hepatocytes as a function of bilirubin concentration. Each point is the average of duplicate determinations after 30 minutes of incubation.

Opportunities also exist for the development of new agents to measure hepatocyte function. Current ^{99m}Tc HB agents either clear the liver too rapidly for adequate multiview imaging of the liver parenchyma or have inadequate target-to-nontarget ratios in jaundiced patients. Possible solutions to these problems include the development of HB agents with long intrahepatic transit times and the use of ^{99m}Tc and ^{131}I-labeled asialo proteins (12,13,39) which bind to the hepatocyte membrane without being excreted in the bile. These agents also would be useful in defining an index of hepatocellular function independent of the function of the biliary system.

The usefulness of these new tracers must be established by clinical trials. These new agents may replace existing diagnostic techniques of they may complement them, but their role in the diagnostic process should be established by comparative modality studies in a variety of disease states.

Current ^{99m}Tc-HB agents have ushered in a renaissance in cholescintigraphy. The current level of enthusiasm generated by these tracers should stimulate researchers to develop new agents with increased specificities, tailored to specific disease states.

REFERENCES

1. Tubis M, Krishnamurthy GT, Endow JS et al: ^{99m}Tc-penicillamine complexes. In Radiopharmaceuticals, Subramanian G, Rhodes BA, Cooper JF, Sodd VJ, eds, New York, The Society of Nuclear Medicine, Inc., 1975, pp 55-62

2. Javitt NB: Hepatic bile formation. N Engl J Med 295:1464-1469, 1511-1516, 1976

3. Fortner EL: Mechanisms of hepatic bile formation. Ann Rev Physiol 39:323-347, 1977

4. Berk PD, Berlin NI, eds: International Symposium on Chemistry and Physiology of Bile Pigments, National Institutes of Health, Bethesda, Maryland, 1977. DHEW Publication No. (NIH) 77-1100

5. Taplin GV, Meredith OM, Kade H: The radioactive I-131 tagged rose bengal uptake excretion test for liver function using external gamma ray scintillation counting techniques. J Lab Clin Med 45:665, 1955

6. Nomoto H, Kato T, Aoyama B: Clinical studies on the visualization of the common bile duct in the sequential scanning with I-131 BSP. In Recent Advances in Nuclear Medicine, Tokyo, First World Congress of Nuclear Medicine, 1974, pp 457-458

7. Butterman G, Wolf I, Pabst HW: Quantitative analysis of hepatograms using a gamma camera and labelled contrast media. In Dynamic Studies with Radioisotopes in Medicine, vol 1, Vienna, IAEA, 1975, pp 137-155

8. Eickenbusch W, Stute A, Chen T: ^{131}I-toluidine blue: A new diagnostic agent for biliary function study and gallbladder scanning. In Recent Advances in Nuclear Medicine, Tokyo, First World Congress of Nuclear Medicine, 1974, pp 459-461

9. Bastomsky CH, Horton PW, Shimmins J: Gamma camera study of the hepatobiliary excretion of ^{131}I-thyroxine glucuronide and ^{131}I-rose bengal in the rat. J Nucl Med 14:34-39, 1973

10. Reiffers S, Lambrecht RM, Wolf AP: Synthesis and structural effects of selective biliary excretion of halogenated indotricarbocyanines. J Nucl Med 17:524,1976

11. Prokopowicz D: Uber die Brauchbarkeit der Indocyaninprobe zur Beurteilung der Leberfunktion bei Infektionskrankheiten. Dtsh Gesundheitsw 46:2188, 1973

12. Van Rijk PP, DeGraff CN: ^{131}I-asialo-α_1-acid-glycoprotein. A new pharmaceutical for dynamic liver function studies. In Dynamic Studies with Radioisotopes in Medicine, vol 1, Vienna, IAEA, 1975, pp 99-110

13. Vera DR, Krohn KA, Stadalnik RC: Liver imaging via labeled hepatic binding protein ligands. J Nucl Med 19:685, 1978

14. Goris ML: ^{123}I-iodo-bromsulphthalein as a liver and biliary scanning agent. J Nucl Med 14:820-825, 1973

15. Christy B, King G, Smook WM: Preparation of iodine-123 labeled rose bengal and its distribution in animals. J Nucl Med 15:484, 1974

16. Ansari AN, Atkins HL, Lambrecht RM: ^{123}I-indocyanine green (^{123}I-ICG) as an agent for dynamic studies of the hepatobiliary system. In Dynamic Studies with Radioisotopes in Medicine, vol 1, Vienna, IAEA, 1975, pp 111-124

17. Tubis M, Krishnamurthy GT, Endow JS: ^{99m}Tc-penicillamine a new cholescintigraphic agent. J Nucl Med 13:652-654, 1972

18. Lin TH, Khentigan A, Winchell HS: A ^{99m}Tc-labeled replacement for ^{131}I-rose bengal in liver and biliary tract studies. J Nucl Med 15:613-615, 1974

19. Tonkin A, DeLand FH: Dihydrothioctic acid: A new polygonal cell imaging agent. J Nucl Med 15:539, 1974

20. Hunt FC, Maddalena DJ, Yeates MG: Technetium-99m-6-mercaptopurine, a new radiopharmaceutical for cholescintigraphy. In Recent Advances in Nuclear Medicine, Tokyo, First World Congress of Nuclear Medicine, 1974, pp 869-870

21. Yokoyama A, Terauchi Y, Horiuchi K: Technetium-99m-kethoxal-bis(thiosemicarbazone), an uncharged complex with a tetravalent ^{99m}Tc state, and its excretion into the bile. J Nucl Med 17:816-819, 1976

22. Heindel ND, Risch VR, Burns HD: Synthesis and tissue distribution of ^{99m}Tc-sulfonylureas. J Pharm Sci 64:687-689, 1975

23. Fritzberg AR, Eshima D, Klingensmith WC: Comparison of ^{99m}Tc-hepatobiliary agents based on naphthalene and similar ring systems. J Nucl Med 19:694-695, 1978

24. Fritzberg AR, Lyster DM, Dolphin DH: ^{99m}Tc-bioquin-7CA, a potential new hepatobiliary scanning agent. J Nucl Med 17:907-910, 1976

25. Baker RJ, Bellen JC, Ronai PM: ^{99m}Tc-pyridoxylideneglutamate: a new rapid cholescintigraphic agent. J Nucl Med 15:476, 1974

26. Fotopoulus A, Chiotelis E, Koutoulidis C: Evaluation of Tc-99m pyridoxal phenylalanine as a hepatobiliary imaging agent. Part 1. Experimental studies. J Nucl Med 18:1189-1193, 1977

27. Baker RJ, Bellen JC: Technetium-99m biliary scanning agents based on pyridoxal: effect of the amino acid component on labelling and biological distribution. Int J of Nucl Med and Biol 4:85-91, 1977

28. Chiotellis E, Subramanian G, McAfee JG: Preparation of Tc-99m labelled pyridoxal amino acid complexes and their evaluation. Int J of Nucl Med and Biol 4:29-41,1977

29. Kato M, Hazue M: Tc-99m (Sn) pyridoxylideneaminates: preparation and biologic evaluation. J Nucl Med 19:397-406, 1978

30. Kohuo A, Maeda T, Shiokawa H: Technetium-99m: 3-hydroxy-4-formyl pyridine: glutamic acid complex. J Nucl Med 18:624, 1977

31. Harvey E, Loberg MD, Cooper M: Tc-99m-HIDA: a new radiopharmaceutical for hepatobiliary imaging. J Nucl Med 16:533, 1975

32. Tjeu HSLM, Van der Pompe WB, Cox PH: Technetium-99m-labelled diethyl-acetanilido-iminodiacetate: a new hepatobiliary agent. Brit J Radiol 50:735-739, 1977

33. Subramanian G, McAfee JG, Henderson RW: The influence of structural changes on biodistribution of Tc-99m labelled N-substituted IDA derivatives. J Nucl Med 18:624, 1977

34. Burns D, Marzilli L, Sowa D: Relationship between molecular structure and biliary excretion of technetium-99m HIDA and HIDA analogs. J Nucl Med 18:624, 1977

35. Wistow BW, Subramanian G, Van Heertum RL: An evaluation of ^{99m}Tc-labeled hepatobiliary agents. J Nucl Med 18:455-461, 1977

36. Saji H, Yokoyama A, Arano Y: Phthalein and fluorescein derivatives with an effective 99m-technetium labeled state for hepatobiliary transport: ^{99m}Tc-PC, a new hepatobiliary radiopharmaceutical. J of Labelled Compounds and Radiopharmaceuticals 16:188-189, 1979

37. Porter DW, Loberg MD, Fields AT: Tc-99m labeled steroids as hepatobiliary radiopharmaceuticals. J Nucl Med 19:687-688, 1978

38. Spencer RP, Miller RE, Antar MA: ^{99m}Tc-protamine complex with biliary excretion. J Nucl Med 15:535, 1974

39. Van Rijk PP, van den Hamer CJ: Labelled asialo-orosomucoid in liver function studies. Nuklearmedizin 14:Supp No 1, 570, 1977

40. Dewanjee MK, Fliegal C, Treves S: ^{99m}Tetracyclines: preparation and biological evaluation. J Nucl Med 15:176-182, 1974

41. Hirom PC, Millburn P, Smith RL, et al: Species variation in the threshold molecular weight factor for the biliary excretion of organic anions. Biochem J 129:1071-1077, 1972

42. Jansholt AL, Krohn KA, Stadalnik RC, et al: Catabolism and protein binding of Tc-99m pyridoxylideneglutamate. J Nucl Med 19: 1036-1044, 1978

43. Chiotellis E, Subramanian G, McAfee JG: ^{99m}Tc-pyridoxylidene glutamate, a gall bladder imaging agent: chemistry of preparation and comparative animal studies with ^{131}I rose bengal. Int J of Nucl Med and Biol 4:21-28, 1977

44. Loberg MD, Cooper M, Harvey E, et al: Development of new radiopharmaceuticals based on N-substitution of iminodiacetic acid. J Nucl Med 17:633-638, 1976

45. Baker RJ, Bellen JC, Ronai PM: Technetium-99m pyridoxylidene glutamate: a new hepatobiliary radiopharmaceutical. I. Experimental aspects. J Nucl Med 16:720-727, 1975

46. Wistow B, Subramanian G, McAfee JG: An evaluation of Tc-99m labeled HIDA derivatives as hepatobiliary agents in experimental animals. J Nucl Med 17:545, 1976

47. Yokoyama A, Saji H, Tanaka H: Preparation of a chemically characterized ^{99m}Tc-penicillamine complex. J Nucl Med 17:810-815, 1976

48. Fields AT, Porter DW, Callery PS, et al: Synthesis and radiolabeling of technetium radiopharmaceuticals based on N-substituted iminodiacetic acid: effect of radiolabeling conditions on radiochemical purity. J of Labelled Compounds and Radiopharmaceuticals 15:387-399, 1978

49. Firnau G: Why do ^{99m}Tc-chelates work for cholescintigraphy? Eur J Nucl Med 1:137-139, 1976

50. Ryan J, Cooper M, Loberg MD, et al: Technetium-99m-labeled N-(2,6-dimethylphenylcarbamoylmethyl) iminodiacetic acid (Tc-99m HIDA): a new radiopharmaceutical for hepatobiliary imaging studies. J Nucl Med 18:995-1002, 1977

51. Burns HD, Worley P, Wagner HN: Design of technetium radiopharmaceuticals. In The Chemistry of Radiopharmaceuticals, New York, Masson Publishing Inc., 1977, p 269

52. Harvey E, Loberg MD, Ryan J, et al: Hepatic clearance mechanism of ^{99m}Tc-HIDA and its effect on quantitation of hepatobiliary function. J Nucl Med (In press)

53. Eikman EA, Ameron JL, Colman M, et al: A test for patency of the cystic duct in cholecystitis. Ann Int Med 82:318-322, 1975

54. Ronai PM, Baker RJ, Bellen JC, et al: Technetium-99m-pyridoxylideneglutamate: a new hepatobiliary radiopharmaceutical. II. Clinical aspects. J Nucl Med 16:728-737, 1975

55. Matolo NM, Stadalnik RC, Wolfman EF: Hepatobiliary scanning using Tc-99m pyridoxylideneglutamate. Am J Surg 133:116-120, 1977

56. Rosenthall L, Shaffer EA, Lisbona R, et al: Diagnosis of hepatobiliary disease by Tc-99m HIDA cholescintigraphy. Radiology 126:467-474, 1978

57. Rosenthall L, Forseca C, Arzoumanian A, et al: Tc-99m IDA hepatobiliary imaging following upper abdominal surgery. Radiology 130:735-739, 1979

58. Reichelt HG, St Stender H, Schmidt FW, et al: Hepatobiliare Sequenzszintigraphie-Fortschritte in der Abklarung von Rest und Wachbeschwerden Cholezystektomieter. Fortschr Roentgenstr 128: 451-459, 1978

59. Poulouse KP, Eckelman WC, Reba RC, et al: Evaluation of Tc-99m pyridoxylideneglutamate for the differential diagnosis of jaundice. Clin Nucl Med 1:70-72, 1976

60. Williams JAR, Baker RJ, Walsh JF, et al: The role of biliary scanning in the investigation of the surgically jaundiced patient. Surg Gyn Obst 114:525-530, 1977

61. Ryan J, Hill M, Isikoff M, et al: An approach to the differential diagnosis of jaundice utilizing Tc-99m HIDA and ultrasound. World Federal of Nuclear Medicine and Biology, Second International Congress, Washington DC, 1978, p 9 (abst)

62. Jenner RW, Howard ER, Clarke MB, et al: Hepatobiliary imaging: The use of Tc-99m pyridoxylidene glutamate scanning in jaundiced adults and infants. Brit J Radiol 51:862-866, 1978

63. Iio M, Kawaguchi S, Murata H, et al: Technetium-99m(Sn) pyridoxine isoleucine analog (PI) as hepatobiliary imaging radiopharmaceutical. World Federation of Nuclear Medicine and Biology Second International Congress, Washington DC, 1978, p 107

64. Klingensmith WC, Fritzberg AR, Koep LJ, et al: A clinical comparison of Tc-99m diethyliminodiacetic acid, Tc-99m diethyliminodiacetic acid, Tc-99m pyridoxylidene glutamate, and I-131 rose bengal in liver transplant patients. Radiology 130:435-441, 1979

65. Fisher R, Malmud LS, Tolin R, et al: Simultaneous imaging and quantitation of gastric and gallbladder emptying. World Federation of Nuclear Medicine and Biology, Second International Congress, Washington DC, 1978, p 125 (abst)

66. Weissmann HS, Frank M, Rosenblatt R, et al: Cholescintigraphy, ultrasonography and computerized tomography in the evaluation of biliary tract disorders. Sem Nuc Med 9:22-35, 1979

67. Isikoff MB, Diaconis JN: Ultrasonography: a new diagnostic approach to the jaundiced patient. JAMA 238:221-223, 1977

68. Biello DR, Levitt RG, Siegel BA, et al: Computed tomography and radionuclide imaging of the liver: a comparative evaluation. Radiology 127:159-163, 1978

69. Lin MS: ^{111}In-phenolphthalexon, a new chelate for extended cholescintigraphy: formulation and preclinical characterization. Radiology 123:783-786, 1977

DEVELOPMENT AND RESULTS OF ROUTINE QUALITY CONTROL PROCEDURES FOR Tc-99m IMINODIACETATE HEPATOBILIARY AGENTS

Alan R. Fritzberg and Donna Huckaby

University of Colorado Medical Center and School of Pharmacy
Denver, Colorado 80262

ABSTRACT

A variety of solvent systems and adsorbent support systems were evaluated for use in the routine quality control of iminodiacetate radiopharmaceuticals. Using systems which gave R_f values between 0 and 1 preparation variables such as time and dilution were studied. Variability in the intensities of multiple peak patterns were found to not significantly alter the rate of biliary excretion as evaluated with Tc-99m-diethyl-IDA. The multiple forms coalesced into narrow single peaks when diluted in saline or in plasma suggesting rapid equilibration in vivo to common chemical forms.

INTRODUCTION

Several Tc-99m labeled hepatobiliary scanning agents based on iminodiacetate analogs of lidocaine are available in Europe or are becoming available in the United States for routine use. These include N-(2,6-diethylacetanilide)-iminodiacetic acid (diethyl-IDA), N-(2,6-dimethylacetanilide)-iminodiacetic acid (HIDA) and N-(*p*-isopropylacetanilide)-iminodiacetic acid (PIPIDA).

The increased lipophilicity of these agents relative to other polar chelates such as Tc-99m-DTPA and Tc-99m phosphonate bone agents allows them to exhibit R_f values between 0 and 1 on paper chromatography (1). These properties also provide the opportunity to observe variations in results due to variables such as time post preparation and dilution. In the case of Tc-99m-diethyl-IDA, components that changed with time were reported to be present as indicated by column gel filtration on G-25 sephadex, although animal evaluations in mice resulted in no significant biological differences (2).

The object of this study was the evaluation of a variety of solvent systems for maximization of convenience and separations and use of them to monitor preparation variables likely to be encountered in routine use.

MATERIALS AND METHODS

Radiopharmaceuticals. Chelating agents, N-(2,6-diethylacetanilide)-iminodiacetic acid and N-(2,6-dimethylacetanilide)-iminodiacetic acid were synthesized from 2,6-diethylaniline* and 2,6-dimethylaniline* by consecutive reactions with chloroacetyl chloride and iminodiacetic acid disodium salt† (3). Analytical data were as expected. Diethyl-IDA kits were formulated with 35 mg of carrier and 0.35 mg $SnCl_2 \cdot 2H_2O$ in 0.8 ml at pH 5.8-6.0. They were stored frozen in evacuated vials. Tagging with Tc-99m-pertechnetate was by the addition of 40 to 60 mCi in 2.2 ml. HIDA kits were formulated with 20 mg of carrier and 0.30 mg $SnCl_2 \cdot 2H_2O$ in 1.8 ml at pH 5.8. Tagging with Tc-99m-pertechnetate was by the addition of 20 to 30 mCi in 1.2 ml. PIPIDA kits were obtained under IND status#. They were prepared by the addition of 20 to 30 mCi in 3.0 ml.

Chromatography. Analyses were carried out on silica gel impregnated glass fiber strips (Gelman ITLC), silica gel on plastic backed sheets (Eastman) and Whatman 3M paper. Solvents used were analytical grade. Chromatographic strips, 2.5 x 10 cm, were dried under a nitrogen stream and developed in ascending fashion. They were scanned on a chromatogram scanner (Packard Model 7200).

Animal Studies. The effect of time post preparation on biliary excretion of Tc-99m-diethyl-IDA was evaluated by monitoring appearance of activity in rat bile. Long-Evans rats weighing 240 to 340 g were used. They were anesthetized with pentobarbital. The initial dose was about 50 mg/kg with further

---

*Aldrich Chem. Co., Milwaukee, Wis.
†Eastman Kodak, Co., Rochester, N.Y.
#Diagnostic Isotopes, Inc., Bloomfield, N.J.

maintainance doses of 10 mg/kg. Internal temperature of the rats was determined with a rectal thermometer and a heat lamp was used to maintain temperature at 36-37°C. A line was inserted into a femoral vein for injection of the radiopharmaceuticals. The common bile duct was cannulated and the tubing was connected to a fraction collector for collection of bile samples. The radiopharmaceutical mixture (1-2 μCi of Tc-99m-diethyl-IDA and 0.5-1 μCi of I-131-Rose Bengal* as a control agent in 0.25 ml) was injected over a several second period. Bile collections were made from 0 to 2 min and then for 4 min periods. Bile samples were then counted in a well counter (Picker Autowell, dual channel) and activity determined for I-131 and Tc-99m. Radioactivity of Tc-99m was corrected for crossover contribution by I-131.

RESULTS AND DISCUSSION

The results of varying the solvent system on 3M paper are shown in Table 1. The studies in the different solvent systems were carried out at times of one to three hours post preparation so that changes in chemical forms would be minimized. The recommendation of saline as a solvent system (1) is useful for the separation of chelated and insoluble technetium and free pertechnetate for Tc-99m-HIDA and -diethyl-IDA but not for Tc-99m-PIPIDA. An R_f of 1 as observed for chelated HIDA and diethyl-IDA, however, does not give information regarding related chemical forms. Acetonitrile:water (3:1) (A-H) also suggested for use (1) resulted in intermediate R_f values for all iminodiacetate chelates. In practice, however, a broad peak for Tc-99m-diethyl-IDA about R_f 0.75, while characteristic, results in a poor separation of chelated technetium and free pertechnetate. This problem appears to be related to concentration which will be discussed later. The other solvents tried did not result in significant advantages to the acetonitrile water mixture.

Table 2 shows the results of different solvent systems on silica gel impregnated in glass fiber strips. The use of saline, which usually showed three peaks, resulted in separation of Tc-99m-PIPIDA from insoluble activity and free pertechnetate. This provides a means for rapid indication of the state of the Tc-99m-PIPIDA preparation. The R_f values of 0 for the iminodiacetates in methyl ethyl ketone (MEK) provide a way of quickly ascertaining the extent of free pertechnetate for all agents. However, routinely Tc-99m-diethyl-IDA has resulted in 5 to 15% at the solvent front in MEK which has not been supported as free pertechnetate by the results on 3M paper or by in vivo evaluation. This suggests possible reduced stability in MEK. The only other solvent showing potential on this adsorbent and support was 1-butanol which gave intermediate R_f values for Tc-99m-diethyl-IDA and -HIDA.

The last adsorbent evaluated was silica gel on plastic backed sheets. The two systems which demonstrated potential included saline for Tc-99m-diethyl-IDA and -PIPIDA and 1-butanol:acetic acid:water (4:1:1) for all three agents. Elution times are longer for this adsorbent system and no real advantages are apparent.

Since others (2) and we had observed time dependent variations in the chromatographic pattern associated with Tc-99m-diethyl-IDA, a more detailed

---

*I-131 Rose Bengal was obtained from Mallinckrodt Nuclear, Inc., St. Louis, Mo. or E.R. Squibb and Sons, Inc., Princeton, N.J.

TABLE 1. BEHAVIOR OF Tc-99m-IMINODIACETATES ON 3M PAPER IN VARIOUS SOLVENTS

Solvent	Diethyl-IDA	PIPIDA	HIDA	Pertechnetate
saline	1.0	0.59 (1.0)	0.93	0.66
MEK	0.0	0.10	0.0	0.73
BEH	0.92	1.0	0.24 (.43) 0.88 (1.0)	0.49
1-butanol	0.0	0.0	0.0	0.0
A-H	0.54 (1.0) 0.75 (1.0)	0.31 (.29) 0.76 (1.0)	0.43 (.49) 0.73 (1.0)	1.0
methanol	0.0 (1.0) 0.69 (.24)	0.0 (.50) 0.37 (1.0)	0.47 (brd)	0.34
BAH	0.84	0.53 (1.0) 0.90 (.59)	0.41 (sh) 0.87 (brd)	0.29
acetone	0.17	0.0	0.0	0.92

Abbreviations are: MEK = methyl ethyl ketone; BEH = 1-butanol:ethanol:water (2:2:1); A-H = acetonitrile:water (3:1); BAH = 1-butanol:acetic acid:water (4:1:1); sh = shoulder; brd = broad. When several peaks were observed, relative peak heights are given in parentheses.

TABLE 2. R_f VALUES OF Tc-99m-IMINODIACETATE BILIARY AGENTS IN VARIOUS SOLVENT SYSTEMS ON SILICA GEL (GLASS FIBER SUPPORT)

Solvent	Diethyl-IDA	PIPIDA	HIDA	Pertechnetate
saline	0.87	0.34 (.62) 0.47 (.88) 0.67 (1.0)	1.0	1.0
MEK*	0.0	0.05	0.0	1.0
BEH*	0.84	1.0	1.0	1.0
1-butanol	0.54 (.38) 0.87 (1.0)	0.0 (1.0) 0.32 (.18) 0.67 (.24)	0.18 (1.0) 0.47 (.50) 0.71 (.78)	1.0
A-H*	0.83	1.0	0.87	1.0
methanol	1.0	1.0	1.0	1.0
BAH*	1.0	1.0	1.0	1.0
acetone	0.0 (.70) 1.0 (1.0)	0.0	0.0 0.27 (sh)	1.0

TABLE 3. R_f VALUES OF Tc-99m-IMINODIACETATE BILIARY AGENTS IN VARIOUS SOLVENT SYSTEMS ON SILICA GEL (PLASTIC SHEET SUPPORT)

Solvent	Diethyl-IDA	PIPIDA	HIDA	Pertechnetate
saline	0.67	0.29 (1.0) 0.77 (.18)	0.91	0.94
MEK*	0.0	0.0	0.0 0.31 (sh)	0.93
BEH*	0.94	0.96	0.89	0.85
1-butanol	0.06	0.0	0.0	0.0
A-H*	0.75 (1.0) 0.91 (.87)	0.40 (.36) 0.92 (1.0)	0.72 (1.0) 0.85 (.85)	1.0
methanol	0.40 (.71) 0.93 (1.0)	0.0 (.22) 0.46 (1.0)	0.40 (1.0) 1.0 (.70)	0.56
BAH*	0.44 (.28) 0.72 (1.0)	0.54 (.35) 0.86 (1.0)	0.54 (sh) 0.85	0.35
acetone	0.0	0.0	0.0	0.95

*For abbreviations see footnote on Table 1.

time course dependency was evaluated with all three agents (Fig. 1). Although some variability is observed, all agents were shifted from higher to lower R_f values in acetonitrile:water (3:1). This was also observed for Tc-99m-PIPIDA in saline on silica gel glass fiber strips. Unfortunately suggestions of general changes in polarity cannot be made since the decreases in R_f values in the A-H system would seem to indicate increased polarity and decreases in R_f values in the saline system would seem to indicate decreased polarity.

An important question is whether or not these variations in peak shapes and intensities are relevant to reproducible biological behavior. To provide an answer, preparations of Tc-99m-diethyl-IDA were injected into rats 5 min post preparation and the rate of appearance in the bile compared to preparations which were injected at 30 min or longer post preparation. The results are shown in Fig. 2. There appears to be a slightly faster rate of appearance in the bile in the 5 min studies although the final amount recovered in the bile differs by less than 3 percent. These results agree with the report that in mice no difference was observed between injections at 2 to 4 min and 20 min post preparation. Thus, it can be concluded that the differences are not large enough to warrant control of time of injection post preparation.

Since these differences in biliary excretion rates are not large, the possibility of rapid equilibration to common forms in vivo was considered. This idea was supported by observations resulting from a study of the effects of dilution in saline and water on the stability of Tc-99m-diethyl-IDA. In each case a single narrow peak resulted from the original broad peak observed with the 3.0 ml preparations. The results for all three agents are shown in Figure 3 for dilutions in saline and in plasma. There is a general result that the R_f values are decreased to varying extents and the multiple peaks are coalesced into relatively narrow peaks in both plasma and saline at ten-fold dilutions. These results also were time independent supporting the suggestion that equilibration occurs in vivo to give a single common chemical form for each agent. Initial preparations at a volume of 8.0 ml instead of 3.0 ml gave essentially the same pattern with a small increase in apparent insoluble activity of R_f value 0.

The shifts from multiple broad peaks to single peaks upon dilution observed in this study complicate the criterion of single components for radiochemical purity indicated by Eckelman (4). To be relevant to biological evaluations it may be necessary to subject complexed Tc-99m samples to chromatographic precedures which include dilution in plasma or saline to an extent that is relevant to the dilution resulting upon injection. In this study the same chromatogram appearance was obtained for Tc-99m-diethyl-IDA at dilutions greater than ten-fold so a complete dilution modeling may not be necessary, however.

It has been shown that iminodiacetate complexes have a 2:1 ratio of ligand to technetium (5). In the earlier study on Tc-99m-diethyl-IDA time dependent changes (2) it was suggested that the variations were due to the passage of the technetium through a 1:1 complex intermediate. That idea does not appear consistent with the formation of a single species within a few minutes upon dilution. It may be that one is seeing greater than 2:1 ratios and/or intermediate states of hydration at higher concentrations instead since technetium has been shown to exhibit higher than six coordination (6).

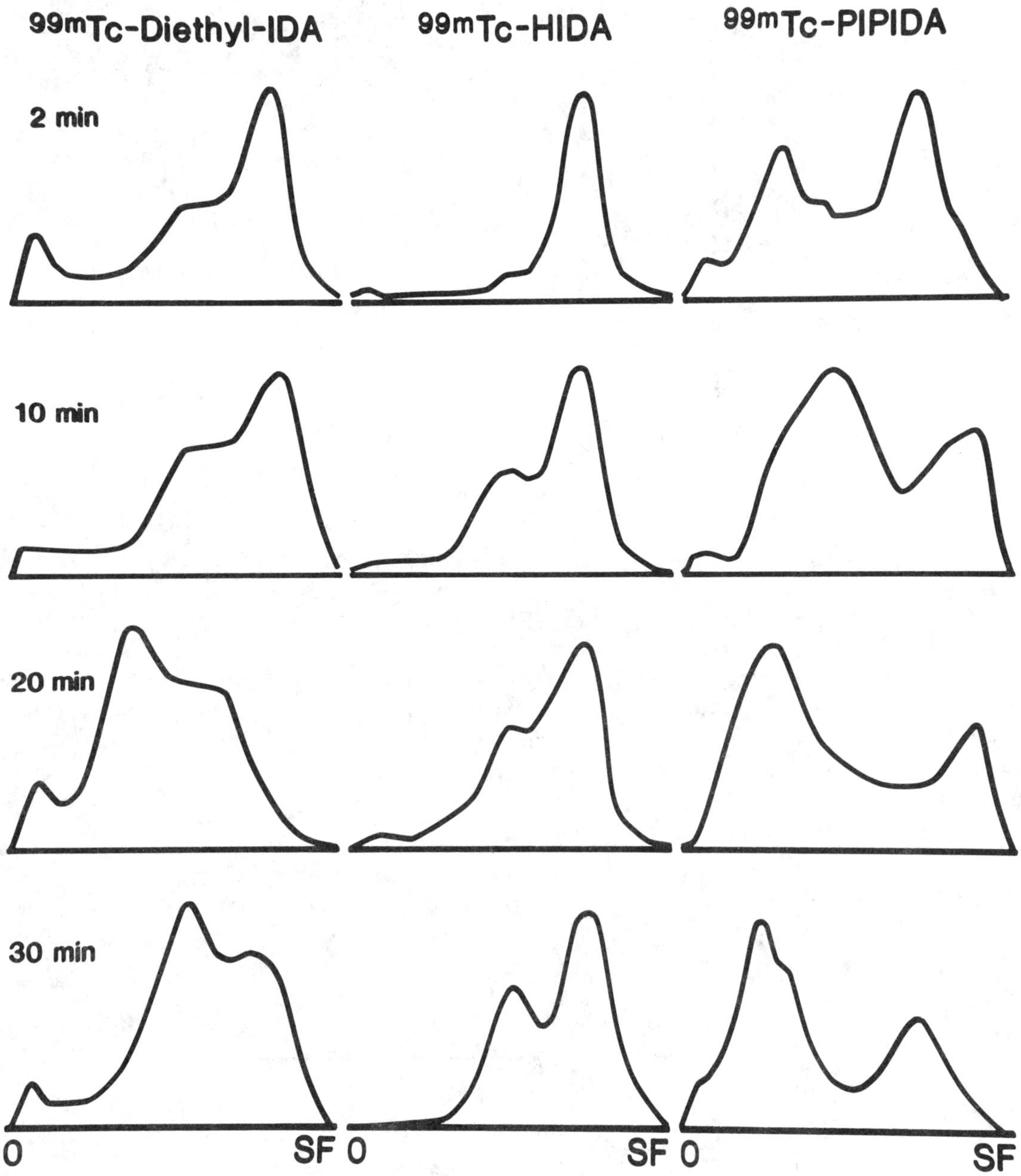

Figure 1.

Chromatogram representations of Tc-99m-iminodiacetate biliary agents at various times post preparation. All agents were compared at 3 ml total volume. Tc-99m-diethyl-IDA and -HIDA were run in acetonitrile-water (3:1) on 3M paper and Tc-99m-PIPIDA was run in saline on silica gel (glass fiber).

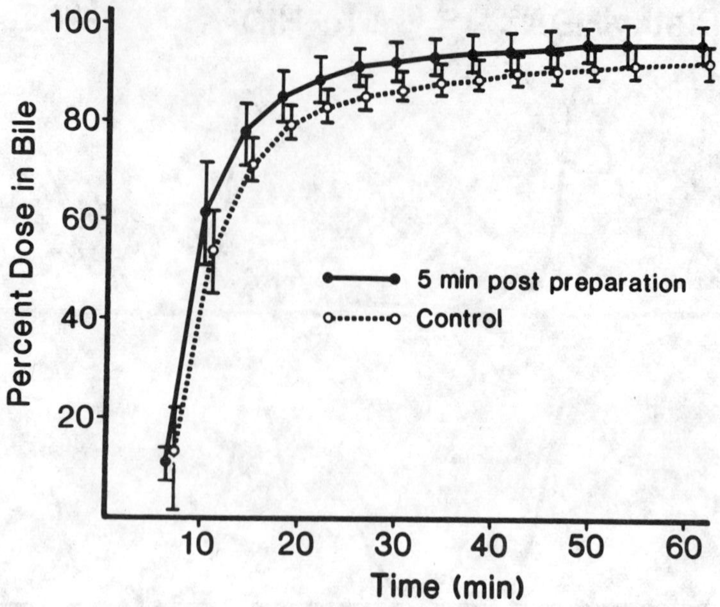

Figure 2.

Comparison of rate of appearance in rat bile of Tc-99m-diethyl-IDA injected 5 min post preparation (●—●) and 30 min or longer post preparation (o--o). Plots indicate average and range for 3 rats for 5 min series and average and range for 4 rats for 30 min or longer series.

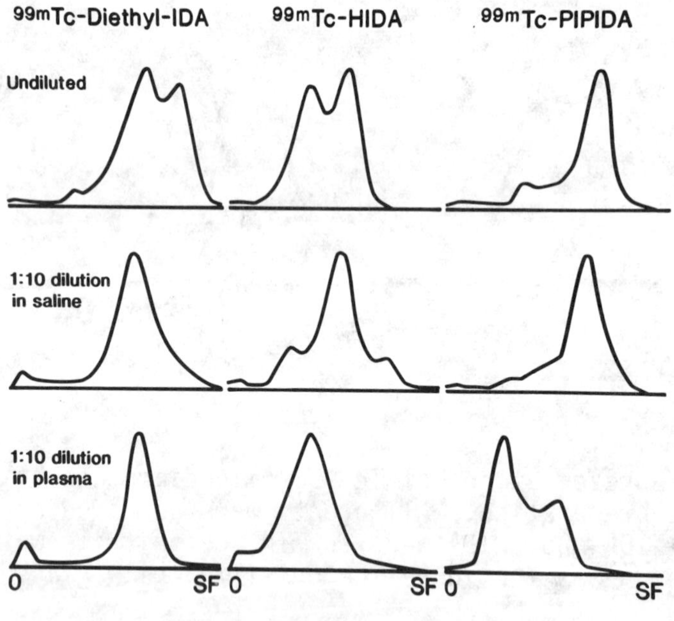

Figure 3.

Chromatogram representations of Tc-99m-iminodiacetates after dilution ten-fold in saline and in plasma. All complexes were prepared at 3 ml total volume and chromatographed in acetonitrile-water (3:1) on 3M paper.

REFERENCES

(1) Loberg MD, Cooper M, Harvey E, et al: Development of new radiopharmaceuticals based on N-substitution of iminodiacetic acid. *J Nucl Med* 17: 633-638, 1976

(2) Fonda U, Pedersen B: ^{99m}Tc-Diethyl-IDA. A contribution to its structure. *Eur J Nucl Med* 3: 87-89, 1978

(3) Callery PS, Faith WC, Loberg MD, et al: Tissue distribution of technetium-99m and carbon-14 labeled N-(2,6-dimethylphenylcarbamoylmethyl)-iminodiacetic acid. *J Med Chem* 19: 962-964, 1976

(4) Eckelman WC: Radiochemical purity of new radiopharmaceuticals. *J Nucl Med* 17: 865, 1976

(5) Loberg MD, Fields AT: Chemical structure of Technetium-99m-labeled N-(2,6-dimethylphenylcarbamoylmethyl)-iminodiacetic acid (Tc-HIDA). *Int J Appl Rad Isot* 29: 167-173, 1978

(6) Deutsch E, Elder RC, Lange BA, et al: Structural characterization of a bridged ^{99}Tc-Sn-dimethylglyoxime complex: Implications for the chemistry of ^{99m}Tc-radiopharmaceuticals prepared by Sn(II) reduction of pertechnetate. *Proc Natl Acad Sci* 73: 4287-4289, 1976

IN VIVO KINETICS OF HEPATOBILIARY AGENTS IN JAUNDICED ANIMALS

Anne-Line Jansholt, David R. Vera
Kenneth A. Krohn, and Robert C. Stadalnik

University of California Davis Medical Center
Department of Nuclear Medicine
Sacramento, California 95817

ABSTRACT

Blood clearance, liver uptake and biliary and urinary excretion kinetics have been measured for I-131 rose bengal, Tc-99m PyG, and several Tc-99m-labeled iminodiacetic acid derivatives in normal rabbits and in rabbits infused with high doses of bilirubin. Serum bilirubin levels of up to 12 mg/dl resulted in some changes in the above parameters, including increased urinary excretion of Tc-99m PyG and delayed liver uptake and clearance of Tc-HIDA and rose bengal. However, the overall picture was one of relatively small changes and poor quantitative correlation with the performance of these hepatobiliary radiopharmaceuticals in jaundiced patients. We conclude that as long as the hepatocytes are healthy, their capacity for excretion of organic anions is relatively insensitive to the amounts of bilirubin used in these experiments. The altered pharmacokinetics of these radiopharmaceuticals in jaundiced patients is thus largely related to impaired hepatocyte function.

INTRODUCTION

Two groups of Tc-99m-labeled radiopharmaceuticals are being actively investigated for their potential for imaging the hepatobiliary system. One group includes Tc-99m pyridoxylideneglutamate (Tc-PyG) and related Schiff's bases and the other includes Tc-99m-labeled N-substituted iminodiacetates (IDA). The properties desirable in a radiopharmaceutical for biliary imaging include (1) rapid extraction from the blood by the liver, (2) rapid excretion into the biliary canaliculi, (3) negligible urinary excretion, (4) no intestinal reabsorption, and (5) simple preparation methods that result in a product of reliably high quality. Numerous comparisons of the various Tc-99m-labeled hepatobiliary agents have been reported, with the general conclusion that, when scored on these characteristics, members of the two groups vary considerably and no single agent has a clear overall advantage. The IDA derivatives clear more rapidly from the blood and are excreted in the urine to a lesser degree than Tc-PyG in experimental animals (1,2), and are therefore favored for hepatobiliary scintigraphy. Liver kinetic measurements in the literature reports are either nonquantitative or are difficult to compare with each other.

In our laboratory Tc-PyG has been used successfully for clinical evaluation of more than 600 patients. Although an improvement as compared with standard radiography and I-131 rose bengal cholescintigraphy (3), its usefulness is limited in jaundiced patients. Using Tc-PyG, Stadalnik (4) studied a group of 90 jaundiced patients with suspected hepatobiliary disease and serum bilirubin levels ranging from 1.6 to 28 mg/dl. The biliary system was consistently visualized when the serum bilirubin was 4.4 mg/dl or less. Even though the biliary system was not visualized above this level, the gastrointestinal tract was visualized at a late time interval (up to 24 hours after injection) in 62 of 64 patients with hepatocellular disease, but in none of 18 patients with proven complete common bile duct obstruction, thereby separating the two conditions. Cholescintigraphy with Tc-PyG can be helpful for evaluation of jaundiced patients. Elevated serum bilirubin similarly limits the usefulness of IDA derivatives (5,6); however, no one has been able to directly correlate pharmacokinetic parameters or degree of biliary tract visualization with the severity of jaundice. From this clinical experience another desirable property of hepatobiliary radiopharmaceuticals has emerged. Their in-vivo distribution and kinetics should be independent of serum bilirubin levels and influenced only by the patency of the hepatobiliary drainage system.

There is widely based clinical experience in jaundiced patients that points toward a common excretory pathway for these radiopharmaceuticals and for bilirubin, similar to that reported for I-131 rose bengal and sodium sulfobromophthalein (BSP) (7,8), a roentgenographic agent excreted by the same mechanism as bilirubin. When bilirubin and BSP are administered simultaneously, BSP is excreted preferentially (7), indicating that the hepatic excretory pathway is quantitatively different for these compounds. That difference is generally thought to relate to different binding affinities for ligandin but may also relate to differences in hepatocyte membrane transport parameters.

Our experiments have been designed to examine the pharmacokinetic implications of the competition that presumably exists between bilirubin and hepatobiliary radiopharmaceuticals for a common excretory pathway. We have measured the blood clearance, liver uptake and biliary and urinary excretion kinetics of several of these radiopharmaceuticals in healthy rabbits and in rabbits infused with high doses of bilirubin.

MATERIALS AND METHODS

Animal model. A jaundiced rabbit model has been developed that allows evaluation of hepatocyte uptake and excretion as well as determination of blood clearance rate and extent of urinary excretion. We chose to use infusions of unconjugated bilirubin to maintain elevated concentrations of serum bilirubin for the duration of our imaging experiment rather than BSP, which has higher binding affinity for ligandin (9).

Bilirubin solutions of up to 8 mg/ml were prepared by quickly dissolving bilirubin in 0.1N NaOH, followed by rapid addition of cold 0.15N sodium carbonate solution. The pH was adjusted to 10.5, and any undissolved material removed by centrifugation. The bilirubin concentration was determined spectrophotometrically (a_s = 1.204 mg/dl·cm at 460 nm) by the method of Jacobsen and Wennberg (10). Elevated serum bilirubin levels were produced by rapid injection of a priming dose of bilirubin followed by constant infusion to maintain the jaundiced state throughout the experiment.

Radiopharmaceuticals. The following were included in this investigation: I-131 rose bengal (RB), Tc-99m PyG, and several derivatives of N-substituted acetanilido-iminodiacetic acid: dimethyl (Tc-HIDA), diethyl (Tc-diethyl IDA), paraisopropyl (Tc-PIPIDA), and parabutyl (Tc-BIDA). Blood clearance rate, urinary excretion and liver kinetic parameters were determined for these agents in normal rabbits and in rabbits whose hepatocytes were otherwise normal but whose serum bilirubin level was elevated.

Pyridoxylideneglutamate kits were prepared as previously described (11). Kits for the preparation of Tc-99m-labeled HIDA*, PIPIDA†, and BIDA§ were supplied by radiopharmaceutical manufacturers, and prepared according to their instructions. The kit for preparing Tc-99m-diethyl IDA was made by combining 35 mg of N-(2,6-diethylacetanilide)-iminodiacetic acid¶ and 0.35 mg $SnCl_2 \cdot 2H_2O$ in 0.8 ml. The pH was adjusted to 6.0 and the kits stored frozen in evacuated vials. Labeling was performed by adding 20 mCi Tc-99m pertechnetate in 1.2 ml sterile saline. RB was obtained commercially. The radiochemical purity of RB was determined by thin-layer chromatography on silicic acid with a solvent system of 9:1 (v/v) chloroform and acetic acid. The amount of free Tc-99m pertechnetate in Tc-PyG was determined by electrophoresis and thin-layer chromatography (11). Radiochemical quality control of each Tc-99m labeled IDA derivative was performed using at least two of the following methods: (1) Electrophoresis at 250V, 5mA, 30 minutes using cellulose acetate strips and a pH 8.8 barbital buffer, (2) Instant thin-layer chromatography using silica gel medium and normal saline, (3) Paper chromatography on Whatman 3MM with 3:1 (v/v) acetonitrile and water as the mobile phase.

---

*Medi-Physics; † Diagnostic Isotopes; § New England Nuclear;
¶ Gift from A.R. Fritzberg, University of Colorado Medical School, Denver.

Data collection and analysis. New Zealand male rabbits weighing 2.6-3.9 kg and maintained on a normal diet were injected I.M. with 5 mg/kg xylazine followed by 50 mg/kg Ketamine for anesthesia. We elected to avoid barbiturate anesthesia because of its known effect on liver physiology, including blood flow and pressure and ligandin activity (12). The amount of bilirubin needed to reach the desired blood level was calculated assuming a first order blood clearance rate of 0.0693 min^{-1} and a blood volume equal to 6% of total body weight. The bilirubin solutions were administered through a surgically exposed leg vein. Infusion was started 5 minutes before injection of the radiopharmaceutical and was continued until the animal was sacrificed. Blood samples were obtained to determine the serum bilirubin level initially, immediately after injection of radiopharmaceutical and 10 minutes later.

The rabbits were positioned under a LFOV gamma camera such that both the heart and the bladder were in the field of view. The camera and computer were started immediately before injection of radiopharmaceutical through an ear vein. Fifteen-second frames were collected in histogram mode for 25 minutes following injection of the radiopharmaceutical. Scintiphotos and blood samples were obtained immediately and at several additional times for up to 30 minutes after injection. At that time the rabbits were sacrificed, the urinary bladder was surgically exposed, and its total content withdrawn.

Plasma samples were analyzed spectrophotometrically to determine the bilirubin level (10). Whole blood and urine samples as well as standards of known activity were counted in a scintillation system to determine blood clearance rate and degree of urinary clearance. The %I.D. in blood was calculated assuming a blood volume of 6% of the body weight. Using the computer-collected data and defining regions of interest (ROI) over different organs, time-activity curves could be generated for the heart, liver (left lobe), gallbladder, left kidney, and urinary bladder. For the liver, the only organ for which data were analyzed in all experiments, a ROI of 8x8 matrix cells was chosen, giving a maximum count rate of between 5 and 30 thousand per frame (average 15K) for all Tc-99m-compounds. For I-131 rose bengal 3 consecutive 15-second frames were summed so that each data point at the maximum of the curves represents 2000 to 5000 counts.

RESULTS

All the Tc-99m-labeled radiopharmaceuticals were simple to prepare and contained <4% free pertechnetate. The serum bilirubin levels indicated on the illustrations are those at the time of injection of the radiopharmaceuticals. Although a constant level was usually maintained, we observed slow increases or decreases in a number of experiments due to variation in clearance rate and blood volume in individual animals. At bilirubin levels of 15 mg/dl or over, the animals reacted to the priming dose so we aimed for levels up to only about 12 mg/dl. Although a large volume of basic carbonate solution was injected, we did not detect changes in blood gas parameters before and shortly after bilirubin infusions.

We have verified differences in the in-vivo kinetics of the Tc-99m-labeled hepatobiliary radiopharmaceuticals and found that their kinetics are altered to some degree by elevated serum bilirubin levels although the effects were variable and not as dramatic as expected based on the behavior of these radiopharmaceuticals in jaundiced patients.

Increased urinary excretion of tracer is a common observation in jaundiced patients, and may cause problems in the interpretation of scans when renal activity remains high. Our study showed Tc-PyG to be affected in this way. At high bilirubin concentrations its urinary excretion was as much as twice the normal amount. Tc-HIDA was moderately affected in some of the animals at levels above 7 mg/dl, but urinary excretion was unchanged for the rest of the IDA compounds and never exceeded 1% for RB and Tc-BIDA, the IDA derivative with the largest non-polar side-chain.

Decreased blood clearance rates were observed in animals with elevated bilirubin levels. These changes were only moderate for diethyl IDA and PIPIDA and were even lower for Tc-HIDA and RB at bilirubin levels as high as 10 mg/dl. At bilirubin levels of 9-11 mg/dl no consistent changes were seen with Tc-PyG or Tc-BIDA, and in general the effect of elevated bilirubin levels on radiopharmaceutical clearance are so small that, although there is a qualitative similarity to what is seen in patients, it is clear that our animal model does not quantitatively reflect what is seen clinically.

Activity in the gastrointestinal tract was evident with all Tc-99m compounds within 5 minutes after injection in normal animals. Some delay in visualization of the biliary and G.I. tracts was observed in jaundiced animals but the G.I. tract was always visualized within 15 minutes, and the gallbladder was usually also seen by that time. At earlier time intervals a clear outline of the gallbladder was often impossible on the scintiphotos because of high liver activity, particularly for Tc-BIDA. The rate of liver uptake in normal animals is very fast for Tc-PyG, considerably slower for RB with the IDA derivatives being intermediate in behavior and following the series HIDA > diethyl IDA ≃ PIPIDA > BIDA. The liver transit time (t_2 of Table 1) follows the same order for the IDA compounds but both RB and PyG are distinctly slower. Again we find that quantitatively the differences associated with infusion of bilirubin are minor compared with the poor liver uptake and nonvisualization of the biliary tract in jaundiced patients.

DISCUSSION

Harvey, et al (13) have measured the pharmacokinetics of Tc-HIDA in dogs after inducing plasma concentrations of BSP of > 10 mg/dl. This resulted in delayed blood clearance, increased urinary excretion and decreased biliary excretion of Tc-HIDA as compared with normal dogs, indicating competition for a common excretory pathway. Since BSP is cleared through the hepatocyte by the same mechanism as bilirubin (7,8), the poor performance of Tc-HIDA and related hepatobiliary radiopharmaceuticals in jaundiced patients is reasonably assumed to be due to competition between these compounds and bilirubin for excretion through the hepatocyte.

The mechanism for clearance of bilirubin and related compounds includes three major steps (8): hepatocyte uptake by a concentrative membrane transport system of high capacity; hepatocyte binding, conjugation and storage; and excretion into the biliary canaliculus by an active membrane transport mechanism of relatively low capacity. The number of plasma protein binding sites available for a given compound is one of the determinants for passage into the hepatocyte as well as for degree of urinary excretion. This variable was measured using dialysis (13) for Tc-HIDA before and after infusion of dogs with BSP. Levels of above 10 mg/dl decreased protein binding of Tc-HIDA minimally, from 64% to 56%. In order for a radiopharmaceutical to compete more effectively with bilirubin for liver clearance, it must be able to displace bilirubin from its transport binding sites. This is not accomplished with present hepatobiliary radiopharmaceuticals. Competition for the very large number of plasma protein binding sites is not a likely explanation for the affects of bilirubin at this concentration.

In contrast to BSP and bilirubin, Tc-PyG and Tc-HIDA are not conjugated or otherwise metabolized by the liver before excretion into the bile (11,14). The quantitative binding affinity of these radiopharmaceuticals for ligandin or other proteins within the hepatocyte is unknown, but the rapid blood clearance rate of Tc-HIDA in both normal and jaundiced rabbits (Fig.1a), combined with decreased liver clearance rate (Fig. 2) in jaundice suggests that a mechanism for its storage exists within the hepatocyte. When the mass of material cleared from the blood by the hepatocyte exceeds its excretory capacity, compounds can be stored within the hepatocyte by ligandin binding, but as the saturation level for storage is approached, material will be increasingly returned to the circulating blood pool, resulting in apparently decreased blood clearance rates. In hepatocellular disease saturation is reached much sooner than with healthy hepatocytes, so that the effect is magnified, leading to poor liver uptake, decelerated blood clearance and increased urinary excretion. Since we worked with an animal model with normal liver function, we did not appreciably saturate the hepatocyte clearance capacity. This explains why we observed qualitatively what is seen clinically in jaundiced patients but our differences were not as dramatic. Interestingly Tc-PyG and Tc-BIDA were influenced much less than other agents by experimentally elevated bilirubin levels. Too little is known about how relatively small structural differences influence hepatic uptake and clearance of chemicals for us to offer a molecular explanation for our results.

Current radiopharmaceuticals for hepatobiliary imaging do not fulfill the criteria of insensitivity to elevated serum bilirubin levels, yet they are still useful in a wide variety of clinical situations. Is one particular hepatobiliary radiopharmaceutical "best"? Tc-BIDA is influenced the least by elevated bilirubin levels in our animal model; however, it reaches a maximum liver uptake later than the other IDA derivatives and the gallbladder cannot be visualized as early because of the longer liver retention time. This inconvenience may be more than compensated for by very low urinary excretion, and Tc-BIDA may be the radiopharmaceutical of choice in jaundiced patients. An alternative solution to the interference of bilirubin with hepatobiliary radiopharmaceuticals would be development of compounds that are being excreted by a mechanism entirely different from that of bilirubin.

REFERENCES

(1) Wistow BW, Subramanian G, Van Heertum RL, et al: An evaluation of Tc-99m-labeled hepatobiliary agents. J Nucl Med 18: 455-461, 1977

(2) Smith RB, Coupal J, DeLand FH, et al: Pharmacokinetics of hepatobiliary imaging agents in rats: Concise communication. J Nucl Med 20:45-49, 1979

(3) Stadalnik RC, Matolo NM, Jansholt A-L, et al: Technetium-99m pyridoxylideneglutamate (PG) cholescintigraphy. Radiology 121: 657-661, 1976

(4) Stadalnik RC and Matolo NM: Value of Technetium-99m-P.G. (pyridoxylideneglutamate) scintigraphy in the evaluation of the jaundiced patient. Int J Nucl Med Biol 5: 176-182, 1978

(5) Ryan J, Cooper M, Loberg M, et al: Technetium-99m-labeled N-(2,6 -dimethylphenylcarbamoylmethyl)iminodiacetic acid (Tc-99m HIDA): A new radiopharmaceutical for hepatobiliary imaging studies. J Nucl Med 18: 997-1004, 1977
(6) Rosenthall L, Shaffer EA, Lisbona R, et al: Diagnosis of hepatobiliary disease by Tc-99m-HIDA cholescintigraphy. Radiology 126:467-474, 1978
(7) Cantarow A, Wirts CW, Snape WJ, et al: Excretion of bilirubin and bromsulfalein in bile. Amer J Physiol 154: 211-219, 1948
(8) Goresky CA: The hepatic uptake and excretion of sulfobromophthalein and bilirubin. Canad Med Ass J 92: 851-857, 1965
(9) Levi AJ, Gatmaitan Z, and Arias IM: Two hepatic cytoplasmic protein fractions, Y and Z, and their possible role in the hepatic uptake of bilirubin, sulfobromophthalein, and other anions. J Clin Invest 48: 2156-2167, 1967
(10) Jacobsen J and Wennberg R: Determination of unbound bilirubin in the serum of newborns. Clin Chrom 20: 783-789, 1974
(11) Jansholt A-L, Krohn KA, Stadalnik RC, et al: Catabolism and protein binding of Tc-99m pyridoxylideneglutamate. J Nucl Med 19: 1036-1044, 1978
(12) Reyes H, Levi AJ, Gatmaitan Z, et al: Organic anion-binding protein in rat liver: drug induction and its physiologic consequence. Proc Nat Acad Sci 64: 168-170, 1969
(13) Harvey E, Loberg M, Ryan J, et al: Hepatic clearance mechanism of Tc-99m HIDA and its effect on quantitation of hepatobiliary function. Accepted for publication, J Nucl Med
(14) Loberg MD, Cooper M, Harvey E, et al: Development of new radiopharmaceuticals based on N-substitution of iminodiacetic acid. J Nucl Med 17: 633-638, 1976

TABLE 1. IN VIVO CLEARANCE RESULTS FOR HEPATOBILIARY RADIOPHARMACEUTICALS

Compound (Dose, mg/kg)	Animal Model*	n	Blood % I.D. 30 min	30 min urine % I.D.± s.d.	Liver uptake† t_1	t_2
Rose bengal	N	8	7	<1	8	15
(0.04-0.8)	J	4	6-15	<1	13	40-60
Tc-PyG	N	5	6-8	18.9 ± 4.8	2	15-20
(0.80-3.7)	J	4	6-8	34.1 ± 6.4	2	15-20
Tc-HIDA	N	4	1.4	15.1 ± 2.3	3.5	5.5
(0.25-0.4)	J	4	1.0-4.1	19.4 ± 5.6	5-6	8-11
Tc-diethyl IDA	N	3	1.2	9.0 ± 1.1	3-4	8
(0.70-0.9)	J	2	3	10.5	4	9
Tc-PIPIDA	N	3	1.3	7.5 ± 3.4	4	8
(0.90-1.1)	J	3	2.5-5	5.6 ± 0.01	4-6	9-15
Tc-BIDA	N	3	1.1	0.7 ± 0.1	5-6	19
(0.35-0.7)	J	3	0.8-1.8	0.6 ± 0.3	6	22

* N = normal animals, J = jaundiced animals; † t_1 = time (min) to reach maximum liver uptake, t_2 = time (min) from t_1 to 50% liver excretion.

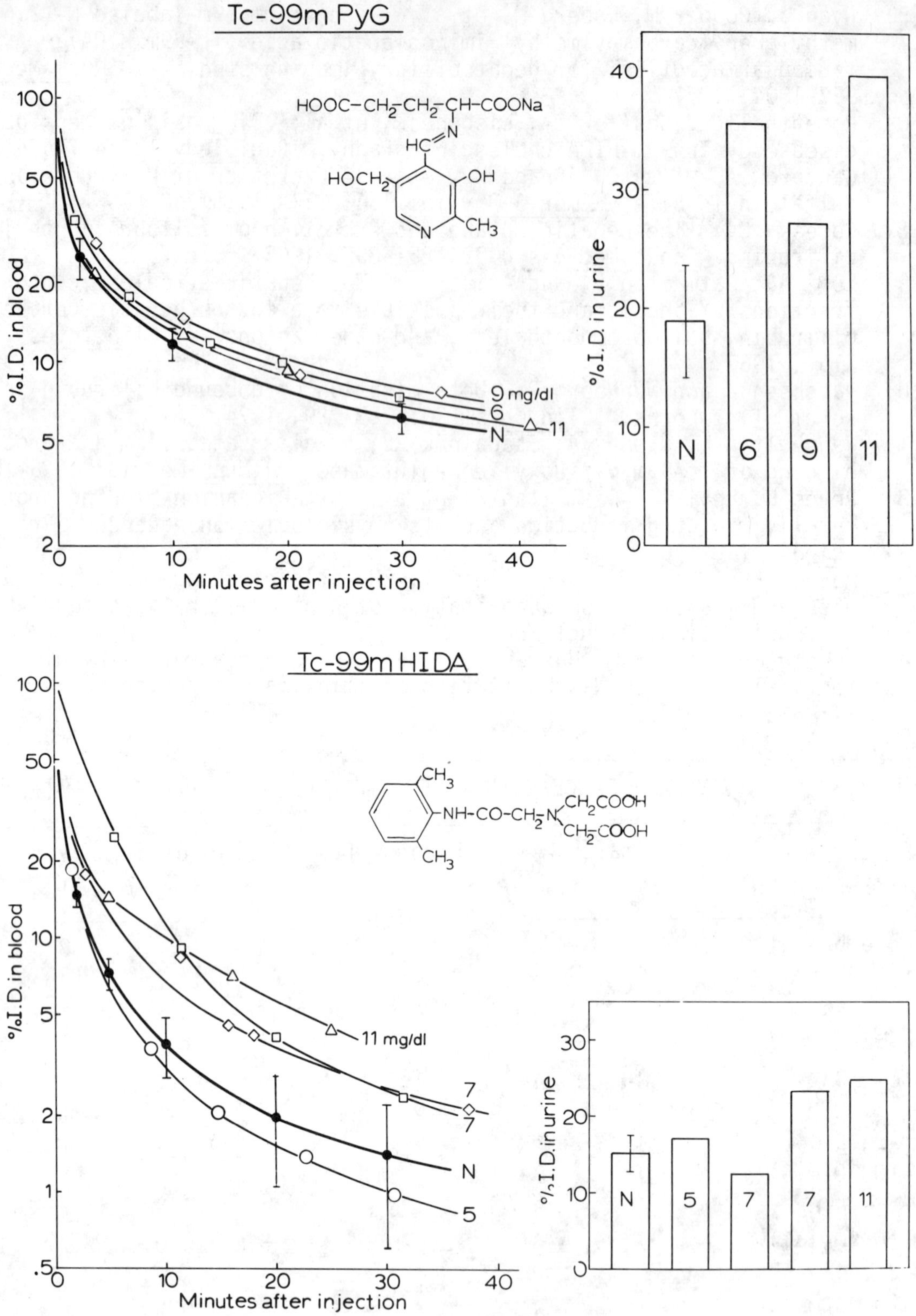

Figure 1a.

Blood and urinary clearance of Tc-99m HIDA and Tc-99m PyG in normal (N) and jaundiced rabbits. The numbers next to the blood clearance curves and on the urine excretion bars indicate serum bilirubin levels in mg/dl.

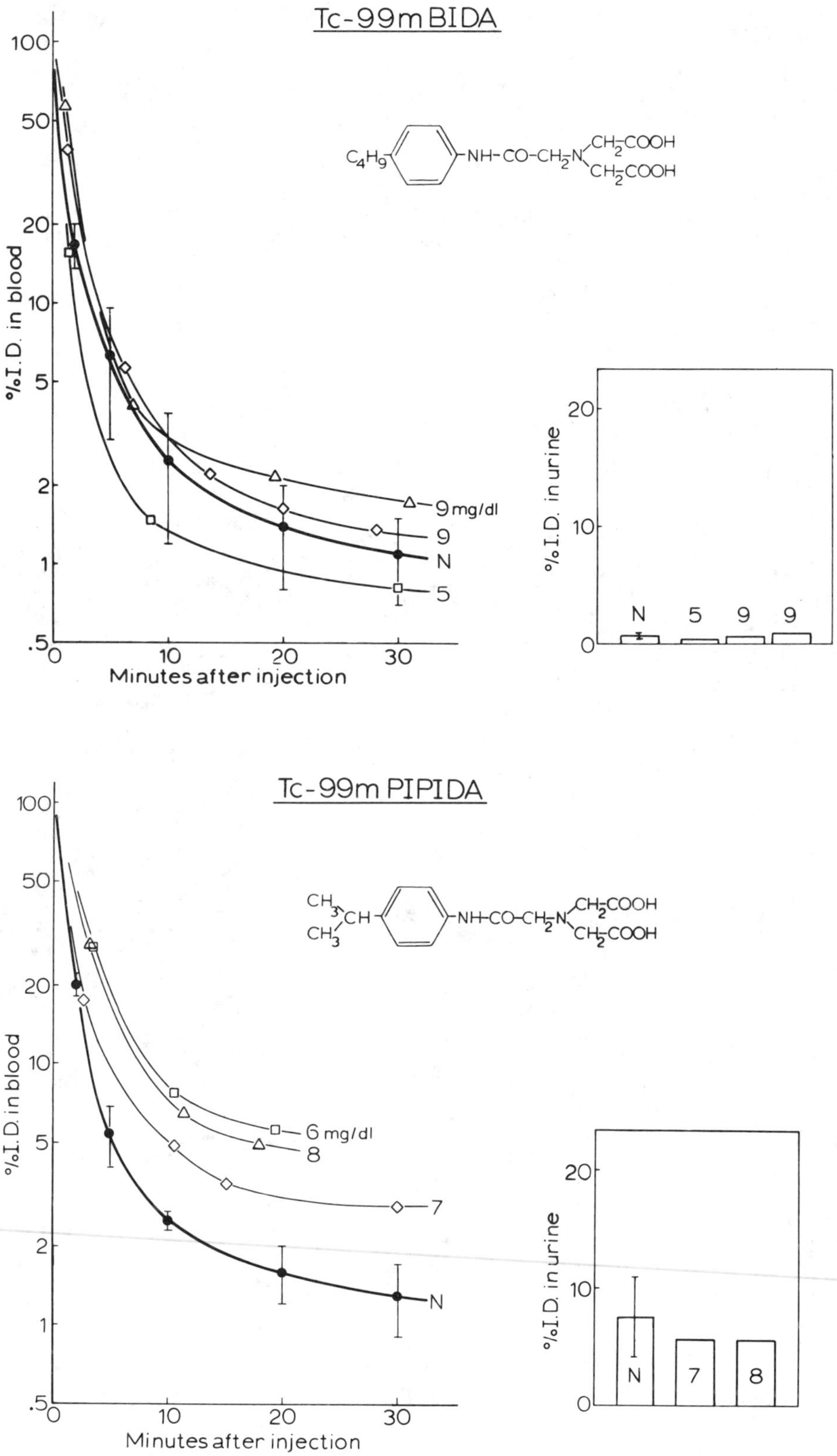

Figure 1b.

Blood and urinary clearance of Tc-99m BIDA and Tc-99m PIPIDA in normal (N) and jaundiced rabbits. The numbers next to the blood clearance curves and on the urine excretion bars indicate serum bilirubin levels in mg/dl.

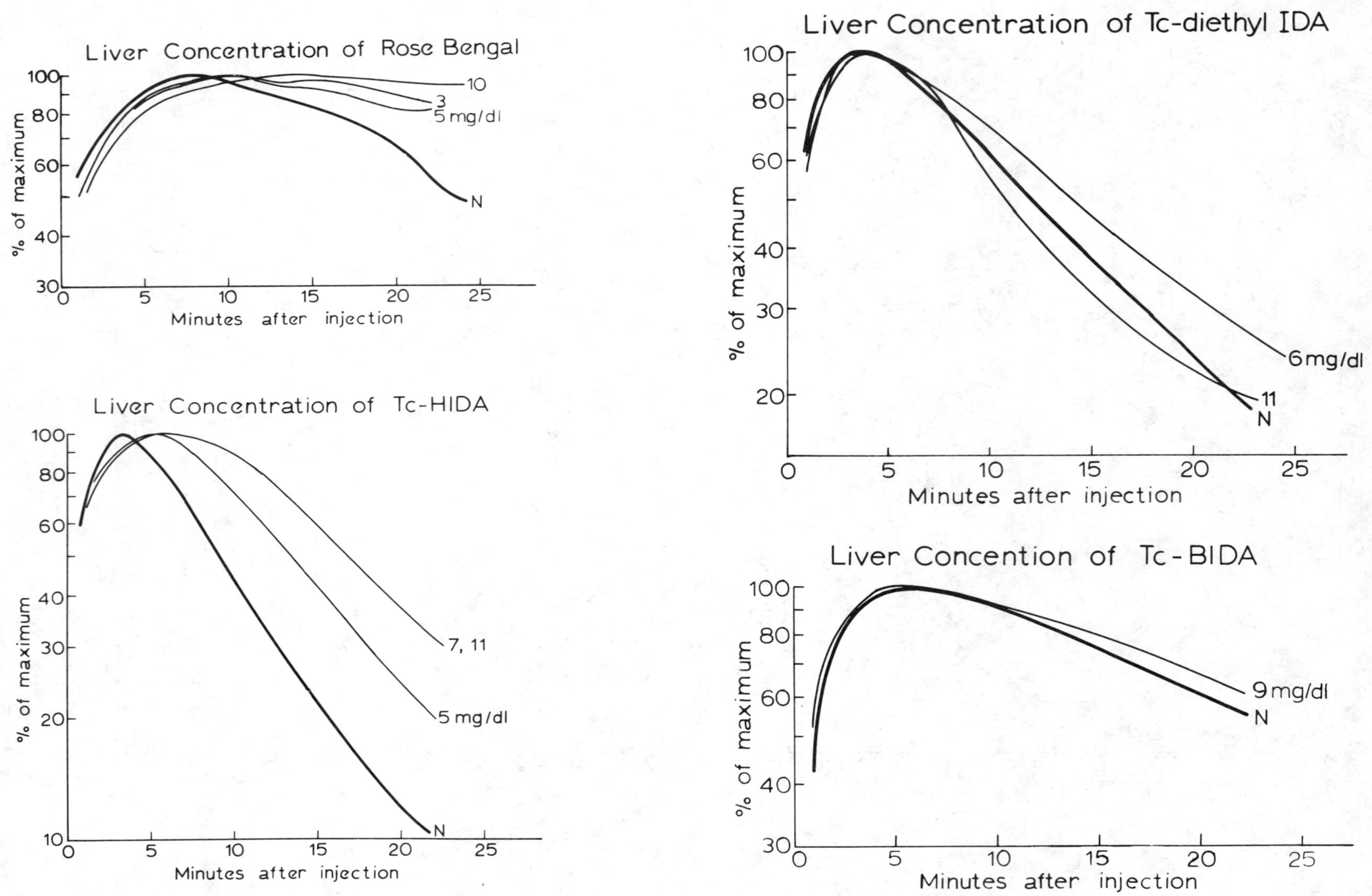

Figure 2.

Liver uptake and excretion of I-131 rose bengal, Tc-99m HIDA, Tc-99m diethyl IDA, and Tc-99m BIDA in normal (N) and jaundiced rabbits. The numbers given next to the curves indicate serum bilirubin levels in mg/dl.

RADIOLIGANDS THAT BIND TO CELL-SPECIFIC RECEPTORS: HEPATIC BINDING PROTEIN LIGANDS FOR HEPATIC SCINTIGRAPHY

David R. Vera, Kenneth A. Krohn,
and Robert C. Stadalnik

University of California Davis Medical Center
Department of Nuclear Medicine
Sacramento, California 95817

ABSTRACT

A conceptual model has been developed for designing and evaluating receptor - specific radiopharmaceuticals and maximizing their potential for scintigraphic evaluation of pathophysiology. Tracer ligands for receptors that serve a significant physiological role may be used to assess the functional state of the target cell. The criteria developed have been used to assess the potential of hepatic binding protein (HBP) ligands, a new class of radiopharmaceuticals based on the molecular biology of a hepatocyte membrane receptor. An analog ligand, technetium-99m-galactosyl neoglycoalbumin has the essential features of isotopic and biologic convenience; high ligand purity, stability and specificity; and the appropriate kinetics to permit differentiation of the hemodynamic, receptor binding and molecular release phases of tracer kinetics.

Recently, considerable interest has been generated in the design of radiopharmaceuticals based on molecular-specific receptor interactions (1,2). These tracer compounds measure the viability and presence of a physiologically significant receptor associated with a specific cell type. The most significant feature of such a labeled compound is that the accumulation of the tracer is the result of normal metabolic processes rather than the result of a unique property of the lesion. There are a few radiopharmaceuticals which exhibit this quality, including iodide ion, rose bengal, and 2-fluoro-2-deoxyglucose, but as molecular biologists develop an increased understanding of membrane biochemistry they will provide the radiopharmaceutical scientist with an increasing number of molecular-receptor specific ligands.

The value of any diagnostic test is proportional to the quantity of information retrieved from the procedure. Hence, the development of any radiodiagnostic procedure must involve an assessment that recognizes the maximum number of alterable parameters of the system under investigation. Such parameters would, as a minimum, include anatomy, blood flow, regional vascularity, vascular permeability and metabolism. In order to derive the maximum amount of benefit from a radiodiagnostic procedure, the radiopharmaceutical should be designed consistent with the following goals:

(1) known molecular biology
(2) tracer kinetics which allow the separation of blood flow, receptor binding and release rates
(3) a set of kinetic parameters which are uniquely altered by each pathological state of diagnostic interest
(4) a convenient and stable biological agent and label

Knowledge of a radiopharmaceutical's behavior at the cellular level is critical for proper diagnostic interpretation. Furthermore, if the substrate-receptor interaction of the native substrate is understood on a molecular level, an analog ligand could be designed to optimize radiopharmaceutical performance. Significant features of ligand bioanalogs might be their increased biological stability, availability and labeling characteristics. Once the physical characteristics of the labeled ligand have been optimized, the biological properties should be modulated to separate the tracer kinetics into three phases: The hemodynamic phase would allow assessment of blood flow and vascularity. The binding phase would permit measurement of regional viability of the target cells based on chemical kinetics analysis. The release phase resulting from cellular metabolism and exocytosis of the labeled products could provide additional functional and anatomic information. Other significant features needed to maximize information should include the presence of the target cell in only one tissue type and the ability to inject the number of ligands needed to produce second order kinetics. The latter situation is required to evaluate the receptor concentration.

How do these principles translate into radiopharmaceutical design? The kinetic distinction between flow and binding can only be accomplished if localization is not blood flow limited. This is contrary to past design practice, where a high extraction coefficient was desirable to measure relative blood flow (3) on first pass through tissue. Consider the kinetic scheme:

$$A_o \xrightarrow{k_1} B \xrightarrow{k_2} C$$

$$B(t) = [k_1/(k_2-k_1)]\ A_o[\exp(-k_1t) - \exp(-k_2t)]$$
$$C(t) = A_o\ [1 - k_2\ \exp(-k_1t) - k_1\exp(-k_2t)]$$

If compartment C has a high extraction coefficient for the tracer, $k_1 << k_2$, C(t) and B(t) reduce to

$$B(t) \simeq (k_1/k_2)\ A_o\ \exp(-k_1t)$$
$$C(t) \simeq A_o[1-\exp(-k_1t)].$$

The time activity curve of compartment C now contains no quantitative information concerning k_2, that compartment's rate constant for extraction of the tracer. This renders the volume of compartment B, the organ blood pool, operationally unaccessible ($k_1/k_2 << 1$). This situation is approached by many of the current radiopharmaceuticals for liver scintigraphy, most notably the colloids, and results in diminished sensitivity to any hemodynamic measurement. Thus, quantitation of ligand binding kinetics can only be accomplished if the ligand-receptor affinity is within a range which produces a moderate extraction coefficient.

Another advantage to a moderate binding affinity is an increase in the number of statistically significant imaging frames available for computer kinetic analysis. This is a desirable feature of rose bengal. This compound yields up to four times more uptake phase data points than the current hepatobiliary agent although the overall information presented by rose bengal is very poor due to low photon flux. Ideally the release of the labeled metabolic product from the target cell should not interfere with the target cell image. Thus, products of exocytosis must be rapidly removed from the vicinity of the target cell. Again the blood flow, the most likely mechanism to achieve the above, must not limit the tracer release dynamics.

In summary, the goal is the design of a radiopharmaceutical by which any of the following physiochemical parameters at the target cell may be measured.

Hemodynamic Phase	Binding Phase	Release Phase
blood flow (Q)	receptor concentration [R]	metabolic rate (k_m)
vascularity (V)	forward rate constant (k_b)	% metabolized (A_m)
permeability (P)	reverse rate constant (k_{-b})	

The result is a set of biologic parameters where an alteration in a single parameter represents two pathological states. The degree of each is measured by the relative increase or decrease from the normal value. The hope is that each disease state of diagnostic interest will uniquely alter the set of physiochemical parameters.

In recent years, researchers have identified a hepatic pathway for the accelerated clearance of a number of desialylated glycoproteins. A common sugar sequence occurring in plasma glycoproteins is sialic acid in the terminal position, followed by galactose then N-acetylglucosamine. When ceruloplasmin which has ten such sequences was treated with the desialylating enzyme, neuraminidase, the resulting asialoceruloplasmin cleared rapidly from the circulation of rats (4,5). This removal was mediated by membrane receptors found only on hepatocytes (6). Studies involving other asialoglycoproteins also revealed substantial decreases in circulatory half-times when compared with the native protein.

The common structural feature for all glycoprotein derivatives which were susceptible to accelerated hepatic clearance was the presence of terminal galactose residues exposed upon treatment of the glycoprotein with neuraminidase. If the asialoprotein was modified further by cleaving the galactosyl residue or converting the aglycon carbon from an alcohol to an aldehyde, the resulting derivative showed circulatory survival times similar to its native counterpart (5). The spectrum of plasma clearance times exhibited by the different desialylated proteins is most likely due to the primary sequence of the different carbohydrate structures of each glycoprotein, and only secondarily to the peptide backbone (7,8). At moderate asialoprotein concentrations (2.0 mg in rats *in vivo*), hepatic binding may be saturated (5). The HBP-asialoglycoprotein binding is operationally irreversible, and is the prelude to endocytosis and lysosomal catabolism (4).

Neuraminidase treatment of HBP on the cellular membrane completely abolishes asialoglycoprotein binding (8) because HBP is also a glycoprotein consisting of many terminal sialic acid and penultimate galactose sequences (9). Thus, sialic acid may play two roles in glycoprotein metabolism (10): Firstly, in the circulating glycoproteins, by masking the galactosyl residues and preventing hepatic clearance, and secondly, in the membrane glycoprotein, HBP, by masking galactosyl residues which, if exposed, would compete with the plasma asialoglycoprotein for the remaining receptor sites. This follows from the proposed physiological role of the hepatic binding protein, namely, the continual removal of proteins flagged by the action of *in vivo* neuraminidase. This catabolism of asialoglycoproteins is essential to prevent the *in vivo* accumulation of denatured plasma proteins (11).

$$\text{Plasma Protein-gal-NANA} \xrightarrow{\text{neuraminidase}} \text{Protein-gal} + \text{NANA (sialic acid)}$$

$$\text{Protein-gal} + \text{HBP-hepatocyte} \longrightarrow \text{Protein-gal-HBP-hepatocyte}$$

$$\text{HBP-hepatocyte-HBP} \xrightarrow{\text{neuraminidase}} \text{gal-hepatocyte-HBP} + \text{NANA}$$

Utilizing the criteria previously set forth, the ideal radiolabeled ligand based upon the molecular properties of HBP would have the following properties: First, its binding constant should be lower than the hepatic blood flow so that membrane binding and not delivery of the labeled ligand is the rate limiting step in the localization process. Second, the molecular processing of the ligand should follow the scheme of delivery to the hepatic sinusoids, binding of labeled ligand to the membrane receptor (HBP) at the rate $k_m \cdot [L] \cdot [HBP]$, where third, the ligand-receptor complex should transfer to the lysosome by endocytosis, resulting in catabolism of the substrate with subsequent release of the label into the blood where it should be rapidly extracted by glomerular filtration. The dynamics of this system can be described by the following kinetic model:

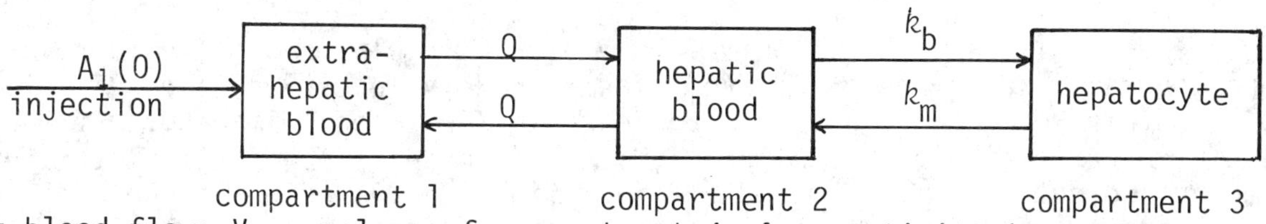

Q = blood flow; V_i = volume of compartment i; A_i = activity in compartment i

The separation of the liver compartment into blood and membrane compartments reflects the design strategy of hemodynamic and molecular differentiation.

Let us now consider four possible tissue conditions: normal, cirrhotic, metastatic and the hepatoma. The hemodymanic determinants of each tissue are well known. The alterations of k_m are based on the general association of decreased lysosomal viability with inflammation and the onset of cellular necrosis. Thus a decrease in k_m with a cirrhotic patient would indicate a late disease stage. Further measurements of the HBP concentration and binding constant (k_b) must be measured for various diseases, but we have made some reasonable assumptions to allow us to continue this discussion. A metastatic tumor residing in the liver will have no hepatic binding protein and hence will exhibit no binding or release phase.

	Q	V	k_b	k_m	[HBP]
normal	-	-	-	-	-
cirrhosis	↓↓	↓↓	(-?)	↓	(↓?)
hepatoma	↑↑	↑↑	(↓?)	↓	(↑?)
metastasis	↓	-	NA	NA	NA

If a region of interest is selected over a given lesion the compartmental model predicts the following time-activity equations resulting from the hepatic uptake of labeled HBP ligand.

$$\frac{dA_1}{dt} = \frac{QA_2}{V_2} - \frac{QA_1}{V_1}$$

$$\frac{dA_2}{dt} = \frac{QA_1}{V_1} - \frac{QA_2}{V_2} + k_m A_3 - \frac{k_b[\mathrm{HBP}]A_2}{V_2}$$

$$\frac{dA_3}{dt} = \frac{k_b[\mathrm{HBP}]A_2}{V_2} - k_m A_3$$

$$A_1(0) = A_1(t) + A_2(t) + A_3(t)$$

$$\text{Region of Interest Activity} = A_2(t) + A_3(t)$$

A computer fit of the above equations to the tracer uptake curves will yield a numerical solution to the kinetic parameters of our model. If each curve is a response to a single set of alterations in model parameters governed by the unique properties of that disease state, a kinetic evaluation of each curve will lead to an unequivocal diagnosis. The specificity of such a procedure will increase with the number of distinguishable and measureable parameters. Thus the design goal is to produce radiopharmacokinetic behavior capable of distinguishing multiple physiological determinants.

The significant physiological role of HBP in plasma protein turnover and the general association of altered plasma protein levels in liver diseases places this receptor as a prime candidate for functional imaging. Of the many native hepatic binding protein ligands that are described in the literature, we have selected three, removed their terminal sialic acid residues and labeled them with the short-lived gamma-emitting nuclides, I-123 or Tc-99m (see protocol Fig. One). Our purpose was to obtain a preliminary evaluation of their liver uptake kinetics in relation to the highly extracted agents (Tc-99m sulfur colloid and the IDA compounds) and the release phase kinetics and distribution of labeled breakdown products.

Table one summarizes the normal rabbit pharmacokinetics of these three compounds; asialoceruloplasmin (AcCp), asialofetuin (AsFet) and asialoorosmucoid (AsOro) using variations in dose, percent sialic acid removed, and label. Two hepatobiliary agents, Tc-99m HIDA and I-131 rose bengal are included for reference. For a preliminary evaluation the functional form $A[\exp(-k_B t) - \exp(-k_A t)]$ was fitted by the algorithm of Marquardt (12) operating under MDOS software (Medical Data Systems, Inc., Ann Arbor, Michigan). The binding phase rate will be estimated by k_A and the release rate by k_B. With the exception of rose bengal, all parameter error estimates were less than 5% with $\chi^2 \leq 300$ (d.f. = 90).

Experiment number	Sialic Acid removed %	Amount injected (mg/kg)	Radio label	k_A (min^{-1})	k_B (min^{-1})
1 AsCp	98	0.33	I-123	0.23	0.023
2 AsCp	98	0.02	I-123	0.29	0.012
3 AsCp	60	0.33	I-123	0.18	0.030
4 AsCp	98	0.50	Tc-99m	0.25	0.005
5 AsFet	98	0.50	Tc-99m	0.14	0.009
6 AsFet	98	0.50	I-123	0.22	0.028
7 AsOro	98	0.33	I-123	0.39	0.048
8 HIDA	-	0.30	Tc-99m	0.74	0.143
9 RB	-	0.05	I-131	0.25	0.039

Experiments 1, 6 and 7 show the variability in liver uptake of different asialoproteins. All have the advantage of a slower hepatic uptake when compared to HIDA. The localization of the latter compound is essentially blood flow limited. The glycoprotein which showed the most promise in this regard is the desialylated sheep submaxillary mucin, asialofetuin. Unfortunately the Tc-99m labeling dramatically altered its uptake rate, probably because fetuin is the 50% carbohydrate and these moieties will favorably compete with the peptide moieties as labeling sites. A decreased HBP affinity could result if the technetium were bound to the carbohydrate of the ligand which is the site of HBP-asialoglycoprotein interaction. Within 10 mins the hepatic uptake peak was reached and it returned to 63% at 30-35 mins. During this same time interval, the plasma activity decreased to less than 10% of the injected dose. Figure Two compares the biodistribution of I-123-AsCp and Tc-99m-HIDA. These curves were generated by computer regions of interest and calibrated by selected tissue sampling. They illustrate two limitations of current hepatobiliary agents for hepatocyte functional imaging: They include a small number of uptake-phase data points, and canalicular flow creates a rapid anatomical shift of activity resulting in impaired functional image.

The technetium labeled compounds exhibited a slower release phase than the iodine tracers, but also showed slight GI activity whereas iodine tracers exhibited none. A small but variable percentage of Tc agents was in the kidney and urinary bladder at 2 minutes post injection and probably represented unbound label. This latter aspect is attributed to the *in vivo* instability of one of two labeling sites encountered during *in vitro* stability measurements. For example the electrolytic labeling of AsCp (80% nuclidic yield after 15 min incubation prior to neutralization, current applied prior to addition of protein) places the reduced technetium at two sites having different stabilities. The more stable site accounts for approximately 60% of the initial activity and exhibits no hydrolysis during 24 hr incubation in HSA (2.5%) and isotonic saline. The other labeling site shows almost complete hydrolysis by 2 hrs when incubated in a similar manner.

Also of interest is the fact that for any given study the binding and release phase as measured by k_A and k_B are homogenous within the entire liver field. This suggests the absence of biliary excretion and is confirmed by the absence of GI and gall bladder activity in four and twelve hour scintiphotos. As a result, the slow release of both labels from the liver is due solely to hepatocyte excretion into the blood followed by rapid filtration by the kidney.

Our preliminary evaluation of labeled native HBP ligand led us to the conclusion that these compounds not only have the superb cellular specificity needed for anatomical imaging, but they also exhibit many of the qualities necessary for successful functional imaging. The rate of binding is sufficient to produce an image within an adequate time but is not too rapid to be blood flow limited. The release phase of both the technetium and iodine labeled metabolic products shows no biliary excretion which would impair the hepatocyte image. Instead this phase is characterized by a slow transport of the label into the circulation where it is rapidly removed by glomerular filtration. This low rate of transfer to the blood insures a dynamic process governed solely by the cell's molecular activity. The predominant factor in release would then be the viability of the lysosomal enzymes. Such a slow exit from the liver field also permits multiple imaging views after the functional data is acquired. This is helpful for the detection of focal lesions deep within the right lobe of the organ.

A special note should be made of a feature of HBP ligands which may be unique among receptor-specific agents, that is their operationally irreversible binding in the normal state. This produces a spectacular liver to background ratio, but unfortunately limits the upper range of possible lesion to normal liver ratios. Another molecular aspect relating to radiopharmaceutical performance is the number of ligands relative to the number of HBP receptors within the liver field. Based on the second order kinetics of a 2 mg dose of AsCp in the rat (5), we might need as much as 200 mg of ligand to give true second order kinetics for functional evaluation of HBP concentration in humans. This might be achieved by simultaneous injection of unlabeled ligands. Without the competitive ligand it would be prudent to expect kinetic alterations produced by HBP concentration changes in only the severest of hepatocellular diseases.

Although we have successfully used Tc-99m-asialoceruloplasmin for hepatic imaging, its routine use is somewhat limited due to a variety of conditions: Ceruloplasmin is a plasma glycoprotein which is expensive, not readily available, and without extensive human use experience. It would also be desirable to have a compound with a better labeling yield and greater biological stability. We therefore concluded that a more clinically acceptable radiopharmaceutical could be achieved by modification of a more advantageous protein. Instead of removing a sugar (sialic acid) to expose the molecular ligand (galactose), an asialoglycoprotein analog may be produced by adding the galactosyl substrate to a protein backbone. The protein of choice would be human serum albumin and the coupled carbohydrate would produce a neoglycoalbumin.

To further investigate the binding properties of HBP, Lee and Krantz (13,15) coupled galactose to bovine albumin and found that the resulting galactosyl-neoglycoalbumin had a high affinity for isolated rabbit HBP receptors as measured by the method of Van Lenten and Ashwell (14). Coupling was achieved by the synthesis of an imidate derivative of galactose: 2-imino-2-methoxyethyl-1-thiogalactoside (IME thiogalactose) (15). The HBP binding affinity depends on the number of sugars attached to each albumin molecule. The binding constants may be altered by as much as 10^4 by controlling the extent of sugar attachment. We are routinely synthesizing IME thiogalactose and coupling this to human albumin. Imidates and the resulting amidine bonds are stable in dilute acid, a condition required for technetium labeling. By using the electrolytic method of technetium labeling (16) we consistently achieve 95% labeling yields of technetium-99m-galactosyl neoglycoalbumin (Tc-99m-gal NGA).

Polyacetate electrophoresis (80 min) will separate reduced Tc (origin), neoglycoalbumin and pertechnetate. Electrophoresis of the unchromatographed product at various times after electrolysis indicates that the product usually loses 5% of its initially bound activity (90-95%) within 20 min. The remaining label is completely stable for several hours.

One concern for this labeled compound as a hepatic radiopharmaceutical was the possibility of product aggregation which would convert the hepatocyte agent into a standard colloidal reticuloendothelial agent. To measure the presence and extent of colloidal formation, the molecular weight distribution of Tc-99m-gal NGA was measured. Over a dozen preparations utilizing the following protocol showed no aggregation or colloid formation. Using the electrolytic method of Benjamin (16), 0.10 mg of gal NGA ($\approx 100 \mu l$) of the same lot of gal NGA (20 gal/HSA, (14)) was added immediately after application of current and 15 minutes later the solution was neutralized. A drop of neutralized product was applied to a 16 x 1 cm column of P-300 Bio Gel. Elution was monitored by absorbance (254 nm) and scintillation detection. The column's void volume was determined by separate elution of dextran blue and ferritin. Prior to elution of the first peak the columns were temporarily removed from the chromatographic system for a rapid standing gel scan.

The elution profile of native HSA and gal NGA reflected the molecular weight shift produced by 20 galactose residues per HSA macromolecule. In all preparations the optical density and activity profiles matched. Two small peaks were present and were interpreted as trimeric and dimeric forms of the basic albumin unit, a common occurrence in albumin chromatography (17). Most important was the consistent absence of activity and absorbance in the void volume and the fact that we have yet to find any retention of activity at the top of the P-300 and desalting (see Protocol, figure one) columns. The standing gel profile confirmed this. Polyacrylamide P-300 elution of Tc-99m-AsCp showed no activity or absorbance at the void volume. From this evidence we conclude that the electrolytic preparation of Tc-99m-gal NGA is free of any product which may localize via recticuloendothelial cells.

With the confidence of high HBP ligand purity, the scintiphotos and time-activity curves of figure three can be regarded solely as the result of hepatocellular function. As with technetium labeled AsCp and AsFet, the Tc-99m-gal NGA exhibited 4-6% ID in the urinary bladder immediately following injection. Within 10 min this number started to rise as labeled lysosomal products were extracted from the blood by glomerular filtration. Removal of the urine from rabbits at 30 minutes showed this activity to be approximately 15% of the injected dose. In all of our *in vivo* rabbit trials the Tc-99m-gal NGA showed a small degree GI activity. In each Tc-99m-gal NGA experiment the gallbladder was removed at 60 min and contained no radioactivity.

We consider technetium-99m-galactosyl neoglycoalbumin to have the features needed for successful cell specific receptor based imaging of the liver. The radiopharmaceutical can be synthesized to produce the desired receptor affinity and its labeling is rapid, efficient, and produces a ligand of high purity. The *in vivo* behavior is specific for the hepatocyte membrane and its uptake rate can be tailored to optimize the functional evaluation of the vascular, receptor binding, and metabolic dynamics. Our plans for the further development of this agent include extensive *in vivo* characterization in normal rabbits. Spontaneous murine and chemically induced rabbit hepatoma models will be used to determine its sensitivity for primary liver tumors. Chemically induced cirrhosis will be used to judge its ability to stage diffuse hepatic disease.

Acknowledgment. We would like to express our appreciation to Gerald L. DeNardo, Edwin C. Glass, and Yuan C. Lee for helpful discussions. This work was supported in part by a Biomedical Research Grant of the University of California, Davis, School of Medicine and by the James Picker Foundation (RCS).

References.

1. Eckelman C, Reba R: The classification of radiotracers. *J Nucl Med* 19:1179-1181, 1977.

2. Taylor D: Biochemical aspects of radiopharmaceutical design. In *Second International Symposium on Radiopharmaceutical Chemistry*, Oxford, 1977, pp 1-3.

3. Saperstein L, Moses, L: Cerebral and cephalic blood flow in man: Basic considerations of the indicator fractionation technique in dynamic clinical studies with isotopes. In *Proceedings of Symposium held at Oak Ridge Institute of Nuclear Studies*, October 21-25, 1963.

4. Gregoriadis G, Morell A, Sternlieb I, Scheinberg I: Catabolism of desialylated ceruloplasmin in the liver. *J Biol Chem* 245:5833-5837, 1970.

5. Morell A, Gregoriadis G, Scheinberg I, Hickman J, Ashwell G: The role of sialic acid in determining the survival of glycoproteins in the circulation. *J. Biol Chem* 246:1461-1467, 1967.

6. Morell A, Irvine R, Sternlieb I, Scheinberg I, Ashwell G: Physical and chemical studies on ceruloplasmin. V. Metabolic studies on sialic acid-free ceruloplasmin in vivo. J Biol Chem 243:155-159, 1968.

7. Van Den Hamer C, Morell A, Scheinberg I, Hickman J, Ashwell G: Physical and chemical studies on ceruloplasmin IX: The role of galactosyl residues in the clearance of ceruloplasmin from the circulation. J Biol Chem 245: 4397-4402, 1970.

8. Hudgin R, Pricer W Jr, Aswell G, Stockert J, Morell G: The isolation and properties of a rabbit liver binding protein specific for asialoglycoproteins. J Biol Chem 249:5536-5543, 1974.

9. Kawasaki T, Ashwell G: Chemical and physical properties of a hepatic membrane protein that specifically binds asialoglycoproteins. J Biol Chem 251:1296-1306, 1976.

10. Stockert R, Morell A, Scheinberg I: Hepatic binding protein: The protective role of its sialic acid residues. Science 197:667-668, 1977.

11. Scheinberg I: Life and death of plasma proteins. In Plasma Protein Turnover, Bianchi, Mariani G, and McFarlane A, eds, Baltimore MD, University Park Press, pp 121-129, 1975.

12. Marquardt D: An algorithm for least-squares estimation of nonlinear parameters. J Soc Indust Appl Math 2:431-441, 1963.

13. Krantz M, Holtzman N, Stowell C, Lee YC: Attachment of thioglycosides to proteins: Enhancement of liver membrane binding. Biochemistry 15: 3963-3978, 1976.

14. Van Lenten L, Ashwell G: The binding of desialylated glycoproteins by plasma membranes of rat liver. J Biol Chem 247:4633-4640, 1972.

15. Lee YC, Stowell C, Krantz M: 2-imino-2-methoxyethyl-1-thioglycosides: New reagents for attaching sugars to proteins. Biochemistry 15:3956-3963, 1976.

16. Benjamin P: A rapid and efficient method of preparing Tc-99m human serum albumin. Its clinical applications. Int J Appl Radiat Isot 20:187-194, 1968.

17. Masinis S, Ott H: In Protides of the Biological Fluids, Proceedings of the 12th Colloquium, Peeters, ed, Elsevier, Amsterdam, p 420, 1965.

purified glycoprotein + neuraminidase → asialoglycoprotein
↓ ↓
purification of asialoglycoprotein — determination of NANA removed
↓
I-123 labeling ← ← ← storage → → → Tc-99m labeling
↓ ↓
↓ desalting ↓
I-123-asialoprotein → polyacrylamide gel ← Tc-99m-asialoprotein
↓
dose/ear i.v.
0.05-0.5 mg/1.0-2.0 mCi
↓
scintiphotos ROI kinetics ← 2.5-3.0 kg anesthesized male rabbit → blood samples % activity protein bound blood clearance

Figure 1. Protocol for In Vivo Studies of Radiolabeled Asialoproteins

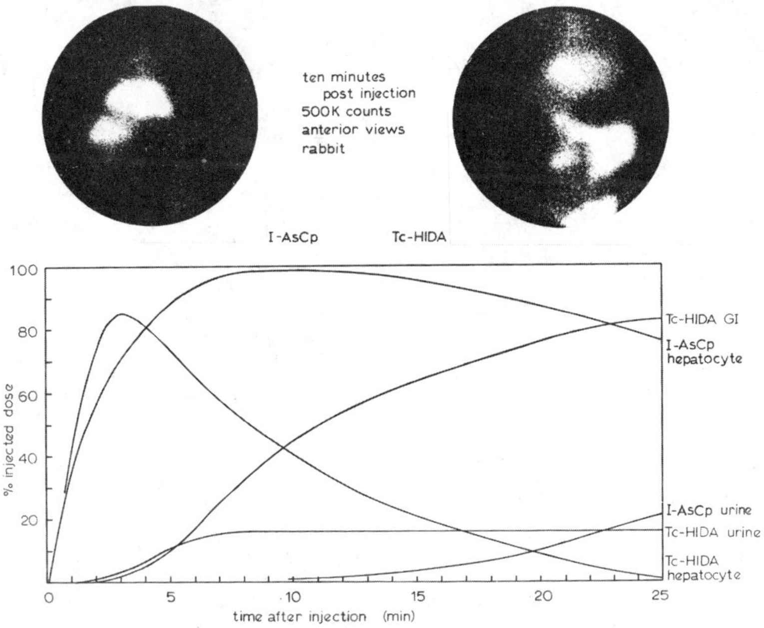

Figure Two. Biological distribution of I-123-asialoceruloplasmin (I-AsCp) and Tc-99m-HIDA.

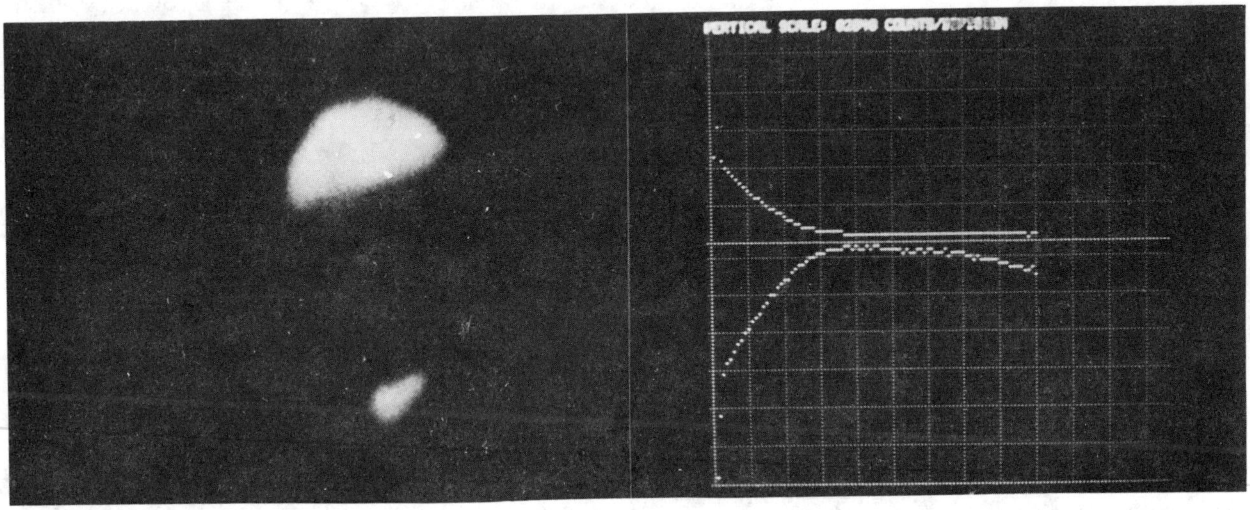

Figure Three. Left: Anterior view of normal rabbit ten minutes after injection of Tc-99m-galactosyl neoglycoalbumin, 750 K counts. Right: Blood clearance (upper) and liver uptake curves, 2048 counts per major division. Liver curve fit to the function $Y(t) = A\exp(-k_B t) - B\exp(-k_A t)$ gave the following values $k_A = 0.19, k_B = 0.015$ (5% error, $X^2 = 220$ d.f. = 90, data was decay corrected).

HEPATOBILIARY TRANSPORT MECHANISM OF Tc-99m-N,α-(2,6-DIETHYLACETANILIDE)-IMINODICARBOXYLIC ACID (Tc-99m-DIETHYL-IDA)

Alan R. Fritzberg, Wayne P. Whitney, and William C. Klingensmith III

University of Colorado Medical Center and
Denver Veterans Administration Hospital
Denver, Colorado 80262

ABSTRACT

The hepatobiliary transport mechanism of ^{99m}Tc-N,α-(2,6-diethylacetanilide)-iminodiacetic acid (^{99m}Tc-diethyl-IDA) was studied by administering saturating doses of representatives of each of the defined transport pathways. The effects on the rate of biliary excretion were monitored by collection of bile from the common bile duct of rats. The administration of sulfobromophthalein decreased the rate of biliary excretion of both ^{99m}Tc-diethyl-IDA and control ^{131}I-Rose Bengal. Taurocholate and procaine amide ethobromide increased the rate of biliary excretion. These results indicate that ^{99m}Tc-diethyl-IDA is transported by the anion dye transport system of BSP and Rose Bengal.

INTRODUCTION

The superior external imaging and absorbed dose radiation properties of Tc-99m over those of I-131 have encouraged attempts to replace ^{131}I-Rose Bengal with a Tc-99m labeled radiopharmaceutical for hepatobiliary imaging studies. The initial report of ^{99m}Tc-N-(2,6-dimethylacetanilide)iminodiacetic acid (^{99m}Tc-HIDA) (1) and then further evaluations of variably substituted phenyl analogs (2) showed this family of agents to exhibit rapid clearance from blood to liver to bile with high specificity. Clinical studies of two agents of this series, ^{99m}Tc-HIDA (3, 4) and the diethyl analog, ^{99m}Tc-diethyl-IDA (5) demonstrated them to be superior to ^{131}I-Rose Bengal. Although animal studies and clinical use have established the pharmacokinetics of these agents, the hepatocyte transport mechanism has not been defined.

Studies of the ^{99m}Tc-HIDA complex (6) postulate a structure which suggests similarities to organic anions and quarternary amines (Fig. 1). The complex bears a net -1 charge indicating it to have anionic properties. However, donation of the imino nitrogen electron pair to the technetium metal ion causes the development of positive charge on the nitrogen atoms and thus structural similarity to quarternary amines such as procaine amide ethobromide (PAEB).

Separate transport pathways have been characterized for organic anions and quarternary amines (7). Therefore, the object of this study was to determine if ^{99m}Tc-diethyl-IDA was transported via one of these systems or via a system unique to metal complexes.

MATERIALS AND METHODS

Materials. Chelating agent, N-(2,6-diethylacetanilide)-iminodiacetic acid was synthesized from 2,6-diethylaniline* by consecutive reactions with chloroacetyl chloride* and iminodiacetic acid disodium salt†. Elemental analysis and nmr spectra of the final compound were as expected. Radiopharmaceutical kits were made by combining 35 mg of the carrier iminodiacetate and 0.35 mg $SnCl_2 \cdot 2H_2O$ in 0.8 ml with final pH adjusted to 5.8 to 6.0. The kits were stored frozen in evacuated vials. Labeling was performed by the addition of 20 μCi of ^{99m}Tc-pertechnetate in 1.2 ml of saline. I-131-Rose Bengal, 10 μCi, was diluted with distilled water to 2.0 ml. The radiopharmaceuticals were combined immediately before injection. I-131-Rose Bengal, sulfobromophthalein and taurocholate were obtained commercially#. Procaine amide ethobromide was kindly supplied by the Squibb Institute for Medical Research.

---

*Aldrich Chemical Co., Inc., Milwaukee, Wis.

†Eastman Kodak, Co., Rochester, N.Y.

#I-131 Rose Bengal was obtained from either Mallinckrodt Nuclear, Inc., St. Louis, Mo. or E.R. Squibb and Sons, Inc., Princeton, N.J. Sulfobromophthalein was obtained from Hynson, Westcott and Dunning, Inc., Baltimore, Md. and sodium taurocholate from Sigma Chemical Co., St. Louis, Mo.

In vivo studies. Male Long-Evans rats weighing 180 to 330 g were used for the experiments. They were anesthetized with pentobarbital. The initial dose was about 50 mg/kg. Further maintenance doses of about 10 mg/kg were given at 45 min intervals or as needed. Internal temperature of the rats was determined with a rectal thermometer and a heat lamp was used to maintain the temperature at 36-37°C.

A. Infusion administration technique. An infusion line for the radiopharmaceutical mixture was placed in one femoral vein and another infusion line for the competitive agent placed in the other. The common bile duct was cannulated and the tubing was connected to a fraction collector for 10 min collection of bile samples. The radiopharmaceutical mixture (15-20 μCi of ^{99m}Tc-diethyl-IDA, 5 μCi of ^{131}I-Rose Bengal in 8.0 ml) was infused at 0.04 ml/min until a constant fraction of the infused dose of both agents appeared in the bile (60-70 min). Then the competitive agent infusion was begun. Both infusions were continued until a total of 150 min was reached. The bile samples were then counted in a well gamma scintillation counter (Picker Autowell, dual channel) and activity determined for I-131 and Tc-99m. Ten minute samples of radiopharmaceutical infusate were used as standards. Radioactivity of Tc-99m was corrected for crossover contribution by I-131.

B. Bolus administration technique. The rats were prepared as described above except that only one femoral vein line was needed. The radiopharmaceutical mixture (1-2 μCi of ^{99m}Tc-diethyl-IDA and 0.5-1 μCi of ^{131}I-Rose Bengal in 0.25 ml) was injected over a several second time period. Bile collections were made from 0-2 min and then for 4 min periods out to 90 min total. For competitive studies with BSP, the same dose used for the infusion study was infused for 15 min before the radiopharmaceutical injection until 25 min after. For the studies with procaine amide ethobromide (PAEB) the renal pedicles were tied off. The dose of 20 mg/kg was given over a 5 min period and the radiopharmaceutical mixture was then injected after a delay of 20 min. Bile samples were counted as above and the total PAEB was determined colorimetrically (8, 9).

For evaluation of the organic anion pathway a dose of 2.5 μmol/min/kg of BSP was infused, approximately twice the Tm of BSP (10). After bile samples were weighed and counted, 5 μl aliquots were diluted with 15.0 ml of 0.1 N NaOH. The concentration was determined by spectrophotometric assay at 580 nm. For evaluation of the bile acid pathway a saturating dose of 13 μmol/min/kg of taurocholate (11) was infused. For the quarternary amine pathway doses of 6, 10 and 20 mg/kg were given. Saturating values were reported above 5 mg/kg by one group (9) while differences in dose responses were reported up to 20 mg/kg by another (12). Saturation kinetics were obtained in this study only at 20 mg/kg so the data are reported only for that dose level.

RESULTS

Saturating doses of sulfobromophthalein (BSP) depressed the biliary output of both ^{99m}Tc-diethyl-IDA and ^{131}I-Rose Bengal under continuous infusion (Fig. 2) and bolus injection (Fig. 3) conditions. The depression of ^{131}I-Rose Bengal was reduced to a greater extent than ^{99m}Tc-diethyl-IDA under both conditions. Since the normal rate of biliary excretion of ^{99m}Tc-diethyl-IDA is much faster than ^{131}I-Rose Bengal (Fig. 3) this difference may mean that

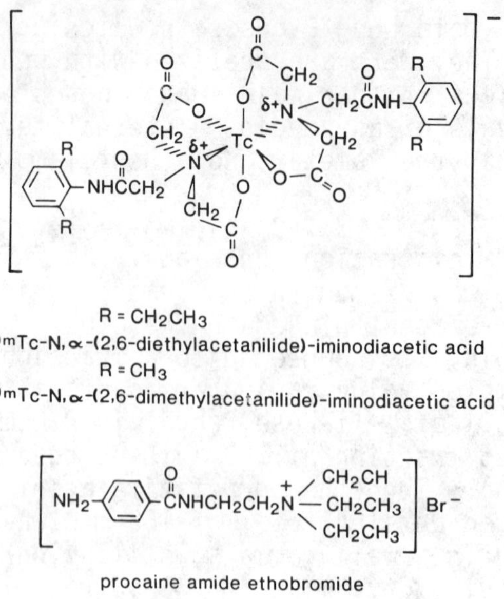

Figure 1.

Structures of Tc-99m iminodiacetate hepatobiliary agents and the quarternary amine, procaine amide ethobromide, for comparison.

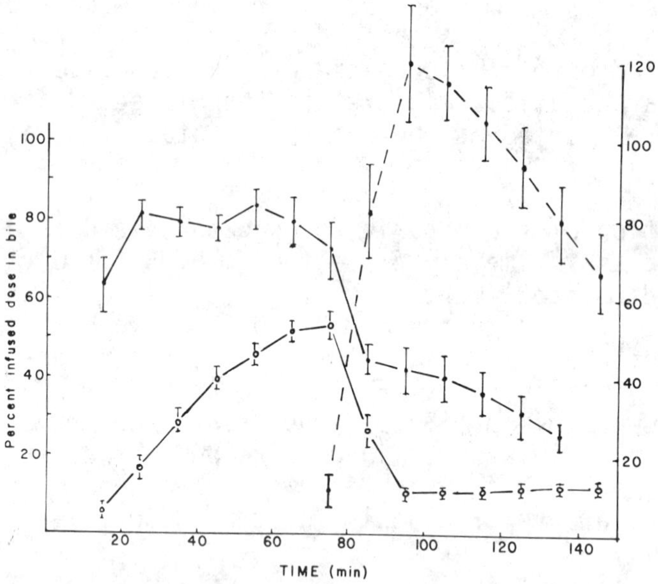

Figure 2.

Effect of BSP on biliary excretion of ^{99m}Tc-diethyl-IDA (●—●) and ^{131}I-Rose Bengal (o--o) under continuous infusion conditions. Values for radiopharmaceuticals represent percentage of 10 min infused dose collected in 10 min from common bile duct. BSP biliary excretion (■--■) is in μg/min/100g rat corrected for bile flow. Values are average of six experiments and bars indicate S.E.M. BSP was infused at 2.5 μmol/min/kg.

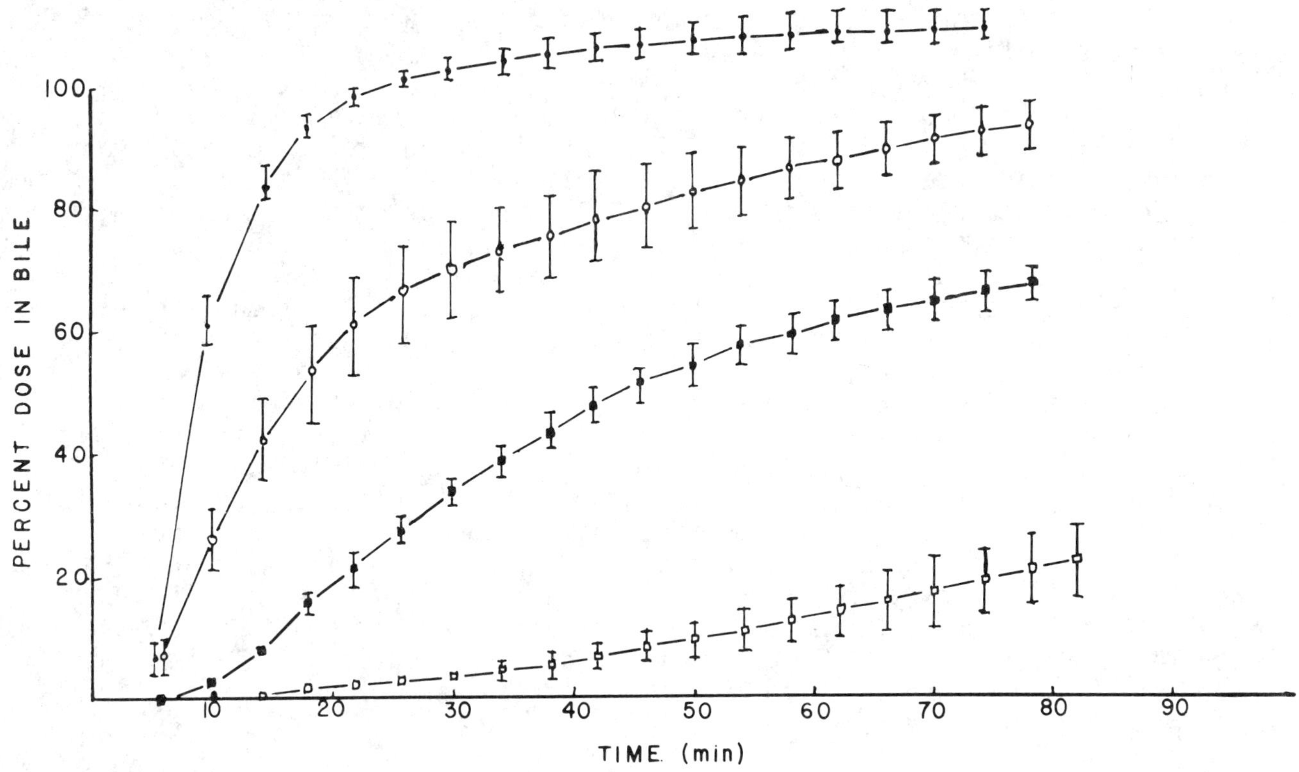

Figure 3.

Effect of BSP on biliary excretion of ^{99m}Tc-diethyl-IDA and ^{131}I-Rose Bengal following bolus injection. Control curves of ^{99m}Tc-diethyl-IDA (●—●) and ^{131}I-Rose Bengal (■—■) are compared with BSP treated curves of ^{99m}Tc-diethyl-IDA (o—o) and ^{131}I-Rose Bengal (□--□). BSP was infused at 2.5 μmol/min/kg for 15 min before radiopharmaceutical injection until 25 min after. Bile samples were collected for 4 min periods after an initial 2 min collection. Values of control ^{99m}Tc-diethyl-IDA are mean and range for three experiments; for ^{131}I-Rose Bengal mean and S.E.M. of four experiments; for BSP experiments, the mean and S.E.M. of four experiments of each agent.

the ^{99m}Tc complex more effectively competes for the transport system under BSP saturating conditions.

The effect of taurocholate on the infused radiopharmaceuticals appeared to be parallel for both (Fig. 4). A rise in bile output followed by a return to previous steady-state output conditions was observed. The result is in agreement with other observed increases in BSP transport maximum caused by taurocholate in perfused rat liver (13), dogs (14), and sheep (15).

The evaluation of the quarternary amine pathway by saturation with procaine amide ethobromide (PAEB) was performed under both constant infusion and bolus injection conditions. Although no inhibition by PAEB was indicated under constant infusion conditions (Fig. 5) alteration of bile output of ^{99m}Tc-diethyl-IDA cannot be inferred from the data since there was variability in the output of ^{99m}Tc-diethyl-IDA before and after the ^{131}I-Rose Bengal reached a plateau in these experiments when the renal pedicles were tied off. The output of ^{131}I-Rose Bengal, however, was increased from a plateau value of 60-65% of the 10 min infused dose appearing in the bile in 10 min to 75-80%. The PAEB inhibition experiments were then repeated using bolus injection techniques (Fig. 6). Under these conditions inhibition of ^{99m}Tc-diethyl-IDA output clearly did not occur. Output of the radiopharmaceutical did appear to be increased in the 2 to 6 and 6 to 10 min collection periods when compared to the control bile output curve, however.

DISCUSSION

The high specificity and very rapid rate of biliary excretion of ^{99m}Tc-diethyl-IDA observed here in rats is in accord with clinical experience in man and suggests transport by a carrier mediated system. The traditional approach to defining the transport of a compound by determining a transport maximum with increasing doses is impractical in this case since the carrier iminodiacetate is excreted by the kidneys (16). This leaves only the tracer amounts of ^{99m}Tc-diethyl-IDA following the pathway of biliary excretion. As an indirect alternative to increasing doses of ^{99m}Tc-diethyl-IDA, saturating doses of compounds representative of potential transport systems were administered. The biliary excretion rate depression of ^{99m}Tc-diethyl-IDA resulting from BSP observed in this study supports the suggestion of active transport.

Of the hepatobiliary transport systems tested, a depression in bile output of ^{99m}Tc-diethyl-IDA was observed only for the dye anion pathway as indicated by saturation with BSP. This result negates the need for consideration of a special pathway for iminodiacetate technetium complexes. The lack of inhibition by PAEB indicates that structural similarity of quarternary ammonium compounds and the complexed imino nitrogen is apparently not a factor in the hepatobiliary transport while the net negative charge on the complex is. This result may have been indirectly predicted since rapid rates of liver uptake and biliary clearance of ^{99m}Tc-diethyl-IDA are seen in the dog (17) while quarternary amines such as PAEB are taken up and cleared very slowly in the dog relative to the rat and rabbit (12). The increased bile output of both ^{99m}Tc-diethyl-IDA and ^{131}I-Rose Bengal after infusion of taurocholate is consistent with reports of similar effects on BSP as mentioned earlier and further supports a common transport system for each.

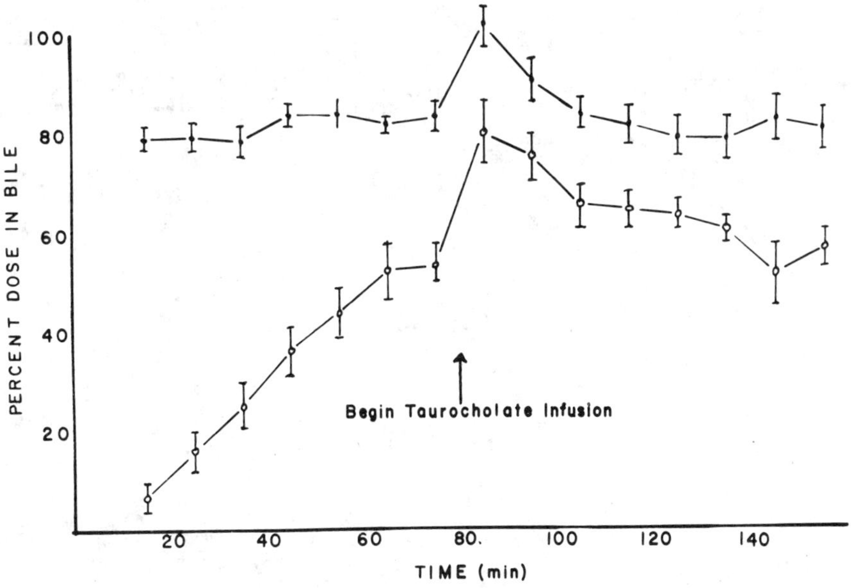

Figure 4.

Plot of effect of sodium taurocholate on bile output of ^{99m}Tc-diethyl-IDA (●—●) and ^{131}I-Rose Bengal (o—o) under constant infusion conditions. Sodium taurocholate was infused at a rate of 13 μmol/min/kg. Values are mean and S.E.M. of five experiments.

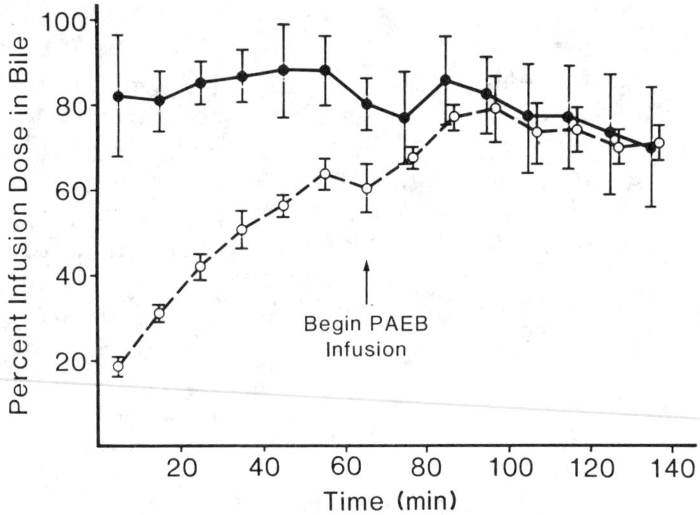

Figure 5.

Effect of procaine amide ethobromide (PAEB) on biliary excretion of ^{99m}Tc-diethyl-IDA and ^{131}I-Rose Bengal under constant infusion conditions. Values are portion of 10 min infused dose collected from the common bile duct in 10 min. PAEB was infused over a five min period at a level of 20 mg/kg. Values are mean of three rats and bars indicate ± S.D. ^{99m}Tc-diethyl-IDA is shown by ●—● and ^{131}I-Rose Bengal by o--o.

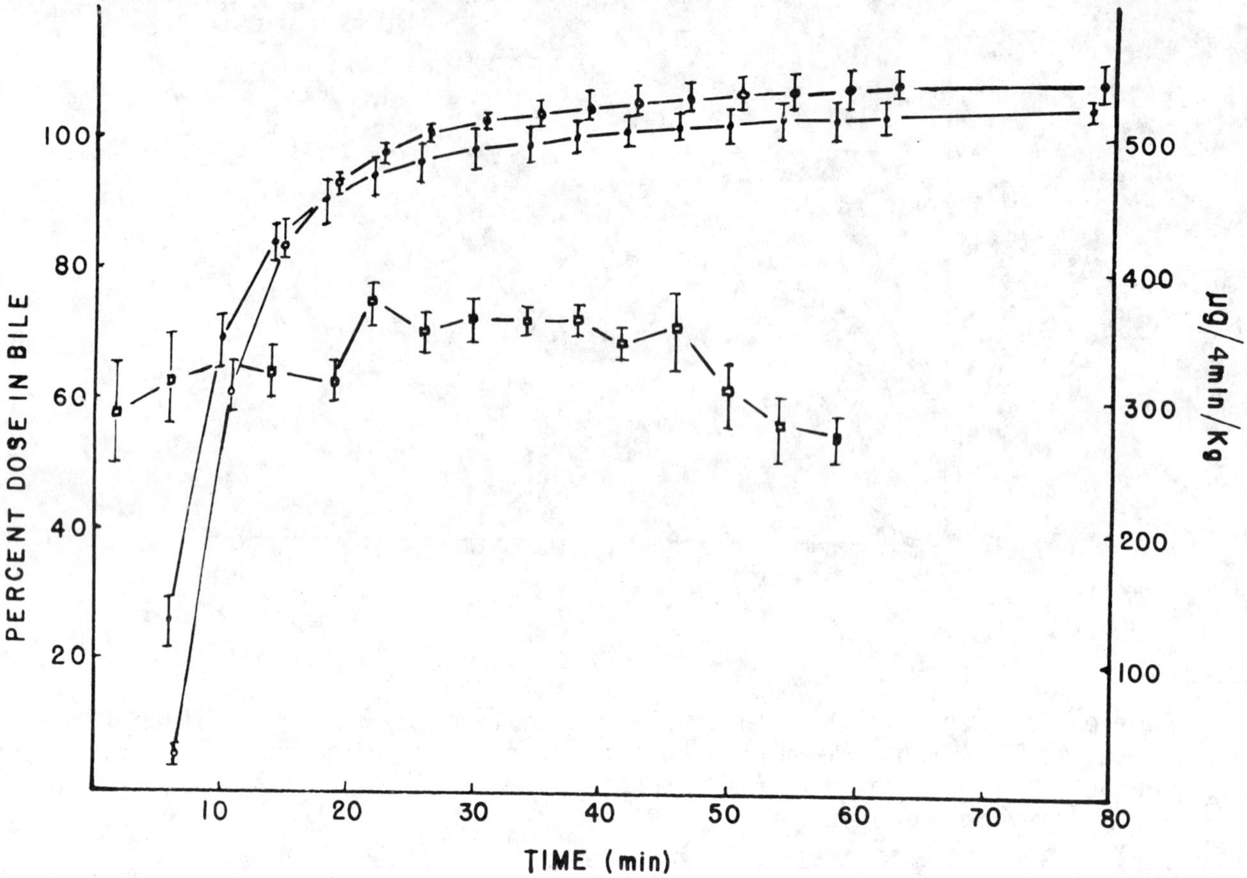

Figure 6.

Effect of procaine amide ethobromide (PAEB) at 20 mg/kg on biliary excretion of ^{99m}Tc-diethyl-IDA under bolus injection conditions. Control ^{99m}Tc-diethyl-IDA values (o—o) are compared to those with added PAEB (●—●). Total PAEB in bile (■--■) is given in µg/min/kg. PAEB was given over a five min period with renal pedicles tied off. Radiopharmaceuticals were injected after an additional 20 min. Bile samples were collected for 4 min after an initial 2 min collection. Control ^{99m}Tc-diethyl-IDA values are mean and range for three experiments; values of ^{99m}Tc-diethyl-IDA and PAEB following injection of PAEB are mean and S.E.M. for five rats.

In terms of clinical significance these results demonstrate that ^{99m}Tc-diethyl-IDA uses the same transport system as BSP, Rose Bengal and hence bilirubin (18). Thus, the high bilirubin levels in the jaundiced patient will cause a competitive situation with ^{99m}Tc-diethyl-IDA and similar agents and will decrease rates of biliary excretion as well as increase renal excretion. These interactions with bilirubin were shown to be the case for ^{99m}Tc-HIDA in dogs by M.D. Loberg (personal communication) while this work was in progress. In addition to direct interactions with bilirubin, the results have implications for patients with Dubin-Johnson syndrome which is characterized by abnormal bilirubin and BSP transport and normal bile acid transport and possibly for patients with Gilbert's syndrome in which hepatic uptake of unconjugated bilirubin is impaired.

ACKNOWLEDGMENTS

The authors wish to thank Professor Franz Simon, M.D., Department of Gastroenterology for help with the animal methods and BSP determinations as well as for helpful discussions.

REFERENCES

(1) Harvey E, Loberg M, Cooper M: Tc-99m-HIDA: A new radiopharmaceutical for hepatobiliary imaging. J Nucl Med 16: 533, 1974

(2) Wistow BW, Subramanian G, Van Heertum RL, et al: An evaluation of ^{99m}Tc-labeled hepatobiliary agents. J Nucl Med 18: 455-461, 1977

(3) Ryan J, Cooper M, Loberg M, et al: Technetium-99m-labeled N-(2,6-dimethylphenylcarbamoylmethyl)-iminodiacetic acid (Tc-99m HIDA): A new radiopharmaceutical for hepatobiliary imaging studies. J Nucl Med 18: 995-1002, 1977

(4) Weissman H, Frank M, Rosenblatt R, et al: Correlation of Tc-99m HIDA with anatomic imaging in patients with hepatobiliary disease. J Nucl Med 19: 738, 1978

(5) Klingensmith WC, Fritzberg AR, Koep LJ, et al: Clinical Comparison of ^{99m}Tc-diethyl-iminodiacetic acid, ^{99m}Tc-pyridoxylideneglutamate, and 131I-Rose Bengal in liver transplant patients. Radiology 130: 435-441, 1979

(6) Loberg MD, Fields AT: Chemical structure of Technetium-99m-labeled N-(2,6-dimethylphenylcarbamoylmethyl)-iminodiacetic acid (Tc-HIDA). Int J Appl Rad Isot 29: 167-173, 1978

(7) Klassen CD: Biliary excretion. In Handbook of Physiology, Section 9: Reactions to Environmental Agents, ed. by DHK Lee, chap 34, pp 537-553, American Physiological Society, Washington, D.C., 1977

(8) Bratton AC, Marshall Jr. EK: A new coupling component for sulfanilanide determination. J Biol Chem 128: 537-550, 1939

(9) Schanker LS, Solomon HM: Active transport of quarternary ammonium compounds into bile. Amer J Physiol 204: 829-832, 1963

(10) Dhumeaux D, Berthelot P, Preaux AM, et al: A critical study of the concept of maximal biliary transport of sulfobromophthalein in the Wistar rat. Rev Europ Etudes - Clin et Biol 15: 279-286, 1970

(11) Paumgartner G, Sauter K, Schwarz HP, et al: Hepatic excretory transport maximum for free and conjugated cholate in the rat. Effect of enzyme induction. In The Liver. Quantitative Aspects of Structure and Function, pp 337-344, Karger, Basel, 1973

(12) Hunter A, Klassen CD: Species difference in the plasma disappearance and biliary excretion of procaine amide ethobromide. Proc Soc Exp Biol Med 139: 1445-1450, 1972

(13) Boyer JL, Scheig RL, Klatskin G: The effect of sodium taurocholate on the hepatic metabolism of sulfobromophthalein sodium (BSP). The role of bile flow. J Clin Invest 49: 206-215, 1970

(14) O'Maille ERL, Richards TG, Short AH: Factors determining the maximal rate of organic anion secretion by the liver and further evidence on the hepatic site of action of the hormone secretion. J Physiol 186: 424-438, 1966

(15) Gronwall R, Cornelius CE: Maximal biliary excretion of sulfobromophthalein sodium in sheep. Amer J Digest Dis 15: 37-47, 1970

(16) Callery PS, Faith WC, Loberg MD, et al: Tissue distribution of Technetium-99m and Carbon-11 labeled N-(2,6-dimethylphenylcarbamoylmethyl)-iminodiacetic acid. J Med Chem 19: 962-964, 1976

(17) Fritzberg AR, Eshima D, Klingensmith WC, et al: Comparison of ^{99m}Tc-hepatobiliary agents based on napthalene and similar ring systems. J Nucl Med 19: 694, 1978

(18) Cantarow A, Wirts CW, Snape WJ, et al: Excretion of bilirubin and bromsulfalein in bile. Amer J Physiol 154: 211-219, 1948

BENZIMIDAZOLYL METHYLIMINODIACETIC ACIDS - NEW BIFUNCTIONAL CHELATORS OF TECHNETIUM FOR HEPATOBILIARY SCINTIGRAPHY

Frederick C. Hunt, J. Gerald Wilson and Desmond J. Maddalena

Isotope Division, Australian Atomic Energy Commission,
Research Establishment, Lucas Heights, N.S.W., Australia

ABSTRACT

Dimethyl- and chloro- substituted benzimidazolyl methyliminodiacetic acids have been synthesised and evaluated as new bifunctional chelators of ^{99m}Tc. Stannous chelates of these compounds were prepared as freeze-dried kits and labeled with ^{99m}Tc. The radiopharmaceuticals thus prepared were rapidly excreted by the hepatobiliary system of rats and rabbits with little urinary excretion. The chloro- compound had a higher biliary and lesser urinary excretion than the dimethyl- however both technetium complexes provided good scintigraphic images of the hepatobiliary system in animals. The compounds behaved similarly to the ^{99m}Tc-lidocaine iminodiacetic acid [HIDA] complexes with respect to their biliary elimination.

INTRODUCTION

The introduction of iminodiacetic acid (IDA) derivatives as bifunctional chelators for technetium-99m has been a recent innovation for the rational design of ^{99m}Tc-radiopharmaceuticals.

These chelators comprise a chelating group to bind technetium and a functional moiety to impart biological specificity to the chelate as a whole. The lidocaine-IDA derivatives (HIDA) introduced by Loberg (1) and extended by Subramanian (2) have been used for hepatobiliary scintigraphy although the biodistribution of the original antiarrythmic drug was substantially changed on conversion to an IDA-derivative and subsequent ^{99m}Tc labeling.

Another moiety with biological action is the benzimidazole group frequently used in the design of anthelmintic drugs. Benzimidazolyl methyliminodiacetic acid, (BIMIDA) [I], a known complexone, chelates a number of metals (3) although no biological use of the metal chelates has been suggested hitherto. The synthesis of two BIMIDA derivatives containing different substituents was therefore undertaken and their potential as technetium radiopharmaceuticals tested.

MATERIALS AND METHODS

Synthesis of BIMIDA Complexones

The dimethyl- and chloro- BIMIDA compounds used in this study were synthesised by alkylation of iminodiacetic acid dimethyl ester with the appropriate chloromethylbenzimidazoles followed by mild alkaline hydrolysis to give the benzimidazolyl methyliminodiacetic acids (3). The chlomethyl intermediates were made from the corresponding dimethyl- and chloro- phenylenediamines.

(a) 5,6-Dimethyl-2-chloromethylbenzimidazole.
To a warm suspension of 4,5-dimethyl-o-phenylenediamine (13.6 g., 0.1 mol) in dry ethanol (40 ml) a suspension of chloracetimino ethyl ether hydrochloride (15.8 g, 0.1 mole) also in dry ethanol (40 m) was added in portions with vigorous agitation. The reaction mixture became very hot and ammonium chloride separated. After further warming for about 15 minutes on a steam bath, the reaction mixture was left at room temperature for 2 hr. then diluted with water (~150 ml). The crystalline product was collected after several hours, washed well with water and dried in air (17.1g. 88%). Recrystallization from hot, dioxan (150 ml) and petroleum ether (bp 60-80°, 25 ml) afforded pale yellow crystals (12.0 g) mp 165-167°dec.

(b) 2-(5-6 Dimethylbenzimidazole)-methylamine - N,N-diacetic acid. [Dimethyl BIMIDA]
5-6 Dimethyl-2-chloromethylbenzimidazole (14.6 g. 0.075 mol) was added in two portions to a stirred warm suspension of iminodiacetic acid dimethyl ester hydrochloride (15.0 g, 1.1 mol equiv.) in ethanol (50 ml) containing triethylamine (21.0 ml, 2.2 mol equiv.). The mixture was stirred for 15 min. and after warming on the steam-bath for 10 min all the solid material dissolved. Gentle heating was continued for 1.5 hr. followed by cooling in the refrigerator. After filtration of the solid (mostly triethylamine hydrochloride) and equal volume of diethyl ether was added to the filtrate and after cooling overnight the second lot of crystals was filtered off. The oil remaining after removal

of the solvent was then hydrolysed in 50% methanol (100 ml) containing sodium hydroxide (7.5 g), and the solution was stirred for 1 hr. Water (50 ml) was added and the solution was extracted with ether. The aqueous layer was warmed to remove dissolved ether and the pH was brought to 3 by the careful addition of conc. hydrochloric acid. The product separated in crystalline form and was collected after cooling overnight. Yield 15.5 g (48%) mp 208 - 210°C dec.

Purification

The product from three such runs (28.3 g) was suspended in water (150 ml) and methanol (50 ml) and a solution of sodium hydroxide (8.0 g) in water (20 ml) was added until solution was obtained after stirring and warming. Acidification of the warm filtered solution to pH 3.0 caused separation of the iminodiacetic acid in a finely granulated form. After 3 hr. in the refrigerator the product was collected and washed in turn with ice-water, cold 50% aqueous methanol and finally methanol, then air dried. Yield 25.4 g. mp 209 - 210° dec. Anal. C,57.72; H, 5.88; N,14.23%. ($C_{14}H_{17}N_3O_4$) requires C,57.73; H,5.84; N,14.43%.

Infra-red (KBr) maxima 1706 (COOH), 1240 cm^{-1} (COOH) obtained were characteristic of the free carboxylic acid group of iminodiacetic acid derivatives. Mass spectrum m/e 273 obtained was the ion of highest mass formed by the thermal loss of water (18 m. u) from the molecular ion.

Technetium-99m Labeling

The dimethyl-and chloro-BIMIDA compounds were labeled with technetium-99m by the stannous chloride reduction method. The compounds were made into lyophilised kits containing 5 mg of the BIMIDA and 50-200 µg of stannous chloride at pH 6-7. Labeling with ^{99m}Tc was achieved by adding 1-2 ml of TcO_4^- to the vial and mixing thoroughly.

Labeling efficiencies and radiochemical purity of BIMIDA compounds were determined by ascending TLC using Gelman ITLC medium and saline solvent. Biodistributions were performed in female albino rats and rabbits fasted overnight. Scintigraphic studies were performed in fasted rabbits. Two millicuries (74 MBq) were injected intravenously and sequential images were taken accumulating 100K counts in the anterior position.

RESULTS AND DISCUSSION

Technetium-99m labeling efficiencies greater than 98% were obtained with dimethyl-BIMIDA and chloro-BIMIDA. With the saline/ITLC system R_F's of 0.5 for the dimethyl-complex and 0.9 for the chloro complex were obtained. There was no colloidal or reduced unbound technetium remaining at the origin. Pertechnetate (if any) appeared at the solvent front.

The biodistributions of the two complexes are shown in Table 1. With both ^{99m}Tc - di CH_3 - BIMIDA and Cl-BIMIDA the predominant excretory route is hepatobiliary with the chloro- complex having the greatest biliary output and the least urinary excretion. With both complexes the spleen and stomach contained less than 1% of the injected dose in each organ indicating absence of colloid and free pertechnetate in the respective organs.

The biodistributions of these compounds are comparable with those of Tc-

TABLE 1. BIODISTRIBUTION OF ^{99m}Tc-LABELLED DIMETHYL- AND CHLORO-BENZIMIDAZOLYL-IDA DERIVATIVES
% DOSE IN ORGAN AT 1 HR (MEANS ± SD OF 6 ANIMALS)

		Blood	Liver	Kidneys	G.I.T.+G.B.	Urine
di-CH_3-BIMIDA	Rats	0.9 ± 0.1	1.7 ± 0.3	0.8 ± 0.08	84.0 ± 2.9	6.0 ± 0.8
	Rabbits	5.1 ± 1.3	7.2 ± 0.9	1.8 ± 0.5	68.6 ± 6.0	6.7 ± 1.0
Cl-BIMIDA	Rats	0.7 ± 0.1	1.3 ± 0.2	0.3 ± 0.0	92.3 ± 2.2	1.7 ± 0.4
	Rabbits	3.8 ± 1.4	3.3 ± 0.3	0.5 ± 0.2	82.5 ± 0.3	1.5 ± 0.7

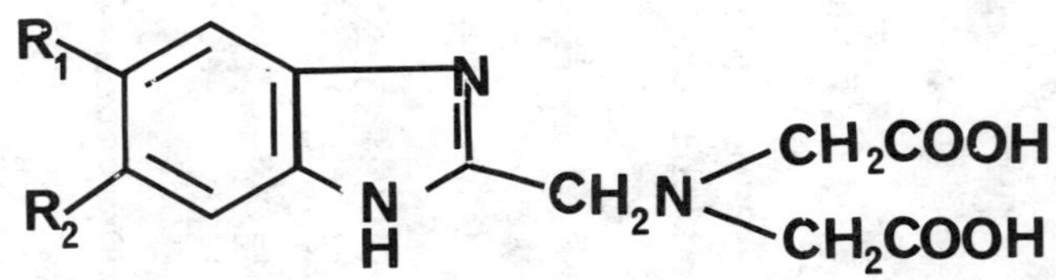

I BIMIDA COMPLEXONE

Dimethyl $R_1 = R_2 = CH_3$

Chloro $R_1 = Cl, R_2 = H$

HIDA (1,2). The one hour G.I.T. excretion of ^{99m}Tc-Cl-BIMIDA in rabbits (82.5%) is comparable with that of ^{99m}Tc-p-isopropyl HIDA (83.0%) however the urinary excretion at one hour (1.5%) is less than that obtained with pip-IDA (4.8%).

Scintigraphic studies of the BIMIDA complexes in rabbits are shown in Fig. 1. With both compounds there is rapid clearance of activity from the blood by the liver after intravenous injection with excretion of activity into the G.I.T. apparent at 10 min. The gallbladder is visualized between 5 and 10 mins. The kidneys are faintly and transiently visualized shortly after injection as the urinary component is excreted. The two complexes give similar images, except the liver remains imaged for longer in the case of the dimethyl-complex and also the urinary bladder is more intensely imaged.

The biliary excretion of the BIMIDA complexes is consistent with the behaviour of lipophilic chelating agents. The biliary excretion of iron (4) and technetium chelates (5) of lipophilic analogues of EDTA have been reported. Benzimidazoles are isosteric with purines and the biliary excretion of ^{99m}Tc-6-mercaptopurine has also been reported. (6)

Loberg has postulated a structure for HIDA-Tc complexes containing two molecules of HIDA per atom of technetium (7). It is possible that a similar structure exists with BIMIDA-Tc, thereby providing high molecular weight complexes facilitating biliary excretion.

To summarise, new iminodiacetic acid complexones, benzimidazolyl methyl-iminodiactic acids have been synthesised, formulated into lyophilised stannous complexes and found to be suitable for hepatobiliary scintigraphy when complexed with technetium-99m.

ACKNOWLEDGEMENT

The authors are grateful for the technical assistance provided by Ruth Potter and Renate Domel.

REFERENCES

1. LOBERG, M.D., COOPER, M, HARVEY, E.,et.al.: Development of new radiopharmaceuticals based on N-substitution of Iminodiacetic acid. J. Nucl. Med. 17: 633-638, 1976.

2. SUBRAMANIAN, G. : The influence of structural changes on biodistribution of Tc-99m labeled N-substituted IDA derivatives. J. Nucl. Med. 18 : 624, 1977.

3. IRVING, H., WEBER, O. : Some potential chelating agents derived from benzimidazole. J. Chem. Soc. 2296-2298, 1959.

4. HADDOCK, E.P., ZAPOLSKI, RUBIN M.,et.al.:Biliary excretion of chelated iron. Proc. Soc. Exptl. Biol. Med. 120 : 663-668, 1965.

5. HUNT, F.C., WILSON, J.G., MADDALENA, D.J. : Technetium chelates of phenolic aminocarboxylic acids for hepatobiliary scintigraphy. Proc. 1st. Asia and Oceania Cong. Nucl. Med. 140-143, 1976.

6. HUNT, F.C., MADDALENA, D.J., YEATES, M.G. : Technetium-99m-6-mercaptopurine, a new radiopharmaceutical for cholescintigraphy. Proc. 1st. World Cong. Nucl. Med. 869-871, 1974.

7. LOBERG, M.D., FIELDS, A.T. : Chemical structure of technetium-99m-labeled N-(2,6-dimethylphenylcarbamolymethyl) iminodiacetic acid. (^{99m}Tc-HIDA) : Int. J. Appl. Radiat. Isot. 29 : 167-173, 1978.

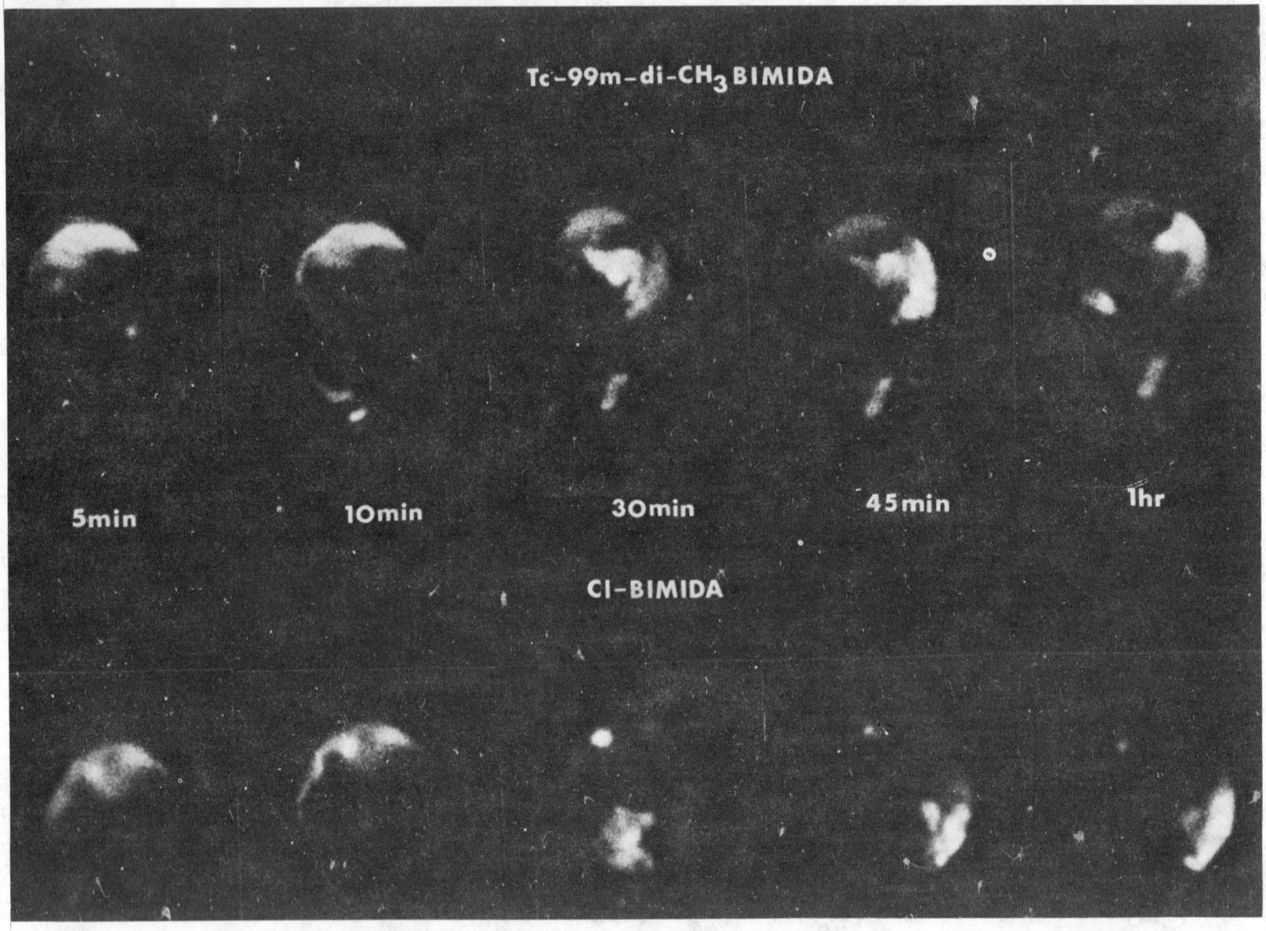

FIGURE 1

Sequential images of two rabbits after the intravenous injection of 2 mCi of ^{99m}Tc-dimethyl- and chloro- BIMIDA. There is less liver retention and urinary excretion with the chloro- complex.

BIOLOGIC EVALUATION OF RADIOCOLLOIDS FOR BONE MARROW SCINTIGRAPHY

Sydney Heyman, Michael A. Davis, Peter M. Shulkin and Salvador Treves

Joint Program in Nuclear Medicine, Harvard Medical School
and Children's Hospital Medical Center
Boston, Massachusetts 02115

The validity of a primate animal model for studying the in vivo distribution of various colloids was established. Computerized images from two adult baboons injected with technetium-99m labeled sulfur colloid, stannous phytate and microaggregated albumin were analyzed to give the relative uptake of radioactivity in the liver, spleen and bone marrow. These values were in good agreement with those previously established in several animal species and in man. Antimony sulfide colloid and minimicroaggregated albumin, each having a significantly smaller particle size than Tc-99m sulfur colloid were evaluated. Compared with sulfur colloid the minimicroaggregated albumin showed three times the bone marrow uptake (15-20%) whereas microaggregated albumin and antimony sulfide gave somewhat lower values (8-12%). The stannous phytate showed no improvement over Tc-99m sulfur colloid (5-7%).

INTRODUCTION

Bone marrow imaging is potentially useful in the diagnosis of both benign and malignant disorders of the skeletal system. It has been shown that normal bone scans and radiographs do not exclude the presence of tumor cells in the bone marrow (1). Since in many instances skeletal metastases involve the marrow cavity before affecting the cortex, marrow imaging may be an earlier indicator of metastatic bone disease than the bone scan. Reticulo-endothelial scanning with Tc-99m-sulfur colloid has demonstrated the replacement of normal bone marrow by metastatic neuroblastoma (2). We have seen recurrent lymphoma with bone involvement produce a defect on bone scanning in the presence of a normal bone scan. Similarly bone marrow scanning may also be an earlier indicator of disease in those conditions which involve the bone marrow primarily, such as leukemia and lymphomas. In eosinophilic granuloma, basically a disease of the reticuloendothelial system, bone marrow scans may be abnormal in those instances where there is expansion of the marrow cavity without cortical involvement (3). The technique has also proved helpful in the diagnosis of bone infarction, especially in patients with sickle cell anemia where bone pain may be due to either infarction or infection (4).

At present, phagocytic bone marrow imaging is performed with Tc-99m-sulfur colloid. Since only about 5% of the injected dose is taken up by reticuloendothelial cells in the marrow, the photon yield is usually low. This means that relatively high administered doses and a prolonged imaging time are necessary to obtain scintigraphs of a suitable information density. This is of particular importance in pediatric nuclear medicine, where it is often difficult for small patients to remain still for a prolonged period of time. The Tc-99m-sulfur colloid particles range in size from about 0.1 μ to 1.0 μ, with a mean of 0.3 μ(5). It has been suggested that the smaller particles tend to be taken up preferentially by the bone marrow although numerous additional factors such as the chemical composition, charge, shape, and surface properties of the particles may also be important (6-10).

The organ distribution of the technetium-99m labeled colloids could be readily determined from computerized images. The actual distributions in liver, spleen and marrow for sulfur colloid, microaggregated albumin and stannous phytate were known from previously published animal and human studies. If the radioactivity in the computerized images were normalized to give the expected organ distribution for sulfur colloid, this should correct for attenuation and geometrical factors, since the same animals were used in each study. We postulated that if the normalizing factors gave the previously established distributions for the microaggregated albumin and the stannous phytate, the validity of the model would be confirmed. With the model in hand, colloids of smaller particle size, such as antimony sulfide and minimicro-aggregated albumin, were then evaluated in the same manner to quantitate differences in bone marrow uptake.

MATERIALS AND METHODS

Studies were performed on two adult female baboons weighing 25 kg and 37 kg. The radiopharmaceuticals tested are listed in Table 1. These were injected intravenously in a sequential manner allowing at least 48 hr between studies, at a dosage of 80 μCi/kg. The syringe activity was determined before and after injection, allowing precise determination of the injected dose. Imaging was commenced 20 minutes after injection, using a gamma camera fitted with a diverging low energy collimator (Ohio Nuclear Sigma 400) and interfaced to a PDP 11/20 computer (Digital Equipment Corp.). Scintiphotos of the liver, spleen and lower spine were obtained in both the anterior and posterior projections, a total of 500 thousand counts being collected. At the same time the images were acquired and stored in the computer.

The posterior computerized images were analyzed as follows. Regions of interest were drawn around the liver and the spleen to determine the total counts within each organ. Additionally, a rectangular region of 28 cells was drawn over the lower spine, the lower limit corresponding to the level of the iliac crest. An adjacent region of similar size representing body background was placed to the right of the spine (Fig. 1). By noting the time taken to collect each image, the activity within the liver, spleen, and the fixed area of bone marrow, was expressed as a percentage of the injected activity. Appropriate corrections for background counts and radioactive decay were made. Based on the known organ distribution of sulfur colloid (liver=75-85%, spleen=5-10%, bone marrow=5-10%), a normalizing factor was selected to give the true distribution within these organs. Thus the liver and spleen percentages were multiplied by 5.0, and the marrow value by 100. These same factors were utilized to determine the true distributions of the other colloids evaluated.

RESULTS

The uptake of the radiocolloids in the three tissues of interest is shown in Table 2. The very low uptakes obtained for Tc-99m-sulfur colloid show the effect of geometry and attenuation factors and demonstrate the need for normalization to the known values. Microaggregated albumin, being of similar size, gave liver and spleen values not dissimilar from sulfur colloid but marrow uptake was unexpectedly increased.

The splenic uptake of Tc-99m-stannous phytate was reduced to about half that of the sulfur colloid value, while the liver and bone marrow uptake remained essentially unchanged. This alteration in the distribution of radioactivity is consistent with previous findings (11).

Both minimicroaggregated albumin and antimony sulfide have a substantially smaller particle size than sulfur colloid. The splenic uptake of both these agents was about half that of sulfur colloid. Liver uptake was similar for minimicroaggregated albumin and somewhat decreased for antimony sulfide. The shift to the bone marrow was marked in the case of minimicroaggregated albumin, showing about a threefold increase. With the antimony sulfide colloid, however, there was only a moderate increase in marrow activity even though the particles were all below 0.012 μ in size.

With these smaller colloids, particularly the antimony sulfide, the blood clearance was slower than with sulfur colloid. Activity in the blood pool was evident in the 20 min images, with further slight organ uptake by 1 hr. There was no advantage in delaying imaging beyond this time, as increasing bladder activity reflected the presence of free pertechnetate resulting presumably from biological degradation.

TABLE 1. LISTING OF RADIOPHARMACEUTICALS
EVALUATED AS BONE MARROW IMAGING AGENTS

Tc-99m Labeled Agent	Particle Size (μ)
Sulfur colloid	0.1 - 1.0 (85%)
Stannous phytate	----
Microaggregated albumin	0.1 - 1.0 (80%)
Antimony sulfur colloid	0.006 - 0.012
Minimicroaggregated albumin	<.08 (>80%)

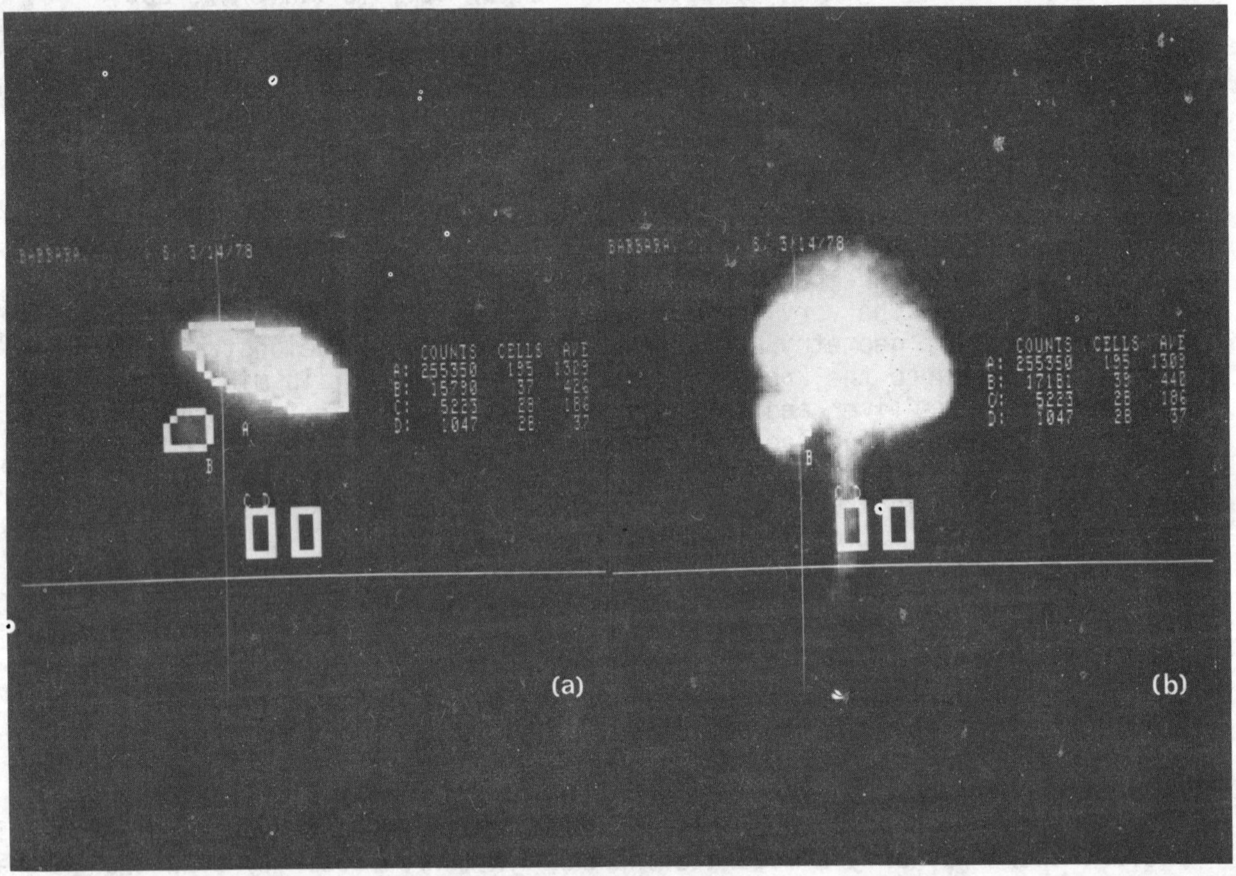

Figure 1.

Irregular regions of interest are placed around the liver and spleen, with regular regions 28 cells in size over the lower spine (marrow) and immediately to the right over soft tissue (background) as shown in (a). Computer thresholding is usually necessary to delineate the marrow activity (b).

TABLE 2. PERCENTAGE UPTAKES IN THE LIVER, SPLEEN, AND BONE MARROW IN PRIMATES AS FOUND WITH FIVE DIFFERENT COLLOID PREPARATIONS

Agent	Time post-injection (mins)	Animal	Percentage Distribution			Normalized Percentage Distribution		
			Liver	Spleen	Marrow	Liver	Spleen	Marrow
Sulfur Colloid 2-22-79	20	W	15.9	1.7	0.061	79.6	8.5	6.1
		B	13.7	1.8	0.068	68.6	9.0	6.8
	60	W	15.5	1.9	0.056	77.6	9.5	5.6
		B	13.3	1.8	0.064	66.7	9.0	6.4
Stannous Phytate 3-23-78	20	W	17.0	0.8	0.062	85.1	4.0	6.2
		B	17.5	0.7	0.036	87.6	3.5	3.6
	60	W	16.1	1.0	0.056	80.6	5.0	5.6
		B	19.4	0.7	0.053	97.1	3.5	5.3
Stannous Phytate 2-12-79	20	B	15.0	0.8	0.054	75.0	4.0	5.4
	60	B	16.9	1.0	0.069	84.5	5.0	6.9
Microagg. Albumin 1-29-79	20	W	17.0	2.0	0.097	85.9	10.0	9.7
		B	14.8	1.9	0.107	74.0	9.5	10.7
	60	W	14.6	2.4	0.103	73.0	12.0	10.3
		B	15.2	1.5	0.117	78.0	7.5	11.7
Antimony Sulfur Colloid 2-14-79	20	W	11.5	0.6	0.084	57.5	3.0	8.4
	60	B	16.1	0.8	0.111	80.5	4.0	11.1

TABLE 2. (Continued)

Agent	Time post-injection (mins)	Animal	Percentage Distribution			Normalized Percentage Distribution		
			Liver	Spleen	Marrow	Liver	Spleen	Marrow
Antimony Sulfur Colloid 2-27-79	20	W	12.0	0.9	0.091	60.0	4.5	9.1
		B	10.0	0.8	0.103	50.0	4.0	10.3
	60	W	12.4	1.0	0.083	62.0	5.0	8.3
		B	12.3	0.9	0.090	61.5	4.5	9.0
Minimicroagg. Albumin 3-14-78	20	W	14.7	0.9	0.181	73.5	4.5	18.1
		B	12.5	0.8	0.219	62.5	4.0	21.9
Minimicroagg. Albumin 3-29-78	20	W	14.7	0.8	0.157	73.5	4.0	15.7
		B	12.1	0.9	0.196	60.5	4.5	19.6
	60	W	15.1	0.8	0.143	75.5	4.0	14.3
		B	12.6	0.9	0.159	63.0	4.5	15.9
Minimicroagg. Albumin 4-6-78	20	W	16.6	1.6	0.156	83.0	8.0	15.6
		B	12.5	1.0	0.175	62.5	5.0	17.5
	60	W	16.3	1.0	0.143	81.5	5.0	14.3
		B	13.3	1.0	0.190	66.5	5.0	19.0
Minimicroagg. Albumin 1-29-79	20	W	16.3	0.8	0.183	81.5	4.0	18.3
	60	W	15.6	0.8	0.187	78.0	4.0	18.7

The computer images were all processed in the same way, with an upper threshold of 15% and a lower threshold of 1%. Since the images all contain a total of 500 thousand counts they are directly comparable. Fig. 2 presents representative posterior images at 1 hr post injection for sulfur colloid, antimony sulfide, stannous phytate, microaggregated albumin and minimicroaggregated albumin. Bone marrow activity was not seen in either the sulfur colloid or phytate image (phytate not shown); some activity was evident with antimony sulfide colloid and microaggregated albumin, and a significantly higher uptake was apparent with minimicroaggregated albumin.

DISCUSSION

Any attempt to influence the tissue distribution of colloids must be directed either against host factors or the colloid particle itself. The localization of colloids in the reticuloendothelial system depends mainly on organ blood flow. This accounts for the high activity which is usually present in the normal liver. Other host factors may also be important. Opsonins immediately form a coating around the injected foreign particle, and this complex interacts with the cell surface where phagocytosis takes place. Different reticuloendothelial cells may respond to different opsonins, thus affecting the localization of particles within the liver, spleen or bone marrow (12). These host factors cannot be manipulated to obtain the desired shift of activity to the bone marrow. The possibility of using "blocking" doses of unlabeled colloid to preload the liver and spleen has been demonstrated in animals. However, the doses needed are not practical for human use (9).

Particle size also influences the organ distribution of the colloids. Smaller particles tend to be cleared more slowly from the blood and to localize in the bone marrow to a greater extent. As expected, there was a shift of activity to the bone marrow with the smaller particles evaluated in this study. Of the agents investigated, the minimicroaggregated albumin, with a particle size ranging from 0.01 to 0.10 μ, showed the greatest effect, with a threefold increase in bone marrow activity (20%). This uptake of about 20% approximates the maximal uptake of sulfur colloid by the marrow observed in severe hepatocellular disease, and possibly represents an upper limit that one can expect to achieve with a colloidal radiotracer specifically designed for bone marrow localization. If the bone marrow uptake varies inversely with particle size, then the uptake of antimony sulfide (0.006 to 0.012 μ) should have been equal to or greater than that of the minimicroaggregated albumin. There are several possible reasons why this did not occur. There may be a lower limit, in fact, to the particle size which the reticuloendothelial system of the bone marrow can efficiently phagocytize. Further, particle size itself does not appear to be the only or even primary determinant in bone marrow localization. The physical characteristics of the particle surface may be equally important. This possibly accounts for the redistribution seen in disease states, when the serum protein charge may be altered. Positively charged colloid particles have been shown to localize in the lungs, presumably because agglomeration occurs as the surface charge is changed to negative on mixing with plasma proteins (13). These properties of the particles were not investigated, but may explain the differences observed between minimicroaggregated albumin and antimony sulfide colloid.

In conclusion, the minimicroaggregated albumin appears to be an improved radiopharmaceutical for bone marrow imaging when compared to sulfur colloid and the other agents evaluated. The optimal imaging time would appear to be from 30-60 min post injection, starting with anterior and posterior images of the pelvis. Liver and spleen activity is still significant and may interfere with images of the anterior ribs and mid-thoracic spine. Species differences are also possible, so that trials in human patients will be necessary to validate the findings in this primate study.

REFERENCES

1. Helson L, Watson RC, Benua RS et al: ^{18}F radioisotope scanning of metastatic bone lesions in children with neuroblastoma. Am J Roentgenol 115: 191-199, 1972

2. Judisch JM, McIntyre PA: Recognition of metastatic neuroblastoma by scanning the reticuloendothelial system (RES). Johns Hopkins Med J 130: 83-86, 1972

3. Heyman S, Treves S: Scintigraphy in pediatric bone tumors. in Bone Tumors in Children, Jaffe N (ed). PSG Publishing Co., Inc., Littleton MA, 1979. pp 91-92

4. DeNardo SJ, Hamel CF, Lewis JP et al: Assessment of the bone and marrow in sickle cell disease. J Nucl Med 13: 425-426, 1975

5. Davis MA, Jones AG, Trindade H: A rapid and accurate method for sizing of radiocolloids. J Nucl Med 15: 923-928, 1974

6. Wilkins DJ: Interaction of charged colloids with the RES. Adv Exp Med Biol 1: 25-33, 1967

7. Scott GBD, Williams HS, Marriott PM: The phagocytosis of colloidal particles of different syzes. Br J Exp Pathol 48: 411-416, 1967

8. Chaudhuri TK, Evans TC, Chandhuri TK: Autoradiographic studies of distribution in the liver of ^{198}Au and ^{99m}Tc-sulfur colloids. Radiology 109: 633-637, 1973

9. Atkins HL, Hauser W, Richards P: Factors affecting distribution of technetium sulphur colloid. J Reticuloendothel Soc 8: 176-184, 1970

10. Dobson EL, Gofman JW, Jones HB et al: Studies with colloids containing radioisotopes of ytrium, zirconium, columbium, and lanthanum: II. The controlled selective localization of radioisotopes of ytrium, zirconium and columbium in the bone marrow, liver and spleen. J Lab Clin Med 34: 305-312, 1949

11. Davis MA, Kaplan ML, Ahnberg DS et al: A modified Tc-99m-phytate colloid for liver-spleen imaging. Int J Appl Radiat Isotop 28: 123-130, 1977

12. Nelp WB: An evaluation of colloids for RES function studies. In Radiopharmaceuticals, Subramian G, Rhodes BA, Cooper JF, Sodd VJ (eds). New York, The Society of Nuclear Medicine, Inc, 1975, pp 354-356

13. Bell EG, McAfee JG: Concepts of colloid chemistry. In Hematopoetic and Gastrointestinal Investigations with Radionuclides, Gilson AJ, Smoak WM, Weinstein MB (eds). Springfield, Ill, Charles C. Thomas, 1972, pp 66-84

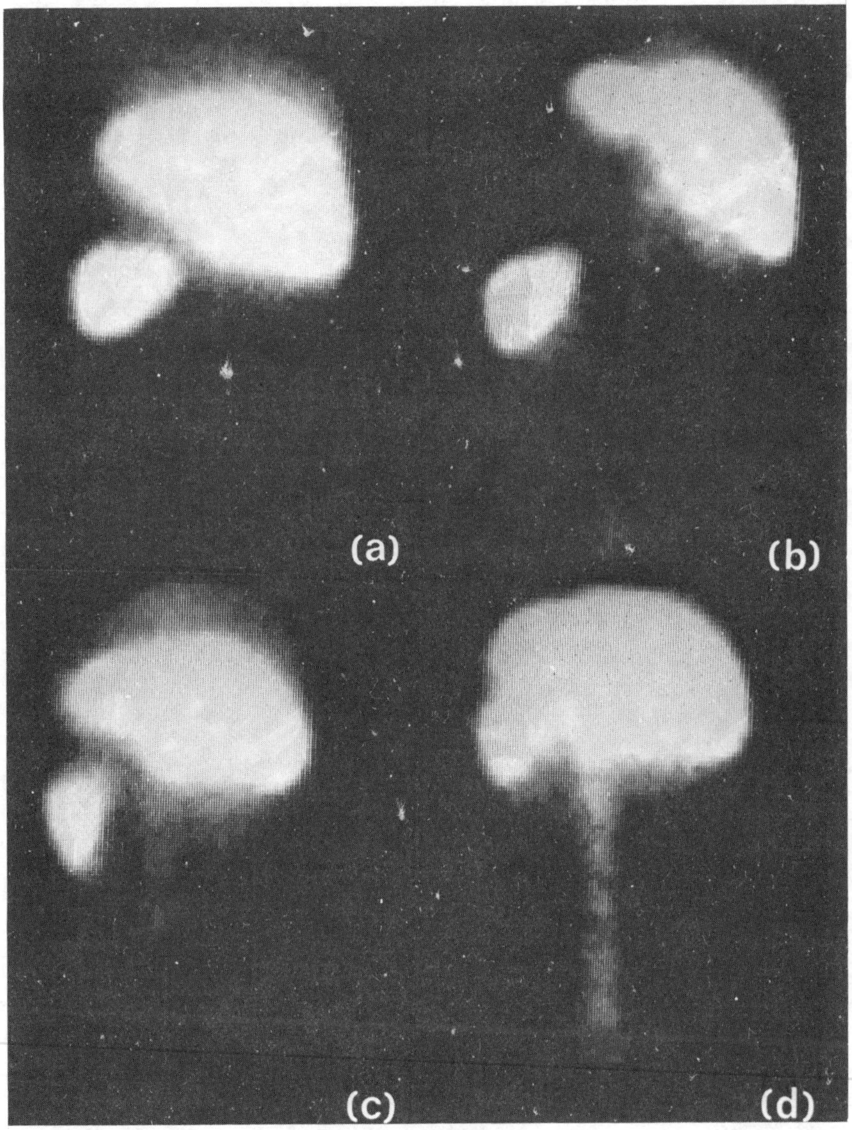

Figure 2.

This shows representative computerized images obtained one hour post injection in the same animal. All have been similarly enhanced with an upper threshold of 15% and a lower threshold of 1%. There is no marrow activity evident in the sulfur colloid image (a), a mild increase with microaggregated albumin (b) and antimony sulfur colloid (c), and a marked increase with minimicroaggregated albumin (d). With stannous phytate there was no marrow activity seen (not shown).

IMAGING THE SKELETAL SYSTEM

Marion D. Francis, Andrew J. Tofe
James J. Benedict and John A. Bevan

Miami Valley Laboratories
The Procter & Gamble Company
Cincinnati, OH 45247

ABSTRACT

The highly vascular system, inorganic calcium phosphates and a supporting fibrous matrix of the skeleton are considered with respect to skeletal imaging agents. The role and interrelations of the structure of the ligands complexing with reduced Tc-99m and their reaction with the organic and inorganic phases of bone are discussed. Two hydroxy-diphosphonates (HEDP and HMDP) and a non-hydroxylated diphosphonate (MDP) were considered on the basis of their structure, calcium salt solubility, and clearance from the mammalian system. Based on specific adsorption properties for normal and active osteogenic bone, HMDP was found to be the most effective technetium complexing ligand.

INTRODUCTION

Bone is a tissue that is dynamically supplied and drained by the cardiovascular system. The administration of materials designed specifically to react with bone substance for skeletal imaging depends upon a viable functioning vascular supply (1) such as is shown in Figure 1 for a typical long bone.

The reactivity of the more recent skeletal imaging agents such as Tc-99m phosphates and Tc-99m phosphonates which are being carried by the vasculature to bone probably depends upon: the composition and surface area of the specific calcium phosphate present in bone where the Tc-99m ligand complex can diffuse; the solubility of the calcium-ligand salt formed on the surfaces of the calcium phosphate in bone; the strength of the Tc-99m ligand complex which determines its stability while circulating in the mammalian system to the bone and the structure of the ligand with which the reduced technetium species reacts to form a soluble complex. Additional factors such as size of the technetium-99m complex, its charge and protein binding potential also can have an effect on its diffusion from the vascular system and its reactivity with bone.

SKELETAL IMAGING AGENTS AND THE COMPOSITION OF BONE

The reactivity of skeletal imaging agents diffusing to the bone depends upon the organic and inorganic composition of bone. The most important organic constituent of bone is the fibrous collagen matrix upon which the inorganic calcium phosphate (70% by weight of bone) is deposited. Some deposition of the Tc-99m bone agents directly on the organic matrix appears to occur in osseous abnormalities such as osteomalacia (2) but Tc-99m deposition is most probably due to nuclei of calcium phosphate present in the uncalcified matrix reacting with the Tc-99m ligand complex rather than to strong specific adsorption or interaction with the organic matrix itself. In order to investigate the relative adsorption of a Tc-99m complexing ligand such as 1-C-14 (1-hydroxyethylidene) diphosphonate (C-14 HEDP) on bone matrix versus bone calcium phosphate, acid insoluble rat bone matrix (organic only) was isolated by the method of Reddi (3). Adsorption of C-14 HEDP on matrix and on solid calcium phosphate was then studied by the method of Francis, et al. (4). The results are shown in Table 1. It is evident that an appreciable adsorption can take place on free matrix (about 17% relative to bone) provided saturation with phosphonate is allowed to occur. It has been shown previously by autoradiography (5), however, that in the absence of complete saturation of the system in vivo, the C-14 HEDP will bind exclusively to the calcium phosphate of bone rather than to the matrix. In addition, malignant osteoid of a Paget's osteogenic sarcoma did not take up Tc-99m HEDP on the uncalcified matrix (6). These data strongly mediate against the Tc-99m diphosphonate reaction with organic bone matrix being a significant contributing factor to scintigraphic visualization unless the matrix is nucleated with calcium phosphate. If this deposition of calcium phosphate is only at the nucleation stages, it will be totally undetectable by insensitive physical techniques such as x-ray, and by chemical techniques such as Von Kossa or alizarin red S staining, but will be detected by the deposition of the Tc-99m radionuclide on the high surface area of the calcium phosphate nuclei.

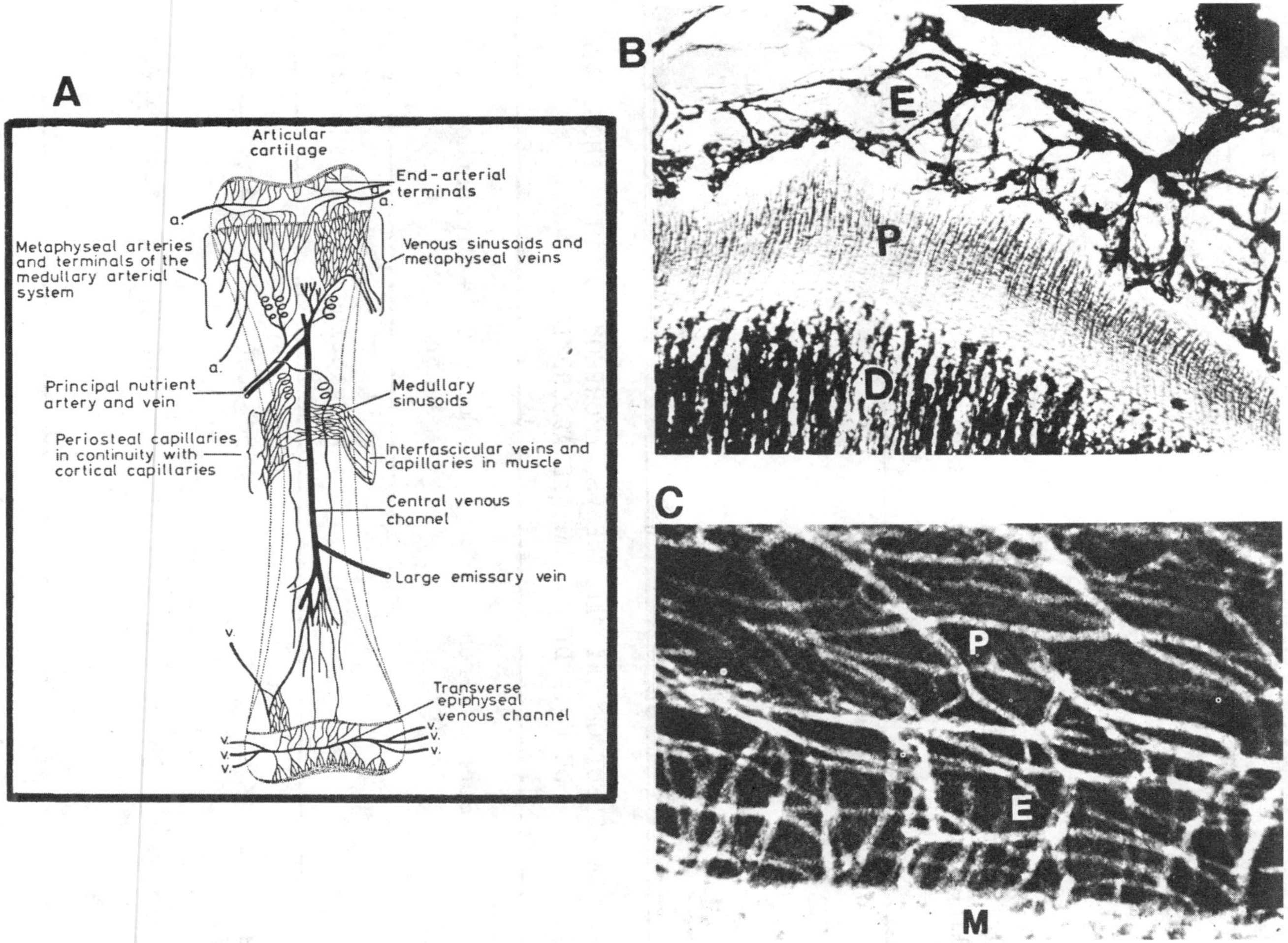

FIG. 1. Vascular organization of a long bone. (A) Schematic of the blood supply to the various segments of a long bone. (B) Growth cartilage of a rabbit: E, epiphyseal subchondral vessles; P, growth plate; D, metaphyseal subchondral vessels (india ink preparation). (C) Microangiograph of rat bone cortex: P, periosteal bone; E, endosteal bone; M, marrow (Reproduced from Brookes M, reference 1).

TABLE 1. THE ADSORPTION EFFICIENCY OF C-14 HEDP FOR ACID INSOLUBLE BONE MATRIX (AIBM) AND FOR CALCIUM PHOSPHATE (Ca-P)

System	Weight of Sample (mg)	Moles of C-14 Adsorbed ($X 10^6$)	% Adsorption Relative to Bone*
Ca-P**	12	1.13 ± 0.05	83.2
AIBM	12	0.53 ± 0.1	16.8

* Assuming 70% inorganic and 30% organic matrix in bone
** Crystalline hydroxyapatite:amorphous calcium phosphate 1:1 (W:W)

TABLE 2. THE EFFECT OF CRYSTALLINE VS. AMORPHOUS CALCIUM PHOSPHATE ON C-14 DIPHOSPHONATE ADSORPTION

System	Surface Area (m^2/g)	Moles C-14 Diphosphonate Adsorbed/Mole Calcium*	
		5 min Reaction	30 min Reaction
Crystalline Hydroxyapatite	60	0.615 ± 0.02	0.723 ± 0.04
Amorphous Calcium Phosphate	35	1.15 ± 0.02	1.471 ± 0.03

* Average of all three diphosphonates (C-14 HEDP, C-14 MDP and C-14 HMDP)

The type or phase of inorganic calcium phosphate present in bone seems to be primarily governed by the rate at which the calcium phosphate is laid down in bone and the "maturity" of the bone. Well crystallized hydroxyapatite [$Ca_{10}(OH)_2(PO_4)_6$] (HA) is found in mature bone areas such as cortical bone as needle or rodlike crystallites which measure 200-600 Å in length (c axis) and 25-75 Å in diameter (7, 8, 9) or as plates 400 Å in length (c axis) and width and about 50 Å in thickness (10). The molar Ca/P ratio of these mature crystals is close to the theoretical 1.67 for hydroxyapatite (11). If, however, bone is examined in an area of rapid metabolic formation such as in embryonic bone (12), newly forming bone (13, 14) or in lesions such as Paget's disease (15), the calcium phosphate present is of low density and high hydration, is amorphous (by x-ray or electron microscopy), and can have a molar Ca/P as low as 1.4 (11, 16). These calcium phosphates are probably mixtures of different hydrated phases (10, 11, 17) such as calcium monohydrogen phosphate [$CaHPO_4$, Ca/P = 1.0], octacalcium phosphate [$Ca_4H(PO_4)_3$, Ca/P = 1.33] or even tricalciumphosphate [$Ca_3(PO_4)_2$, Ca/P = 1.5] (18) along with hydroxyyapatite (HA). The probable implication of $CaHPO_4 \cdot 2H_2O$ as a precursor to the formation of HA has been reported previously (17, 19) and can account for all the variations in Ca/P found in mammalian HA formation. The reactivity of the ligand (diphosphonate, pyrophosphate, etc.) or of the Tc-99m labeled ligand complexes with each of these specific phases has not at this time been investigated but the composition of the phases will undoubtedly alter specific reactivity. An additional factor in the ligand reactivity with apatite just nucleated versus long needle-like crystals or ribbons of apatite may be the ratio of the growing face (001 face) to the other faces of HA. HA grows very rapidly along the (c) axis first and only slowly along the (b) and (a) axis. Thus any selective adsorption of ligand or Tc-99m ligand complex, to a given face (configuration of calcium atoms) could markedly alter the adsorption as the calcium phosphate changes from initial formation at a calcifying center (nuclei) to mature HA crystals as in cortical bone. The final factor is the surface area available for adsorption: the higher the surface area, the higher should be the adsorption.

The remarkable difference in reactivity of three diphosphonates [C-14 labeled HEDP, C-14 labeled methylenediphosphonate (MDP) and C-14 labeled hydroxymethylenediphosphonate (HMDP)] for crystalline HA versus newly formed or amorphous calcium phosphate (Figure 2) is shown in Table 2. A small but significant increase in adsorption occurs from 5 to 30 minutes reaction indicating the initial adsorption reaction is fairly rapid. By far the major effect, however, is the increase in adsorption per mole of calcium when the diphosphonates are reacted with amorphous calcium phosphate rather than with crystalline. This high adsorption effect takes place despite the inverse relation in surface area measurement (i.e., 35 m^2/g for amorphous versus 60 m^2/g for crystalline). Through their ligands all the Tc-99m labeled phosphorus skeletal imaging agents exhibit their affinity for rapid metabolic bone activity (osteogenesis). The fundamental difference in effectiveness among these agents is the degree of reactivity of the ligand for the lesion/normal bone site or for the heterotopic site of calcification such as in a myocardial infarct lesion.

MECHANISMS OF SKELETAL REACTIVITY WITH COMPOUNDS APPLICABLE TO NUCLEAR MEDICINE

The metal ions (cations) frequently encountered in nuclear medicine for diagnostic imaging, such as stannous tin [Sn(II)], indium [In(III)], gallium [Ga(III)] and rare earths, can adsorb or interact with bone by direct reaction with the phosphate anion of bone to yield very thin coatings of insoluble phosphates tightly adhering to the inorganic phase of bone (20). The anions of interest such as fluoride, pyrophosphate, polyphosphates and diphosphonates adsorb or interact with bone by direct reaction with calcium hydroxyapatite of bone to yield very thin coatings of insoluble calcium salts also tightly adhering to the inorganic phase of bone (21, 22). The fluoride anion reacts to form fluorapatite [$Ca_{10}F_2(PO_4)_6$] by exchange with the OH of calcium hydroxyapatite or can form calcium fluoride (CaF_2) at the surface if certain conditions of pH and ion concentrations are met (20). The diphosphonates and pyrophosphate (P_2O_7) form calcium salts of widely varying solubility which in most cases are thought to be the dicalcium diphosphonate and calcium pyrophosphate salts respectively. The relative solubility of the calcium salts of several diphosphonates [as determined by solubility studies (23)] are in decreasing order of solubility MDP > HEDP > HMDP $\gg$ P_2O_7. This insolubility of the calcium salts is one of the several factors of importance in effectiveness of deposition of the ligand in Tc-99m bone scanning as well as infarct scanning.

The ultimate key to the most sensitive diagnostic bone agent is the ability to detect abnormal sites of osteogenesis, e.g., metastatic lesions, in a milieu of normal bone (target to non-target ratio) while optimizing the visualization of the normal skeleton. Selectivity is not only based on the phase of calcium phosphate in the lesion, but also on the specific structure of the ligand carrying the Tc-99m radioisotope.

The adsorption of geminal diphosphonates to the crystallites of bone is presumed to be either bidentate (through both geminal diphosphonate groups to the calcium of the crystal as in MDP) or tridentate [through the hydroxy group and both geminal diphosphonate groups to the calcium as in HEDP or HMDP (24)]. It is obvious that diphosphonates containing no hydroxy groups can only bind with bidentate addition while molecules containing hydroxy groups should be able to effect both bidentate and tridentate addition to calcium phosphate solids. Thus the presence of the hydroxy group, the solubility of the calcium salt of the phosphonate, and the size of the molecule may be the factors most important for superior bone scanning ligand requirements along with the stability of the Tc-99m diphosphonate-soluble complex. The reactivity of three diphosphonates C-14 HEDP, C-14 MDP and C-14 HMDP at typical bone imaging agent concentrations with crystalline HA is shown in Figure 3. The C-14 HMDP was adsorbed on the crystalline apatite at a significantly higher level than HEDP which in turn was adsorbed more than MDP. This type of adsorption is probably similar to the "irreversibly bound compartment" of Arnold (25) and would suggest a superiority could be obtained for HMDP for total skeletal imaging.

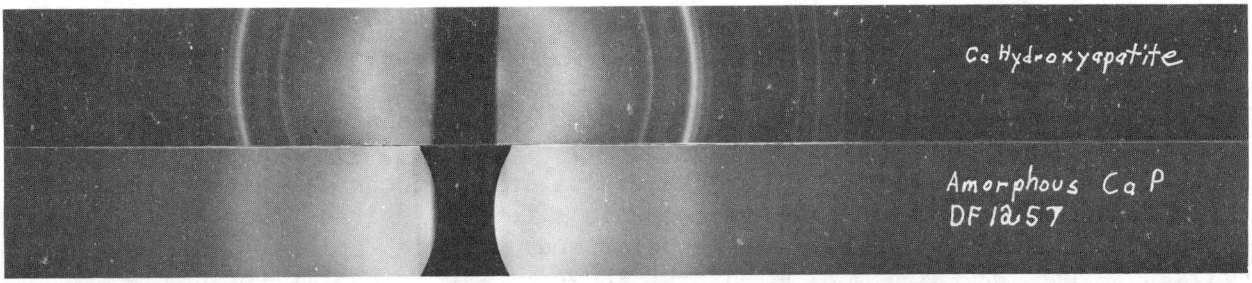

FIG. 2. Cylindrical x-ray diffraction patterns of crystalline hydroxyapatite (top) and amorphous calcium phosphate (bottom).

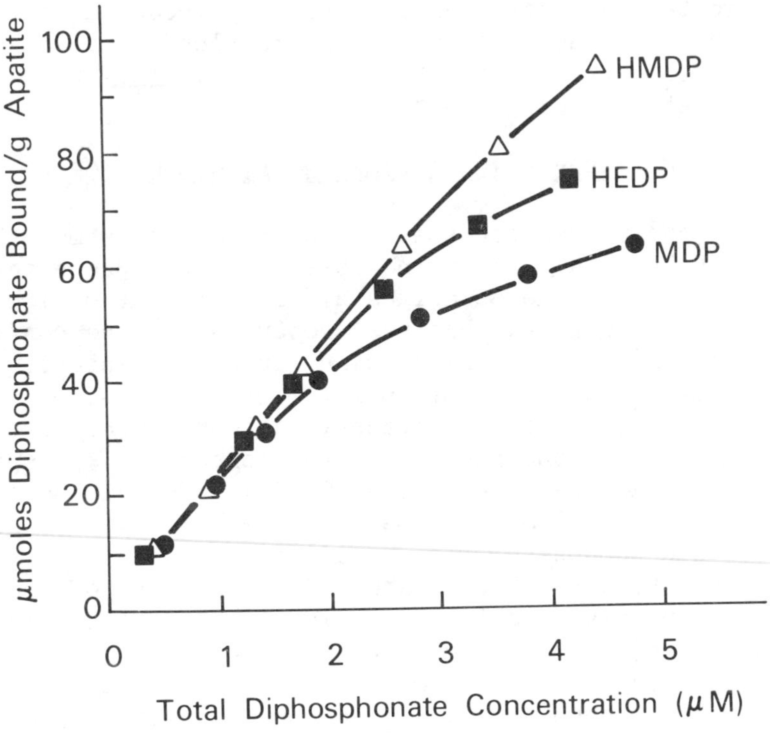

FIG. 3. Adsorption of C-14 methylenediphosphonate (MDP), 1-C-14 (1-hydroxyethylidene) diphosphonate (HEDP) and C-14 hydroxymethylene-diphosphonate (HMDP) on crystalline hydroxyapatite.

The factors sited above for ligand superiority for crystalline apatite, pertain even more to the sites of osteogenesis. At these sites, the 001 face of the HA surface is presumably high with respect to the other faces of apatite when significant crystal growth has not yet taken place and in addition the early non apatite phases of calcium phosphate are present. Confirmation of these differences among the diphosphonates on this type of calcium phosphate was studied by the adsorption of C-14 labeled HEDP, MDP and HMDP on amorphous calcium phosphate (Figure 2) synthesized by the method of Termine and Eanes (26) . In Table 3 is shown the significantly greater uptake of C-14 HMDP on amorphous calcium phosphate than of C-14 labeled MDP or HEDP. These latter two were not different from each other.

Selectivity of the ligand with Tc-99m for lesion scanning was also studied in an in vivo system of isolated osteogenesis. This involves the formation of an osteogenic nodule or pellet in an isolated focus by the use of an implanted acid insoluble bone matrix protein (3, 4) in a soft tissue site. The in vivo model thus allows evaluation of technetium uptake from bone scanning agents in a pure osteogenic site (3) separate from normal bone (4). The results of uptake studies are shown in Table 4. As with the ligand alone, the Tc-99m labeled HMDP showed a significantly ($p<.05$) higher uptake/mg calcium in the osteogenic site than either Tc-99m labeled MDP or HEDP. The combination of increased adsorption of HMDP for crystalline hydroxyapatite with the even more selective adsorption of HMDP on amorphous calcium phosphate compared to HEDP and MDP probably explains why HMDP seems to indicate advantages for skeletal or myocardial imaging (27). These factors when combined with a more rapid blood clearance of Tc-99m HMDP than the other three agents (MDP, HEDP, P_2O_7) (27) places HMDP as a preferred skeletal imaging agent.

MECHANISM OF DEPOSITION OF Tc-99m SKELETAL IMAGING AGENTS ON BONE

The necessity of reducing pertechnetate with tin(II) or some other reducing agent such as $NaBH_4$ (28) and the necessity for the presence of the bone seeking ligand to give the scintigraphic specificity of low soft tissue retention and high bone uptake has been reported (5) and confirmed (29). Exactly what happens at the intracellular fluid-bone interface, however, is still not clarified. The possibility that the Tc-99m phosphorus complex travels to the bone surface intact and then dissociates because of the high affinity of the phosphorus (phosphonate or phosphate) compound for calcium phosphate has been sited (5, 29). The dissociation would then allow the reduced Tc-99m to deposit as the hydrolyzed oxide (5) or as the slightly soluble phosphate. The requirement of the tin (II) for this Tc-99m deposition on bone has been disproved by the work of Deutsch, et al. (28) who have shown that excellent bone imaging can be obtained in the complete absence of tin(II) using substitution reactions or borohydride reductions for the formation of the Tc-99m HEDP bone seeking complex. Efforts to detect adsorbed technetium-99 dioxide or technetium-99 phosphate on apatite or amorphous calcium phosphate by physical techniques such as electron diffraction have failed, however, probably because of the fine amorphous character of the deposit. The alternative possibility exists that the technetium-ligand adsorbs as an entity. This aspect of bone scanning, then, is as yet an unsolved problem.

TABLE 3. THE ADSORPTION EFFICIENCY OF THREE C-14 LABELED DIPHOSPHONATES FOR AMORPHOUS CALCIUM PHOSPHATE (ACP)

C-14 Diphosphonate*	Moles Diphosphonate Adsorbed/Mole Calcium in ACP	
	5 min Reaction	30 min Reaction
HEDP	0.907 ± 0.027	1.43 ± 0.047
MDP	0.868 ± 0.051	1.33 ± 0.031
HMDP	1.68 ± 0.020	1.65 ± 0.023

* All systems were treated at 1.6 X 10^{-4}M, pH = 7.4.

TABLE 4. THE RETENTION OF Tc-99m FROM THREE SKELETAL IMAGING AGENTS IN AN ISOLATED SITE OF OSTEOGENESIS IN RATS

Tc-99m Diphosphonate	n	Percent Dose Per Mg Calcium	
HEDP	30	0.326	
MDP	32	0.254	$p < 0.05$
HMDP	34	0.804	

PHOSPHATASE ENZYME SYSTEMS AND Tc-99m SKELETAL IMAGING AGENTS

Phosphatase enzymes, particularly alkaline phosphatase (pyrophosphatase) enzyme has been suggested as the localizing factor for Tc-99m from bone agents. Recently Rohlin, et al (30) have shown that soft tissues such as intestinal mucosa, stratum intermedium, and subodontoblastic layers which show high pyrophosphatase activity did not accumulate Tc-99m labeled pyrophosphate. Other observations also tended to suggest that bone agents are not directed by enzyme molecules. Zimmer, et al, measured enzyme inhibition using solution concentrations of diphosphonates in the range of 0.7-2.6 $X10^{-3}$M (31). This is known to be above the substrate concentration range and therefore is not indicative of a selective interraction with the enzyme but rather only competitive inhibition by substrate competition. Furthermore, only 1-3 minutes after injection of the phosphonate skeletal agent into man the concentration of free diphosphonate is only in the 10^{-6}M range - three orders of magnitude lower than those indicating inhibition by Zimmer. The concentration of Tc-99m diphosphonate is, of course, orders of magnitude below this. It is to be noted additionally that Zimmer used stannous phosphonate in his studies, the concentration of Sn(II) then would have been two times the concentration of diphosphonate (1.4-5.2 X 10^{-3}M) and could easily have accounted for inactivation of the enzyme through the well known heavy metal-protein reaction. Evidence, therefore, for phosphatase enzyme being a significant factor in Tc-99m localization from bone agents in significantly lacking. It is to be remembered, however, that phosphatase enzyme is an important factor in the calcification process and deposited calcium is certainly involved in localization of bone agents.

REFERENCES

1. Brookes M: *The Blood Supply of Bone*. London, Butterworths, 1971

2. Kaye M, Silverton S, Rosenthall L: Technetium-99m-pyrophosphate: studies in vivo and in vitro. *J Nucl Med* 16: 40-45, 1975

3. Reddi AH, Huggins C: Biochemical sequences in the transformation of normal fibroblasts in adolescent rats. *Proc Nat Acad Sci* USA 69: 1601-1605, 1972

4. Francis MD, Ferguson DL, Tofe AJ, et al: Comparative evaluation of three diphosphonates: in vitro adsorption (C-14 labeled) and in vivo osteogenic uptake (Tc-99m complexed). *J Nucl Med* (submitted)

5. Jones AG, Francis MD, Davis MA: Bone scanning: radionuclidic reaction mechanisms. *Seminars Nucl Med* 6: 3-18, 1976

6. McKillop JH, Fogelman I, Boyle IT, et al: Bone scan appearance of a Paget's osteosarcoma: Failure to concentrate HEDP. *J Nucl Med* 18: 1039, 1977

7. McLean FC, Urist MR: *Bone: Fundamentals of the Physiology of Skeletal Tissue*. Chicago, University of Chicago Press, 1968, p. 57

8. Suzuki M: Studies on the physicochemical nature of hard tissue. In *Physico-Chimie et Crystallographie des Apatites d'Interet Biologique*. Paris, Centre National de la Recherche Scientifique, 1975, p. 82

9. Bourne GH: *The Biochemistry and Physiology of Bone*. New York, Academic Pres, 1972, p. 5

10. Brown WE, Chow LC: Chemical properties of bone mineral. *Ann Rev Mater Sci* 6: 213-236, 1976

11. Muenzenberg KJ, Gebhardt M: Brushite, octacalcium phosphate, and carbonate-containing apatite in bone. *Clin Orthop Rel Res* 90: 271-273, 1973

12. Landis WJ, Hauschka BT, Paine MC: Orthopaedic Research Society, San Francisco, abstract #43, 1975

13. Robinson RA, Watson ML: Crystal-collagen relationships in bone as observed in the electron microscope. III. Crystal and collagen morphology as a function of age. *Ann NY Acad Sci* 60: 596-628, 1955

14. Miller AL, Schroer H: Ultrastructural observations of amorphous bone mineral in avian bone. *Calcif Tissue Res* 18: 311-324, 1975

15. Posner AS, Betts F: Synthetic amorphous calcium phosphate and its relation to bone mineral structure. *Accounts Chem Res* 8: 273-281, 1975

16. Termine JD, Posner AS: Infrared analysis of rat bone: Age dependency of amorphous and crystalline mineral fractions. *Science* 153: 1523-1525, 1966

17. Francis MD, Webb NC: Hydroxyapatite formation from a hydrated calcium monohydrogenphosphate precursor. *Calcif Tissue Res* 6: 335-342, 1971

18. Eanes ED: Amorphous intermediates in the formatin of biological apatites. See reference 6, pp. 295-301

19. Francis, MD, Briner WW, Gray JA: Chemical agents in the control of calcification processes in biological systems. *Hard Tissue Growth, Repair and Remineralization*. Amsterdam, Elsenier, 1973, pp. 57-90

20. Francis MD, Gray JA, Griebstein WJ: The formation and influence of surface phases on calcium phosphate solids. *Adv Oral Biol* 3: 83-120, 1968

21. Francis MD: The inhibition of calcium hydroxyapatite crystal growth by polyphosphonates and polyphosphates. *Calcif Tissue Res* 3: 151-162, 1969

22. Francis MD, Briner WW: The effect of phosphonates on dental enamel in vitro and calculus formation in vivo. *Calcif Tissue Res* 11: 1-9, 1973

23. Russell RGG, Muehlbauer RC, Bisaz S, et al: The influence of pyrophosphate, condensed phosphates, phosphonates and other phosphate compounds on the dissolution of hydroxyapatite in vitro and on bone resorption induced by parathyroid hormone in tissue culture and in thyroparathyroidectomized rats. Calcif Tissue Res 6: 183-196, 1970

24. Barnett BL, Strickland, LC: The crystal structure of disodium dihydrogen-1-hydroxyethylidene diphosphonate tetrahydrate: a bone growth regulator. Acta Cryst B35, 1979 (in press)

25. Arnold JS: Kinetic analysis of bone. Proc First Internat Symp on Radiopharm, Innsbruck, Austria, in press

26. Termine JD, Eanes ED: Comparative chemistry of amorphous and apatitic calcium phosphate preparations. Calcif Tissue Res 10: 171-197, 1972

27. Bevan JA, Tofe AJ, Francis MD, et al: Tc-99m hydroxymethylene-diphosphonate (HMDP): A new skeletal imaging agent. Second Internat Symposium on Radiopharmaceuticals Seattle, WA, 1979

28. Deutsch EA, Libson KF, Becker CB, et al: Preparation, chemistry and biological distribution of technetium phosphonate radiopharmaceuticals synthesized without stannous ion. Soc Nucl Med, Third Annual Western Reg. Meeting, Oct. 13-15, Abstract #E3, 1978

29. van Langevelde A, Driessen OMJ, Pauwels EKJ, et al: Aspects of technetium-99m binding from an ethane-1-hydroxy-1,1-diphosphonate-Tc-99m complex to bone. Eur J Nucl Med 2: 47-51, 1977

30. Rohlin M, Larsson Å, Hammerstrom L: In vitro interaction between Tc-99m-labeled pyrophosphate, P-32-labeled pyrophosphate and rat tissues. Eur J Nucl Med 3: 249-255, 1978

31. Zimmer AM, Isitman AT, Holmes RA: Enzymatic inhibition of diphosphonate: a proposed mechanism of tissue uptake. J Nucl Med 16: 352-356, 1975

AN IMPROVED MODEL FOR CALCIUM-HEDP SOLUTION EQUILIBRIA

Myles L. Lamson, Jeffrey L. Fox, and William I. Higuchi

College of Pharmacy
The University of Michigan
Ann Arbor, Michigan, 48109

ABSTRACT

Since hydroxyethylidene-1,1-diphosphonic acid has been reported to form polynuclear complexes with calcium ion, an investigation of the in vivo behavior of this bone agent should consider the equilibria of the Ca-HEDP species. This study employs TMAOH titration data and published formation curve data for non-linear least squares parameter estimation based on a solution model which includes several polynuclear species. Fittings of both types of experimental data, and computer experiments varying sets of complexes and individual constants, predict that large polynuclear species may exist near neutral pH and that these species would have an extreme pH dependency in this region.

INTRODUCTION

Rarely in medical science does any single pharmaceutical compound have such diverse application as ethane-1-hydroxy-1,1-diphosphonic acid (EHDP). During the last decade this material has been established as an important agent both for diagnosis and treatment of various diseases, and for fundamental biomedical research (1). Its most important use is as a diagnostic radiotracer in bone imaging (2). During the period 1970-73 bone scanning rapidly progressed through Sr-85 and F-18 to Tc-99m phosphates and phosphonates. During that short span of time radiopharmaceutical efficacy progressed to such an extent that scanning became a screening procedure for metastatic bone disease, with the number of studies approaching 20-30 times the number done with Sr-85. Diagnostic utility would be increased if the localization mechanism were accurately understood. Toward that objective, and since all applications of HEDP are related in one way or another to the physiologic role of calcium (3-5), the present study was undertaken in an attempt to understand Ca-HEDP solution equilibria.

A number of studies have reported determinations of the HEDP acid dissociation constants (6-10); these values are summarized in Table I. Since the first and second protons are so easily

TABLE I. HEDP Acid Dissociation Constants.

ionic strength	pK1	pK2	pK3	pK4	pK5	ref.
0.0		3.03	7.31	11.52		6
0.1		2.80	7.00	11.16		6
0.1	1.7	2.47	7.28	10.29	11.13	9
0.5		2.54	6.97	11.41		5
0.1		2.31	6.99	10.93		8
0.0	2.01	3.08	7.60	11.96	13.63	10
*	2.00	2.90	7.23	11.37	11.96	10
	(.01)	(.02)	(.04)	(.06)	(.07)	

*ionic strength varies (noted in parentheses).

ionized, and have no independent inflection points, a measurement of pK1 is difficult and has often been ignored. At the other extreme of pH the hydroxyl proton is very difficult to remove and has generally been neglected in considering the acid dissociation equilibria . Recently, however, Collins has presented evidence that this proton also is ionizable (10). Kabachnik had noted earlier that the two closely neighboring phosphonate groups cause a significant increase in the acidity of the hydroxyl group (9).

Carroll also studied the acidity of substituted gem-diphosphonates and noted a correlation with electron donor ability on the bridging carbon atom (11). This study seemed to indicate intramolecular participation of the -OH group in both proton and metal ion binding. Complex formation involves the replacement of molecules in the solvation shell by coordinating atoms of the ligand, usually O, N or S. If the ligand contains more than three coordinating atoms, or when the chelate rings are strained or many membered, hydrogen complexes and polymetallic species may also be formed. Hydroxo complexes and metal ion hydrolysis can occur when working above a critical hydrolysis pH. Constant ionic strength must be maintained during experimental studies since concentration constants for equilibria involving highly charged metal ions and ligands vary strongly with the salt concentration (16). Several authors have determined values for Ca-HEDP complex constants, as shown in Table II. Most studies have not considered the presence of polynuclear species.

TABLE II. Calcium-HEDP Equilibrium Constants.*

log of association constant				reference
K(ml)	K(m-hl)	K(1-ml)	K(m21)	
5.74	3.58			11(u=0.5)
6.04			15.59	9
7.09		6.20(50o)		12
6.40	3.53		12.2	8
5.52				7

*Determined at u=0.1, 25o, except as noted.

Kabachnik (9) determined formation constants for a number of alkaline earth and transition metals using a potentiometric technique, and claims to have found 1:1, 2:1, and 1:2 (metal:ligand) complexes. It was noted that HEDP forms complexes of the composition MHL with all cations, and that the hydroxyl group enables an additional ring to be formed which increases overall complex stability. Carroll also suggested this -OH group stabilization and noted that the role of the hydroxyl group may increase with decreasing charge on the phosphonate groups (11).

Callis studied the Ca-HEDP system using a nephalometric method at pH 11 (12). This technique was later modified by Grabenstetter (7), who interpreted formation curve results as indicative of polynuclear complex formation. These formation curves were also determined at pH 11. Since the Grabenstetter data indicated a clear dependence of formation number on total diphosphonate, the data were analyzed for average composition of complexes by the method of Sillen (16). Average numbers of Ca and DP bound per complex are linearly related with no grouping of data (which would indicate that a certain complex is preferred). Another group has examined the Ca-HEDP complex formation equilibria potentiometrically based on proton displacement, and has interpreted their results as 1:1, 2:1 equilibria (8). Carroll (11) has reported a value of 0.52 Kcal/mole for the enthalpy of complexation of the Ca-HEDP complex. Their study suggests that binding is electrostatic.

Wiers (14) has studied precipitation boundaries of Ca-HEDP species at pH 11. Two types of precipitate appeared to form depending on the value of metal ion to ligand ratio, and analysis revealed that the precursors were polynuclear, macromolecular species. The author noted that the boundary data could not be fit well by assuming the presence of a strong Ca2DP species. In another study Wiers used classical macromolecular techniques which demonstrated that in relatively concentrated solutions at pH 11 very large aggregates were formed which eventually develop into a crystalline precipitate(13). Another Proctor & Gamble worker studied the x-ray crystal structure of Ca-HEDP, and noted that only those gem-diphosphonates which contain a ligand atom in one of the -R groups would be capable of significant polynuclear aggregate formation (15). Two different ring structures may be formed with calcium ion: one six membered ring which includes both phosphonate groups, and a five membered ring involving the methylene -OH group. It is this capability, the author feels, which allows polymerization at excess molar ratios of Ca++.

THEORETICAL

The model used for fitting the formation and titration data is described by the set of equilibrium expressions presented in Table III. All constants employed in these expressions are concentration constants valid at an ionic strength of 0.1 M. No species activity corrections were attempted, with the exception of converting measured pH to hydrogen ion concentration using an activity coefficient of 0.826. Neither the acid dissociation

TABLE III. Constants used in or Determined From Simultaneous Parameter Estimation.# (fitted parameters noted by asterisk.)

reaction	log of association constant (or pKa for acids)
H4DP ⟷ H + H3DP	1.7
H3DP ⟷ H + H2DP	2.7
H2DP ⟷ H + HDP	7.0
HDP ⟷ H + DP	11.0
Ca + DP ⟷ CaDP	5.94*
Ca + H + DP ⟷ CaHDP	14.54*
Ca + OH ⟷ CaOH	1.4
Ca + 2OH ⟷ Ca(OH)2	3.8
3Ca + 2DP ⟷ Ca3DP2	18.75*
4Ca + 3DP ⟷ Ca4DP3	30.14*
7Ca + 4DP ⟷ Ca7DP4	46.20*
2(Ca7DP4) ⟷ (Ca7DP4)2	9.84*
(Ca7DP4)2+Ca7DP4 ⟷ (Ca7DP4)3	9.84*
H + OH ⟷ H2O	14.0

#Ca2DP used in some trials (log K=12.2); species charges omitted for brevity; double arrows denote equilibrium.

constants nor the calcium hydroxide constants were fitted in the computations. pKa's were rounded average values based on the several sets previously reported. Calcium hydrolysis constants are established values which were independently determined.

The determination of formation curves provides a sensitive experminetal method for establishing the type of complexes formed between a metal ion and ligand in solution. The theory has been described in detail by Sillen and others (e.g.16). In this case the formation number, Z, is the ratio of bound calcium to total diphosphonate (DPTOT). $Z=(CATOT-Ca^{++})/DPTOT$. Experimentally, if calcium ion is measured for various CATOT at several concentrations of DPTOT then polynuclear species are indicated if multiple curves are obtained corresponding to the different concentrations of total diphosphonate. Simple mononuclear complexes would produce a single formation curve independent of DPTOT. Further, it is possible to determine average values of calcium and diphosphonate per complex. (7).

EXPERIMENTAL

Several pH titrations were performed using the disodium salt of ethane-1-hydroxy-1,1-diphosphonic acid supplied by Proctor & Gamble Co. The experiments were carried out at 25.0oC and moderate precautions were taken to exclude CO_2 from the solution. Three solutions were titrated in both directions starting at the second equivalence point with either HCl or tetramethyl ammonium hydroxide, prepared at approximately 10x the concentration of the HEDP solution. The solutions titrated were: 1) 0.5 mM HEDP, 2) 0.5 mM HEDP with 1.0 mM calcium, and 3) 1.0 mM HEDP with 2.0 mM calcium. Ionic strength was maintained at 0.1 M with the non-complexing salt tetramethyl ammonium chloride. Sodium forms a very weak complex with HEDP but there was no apparent affect on the titration curve from the sodium initially present as the HEDP salt. Resulting data were corrected for volume changes and expressed as pH as a function of moles of base added per mole of HEDP.

Values for the various parameters in the model were determined using a non-linear least squares procedure. Basically, the nonlinear system solver described by Powell (17) was modified to handle the nonlinear least squares problem. For a given set of parameter values, a given point on either one of the pH titration curves or one of the formation curves was determined by solving the appropriate chemical equilibrium problem. The problem was formulated as a nonlinear system of three equations (for proton, Ca, and diphosphonate mass balances) in three unknowns (hydrogen ion, calcium ion, and the unprotonated diphosphonate species) and Powell's program in it's original form was used. The high order of the equations required that much effort be expended to ensure that the nonlinear system solver was given reasonable approximations to the solution before beginning its iterations.

After solving the basic system of equations it was possible to express the computational results in a variety of formats, as shown graphically in Fig. 1-6. The computation of formation number as a function of free calcium ion was a relatively easy problem since hydrogen ion concentration was specified and CATOT unconstrained. The titration curves, however, were considerably more difficult to calculate and fit since 1) the system required three variables, 2) CATOT and DPTOT were fixed, and 3) the order of certain species resulted in extreme pH dependencies.

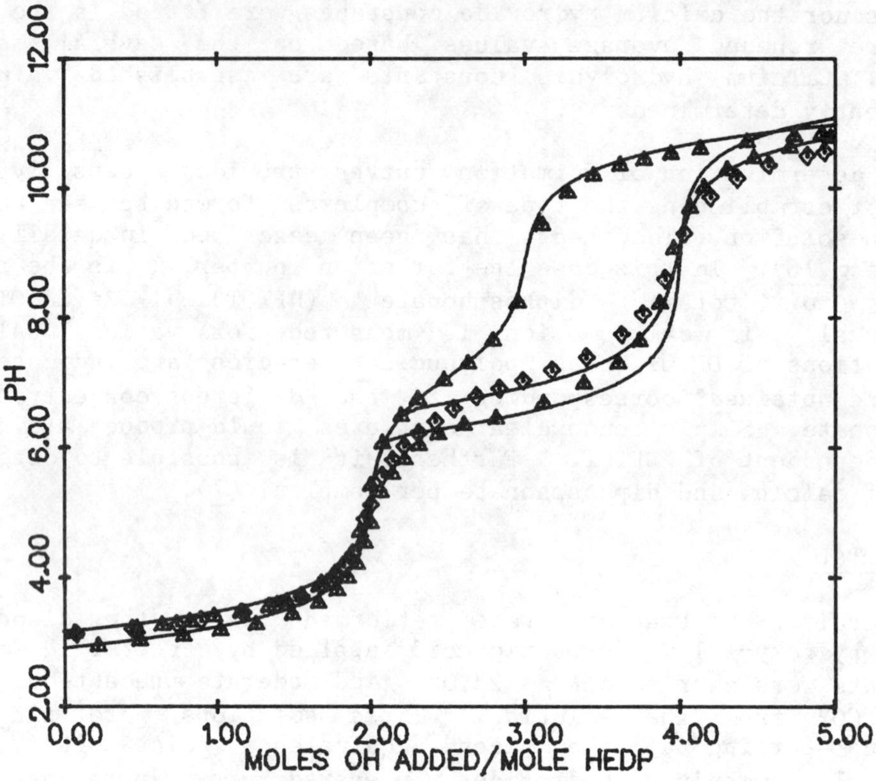

Fig. 1. pH titration curves calculated using best constants (set 10) from ref. (7).

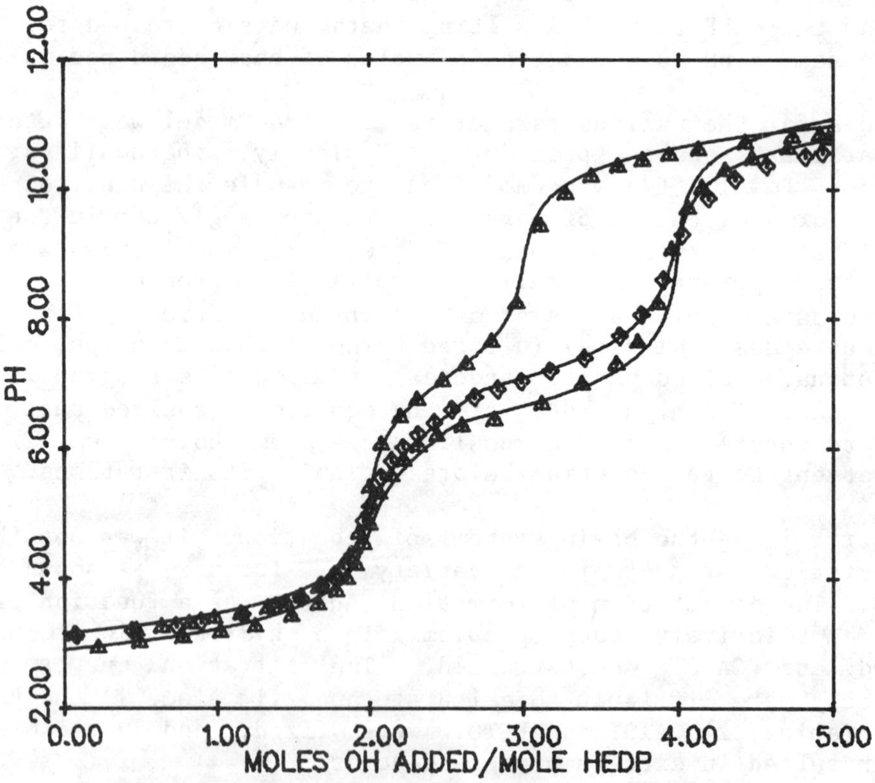

Fig. 2. Titration curves calculated using best constants for above model plus CaHDP.

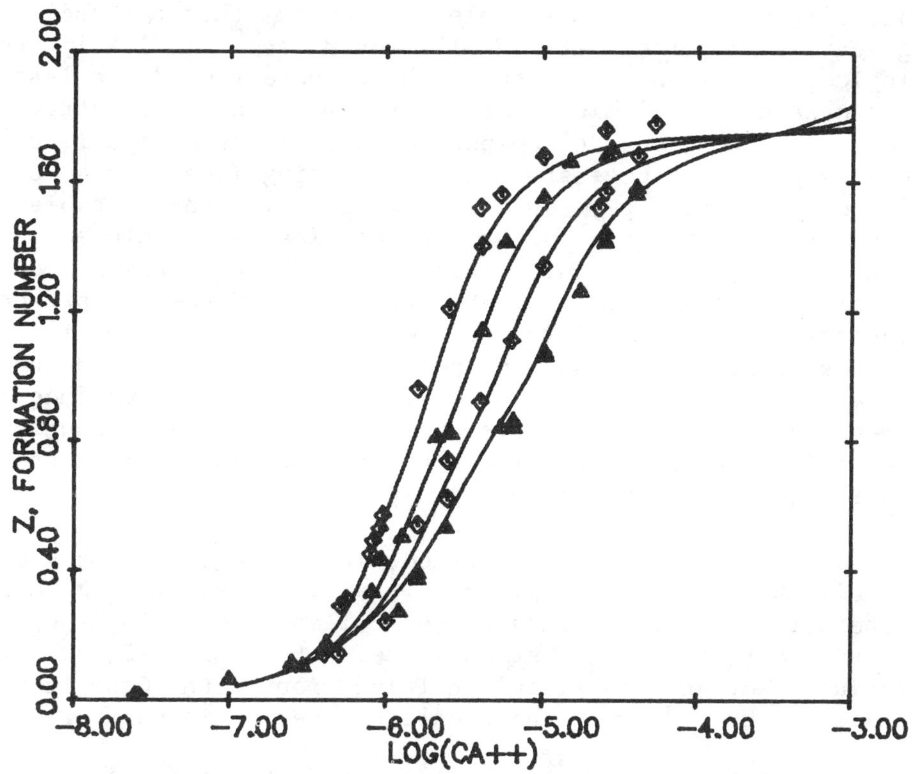

Fig. 3. Formation curves, pH 11. Simultaneous best fit with titration curves in Fig. 2.

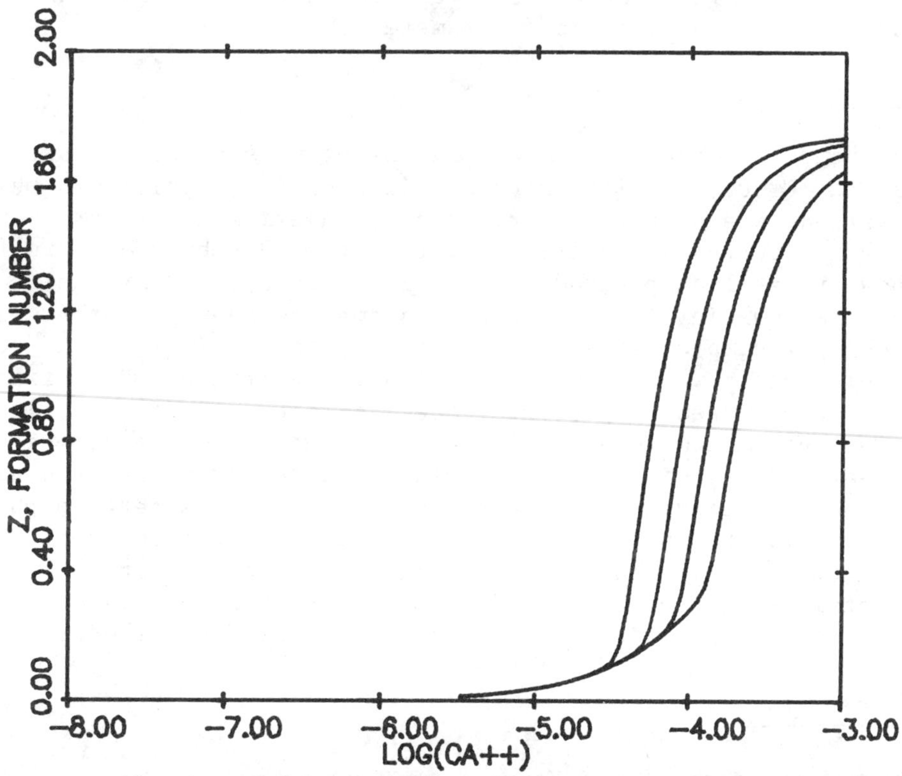

Fig. 4. Model-predicted formation curves, pH 8.

Initially it seemed reasonable to expect that polynuclear complexation and aggregation reported at pH 11 would not be likely to occur in a physiologic range of pH where the diphosphonate ligand has less of a tendency to form complexes. Preliminary computer experiments then attempted to fit the titration curves with a model composed only of the mononuclear species. This model was marginally successful in predicting the spread of the titration curves at near neutral pH, but was completely unacceptable for the pH 11 formation curves. When the polynuclear species were included in the model the formation curves were fit quite successfully but titration curves were not, particularly in the neutral pH region. Since a protonated calcium diphosphonate complex is likely to be an important species in the neutral region, this complex was also included in the model and the entire set of equilibria were fitted to the titration data. This produced a significant improvement in the titration curve model, without significantly affecting the pH 11 formation curves. Titration curves with and without the CaHDP complex are seen in Figures 2 and 1 respectively.

The Ca2DP complex was also tried in the equilibrium model with little success. Inclusion of this species with an overall log K greater than 7 causes the set of formation curves to collapse to a single curve. Since the value of the CaDP constant is approximately 6.0 this indicates that the log K for association of an additional calcium ion with CaDP is no greater than approximately 1.0, and hence the dicalcium species is unimportant.

Subsequently a program was developed which allowed us to fit parameters to the model for both types of experimental results simultaneously. The best values of constants determined from both titration and formation curve data are presented in Table III. Figures 2 and 3 are plots of the data and calculated curves based on on their best fitting parameters for the common model. The remaining Figures 4, 5 and 6 are predictions which are based on this model and on the optimized parameters.

DISCUSSION

The computer model, based on Grabenstetter's suggested set of polynuclear species plus CaHDP, permits several interesting predictions which, it should be emphasized, are as yet experimentally unverified. Figure 4 indicates that the formation curves are still separated at pH 8 (the model, in fact, predicts that they are still separated at pH close to 6), and are shifted to a higher range of free calcium ion. Figure 5 shows the average number of Ca (red) and DP (black) per complex as a function of pH; the maximum, it is noted, is close to physiological pH. A complete species concentration pH profile is presented in Fig. 6, where red represents free calcium and simple calcium complexes, black is the uncomplexed DP species, and blue represents the various polynuclear CaDP species. A particularly interesting feature of this plot is the extreme pH dependency of the polynuclear species between pH 6 and 7. Our purpose thus far has been primarily to look at computer designed experiments and model optimization based on a variety of input. It is hoped that future experimental aspects of this project will be able to relate these predictions to bone agent localization (18) and formation of Tc-HEDP complexes.

ACKNOWLEDGEMENT

Supported in part by NIH grant #DE01830.

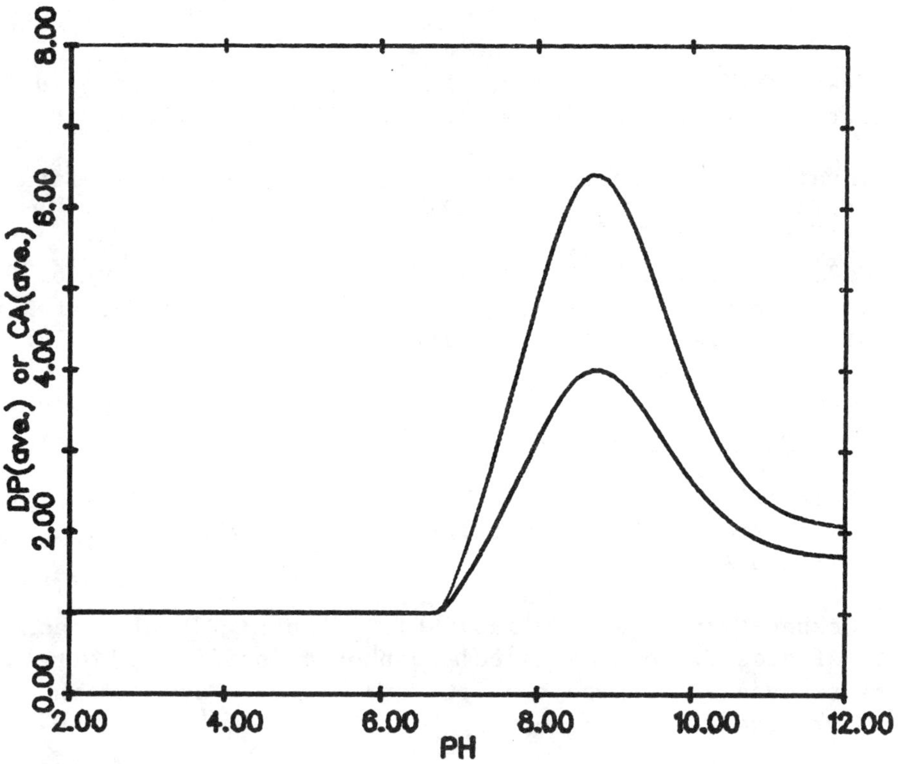

Fig. 5. pH profile: average Ca and DP per complex.

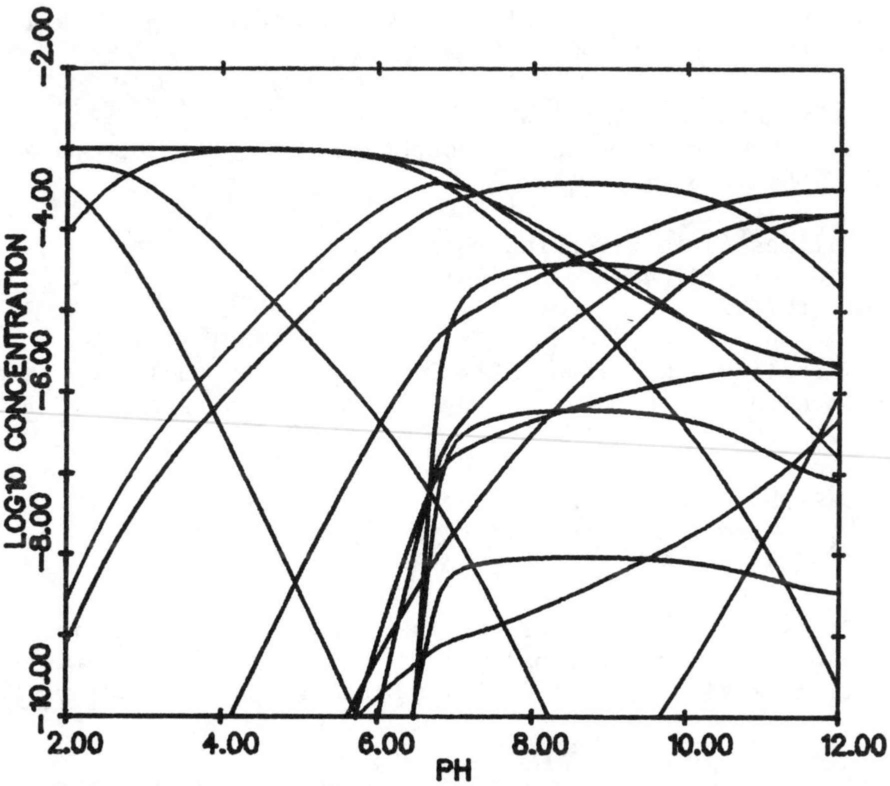

Fig. 6. pH profile: species concentrations.

REFERENCES

1. Francis MD, and Centner RL: The development of diphosphonates as significant health care products. J Chem Ed 55:760-766 (1978).

2. Thrall JH: Technetium-99m labeled agents for skeletal imaging. CRC Crit Rev Clin Rad Nucl Med, Aug. 1976, pp. 1-31.

3. McRae J, Hambright P, Valk PE, et al: Chemistry of 99m-Tc tracers. II. In vitro conversion of tagged HEDP and pyrophosphate into gluconate. Effects of Ca and FeII on in vivo distribution. J Nucl Med 17:208-216 (1976).

4. Jones AG, Francis MD, Davis MA: Bone scanning: radionuclide reaction mechanisms. Semin Nucl Med 6:3-18 (1976).

5. Carroll RL, Irani RR: On the acidity of substituted methylenediphosphonates and their interaction with alkali metal ions. Inorg Chem 6:1994-1998 (1967).

6. Grabenstetter RJ, Quimby CT, Flautt TJ: The acid dissociation constants of substituted methanediphosphonic acids: A correlation with P-31 magnetic resonance chemical shift and with Taft sigma. J Phys Chem 71:4194-4202 (1967).

7. Grabenstetter RJ, and Cilley WA: Polynuclear complex formation in solutions of calcium ion and ethane-1-hydroxy-1,1-diphosphonic acid. I. Complexometric and pH titrations. J Phys Chem 75:676-682 (1972).

8. Wada H, Fernando Q: Interaction of methane-hydroxyphosphonic acid and ethane-1-hydroxy-1,1-diphosphonic acid with alkali and alkaline earth metal ions. Anal Chem 44:1640-1643 (1972).

9. Kabachnik MI, Latovski RP, Medved TY, et al: Complex forming properties of hydroxyethylidenediphosphonic acid in aqueous solutions. Dokl Acad Nauk USSR 177:582-585 (1967).

10. Collins AJ, and Perkins PG: Acidity constants and salt formation by 1-hydroxyethyl-1,1-diphosphonic acid (HEDPA). J Appl Chem Biotechnol 27:651-661 (1977).

11. Carroll RL, and Irani RR: On the delta H and delta S of metal ion interaction with substituted methylene diphosphonates. J Inorg Nucl Chem 30:2971-2976 (1968).

12. Callis CF, Kerst AF, Lyons JW: "Calcium complexing by phosphorous compounds", in Coordination Chemistry, Stanley Kirschner, ed., Plenum Press, New York (1969), pp. 223-240.

13. Wiers BH: Polynuclear complex formation in solutions of calcium ion and ethane-1-hydroxy-1,1-diphosphonic acid. II. Light scattering, sedimentation, mobility, and dialysis measurement. J Phys Chem 75:682-687 (1971).

14. Wiers BH: Precipitation boundaries in calcium pyrophosphate and

calcium ethane-1-hydroxy-1,1-diphosphonate systems. Inorg Chem 10:2581-2584 (1971).

15. Uchtman VA: Structural investigations of calcium binding molecules. I. The crystal and molecular structures of calcium dihydrogen ethane-1-hydroxy-1,1-diphosphonate dihydrate: Implications for polynuclear complex formation. J Phys Chem 76:1304-1310 (1972).

16. Sillen LG: "Polynuclear complexes in solution", in Coordination Chemistry, Vol. I, ACS Monograph 168. Arthur E. Martel, ed. Van Nostrand Reinhold Co., New York (1971), pp. 491-541.

17. Powell MJD: "A FORTRAN subroutine for solving systems of nonlinear algebraic equations", in Numerical Methods for Nonlinear Algebraic Equations, P. Rabinowitz, ed., Gordon and Breach Science Publishers, London (1970), pp. 115-161.

18. van Langeveld A, Driessen OMJ, Pauwels EKJ, et al: Aspects of 99m-Technetium binding from an Ethane-1-hydroxy-1,1-diphosphonate-99m-Tc complex to bone. Eur J Nucl Med 2:47-51 (1977).

OXIDATION STATE OF TECHNETIUM IN BONE SCANNING AGENTS

Charles D. Russell
and
Anna G. Cash

VETERANS ADMINISTRATION AND UNIVERSITY MEDICAL CENTERS
Birmingham, AL

We have identified the oxidation state of technetium by polarography after electrolytic reduction of pertechnetate in complexing media of pyrophosphate, methylene diphosphonate (MDP), and ethane-1, hydroxy-1, 1-diphosphonate (EHDP). In the same media we have identified the oxidation state resulting from reduction with Sn(II) by means of amperometric titration. The stable states resulting from electrolytic reduction were found to be Tc(III) and Tc(IV) below pH 6, and Tc(IV) and Tc(V) above pH 10,for all three ligands. Between pH 6 and 10 the results depended on the ligand present. Tc(III) and Tc(IV) were identified in pyrophosphate and MDP, while EHDP Tc(II) and Tc(V) were identified with certainty, Tc(IV) questionably. Though two stable products thus resulted from electrolytic reduction at each pH, tin reduction gave only a single product, which was always Tc(IV) except in EHDP, where Tc(V) was formed in neutral or alkaline solution.

Technetium is no exception to the rule that transition metals exhibit a multiplicity of oxidation states; compounds of technetium have been reported in all oxidation states from 0 to +7 (1). In technetium pharmaceuticals the oxidation state has seldom been identified; usual analytical methods are generally incapable even of telling whether the agent is a single species or a mixture. In an increasing number of examples careful study shows the presence of multiple components (2-7); how these differ in oxidation state, coordinate bonding, or polynuclear aggregation is obscure. In this report we shall address the question of oxidation state.

Various approaches have been used to identify the oxidation state of technetium in pharmaceuticals. Most of these agents are prepared by reducing TcO_4^- with Sn(II) in the presence of a chelating agent. Since tin has only two stable oxidation states in aqueous solution, the change in oxidation state of the technetium can be calculated from the excess of TcO_4^- or of Sn(II) remaining after reaction. To measure excess Sn(II) two methods have been used: titration with I_3^- to a starch (8) or potentiometric (9, 10) endpoint, or direct polarographic measurement of the diffusion current for Sn(II) (9, 10). These methods have certain shortcomings. In the iodometric titration one cannot be certain that reduced technetium species do not react as well as the Sn(II) (9): to repeat with excess TcO_4^- (8) is inconclusive, since TcO_4^- itself could oxidize lower states to a higher one. Indeed, an example of such a reaction is given below. In the direct polarographic determination of excess Sn(II), on the other hand, it has been assumed that the background current does not change in the presence of technetium complexes, *i.e.*, that the sole significant reaction at the dropping mercury electrode is the reduction of tin. This assumption is not always warranted; in coulometric studies we have often found the background current to increase in the presence of technetium complexes, presumably from the catalytic reduction of water.

Another variant of this approach entails using an excess of TcO_4^- and measuring the excess by thin layer chromatography (2). This appears reliable for chelates stable enough not to be reoxidized to pertechnetate during chromatography, but may not detect all stable products; for example, if reduction occurs first to Tc(IV) and then to Tc(III), this method will give no hint of the second step.

The oxidation state can be inferred from x-ray crystallography (11). that most powerful of structural methods. However, correct results depend on locating all hydrogen atoms present, which may not be easy. The difficulty of obtaining single pure crystals, and the complexity of data analysis, will always restrict use of this method to a limited number of compounds of special interest.

Pertechnetate has also been directly titrated with Sn(II) to a potentiometric endpoint (2, 9, 12). The Tc(III)/Tc(IV) redox couple appears to be fairly reversible at mercury electrodes in a variety of media, and the Sn(II)/Sn(IV) couple not too irreversible in strongly acidic chloride, so that potentiometric titration should be suitable in selected cases. In the absence of such reversible couples the potential of the indicator electrode will be poised by a balance of exchange currents interpretable only with detailed knowledge of the corresponding electrochemical kinetics (9). If the indicator electrode gives stable readings and the endpoint is sharp this can be a reliable method; but even then it may not reveal every step of a stepwise reaction, or all stable products.

Polarography has also been used not only to measure excess Sn(II) after reduction of TcO_4^- by Sn(II), but also to identify the oxidation states resulting from direct electrolytic reduction of TcO_4^- at the dropping mercury electrode. From the polarographic limiting current and the diffusion coefficient for pertechnetate, the number of electrons transferred per mole of pertechnetate can be calculated using the Ilkovic or Koutecky equations. However, this requires accurate knowledge of the diffusion coefficient, and requires that the assumptions of the Ilkovic or Koutecky equation hold; it has led in the past to discrepant results (13). We have used a different polarographic approach, based on reliably determining the product in a single medium, and then comparing other polarographic measurements to that medium as a standard (13). Not only is this more reliable, as we believe, but it is technically easier as well, and can be applied to technetium chemistry with very little handling of the radioactive solutions. We have applied it to over twenty media ((13, 14) and unpublished work), and present below a study of the bone scanning agents by this means. A further virtue of this technique is that it has a synthetic counterpart; reactions identified by polarography can be used for radiopharmaceutical synthesis by controlled-potential electrolysis (6, 10, 15, 16).

Figure 1 summarizes the polarographic properties of pertechnetate in 0.1 M MDP (methylene diphosphonate) as function of pH (14). The number n of electrons transferred for each wave was measured by comparison with the known 4-electron wave in pH 3 citrate, as described elsewhere (13). Below pH 6 the first step in electrolytic reduction is to Tc(III); this can be reduced further at more cathodic electrode potentials but the products are unstable, being strong enough reducing agents to reduce water. Between pH 6 and pH 10 two stable oxidation states are identified, Tc(IV) and Tc(III), formed successively as the electrode potential is made more cathodic. Above pH 10 Tc(V) becomes stable, and Tc(III) disappears.

It is illuminating to examine the electrochemical behavior not only of pertechnetate, as in the foregoing paragraph, but also of the products of pertechnetate reduction. A convenient way to do this is by pulse polarography, holding the electrode at a potential on the plateau of one of the pertechnetate reduction waves, but applying an anodic voltage pulse so that the products, if electroactive, will be reoxidized. The results of doing this in MDP solutions are shown in Figure 2. Tc(III) is reoxidized to Tc(IV) at an electrode potential which varies with pH. Further reoxidation to TcO_4^- occurs with more anodic pulses; the wave for oxidation of Tc(IV) merges with that for oxidation of Tc(V).

We have obtained results for pyrophosphate and EHDP (ethane-1, hydroxy-1, 1-diphosphonate) similar to those shown for MDP. These data, together with procedural details and quantitative measurements, will be described in a forthcoming communication (14). The other two agents behaved essentially the same as MDP, except that a Tc(V) complex of EHDP was formed in the pH range from 6 to 10. The existence of multiple stable

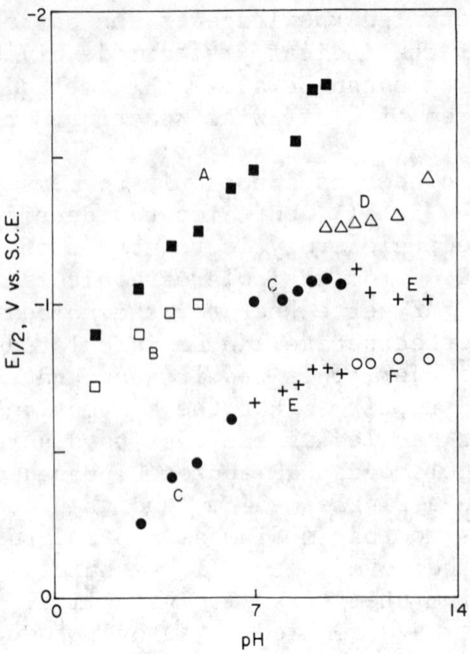

FIG. 1. Half-wave potentials for reduction of 0.14 mM TcO_4^- in 0.1 M MDP. A, n = 6 or 7; B, 5 or 6; C, 4; D, 4, 5, or 6; E, 3; F, 2 (14, by permission).

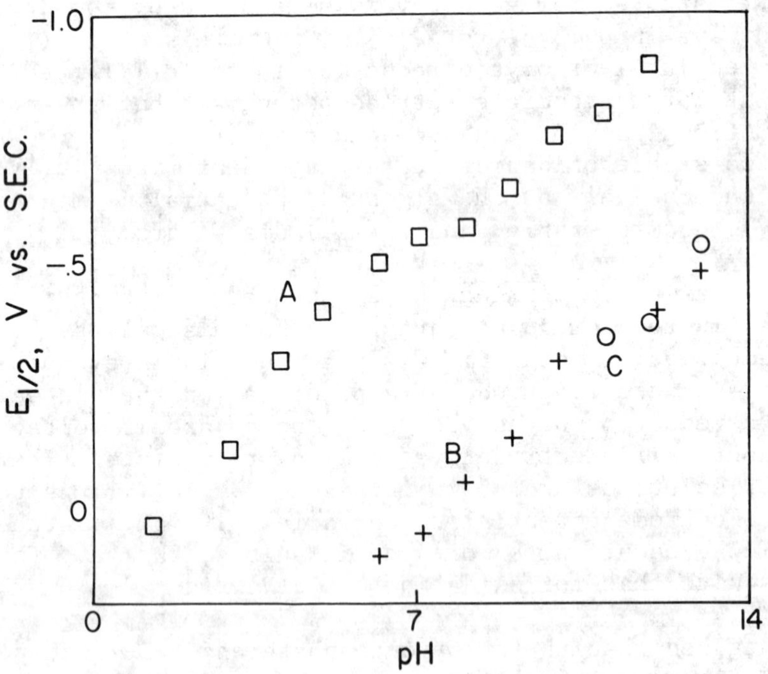

FIG. 2. Half-wave potentials of anodic waves for products of reduction of TcO_4^- in 0.1 M MDP. The products of cathodic TcO_4^- reduction are detected by anodic pulses, using normal pulse polarography and sweeping in the anodic direction from a resting point on the plateau of a cathodic wave. A, Tc(III) → Tc(IV); B, Tc(IV) → TcO_4^-; C, Tc(V) → TcO_4^- (14 , by permission).

states in the pH range used clinically raises the interesting question: which state is present in clinical preparations, or are both?

The foregoing method is eminently suited to the rapid survey of a large number of media, and can be quite informative about the solution chemistry of technetium. From study of over twenty media ((13, 14) and unpublished work), we can venture the following generalizations. The stable oxidation states of technetium complexes in aqueous media are III, IV, and V. Stability depends on pH; Tc(IV) is found at any pH, but Tc(III) is more characteristic of acidic media and Tc(V) of alkaline. There are often several stable species at any given pH. Exceptions will no doubt be found, (e.g., the cyanide complex of Tc(I)) but we have found none so far in our studies. This provides more direct means for studying the species in solution than previous methods that depend on isolating solid compounds, which do not necessarily reflect the species in solution.

Coulometry is also a powerful method for determining the oxidation state resulting from redox reactions, but we have not found it easily applied to technetium compounds. The results are quite sensitive to electrode potential, as pointed out by Salaria, Rulfs, and Elving (17), so that cell design and solution composition must be so chosen to insure uniform conditions over the electrode surface; otherwise different products may be formed at different parts of the electrode. We have found that the background current often increases in the presence of technetium complexes, presumably as the result of catalytic water reduction, so that the background correction tends to be large, uncertain, and not overcome by working at high concentrations. The catalytic current has the further disadvantage of requiring the cell to be cleaned after each measurement, so that replication is tedious. Current-time curves often indicate complex processes. For example, a maximum is observed in the current-time curve for pertechnetate reduction in 0.1 M pyrophosphate at pH 8.15, establishing that the mechanism is autocatalytic. In light of the polarographic data on this system, a plausible mechanism is:

$$TcO_4^- + 3e^- \rightarrow Tc(IV)$$

$$2Tc(IV) + TcO_4^- \rightarrow 3Tc(V)$$

$$Tc(V) + e^- \rightarrow Tc(IV)$$

Another peculiar current-time curve is found on 0.1 M EHDP at pH 3.0, for which the log current-time curve shows continuous and dramatic increase in slope throughout the course of electrolysis, eventually exceeding the maximum diffusion-limited slope. Again using the polarographic data as guide, a plausible mechanism is:

$$TcO_4^- + 4e^- \rightarrow Tc(III)$$

$$3Tc(III) + TcO_4^- \rightarrow 4Tc(IV)$$

Even these mechanisms are probably oversimplified. Present information, though limited, leads us to suspect that the reduction of pertechnetate is messy even in the absence of tin, leading by complex mechanisms to multiple products. The simplicity of the polarographic results may be deceptive, side reaction being limited by the short reaction time (we used two second drops). A properly done coulometric study is thus a tedious job. An interesting example of the problems one can encounter in careful coulometric study of a transition metal is given by the thorough study of the reduction of vanadate by Israel and Meites (18). We believe coulometric data on technetium complexes to be trustworthy only if presented with full information on the precision of the results and on their variation with electrode potential. We consider it meaningless to report an integral number of electrons transferred at some arbitrary potential, as has often been done, without demonstrating that a plot of number of electrons versus potential levels off in the potential range selected. Though good coulometric studies are needed, they are not easy. Polarography leads more quickly to reliable and useful results, which are of great value in interpreting coulometric measurements, and in our opinion polarography should be done first.

Although polarographic studies of the electrolytic reduction of pertechnetate can identify stable and unstable species in a given complexing medium, it does not answer the question of what product is formed upon chemical reduction by Sn(II). To answer this question we have applied the technique of amperometric titration. Both the tin wave and the pertechnetate wave can generally be followed through the titration of pertechnetate with Sn(II), using a dropping mercury electrode at controlled potential as the indicator electrode. This is related to the previously described method, the polarographic determination of excess tin (9, 10), except that to obtain sequential measurements throughout the titration conveys certain advantages. Both the Sn(II) and TcO_4^- can be monitored simultaneously by recording the current at two different potentials after each addition of titrant, choosing the potentials from polarographic data to correspond to diffusion currents for each species. When the entire titration curve is available, stepwise reactions can be identified by kinks, and slow reactions by curvature near the endpoint. But most importantly, changes in the background current during the titration have no effect on the position of the endpoint, though they will affect the slope of the titration curve and will cause an error in the method previously employed. The endpoint will also be insensitive to the reported time-dependence of the diffusion current for Sn(II) (10).

We have conducted amperometric titrations of pertechnetate with Sn(II) in pyrophosphate, MDP, and EHDP. These were carried out in each of the three pH ranges for which polarography (above) showed differences in behavior to be likely. The results are summarized in Table 1; procedural details and numerical data will be presented in a forthcoming communication (19). In every case the oxidation state identified by amperometric titration as the reaction product was one which had been previously identified as stable by polarography (14). Again, the difference between EHDP and the other agents is noteworthy. The titration curves were linear, with no evidence of stepwise reaction. There was little or no curvature in the region of the equivalence point, though the solution was stirred for only 60 seconds after each addition of titrant, showing the reaction of TcO_4^- with Sn(II) to be rapid and complete. This method appears reliable for identifying the oxidation state formed in the reduction of pertechnetate by tin at carrier concentrations. It is capable of disclosing stepwise reactions, though none have been found in the cases studied so far.

Determination of oxidation states does not appear, unfortunately, to tell the whole story. Evidence suggests that each oxidation state can form several stable species (2, 3). We shall review here such evidence for the bone scanning agents. 1) Tissue distribution in mice: We have reported data for technetium pyrophosphate labelled electrolytically at two different voltages (6). If only two pyrophosphate species can be formed, then the results of labelling at a third voltage should be a linear combination of the results of the other two. We have done this, and cannot explain the quantitative results without hypothesizing more than two species. 2) Coulometry: In the presence of pyrophosphate, we have shown that (in neutral solution) pertechnetate can be reduced electrolytically to Tc(III), and the Tc(III) complex can then be oxidized at more anodic electrode potential to Tc(IV) (6). By cyclically changing the electrode potential, the technetium can be converted from one oxidation state to the other, and back again. When this is done, however, the coulombs of electricity required decreases with each successive cycle, showing partial conversion into technetium species which are unreactive at the electrode. This cannot be accounted for by electrodeposition, since the total radioactivity in solution does not decrease, nor by the formation (except at high technetium concentrations) of "hydrolyzed-reduced" technetium, since no significant amount of that component was found by paper chromatography. Thus we have evidence for soluble electroinert species, in addition to the soluble electroactive species hitherto demonstrated. 3) Paper Chromatography: At high technetium concentrations (above 1mM) "hydrolyzed-reduced" technetium is found in electrolytically labelled preparations, and the solutions, mostly colored, tend to look muddy. The isolation by Deutsch, Barnett, and Libson of a polymeric technetium-MDP complex suggests an explanation (20). Even at low concentrations soluble polynuclear complexes are likely, since this is common with transition metals (21), and even the calcium complex of EHDP forms polynuclear aggregates (22). 4) Column Chromatography: The most direct evidence for a multiplicity of species consists in their separation by column chromatography. Using the ion-exchange method described previously (4), multiple peaks can be demonstrated in technetium pyrophosphate preparations for which polarography reveals only a single oxidation state.

To define a technetium chelate, it is therefore insufficient to identify only its oxidation state. Radioanalytical methods are needed capable of separating closely related species, and should be correlated with studies of solution chemistry by electrochemical and other means, and with structure studies by spectroscopy and x-ray diffraction. The complexity of this problem is greater than generally appreciated and will no doubt require the contribution of a number of laboratories, using different tools, for its solution.

ACKNOWLEDGEMENT

We are grateful to Mrs. Darrow Davis for helping to prepare the manuscript. This work was supported by the Veterans Administration Medical Center, Birmingham, Alabama, and the Department of Radiology, University of Alabama in Birmingham.

REFERENCES

1. Schwochau K: Technetium. Chemiker-Zeritung 102: 329-331, 1978

2. Ikeda I, Inoue O, Kuraja K: Chemical and biological studies on Tc-99m-DMS II. Int J Appl Radiat Isotopes 27: 681-688, 1976

3. Yokoyama A, Saji G, Tanaka H, Odori T, Morita R, Mori T, Torizuka K: Preparation of a chemically characterized Tc-99m penicillamine complex. J Nucl Med 17: 810-815, 1976

4. Russell CD, Majerik JE: Radiochemical purity of technetium phosphate. J Nucl Med 19: 1270, 1978

5. Jansholt AL. Krohn KA, Stadalnik RC, Matolo NM, DeNardo GL: Catabolism and protein binding of Tc-99m pyridoxylidene-glutamate J Nucl Med 19: 1036-1044, 1978

6. Russell CD, Majerik JE, Cash AG, Lindsay RH: Technetium pyrophosphate, a mixture? - Preparation and comparative biologic properties of Tc(III) and Tc(IV) pyrophosphate. Int J Nucl Med Biol 5: 190-194, 1978

7. Deutsch EA, Libson KF, Becker CB, Francis MF, Tofe AJ, Ferguson DL, McCreary LD: Preparation, chemistry and biological distribution of technetium phosphonate pharmaceuticals synthesized without stannous ion. Presented at the Western Regional Meeting, Society of Nuclear Medicine, Vancouver, B.C., Canada, October 1978

8. Hambright P, McRae J, Valk P, Bearden AJ, Shirley BA: Chemistry of technetium radiopharmaceuticals I. J Nucl Med 16: 478-482, 1975

9. Steigman J, Meinken G, Richards P: The reduction of pertechnetate-99 by stannous chloride I. Int J Appl Radiat Isotopes 26: 601-609, 1975

10. Steigman J, Meinken G, Richards P: The reduction of pertechnetate-99 by stannous chloride II: Int J Appl Radiat Isotopes 29: 653-660, 1978

11. Byrne EF, Smith JE: Synthesis, structure elucidation and bio-distribution of technetium-99 and technetium-99m dithiolates. J Nucl Med 19: 695, 1978

12. Crouthamel CE: Thiocyanate spectrophotometric determination of technetium: Anal Chem 29: 1756-1760, 1957

13. Russell CD, Cash AG: Polarographic reduction of pertechnetate. J Electroanal Chem 92: 85-99, 1978

14. Russell CD, Cash AG: Complexes of technetium with pyrophosphate, etidronate, and medronate. J Nucl Med, in press

15. Russell CD: Carrier electrochemistry of pertechnetate. Int J Appl Radiat Isotopes 28: 241-249, 1977

16. Russell CD, Majerik J: Tracer electrochemistry of pertechnetate. Int J Appl Radiat Isotopes 29: 109-114, 1978

17. Salaria GBS, Rulfs, Elving PG: Polarographic behavior of technetium. J Chem Soc pp 2479-2484, 1963

18. Israel Y, Meites L: The reduction and oxidation of vanadium in acidic aqueous sulfate solutions at mercury electrodes. J Electroanal Chem 8: 99-119, 1964

19. Russell CD, Cash AG: Oxidation state of technetium in bone scanning agents as determined at carrier concentration by amperometric titration. Int J Appl Radiat Isotopes, in press

20. Deutsch EA: Private Communication

21. Sillen L: Polynuclear complexes in solution. In Coordination Chemistry, Martell AE, ed, New York, Van Nostrand-Reinhold, 1971, Vol. I, pp 492-500

22. Grabenstetter RJ, Cilley WA: Polynuclear complex formation in solutions of calcium ion and ethane-1, hydroxy-1, 1-diphosphonic acid I. J Phys Chem 75: 676-682, 1971

TABLE 1. OXIDATION STATE RESULTING FROM REDUCTION OF TcO_4^- WITH SN(II), AS MEASURED AT CARRIER CONCENTRATIONS BY AMPEROMETRIC TITRATION

pH	MDP	$Na_4P_2O_7$	EHDP
acidic	IV	IV	IV
neutral	IV	IV	V
basic	IV	IV	V

ANTIOXIDANT STABILIZATION OF BONE AGENTS

Andrew J. Tofe, John A. Bevan, Mahdi B. Fawzi
Harry S. Whitehouse and Marion D. Francis

Miami Valley Laboratories
The Procter & Gamble Company
Cincinnati, OH 45247

ABSTRACT

In vitro instability associated with Tc-99m phosphonate skeletal imaging agents whose formulations contain a low tin(II) concentration is a direct result of the presence of oxidants. These contaminants are either radiolytically induced or introduced through chemical contamination. Stabilization has been demonstrated with ascorbic acid and in this paper, gentisic acid (2,5-dihydroxybenzoic acid) is shown to be an equally effective antioxidant for the (1-hydroxyethylidene) diphosphonate (HEDP) and hydroxymethylenediphosphonate (HMDP) skeletal imaging agents. In vitro studies in the presence of oxygen show less than 2% free [^{99m}Tc] sodium pertechnetate at 24 hours. Biodistribution studies in guinea pigs show no effect of the antioxidant upon the osseous specificity and suggest the sole function of gentisic acid is the interruption of radical chain propagated reactions.

INTRODUCTION

Diagnostic skeletal imaging procedures in nuclear medicine continue to grow such that in 1980 an estimated 1.3 million procedures utilizing the technetium-99m phosphorus agents will be performed. Its growth, as well as the growth of other diagnostic procedures using technetium-99m labeled agents, continues to create a demand for higher daily levels of [^{99m}Tc] sodium pertechnetate. This increased demand creates problems for manufacturers because of the formation of radiolysis by-products that are associated with these higher specific activity sources. These radiolysis effects necessitate the use of oxidants to maintain high [^{99m}Tc] sodium pertechnetate extraction or elution yields. It is these oxidants which are added during the processing of ^{99}Mo for [^{99m}Tc] sodium pertechnetate sources, as well as ambient oxygen and radiolysis by-products that have been indicted as major causes of in vitro instability in diagnostic kits containing Sn(II) as the reducing agent. Analyses of both saline eluants and first eluates from commercially available high activity generators have detected high levels of oxidants (1). Thus, diagnostic kits of low Sn(II) reductant are especially susceptible to instability and clinically this instability manifests itself by visualization of activity in the gut and thyroid on the skeletal image, indicating the presence of free [^{99m}Tc] sodium pertechnetate.

Two skeletal imaging agents, Tc-99m labeled (1-hydroxyethylidene) diphosphonate (Tc-99m HEDP) and Tc-99m labeled hydroxymethylenediphosphonate (Tc-99m HMDP), whose formulations contain 5.0 mg and 2.0 mg respectively of the phosphonic acid, each utilize a low level of stannous tin (0.10 mg per vial). The formulations contain far in excess of the Sn(II) concentration required for both stoichiometric reduction Tc(VII) of and subsequent formation of the Tc-99m complex with the diphosphonate ligand. However, the concentration level is not sufficient to prevent temporal decomposition in either the presence of radiolytically induced or extraneously introduced oxidants and oxygen.

The in vitro instability of low-Sn(II) diphosphonate complexes with respect to the formation of free [^{99m}Tc] sodium pertechnetate can be minimized by a number of techniques which will be discussed.

OXYGEN/NITROGEN EFFECT

The reaction of oxygen with the oxidants/radiolysis products in the [^{99m}Tc] sodium pertechnetate eluate is believed to be responsible for the in vitro instability (1). This has recently been confirmed by Billinghurst, et al. (2) who showed that radiation-induced decomposition is catalyzed by the presence of dissolved oxygen.

Vials of the HEDP and HMDP formulations containing no stabilizer under either nitrogen (<0.5% oxygen) or air were prepared with 75-100 mCi obtained from high specific activity generators. Regeneration of free [^{99m}Tc] sodium pertechnetate levels of the unstabilized vials ranged from 25% to 38% for vials vented to the air and less than 3% for nitrogen vials at 8 hours post preparation (Figure 1 - shaded areas). Thus, the inhibition of instability by maintenance of a nitrogen atmosphere is clearly demonstrated.

Similarly, technetium-99m pyrophosphate preparations are also effectively stabilized by nitrogen (2), and nitrogen purging of the generator saline eluant and the diagnostic kit vial is also reportedly an effective technique of minimizing instability for low Sn(II) formulations (2, 4).

TIN CONCENTRATION EFFECT

Another method to overcome the stability problem is to increase the level of Sn(II) to counteract the effects of oxygen and oxidants (2, 3). Distribution studies conducted in our laboratories on the retention of Tc-99m in stomachs (indicative of free [^{99m}Tc] sodium pertechnetate) in Sprague-Dawley rats as a function of increasing stannous content, biologically illustrate the stabilizing effect of higher concentrations of tin (Figure 2). Increasing the tin concentration doses effectively inhibit the formation of free [^{99m}Tc] sodium pertechnetate. However, the higher tin levels in skeletal imaging agents have two detrimental effects. Reports in the literature (5-9) indicate that red blood cell labeling from bone seeking agents containing high levels of tin are responsible for the altered activity distribution found when [^{99m}Tc] sodium pertechnetate is employed as a brain scanning agent. Additionally, and perhaps more common, is hydrolysis of the excess tin(II). The result of this hydrolysis is localization of a technetium-99m labeled stannous colloid in the liver which consequently masks portions of the skeleton.

Mechanistically, the most likely explanation of the role of stannous tin is that the reduction of technetium(VII) oxidation state results in competing rates of formation between complexation with the phosphorus ligand and hydrolysis to TcO_2 as shown for the latter below:

$$2TcO_4^- + 3Sn^{++} + 2H_2O \rightleftharpoons 2TcO_2\downarrow + 3SnO_2\downarrow + 4H^+ \qquad (1)$$

In the case of low tin diagnostic skeletal imaging agents, the oxidation of TcO_2 with no excess Sn^{++} would rapidly yield free [^{99m}Tc] sodium pertechnetate. Excess Sn^{++} would inhibit oxidation of TcO_2 but would be expected with progressive oxidation to form SnO_2, which when coordinated with Tc-99m would be visualized in the reticuloendothelial system.

ANTIOXIDANT EFFECT

A solution to the nearly impossible problem of totally eliminating oxygen is to protect the low tin radiopharmaceutical formulations through the use of antioxidant stabilizers. The introduction of one of these compounds into the existing formulations offers a variable alternative to increasing the concentration of tin to inhibit instability.

Ascorbic acid and now gentisic acid are the only antioxidants presently being utilized in diagnostic nuclear medicine to inhibit the effects of both oxygen and oxidants. Ascorbic acid is an extremely safe and nontoxic compound and has been found to be extremely effective in inhibiting the formation of free [^{99m}Tc] sodium pertechnetate. Studies with the Tc-99m

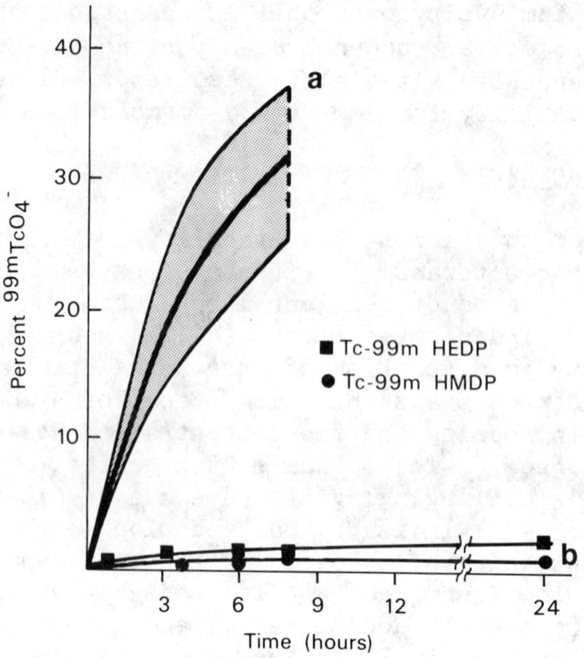

FIG. 1. In vitro stability of Tc-99m labeled HEDP and HMDP under (a) oxygen (no stabilizer) or (b) nitrogen or oxygen (with stabilizer). Stabilizer was 0.5 mg of antioxidants (ascorbic acid or gentisic acid).

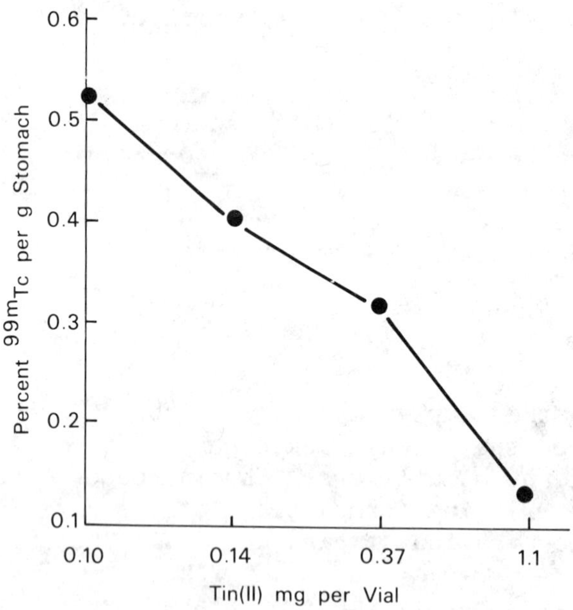

FIG. 2. Biological retention of ^{99m}Tc in stomachs of Sprague-Dawley rats as function of increasing stannous concentration for HEDP (5 mg) skeletal imaging formulation.

HEDP and low Sn(II) containing 0.5 mg of ascorbic acid have consistently shown free [^{99m}Tc] sodium pertechnetate values of approximately 2% or less in the presence of either oxygen or nitrogen over a 6 hour in vitro period (1). Ascorbic acid's sole function as an in vitro stabilizer was deduced from observing similar animal biodistribution patterns regardless of the presence and absence of ascorbic acid in the HEDP formulation. In addition, the lack of any increased renal uptake when the bone agent contains ascorbic acid stabilizer biologically indicates a negligible, if any, formation of a Tc(IV)-ascorbate complex. The use of ascorbic acid is not limited to the stannous phosphonate bone agents as Winchell (10) has shown that ascorbic acid also effectively stabilizes dimercaptosuccinic acid as well as DTPA technetium-99m labeled stannous complexes.

The polyhydroxy phenolic compounds, particularly gentisic acid (2,5-dihydroxybenzoic acid), have been shown to also be effective antioxidants (11). Like ascorbic acid, gentisic acid's toxicological and chemical properties are most favorable for human use (12-13). Its effectiveness in stabilizing the low tin HEDP and HMDP skeletal imaging formulations with a level of 0.5 mg is illustrated at the bottom of Figure 1. Free [^{99m}Tc] sodium pertechnetate values of approximately 2% or less in the presence of either oxygen or nitrogen were obtained over an 8 hour in vitro period. Even at 24 hours there was still less than 2% free [^{99m}Tc] sodium pertechnetate in the stabilized vials prepared under nitrogen.

Similar to ascorbic acid, the addition of gentisic acid to the stannous reductant and the two hydroxydiphosphonates does not alter the expected osseous specificity of the skeletal imaging agents. The biological distribution data of C-14 labeled gentisic acid with and without Tc-99m HEDP, and H-3 labeled gentisic acid with and without Tc-99m HMDP have recently been investigated (14) and are summarized in Tables 1 and 2 respectively. The similar distribution pattern of C-14 labeled gentisic acid alone to the Tc-99m HEDP and C-14 labeled gentisic acid uptake, and the H-3 labeled gentisic acid alone to the Tc-99m HMDP with H-3 gentisic acid uptake, indicate that gentisic acid does not to any extent appreciably complex with the hydroxydiphosphonate formulations and acts solely as an excipient antioxidant to maintain in vitro stability.

The role of oxygen in accelerating in vitro instability and the antioxidant effectiveness of both ascorbic acid and gentisic acid supports the contention of two oxidation intermediates which act as reducible contaminants to promote in vitro instability. As previously indicated (1), the first are peroxides in whose absence atmospheric oxidation is not normally possible (15). The scission of the oxygen-oxygen bond of peroxides forms powerful oxidizing alkoxy or hydroxy radicals (Eq. 2). The other intermediate are the powerful peroxy radicals which are formed from oxygen and radiolysis by-products (Eq. 3). The antioxidant's primary function is the interruption of these radical-chain-propagated reactions. Both ascorbic acid (Eq. 4) and gentisic acid (Eq. 5) function by donating reactive hydrogen atoms which in turn are resonance stabilized and hence cannot propagate the chain reaction.

$$RO\text{-}OR' \longrightarrow RO\cdot + \cdot OR' \quad (2)$$

$$R\cdot + O_2 \longrightarrow RO_2 \quad (3)$$

TABLE 1. TISSUE DISTRIBUTION OF Tc-99m HEDP, Tc-99m HEDP WITH ^{14}C-GENTISIC ACID, AND ^{14}C-GENTISIC ACID ALONE IN GUINEA PIGS AT 3 HOURS POST INTRAVENOUS DOSE

Organ	Percent Dose Per Gram			
	Tc-99m HEDP	Tc-99m HEDP with ^{14}C-GA Tc-99m	^{14}C-GA	^{14}C-GA
Femur	1.66 (1.60-1.71)*	1.62 (1.49-1.71)	0.005 (0.004-0.007)	0.007 (0.005-0.011)
Liver	0.016 (0.014-0.017)	0.017 (0.015-0.022)	0.005 (0.004-0.005)	0.006 (0.005-0.006)
Kidney	0.150 (0.117-0.176)	0.139 (0.117-0.157)	0.017 (0.008-0.026)	0.009 (0.007-0.011)

* Values in parentheses indicate range for n=3

TABLE 2. TISSUE DISTRIBUTION OF Tc-99m-HMDP, Tc-99m-HMDP WITH ^{3}H-GENTISIC ACID AND ^{3}H-GENTISIC ACID ALONE IN GUINEA PIGS AT 3 HOURS POST INTRAVENOUS DOSE

Organ	Percent Dose Per Gram			
	Tc-99m HMDP	Tc-99m HMDP with ^{3}H-GA Tc-99m	^{3}H-GA	^{3}H-GA
Femur	1.91 (1.79-2.03)*	2.00 (1.93-2.07)	0.005 (0.005-0.006)	0.006 (0.005-0.007)
Liver	0.043 (0.022-0.063)	0.087 (0.077-0.098)	0.010 (0.010-0.011)	0.011 (0.008-0.015)
Kidney	0.42 (0.34-0.49)	0.23 (0.20-0.25)	0.020 (0.020-0.020)	0.022 (0.020-0.024)

* Values in parentheses indicate range for n=3

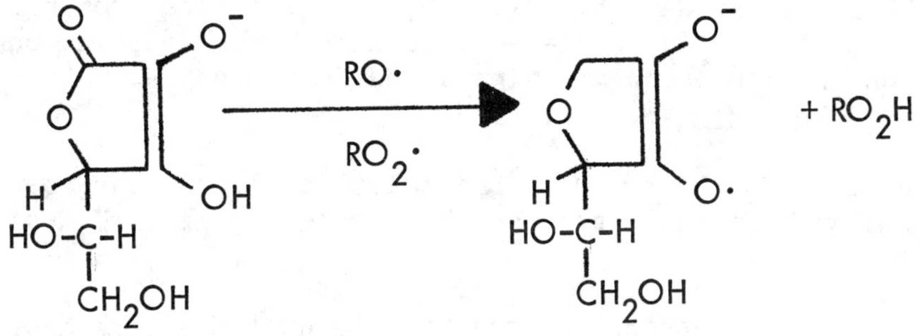

(4)

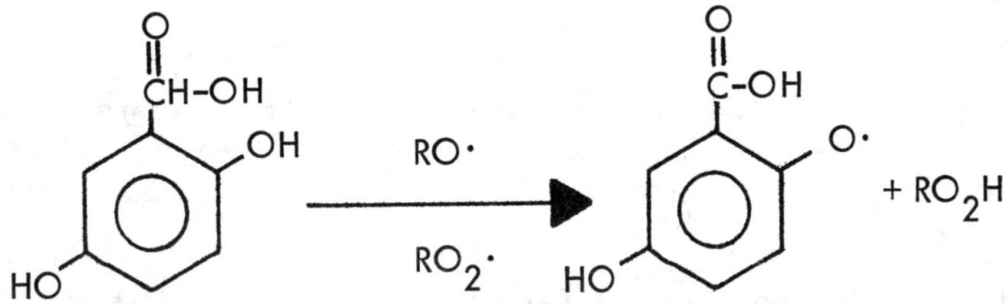

(5)

The elimination of the intermediates by both antioxidants provides the in vitro stability by inhibiting the slow oxidation of TcO_2 which is responsible for the production of Tc(VII) (4).

In conclusion, both ascorbic acid and gentisic acid are safe and effective agents which inhibit the effects of reducible contaminants and provide a means of continued use of diagnostic low tin formulations.

REFERENCES

1. Tofe AJ, Francis MD: In vitro stabilization of a low tin bone imaging agent (Tc-99m-Sn EHDP) by ascorbic acid. J Nucl Med, 17: 726, 1976

2. Billinghurst NW, Remple S, Westendorf BA: Radiation decomposition of technetium-99m radiopharmaceuticals. J Nucl Med 20: 138-143, 1978

3. Owunwanne A, Church LB, Blau M: The effect of oxygen on the reduction of pertechnetate ion by stannous ion. J Nucl Med 15: 521, 1974

4. Merlin L, Besnard M, Cohen Y: Chemistry of technetium: Effect of oxidoreduction systems on the stability of complexes used as radiopharmaceuticals. In New Developments in Radiopharmaceuticals and Labeled Compounds, vol. 1 Copenhagen, IAEA, 1974, pp. 63-70

5. McRae J, Sugar RM, Shipley B, et. al: Alterations in tissue distribution of Tc-99m pertechnetate in rats given stannous tin. J Nucl Med, 15: 151-155, 1974

6. Khentigan A, Garrett M, Lum D, et al.: Effect of prior administration of Sn(II) complexes used in nuclear medicine on in vivo distribution of subsequently administered Tc-99m pertechnetate and Tc-99m compounds. J Nucl Med 16: 541, 1975

7. Chandler WM, Shuck LD: Abnormal technetium-99m pertechnetate imaging following stannous pyrophosphate bone imaging. J Nucl Med 16: 518-519, 1975

8. Shuck LD: Case of the quarter. J Nucl Med Techol 3: 91, 1975

9. Pavel OG, Zimmer AM, Patterson VN: In vitro labeling of red blood cells with Tc-99m: A new approach to blood pool visualization. J Nucl Med 18: 305-308, 1977

10. Winchell HS, Khentigan A, Garrett M, et al: Instability of Sn(II) reagents to oxidation and hydrolysis: stabilization with ascorbic acid and erythorbic acid. Southern and Northern California Chapter Meeting S.N.M., C4, 1975

11. Martin AN, Swarbrick J, Cammarata A, et al: Physical Pharmacy. Philadelphia, Lea and Febiger, 197X, p. 391

12. Nash JF, Bope FW, Christensen BV: Esters of gentisic acid and their toxicities. Sci Ed, 42: 254, 1953

13. Clarke NE: Phenolic compounds in the treatment of rheumatic fever I. A study of gentisic acid derivatives. Cit, 7: 247, 1953

14. Tofe AJ, Bevan JA, Fawzi MB, et al: Gentisic acid: A new stabilizer for low tin skeletal imaging formulations. Submitted to J Nucl Med

15. Scott G: Atmospheric Oxidation and Antioxidants. New York, Elsevier, 1965 p. 14

Tc-99m HYDROXYMETHYLENE DIPHOSPHONATE (HMDP): A NEW SKELETAL IMAGING AGENT

John A. Bevan, Andrew J. Tofe, Marion D. Francis,
Bobby L. Barnett and James J. Benedict

Miami Valley Laboratories
The Procter & Gamble Company
Cincinnati, OH 45247

ABSTRACT

The hydroxymethylenediphosphonate ligand (HMDP) is a homolog of (1-hydroxyethylidene) diphosphonate (HEDP). Crystallographic studies with HEDP suggest the importance of the hydroxyl group in binding of the ligand to the growing surface of hydroxyapatite. Tc-99m HMDP has been investigated for use as a skeletal imaging agent. Animal biodistribution studies indicated excellent potential for skeletal and acute myocardial infarct imaging. In rats, femur/muscle uptake ratio averaged 638. In dogs, three hour whole blood retention averaged 1.3% of the dose, and the biodistribution of Tc-99m HMDP was significantly superior to that of Tc-99m MDP (methylenediphosphonate) and Tc-99m HEDP for skeletal imaging. In a dog myocardial infarct model, Tc-99m HMDP was shown to be equal to Tc-99m pyrophosphate for infarct imaging.

INTRODUCTION

A number of the technetium-99m phosphate agents (including diphosphonates and imidodiphosphate) have been compared for use as skeletal imaging agents, especially by Subramanian, et al (1), and Davis and Jones (2). One of the conclusions was that, at least among the complexes of interest, the primary factor affecting scintigraphic image quality was the rate of renal clearance, rather than the bone uptake, *per se* (1). There are, undoubtedly, many circumstances when this is a reasonable conclusion, since image quality can be drastically reduced (lower bone:soft tissue ratios) by impaired renal clearance or obesity, i.e., by delayed soft tissue clearance.

However, it is important not to let this divert us from considering the implications of structural variations among the diphosphonates and their relative affinity for the surface of calcium phosphate in its various forms. At this time, not much is known about the structural details of the technetium complexes of radiopharmaceutical interest although this is changing (3, 4) as crystal structure analysis is at last being brought to bear on the problem. On the other hand, we do know the structural configurations of some of the sodium and calcium diphosphonates (5-7) and it is not inappropriate to make use of those insights when considering the technetium complexes. For example, all the diphosphonate structures which we have studied in our laboratory include a rigid planar conformation, O-P-C-P-O, in the shape of the letter W. This is illustrated in Fig. 1. We believe that this is a characteristic of the diphosphonate ligand which will be maintained, not only in the solid state, but also in aqueous solution at tracer concentration levels, such as in radiopharmaceutical systems. Furthermore, the planar, W-shaped conformation defines the configuration of the remaining phosphate oxygen atoms. In the case of the methylenediphosphonate (MDP) structure (see Fig. 1), for example, this results in the formation of a bidentate ligand. However, in the case of the hydroxydiphosphonates, such as (1-hydroxyethylidene) diphosphonate (HEDP) and hydroxymethylenediphosphonate (HMDP), the hydroxyl group falls naturally into a triangular arrangement of oxygen atoms to form a tridentate ligand. This triangular oxygen grouping is strikingly similar to the oxygen triad which coordinates to calcium ions in the 001 face of the hydroxyapatite (HA) structure (5). Since the 001 face is the most rapidly growing face of HA in vitro and in vivo, this scheme offers an explanation for the therapeutic ability of HEDP to control the rate of growth of HA crystals.

Notice in Fig. 1 that both bidentate and tridentate binding are not only theoretically possible in the hydroxydiphosphonate ligand, but also have, in fact, been observed (5, 6). One might expect that this multi-coordination feature would enhance the affinity (i.e., increase the uptake) of hydroxydiphosphonate complexes for the various biological phases of calcium phosphate, compared to diphosphonates which lack the hydroxy group (such as MDP).

Recent clinical data (8) show significantly higher tumor-to-bone ratios with Tc-99m HEDP compared to Tc-99m MDP. Since active bone tumors (lesions) are regions of increased bone turnover, they will include a relatively increased proportion of growing HA crystals. Therefore, the tridentate coordinating affinity of Tc-99m HEDP could be responsible for enhanced lesion uptake. On the other hand, Arnold (9) has determined, also in patients, that the higher lesion-to-normal bone uptake of Tc-99m HEDP compared to Tc-99m MDP is due to the lower uptake of the HEDP complex in normal bone (i.e., a smaller denominator, rather than a larger numerator).

In the light of this, there is the possibility that the CH_3 group of the HEDP structure (see Fig. 1) sterically hinders the bidentate function of the ligand, and thereby reduces its binding with mature bone mineral, but not impairing its uptake in abnormal bone. If this is the case, then it follows that the HMDP structure (see Fig. 1) should be free of such hindrance and its coordination potential, and therefore its uptake on all phases of HA, should be maximized.

We have studied the biodistribution properties of Tc-99m HMDP in animals and found it to show exceptional promise as a skeletal imaging agent.

SYNTHESIS OF HMDP AND COMPLEXATION WITH TECHNETIUM

The Na_2HMDP was prepared by a modification of the method of Quimby, Prentice and Nicholson (10). Dichloromethylene diphosphonic acid was hydrolyzed to the tetrasodium carbonyldiphosphonate and then hydrogenated at high pressure and temperature to the hydroxymethylene diphosphonate salt. Chemical identity was confirmed by ^{31}P-NMR spectrum, S = 15.1 ppm (J = 14Hz); ^{1}H NMR spectrum, S = 3.88 ppm (J = 14Hz). Purity was determined by chemical analysis as the stable hemihydrate $HC(OH)(PO_3HNa)_2 \cdot 1/2\ H_2O$ (6).

The technetium-99m complex of HMDP was prepared in the familiar manner by the reduction of [^{99m}Tc] sodium pertechnetate with stannous ion (Sn^{++}) in the presence of an excess of ligand in aqueous solution. The initial experiments were performed with *ad hoc* liquid preparations but, once the formulation had been chosen, a conventional lyophilized kit in standard dosing vials was prepared. The formulation consisted of 2.0 mg Na_2HMDP, 0.16 mg $SnCl_2$ and 0.56 mg gentisic acid. Thus, a single adult human dose (e.g., one-fifth of the lyophilized formulation) would amount to a level of 0.006 mg/kg of the active diphosphonate in a 70 kg patient.

A low-tin formula was chosen in order to minimize the possibility of interference with subsequent [^{99m}Tc] sodium pertechnetate diagnostic procedures in the event of human clinical practice. As a result, we have added an antioxidant in the form of gentisic acid (2,5-dihydroxybenzoic acid) to prolong the *in vitro* stability of the complex by preventing the reformation of free [^{99m}Tc] sodium pertechnetate. Gentisic acid is safe and effective at this task and has been shown bio-equivalent to ascorbic acid (11).

BIODISTRIBUTION IN ANIMALS

Preliminary animal studies showed that above a certain dose level (0.1 mg/kg Na_2HMDP) there was increasing visualization of liver. Below this threshhold, Tc-99m HMDP showed excellent promise and consequently experimental doses were limited to below this level in the biodistribution studies.

Table 1 contains data from a biodistribution study in Sprague-Dawley rats (195 ± 7 gm females) following intravenous dosing of the Tc-99m labeled formulation at Na_2HMDP levels of 0.02, 0.05 and 0.10 mg/kg. The data are indicative of excellent scintigraphic imaging potential with high uptake in bone, moderate activity in kidney during renal clearance of the excreted dose fraction, and low retention in the other sample tissues. The average bone/blood values were 167, 164 and 156, and average bone/muscle values were 627, 672 and 619, for the three dose groups, respectively. There were no significant differences in biodistribution over this dose range. Gamma camera scintiphotos visually confirmed the expectations of the radioassay data.

For the technetium-99m phosphate skeletal imaging agents, the plasma clearance is generally indicative of the rate of soft-tissue clearance, an important determinant of scintigraphic image quality. As far as animal models are concerned, the dog is probably the most useful for predicting the human case (1, 12). The blood clearance of Tc-99m HMDP was compared to two commercial bone scanning kits of Tc-99m MDP and Tc-99m HEDP by radioassay of serial venous blood samples in the period which is most pertinent to imaging, 0.5-4.0 hours post dose. Sixteen adult beagle dogs were divided into four equal groups and given single cephalic vein doses as follows: 0.1 mg Tc-99m HMDP, 0.5 mg Tc-99m HMDP, 0.5 mg Tc-99m MDP, and 0.5 mg Tc-99m HEDP. In each instance, dose levels are calculated in terms of the disodium salt. The measured clearance data are shown in Fig. 2 in which the two Tc-99m HMDP groups have been combined, as they were indistinguishable. The Tc-99m HMDP and Tc-99m MDP clearances are closely similar through two hours post dose after which Tc-99m HMDP appears to clear at a slightly faster rate ($p<0.05$ at 4h). Both Tc-99m HMDP and Tc-99m MDP are significantly faster $p<0.05$) than Tc-99m HEDP from one hour post dosing.

The cumulative urinary excretion of Tc-99m HMDP in six beagle dogs was determined 24 and 48 hours post dose. The results are given in Table 2. Comparison with other measurements (not included here) in the same colony of dogs at our laboratory indicates that less Tc-99m HMDP is excreted than for the Tc-99m MDP and Tc-99m HEDP. Since the soft tissue clearance is rapid, this implies a higher degree of bone uptake for the new agent, which should result in high scintigraphic count rates and shorter scan times than Tc-99m MDP or Tc-99m HEDP.

COMPARATIVE STUDIES IN A CANINE ACUTE MYOCARDIAL INFARCT MODEL

The canine acute myocardial infarct (AMI) model has been used to evaluate potential AMI imaging agents (13, 14). Here, we have used this animal model to evaluate Tc-99m HMDP as an AMI imaging agent compared to the widely used Tc-99m pyrophosphate (PPi), and also, for its skeletal imaging potential compared to two diphosphonate agents, Tc-99m MDP and Tc-99m HEDP. The high degree of variability in the canine infarct model creates the need for some type of internal standard. In previous comparative studies of

TABLE 1. BIODISTRIBUTION OF Tc-99m HMDP IN SPRAGUE-DAWLEY RATS AT THREE HOURS POST INTRAVENOUS DOSING

	Tc-99m content in percent dose per gram		
Tissue	0.02 mg HMDP/kg	0.05 mg HMDP/kg	0.10 mg HMDP/kg
Femur	3.01 (2.81-3.25)	2.62 (2.37-2.83)	2.97 (2.44-3.55)
Blood	0.018 (0.016-0.021)	0.016 (0.014-0.018)	0.019 (0.012-0.023)
Muscle	0.0048 (0.0043-0.0053)	0.0039 (0.0032-0.0044)	0.0048 (0.0034-0.00066)
Kidney	0.276 (0.221-0.329)	0.306 (0.230-0.471)	0.414 (0.297-0.566)
Liver	0.046 (0.023-0.101)	0.041 (0.020-0.077)	0.073 (0.039-0.119)
Heart	0.014 (0.013-0.015)	0.011 (0.010-0.012)	0.013 (0.010-0.015)

* Values in parentheses are the range of values per group (n=4 rats/group).

TABLE 2. URINARY EXCRETION OF Tc-99m HMDP IN BEAGLE DOGS

	Percent Dose in Urine	
Dog #	0-24h	24-48h
1	53.1	3.1
2	61.8	5.0
3	49.5	7.6
4	59.8	2.3
5	78.8	1.6
6	48.8	1.3
Mean ± s.d.	58.6 ± 11.2	3.55 ± 2.4

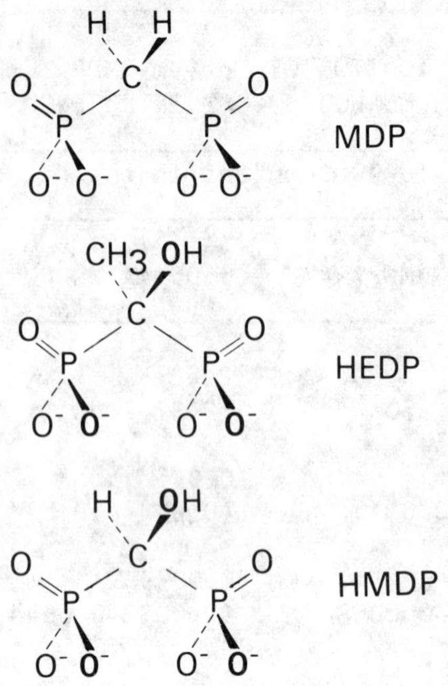

FIG. 1. Structures of HMDP compared with MDP and HEDP. The phosphate oxygen atoms are shown unprotonated. The tridentate binding potential of the hydroxylated molecules is emphasized in heavy print.

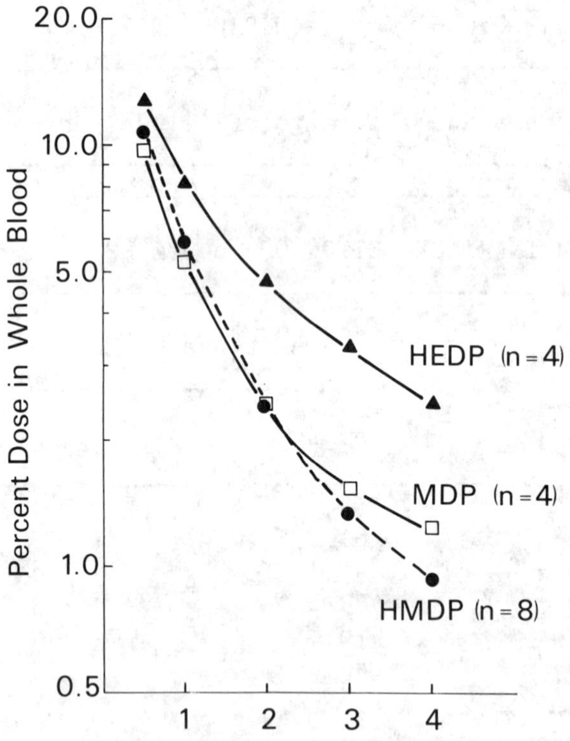

FIG. 2. Blood clearance of Tc-99m HMDP in dogs compared with Tc-99m MDP and Tc-99m HEDP assuming blood volume was 8.3% of body weight.

Tc-99m phosphorus agents for skeletal imaging, Subramanian (1) used Sr-85 as an internal standard. However, Sr-85 does not concentrate in AMI tissue. Following the observation of Buja, et al. (15) that labeled HEDP is highly correlated with Tc-99m PPi in AMI, we decided to administer concomitant doses of C-14 labeled HEDP with the test agents. This standard has the dual advantage of normalizing for both bone and infarcted myocardium.

In a series of individual studies, adult purebred beagle dogs were placed under general anesthesia and the left anterior descending coronary artery (LAD) permanently ligated. Forty-eight hours post occlusion, each animal received single intravenous doses of 10-15 mCi of the Tc-99m complex of one of test compounds, followed in each instance, 15 minutes later by 35-40 μCi of C-14 HEDP (1 mg HEDP) via the same route. Ninety minutes after Tc-99m dosing, scintiphotos were obtained with a gamma camera, a blood sample taken, and the animals sacrificed with lethal doses of pentobarbitol. Multiple tissue samples were taken and radioassayed for Tc-99m and C-14 content.

The superiority of the Tc-99m HMDP agent for bone uptake and skeletal contrast (with soft tissue) is evident in the Tc-99m results and is made particularly clear in the Tc-99m/C-14 concentration ratios (Table 3). Higher bone uptake is strongly indicative of a higher extraction rate onto bone for Tc-99m HMDP compared to the other agents. It should be noted that independent studies with normal (non-infarcted) beagle dogs show no significant difference in normal osseous and non-osseous tissue distribution compared to the infarcted dogs. The concentration ratios in muscle appear to be at odds with the blood clearance data (Fig. 2) but this should be treated with caution. It is logical to normalize for tissues in which substantial uptake is to be expected but, in muscle, we are calculating the ratio of two small numbers which is meaningless, from the scintigraphic viewpoint. (Note the large standard errors in the Tc-99m/C-14 ratios for muscle and blood).

The myocardial uptake for Tc-99m HMDP and Tc-99m PPi are compared in Table 4. The Tc-99m/C-14 concentration ratios are not significantly different from each other and suggest that Tc-99m HMDP is as effective as Tc-99m PPi in AMI scintigraphy.

CONCLUSIONS

Tc-99m hydroxymethylenediphosphonate (Tc-99m HMDP) is a new skeletal imaging agent which appears to have intrinsic superiority to the Tc-99m complexes of MDP and HEDP for skeletal imaging, and equivalence to Tc-99m PPi for the imaging of acute myocardial infarct. These superior properties are due to a more rapid uptake on bone mineral surface (hydroxyapatite and amorphous calcium phosphate) which, we believe, is related to structure. In this new agent, the enhanced scintigraphic potential, compared to other agents, is clearly due to a higher affinity for the biological phases of calcium phosphate, rather than to the renal clearance rate.

TABLE 3. ABSOLUTE AND NORMALIZED (WITH C-14 HEDP) BIODISTRIBUTION OF Tc-99m DIPHOSPHONATE AGENTS IN BEAGLE DOGS

Tissue	HMDP (n=8)	MDP (n=3)	HEDP (n=4)
	Percent dose per gram tissue X 10^2 (mean ± s.e.)		
Rib	5.37 ± 0.44	4.32 ± 0.86	2.27 ± 0.26*
Femur	1.76 ± 0.13	1.30 ± 0.44	0.42 ± 0.02**
Muscle	0.07 ± 0.01	0.10 ± 0.02	0.07 ± 0.01
Blood	0.49 ± 0.06	0.74 ± 0.11	0.56 ± 0.14
	Tc-99m/C-14 HEDP Concentration Ratios		
Rib	1.00 ± 0.03	0.69 ± 0.03*	0.78 ± 0.10*
Femur	1.09 ± 0.08	0.64 ± 0.05*	0.56 ± 0.05**
Muscle	1.83 ± 0.33	1.52 ± 0.51	1.30 ± 0.10
Blood	1.96 ± 0.33	2.17 ± 0.81	2.92 ± 0.54

Significant difference compared to Tc-99m HMDP as follows: * $p<0.01$
** $p<0.001$

REFERENCES

1. Subramanian G, McAfee JG, Blair RJ, et al.: Technetium-99m-methylene diphosphonate - a superior agent for skeletal imaging: comparison with other technetium complexes. *J Nucl Med* 16: 744-755, 1975

2. Davis MA, Jones AG: Comparison of ^{99m}Tc-labeled phosphate and phosphonate agents for skeletal imaging. *Sem Nucl Med* 6: 19-31, 1976

3. Deutsch E, Elder RC, Lange BA, et al: Structural characterization of a bridged ^{99m}Tc-Sn-dimethylglyoxime complex: Implications for the chemistry of ^{99m}Tc-radiopharmaceuticals prepared by the Sn(II) reduction of pertechnetate. *Proc Natl Acac Sci USA* 73: 4287-4289, 1976

4. Libson K, Barnett BL, Deutsch E: Structural Characterization of a diphosphonate complex of technetium synthesized via a substitution procedure. *Proc Natl Acac Sci USA* (to be submitted)

5. Barnett BL, Strickland LC: The crsytal structure of disodium dihydrogen-1-hydroxyethylidene diphosphonate tetrahydrate: a bone growth regulator. *Acta Cryst* B35, 1979 (in press)

6. Barnett BL, Strickland LC: Structure of disodium dihydrogen hydroxymethylenediphosphonate hemi ydrate. *Acta Cryst* (to be submitted)

7. Uchtman VA: Structural investigations of calcium binding molecules. II. The crystal and molecular structures of calcium dihydrogen ethane-1-hydroxy-1,1-diphosphonate dihydrate, $CaC(CH_3)(OH)(PO_3H)_2 \cdot 2H_2O$; Implications for polynuclear complex formation. *J Phys Chem* 76: 1304-1310, 1972

8. Fogelman I, Citrin DL, McKillop JH, et al.: A clinical comparison of Tc-99m HEDP and Tc-99m MDP in the detection of bone metastases: concise communication. *J Nucl Med* 20: 98-101, 1979

9. Arnold JS: Kinetic analysis of bone. *Proc First Internat Symp on Radiopharm*, Innsbruck, Austria, in press

10. Quimby OT, Prentice JB, Nicholson DA: *J Org Chem* 32: 4111-4114, 1967

11. Tofe AJ, Bevan JA, Fawzi MB, et al: Antioxidant stabilization of bone agents. *Second Internat Symp on Radiopharmaceuticals*, Seattle, WA, 1979

12. Rosenthall L, Arzoumanian A, Lisbona R, et al.: A longitudinal comparison of the kinetics of ^{99m}Tc-MDP and ^{99m}Tc-HEDP in humans. *Clin Nucl Med* 2: 2332-234, 1977

13. Zweiman FG, Holman BL, O'Keefe A, et al: Selective uptake of ^{99m}Tc complexes and ^{67}Ga in acutely infarcted myocardium. *J Nucl Med* 16: 975-979, 1975

14. Kulkarni PV, Parkey RW, Buja LM, et al: Technetium-labeled heparin: Preliminary report of a new radiopharmaceutical with potential for imaging damaged coronary arteries and myocardium. *J Nucl Med* 19: 810-815, 1978

15. Buja LM, Tofe AJ, Parkey RW, et al: Sites and mechanisms of localization of technetium-99m phosphorus radiopharmaceuticals in acute myocardial infarcts and other tissues. *J Clin Invest* 60: 724-740, 1977

TABLE 4. ABSOLUTE AND NORMALIZED (WITH C-14 HEDP) MYOCARDIAL DISTRIBUTION OF Tc-99m HMDP AND Tc-99m PPi IN INFARCTED BEAGLE DOGS

Tissue	Percent dose per gram tissue X 10^2 (mean ± s.e.) Tc-99m HMDP (n=8)	Tc-99m PPi (n=5)
Myocardial infarct	5.05 ± 1.13	4.44 ± 1.39
Normal myocardium	0.15 ± 0.02	0.31 ± 0.09
Blood	0.49 ± 0.06	0.78 ± 0.06
	Tc-99m/C-14 HEDP Concentration Ratios	
Myocardial infarct	1.18 ± 0.07	1.27 ± 0.12
Normal myocardium	1.68 ± 0.32	1.62 ± 0.22
Blood	1.96 ± 0.33	2.07 ± 0.10

RADIOPHARMACEUTICALS FOR THYROID IMAGING: A REVIEW

Hiroshi Nishiyama, M.D.

Bureau of Radiological Health, FDA
Nuclear Medicine Laboratory/
Radioisotope Laboratory
University of Cincinnati Medical Center
Cincinnati, Ohio 45267

ABSTRACT

A review of radiopharmaceuticals which have been used for thyroid imaging was made with special emphasis on palpable thyroid nodules. An attempt was made to evaluate cold nodules derived from imaging methods using radioiodine or Tc-99m pertechnetate, followed by a successive application of another radiopharmaceutical. An attempt was also made to understand the patho-physiology of various thyroid disorders. The latter was based on the accumulated cases with discordant images between radioiodine and Tc-99m pertechnetate, and also on the iodine content within the gland by means of fluorescent techniques. Better radiopharmaceuticals are anxiously awaited in order to realize the distinction between benign and malignant thyroid disorders at the preoperative decision-making stage.

INTRODUCTION

One goal of thyroid imaging can be defined as an absolute differentiation between benign and malignant disorders. Various radiopharmaceuticals have been investigated for this purpose, in addition to fluorescent techniques. The goal appears to still be beyond our reach; however, a better understanding of the patho-physiology of the thyroid gland has been defined. The purpose of this review is to evaluate the accumulated experiences appearing in the literature, and, hopefully, to elucidate a logical and judicious use of radiopharmaceuticals for thyroid study. The search for new radiopharmaceuticals certainly continues.

DATA COLLECTION AND ANALYSES

Cases with thyroid disorders were collected from the literature and attention was focused on the solitary thyroid nodule, especially the so-called cold nodule, defined by either radioiodine or Tc-99m-pertechnetate (Tc-99m) imaging. The literature search is by no means complete, and certain omissions are expected. The patient population was not uniform, and all were highly selected groups in various stages of evaluation. Therefore, the analyses based on the compiled results are applicable, in a strict sense, only to those selected populations, and caution must be exercised when the analyzed data are applied to the population as a whole.

A good test should indicate with high probability whether the patient does or does not have a disease (posterior probability)(1). For this assessment, the tabulated data under each radiopharmaceutical were analyzed by means of sensitivity, specificity, predictive value of positive test [PV(+)] and predictive value of negative test [PV(-)]. PV(+) and PV(-) are given as posterior probability of having, and posterior probability of not having, malignant thyroid disease, respectively. These terms were calculated using the following defined subgroups: 1) disease present with positive test (TP), 2) disease present with negative test (FN), 3) disease absent with positive test (FP), and 4) disease absent with negative test (TN).

$$\begin{aligned} \text{Sensitivity} &= TP / (TP + FN) \\ \text{Specificity} &= TN / (TN + FP) \\ PV(+) &= TP / (TP + FP) \\ PV(-) &= TN / (TN + FN) \\ \text{Overall Accuracy} &= (TP + TN) / (TP + FN + TN + FP) \end{aligned}$$

RADIOPHARMACEUTICALS

Table 1 is a summary of the radiopharmaceuticals used for thyroid study (2-10). Limited trials using Tc-99m-bleomycin, Tc-99m-solcocitran, and Hg-197 were also made; however, these will not be elaborated on.

THYROID NODULES

Palpable thyroid nodules are a prime reason for patient referral to the nuclear medicine laboratory, and diagnosis and management of the so-called cold nodule is of utmost importance. In addition, patients with a past history of therapeutic radiation exposure to the neck have been considered by some to need thyroid imaging studies, regardless of the presence or absence of palpable nodules.

The results of a review of the literature on the importance of finding a solitary cold nodule are shown in Table 2 (11-25). Although the population and the imaging technique differed somewhat, an attempt to further subdivide palpable nodules based on imaging and pathological findings can be considered on the basis of statistical representations. Scintigraphic classification of cold, warm, and hot nodules was uniform among authors. The warm nodule was defined as equal radionuclide concentration in the nodule compared with the surrounding non-nodular thyroid tissue; the hypofunctioning nodule was classified as a cold nodule. A recent practice of the oblique view utilizing the versatility of the gamma camera will, to an extent, decrease the warm nodule incidence and increase the cold nodule incidence.

The overall probability of malignancy among patients with palpable solitary nodules is 13.6% [probability of disease present in a given population: P(D+)]. The probability of malignancy among patients with a cold nodule is 16.2% [PV(+)]; the probability of warm and hot nodules that turn out to be malignant is 6.2% and 3.7%, respectively. However, patients with malignancies in these two categories comprise only 1.2% of the population cited herein. Thus, 12.4% of the total population is found to have a malignancy and come from patients with nonfunctioning and hypofunctioning nodules.

Cs-131-CHLORIDE

The mechanism of cesium concentration in tumor tissue is not well understood; however, it appears similar to the uptake of potassium and rubidium, which are known to concentrate in certain tumors (26).

The collected cases, all of which were initially studied with conventional thyroid imaging methods as having cold nodules, are tabulated and shown in Table 3 (25,27-30). The employed activity ranged from 0.5 to 1.5 mCi i.v., and the image was usually obtained about 2-h post-dose. As noted in Table 7, a good sensitivity of 77% can be seen; however, the PV(-) is the most interesting one and was calculated to be 95%. That is, if a patient with a cold nodule is studied with Cs-131 and shows no appreciable uptake in the lesion, the probability of malignancy in this patient is down to 5%; a good agent if the test is negative. On the other hand, the PV(+) is only 22%, and the probability of benignancy with a positive Cs-131 uptake is still as high as 78%; a poor agent to use for the judgment of the posterior priority of having thyroid malignancy when the test is positive. It is concluded that Cs-131 will give additional information in conjunction with routine thyroid scintigraphy, especially when the test is negative, but any added advantage of this test in a given patient can be argued. It would be interesting to see the detailed analyses of a Cs-131 study in conjunction with its uptake and cellular types of the lesion in a large series as predicted (27).

Tl-201-CHLORIDE

Tl-201 is an alkali congener, and its biological behavior is similar to cesium, rubidium, and potassium. The mechanism of its concentration in the thyroid tissue is not known. Its thyroid uptake in terms of the total administered dose in the euthyroid individual is 0.2% in four patients studied (9), and a similar study indicated the thyroid uptake as being 0.14 ± 0.05% based on 40 euthyroid patients (31). A slightly higher thyroid uptake in rats was reported as 0.40 ± 0.17% (32). An interesting observation among the rats treated with T_3 for 3 d is that Tl-201 thyroid concentration is markedly decreased to

TABLE 1. RADIOPHARMACEUTICALS FOR THYROID IMAGING

Radio-pharmaceutical	Physical Half-Life	Decay Mode	Principal Energy and Abundance(%)	Thyroid Uptake(%)	Absorbed Dose(rads/mCi) Thyroid	Absorbed Dose(rads/mCi) Whole Body	Ref.
I-123	13 h	EC	159 keV (83)	5 15 25	2.4 7.5 13.0	0.025 0.027 0.029	(2)
I-125	60 d	EC	27.5 keV (x-ray)	5 15 25	140 450 790	0.11 0.29 0.49	(2)
Am-241 - I-127	(Stable)	--	28.6 keV (x-ray)	--	0.015-0.050*	--*	(3,4)
I-131	8 d	β^-	364 keV (82)	5 15 25	260 800 1300	0.24 0.47 0.71	(2)
Tc-99m	6 h	IT	140 keV (90)	--	0.13	0.011	(5)
Ga-67	78 h	EC	93 keV (38) 185 keV (24) 300 keV (16)	--	0.3 - 0.4	0.26	(6,7)
Se-75 seleno-methionine	120 d	EC	136 keV (56) 264 keV (57) 280 keV (25)	--	6.0	8.0	(8)
Tl-201	73 h	EC	71 keV (95) 167 keV (10)	--	1.03	0.21	(9)
Cs-131	9.7 d	EC	30 keV (x-ray)	--	--†	0.21	(10)

*These doses are in rads per study. No radionuclide is administered internally, so the dose is limited to the neck region. †Data not readily available.

TABLE 2. SOLITARY THYROID NODULES IN RELATION TO FUNCTIONAL STATES AND PATHOLOGY

Author (Year)	Cold Benign	Cold Malignant	Warm Benign	Warm Malignant	Hot Benign	Hot Malignant	# of Cases
Johnson (1955)	18	14	9	1	2	0	44
Perlmutter (1956)	76	23	0	0	39	0	138
Groesbeck (1959)	94	16	51	2	11	0	174
Attie (1960)	48	15	27	1	14	3	108
Meadows (1961)	51	16	14	1	20	1	103
Grant (1963)	71	18	24	2	2	2	119
Dische (1964)	45	4	9	0	4	1	63
Shimaoka (1964)	17	3	14	3	3	0	40
Robinson (1966)	50	23	41	3	29	2	148
Shukri (1967)	63	4	21	0	12	0	100
Psarras (1972)	529	78	42	3	27	0	679
Messaris (1974)	410	56	30	2	70	0	568
Alderson (1976)	62	13	4	0	4	0	83
Thomas (1976)	26	16	3	1	8	1	55
Koutras (1976)	47	11	0	0	12	0	70
Total	1607	310	289	19	257	10	2492

Probability of Malignancy Present in Solitary Nodules or P(D+): (339/2492) = 13.6%
Probability of Malignancy Present in Cold Nodules or PV(+) : (310/1917) = 16.2%
Probability of Malignancy Present in Warm & Hot Nodules or 1.00-PV(-): (29/575) = 5.0%

0.16 ± 0.3%. It is suggested that the thyroid uptake of Tl-201 may reflect the cationic side of the iodide pump (iodide trap) or possible thyroidal blood flow (32). Regardless of the functional state of the thyroid gland, a markedly increased Tl-201 uptake (1.07 - 2.11%) was noted in four patients with goiter: two with Hashimoto's thyroiditis (31). These available data suggest that its uptake mechanism is difficult to explain by either trapping or thyroid blood flow alone. Instruments used varied, but the activity employed and the initiation of imaging were the same as those used in the cardiac study which employed Tl-201. Although the total number is small, the reported cases studied with Tl-201 are tabulated and shown in Table 4 (29,31,33,34). Because of only one false negative case out of a total of 24 cases with thyroid malignancy, the sensitivity is 96%. The PV(-) is also 96%, but both specificity and the PV(+) are not encouraging (see Table 7). If the trend remains similar to that shown in Table 4, Tl-201 will fall into a category similar to Cs-131; a good agent to predict a benign thyroid lesion when no Tl-201 accumulation is seen in the already known cold lesion.

Se-75-SELENOMETHIONINE

As in other organs or tumors where a high rate of protein synthesis is expected, methionine, whose sulfur was replaced by Se-75, is a logical choice for the study of cold nodules. However, the majority of thyroid tumors of both benign and malignant nature are slow growing; thus, the difference in the metabolic state of the tumor tissue as compared to the normal surrounding tissue may not be easy to realize. An exception to the rule would be anaplastic carcinoma of the thyroid.

Cases with cold nodules studied by radioiodines and imaged with Se-75-selenomethionine are shown in Table 5 (24,35-37). Sensitivity is 64% and specificity is 83%. The PV(-) is less specific when Se-75-selenomethionine is used as compared with Cs-131 or Tl-201, but the PV(+) is better specified. (See Table 7.)

The activity administered by various investigators varied from 75 μCi to 250 μCi, and imaging was initiated within 40-m post-dose. The relatively low activity administered, the low target-to-nontarget ratio, which ranged from 1.3 to 2.2 in the thyroid cancer (37), and also probably the high background activity, preclude obtaining good quality images. Because of the long residence time of the Se-75-selenomethionine in malignant tumors (biological half-life of approximately 24 d)(35), the image quality could be improved if the scintigraphy is obtained a day or two later, as in the case of radioiodine, in addition to using higher initial activity. However, neither was carried out by any investigator cited herein, and a higher activity employment is difficult to justify since the absorbed dose to the whole body is already quite high within the administered activities employed (see Table 1).

Ga-67-CITRATE

The mechanism of Ga-67 concentration is not known; however protein binding seems to play a role in its transport and tissue localization. The former is binding to transferrin, and the latter, to a variety of organelles in cytoplasm (38). The activity employed for the thyroid study ranged from 1.5 to 2.5 mCi, and imaging was usually obtained 48- to 78-h post-dose. The majority of cases reported were studied using the rectilinear scanner, and no series cited herein was imaged using a camera with dual or triple window settings.

TABLE 3. Cs-131 SCINTIGRAPHY IN NON-FUNCTIONING NODULES

Pathology	Cs-131 Uptake: Positive	Cs-131 Uptake: Negative	# of Cases
Benign:			
Follicular Adenoma	28	33	61
Hürthle Cell Tumor	1	0	1
Colloid Nodule	1	2	3
Multinodular Gland	0	2	2
Subacute Thyroiditis	1	0	1
Unclassified	131	231	362
(Subtotal)	(162)	(268)	(430)
Malignant:			
Follicular Ca	2	2	4
Papillary Ca	1	0	1
Papillary-Follicular Ca	1	0	1
Anaplastic Ca	2	0	2
Metastatic Ca	1	0	1
Unclassified	39	12	51
(Subtotal)	(46)	(14)	(60)
Total	208	282	490

TABLE 4. Tl-201 SCINTIGRAPHY IN NON-FUNCTIONING THYROID NODULES

Pathology	Tl-201 Uptake: Positive	Tl-201 Uptake: Negative	# of Cases
Benign:			
Adenoma*	11	13	24
"Thyroiditis"	0	2	2
Subacute Thyroiditis	0	2	2
Chronic Thyroiditis	5	0	5
Unclassified	0	8	8
(Subtotal)	(16)	(25)	(41)
Malignant:			
Papillary Ca	15	1	16
Follicular Ca	3	0	3
Papillary-Follicular Ca	2	0	2
Anaplastic Ca	3	0	3
(Subtotal)	(23)	(1)	(24)
Total	39	26	65

*20 adenoma, 2 adenomatous goiter, 1 follicular adenoma and 1 nodular goiter.

Table 6 shows cases with thyroid disorders imaged with Ga-67 (25,39-47). Sensitivity is 44% and specificity is 93%, which is somewhat similar to Se-75-selenomethionine. Similarly, posterior probability of either having, or not having, thyroid malignancy is almost identical to Se-75-selenomethionine, but Ga-67 is slightly better than Se-75-selenomethionine in the PV(+) (see Table 7). The role of Ga-67 imaging in thyroid disorders may contribute to the preoperative identification of malignant thyroid lesions, but this contribution in a given patient is a questionable one.

Aside from the diagnostic interest, four cases of subacute thyroiditis showed a markedly increased Ga-67 uptake, and the decreased degree of Ga-67 uptake was noted along with the recovery phase of thyroiditis (41). An evaluation of the target-to-nontarget ratio of Ga-67 concentration in the neck demonstrated a high ratio in cases of Hashimoto's thyroiditis, and anaplastic carcinoma (39). Similarly, a high Ga-67 concentration in anaplastic carcinoma is noted during the imaging procedure (25,42), but this was not a consistant finding (47). Tissue distribution study, however, showed the highest concentration in nodules of Hashimoto's thyroiditis and anaplastic carcinoma (48).

FURTHER ANALYTICAL CONSIDERATION OF COLD NODULES

Palpable thyroid nodules were evaluated in 2492 patients using either radioiodine or Tc-99m imaging, and the results are shown in Table 2. Among them, 1917 had cold nodules, and this group has the highest risk of having thyroid malignancy. However, the PV(+) is merely 16.2%, and if a patient is sent for surgical intervention at this stage, the probability of having a benign lesion is high. This is the prime reason that many investigators have sought additional test results for those patients who have shown cold nodules.

If a given population of 2492 is studied with radioiodine (or Tc-99m), and if those 1917 patients with cold nodules are subsequently studied with a second radiopharmaceutical whose sensitivity and specificity are already known, the predictive values should reflect the results of two successive diagnostic tests in a given population. The diagram below shows hypothetical results if Cs-131 were employed as the second radiopharmaceutical in those 1917 patients with cold nodules; the subsequent numbers in each category are based on the sensitivity and specificity of Cs-131 derived from the data shown in Table 3.

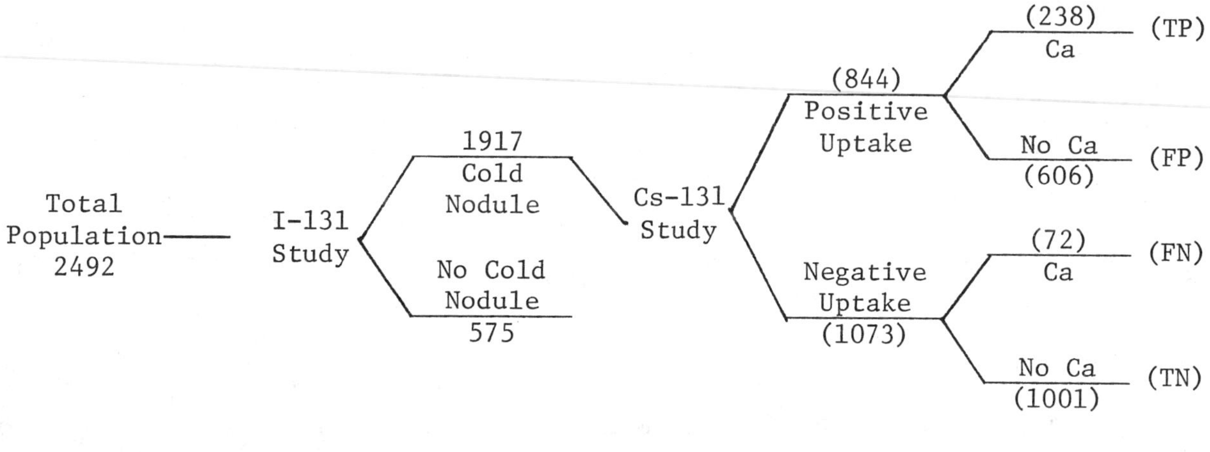

TABLE 5. Se-75 SELENOMETHIONINE SCINTIGRAPHY IN NON-FUNCTIONING THYROID NODULES

Pathology	Se-75 Uptake		# of Cases
	Positive	Negative	
Benign:			
Follicular Adenoma	3	27	30
Papillary Adenoma	0	1	1
Hürthle Cell Adenoma	1	0	1
Other Adenoma*	3	15	18
Cyst	2	7	9
Nodular Goiter	3	18	21
Thyroiditis†	3	4	7
Unclassified	0	2	2
(Subtotal)	(15)	(74)	(89)
Malignant:			
Follicular Ca	5	4	9
Papillary Ca	5	2	7
Papillary-Follicular Ca	3	3	6
Anaplastic Ca	6	2	8
Medullary Ca	1	1	2
Unclassified	9	8	17
Other Primary Thyroid Malignancy‡	7	0	7
Metastatic Ca	5	3	8
(Subtotal)	(41)	(23)	(64)
Total	56	97	153

*12 adenomatous goiter, 1 colloid adenoma and 5 "adenoma".

†1 Hashimoto's, 1 Riedel's, 1 chronic granulomatous thyroiditis and 4 "thyroiditis".

‡2 lymphoma, 4 sarcoma and 1 squamous cell carcinoma.

TABLE 6. Ga-67 SCINTIGRAPHY IN NODULAR THYROID DISORDERS*

Pathology	Ga-67 Uptake		# of Cases
	Positive	Negative	
Benign:			
Adenoma†	0	22	22
Autonomous/Toxic Adenoma	0	3	3
Nodular Goiter	1	22	23
Cyst	0	3	3
Thyroiditis‡	6	0	5
Unclassified	1	62	63
(Subtotal)	(8)	(112)	(120)
Thyroid Dysplasia	0	2	2
Malignant:			
Follicular Ca	4	3	7
Papillary Ca	1	7	8
Papillary-Follicular Ca	1	2	3
Anaplastic Ca	7	3	10
Medullary Ca	0	4	4
Adenocarcinoma	1	0	1
Leiomyosarcoma	1	0	1
Unclassified	10	15§	25
Metastatic Ca	2	0	2
(Subtotal)	(27)	(34)	(61)
Total	35	148	183

*Functional aspects of the nodules by means of radioiodine or Tc-99m were not specified in one-third of the cases. Tabulation of non-functioning nodules alone, however, shows almost identical figures as those shown in this table.

†Includes 6 colloid nodules.

‡2 "thyroiditis" and 4 subacute thyroiditis.

§Includes 4 cases classified follicular and/or papillary carcinoma.

Thus, the PV(+) has changed from 16.2% when using radioiodine alone to 28.2% when using a combination of two tests. However, the number of patients sent to surgery has decreased from 1917 to 844. The number of missed malignancies was 101 (29 from the radioiodine study and 72 from the cesium study) and the number of patients without malignancy who underwent surgery decreased from 1607 to 606. Similar calculations are made using Tl-201, Se-75, and Ga-67, and these are shown in Table 7. Provided that our assumption that both sensitivity and specificity of each second radiopharmaceutical are kept constant is correct, none of the radiopharmaceuticals employed for the second study change the predictive values drastically. However, the combined two tests could only be of value, depending upon which posterior probability is more important, in preoperative decision making.

PHYSIOLOGICAL CONSIDERATION OF RADIOIODINE AND Tc-99m

Although the majority of tissues do not concentrate iodide ions above the level found in the plasma or surrounding media, several non-thyroidal tissues do exhibit iodide-concentrating ability. These include gastric mucosa, salivary glands, mammary glands, choroid plexus, sweat glands and hair. However, none of these tissues exhibit iodide ion concentrating ability as great as the thyroid gland (49). Thyroid metabolism of iodine is roughly divided into iodide ion trapping, organic binding of iodine, storage of thyroid hormone and release of the hormone. Almost one-fifth of the iodide in the blood that perfuses the thyroid is extracted (trapping) in a single passage through the gland, and the iodide trapping follows saturation kinetics (50). After trapping iodide, a rapid succession of thyroid hormone synthesis occurs. Organic radioactivity is already demonstrated 5 s after intravenous administration of I-125 in the experimental animals (51), and secretion of radioiodine-labeled thyroxine begins within 2 h after intravenous injection in humans (50). Because of continuous trapping, organification, storage and release functions, the saturation kinetics of the iodide continue to change with an overall increase in radioiodine in the gland until the rate of hormone release exceeds the other functions, at which time the peak thyroid uptake is seen.

The tetrahedral anions such as ClO_4^- (perchlorate) and TcO_4^- (pertechnetate) are also concentrated by the thyroid gland. Perchlorate is a potent inhibitor of iodide trapping, while pertechnetate is less potent as an inhibitor (49,52,53). The saturation kinetics of pertechnetate differ from those of iodide because no significant organification and storage occur (54), and saturation occurs within a short time after injection.

Thus, any thyroid tissues with trapping function can potentially be visualized by either radioiodine or Tc-99m imaging during the trapping phase. Iodide ions, however, can visualize the thyroid tissue at any mixed functional phases of trapping, organification and storage. Follicular adenoma and follicular carcinoma can, at times, be visualized during primarily the trapping phase, but those lesions are usually seen as cold nodules at a later stage (55,56) because little to no organifying function remains. This is presumably due to nonorganified iodine being released from the lesion rather rapidly. The autonomous adenoma, if succession of thyroid functions is extremely rapid, can be seen as cold nodules as early as 4 h after oral radioiodine administration (57). Therefore, an advantage of radioiodine for thyroid imaging is an ability to visualize the patho-physiological sequences as opposed to Tc-99m, whose sole known function is confined to the trapping phase.

DISCORDANT SCINTIGRAPHIC FINDINGS BETWEEN Tc-99m AND RADIOIODINE

The incidence of discordant cases varies widely from series to series. In this country, its incidence ranges from 0% to 22% (58,59). European counterparts reported higher incidences ranging from 15.7% to as high as 31% (55,60). These differences are presumably due to the selecting procedures of patients, the imaging procedures, and criteria set forth for evaluating images, rather than different population groups as a whole. Dige-Petersen's study (60) is unique in that 500 μCi of I-123 was given intravenously and pinhole images were obtained 1-h post-dose rather than at a later time as has been customarily done in radioiodine imaging.

Reported cases with discordant scintigraphic findings are tabulated in Table 8 (56-58,60-67). All 48 cases had histological confirmation. A notable finding here is 14 cases with follicular carcinoma, all of which showed positive Tc-99m uptake, but non-positive radioiodine uptake. Two cases listed under papillary carcinoma showed follicular components, and one case under unclassified showed follicles within the tumor though these may not necessarily be functioning. These comprise 17 cases out of 19 malignant cases with various degrees of follicular components. Similarly, follicular adenoma under benign disease is also suggestive of the function related with the follicle.

Autonomous adenoma is another interesting entity (57,60). All five cases could have shown positive if the image had been obtained earlier. Four cases showed hot nodules with I-123 after T_3 suppressions and at least one case investigated showed a hot nodule at 1- to 20-m post-injection of I-123 (57). These cases are probably not only related with follicular component as discussed earlier, but also related to the high functional state; thus, markedly accelerated iodine turnover within the autonomous nodule had presumably taken place.

FLUORESCENT IMAGING

The amount of iodine content within the thyroid gland has been determined by means of fluorescent techniques. The clinical role of this technique in the evaluation of solitary nodules has yet to be determined in terms of resolution (68), and further understanding of functional states of nodules in terms of iodine content is in order. However, available data have clarified certain issues directly related with the conventional imaging findings in functioning and nonfunctioning thyroid nodules despite the fact that there are certain overlaps of iodine content between nodules of various pathological states and the surrounding normal tissue (4,67-71).

Nonfunctioning nodules defined by either radioiodine or Tc-99m imaging have, most of the time, shown low iodine content; the lowest being found in cold nodules of malignant nature (4,69); the higher iodine content in either cystic or solid benign cold nodules (4). The considerable overlap compared with the normal tissue was observed in the latter groups (68,69). The work published by Patton (4) is the most interesting one to date; the ratios of iodine content in malignant cold nodules, benign cystic nodules, and benign cold nodules to that in the corresponding normal area are reported to be 0.41 ± 0.13 (n=15), 0.91 ± 0.36 (n=10), and 0.78 ± 0.29 (n=17), respectively. No such demarcation between benign and malignant solitary nodules was shown by others. Autonomous thyroid nodules defined by radioiodine or Tc-99m

TABLE 7. DIAGNOSTIC VALUES OF VARIOUS RADIOPHARMACEUTICALS*

Analytical Method	Radiopharmaceutical				
	I-131 (Tc-99m)	Cs-131	Tl-201	Se-75	Ga-67
P(D+)†	13.6%	12.2%	36.9%	41.8%	33.3%
Sensitivity	91.4%	76.7%	95.8%	64.1%	44.3%
Specificity	25.4%	62.3%	61.0%	83.1%	93.4%
PV(+)‡	16.2%	22.1% (28.2)	59.0% (32.1)	73.2% (41.9)	77.1% (56.4)
PV(-)‡	95.0%	95.0% (93.3)	96.2% (98.7)	76.2% (92.3)	77.0% (89.7)
Accuracy‡	34.3%	64.0% (64.6)	73.8% (66.6)	75.2% (80.0)	76.0% (85.5)

*Values calculated are individual sets of data corresponding to Tables 2-6.

†Probability of disease present in a given population (prior probability).

‡Numbers in parentheses are derived from the hypothetical setting as discussed in the text when both sensitivity and specificity are set constant for each radiopharmaceutical. For the combined test analyses, the prior probability is now set to 16.2%.

TABLE 8. FORTY-EIGHT REPORTED CASES WITH DISCORDANT SCINTIGRAPHIC FINDINGS BETWEEN RADIOIODINE AND Tc-99m

Pathology	Tc-99m Uptake (+) Iodine Uptake (-)	Tc-99m Uptake (-) Iodine Uptake (+)	# of Cases
Benign:			
Follicular Adenoma	6	1	7
Goiter*	6	4	10
Thyroiditis†	4	0	4
Autonomous Adenoma‡	5	0	5
Other Adenoma§	0	3	3
Malignant:			
Follicular Ca	14	0	14
Papillary Ca¶	2	0	2
Unclassified⌈	3	0	3
Total	40	8	48

*3 colloid goiter, 7 nodular and/or cystic goiter.

†1 Riedel's, 2 de Quervain's and 1 subacute thyroiditis.

‡1 case with IV injection of I-123 and imaged 1 h later; 4 cases with p.o. administration of radioiodine and imaged 4-24 h later.

§trabecular adenoma, Hürthle cell adenoma and adenomatous dysplasia.

¶2 cases showed follicular component.

⌈1 case showed follicles in compactly arranged tumor cells.

imaging, in addition to T_3 suppression test, showed slightly higher iodine content within the nodules (70), but some cases showed no more iodine content than the surrounding normal tissue (69). This finding was supported by tissue assay with double-isotope labeling before operation of the autonomous nodule (72). Dissociation of high Tc-99m trapping function and relatively low iodine content within the autonomous nodule is well documented (70), suggesting a caution of image interpretation. Different degrees of iodine content in nodules of various pathological states are not clearly understood in terms of functions, but, at least, a low iodine content could be seen in cases of extremely high functional states and conversely low functional states. The former includes autonomous nodule as in certain cases of hyperthyroidism (68, 71) and the latter includes benign and malignant nodules.

Further study of fluorescent imagings in conjunction with the functional state of the nodule will enhance our knowledge and give us a clue for better utilization of conventional imaging techniques. For that matter, the mechanism of Tc-99m release from the thyroid gland after its trapping in various functional states is hardly known at present.

SUMMARY AND CONCLUSION

A review of the literature for thyroid imaging was done with special emphasis on nonfunctioning thyroid nodules as defined by conventional imaging methods. No radiopharmaceutical, as yet, is available to differentiate cold nodules of benign and malignant nature; however, extensive investigation using Cs-131 has been carried out, as well as the use of Ga-67, Se-75-selenomethionine and, more recently, Tl-201 among those patients who have shown a variety of thyroid disorders. It appears that none of these latter radiopharmaceuticals show a clear cut affinity to a cell type regardless of benign or malignant disorders. The question remains as to whether the aforementioned second radiopharmaceuticals can help establish a more accurate diagnosis. Although populations cited herein were all highly selected groups, the compiled data were analyzed with an attempt to answer this question. The posterior probabilities change somewhat as shown in Table 7, and the result suggests that certain merits can be seen depending upon whether the PV(+) or PV(-) is looked for in a given patient.

However, a question remains— What can be done at the current state of art? Our knowledge of iodine moiety within the thyroid gland has increased based on the accumulated cases of disparate images between radioiodine and Tc-99m. Adding to this is the knowledge of iodine content in the gland as determined by fluorescent techniques. A single best procedure appears to be the administration of radioiodine intravenously, preferably I-123 in high activity, and following sequential patho-physiological changes by means of imaging technique. However, functional similarities between benign and malignant diseases are certainly impossible to differentiate when viewed as images. Technetium pertechnetate is the single best agent in terms of economy and absorbed radiation dose, but its obvious disadvantage as an imaging agent is its limited functional relationship to the complex thyroidal functions of normal and abnormal states. It makes sense to employ both radioiodine and Tc-99m in a questionable case in order to minimize time economy required from the patient and technician until better radiopharmaceuticals are available. However, the employment sequence of these two radiopharmaceuticals cannot be easily realized from the available data.

ACKNOWLEDGEMENT

I wish to thank Alan B. Ashare, M.D. for his analyses of the data presented here. Also, I would like to thank Paul A. Feller and Kenneth L. Scholz for their help in preparing Table 1.

REFERENCES

1. FEINSTEIN AR: On the sensitivity, specificity, and discrimination of diagnostic tests. Clin Pharm Therap 17: 104-116, 1975
2. MIRD/Dose Estimate Report No. 5: Summary of current radiation dose estimates to humans from ^{123}I, ^{124}I, ^{125}I, ^{126}I, ^{130}I, ^{131}I and ^{132}I as sodium iodide. J Nucl Med 16: 857-860, 1975
3. HOFFER PB: Clinical results in fluorescent thyroid scanning. In Semiconductor Detection in the Future of Nuclear Medicine, Hoffer PB, Beck RN and Gottschalk A, eds., New York, Society of Nuclear Medicine, 1971, pp 230-239
4. PATTON JA et al: Differentiation between malignant and benign solitary thyroid nodules by fluorescent scanning. J Nucl Med 17: 17-21, 1976
5. MIRD/Dose Estimate Report No. 8: Summary of current radiation dose estimates to normal humans from ^{99m}Tc as sodium pertechnetate. J Nucl Med 17: 74-77, 1976
6. MIRD/Dose Estimate Report No. 2: Summary of current radiation dose estimates to humans from ^{66}Ga-, ^{67}Ga-, ^{68}Ga-, and ^{72}Ga-citrate. J Nucl Med 14: 755-756, 1973
7. FELLER PA: Unpublished data
8. MIRD/Dose Estimate Report No. 1: Summary of current radiation dose estimates to humans from ^{75}Se-L-selenomethionine. J Nucl Med 14: 49-50, 1973
9. ATKINS HL et al: Thallium-201 for medical use. Part 3: Human distribution and physical imaging properties. J Nucl Med 18: 133-140, 1977
10. LEBOWITZ E et al: Thallium-201 for medical use. I. J Nucl Med 16: 151-155, 1975
11. JOHNSON PC et al: Reliability of scintiscanning nodular goiters in judging the presence or absence of carcinoma. J Clin End 15: 865, 1955
12. PERLMUTTER M et al: Which nodular goiters should be removed. N Eng J Med 255: 65-71, 1956
13. GROESBECK HP: Evaluation of routine scintiscanning of nontoxic thyroid nodules. 1. The preoperative diagnosis of thyroid carcinoma. Cancer 12: 1-5, 1959
14. ATTIE JN: The use of radioactive iodine in the evaluation of thyroid nodules. Surg 47: 611-614, 1960
15. MEADOWS PM: Scintillation scanning in the management of the clinically single thyroid nodule. JAMA 177: 229-234, 1976
16. GRANT HTH et al: The enigmatic thyroid nodule. Ann Surg 158: 205-210, 1963
17. DISCHE S: The radioisotope scan applied to the detection of carcinoma in thyroid swelling. Cancer 17: 473-479, 1964
18. SHIMAOKA K et al: Differentiation of benign and malignant thyroid nodules by scintiscan. Arch Int Med 114: 36-39, 1964
19. ROBINSON E et al: Incidence of cancer in thyroid nodules. Surg Gyn Obst 123: 1024-1026, 1966
20. SHUKRI AM: The solitary thyroid nodule in Iraq. Brit J Clin Pract 21: 75-80, 1967
21. PSARRAS A: The single thyroid nodule. Brit J Surg 59: 545-548, 1972
22. MESSARIS G et al: The single thyroid nodule and carcinoma. Brit J Surg 61: 943-944, 1974
23. ALDERSON PO et al: The single palpable thyroid nodule. Cancer 37: 258-265, 1976

24. THOMAS CG et al: Evaluation of dominant thyroid masses. Ann Surg 183: 463-469, 1976

25. KOUTRAS DA et al: Thyroid scanning with gallium-67 and cesium-131. J Nucl Med 17: 268-271, 1976

26. CHARKS ND et al: Tumor scanning with radioactive Cs^{131}. In Recent Advances in Nuclear Medicine, Croll MN, Brady LW, eds., New York, Appleton-Century-Crofts, 1966, pp 235-245

27. BURKE J et al: Thyroid scanning with Cesium-131. J Nucl Med 18: 97, 1977

28. CROCKER EF et al: Comparison of cesium scintiscanning with grey-scale echography in investigation of solitary, non-functioning thyroid nodules. In Ultrasonics in Medicine, Kazner E et al., eds., Amsterdam, Excerpta Medica, 1975, pp 207-212

29. PALERMO F et al: ^{201}Tl for the differential diagnosis of cold thyroid nodules. La Riceroa Clin Lab 7: 289-295, 1977

30. MURRAY IPC et al: Thyroid scanning with ^{131}Cs. Brit Med J 2: 653-656, 1970

31. FUKUCHI M et al: Marked thyroid uptake of thallium-201 in patients with goiter: Case Report. J Nucl Med 18: 1199-1201, 1977

32. OSTER Z et al: Thallium-201 distribution in the thyroid: Relationship to thyroidal trapping function. Radiol 126: 733-734, 1978

33. TONAMI N et al: Clinical application of ^{201}Tl scintigraphy in patients with cold thyroid nodules. Clin Nucl Med 3: 217-221, 1978

34. FUKUCHI M et al: Thallium-201 imaging in thyroid carcinoma - Appearance of a lymphnode metastasis. J Nucl Med 19: 195-196, 1978

35. WEINSTEIN MB et al: ^{75}Se selenomethionine as a tumor scanning agent for the differential diagnosis of the cold thyroid nodule. Sem Nucl Med 1: 390-396, 1971

36. SPENCER R et al: The value of ^{75}Se-selenomethionine scanning in solitary nodules of the thyroid gland. Brit J Radiol 47: 457-463, 1974

37. THOMAS CG et al: Differentiation of malignant from benign lesions of the thyroid gland using complementary scanning with ^{75}Selenomethionine and radioiodine. Ann Surg 170: 396-408, 1969

38. LARSON SM: Mechanisms of localization of gallium-67 in tumors. Sem Nucl Med 8: 193-203, 1978

39. ERJAVEC M et al: Computer-assisted scanning in evaluation ^{67}Ga-citrate uptake in thyroid disease. J Nucl Med 15: 810-813, 1974

40. EDWARDS CL et al: Scanning malignant neoplasms with gallium-67. JAMA 212: 1182-1190, 1970

41. GROVE RB et al: Uptake of ^{67}Ga-citrate in subacute thyroiditis. J Nucl Med 14: 403, 1973

42. LANGHAMMER H et al: ^{67}Ga for tumor scanning. J Nucl Med 13: 25-30, 1972

43. VAN DER SCHOOT JB: Gallium-67, its tumor affinity and tumor specificity. Radiol Clin Biol 41: 371-377, 1972

44. KAPLAN WD et al: ^{67}Ga-citrate and the nonfunctioning thyroid nodule. J Nucl Med 15: 424, 1974

45. HIGASHI T et al: Clinical evaluation of ^{67}Ga-citrate scanning. J Nucl Med 13: 196-201, 1972

46. ROOS J et al: The uptake of gallium-67 in euthyroid patients with multinodular goiter. Acta Med Scan 194: 225-228, 1973

47. HEIDENDAL GAK et al: Evaluation of cold areas on the thyroid scan with ^{67}Ga-citrate. J Nucl Med 16: 793-794, 1975

48. SOIN JS: Quantitative estimation of ^{67}Ga citrate in hypofunctioning thyroid nodules. Radiol 122: 243-245, 1977

49. WOLF J: Transport of iodide and other anions in the thyroid gland. Physiol Rev 44: 45-90, 1964

50. BEIERWALTES WH: Physiology of the thyroid gland. In Principles of Nuclear Medicine, Wagner HN, ed., Philadelphia, N.B. Saunders Comp., 1968, pp 308-313

51. PITT-RIVERS R: Some biological reactions of iodine. In Further Advances in Thyroid Research, Vol. 1, Fellinger K, Hoffer R, eds., Wein, Verlarg der Wiener Mediziniochen Akademie, 1971, pp 15-22

52. DEGROOT LJ: Current views on formation of thyroid hormones. N Eng J Med 272: 243-250, 1965

53. HAYS MT et al: Simultaneous measurement of thyroidal trapping ($^{99m}TcO_4^-$) and binding ($^{131}I^-$): Clinical and experimental studies in man. J Nucl Med 14: 785-791, 1973

54. HAYS MT: Kinetics of the human thyroid trap: A compartmental model. J Nucl Med 19: 789-795, 1978

55. ERJAVEC M et al: Comparative accumulation of ^{99m}Tc and ^{131}I in thyroid nodules: Case Report. J Nucl Med 18: 346-347, 1977

56. O'CONNOR MK et al: A kinetic study of (^{131}I) iodine and (^{99m}Tc) pertechnetate in thyroid carcinoma to explain a scan discrepancy: Case Report. J Nucl Med 18: 796-798, 1977

57. MILLER JM et al: Disparate imaging of the autonomous functioning thyroid nodule with ^{99m}Tc pertechnetate and radioiodine. J Nucl Med 17: 532-533, 1976

58. TURNER JW et al: Thyroid carcinoma presenting as a pertechnetate "hot" nodule, but without ^{131}I uptake: Case Report. J Nucl Med 17: 22-23, 1976

59. ARNOLD JE et al: Comparison of ^{99m}Tc and ^{123}I for thyroid imaging. J Nucl Med 17: 261-267, 1976

60. DIGE-PETERSEN et al: A comparison of ^{99m}Tc and ^{123}I scintigraphy in nodular thyroid disorders. Eur J Nucl Med 3: 1-4, 1978

61. GOROWSKI T et al: Some comparative studies using ^{99m}Tc and ^{131}I in thyroid scanning. Nuklear Medizin 15: 268-272, 1978

62. STEINBERG M et al: Uptake of technetium-99-pertechnetate in a primary thyroid carcinoma: Need for caution in evaluating nodules. J Clin Endocr 31: 81-84, 1970

63. BARTH JD et al: Discrepancy between iodine and technetium thyroid scintigraphy. JAMA 240: 463-464, 1978

64. USHER MS et al: Thyroid nodule scans made with pertechnetate and iodine may give inconsistent results. J Nucl Med 12: 136-137, 1971

65. SHAMBAUGH GE III et al: Disparate thyroid imaging. Combined studies with sodium pertechnetate Tc-99m and radioactive iodine. JAMA 228: 866-869, 1974

66. HIRABAYASHI S et al: Inconsistent images of thyroid nodule scintigrams made with iodide and pertechnetate: Case Report. J Nucl Med 16: 918, 1975

67. THRALL JH et al: Discordant imaging of a thyroid nodule with ^{131}I and ^{99m}Tc: Concordance of ^{131}I and fluorescent scans. Radiol 128: 705-706, 1978

68. WAHNER HW et al: Fluorescent thyroid scanning. A method based on stable iodine measurements. Mayo Clin Proc 53: 151-156, 1978

69. LEBLANC AD et al: Measurement of ^{127}I concentration in thyroid tissue by x-ray fluorescence. J Nucl Med 14: 816-819, 1973

70. THRALL JH et al: Solitary autonomous thyroid nodules: Comparison of fluorescent and pertechnetate imaging. J Nucl Med 18: 1064-1068, 1977

71. HOFFER PB et al: Fluorescent technique in thyroid imaging. Sem Nucl Med 1: 379-389, 1971

72. MEDEIROS-NETO GA et al: Autonomous thyroid nodules. II. Double labeling with iodide isotopes and study of biochemical constituents of nodular and paranodular tissue. J Nucl Med 13: 738-743, 1972

RADIOPHARMACEUTICALS FOR PANCREAS, PROSTATE AND ADRENAL IMAGING

Monte Blau and Hank F. Kung

Department of Nuclear Medicine
State University of New York at Buffalo
Buffalo, New York 14215

ABSTRACT

A large number of compounds have been investigated as possible agents for pancreas, prostate and adrenal imaging. These have been based on synthetic activity of the organs, organ control hormones, or drugs known to affect organ function. This effort has been only moderately successful. Clinically useful radiopharmaceuticals have been developed for pancreas (Se^{75}-L-selenomethionine) and for adrenals (I-131-6-iodomethylnorcholesterol) based on the synthesis of the digestive enzymes and the adrenal cortical steroid hormones, respectively. Neither of these meets the need for a low radiation dose, high photon flux, readily available agent. The record for the prostate is even more dismal. No clinically useful imaging compounds have been developed to date.

INTRODUCTION

The development of radiopharmaceuticals for pancreas, prostate and adrenal imaging has been the subject of active research for the past quarter century. In spite of this effort, much of it fairly sophisticated, it has not been possible to develop generally applicable, high photon flux, low radiation dose agents for imaging any of these organs. Prostate imaging in humans has been essentially unsuccessful and the clinically useful agents for pancreas and adrenal leave much to be desired.

This paper is a brief overview of the development of various localizing agents based on metabolic, hormonal or pharmacologic activity. Excellent detailed reviews of the pancreas (1) and adrenal (2) radiopharmaceutical literature are available.

PANCREAS

Se-75-L-selenomethionine.

The rapid synthesis of the digestive enzymes following emptying of the pancreas at each meal results in a high uptake of amino acids. While these enzymes do not represent a large portion of the total protein in the body at one time, the short turnover time compared to the blood proteins and tissue proteins accounts for the high uptake in pancreas compared to other tissues. Several successful agents have been based on this mechanism.

The failure of externally labeled amino acids (I-131-iodotyrosine and N-iodoacetyl tryptophan) to localize in the pancreas led Blau and Manske (3) to prepare and test Se-75-L-selenomethionine. This agent showed the high pancreas uptake characteristic of natural amino acids (6% in rats at 1 hour) and also had high tissue specificity (concentration ratios; pancreas:liver = 8, pancreas:blood = 31, pancreas:spleen = 29). Subsequent studies demonstrated that selenomethionine displayed many biochemical properties of natural amino acids. It is actively transported across the gut wall (4), synthesized into blood proteins (5), and into the digestive enzymes (6).

In the almost two decades since the introduction of selenomethionine there have been many clinical studies of the efficacy of this agent. The results reported have varied from unbelievably good to totally negative. In part, this is due to the unusual variability in the size, location and function of the normal pancreas, but the principal problem is uptake in liver. Although the concentration of selenomethionine is higher in pancreas than in liver, the large size of the liver compared to the pancreas and its location near or over the pancreas leads to substantial confusion on routine scans. The general conclusion is that while pancreas scanning can be a useful procedure, Se-75-selenomethionine is far from an ideal agent. This conclusion is strengthened by the relatively high radiation dose (1.6 rad to total body for the routine clinical dose) from this agent because of the combination of the long physical (120 days) and biological (220 days) half lives.

In order to improve the pancreas uptake and retention of selenomethionine various patient protocols have been introduced. Emptying the pancreas shortly before the study with special meals (7) or pancreas stimulating hormones (8) can increase uptake. The release of digestive enzymes labeled with Se-75 during the study can be slowed down with morphine (9) or probanthene (8). However, the usefulness of these various procedures has never been established with controlled studies in patients.

The deficiencies of selenomethionine as a pancreas scanning agent have resulted in the continuing search for better radiopharmaceuticals.

Other amino acids.

1. C-11 and N-13 natural amino acids. Since the uptake of amino acids in the pancreas is quite prompt (peak reached in 15-30 minutes in normals), it is advantageous to use nuclides with short half-lives. Natural amino acids labeled with C-11 (20 min) or N-13 (10 min) have been prepared and evaluated as pancreas scanning agents. The C-11 amino acids include DL-valine (10), phenylglycine (11), phenylalanine (11) and DL-tryptophan (12) which have been used as pancreas agents and many which have been synthesized but not tested. Nitrogen 13-L-alanine has been used for pancreas imaging (13) as has N-13-L-glutamic acid (14, 15). Many other N-13 amino acids have been prepared but are as yet unevaluated for pancreas specificity.

When enzymatic or biosynthetic methods are used for preparation of labeled amino acids the biologically active L- form is obtained. Chemical synthesis yielding the racemic mixture produces agents with diminished pancreas uptake (16). The full potential of the C-11 and N-13 amino acids has not been realized because of the limited availability of these short-lived nuclides.

2. Halogen labeled amino acids. The F for H substitution, particularly on ring structures, often produces compounds which retain biological activity (17). F-18 labeled tyrosines (18), phenylalanines (18) and tryptophan (18, 19) have been proposed as pancreas localizing agents. Of these only F-18-6-fluoro-tryptophan has shown useful levels of uptake. Limitations on the F-18 amino acids include rapid dehalogenation *in vivo* (20) and large specie to specie differences in pancreas uptake (19). Fluorine labeled m fluoro-L-phenylalanine and F-18-5 fluoro-L-tryptophan are superior to selenomethionine in rats (21).

Another attractive isosteric substitution is Br or I for methyl. Several aliphatic amino acids have been prepared (22) but the resulting compounds either decomposed rapidly or had no pancreas specificity.

Externally halogenated amino acids have uniformly low pancreas specificity with the possible exception of iodo phenylalanine which has some specificity in mice but not in larger animals (23).

3. Other seleno-amino acids. In the search for biologically active amino acids which have high pancreas uptake but a more rapid clearance from the body, a series of Se-75 labeled amino acids have been synthesized and evaluated. These include L-selenocystine (24), DL-meso selenocystine (24), 2-aminoethyl L-selenocystine (25, 26), methyl-L-selenocysteine (25, 26), L-selenoethionine (25, 26) and DL-selenotryptophan (25). None of these compounds proved to be more useful than selenomethionine.

Selenomethionine has been prepared with the 7.2 hour positron emitting isotope Se-73 by both chemical (27) and biological (28) synthesis. Because of its limited availability, this agent has not had widespread clinical trials.

4. Tellurium-123m amino acids. The 120-day half-life Te-123m has been used for the preparation of telluromethionine (29). No biological data have been published for this compound. The lower energy photons (.159MeV) emitted by this nuclide are desirable for gamma camera imaging. However, the equally long half-life of this nuclide and the total internal conversion of the 85 keV gamma photons make it even more undesirable than Se-75 from a radiation dose standpoint.

Pancreas control hormones.

The pancreas stimulating hormone secretin is released from the gut wall when acid chyme enters the upper small intestine. Secretin is a polypeptide containing 27 amino acids (MW $\sim$ 3400). Attempts at direct iodination were unsuccessful but a radioactive derivative prepared by coupling I-131 p-iodobenzoate groups did show high pancreas uptake (8). Unfortunately, secretin is quite labile and the various preparations did not show uniform ability to localize in pancreas. It is possible that secretin labeled with other groups would be a useful imaging agent.

Drugs with pancreas activity.

1. Sulfonylureas. Several of the hypoglycemic sulfonylureas have been labeled and studied. The iodo analog of chlorpropamide (30) and several Tc-99m labeled derivatives of tolbutamide (31) all failed to localize sufficiently for pancreas imaging. This is not surprising since the site of pharmacologic action is the β-cells which represent only a few percent of the total pancreas mass.

2. Drugs with pancreas toxicity. Alloxan (32) and streptozotocin (33) are diabetogenic drugs which cause extensive β-cell damage. The C-14 labeled drugs show no pancreas specificity, however. Diphenylhydantoin is a hyperglycemic agent but neither the C-11 compound (34) nor iodo derivatives (35) showed significant pancreas uptake.

Miscelleneous unsuccessful agents.

1. Zinc. The relatively high zinc content of pancreatic tissue has stimulated many groups to investigate ionic zinc and many zinc chelates as pancreas localizing agents. In spite of negative results over the past 20 years (36-40), there seems to be a need to repeatedly confirm that the specificity of zinc for the pancreas is inadequate for imaging studies. This fact is now well established and the radiopharmaceutical community could profitably turn its attention to other approaches.

2. Fluorescent compounds. Following reports (41) that the fluorescent alkaloid berberine localized in pancreas, I-131 labeled derivatives were prepared and studied (42). These proved to have no pancreas specificity and it was demonstrated that the apparent uptake of berberine was an illusion created by the difference in quenching of fluorescence in all abdominal organs except the pancreas (which is relatively transparant to ultraviolet light). A similar effect is probably the basis for the often observed (but rarely measured) uptake of fluorescent compounds in tumors. Toluidine Blue has been investigated as a pancreas specific agent (43, 44) because of visual estimations of its uptake in the pancreas. Isotope studies show no specificity.

3. Anti-tissue antibodies. Early success with high tissue specificity of anti-kidney antibodies (45) led to attempts to prepare anti-pancreas antibodies for pancreas visualization. In spite of high *in vitro* specificity for pancreas tissue, these labeled antibodies cross-reacted with liver and kidney antigens *in vivo* and were useless for pancreas imaging (8).

4. Thioglucose. Following the observation that Pb-203 thioglucose (46) has a high pancreas uptake upon intraperitoneal injection, Tc-99m thioglucose was prepared and studied (47). This agent has no pancreas specificity when given intravenously but has some on intraperitoneal injection. One can't help but wonder whether agents administered intraperitoneally are *on* the pancreas rather than *in* it.

PROSTATE

Radiopharmaceuticals have been proposed for prostate imaging based on organ synthetic activity, hormonal control and drugs. None have been developed to the point of routine clinical usefulness.

Putrescine.

Although their function is not well established (48), it is known that human semen is rich in polyamines. Spermine, for example, is present in concentrations of the order of 10 mM (49). Putrescine [$H_2N-(CH_2)_4-NH_2$] has been shown to be a precursor of spermine [$H_2N-(CH_2)_3-NH-(CH_2)_4-NH-(CH_2)_3-NH_2$] (50).

Clark and Fair (51) were able to demonstrate moderate uptake of H-3 labeled putrescine in the rat prostate. Welsh, et al (52) prepared C-11 labeled methyl putrescine [$NH_2-(CH_2)_4-NH^{11}CH_3$] and demonstrated prostate uptake comparable to the H-3 labeled compound. They were able to visualize the dog prostate using a positron emission tomographic scanner. The limited availability of this compound has delayed its general clinical application.

Prostate control hormones.

The prostate gland is responsive to both androgens and estrogens with specific binding sites in both the cytoplasmic and nuclear fractions (53-56). A variety of steroids have been investigated with relatively little success. These have included various H-3 (57-60), C-14 (60), I-131 (61), I-123 (62), Br-77 (63)and F-18 (64, 65) labeled androgens and estrogens.

While there are significant variations between compounds, the prostate uptake rarely exceeds a few tenths of a percent of the injected dose. Where measurements have been made on human tissues removed at surgery even 0.1% uptake is considered high (63). In spite of this, many "images" of animal and even human prostates have been published. On close examination (62) these prove to be due to urine in the bladder or radioactivity in the bowel.

The failure of these agents to be taken up in significant quantity in the prostate is in part due to the very limited number of binding sites available. There have been many measurements of the binding sites in prostatic tissue (53-56) and substantial controversy remains (56). However, for androgens (dihydrotestosterone) the total number of sites of all kinds (nuclear and cytoplasmic; bound and free) is roughly 0.5 nanomoles in the total human prostate.

If we assume that all of these sites can be occupied by the labeled hormone (a very unrealistic assumption) and that 1% of an injected dose of 1 mCi of radiopharmaceutical localizes in the prostate, it would require a specific activity of 20 Ci per millimole not to saturate the sites. Many, if not all, of the studies with gamma emitting derivatives have been done at much lower specific activities. Sturman, et al (60) have pointed out this effect when measuring the uptake of low specific activity C-14 steroids compared to the high specific activity H-3 labeled compounds. The number of binding sites available for estrogens may be much higher than for androgens (56).

Prostate active drugs.

1. 'Estracyt'. Estradiol-3-N-bis-2-chlorethyl-carbamate-17-phosphate (estramustine or Estracyt) is a drug which has been used to treat prostatic cancer. The I-131 labeled compound has been claimed to localize in prostate (66) but this remains to be confirmed in detailed studies.

2. 'Flutamide'. 3'-trifluoromethyl-4'nitro-isobutyranalide (Flutamide) is a non-steroid antiandrogen. An I-131 derivative (3'iodo-4'nitro-isobutyranilide) has been prepared and the distribution in rats and dogs has been measured (59). No significant prostate uptake was observed.

Zinc.

As with the pancreas, the high content of zinc in the prostate and in semen has led to many attempts to use this element for prostatic imaging (40, 67-73). The chemical forms studied include the chloride, gluconate, citrate and a variety of amino acid complexes. There are some changes in distribution with chemical form but prostate uptake is between 0.2 and 0.3% of the dose in the total organ (72, 74). Since excretion of non-prostate localized zinc from the body is very slow, it is unlikely that such low uptakes will yield satisfactory scans. Zinc-69m (70) (14 hr, 440 keV) and Zn-62 (40) (9.2 hr positron emitter) have been suggested as useful isotopes, but since the prostatic uptake of zinc is so low it is unlikely that these relatively short-lived nuclides will find much clinical application.

One of the most important limiting factors in the localization of all prostatic imaging agents is the low blood flow through this organ. Using Xe-133 washout in baboons, Schoonees, et al (71) have measured the flow as 0.2 ml per minute per gram. This corresponds to less than 0.1% of the cardiac output passing thru the prostate. Even if an agent were completely removed from the blood as it passed thru the prostate, it would concentrate very slowly.

ADRENALS

The adrenal glands are each composed of two physiologically and metabolically distinct parts. The adrenal cortex secretes a group of hormones, the corticosteroids, synthesized from cholesterol. The adrenal medulla, on the other hand, is functionally related to the sympathetic nervous system and secretes the catecholamine hormones epinepherine and norepinepherine. Cortical and medullary localizing agents will be discussed separately.

Adrenal cortex agents based on steroid synthesis.

1. Iodocholesterol. Following early attempts with labeled sterols (75), the group at the University of Michigan led by Beierwaltes and Counsell prepared the first clinically useful adrenal agent, I-131 19-iodocholesterol (76-78). In spite of the ubiquitous presence of cholesterol in body tissues, the adrenal cortex is more or less alone in its high uptake and long term retention (79, 80). It is likely that the iodo derivative, which is taken up by adrenal cortex but cannot be used for the synthesis of the adrenal steroids, has even better retention than cholesterol itself.

The uptake of 19-iodocholesterol in the adrenals is quite low; less than 0.4% dose in the total organ. However, the long term retention permits the clearance from other tissues and the attainment of very high adrenal concentration ratios (adrenal:kidney = 300, adrenal:liver = 170) (81). With 19-iodocholesterol the clearance from other organs takes about a week, precluding the use of short-lived isotopes of iodine (or other elements) as labels.

Kojima (82) discovered that 19-iodocholesterol (19-iodocholest-5(6)-en-3β-ol) spontaneously undergoes homoallyic rearrangement to 6-iodomethylnorcholesterol (6β-iodomethyl-19-norcholest-5(10)-en-3β-ol). The 6-iodomethyl compound proved to have significantly higher uptake than the original 19-iodocholesterol (83-85). Of even more importance is the more rapid clearance of background radioactivity in other organs permitting scanning at 2-4 days.

It has been suggested that most or all of the adrenal uptake observed with 19-iodocholesterol is the result of the 6-iodomethyl impurity (86). This conclusion has been contested, however (87, 88). In any case, it is clear that the current agent of choice for adrenal imaging is 6-iodomethylnorcholesterol.

2. Selenium and tellurium cholesterols. Selenium-75 labeled 19-methylselenocholesterol (89) and 6-selenocholesterol (90) have been prepared and evaluated. The 19-seleno compound has higher adrenal uptake than the iodo analog, but the slow clearance from blood and liver is a disadvantage (89). No data are available for the 6-seleno derivative.

The advantages of Se-75 cholesterol over I-131 cholesterol include longer shelf life, higher stability and better gamma ray energies (Se-75; 260 and 280 keV, I-131; 364 keV) for imaging. The very low particulate radiation from Se-75 compared to I-131 should result in a lower radiation dose to a small organ such as the adrenal. This advantage is partly lost because of the long physical half-life (120 days) and the long biological retention in the adrenal. A further radiation dose advantage of Se-75 labeled compounds is the absence of thyroid dose from I-131 released from the iodocholesterol.

One peculiar finding with selenocholesterol (89) is the equal uptake in adrenal medulla and cortex. To some extent the same is true for 6-iodomethylcholesterol (85). It is possible that synthesis of steroid hormones is not the only mechanism of uptake in adrenals.

A tellurium-123m labeled cholesterol (24-nor-23-(isopropyltellura)-5α-cholan-3β-ol) has been shown to localize sufficiently to image rat adrenals (91). Although the gamma ray energy (160 keV) of this nuclide is more favorable for imaging with Anger cameras, it is not clear that this advantage can overcome the high cost. The physical half-life is as long as Se-75 and the higher particulate radiation would deliver a larger radiation dose to the adrenal.

Adrenal cortex agents based on drugs.

Non-steroid drugs have been developed which provide specific inhibition of many of the enzymes involved in steroid hormone synthesis in the adrenal cortex. Beierwaltes, et al have prepared H-3 and I-125 derivatives of several of these and found high adrenal uptake and good concentration ratios (92). Based on these preliminary studies, metyrapol and iodo SKF-12185 (β-(4-amino-3-iodo-phenyl)-phenylethylamine) were investigated in humans (93). Both showed uptake adequate for routine scanning. Since the uptake of these agents is quite fast, it is possible that I-123 labeling will be practical. Although the concentration ratios to other organs with these agents is not as good as iodocholesterol, the possibility of labeling with short-lived nuclides makes the enzyme inhibitors attractive as possible adrenal imaging agents.

Stereospecificity is important in the activity of these enzyme inhibitors. When the (-)-iodo SKF-12185 was tested it showed a significant increase in adrenal uptake and specificity over the ± racemic mixture (94).

Adrenal medulla agents based on catecholamine synthesis.

The adrenal medullary hormones and their precursors have been investigated as possible imaging agents (95). Dopamine proved to have the highest uptake (96). Attempts to prepare gamma emitting derivatives which retained activity were largely unsuccessful (96). However, 6-iodo dopamine has been prepared and shows some medullary uptake (97).

Carbon-11 dopamine and norepinepherine have been prepared in the carrier-free state (98). As expected, the adrenal uptake was high.

Adrenergic blocking agents.

Derivatives of the quaternary amine bretylium, an antiadrenergic drug, have been proposed as medullary imaging agents. Iodine-125 labeled p-iodo-benzyltrimethyl ammonium has a high medullary concentration and very low cortical uptake (99). In a study of a series of bretylium derivatives significantly higher uptake was found for o-iodobenzyl dimethyl ethyl ammonium or the corresponding ethylene hydroxy derivative (100). These agents have not been tested in humans.

REFERENCES

1. Risch VR: Radiopharmaceuticals in pancreatic imaging. In The Chemistry of Radiopharmaceuticals ed by Heindel ND, Burns HD, Honda T, et al. Masson Publishing Inc., New York, 1978, pp53-73

2. Beierwaltes WH, Wieland DM, Yu T et al: Adrenal imaging agents: Rationale, synthesis, formulation and metabolism. Seminars in Nucl Med 8:5-18, 1978

3. Blau M, Manske RF: The pancreas specificity of Se^{75}-selenomethionine. J Nucl Med 2:102-105, 1961

4. Spencer RP, Blau M: Intestinal transport of selenium-75 selenomethionine. Science 136:155-156, 1962

5. Awwad HK, Potchen EJ, Adelstein SJ, et al: Se-75 selenomethionine incorporation into human plasma proteins and erythrocytes. Metabolism 15: 626-640, 1966

6. Hansson E, Blau M: Incorporation of Se^{75}-selenomethionine into pancreatic juice protein in vivo. Biochem Biophys Res Comm 13:71-74, 1963

7. Blau M, Bender MA: ^{75}Se-selenomethionine for visualization of the pancreas by isotope scanning. Radiology 78:974, 1962

8. Blau M: Pancreas scanning with Se-75 selenomethionine. In Medical Radioisotope Scanning Vol 2, IEAE, Vienna, 1964, pp275-287

9. Rodriguez-Antunez A: Pancreatic scanning with selenium-75 methionine, utilizing morphine to enhance contrast. Cleve Clin Q 31:213-218, 1964

10. Washburn LC, Wieland BW, Sun TT, et al: [1-^{11}C] DL-Valine, a potential pancreas-imaging agent. J Nucl Med 19:77-83, 1978

11. Vaalburg W, Beerling-van der Molen HD, Woldring MG: Evaluation of carbon-11 labelled phenylglycine and phenylalanine for pancreas scintigraphy. Nucl Med (Stutt) 14:60-66, 1975

12. Hubner KF, Hayes RL, Washburn LC, et al: Scanning of the human pancreas with DL-valine-1-C-11 and DL-tryptophan-1-C-11. J Nucl Med 19:686, 1978

13. Spolter L, Cohen MB, McDonald N, et al: Synthesis and evaluation of N-13-L-alanine for pancreatic imaging. J Nucl Med 15:535, 1974

14. Lembares N, Dinwoodie R, Gloria I, et al: A rapid enzymatic synthesis of 10 min ^{13}N-glutamate and its pancreatic localization. J Nucl Med 13:786, 1972

15. Lathrop KA, Dinwoodie R, Gloria I, et al: Comparison of ^{13}N-glutamate and ^{75}Se-selenomethionine as imaging agents for the pancreas. J Nucl Med 14: 421, 1973

16. Stadalnik RC, Matolo NM, Hein LJ: L or D/L tryptophan for pancreas scintigraphy? J Nucl Med 19:689, 1978

17. Robinson GD: Prospects for ^{18}F radiopharmaceuticals. In Radiopharmaceuticals ed by Subramanian G, Rhodes BA, Cooper JF, et al, The Society of Nuclear Medicine, New York, 1975, pp141-148

18. Taylor DM, Cottrall MF: Evaluation of amino acids labelled with ^{18}F for pancreas scanning. In Radiopharmaceuticals and Labelled Compounds Vol 1, Vienna, IAEA, 1973, pp433-441

19. Atkins HL, Christman DR, Fowler JS, et al: Organic radiopharmaceuticals labeled with isotopes of short half-life. V ^{18}F-labeled 5- and 6-fluorotryptophan. J Nucl Med 13:713-719, 1972

20. Cottrall MF, Taylor MD, McElwain TJ: Investigations of ^{18}F-p-fluorophenylalanine for pancreas scanning. Brit J Radiol 46:277-288, 1973

21. Eakins MN, Somaia S: ^{18}F-labelled L-amino acids as pancreatic scanning agents. In Second International Symposium on Radiopharmaceutical Chemistry, 1978, pp148

22. Kung HF, Gilani S, Blau M: Pancreas specificity of iodo and bromo aliphatic amino acid analogs. J Nucl Med 19:393-396, 1978

23. DiGuillo W, Counsell RE, Skinner RWS: Radioiodinated phenylalanine analogs for parathyroid and pancreas scanning. J Nucl Med 10:398-399, 1969

24. Melmed RN, Agnew JE, Bouchier IAD: Studies on the metabolism of ^{75}Se-L-selenocysteine and ^{75}Se-D-L-meso-selenocysteine with particular reference to their use as pancreatic scanning agents. J Nucl Med 10:580-582, 1969

25. Agnew JE, Maze M: Clinical trial of four pancreatic scanning agents. Brit J Radiol 51:206-209, 1978

26. Tothill P, Heading RC: The fate of four ^{75}Se-labeled amino acids: studies of some unsuccessful pancreas-scanning agents. J Nucl Med 16:933-937, 1975

27. Guillaume M, Lambrecht RM, Christiaens L, et al: Carrier-free selenium-73 cyclotron production. In Second International Symposium on Radiopharmaceutical Chemistry, 1978, pp126

28. Ogawa K, Taki K, Nozaki, et al: Synthesis and application of ^{73}Se-selenomethionine. In First International Symposium on Radiopharmaceutical Chemistry, 1976, pp48

29. Knapp FF, Ambrose KR, Callahan AP: Telluro amino acid-synthesis, characterization and properties of a new and potentially useful class of compounds. In Second International Symposium on Radiopharmaceutical Chemistry, 1978, pp157

30. Boyd CM, Holcomb GN, Counsell RE, et al: Studies on the tissue distribution of ^{125}I-iodopropamide in the dog. J Nucl Med 12:117-118, 1971

31. Heindel ND, Risch VR, Burns HD, et al: Synthesis and tissue distribution of ^{99m}Tc-sulfonylureas. J Pharm Sci 64:687-689, 1975

32. Bekdik FC, Farmelant MH, Tyson I: Studies of tissue alloxan uptake. J Nucl Med 9:31-34, 1968

33. Heindel ND, Burns HD, Risch VR, et al: I-131 streptozotocin analogs: Tissue and tumor distribution studies. J Nucl Med 16:535, 1975

34. Winstead MB, Parr S, Lin TH, et al: Synthesis and in vivo distribution of ^{11}C-hydantoins. J Nucl Med 13:479, 1972

35. Balachandran S, Beierwaltes WH, Ice RD, et al: Tissue distribution of ^{14}C-, ^{125}I- and ^{131}I-diphenylhydantoin in the toadfish, rat and human with insulinomas. J Nucl Med 16:775-779, 1975

36. Meschan I, Quinn JL, Witcofski RL, et al: The utilization of radioactive zinc and manganese in an effort to visualize the pancreas. Radiology 73:62-70, 1959

37. Wakeley JCN, Moffatt B, Crook A, et al: The distribution and radiation dosimetry of zinc-65 in the rat. Intl J Appl Rad Isot 7:225-232, 1960

38. Greenlaw RH, Strain WH, Callear TE, et al: Experimental studies for scintillation scanning of the pancreas. J Nucl Med 3:47-50, 1962

39. Cottrall MF, Taylor MD: Investigations relating to the use of zinc-62 for pancreas scanning. Radioaktive Isotope in Klinik und Forschung IX:412-420, 1970

40. Yano Y, Budinger TF: Cyclotron-produced Zn-62: Its possible use in prostate and pancreas scanning as a Zn-62 amino acid chelate. J Nucl Med 18:815-821, 1977

41. Nardi GL, Seipel JH: The selective localization of a alkaloid in pancreatic tissue. Surg Forum 6:381-385, 1955

42. Blau M, Bender MA: Does berberine localize in the pancreas? Gastroenterology 38:217-218, 1960

43. Larose JH, Whitaker RH, Reba RC: Radioiodinated toluidine blue: An unsatisfactory scanning agent. J Nucl Med 11:731-732, 1970

44. Archer EG, Potchen EJ, Studer R, et al: Tissue distribution of ^{125}I-toluidine blue in the rat. J Nucl Med 13:85-91, 1972

45. Blau M, Day ED, Pressman D: Rate of localization of anti-rat kidney antibodies. J Immunol 79:330, 1957

46. Atkins HL, Lambrecht RM, Wolf AP, et al: Distribution of various labeled thioglucoses in rodents. Radiology 115:723-725, 1975

47. Risch VR, Heindel ND, Markoe AM, et al: Distribution of ^{99m}Tc-1-thioglucose in rats: Effect of administration route on pancreatic specificity. Radiology 124:837-838, 1977

48. Williams-Ashman HG, Lockwood DH: Role of polyamines in reproductive physiology and sex hormone action. Ann NY Acad Sci 171:882-894, 1970

49. Tabor H, Tabor CW: Spermidine, spermine and related amines. Pharmacol Rev 16:245-300, 1964

50. Russell DH, Medina VJ, Snyder SH: The dynamics of synthesis and degradation of polyamines in normal and regenerating rat liver and brain. J Biol Chem 245:6732-6738, 1970

51. Clark RB, Fair WR: The selective in vivo incorporation and metabolism of radioactive putrescine in the adult male rat. J Nucl Med 16:337-342, 1975

52. Welch MJ, Coleman RE, Straatmann MG, et al: Carbon-11-labeled methylated polyamine analogs: uptake in prostate and tumor in animal models. J Nucl Med 18:74-78, 1977

53. Ghanadian R, Auf G, Chaloner PJ, et al: The use of methyltrienolone in the measurement of the free and bound cytoplasmic receptors for DHT in benign hypertrophied human prostate. J Steroid Biochem 9:325-330, 1978

54. Shain SA, Boesel RW, Lamm DL, et al: Characterization of unoccupied (R) and occupied (RA) androgen binding components of the hyperplastic human prostate. Steroids 31:541-556, 1978

55. Shain SA, Boesel RW: Androgen receptor content of the normal and hyperplastic canine prostate. J Clin Invest 61:654-660, 1978

56. deVoogt HJ, Dingian P: Steroid receptors in human prostatic cancer, a preliminary evaluation. Urol Res 6:151-158, 1978

57. Bruchovsky N, Wilson JD: The conversion of testosterone to 5α-Androstan-17β-ol-3-one by rat prostate in vivo and in vitro. J Biol Chem 243:2012-2021, 1968

58. Hansson V, Tveter KJ, Attramadal A: Uptake and binding of 5α-dihydrotestosterone in rat ventral prostate in vivo. Acta Endocrinologica 67:384-392, 1971

59. Berger B, Coffey DS, Scott WW: Concepts and limitations in the application of radiolabeled antiandrogens, estrogens or androgens as isotopic scanning agents for the prostate. Invest Urol 13:10-16, 1975

60. Sturman MF, Beierwaltes WH, Prakash S, et al: Uptake of radiolabeled testosterone, 5-α-dihydrotestosterone, estradiol and pregnenolone by canine prostate. J Nucl Med 15:94-97, 1974

61. Shida K, Shimazaki J, Kurihara H, et al: Uptake and scintiscanning of the prostate with ^{125}I-labeled estradiol phosphate. Intl J Nucl Med Biol 3:86-89, 1976

62. Ghanadian R, Waters SL, Williams G, et al: The use of ^{123}I labelled oestradiol in detecting the human prostate using a gamma camera. Eur J Nucl Med 1:159-160, 1976

63. Ghanadian R, Waters SL, Chisholm GD: Investigations into the use of ^{77}Br labelled 5α-dihydrotestosterone for scanning the prostate. Eur J Nucl Med 2:155-157, 1977

64. Spitzangle LA, Marino CA, Rapacz P: Evaluation of the radiopharmaceutical potential of an analog of dihydrotestosterone, labeled with F-18. J Nucl Med 19:702, 1978

65. Potential organ or tumor imaging agents 16. Fluorinated androstanes and the prostate. J Med Chem 21:391-393, 1978

66. Szendroi Z, Kocsar L, Karika Z, et al: Isotope scanning of the prostate. Lancet 1:1252-1253, 1973

67. Mirand EA, Bender M: The selective uptake of zinc^{65} in normal and carcinomatous prostate glands of the rat. Anat Rec 125:618, 1956

68. Daniel O, Haddad F, Prout G, et al: Some observations on the distribution of radioactive zinc in prostatic and other human tissues. Brit J Urol 28:271-278, 1956

69. Johnston GS, Wade JC, Murphy GP, et al: ^{65}Zn and ^{69m}Zn studies in the dog, monkey and man. J Surg Res 8:528-534, 1968

70. Gold FM, Lorber SA: Radioisotope ^{69m}Zn chloride prostate gland scan. Invest Urol 8:231-238, 1970

71. Schoonees R, deKlerk JN, Mirand EA, et al: The effect of bovine growth hormone on zinc-65 metabolism and prostatic blood flow of intact, testosterone treated, and castrated adult male chacma baboons. Invest Urol 8:103-115, 1970

72. Chisholm GD, Short MD, Ghanadian R, et al: Radiozinc uptake and scintiscanning in prostatic disease. J Nucl Med 15:739-742, 1974

73. Fruhling J, Coune A: Radiozinc as a scintigraphic agent for the human prostate. J Nucl Med 16:495-497, 1975

74. Schoonees R, Bender MA, Schalch DS, et al: Correlation of androgenic stimulation with prostatic function in aging males. In Current Topics in Surgical Research Vol 3, ed by Skinner DB, Ebert PA, Academic Press, New York, 1971, pp233-239

75. Nagai T, Sotis BA, Koh CS: An approach to developing adrenal-gland scanning. J Nucl Med 9:576-581, 1968

76. Counsell RE, Ranade JJ, Blair RJ, et al: IX Radioiodinated cholesterol. Steroids 16:317-328, 1970

77. Beierwaltes WH, Lieberman LM, Ansari AN, et al: Visualization of human adrenal glands in vivo by scintillation scanning. J Am Med Ass 216:275-277, 1971

78. Lieberman LM, Beierwaltes WH, Conn JW, et al: Diagnosis of adrenal disease by visualization of human adrenal glands with I-131-19-iodocholesterol. N Eng J Med 285:1387-1393, 1971

79. Applegren LE: Sites of steroid hormone formation. Autoradiographic studies using labeled precursors. Acta Physiol Scand Suppl 301:1-108, 1967

80. Beierwaltes WH, Varma VM, Lieberman LM, et al: Percent uptake of labeled cholesterol in adrenal cortex. J Nucl Med 10:387, 1969

81. Blair RS, Beierwaltes WH, Lieberman LM, et al: Radiolabeled cholesterol as an adrenal scanning agent. J Nucl Med 12:176-182, 1971

82. Maeda M, Kojima M, Ogawa H, et al: Homoallylic rearrangement of 19-iodocholest-5-en-3β-ol: New adrenal scanning agent. Steroids 26:241-250, 1975

83. Kojima M, Maeda M, Ogawa H, et al: New adrenal scanning agent. J Nucl Med 16:666-668, 1975

84. Sarkar SD, Ice RD, Beierwaltes WH, et al: 6β-I-131-iodomethyl-19-norcholest-5(10)-en-3-ol (NP-59) concentrates in the adrenal gland better than I-131-19-iodocholesterol. J Nucl Med 16:565, 1975

85. Sarkar SD, Beierwaltes WH, Ice RD, et al: A new and superior adrenal scanning agent: NP-59. J Nucl Med 16:1038-1042, 1975

86. Couch MW, Williams CM: Comparison of 19-iodocholesterol and 6-iodomethylnorcholesterol as adrenal-scanning agent. J Nucl Med 18:724-727, 1977

87. Kraychy S, Leeper MA: Comparison of 19-iodocholesterol and 6-iodomethylnorcholesterol as adrenal scanning agents. J Nucl Med 19:333-334, 1978

88. Scott KN, Couch MW, Mareci TH, et al: Purity of the adrenal-scanning agents, 19-iodocholesterol and 6-iodomethylnorcholesterol. J Nucl Med 18:492-493, 1977

89. Sarkar SD, Ice RD, Beierwaltes WH, et al: Selenium-75-19-selenocholesterol - a new adrenal scanning agent with high concentration in the adrenal medulla. J Nucl Med 17:212-217, 1976

90. The Radiochemical Center, Amersham, England

91. Knapp FF, Ambrose KR: Tellurium-123m labeled 24-nor-23-(isopropyl-tellura)-5α-cholan-3β-ol: A new potential adrenal imaging agent. J Nucl Med 16:600, 1977

92. Beierwaltes WH, Wieland DM, Ice RD, et al: Localization of radio-labeled enzyme inhibitors in the adrenal gland. J Nucl Med 17:998-1002, 1976

93. Beierwaltes WH, Wieland DM, Mosley ST, et al: Imaging the adrenal glands with radiolabeled inhibitors of enzymes. J Nucl Med 19:200-203, 1978

94. Wu JL, Wieland DM, Beierwaltes WH, et al: Radiolabeled enzyme inhibitors - enhanced localization following enatiomeric purification. J Nucl Med 19:677, 1978

95. Morales JO, Beierwaltes WH, Counsell RE, et al: The concentration of radioactivity from labeled epinephrine and its precursors in the dog adrenal medulla. J Nucl Med 8:800-809, 1967

96. Ice RD, Wieland DM, Beierwaltes WH, et al: Concentration of dopamine analogs in the adrenal medulla. J Nucl Med 16:1147-1151, 1975

97. Fowler JS, MacGregor RR, Wolf AP: Radiopharmaceuticals. Halogenated dopamine analogs: Synthesis and radiolabeling of 6-iododopamine and tissue distribution studies in animals. J Med Chem 19:356-360, 1976

98. Fowler JS, Wolf AP, Christman DR, et al: Carrier-free ^{11}C-labeled catecholamines. In Radiopharmaceuticals ed by Subramanian G, Rhodes BA, Cooper JF, et al: The Society of Nuclear Medicine, New York, 1975, pp196-204

99. Korn N, Buswink A, Yu T, et al: A radioiodinated bretylium analog as a potential agent for scanning the adrenal medulla. J Nucl Med 18:87-89, 1977.

100. Wieland DM, Swanson DP, Brown LE, et al: Imaging the adrenal medulla with an I-131-labeled antiadrenergic agent. J Nucl Med 20:155-158, 1979

DESIGN OF RADIOPHARMACEUTICALS FOR ADRENAL IMAGING: EFFECT OF STRUCTURE AND METABOLISM ON THE DISTRIBUTION OF H-3 AND F-18 LABELED PREGNENOLONES

Larry A. Spitznagle, Robert R. Eng, Carol A. Marino

Department of Nuclear Medicine
University of Connecticut Health Center
Farmington, CT 06032

William F. Trager

Department of Pharmaceutical Sciences
University of Washington
Seattle, WA 98195

The kinetics of the tissue distribution of pregnenolone and pregnenolone-3-acetate labeled with ^{3}H and of 21-fluoropregnenolone and 21-fluoropregnenolone-3-acetate labeled with ^{3}H and with ^{18}F were studied in the rat. The maximum tissue concentration was 4.1±0.5 %D/g in the adrenal glands, obtained 5 minutes after the administration of 21-fluoropregnenolone labeled with ^{18}F. The presence of the 3-acetate group reduced the maximum adrenal concentration to 2.4 ±0.4 %D/g. The ratio of $^{18}F/^{3}H$ concentrations were high in the adrenal glands. The data indicate that the adrenal glands contain a high concentration of ^{18}F activity not associated with the steroid nucleus. Both ^{18}F-labeled pregnenolones are potentially useful radiopharmaceuticals for imaging the adrenal glands and/or measuring adrenal metabolism rates.

INTRODUCTION

The use of the radiopharmaceuticals, 19-iodocholesterol-^{131}I and 6β-iodomethylnorcholesterol-^{131}I for adrenal imaging has clearly established the clinical value of this nuclear medicine procedure (1,2). However, the same facts which make the cholesterol derivatives useful imaging agents also make them less than the "ideal" adrenal imaging agent. The adrenals cannot be imaged until several days after administration of the radiopharmaceutical (2). It is necessary to use a long-lived radionuclide to label the radiopharmaceuticals, producing a high radiation dose to the adrenal glands and other organs (3,4). As with other iodinated radiopharmaceuticals, these compounds are prone to deiodination and loss of the radioactive label both *in vitro* and *in vivo* (5).

This laboratory has been involved for some time in the design of radiopharmaceuticals, labeled with short-lived radionuclides, for adrenal imaging. Our approach has been to synthesize fluorinated analogs of steroids which are known to concentrate in or to interact with the adrenal glands in some specific manner. We have chosen to use fluorinated analogs for several reasons: (1) the substitution of a fluorine for a hydrogen has been found to enhance the biological activity of several steroids (6); (2) the strength of the carbon-fluorine bond, 108 kCal/mole (7), is such that *in vitro* decomposition and loss of the label is unlikely; (3) the strength of the carbon-fluorine bond is such that most metabolic reactions are prevented (8), thus decreasing the likelihood of *in vivo* loss of the label; (4) the half-life of the only useful radioisotope of fluorine, ^{18}F, is 110 minutes which is short enough to avoid large radiation doses to the patient; (5) ^{18}F decays by positron emission and thus changes in its tissue concentration can be measured with a positron emission tomograph (9).

We chose to begin the design of an adrenal imaging agent with fluorinated analogs of pregnenolone. Pregnenolone, a naturally occuring precursor to the adrenal cortical hormones, has been shown to exhibit a rapid concentration in and clearance from the adrenal glands when the ^{14}C-labeled steroid was injected intravenously (10). We compared the kinetics of the tissue distribution of pregnenolone-7-^{3}H with that of its acetylated derivative, pregnenolone-3-acetate-7-^{3}H. To determine the effects of fluorination on the kinetics of tissue distribution, we also studied 21-fluoropregnenolone-7-^{3}H and 21-fluoropregnenolone-3-acetate-7-^{3}H. We have successfully developed a technique for the synthesis of ^{18}F-labeled analogs of pregnenolone (11). The ^{18}F-labeled pregnenolones were used to determine the effects of label position on the kinetics of tissue distribution by comparing their distribution with that of the corresponding ^{3}H-labeled steroid.

MATERIALS AND METHODS

Pregnenolone-7-^{3}H was obtained from New England Nuclear with a radiochemical purity of greater than 99%. Pregnenolone-3-acetate-7-^{3}H was prepared by reacting pregnenolone-7-^{3}H with acetic anhydride (12). The product was purified to greater than 98% radiochemical purity by preparative thin-layer chromatography on silica gel using dichloroethane-ethyl acetate (19:1) as the solvent system. The ^{3}H-labeled fluorinated steroids, 21-fluoropregnenolone-7-^{3}H and 21-fluoropregnenolone-3-acetate-7-^{3}H, were prepared from pregnenolone-7-^{3}H using perchloryl fluoride as the fluorinating agent (12). The ^{18}F-labeled fluorinated steroids, 21-fluoropregnenolone-21-^{18}F and 21-fluoropregnenolone-3-acetate-21-^{18}F, were prepared from the corresponding iodinated steroid using potassium fluoride-^{18}F and the crown ether 18-Crown-6 dissolved in acetonitrile as the fluorinating agent (12). The radiochemical purity of the labeled steroids was determined by thin-layer chromatography on silica gel using either solvent system A, dichloroethane-ethyl acetate (19:1), or solvent system B, chloroform-methanol (37:4). Unlabeled standards were used to locate the radioactive compounds which were then assayed by either liquid scintillation counting (^{3}H-labeled steroids) or sodium iodide scintillation counting (^{18}F-labeled steroids). The following radiochemical purities (RCP) were obtained: pregnenolone-3-acetate-7-^{3}H, R_f = 0.260-0.32 in solvent system A, RCP > 99%; 21--fluoropregnenolone-7-^{3}H, R_f = 0.07-0.12 in solvent system A, RCP > 97%; 21--fluoropregnenolone-3-acetate-7-^{3}H, R_f = 0.33-0.41 in solvent system A, RCP >95%; 21-fluoropregnenolone-21-^{18}F, R_f = 0.58 in solvent system B, RCP > 97%; 21--fluoropregnenolone-3-acetate-21-^{18}F, R_f = 0.4 in solvent system A, RCP > 95%.

For animal distribution studies, 15-22 μCi of ^{3}H-labeled steroids and/or 16-32 μCi of ^{18}F-labeled steroids were dissolved in ethanol and injected into the femoral vein of female Sprague-Dawley rats weighing 250 g. The animals were anesthetized with intramuscular injections of Ketaset® containing acepromazine. For each study, at least 3 anesthetized animals were sacrificed by exsanguination through the heart at 5, 10, 15, 30, 60, and 120 minutes after injection of the labeled steroid. The following organs and tissues were removed from the animals, blotted and weighed: brain, kidney, liver, lung, muscle, ovaries, pituitary, uterus, and heart. Urine was obtained where available. The ^{18}F-activity of the samples was measured in a sodium iodide (Tl) scintillation detector and compared to a standard containing 1% of the dose. The ^{3}H-activity of the samples was determined by liquid scintillation counting. Adrenals, muscle, ovaries, pituitary, and uterus were dissolved in 0.5 ml of Protosol® at 55°. The brain, heart, kidney, liver and lung were added to 5 ml or an equal weight of water and homogenized with a Tekmar® homogenizer. Aliquots of the tissue suspensions were then dissolved in 0.5 ml of Protosol® at 55°. All of the dissolved tissue samples and aliquots were then dissolved in 15 ml of Aquasol II® for liquid scintillation counting. To each 0.1-0.2 g of blood was added 0.5 ml of Protosol®-ethanol (1:2). The resulting blood samples were shaken at 55° for 1 hour, and cooled, and 0.5 ml of 30% hydrogen peroxide was added to each sample. The samples were shaken at 55° for 30 minutes, and dissolved in 15 ml of Aquasol II® and 0.5 ml of 0.5 ml of 0.5 N HCl. The ^{3}H-counts were corrected for quenching by external standardization.

Urine samples were analyzed for exchangable ^{3}H-activity by evaporation, reconstitution and recounting. Less than 5% of the ^{3}H-activity in the urine was exchangable with water.

RESULTS

The data are presented as percent of administered dose per gram of tissue (%D/g), in Table 1, and in most cases as a ratio of the concentration in the tissue, %D/g, to that in the blood, %D/g, see discussion. The rate of clearance of the radioactivity from blood was determined from the least squares line fit to the logarithm of the %D/g versus time data. The results are given in Table 2.

TABLE 1. TISSUE DISTRIBUTION OF LABELED PREGNENOLONE DERIVATIVES IN THE RAT (%D/g)

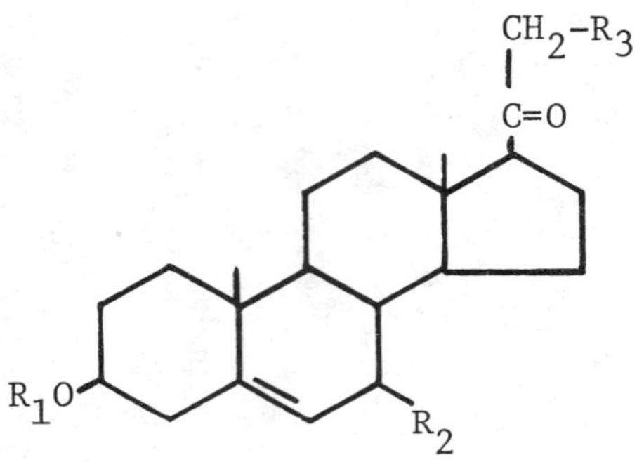

Compound	R_1	R_2	R_3
Pregnenolone-7-^{3}H, I	H	^{3}H	H
Pregnenolone-3-Acetate-7-^{3}H, II	CH_3CO-	^{3}H	H
21-Fluoropregnenolone-7-^{3}H, III	H	^{3}H	F
21-Fluoropregnenolone-3-Acetate-7-^{3}H, IV	CH_3CO-	^{3}H	F
21-Fluoropregnenolone-21-^{18}F, V	H	H	^{18}F
21-Fluoropregnenolone-3-Acetate-21-^{18}F, VI	CH_3CO-	H	^{18}F

Tissue	Compound	Time after injection (min) 5	10	15	30	60	120
Adrenal	I	0.50±0.07	0.40±0.03	0.38±0.03	0.19±0.02	0.14±0.03	0.08±0.02
	II	0.78±0.22	0.59±0.02	0.37±0.02	0.34±0.04	0.32±0.08	0.17±0.01
	III	0.72±0.05	0.47±0.09	0.34±0.03	0.17±0.04	0.12±0.04	0.04±0.00
	IV	0.73±0.07	0.46±0.03	0.37±0.10	0.25±0.05	0.08±0.03	0.07±0.02
	V	4.10±0.47	2.92±0.59	2.31±0.41	0.68±0.15	0.59±0.35	0.18±0.04
	VI	2.37±0.38	1.61±0.14	1.64±0.67	0.70±0.08	0.20±0.10	0.07±0.02
Blood	I	0.15±0.02	0.15±0.03	0.14±0.02	0.15±0.02	0.09±0.01	0.06±0.00
	II	0.24±0.04	0.20±0.03	0.14±0.01	0.12±0.01	0.10±0.02	0.05±0.01
	III	0.14±0.02	0.11±0.01	0.12±0.03	0.06±0.01	0.04±0.01	0.02±0.00
	IV	0.12±0.01	0.09±0.01	0.08±0.01	0.04±0.01	0.02±0.01	0.02±0.00
	V	0.29±0.04	0.23±0.03	0.24±0.01	0.11±0.09	0.09±0.03	0.04±0.00
	VI	0.17±0.02	0.16±0.01	0.15±0.02	0.13±0.01	0.05±0.02	0.02±0.01
Brain	I	0.52±0.02	0.39±0.03	0.49±0.08	0.16±0.03	0.09±0.02	0.04±0.01
	II	0.31±0.02	0.31±0.01	0.26±0.02	0.15±0.03	0.10±0.02	0.04±0.01
	III	0.67±0.02	0.49±0.08	0.32±0.03	0.13±0.01	0.07±0.02	0.02±0.00
	IV	0.36±0.07	0.32±0.04	0.29±0.10	0.11±0/00	0.04±0.01	0.02±0.01
	V	0.82±0.02	0.60±0.07	0.44±0.04	0.18±0.03	0.08±0.04	0.02±0.00
	VI	0.49±0.05	0.36±0.01	0.31±0.13	0.12±0.00	0.04±0.01	0.02±0.00
Heart	III	0.31±0.04	0.20±0.02	0.13±0.02	0.05±0.01	0.04±0.01	0.02±0.00
	IV	0.24±0.04	0.15±0.01	0.13±0.04	0.06±0.01	0.02±0101	0.02±0.00

TABLE 1. Continued

Tissue	Compound	Time after injection (min)					
		5	10	15	30	60	120
Heart	V	0.39±0.05	0.27±0.03	0.20±0.02	0.09±0.02	0.06±0.02	0.02±0.00
	VI	0.27±0.02	0.19±0.02	0.18±0.02	0.10±0.01	0.03±0.01	0.01±0.00
Kidney	I	0.57±0.07	0.53±0.11	0.51±0.07	0.51±0.08	0.36±0.05	0.24±0.04
	II	0.56±0.05	0.49±0.02	0.44±0.07	0.44±0.10	0.34±0.02	0.21±0.04
	III	0.54±0.05	0.35±0.04	0.30±0.02	0.17±0.01	0.08±0.07	0.06±0.01
	IV	0.44±0.04	0.29±0.02	0.28±0.06	0.17±0.04	0.09±0.03	0.05±0.01
	V	0.69±0.05	0.50±0.02	0.49±0.02	0.31±0.05	0.17±0.07	0.06±0.01
	VI	0.51±0.02	0.38±0.00	0.39±0.05	0.30±0.01	0.10±0.04	0.04±0.01
Liver	I	2.39±0.45	3.09±0.80	3.23±0.15	2.12±0.14	0.54±0.13	1.07±0.19
	II	2.90±0.34	3.17±0.25	2.90±0.25	1.83±0.42	1.72±0.38	0.91±0.06
	III	3.29±0.29	2.82±0.39	2.91±0.08	1.60±0.06	1.50±0.10	0.67±0.12
	IV	2.20±0.25	2.03±0.10	1.83±0.15	1.11±0.18	0.46±0.08	0.28±0.05
	V	3.53±0.04	2.60±1.09	2.90±0.15	1.38±0.23	0.98±0.47	0.44±0.05
	VI	2.11±0.26	2.00±0.05	1.58±0.22	1.00±0.13	0.29±0.10	0.16±0.05
Muscle	I	0.18±0.05	0.15±0.03	0.14±0.01	0.08±0.01	0.06±0.01	0.05±0.00
	II	0.17±0.02	0.24±0.07	0.13±0.02	0.08±0.01	0.06±0.02	0.03±0.01
	III	0.18±0.05	0.14±0.05	0.10±0.01	0.17±0.04	0.03±0.01	0.01±0.01
	IV	0.15±0.00	0.12±0.03	0.09±0.03	0.05±0.01	0.02±0.00	0.02±0.00
	V	0.20±0.05	0.17±0.04	0.12±0.01	0.07±0.03	0.06±0.02	0.01±0.01
	VI	0.15±0.00	0.12±0.02	0.11±0.02	0.09±0.02	0.04±0.01	0.03±0.01
Ovaries	I	0.30±0.07	0.25±0.05	0.22±0.01	0.13±0.01	0.10±0.02	0.06±0.01
	II	0.32±0.03	0.30±0.05	0.21±0.04	0.14±0.02	0.16±0.05	0.08±0.01
	III	0.43±0.03	0.24±0.04	0.17±0.04	0.11±0.03	0.10±0.05	0.04±0.01
	IV	0.36±0.08	0.22±0.02	0.17±0.07	0.10±0.02	0.08±0.03	0.05±0.01
	V	0.42±0.13	0.30±0.03	0.25±0.02	0.17±0.02	0.15±0.05	0.06±0.02
	VI	0.40±0.06	0.25±0.02	0.22±0.06	0.13±0.01	0.08±0.03	0.04±0.01

TABLE 1. Continued

Tissue	Compound	Time after injection (min)					
		5	10	15	30	60	120
Pituitary	I	0.54±0.05	0.49±0.03	0.49±0.05	0.19±0.02	0.09±0.03	0.05±0.01
	II	0.38±0.09	0.31±0.05	0.33±0.05	0.16±0.06	0.11±0.03	0.05±0.01
	III	0.71±0.05	0.56±0.11	0.42±0.03	0.14±0.03	0.06±0.02	0.02±0.01
	IV	0.47±0.05	0.31±0.01	0.31±0.12	0.13±0.01	0.04±0.01	0.03±0.01
	V	0.83±0.13	0.58±0.11	0.63±0.08	0.20±0.12	0.07±0.05	0.03±0.03
	VI	0.45±0.04	0.33±0.01	0.30±0.12	0.13±0.02	0.05±0.02	0.04±0.01
Uteri	I	0.20±0.06	0.19±0.03	0.13±0.06	0.09±0.01	0.07±0.01	0.05±0.01
	II	0.22±0.03	0.22±0.04	0.14±0.02	0.09±0.01	0.07±0.01	0.04±0.01
	III	0.23±0.02	0.16±0.02	0.11±0.01	0.06±0.01	0.07±0.04	0.03±0.01
	IV	0.20±0.02	0.14±0.01	0.11±0.04	0.06±0.02	0.04±0.02	0.03±0.01
	V	0.28±0.04	0.22±0.02	0.19±0.02	0.29±0.27	0.11±0.06	0.03±0.01
	VI	0.23±0.03	0.16±0.02	0.17±0.02	0.11±0.02	0.05±0.03	0.03±0.01

Values are the means ± standard deviation for 3 or more 250 g rats at each time period.

Pregnenolone-7-^{3}H, (I). The maximum adrenal uptake of I, 0.5 %D/g, was found at 5 minutes, after which it declined rapidly to 0.08 %D/g at 2 hours. The maximum blood concentration of I was 0.15%D/g at 5 minutes, with a blood clearance half-time of 82 minutes. The maximum adrenal/blood ratio, 3.4, was found at 5 minutes. The adrenal/blood ratio decreased to 1.3 at 2 hours indicating that I was cleared from the adrenals faster than from the blood. The maximum liver uptake of I, 2.4 %D/g, was found at 15 minutes after which it decreased slowly to 1.1 %D/g at 2 hours. The maximum liver/blood ratio, 23, was found at 15 minutes. The liver/blood ratio declined to 16 where it remained from 30 to 120 minutes. The maximum kidney uptake of I, 0.57%D/g, was found at 5 minutes, decreasing slowly to 0.24 %D/g at 2 hours. The kidney/blood ratio ranged from 3.5 at 5 minutes to 3.9 at 2 hours.

Pregnenolone-3-acetate-7-^{3}H, (II). The maximum adrenal uptake of II, 0.78 %D/g, was found at 5 minutes, after which it declined to 0.17 %D/g at 2 hours. The maximum blood concentration of II was 0.24 %D/g at 5 minutes with a blood clearance half-time of 57 minutes. The maximum adrenal/blood ratio was 3.3 at 5 minutes and it did not differ significantly at any other time, indicating that II was cleared from the adrenals at the same rate as from blood. The maximum liver uptake of II, 3.2 %D/g, was found at 10 minutes after which it decreased to 0.9 %D/g at 2 hours. The maximum liver/blood ratio, 20, was found at 15 minutes, decreasing to 15 and remaining there from 30 to 120 minutes. The maximum kidney uptake of II, 0.5 %D/g, was found at 5 minutes after which it decreased to 0.2 %D/g at 2 hours. The kidney/blood ratio increased from 2.4 at 5 minutes to 3.8 at 2 hours indicating that II was cleared more rapidly from blood than from the kidneys.

21-Fluoropregnenolone-7-^{3}H, (III). The maximum adrenal uptake of III, 0.7 %D/g, was found at 5 minutes, decreasing to 0.04 %D/g at 2 hours. The maximum blood concentration of III was 0.14 %D/g at 5 minutes, with a blood clearance half-time of 41 minutes. The maximum adrenal/blood ratio was 5 at 5 minutes. The adrenal/blood ratio decreased to 2 at 2 hours indicating that III was cleared from the adrenals more rapidly than from blood. The maximum liver uptake of III, 3.3 %D/g, was found at 5 minutes after which it declined to 0.7 %D/g at 2 hours. The liver/blood ratio increased from 23 at 5 minutes, to 32 at 2 hours, indicating that III was cleared more rapidly from the blood than from the liver. The maximum kidney uptake of III, 0.5 %D/g, was found at 5 minutes after which it declined to 0.06 %D/g at 2 hours. The kidney/blood ratio decreased from 3.8 at 5 minutes to 2 at 2 hours, indicating that III was cleared from the kidneys more rapidly than from blood.

21-Fluoropregnenolone-3-acetate-7-^{3}H, (IV). The maximum adrenal uptake of IV, 0.73 %D/g, was found at 5 minutes, after which it declined to 0.07 %D/g at 2 hours. The maximum blood concentration of IV was 0.12 %D/g at 5 minutes, with a blood clearance half-time of 44 minutes. The maximum adrenal/blood ratio was 5.9 at 5 minutes. The adrenal/blood ratio decreased to 5 at 30 minutes and to 4.3 at 2 hours, indicating that IV was cleared from the adrenals slightly more rapidly than from the blood. The maximum liver uptake of IV, 2.2 %D/g, was found at 5 minutes after which it decreased to 0.28 %D/g at 2 hours. The liver/blood ratio, 21, remained fairly constant from 5 to 30 minutes. It decreased to 17 at 2 hours indicating that IV was cleared from the liver slightly more rapidly than from the blood. The maximum kidney uptake of IV, 0.4 %D/g was found at 5 minutes after which it declined to 0.05 %D/g at 2 hours.

TABLE 2. PARAMETERS FOR LINEAR REGRESSION FIT OF LOGARITHM BLOOD ACTIVITY-TIME DATA

COMPOUND	r	λ (MIN^{-1})	HALF-TIME (MIN)	FRACTION OF DOSE
I	0.973	0.0085	82	0.164
II	0.958	0.0121	57	0.205
III	0.969	0.0167	41	0.103
IV	0.87	0.0155	44	0.091
V	0.962	0.0166	41	0.265
VI	0.99	0.0195	35	0.196

The kidney/blood ratio increased from 3.5 at 5 minutes to 4 at 30 minutes and then decreased to 2.9 at 2 hours indicating that IV was cleared from the kidneys at about the same rate as from blood.

21-Fluoropregnenolone-21-^{18}F, (V). The maximum adrenal uptake of V, 4.1 %D/g, was found at 5 minutes after which it decreased rapidly to 0.18 %D/g at 2 hours. The maximum blood concentration of V was 0.29 %D/g at 5 minutes, with a blood clearance half-time of 41 minutes. The adrenal/blood ratio was 14.5 at 5 minutes after which it decreased to 4 at 2 hours, indicating that V was cleared from the adrenals more rapidly than from blood. The maximum liver uptake of V, 3.5 %D/g was found at 5 minutes after which it decreased to 0.4 %D/g at 2 hours. The liver/blood ratio did not vary significantly, as it was 12 at 5 minutes and 11.8 at 2 hours, indicating that V was cleared from the liver at the same rate as from the blood. The maximum kidney uptake of V, 0.7 %D/g, was found at 5 minutes after which it decreased to 0.06 %D/g at 2 hours. The kidney/blood ratio decreased from 2.4 at 5 minutes to 1.6 at 2 hours indicating that V was cleared from the kidneys slightly more rapidly than from blood.

21-Fluoropregnenolone-3-acetate-21-^{18}F, (VI). The maximum adrenal uptake of VI, 2.4 %D/g, was found at 5 minutes, after which it decreased to 0.07 %D/g at 2 hours. The maximum blood concentration of VI was 0.17 %D/g at 5 minutes, with a blood clearance half-time of 35 minutes. The maximum adrenal/blood ratio was 13.8 at 5 minutes after which it decreased to 3.3 at 2 hours, indicating that VI was cleared from the adrenals more rapidly than form the blood. The maximum liver uptake of VI, 2.1 %D/g, was found at 5 minutes after which it decreased to 0.2 %D/g at 2 hours. The maximum liver/blood ratio was 12 at 5 minutes, after which it decreased to 7 at 2 hours indicating that VI was cleared more rapidly from the liver than from the blood. The maximum kidney uptake of VI, 0.5 %D/g, was found at 5 minutes, it decreased to 0.04 %D/g at 2 hours. The kidney/blood ratio decreased from 3.0 at 5 minutes to 2.0 at 2 hours indicating that VI was cleared from the kidneys slightly faster than from the blood.

DISCUSSION

The data obtained in this investigation may be used to answer three questions which will be helpful for designing an adrenal imaging agent based on pregnenolone. (1) Is a free hydroxyl group in the 3-position essential for adrenal uptake? (2) Does fluorination alter the biological distribution of pregnenolone? (3) Does the position of the radioactive label alter the target and non-target tissue concentrations of these pregnenolone analogs?

(1) Effect of acetylation on adrenal uptake. Since Counsell reported that the adrenal concentration of 19-iodocholesterol-^{125}I was 300 times greater than that obtained from 19-iodocholesterol-3-acetate-^{125}I (13), we wished to determine if such an effect existed with pregnenolone analogs. Our data indicate that a free hydroxyl group in the 3-position is not essential for adrenal uptake. In fact, only with the analogs of pregnenolone labeled with ^{18}F was the adrenal concentration of the pregnenolone analog containing a free hydroxyl group in the 3-position, V, greater that that for the acetylated analog VI. The magnitude of this effect was much less than in the cholesterol analogs since the adrenal concentration of V was only 1.7 times greater than that of VI. In

addition, the blood concentration of VI was also reduced so that both V and VI gave similar adrenal/blood ratios.

(2) Effect of fluorination on tissue distribution. While it has been reported that fluorination may increase the biological activity of a steroid, little is known about the effects of fluorination on tissue distribution. In order to determine what effect the fluorination of pregnenolone in the 21-position had on the kinetics of tissue distribution, we compared the results obtained for pregnenolone, I, with those obtained for 21-fluoropregnenolone, III. Fluorination caused a decrease in the blood clearance half-time from 82 minutes for I to 41 minutes for III. Fluorination increased both the adrenal uptake and the adrenal/blood ratios of III by 1.5 times the corresponding values for I. Fluorination had little effect on the uptake of label by the liver and kidneys. However, the fluorinated molecule was cleared much more rapidly from the kidneys, thus the kidneys retained 4 times more I than III.

In a similar study, we compared the corresponding acetylated derivatives II and IV. Fluorination also caused a decrease in the blood clearance half-time from 57 minutes for II to 44 minutes for IV. In addition, the quantity of IV remaining in the blood at 5 minutes was one-half that of II. Fluorination had no effect on the adrenal uptake in this series. However, since there was less of IV in the blood, the adrenal/blood ratio increased from 3.3 for II to 5.9 for IV. Fluorination had little effect on theliver and kidney uptake. However, the fluorinated molecule was cleared more rapidly from both the liver and kidneys so that by 2 hours the liver retained 3 times more II than IV, and the kidneys retained 4 times more II than IV.

(3) Effect of label position on tissue distribution. As a result of the above studies we would have concluded that 21-fluoropregnenolone or the acetylated analog would not reach high enough adrenal concentrations or adrenal /blood ratios to warrant further investigation as adrenal imaging agents. However, a preliminary investigation in this laboratory suggested that these ^{18}F-labeled steroids reached higher adrenal/blood ratios than those obtained with the ^{3}H-labeled steroids (11). Therefore, we compared the tissue distribution kinetics of 21-fluoropregnenolone labeled with ^{3}H, III, and with ^{18}F, V, when administered simultaneously to the same group of animals. The effect of label position on the adrenal uptake was dramatic. At 5 minutes the adrenals contained about 6 times more ^{18}F than ^{3}H from 21-fluoropregnenolone. Similar though less dramatic results were obtained from 21-fluoropregnenolone-3-acetate. At 5 minutes the adrenals contained 3 times more ^{18}F than ^{3}H.

Blood clearance rates of the ^{18}F and ^{3}H activities from 21-fluoropregnenolone were identical from 5 minutes to 2 hours. However, there was twice as much ^{18}F as ^{3}H remaining in the blood at 5 minutes. The ^{18}F/^{3}H ratio in the adrenals after the administration of 21-fluoropregnenolone decreased from 5.7 at 5 minutes to 4.5 at 2 hours. The ^{18}F/^{3}H ratio in the adrenals after the administration of 21-fluoropregnenolone-3-acetate decreased from 3 at 5 minutes to 1.0 at 2 hours indicating that ^{18}F was being cleared more rapidly from the adrenals than was ^{3}H.

In most other tissues there was no significant difference between ^{18}F and ^{3}H concentrations resulting from the administration of 21-fluoropregnenolone or 21-fluoropregnenolone-3-acetate except in the liver which contained approximately 1.5 times more ^{3}H-activity than ^{18}F-activity at 1 and 2 hours after the administration of both compounds.

These results strongly suggest that the high adrenal concentrations of ^{18}F-activity are a result of adrenal metabolism of the 21-fluoropregnenolone in such a manner that the ^{18}F-activity is removed from the steroid nucleus. Pregnenolone is a precursor to the adrenal cortical hormones, which contain a 21-hydroxyl group, and to the androgens and estrogens which do not contain the 17β-acetyl group. If 21-fluoropregnenolone is a substrate for the enzymes which produce the adrenal cortical hormones then this study indicates the possibility that the ^{18}F-label is removed from the steroid nucleus in the form of fluoride ion. If 21-fluoropregnenolone is a substrate for the enzymes which produce the androgens and estrogens then this study indicates the possibility that the ^{18}F-label is removed from the steroid nucleus in the form of an organically bound fluorine.

CONCLUSIONS

(1) A free hydroxyl group in the 3-position is not required for adrenal uptake of pregnenolone analogs.

(2) Fluorination in the 21-position has little effect on the adrenal concentrations of the steroid nucleus of pregnenolone analogs. The fluorination does cause a more rapid clearance of the steroid nucleus from the blood, liver, and kidneys.

(3) The large differences in ^{3}H and ^{18}F-activities found in the adrenal glands result from the metabolic removal of ^{18}F-activity from the steroid nucleus.

(4) Since the rat is capable of metabolically removing fluorine in some form from 21-fluoropregnenolone, future radiopharmaceutical studies of these or similar compounds should include an analysis of bone uptake.

(5) Both ^{18}F-labeled pregnenolones reported here are potentially useful for imaging the adrenal glands and/or measuring adrenal metabolism rates.

(6) These results demonstrate that radiopharmaceutical studies should be thorough and that one must not lose sight of the fact that the radioactive label may no longer be attached to the original molecule.

ACKNOWLEDGMENTS

This research was supported in part by NIH Grants GM-22973 and CA-25739.

REFERENCES

(1) Conn JW, Cohen E., Hewig KR: The dexamethasone-modified adrenal scintiscan in hyporeninemic aldosteronism (tumor versus hyperplasia). A comparison with adrenal venography and adrenal venous aldosterone. J Lab Clin Med 88: 841-856, 1976

(2) Freitas JE, Grekin RJ, Thrall JH, Gross MD, Swanson DP, Beierwaltes WH: Adrenal Imaging with Iodomethyl-Norcholesterol (I-131) in Primary Aldosteronism.

J Nucl Med 20: 7-10, 1979

(3) Kirschner AS, Ice RD, Beierwaltes WH: Radiation Dosimetry of ^{131}I-19-Iodocholesterol. J Nucl Med 14: 713-717, 1973

(4) Ice RD, Kircos LT, Coffey JL, Watson E, Sarkar SD, Beierwaltes WH: Radiation Dosimetry of I-131-6-Iodomethylnorcholesterol. J Nucl Med 17: 540, 1976

(5) Counsell RE, Ice RD: The Design of Organ Imaging Radiopharmaceuticals. In Drug Design, vol 5, Ariens EJ, ed, New York, Academic Press, 1975, pp 171-259

(6) Haynes RC, Larner J: Adrenocorticotropic Hormone; Adrenocortical Steroids and their Synthetic Analogs; Inhibitors of Adrenocortical Steroid Biosynthesis. In The Pharmacological Basis of Therapeutics, 5th Ed, Goodman LS, Gilman A, eds, New York, Macmillan, 1975, pp 1472-1506

(7) Cottrell TL: The Strength of Chemical Bonds, London, Buttersworths, 1958

(8) Goldman P: The Carbon-Fluorine Bond in Compounds of Biological Interest. Science 164: 1123-1130, 1969

(9) Kuhl D, Reivich M, Wolf A, Greenburg J, Phelps M, Ido T, Casella V, Fowler J, Gallagher B, Hoffman E, Alavi A, Sokoloff L: Determination of Local Cerebral Glucose Utilization by Means of Radionuclide Computed Tomography and (F-18)-2-Fluoro-2-Deoxy-D-Glucose. J Nucl Med 18: 614, 1977

(10) Appelgren LE: Sites of Steroid Hormone Formation Autoradiographic Studies Using Labelled Precursors. Acta Physiol Scand Suppl 301, pp 1-108, 1967

(11) Eng R, Hinn G, Spitznagle L: F-18-21-Fluoropregnenolone-3-Acetate As An Adrenal Scanning Agent J Nucl Med 16: 526, 1975

(12) Eng RR: I. Radiopharmaceutical Potential Of ^{18}F-Labeled Pregnenolone Analogs. II. Carbon-13 NMR of Some Androstene, Pregnene, and Cholestene Derivatives, PhD Thesis, University of Washington, Seattle, 1979

(13) Counsell RE, Ranada VV, Kulkarni Pg, Afiatpour P: Potential Organ or Tumor-Imaging Agents. 12. Esters of 19-Radioiodinated Cholesterol. J Nucl Med 14: 777-780, 1973

SYNTHESIS AND BIOLOGICAL STUDIES ON AN ^{125}I-ANALOG OF DIAZOXIDE

Ned D. Heindel, Victor R. Risch, H. Donald Burns, Edward G. Corley, Edward A. Michener, and Takashi Honda

Department of Radiation Therapy and Nuclear Medicine
Hahnemann Medical College
Philadelphia, Pa. 19102

ABSTRACT

An ^{125}I-labeled diazoxide analog (7-iodo-3-methyl-2H-1,2,4-benzothiadiazine 1,1-dioxide) was synthesized and studied with respect to localization in a β-cell adenocarcinoma of the pancreas in hamsters. The iododiazoxide is stable in vivo and exhibits pharmacokinetic and excretion patterns similar to ^{3}H-diazoxide. Tumor to pancreas activity ratios demonstrated a progressive increase from 0.5 at 5 min to 3.8 at 6 hr. However, tumor to blood ratios never exceeded one at all time periods studied. The failure to achieve elevated tumor concentration of radioactivity may be due to the relatively low specific activity of the ^{125}I-iododiazoxide used in these studies.

Diazoxide (7-chloro-3-methyl-2H-1,2,4-benzothiadiazine 1,1-dioxide)(1) is a non-diuretic thiazide with antihypertensive activity (1) which is currently the drug of choice for the treatment of hypertensive crises (2). Soon after introduction of this drug, however, reports appeared on its marked hyperglycemic activity (3,4). Diazoxide is currently "the most important and most widely employed drug in the medical management of spontaneous hypoglycemias" (5), and its use in the treatment of hypoglycemia has been the subject of two major international conferences (6,7). Nevertheless, theories on the mechanism of diazoxide-induced hyperglycemia are still controversial, primarily due to the complex modes of action of this drug. It is generally agreed that the major mechanism of diazoxide-induced hyperglycemia is via a direct action on the pancreas (5,8,9) which inhibits insulin release in both normal and neoplastic islets. Extrapancreatic mechanisms involving catecholamine release from the adrenals (10,11) and stimulation of liver glycogenolysis (12,13) also appear to be involved, although it is unlikely that these mechanisms contribute significantly to the hyperglycemic action of diazoxide in normal therapeutic dosages (5). Detailed investigations as to the nature of the direct pancreatic effect of diazoxide indicate that the drug interacts with a receptor on β-cells in the pancreatic islet. The exact nature of this receptor is not known, although there is evidence to both suggest and contradict that it is either an α-adrenergic receptor (14,15)(currently the most popular theory (9)), a β-adrenergic receptor (16,17) or a non-adrenergic receptor which is affected antagonistically by diazoxide and the hypoglycemic sulfonylurea agents (18,19).

Our continuing interests in pancreatic radiopharmaceuticals based upon receptor-specific agents for pancreatic β-cells (20,21) have prompted us to investigate iodinated analogs of diazoxide to determine if specific receptor interaction *in vivo* could result in an elevated concentration of these analogs in pancreatic β-cells. Should this be the case, a γ-emitting radioiodinated analog offers the potential of preoperative localization of insulinomas within pancreatic tissue.

In an attempt to prepare 5-iododiazoxide via direct electrophilic iodination, diazoxide (1) was refluxed with iodine monochloride in acetic acid for four hours. However, even though diazoxide can be nitrated at the 5 position (22)(indicating that this position is susceptable to electrophilic attack) iodine monochloride was not sufficiently electrophilic to bring about the desired reaction. An alternative route which would be expected to yield 5-iododiazoxide is via the diazotization/ iodination of 7-chloro-5-amino-3-methyl-2H-1,2,4-benzothiadiazine 1,1-dioxide (2). An attempt to prepare 2 by the hydrazine reduction of 7-chloro-5-nitro-3-methyl-2H-1,2,4-benzothiadiazine 1,1-dioxide (3) led, through hydrogenolysis, to the dechlorinated 5-amino-3-methyl-2H-1,2,4-benzothiadiazine 1,1-dioxide (4). The desired 5-amino compound (2) was obtained by the reduction of 3 with activated iron powder and hydrochloric acid.

While diazotization of both of these amines (2,4) with nitrous acid appeared to proceed normally, treatment with sodium iodide gave products which did not contain any iodine. Elemental analyses indicated the retention of the molecular equivalents of a diazonium function and a molecule of water. The infrared spectra show strong absorptions at 1725 cm^{-1} which are in the expected region for N-acetylbenzotriazoles. Although amide carbonyls normally appear at less than 1700 cm^{-1}, N-acetylbenzotriazoles display carbonyl absorptions as high as 1744 cm^{-1} due to poor resonance overlap with the triazole ring (23). Mass spectrophotometric studies revealed parent ions at 240 and 275 amu for the compounds derived from the dechlorinated and chlorine-containing amines respectively. Based on this data, we have assigned the structures shown in figures 5 and 6 (Scheme 1) to the products obtained from these diazotization reactions. Although interesting and unexpected, this novel chemistry prevented the preparation of 5-iododiazoxide by this route.

In view of the difficulties encountered in the attempted synthesis of 5-iododiazoxide, we redirected our efforts toward the synthesis of the 7-iodo analog of diazoxide (7, Scheme 2). Initial attempts to prepare 7 from diazoxide by halogen exchange using either red phosphorous and hydriodic acid (24) or sodium iodide in refluxing ethylene glycol (25) were unsuccessful. The desired compound was obtained from o-aminobenzenesulfonamide as outlined in Scheme 2. Exchange labeling of this diazoxide analog with I-131 sodium iodide in propylene glycol did give a low yield of radioiodinated 7-iododiazoxide. However, the radiochemical yield was too low to permit preparation of the labeled compound in practical quantities.

An alternative synthesis of 7 was developed which involved the diazotization/iodination of 7-amino-3-methyl-2H-1,2,4-benzothiadiazine 1,1-dioxide (9) as shown in Scheme 3. Reduction of the 7-nitro precursor by palladium and hydrazine gave the required amine in 90% yield. Diazotization and treatment with sodium iodide resulted in formation of 7 in a 60% yield. The I-125 labeled compound was prepared in 10% radiochemical yield by the same route substituting I-125 sodium iodide for the stable sodium iodide. TLC on silica gel plates with 1:1 $CHCl_3$:HOAc eluted a single radioactive spot with R_f = 0.56 identical to that of authentic I-127 7-iodo-3-methyl-2H-1,2,4-benzothiadiazine 1,1-dioxide.

Biologic studies on ^{125}I-7-iododiazoxide were performed in an inbred strain of golden Syrian hamsters bearing a transplantable pancreatic β-cell adenocarcinoma (21). The radiochemical was administered i.v. into a femoral vein (26) in both a 50:50 ethanol:water mixture and dissolved in physiologic saline with the pH of the solution adjusted to 11.6 with sodium hydroxide in order to determine any effect of the injection vehicle on biological behavior. Furthermore, clinical evidence suggests that the hypotensive effect of the drug is dependent on a rapid bolus injection to minimize protein binding and allow for more free drug for receptor interaction (2), whereas the hyperglycemic effect is independent on speed or route of administration. Injections were therefore performed both as a bolus as well as a slow infusion to determine if receptor localization in pancreatic β-cells could be enhanced by bolus injection. Results from the biodistribution data demonstrated no effect of either injection vehicle or rapidity of administration on biological behavior. All results reported therefore represent pooled data.

EXPERIMENTAL SECTION

Melting points were taken on a Thomas-Hoover melting point apparatus

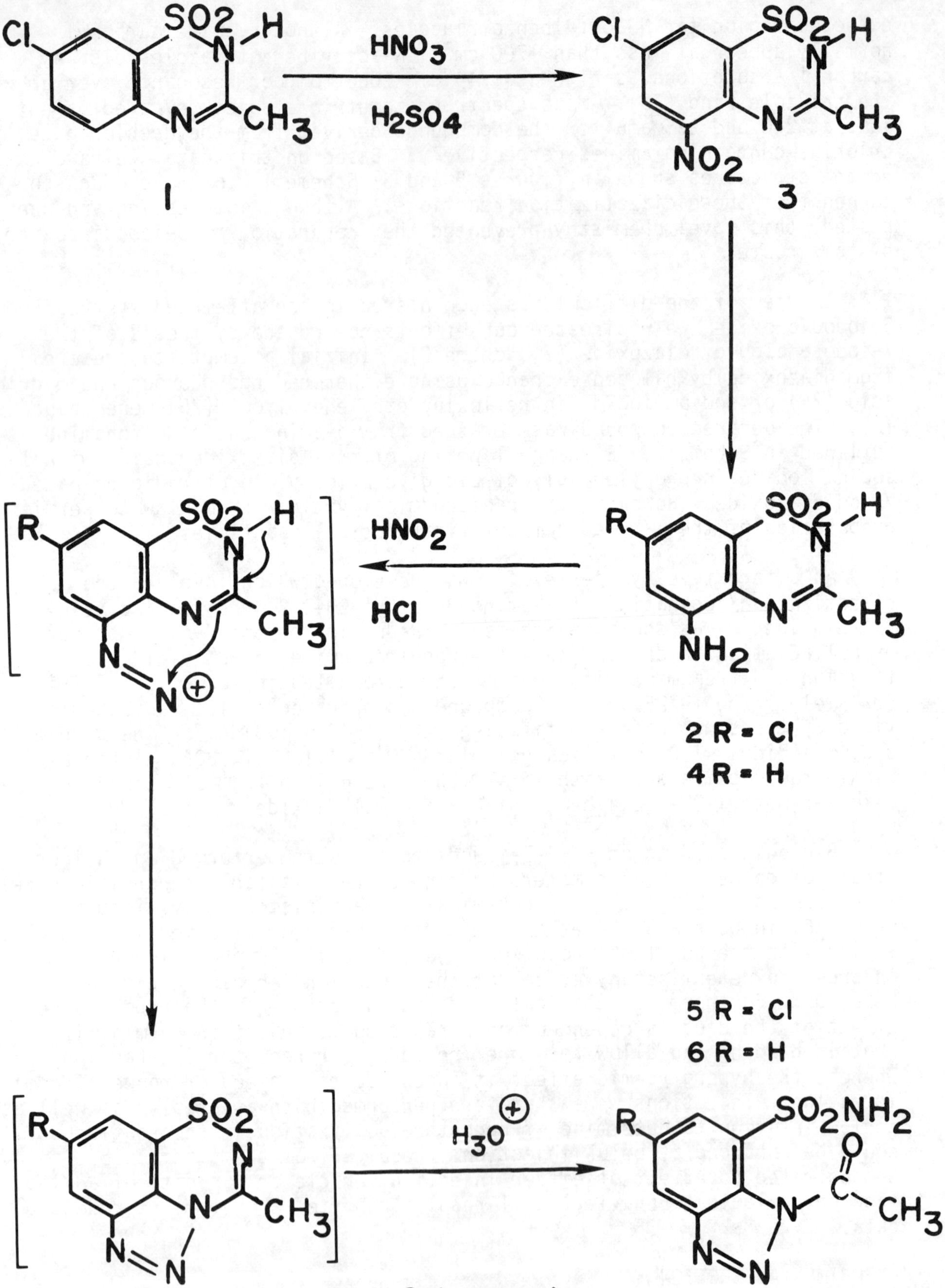

Scheme 1

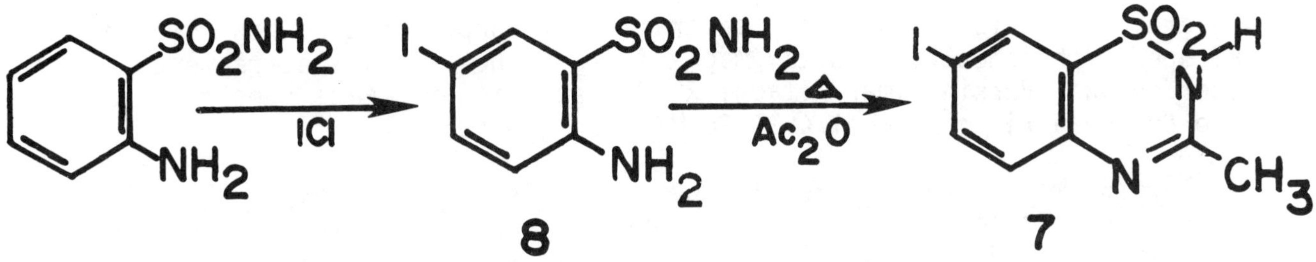

Scheme 2

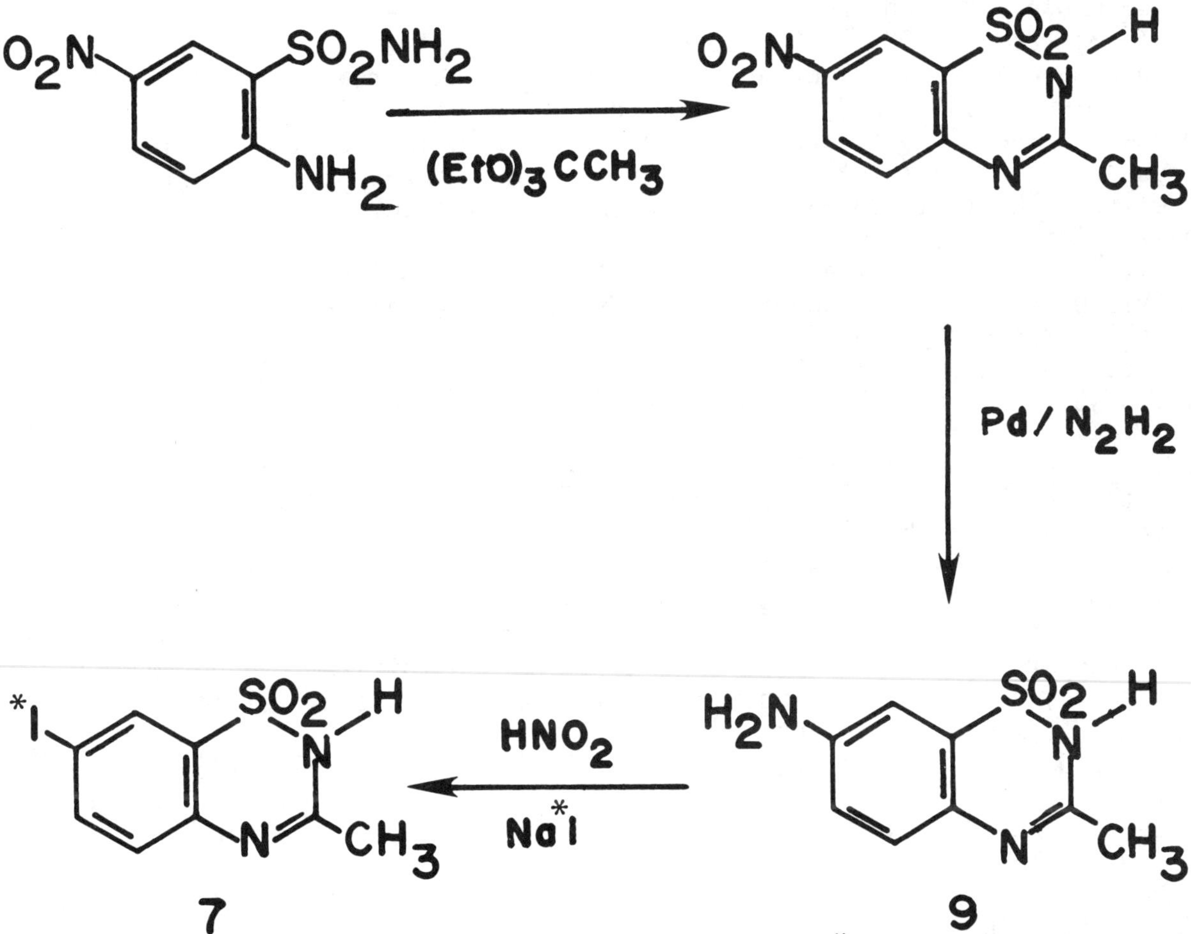

*I = I-125 or I-127

Scheme 3

and are uncorrected. Infrared spectra were recorded on a Perkin-Elmer Model 283 spectrophotometer. Nuclear magnetic resonance (NMR) spectra were recorded on a Perkin-Elmer Hitachi R-20 spectrometer. Mass spectra were recorded on a Finegan Model 4021 GC-MS mass spectrometer.

7-Chloro-5-amino-3-methyl-2H-1,2,4-benzothiadiazine 1,1-dioxide (2). A stirred suspension of 2.00 g of 7-chloro-5-nitro-3-methyl-2H-1,2,4-benzothiadiazine 1,1-dioxide (3) (22)(7.25 mmole) was treated with 2.50 g (47 mmole) of ammonium chloride. The mixture was heated to reflux. Iron powder, 2.50 g (43 mmole), was added in small portions over 30 min to the refluxing mixture. The reflux was continued for 8 hours with constant mechanical stirring. The mixture was filtered hot through Celite. The filter cake was washed with 150 ml of methanol. The combined filtrates were treated with 275 ml of water and heated to boiling. After boiling for 10 min the mixture was again filtered hot through Celite. On cooling the filtrate yielded 1.30 g of crude product which was recrystallized from ethanol/water (Norit) to yield 0.90 g colorless needles, mp 316-18°C. Anal. calcd. for $C_8H_8ClN_3O_2S$: C 39.11 ; H 3.28 ; N 17.10. Found: C 39.01 ; H 3.35; N 17.03.

5-Amino-3-methyl-2H-1,2,4-benzothiadiazine 1,1-dioxide (4). A solution of 0.67 g (2.43 mmole) of 7-chloro-5-nitro-3-methyl-2H-1,2,4-benzothiadiazine 1,1-dioxide (3) (22) in 100 ml 95% ethanol at 50°C was treated with 0.10 g of 10% Pd on charcoal in 15 ml ethanol. The mixture was treated to the dropwise addition of 14 ml hydrazine hydrate over 15 min. After the addition was complete, another 0.10 g Pd/C was added. The solution was then refluxed for two hours. At the completion of reflux the mixture was filtered hot through Celite. The filtrate was then concentrated *in vacuo* to approximately one-half volume. 100 ml of water was then added and the solution taken to neutrality with 6N HCl. The neutral solution was then concentrated to 100 ml to yield crystals of product which were collected and recrystallized from methanol/water. A yield of 0.42 g (82%) of colorless needles, mp 229-31°C was obtained: IR (Nujol) 3435, 3333, 3285 (sh , NH). Anal. Calcd. for $C_8H_9N_3O_2S$: C 45.48 ; H 4.29 ; N 19.89. Found: C 45.55 ; H 4.38 ; N 20.20.

Preparation of 1-acetyl-5-R-7-sulfonamido-1H-benzotriazoles. To a solution of 30 ml water and 5 ml of conc hydrochloric acid was added 1.0 mmole of the appropriate 5-amino-3-methyl-2H-1,2,4-benzothiadiazine 1,1-dioxide (compound 2 or 4). Although a homogeneous solution did not result, the reaction proceeded smoothly. After chilling the mixture to 0°C, it was treated with 1.1 equivalents of $NaNO_2$ in 1.0 ml of water, stirred in the ice bath for two hr and then at room temperature for 1 hr. The precipitated product was filtered directly from the reaction mixture and recrystallized from 1:1 ethanol:water. Compound 6 was obtained in a 76% yield, mp 263-264°C: ^{1}H NMR (DMSO-d_6) δ 1.99 (s, 3, CH_3CO), 7.57-8.45 (m, 3, ArH); IR (Nujol) 3276, 3090 (sh , NH), 1718 (C=O). Anal. Calcd. for $C_8H_8N_4O_3S$: C 39.99; H 3.36 ; N 23.32. Found: C 39.51 ; H 3.33 ; N 23.39. By the same experimental procedure an 85% yield of 5 was obtained, mp 261-263°C: ^{1}H NMR (DMSO-d_6) δ 1.98 (s, 3, CH_3CO), 7.95 (d, J= 3 Hz, 1, ArH), 8.52 (d, J = 3 Hz, 1, ArH); IR (Nujol) 3312, 3092 (NH), 1728 (C=O). Anal. Calcd. for $C_8H_7ClN_4O_3S$: C 34.98 ; H 2.57 ; N 20.39. Found: C 34.75 ; H 2.73 ; N 20.24.

2-Amino-5-iodobenzenesulfonamide (8). A solution of 3.00 g (17.4 mmole) of 2-aminobenzenesulfonamide in 25 ml of glacial acetic acid was treated to the dropwise addition of 2.82 g (17.4 mmole) of iodine monochloride in 25 ml of glacial acetic acid. After addition had been completed the solution was warmed to 65-75°C and maintained at that temperature for 8 hr. Upon refrigeration a mass of pink micro-needles separated and was removed by filtration, washed with cold water, and recrystallized from hot water to yield 3.92 g (62%) of the title compound, mp 203-204°C; IR (Nujol) 3454,3401, 3369, 3293 (NH), 1380, 1311 (SO). Anal. Calcd. for $C_6H_7IN_2O_2S$: C 24.16 ; H 2.35 ; N 9.29. Found: C 24.24 ; H 2.55 ; N 8.86.

7-Iodo-3-methyl-2H-1,2,4-benzothiadiazine 1,1-dioxide (7). A solution prepared from 10 ml acetic anhydride, 5.0 ml of anhydrous pyridine and 2.50 g (8.38 mmole) of 8 was allowed to stand overnight, chilled, and the precipitated crystals collected, washed with cold water and dried *in vacuo*. These crude crystals were fused at 240-250°C and held at that temperature for 5 min. The crude mass solidified on cooling and was recrystallized from 1:1 $MeOH:H_2O$ to give pale yellow crystals of 7 (1.16 g, 43%) mp 325-330°C (lit 331-332°C (22) or 318-320°C (27). Anal. Calcd. for $C_8H_7IN_2O_2S$: C 29.83 ; H 2.19 ; N 8.69. Found: C 30.02 ; H 2.38 ; N 8.65.

7-Iodo-3-methyl-2H-1,2,4-benzothiadiazine 1,1-dioxide (7). A solution of 7-amino-3-methyl-2H-1,2,4-benzothiadiazine 1,1-dioxide (9) (28),105 mg (0.5 mmole), in 1.5 ml conc hydrochloric acid and 10 ml water at 0°C was treated with 40 mg (0.58 mmole) $NaNO_2$ in 1 ml water. The solution was stirred in an ice bath for 30 min. Urea was then added, portionwise, until the excess nitrous acid was destroyed as indicated by a negative test on starch iodide paper. Then 87.5 mg NaI (0.625 mmole) in 2.5 ml water was added, dropwise, over 15 min. The mixture was stirred in an ice bath for an additional 2 hr, then gradually warmed to 40°C over a 90 min period. A dilute bisulfite solution was then added to reduce any iodine that may have formed by oxidation of the iodide. The product was then filtered and recrystallized from methanol/water (Norit) to yield 90 mg (56%) of (7), mp 325-328°C. TLC on silica gel plates with 1:1 $CHCl_3$:HOAc gave one spot, R_f 0.57.

The above procedure was followed exactly as stated in the preparation of the ^{125}I-iododiazoxide except that Na^{125}I was added along with the Na^{127}I to the diazonium salt. Radiochemical yields averaged about 15%. TLC of the ^{125}I-iododiazoxide showed only one radiochemical with activity at an R_f matching that of the authentic iodo-compound. The product was obtained at a specific activity of about 10μCi/mg.

Tissue Distribution Studies. Male animals of an inbred strain of golden Syrian hamsters (LHC/Lak) were used throughout these studies. The animals were 81-158 days old at time of transplantation of the β-cell tumor (Kirkman pancreatic β-cell adenocarcinoma 2309V) (21). The tumor was 35-55 days old at the time of study. Tissue distribution studies were performed by our previously published methods (29), and all values were normalized to 100-g hamsters. Each animal received 0.3 μCi of ^{125}I-7-iododiazoxide dissolved in either saline at pH 11.6 or in a 50:50 $EtOH:H_2O$ mixture i.v. into a femoral vein. Tissues assayed for radioactivity concentration are listed in Table 1. In all cases, both kidneys were taken, and stomach activity refers both to parenchymal uptake as well as any activity in the gastric lumen. Femur

TABLE 1. TISSUE DISTRIBUTION OF ^{125}I-7-IODODIAZOXIDE IN HAMSTERS BEARING THE β-CELL ADENOCARCINOMA*

Tissue	5 min	15 min	1 hr §	2 hr	6 hr φ	24 hr
Brain	0.13 ± 0.05¶	0.11 ± 0.02	0.06 ± 0.008	0.03 ± 0.01	0.01 ± 0.003	0.002 ± 0.001
Heart	2.03 ± 0.42	1.44 ± 0.19	0.79 ± 0.04	0.36 ± 0.11	0.06 ± 0.03	0.006 ± 0.002
Lung	2.18 ± 0.48	1.64 ± 0.20	0.84 ± 0.02	0.47 ± 0.17	0.09 ± 0.05	0.006 ± 0.002
Liver	3.27 ± 0.78	2.59 ± 0.43	1.51 ± 0.01	0.87 ± 0.29	0.19 ± 0.12	0.004 ± 0.002
Spleen	1.12 ± 0.29	0.75 ± 0.14	0.30 ± 0.03	0.16 ± 0.08	0.04 ± 0.03	0.018 ± 0.008
Pancreas	1.69 ± 0.58	1.18 ± 0.22	0.46 ± 0.11	0.22 ± 0.11	0.05 ± 0.04	0.010 ± 0.004
Kidneys	10.66 ± 2.75	8.91 ± 2.39	2.95 ± 0.69	1.74 ± 0.61	0.52 ± 0.39	0.011 ± 0.009
Stomach	0.38 ± 0.06	0.33 ± 0.07	0.29 ± 0.02	0.26 ± 0.06	0.24 ± 0.13	0.29 ± 0.24
Muscle	0.76 ± 0.21	0.75 ± 0.18	0.22 ± 0.07	0.15 ± 0.08	0.03 ± 0.01	0.006 ± 0.001
Femur	0.37 ± 0.19	0.23 ± 0.06	0.07 ± 0.04	0.08 ± 0.02	0.03 ± 0.01	0.014 ± 0.007
Tumor	0.86 ± 0.37	0.78 ± 0.26	0.67 ± 0.06	0.27 ± 0.20	0.18 ± 0.12	0.001 ± 0.001
Blood	2.82 ± 0.44	2.16 ± 0.22	1.87 ± 0.25	0.98 ± 0.21	0.22 ± 0.12	0.007 ± 0.006
Urine	2.49 ± 0.90	17.12 ± 7.27	154.2†	276.9 ± 84.5	91.7 ± 56.4¶	1.61 ± 0.91

* Data represent normalized mean % dose/g values ± standard deviation
¶ 4 animals per data point unless otherwise specified
§ 2 animals per data point
φ 6 animals per data point
† one value only

activity represents both bone and bone marrow, and the gallbladder was assayed with the liver. Blood and urine data are expressed as % dose/1.0 ml.

RESULTS AND DISCUSSION

The tissue distribution of ^{125}I-7-iododiazoxide is given in Table 1. An analysis of the blood distribution data demonstrates that clearance of 7-iododiazoxide follows first order kinetics with a blood half-life of 1-2 hr. This is in agreement with the published plasma clearance data on ^{3}H-diazoxide in rats (30). However, unlike the ^{3}H-compound, ^{125}I-diazoxide is not concentrated against the blood gradient in any tissue except in the kidneys, and in the liver at times less than 1 hr. Activity in these tissues presumably represents excretion of the radiochemical as evidenced by the elevated urinary activity levels. Urinary excretion is known to be the principal route of elimination of the parent drug (31), and biliary excretion is known to occur as well, becoming a more important route in conditions of kidney disease (32). Moreover, activity levels in the stomach as a measure of free ^{125}I-iodide (33) demonstrate that ^{125}I-iododiazoxide does not undergo significant *in vivo* deiodination up to 6 hr postadministration.

Thus, we have synthesized and characterized an iodinated analog of diazoxide which is stable *in vivo* and which demonstrates a similar pharmacokinetic and excretion pattern to the parent compound. We are currently investigating retention of hyperglycemic properties of stable iododiazoxide by our previously published method (34).

Although activity concentration of the ^{125}I-iododiazoxide in the β-cell tumor never exceeds levels in the blood, a comparison of tumor/pancreas ratios demonstrates a progressive rise from 0.5 at 5 min to 3.8 at 6 hr. This prolonged activity retention in the tumor may be due to receptor interaction with the neoplastic β-cells, and is therefore viewed as encouraging. If this is indeed the case, the reason that higher activity levels are not seen in the tumor is likely due to the low specific activity (∿10μCi/mg) of the ^{125}I-iododiazoxide used in these studies. The relatively high concentration of carrier iododiazoxide would result in competition of labeled and unlabeled material for the receptors, with resultant low activity concentration in the tumor. We are currently attempting to prepare ^{125}I-iododiazoxide of higher specific activity to determine if higher tumor activity levels can indeed be achieved. Such higher specific activity material, furthermore, should also result in higher tumor/blood ratios. Although diazoxide is known to be highly protein bound in both rodents and humans (30), ^{3}H-diazoxide was nevertheless selectively concentrated in a variety of animal tissues compared to plasma (30). Thus, it should be possible to achieve elevated tumor/blood ratios in likewise fashion with ^{125}I-iododiazoxide.

REFERENCES

1. Rubin AA, Roth FE, Winbury MM, et al: New class of antihypertensive agents. Science 133: 2067, 1961
2. Koch-Weser J: Diazoxide. N Eng J Med 294: 1271-1273, 1976
3. Dollery CT, Pentecost BL, Samann NA: Drug induced diabetes. Lancet 2: 735-737, 1962
4. Wolff FW, Langdon RG, Ruebner BH, et al: A new form of experimental

diabetes. Diabetes 12: 335-338, 1963
5. Scandellari C, Zaccaria M, Sicolo N, et al: Medical treatment of endogenous organic hyperinsulinism. Horm Metab Res: Suppl 6, 45-54, 1976
6. Smith HM, ed: Diazoxide and the Treatment of Hypoglycemia, Ann NY Acad Sci 150: 191-467, 1968
7. Austoni M, Scandellari C, Trisotto A, et al: Hypoglycemia and Diazoxide, Padua (Italy), Cedam, 1972
8. Unauthored Review: Diazoxide and the secretion of insulin. Nutr Rev 30: 194-198, 1972
9. Unauthored Review: Diazoxide - a review of its pharmacological properties and therapeutic use in hypertensive crises. Drugs 2: 78-137, 1971
10. Loubatieres A, Mariani MM, Alric R: The action of diazoxide on insulin secretion, medullo-adrenal secretion, and the liberation of catecholamines. Ann NY Acad Sci 150: 226-241, 1968
11. Graber AL, Porte D Jr, Williams RH: Clinical use of diazoxide and studies of the mechanism of its hyperglycemic effects in man. Ann NY Acad Sci 150:303-308, 1968
12. Senft G: Biochemical aspects of the hyperglycemic action of diazoxide. Ann NY Acad Sci 150: 242-255, 1968
13. Bleicher SJ, Chowdhury F, Podolsky S, et al: Studies on diazoxide-induced hyperglycemia: and extrapancreatic mechanism. Ann NY Acad Sci 150: 294-302, 1968
14. Mariani MM, Loubatieres A, Chapal J: Etude de l´action inhibitrice du diazoxide sur l´insulino-secretion du pancreas isole et perfuse du rat - Absence d´implication des recepteurs alpha-adenergiques. C R Soc Biol (Paris) 165: 1398-1401, 1971
15. Unauthored Review: Diazoxide. Br Med J 4: 417-418, 1972
16. Burr IM, Marliss EB, Stauffacher W, et al: Diazoxide effects on biphasic insulin release - adenergic suppression and enhancement in the perfused rat pancreas. J Clin Invest 50: 1444-1450, 1971
17. Lefebvre P, Luyckx A, Cessian-Fossion A: Effect of diazoxide on golden hamsters with chronic hyperglycemia due to a transplantable islet-cell tumor. Horm Metab Res 4: 355-360, 1972
18. Ariens EJ: Oral antidiabetics. Dose, plasma concentration and effect. Acta Diabet Lat 6: Suppl 1, 143-176, 1969
19. Mariani MM, Loubatieres A: Recentes experiences concernant l´antagonisme entre certains sulfamides hypoglycemiants et le diazoxide. In Hypoglycemia and Diazoxide, Austoni M, Scandellari C, Trisotto A, et al, eds, Padua,Cedam, 1972, pp 33-53
20. Heindel ND, Risch VR, Burns HD, et al: Syntheses and tissue distribution of ^{99m}Tc-sulfonylureas. J Pharm Sci 64: 687-689, 1975
21. Honda T, Risch VR, Heindel ND, et al: Radiopharmaceuticals for pancreatic tumors. In Radioaktive Isotope in Klinik und Forschung XII. Hofer R, ed, Vienna, H. Egermann, 1976, pp 783-803
22. Monzani A, Di Bella M, Fabio U, et al: Antimicrobial action of 1,2,4-benzothiadiazine 1,1-dioxide derivatives. Farmaco, Ed Sci 28: 149-156, 1973
23. O'Sullivan DG: Vibrational frequency correlations in heterocyclic molecules. J Chem Soc: 3653-3658, 1960
24. Beyer H, Volcker C: Uber die synthese von pyridazino (3.2-b) chinazolonen-10. Chem Ber 97: 390-395, 1964
25. Counsell RE, Ranade V, Kline W, et al: Potential organ or tumor imaging agents XV: Radioiodinated phenytoin derivatives. J Pharm Sci 65: 285-287, 1976

26. Risch VR, Markoe AM, Honda T: The pharmacology and preclinical evaluation of new radiopharmaceuticals. In The Chemistry of Radiopharmaceuticals, Heindel ND, Burns HD, Honda T, et al, eds, New York, Masson Publishing USA Inc., 1978, pp 123-154
27. Topliss JG, Yudis M: Pi-substituent constants for the 2H-1,2,4-benzothiadiazine 1,1-dioxide system. J Med Chem 15: 400-403, 1972
28. Topliss JG, Sherlock MH, Reimann H, et al: Antihypertensive agents I. Non-diuretic 2H-1,2,4-benzothiadiazine 1,1-dioxides. J Med Chem 6: 122-127, 1963
29. Risch VR, Honda T, Heindel ND, et al: Distribution of ^{99m}Tc-1-thioglucose in rats: Effect of administration route on pancreatic specificity. Radiology 124: 837-838, 1977
30. Pruitt AW, Dayton PG, Patterson JH: Disposition of diazoxide in children. Clin Pharm Ther 14: 73-82, 1973
31. Pruitt AW, Faraj BA, Dayton PG: Metabolism of diazoxide in man and experimental animals. J Pharm Exp Ther 188: 248-256, 1974
32. Pearson RM: Pharmacokinetics and response to diazoxide in renal failure. Clin Pharmacokinet 2: 198-204, 1977
33. Logothetopoulos JH, Myant NB: Concentration of radio-iodide and ^{35}S-labeled thiocyanate by the stomach of the hamster. J Physiol 133: 213-219, 1966
34. Tutwiler GF, Bridi GJ, Kirsch TJ, et al: Hyperglycemic activity of some non-nitrosated streptozotocin analogs. Proc Soc Exp Biol Med 152: 195-198, 1976

ACKNOWLEDGMENT

This work was supported by grant CA 14043 from NIH/NCI

RADIOLABELED ENZYME INHIBITORS AS PANCREATIC SCANNING AGENTS

Stephen T. Mosley, Padmakar V. Kulkarni, and Robert W. Parkey

Department of Radiology
The University of Texas Health Science Center at Dallas
Dallas, Texas 75235

ABSTRACT

Radiolabeled inhibitors of the pancreatic protease enzymes trypsin and chymotrypsin: I-125-soybean inhibitor (SBI), I-125-human antitrypsin (HAT), and ^{3}H-diisopropylfluorophosphate (DFP) were studied in rats for their uptake in the pancreas. The protease enzyme inhibitor, SBI, was found to retain its enzymatic activity after radio-iodine labeling by iodine-monochloride method. Assay of tissue radioactivity at 5, 15, 30, 60 min and 24 hr after iv injection in rats showed pancreas:liver (P:L) ratios of < 1.4 and pancreas:blood (P:B) ratios of < 1 at all time intervals for each compound. However, ip injection of I-125 SBI showed preferential uptake in pancreas, P:L mean ratios [$\bar{x}$ (range)] of 6.0 (5.5-6.3) and P:B mean ratios of [$\bar{x}$ (range)] 2.60 (1.8-3.4), up to 60 minutes. These ratios resemble those of Se-75-selenomethionine except that kidney values are higher. As a control, when the nonspecific I-131 IgG was given ip, similar uptake in the pancreas was noted at 15 min, P:L = 6.2 and P:B = 2.0. Most amino acid analogs given iv can pass relatively intact through the liver to the pancreas with good P:L ratios based on the higher metabolic rate of the pancreas. Our data suggest that radiolabeled proteins can also show similar good early pancreatic uptake when given ip by escaping immediate liver metabolism. Proteins can readily be tagged with desirable gamma-emitting radionuclides (viz. I-131, I-123, Tc-99m) compared to amino acids and may potentially be suitable for imaging the pancreas. It may be important to study both iv and ip routes of administration in evaluating the specificity/affinity of potential pancreatic radiopharmaceuticals.

INTRODUCTION

The enzymes within an organ or tissue are a reflection of that organ's or tissue's biological function. Thus, it is logical that an enzyme system unique to a particular organ could provide a localized substrate that would bind to a specific radiopharmaceutical. Beierwaltes (1,2) has suggested that enzyme inhibitors that bind reversibly or irreversibly to enzymes that are unique to a target area might serve as a starting point in the development of potential radiopharmaceuticals. His group has examined a group of radiolabeled reversible inhibitors of enzymes of the adrenal cortex and found significant localization of radioactivity with good target-to-nontarget ratios and have produced diagnostic quality adrenal scans with a gamma-labeled enzyme inhibitor analog, I-131-SKF-12185. Recently, they have extended this approach to the imaging of the adrenal medulla and have selectively imaged the medulla with a radio-iodinated bretylium analog (3).

We postulated that the enzyme inhibitors of the pancreatic enzymes trypsin and chymotrypsin may localize in the pancreas. We evaluated three different radio-labeled enzyme inhibitors of trypsin and chymotrypsin: soybean inhibitor (SBI), alpha-1-human antitrypsin (HAT), and diisopropylfluorophosphate (DFP). SBI and HAT are moderately large size proteins (molecular weights: 21K and 56K Daltons respectively) whereas DFP is a small nonprotein, lipid soluble organic compound with molecular weight of 184 Daltons.

MATERIALS AND METHODS

Radiolabeled Compounds

SBI and HAT were purchased from Worthington Biochemicals, Inc., Freehold, N. J., and labeled with I-125 by a modification of the iodine monochloride method of McFarlane (4). The labeled proteins were purified on an anion exchange column (Dowex 1x8 resin). The radiochemical purity was assessed by ascending thin layer (ITLC-SG)* or paper chromatography (Whatman #1 paper) using 70 percent methanol as the solvent. The radiochromatograms were scanned using a radiochromatogram scanner. Gel electrophoresis of the labeled and unlabeled proteins was also carried out, followed by autoradiography of the labeled products. The inhibitory activity of the SBI was determined by inhibition of trypsin hydrolysis by benzoyl-L-arginine ethyl ester (BAEE) (5). H-3-DFP was purchased from New England Nuclear, Inc., Boston, Massachusetts. Nonspecific IgG was purified from normal human serum by 50 percent ammonium sulfate precipitation and DEAE cellulose column chromatography and subsequently labeled with I-125 and purified as mentioned above.

Animal Studies

Male Sprague-Dawley rats (200-300 gm) were used. Each animal received 10-50 μCi either iv via tail vein injection or ip near the umbilicus.

---

*Gelman Co., Ann Arbor, Michigan

The rats were sacrificed in groups of 2 to 15 at various time intervals between 5 minutes and 24 hours after the dose. Samples (at least 2 to 3) of pancreas, liver, kidney, spleen, and blood were taken from each rat, cleaned of connnective tissue and fat, weighed, and counted in a gamma well counter. The data were expressed as percent of injected dose per gram $\pm$ S.E.M. (% I.D./gm $\pm$ S.E.M.).

RESULTS

Radiochemical analysis showed that more than 95 percent of the radioactivity stayed at the origin, and all the preparations had less than 5 percent of free iodide. Gel electrophoresis showed approximately 90 percent of the radioactivity was in a single band corresponding to the proper molecular weight of each compound. The enzyme inhibition activity of the labeled SBI was not diminished compared to the native protein.

The animal tissue distribution showed that the maximum pancreas to liver (P:L) ratios in the intravenous series was 1.4 for SBI at 30 min. The P:L ratio for HAT and DFP was less than 1.0 for all the time intervals examined (Tables 1,2,3).

In the intraperitoneal series, consisting of SBI, IgG, and DFP, the proteins SBI and IgG showed maximum P:L ratios of 6.2 (range 4.6 - 9.3, at 5, 15, 30, and 60 minutes after injection). DFP, the only nonprotein in this series, showed P:L of less than 1.0 following ip injection (Table 4).

DISCUSSION

The high affinity of amino acids for the pancreas was first observed in 1949 (6) and led to development of selenium-75 (Se-75) selenomethionine by Blau and Manske in 1961 (7). Since then, scores of amino acid analogs and other organic type compounds have been carefully evaluated as potential pancreatic radiopharmaceuticals (8-10). While some have shown promise (9,10), as a whole, P:L ratios for all compounds tested have been 10 to 20 or less, roughly the same order of magnitude as Se-75-selenomethionine.

The P:L values for Se-75-selenomethionine following subcutaneous administration are also less than 10 (7), and SBI and nonspecific IgG approach the same value (approximately 6, range 4.6-9.3) when given ip. When injected iv, the P:L ratio of Se-75-selenomethionine still approaches 10, but that of SBI and IgG fall to around 1. This suggests that the decrease in P:L ratio is a function of the normal metabolic processing of proteins by the liver which is circumvented at early time intervals when the proteins are given ip. DFP is a nonprotein, irreversible inhibitor of trypsin and other serum proteases, has a low molecular weight (184 Daltons) corresponding to the size of a single amino acid, yet shows no pancreatic specificity (P:L < 1) after either iv or ip administration. This might be due to labeling of red cells prior to its localization in pancreatic tissue resulting in a high blood background. Thus, our data suggest that the pancreatic uptake has a distinct but broad-based specificity, and/or low P:L ratios may be due either to lack of pancreatic affinity/specificity or to liver metabolism or both. It

TABLE 1

I-125-α-Human Anti-Trypsin

Tissue Distribution in Sprague-Dawley Rats

% Injected Dose/gm ± SEM*

Dose: 10 μCi IV

	5 min 3 rats	30 min 3 rats	60 min 3 rats	24 Hr 3 rats
Pancreas	.045 ±.012	.037 ±.007	.033 ±.004	.014 ±.002
Liver	.157 ±.020	.159 ±.009	.130 ±.013	.028 ±.005
Kidney	.213 ±.042	.183 ±.009	.186 ±.016	.036 ±.005
Spleen	.076 ±.011	.101 ±.004	.132 ±.035	.020 ±.002
Heart	.088 ±.022	.227 ±.060	.104 ±.014	.032 ±.008
Lung	.142 ±.080	.136 ±.026	.182 ±.022	.041 ±.010
Thyroid	1.213 ± .445	2.932 ± .558	20.592 ±13.118	132.990 ± 19.052
Blood	.839 ± .108	.629 ±.071	.665 ±.096	.119 ±.106

*2-4 samples of each tissue from each rat

TABLE 2

H-3-Diisopropylfluorophosphate
Tissue Distribution in Sprague-Dawley Rats
% Injected Dose/gm ± SEM*
Dose:10μCi IV

	5 min 3 rats	15 min 3 rats	30 min 3 rats	60 min 3 rats	24 hr 3 rats
Pancreas	.177 ±.015	.192 ±.022	.125 ±.053	.211 ±.023	.129 ±.012
Liver	.285 ±.040	.364 ±.090	.262 ±.039	.308 ±.033	.155 ±.032
Kidney	.910 ±.043	.786 ±.078	.695 ± .045	.532 ±.015	.214 ±.024
Spleen	.287 ±.025	.348 ±.043	.330 ± .021	.297 ±.021	.176 ±.042
Heart	.600 ±.143	.472 ±.041	.689 ± .343	.348 ±.008	.141 ±.007
Lung	1.743 ± .319	3.822 ±1.937	1.004 ± .282	1.545 ± .121	.421 ±.044
Blood	2.254 ± .140	1.918 ± .174	3.452 ± .830	1.530 ± .149	6.35 ±2.027

*2-4 samples of each tissue from each rat

may be important to study both iv and ip routes of administration in evaluating the specificity/affinity of potential pancreatic radiopharmaceuticals.

ACKNOWLEDGMENTS

The authors thank Ms. Rose Sullivan for excellent technical assistance and Ms. Gayle Blust for excellent secretarial assistance.

This work was supported in part by the Education and Research Foundation of the Society of Nuclear Medicine.

REFERENCES

1. Beierwaltes WH, Wieland DM, Ice RD, et al.: Localization of radiolabeled enzyme inhibitors in the adrenal gland. J Nucl Med 17:998-104, 1976.

2. Beierwaltes WH, Wieland DM, Yu T, Swanson DP, Mosley ST: Adrenal imaging agents: Rationale, synthesis, formulation, and metabolism. Semin Nucl Med 8:23-42, 1978.

3. Wieland DM, Swanson DP, Brown LE, Beierwaltes WH: Imaging the adrenal medulla with an I-131-labeled antiadrenergic agent. J Nucl Med 20:155-158, 1979.

4. McFarland AS: Labeling of plasma proteins with radioactive iodine. Biochem J 62:135, 1956.

5. Schwert GW and Takenaka YA: A spectrophotometric determination of trypsin and chymotrypsin. Biochim et Biophys Acta 16:570, 1955.

6. Wheeler JE, Lukens FDW, Gyorgy P: Studies on the localization of tagged methionine within the pancreas. Proc Soc Exp Biol Med 70: 187, 1949.

7. Blau M, Manske RF: The pancreas specificity of Se-75-selenomethionine. J Nucl Med 2:102-109, 1961.

8. Busch H, Davis JR, Honig GR, et al.: The uptake of a variety of amino acids into nuclear proteins of tumors and other tissues. Cancer Res 19:1030-1039, 1959.

9. Washburn LC, Sun TT, Rafter JJ, et al.: C-11-Labeled amino acids for pancreas visualization. J Nucl Med 17:557-558, 1976 (abst).

10. Washburn LC, Wieland BW, Sun TT, et al.: 1-C-11-DL-Valine, a potential pancreas imaging agent. J Nucl Med 19:77-83, 1978.

TABLE 3

I-125-Soybean Inhibitor
Tissue Distribution in Sprague-Dawley Rats
% Injected Dose/gm ± SEM*
Dose: 10 μCi IV

	5 min 3 rats	30 min 3 rats	60 min 3 rats
Pancreas	.213 ± .027	.093 ± .010	.047 ± .002
Liver	.158 ± .017	.070 ± .012	.407 ± .203
Kidney	4.532 ± .712	4.415 ± .442	2.180 ± .179
Heart	.094 ± .001	.039 ± .011	.025 ± .002
Lung	.220 ± .003	.130 ± .013	.097 ± .007
Thyroid	2.365 ± .197	7.587 ± 1.114	21.739 ± 2.753
Blood	.467 ± .047	.097 ± .003	.098 ± .001

*2-4 samples of each tissue from each rat

TABLE 4

Tissue Distribution in Sprague-Dawley Rats

% Injected Dose/gm ± SEM*

Dose: 10 µCi IP

	I-125-Soybean Inhibitor				H-3-DFP	I-131-IgG
	5 MIN 15 rats	15 MIN 4 rats	30 MIN 4 rats	60 MIN 3 rats	15 MIN 3 rats	15 MIN 4 rats
Pancreas	.143 ±.013	.159 ± .014	.123 ±.052	.172 ±.020	.556 ±.114	.145 ±.007
Liver	.024 ±.003	.029 ± .003	.025 ±.004	.031 ±.002	.574 ±.069	.039 ±.0.8
Kidney		1.361 ± .73	1.632 ± .975	.444 ±.147	1.180 ±.164	--
Spleen		.076 ± .016	.137 ±.010	.081 ±.005	.421 ±.087	
Blood	.043 ±.006	.053 ± .003	.067 ±.010	.083 ±.017	2.106 .478	.121 ±.031

*2 - 3 samples from each tissue from each rat

RADIOBROMINATED ANDROGENS AS POTENTIAL PROSTATIC IMAGING AGENTS

Raymond E. Counsell, William H. Klausmeier and Andre Castonguay

Department of Pharmacology
and Laboratory of Medicinal Chemistry
The University of Michigan
Ann Arbor, Michigan 48109

and

Rodney V. Pozderac, Paul A. Weinhold and R.W. Scot Skinner

The Nuclear Medicine Unit
Veterans Administration Hospital
Ann Arbor, Michigan 48105

ABSTRACT

A variety of analogs of testosterone were synthesized containing a bromine atom at different positions on the steroid nucleus. Analysis of these brominated androgens for their ability to compete with tritiated 5α-dihydrotestosterone in a rat prostate receptor protein assay revealed several brominated analogs to have an affinity in the order of testosterone. Of these analogs, 2-bromo-17β-hydroxy-5α-androst-1-en-3-one (I) was selected for further study. A synthetic route to this compound was devised that involved bromination in the presence of silver tetrafluoroborate as catalyst. Despite the stability of compound I under *in vitro* conditions, labeling with bromine-82 showed considerable debromination to occur *in vivo*. Such rapid debromination of radiobrominated steroids represents a serious constraint upon their potential use as radiodiagnostic agents.

INTRODUCTION

Within the past fifteen years a class of biological structures called receptors has been identified and implicated in a wide range of hormone actions. In a general sense, these receptors are "locks" within which the appropriate hormone "key" can fit and, thereby, exert its action. A particular hormone receptor is usually only present in those tissues whose activities are regulated by that hormone. This selective distribution has long been considered a desirable means of concentrating a radionuclide-containing hormone or hormone analog in a particular tissue containing that hormone's receptor. Receptors have been identified for the sex steroid hormones testosterone (and 5α-dihydrotestosterone or DHT), estradiol and progesterone. The intense study of these receptor-hormone systems has led to detailed descriptions of the structure-activity relationships governing the binding of a hormone analog to a particular receptor (1). Many of the target organs which contain one or more of the sex steroid hormone receptors are among those for which suitable imaging agents have not yet been designed. One of these is represented by the prostate gland.

PROSTATE RECEPTORS AND HORMONE ACTION

The prostate contains a receptor protein which specifically binds testosterone and DHT (2). Testosterone is delivered by the circulatory system to the target cells either bound to sex steroid binding globulin (SSBG) or in the free form. The free testosterone is then transported across the cell membrane into the cell where it may then undergo reduction by cytoplasmic testosterone-5α-reductase to DHT. DHT then combines with the androgen receptor complex and becomes translocated to the nucleus whereupon the hormone exerts its influence on genetic expression and increased RNA and protein synthesis (3).

PERCEIVED REQUIREMENTS FOR IMAGING

In order for binding to a hormone receptor protein to qualify as a potential means of selectively concentrating radioactivity, that receptor must fulfill the following additional constraints:

1. The receptor must bind the radiolabelled hormone to a significant extent at very low concentrations ($K_D \leq 10^{-9}$M.).

2. There must be enough receptors per unit weight of target tissue to allow adequate uptake of radiolabel ($\geq 10^5$ cpm/gm tissue).

3. There must be a minimal number of tissues with a high receptor density.

4. There must be low tissue levels of endogenous hormones in order for injected radiolabelled hormones to compete successfully for receptor sites.

There are, as well, quite strict requirements that a radiolabelled hormone should fulfill if diagnostic imaging is to be achieved:

1. The radiolabelled hormone must have a receptor binding affinity comparable to the endogenous hormone.

2. The compound must have low affinity for other biological structures.

3. The compound must have ready access to the receptor-containing tissues.

4. Metabolic inactivation and/or loss of radiolabel must be minimal.

5. The radioisotope employed must be nearly carrier-free and have a short enough half-life for the radiolabelled hormone to have sufficiently high specific activity.

6. Radiolabel incorporation must occur readily at low reactant concentrations with minimal side-product formation.

7. The radioisotope employed must have the proper emission characteristics (γ-rays of 100-510 kev).

The further that any one of these characteristics deviates from the ideal, the greater the imperative that the remaining properties be optimized.

TISSUE DISTRIBUTION OF TRITIATED ANDROGENS

The biodistribution of tritiated androgens did not show specificity for androgen receptor-containing tissues such as the prostate until the labelled hormones were available with sufficiently high specific activity ($\geq 10^4$ Ci/mole) (4). This is due to the fact that there are only a small number of receptors in these tissues to which the tritiated hormones can bind. The uptake of the hormone by the prostate can be enhanced by prior castration of the animals which serves to diminish the level of endogenous hormone and increase the availability of unfilled receptor sites. The rather modest prostatic uptake of [^{3}H]-DHT noted in whole animals suggests that the variables involved cannot deviate much from optimal conditions if imaging is to be achieved (5). Moreover, most studies to date have been conducted in rats and it is currently unclear what role androgen receptors play in pathological disorders of the prostate in man.

RESEARCH PLAN

Despite the complexity of the problem, it was viewed that the androgens could serve as a carrier for a γ-emitting nuclide for use in prostate imaging. Certainly this approach had as much chance of success as any others that have been attempted such as radiolabelled estrogens (6), antiandrogens (5) and polyamines (7). Moreover, we were intrigued by the possible use of a γ-emitting androgen as a tool in endocrinology. Accordingly, our research plan took the following course:

1. A variety of androgens containing potential γ-emitting nuclides were synthesized.

2. The ability of these compounds to compete with ^{3}H-DHT for the binding to receptor protein in both the supernatant and nuclear fractions of rat prostate minces was evaluated.

3. Those analogs showing favorable inhibition *in vitro* were then evaluated *in vivo* for their ability to prevent the selective uptake of ^{3}H-DHT in the ventral prostate.

4. Once the appropriate compound was identified, a product with low specific activity was synthesized in order to evaluate *in vivo* stability.

It was perceived that a compound surviving these assays would be an excellent candidate for labelling with high specific activity for tissue distribution studies.

STRUCTURE-ACTIVITY RELATIONSHIPS

On the basis of an evaluation of over 140 androgen analogs for their ability to bind to the rat prostate androgen receptor, the following structural features were found to be compatible with high affinity:

1. The C-3 carbonyl group and the 17β-hydroxyl group are important.

2. Analogs lacking a 4-5 double bond must have the C-5 proton in the α-configuration.

3. Introduction of unsaturation in the A and B rings.

4. Substitution of non-bulky groups at 2,6, 7α and 17α-positions.

Conceptually then, we looked upon the DHT molecule being bound by a receptor protein wherein interaction with the 17β-OH was crucial, but there was some latitude with respect to substitution at other positions.

SELECTION OF RADIONUCLIDE AND TARGET COMPOUND

Since our previous studies with selenium-labelled androgens (8) had to be terminated because of an inability to obtain selenium-75 with sufficiently high specific activity, our attention turned to radiohalogens as potential labels. The short-lived isotopes of iodine-123, bromine-76 and bromine-77 can generally be readily incorporated into carrier molecules and can be produced carrier-free with a theoretical maximum specific activity of approximately 10^7 Ci/mole.

Radioisotopes of bromine rather than iodine were chosen as the foreign label for this study for the following reasons:

1. Radioisotopes of bromine suitable for preliminary radiolabelling and for carrier-free radiolabelling are readily available at low cost.

2. In general, bromine chemistry can be better controlled than iodine chemistry.

3. Vinyl bromides are more stable chemically than vinyl iodides.

The next step was selection of those brominated androgens that showed high affinity in both the *in vitro* and *in vivo* binding assays. Among those showing the highest affinity were 2-bromo-DHT and the corresponding 1-dehydro analog (I, Fig. 1). Subsequent studies focused on the latter series because of the known chemical reactivity of α-haloketones and their expected instability *in vivo*. The corresponding 17α-methyl analog (II, Fig. 1) was also selected as an appropriate target compound since the 17α-substituent would retard metabolic oxidation of the essential 17β-alcohol and at the same time not interfere with binding to the receptor.

SYNTHESIS OF TARGET COMPOUNDS

Although the target compounds were prepared by one of us over fifteen years ago (9), the method of synthesis was not amenable to incorporation of radiobromine with high specific activity. Consequently, a new synthetic route suitable for such a purpose had to be devised which was rapid and free of side-product formation.

The most obvious means of catalyzing the interaction between bromine and a steroidal olefin is by converting one of the two into a charged species. The most established means to accomplish this is to generate a "positive" halogen by metathetic interaction of bromine with silver cation. Such approaches, employing bromine and silver trifluoroacetate have been utilized to catalyze both aromatic halogenation (10) and bromine additions to steroidal double bonds (11).

Accordingly, treatment of 17β-hydroxy-5α-androst-1-en-3-one with equimolar amounts of bromine and silver tetrafluoroborate ($AgBF_4$) at a concentration of 10^{-4} M followed by dehydrohalogenation with 1,5-diazabicyclo-[4,3,0]-non-5-ene (DBN) and quenching with water afforded the desired bromosteroid in excellent yield with no side-product formation (Fig. 2).

The same sequence applied to the 17α-methyl androgen required initial protection of the 17β-OH as the trifluoroacetate in order to prevent dehydration and rearrangement during the course of the reaction (Fig. 3).

These procedures were subsequently employed to label the target compounds with bromine-82.

BIOLOGICAL STUDIES WITH RADIOBROMINATED ANDROGENS

In vitro and *in vivo* stability studies were undertaken with compound I labelled with bromine-82. Samples of prostate, liver and whole blood were obtained from castrated rats at 60 minutes following I.V. administration of the radiobrominated androgen. The tissues were extracted by the Folch method (12) to afford a lipid phase and a water-soluble phase. These studies (Table I) indicated almost complete debromination *in vivo*. Debromination was also observed when the compound was incubated *in vitro* with whole blood and homogenates of prostate and liver (Table I), but the extent of debromination was much less. Subsequent thin layer chromatography of the two phases demonstrated the radioactivity in the aqueous phase to be bromide, while that in the lipid phase migrated coincident with compound I.

Thus, within 1 hour following administration over 90% of the injected dose was debrominated. Moreover, based on the *in vitro* studies, this capacity to debrominate compound I resides in several tissues such as prostate, liver and blood. Subsequent experiments indicated that the debromination in the blood

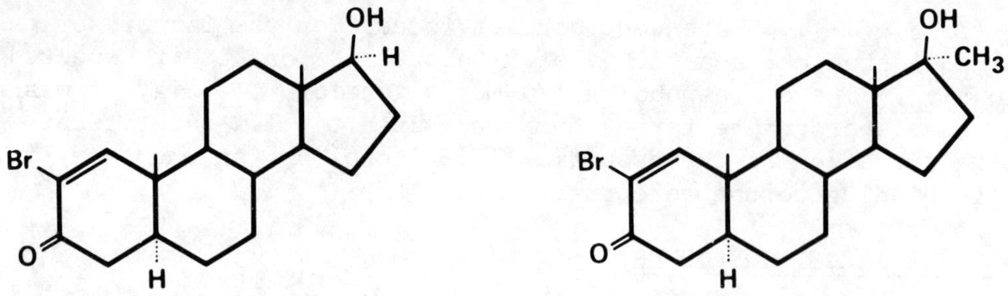

I II

Figure 1.

Brominated androgens selected as target compounds.

Table I

In vitro

	Methanol-water	Chloroform
Prostate	35.6%	64.4%
Liver	55.6%	44.4%
Whole Blood	44.8%	55.2%

In vivo

	Methanol-water	Chloroform
Prostate	95.3%	4.7%
Liver	86.2%	13.8%
Whole Blood	98.7%	1.3%

Biphasic distribution of radioactivity extracted from rat tissue and blood samples. Samples were either incubated for 60 minutes or obtained 60 minutes post I.V. injection.

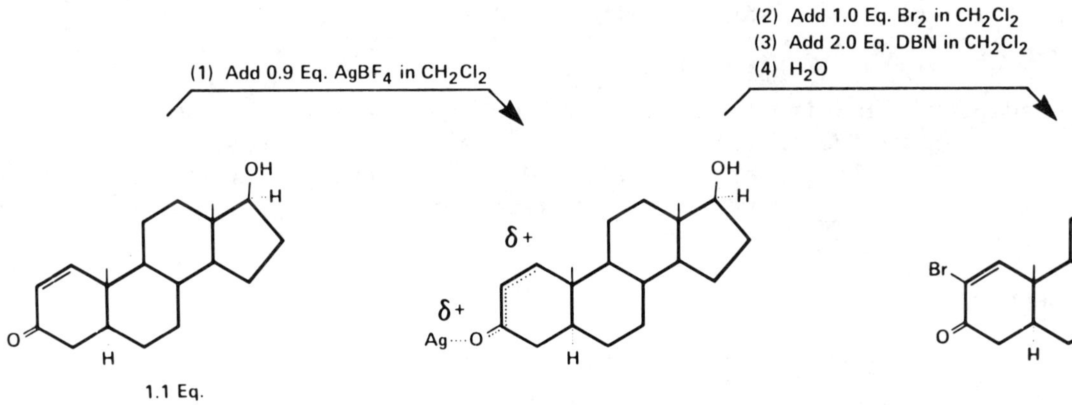

Figure 2.

Bromination sequence employing silver tetrafluoroborate as catalyst.

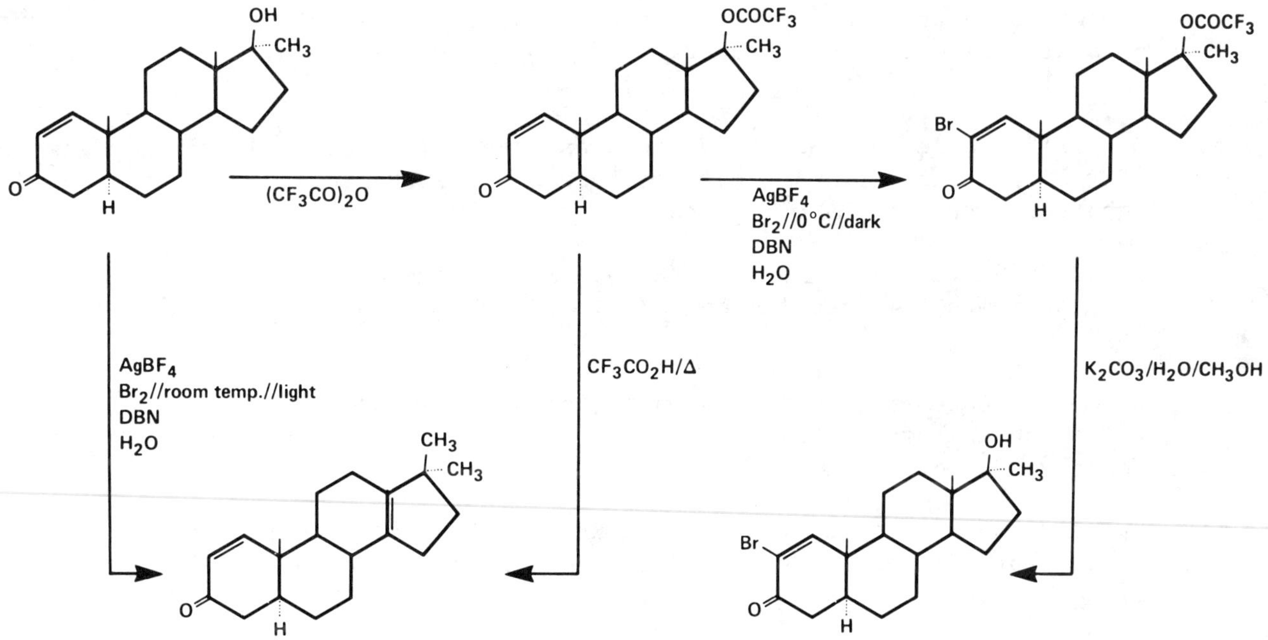

Figure 3.

Bromination reactions with 17α-methyl substituted androgens.

required cellular elements, since no debromination was observed when compound I was incubated with serum.

The unsuspected *in vivo* instability of compound I does not auger well for the future of such radiobrominated androgens as potential prostatic imaging agents. Perhaps labelling such steroids in high specific activity with bromine-77 will circumvent this *in vivo* instability to some extent and such studies are now in progress.

ACKNOWLEDGEMENTS

This research was supported by USPHS grant CA-08349 and Veterans Administration project number 7169:02. The authors are grateful to Searle Laboratories for their generous gift of steroid starting materials.

REFERENCES

1. Counsell, RE, Klausmeier WH: Radiotracer interactions with sex steroid receptor proteins and receptor mapping. In *Principles of Radiopharmacology*, Colombetti, L, Eds, CRC Press, in press.

2. Verhoeven G, Heyns W, de Moor P: Testosterone receptors in the prostate and other tissues. *Vitam Horm* 33: 265-281, 1975

3. Chan L, O'Malley BW: Mechanism of action of the sex steroid hormones. *New Eng J Med* 294: 1322-1328, 1976

4. Treter KJ, Attramadahl A: Selective uptake of radioactivity in rat ventral prostate following administration of testosterone-1,2-^{3}H. *Acta Endocrinol* 59: 218-226, 1968

5. Berger B, Coffey DS, Scott WW: Concepts and limitations in the application of radiolabeled antiandrogens, estrogens, or androgens as isotopic scanning agents for the prostate. *Invest Urol* 13: 10-16, 1975

6. Shida K, Shimazaki J, Kurihara H, *et al*: Uptake and scintiscanning of the prostate with ^{131}I-labeled estradiol phsophate. *Int J Nucl Med Biol* 3: 86-89, 1976

7. Welch MJ, Coleman RE, Straatmann MG, *et al*: Carbon-11-labeled methylated polyamine analogs: uptake in prostate and tumor in animal models. *J Nucl Med* 18: 74-78, 1977

8. Skinner RWS, Pozderac RV, Counsell RE, *et al*: Androgen receptor protein binding properties and tissue distribution of 2-selena-A-nor-5α-androstan-17β-ol in the rat. *Steroids* 30: 15-23, 1977

9. Counsell RE, Klimstra PD: Anabolic agents: derivatives of 2-halo-5α-androst-1-ene. *J Med Pharm Chem* 5: 477-483, 1962

10. Bergmann ED, Shahak I: Organic fluorine compounds. Part X. Reaction of organic compounds with iodine in the presence of silver fluoride and trifluoroacetate. *J Chem Soc*: 1418-1422, 1959

11. Hey DG, Meakins GD, Pemberton MW: Hydroxysteroids. VII. Preparation and reactions of 3α-iodo-5α-cholestan-2β-yl trifluoroacetate. _J Chem Soc C Org_: 1331-1334, 1966

12. Folch J, Lees M, Stanley GHS: A simple method for the isolation and purification of total lipids from animal tissues. _J Biol Chem_ 226: 497-509, 1957

RADIONUCLIDE PRODUCTION

David J. Silvester and Stephen L. Waters

Medical Research Council, Cyclotron Unit,
Hammersmith Hospital
London W12 OHS, U.K.

ABSTRACT

The current programmes of 51 accelerators (all but 2 of which are cyclotrons) producing radionuclides for medical use have been tabulated. With very few exceptions, notably ^{123}I and ^{201}Tl, only those radionuclides which are not available from the 9 commercial sources are made on the 42 non-commercial machines. Development of "new" radionuclides is mostly focused on short-lived positron emitters, or on generator systems, and some of the more innovative production methods are discussed. The cyclotrons now offered by the major manufacturers are reviewed in the light of present trends in radionuclide production and other influencing factors.

INTRODUCTION

Computerised information retrieval systems are widely available nowadays, and, unlike conventional review papers, they can be tailored to meet the needs of individual investigators, thereby eliminating much redundant information. Furthermore, they can be revised at will, whereas a review paper may be out-of-date in some respects even before it appears in print. In this paper, therefore, we have attempted to gather together some information which, as yet, is not stored in computer data banks: this a compilation of the current radionuclide production programmes of all the accelerators contributing to the medical field. We believe this will be of interest to those already working with accelerators, and of considerable value to those thinking of acquiring one.

From our own literature search, we have compiled a list of some 140 references to radionuclide production methods published in the past 2 or 3 years which we will gladly distribute on request. Our purpose here, however, is just to draw attention to current trends, and to highlight some of the more innovative production techniques that have been developed recently.

Finally, because an appreciation of the value of an in-house medical cyclotron is still growing, we include a brief survey of the machines currently on offer by the major commercial manufacturers.

CURRENT PRODUCTION PROGRAMMES

The production of radionuclides to meet medical needs has been a major, and sometimes the only, justification for the construction of 20 or more cyclotrons in different parts of the world in the past 15 years, and for the continued operation of many more that might otherwise have been shut down. Furthermore, some of the world's biggest accelerators have also found it expedient to add radionuclide production programmes to their main objectives in nuclear physics research. Since it is still reactor-produced radionuclides (especially ^{99m}Tc, of course) that play the biggest role in nuclear medicine, the question arises: what contribution are all these machines making?

In the hope of answering that question without guessing, we recently sent out a questionnaire to more than 60 institutes operating accelerators, requesting information on their radionuclide production programmes. The response was gratifying in that 48 replies, relating to 55 accelerators were received, though unfortunately we have no information from Eastern Europe, South Africa or South America. Several respondents provided much additional material, to which full justice cannot be done in this paper; however, in Table 1 we have summarised the main findings. Part A comprises the "non-commercial" accelerators (40 cyclotrons and 2 linacs) and lists, for each of them, those radionuclides (a) which are currently being produced at least once a week, (b) those which are produced less frequently, and (c) those which are expected to be in production by 1980. Part B of the table lists the world's 9 commercially operated cyclotrons and their products.

A few immediate conclusions can be drawn from this information. First, to state the obvious, the diversity of radionuclides being produced indicates the continuing need for accelerators to augment the more readily available reactor products. Second, very few non-commercial cyclotrons find it worthwhile making commercially-produced radionuclides such as ^{67}Ga and ^{111}In, unless their geographical location makes supplies difficult to obtain. However, it is equally evident that demand for two products, ^{123}I and ^{201}Tl, far exceeds the commercial supply, especially in Europe. Third, recognition of the enormous clinical value of ^{81m}Kr (1) has put the production of ^{81}Rb-^{81m}Kr generators high on many priority lists. Finally, the long-awaited availability of commercial positron-imaging devices is stimulating interest in the development of 'new' short-lived positron-emitting nuclides, e.g. ^{21}Na, ^{30}P, ^{38}K etc., as well as boosting the production of the well-known ones, ^{11}C, ^{13}N, ^{15}O, ^{18}F and ^{68}Ga.

NEW INNOVATIONS

Such is the demand for ^{123}I that even now, 10 years after Blue and Sodd (2) first reported its production in ultra-high purity, the challenge of making it is still bringing about technological advances which must influence the whole field of radionuclide production. Replies to our questionnaire show that the $^{124}Te(p,2n)^{123}I$ route, using relatively inexpensive isotopically enriched ^{124}Te target material, has now been adopted at most institutes with machines providing proton beams of up to about 30 MeV, as well as some bigger ones. Kondo et al (3) provide a useful review of the yields obtainable by this method, the highest reported being 45 μCi/μAh (4) for 28 MeV incident protons, though 34 μCi/μAh (3) may be a more reliable figure.

Directly produced ^{123}I is invariably contaminated by as much as 1% ^{124}I at end-of-bombardment. This severe disadvantage is avoided by the more powerful cyclotrons with proton or deuteron beams of over 50 MeV, which can produce ^{123}I indirectly via its ^{123}Xe parent, and it is here that we find almost heroic advances have been made in concepts of target design and product recovery. Until recently, except in the production of aqueous ^{18}F solutions, liquid phase targets had found very limited application, being regarded as potentially too hazardous to the health of the cyclotron ever to be practicable. Now, however, this has been proved untrue, and Table 2 lists some of the liquid phase targets which have been developed for the production of ^{123}I. Although most have been used on only a small scale, it is worthy of note that the liquid NaI target at Davis (6) and the CH_2I_2/I_2 target at Harwell (9) are between them currently producing curie-quantities of ^{123}I each week.

Novel methods for the preparation of anhydrous ^{18}F-fluorinating intermediates have been discussed (12) and the "in-target" preparation of $SF_3{}^{18}F$ (13) and $H^{18}F$ (14) have also been described. The production of new inorganic radionuclides for positron emission tomography is the subject of another paper (15) in this Symposium where special emphasis will be placed on ^{38}K. However, two others merit special mention, because of their potential as physiological tracers: ^{21}Na, with a half-life of only 23 s, has been produced in sufficient quantities for clinical investigations by the $^{20}Ne(d,n)$ reaction (16); and the preferred route by which to make the 2.5 m ^{30}P has been found to be the $^{27}Al(\alpha,n)$ reaction (17). Some particularly interesting results have been achieved with a 12 MeV tandem accelerator at the University of Wisconsin, where, in addition to ^{11}C, ^{13}N and ^{15}O, $^{14}O(t_{\frac{1}{2}} = 71s)$ and $^{17}F(t_{\frac{1}{2}} = 65s)$ have been made, incorporated in simple molecules, and used in cerebral blood flow studies (18).

Despite its limitations, ^{111}In-labelled Bleomycin is still widely used in the diagnosis of soft-tissue tumours, and so the development of a positron-emitting cobalt label for this drug must be regarded as innovative. Cobalt-55 ($t\frac{1}{2}$ = 18h) has been made and used for this purpose by the group at Davis (19).

Amongst new generator systems which are in the process of development, two, those for ^{52m}Mn and ^{178}Ta, will be discussed in later papers (20,21) in this Symposium, as will improvements to a third: the ^{68}Ga generator (22). One of the simplest, yet most effective, innovations in generator systems, however, must be the gas-phase ^{81m}Kr generator described by Fremlin et al (23). The ^{81}Rb parent is made by α-bombardment of sodium bromide, and to reduce handling problems this is dissolved in situ. The resulting solution is easily transferred to the generator system, where it is simply soaked up on a coil of filter paper which takes the place of the ion-exchange medium used in conventional generators.

In order to make more efficient use of their cyclotrons several laboratories, including our own, are working on "tandem" target systems. Naturally, this approach is highly appropriate for really high-energy machines, such as the 200 MeV BLIP (24), but an ingenious three-tier target has been described (25) by means of which a relatively modest 60 MeV proton beam can produce simultaneously ^{123}I, ^{62}Zn and ^{13}N.

CHOOSING A CYCLOTRON

In Table 3 we have listed the cyclotrons currently on offer by the major manufacturers, together with their basic specifications.

Before deciding to install a cyclotron for radionuclide production, any institute would do well to consider the possible impact on its programme both of further development of commercial sources in its region (there are already 6 commercially operated cyclotrons in the U.S.A.) and of the very large, but as yet not fully realised, potential output of radionuclides from the "giant" machines, (e.g. BLIP, LAMPF and TRIUMF in N. America, and in Europe, SIN and possibly the machine being rebuilt at Uppsala). If the problems of distribution can be overcome (and this should not be difficult, as the commercial firms have shown) these big machines could become very important indeed as radionuclide producers.

However, in order fully to exploit the advances in research, and perhaps diagnosis, made possible by the newly developed positron-imaging devices, it will always be vital to have access to a cyclotron capable of producing at least the four "basic" short-lived radionuclides: ^{11}C, ^{13}N, ^{15}O and ^{18}F. Fortunately, these can be made in good yield with relatively low energy proton or deuteron beams (26), and, realising this, cyclotron manufacturers are now competing in their offers of small machines (with proton or deuteron beams of up to 10-16 MeV) designed to meet just this basic need. Although the Japanese "baby" cyclotron must be the closest to being commissioned, the others being essentially still on the drawing board, even this one has yet to be fully tested in a clinical environment, and so it is not possible to forecast which may prove the most successful in the long run.

A programme calling for the production of a wider range of radionuclides will require a bigger cyclotron. A rough guide to the capabilities of some machines of different sizes can be obtained from Table 1, though it must be borne in mind that for various reasons only a few institutes use their machines to anything approaching the limits of their production capacity.

Many factors will influence the final decision as to what will be the most suitable machine for a given institute. Ultimately, the most important of these may be the total cost of the installation, which will in turn be affected by the nature of the site (greenfield, or otherwise?), proximity to the manufacturer, and, in some countries perhaps, national policy. However, the technical factors, too, need careful consideration.

First it must be decided whether to buy a cyclotron which can accelerate ^{3}He- and ^{4}He- particles as well as protons and deuterons. The recent development of cyclotrons which accelerate negative ions is of great interest, since beam extraction is much simpler to achieve than with conventional positive ion machines. In principle, beams of different energies can be extracted to bombard two or three targets simultaneously, so such a machine could be used very efficiently. Furthermore, they should be cheaper than positive-ion cyclotrons of similar performance. A major drawback to such a machine, however, is the fact that only singly charged particles (H^- or D^-) can be accelerated and so ^{3}He or ^{4}He beams will not be available. Whilst, again in principle, most radionuclides of medical interest can be made with proton or deuteron beams of sufficient energy, there are nevertheless some (e.g. ^{43}K, ^{52}Fe) for which is is scarcely practicable. A positive ion accelerator therefore offers greater flexibility of output.

There are now more cyclotrons on offer with variable- than with fixed-energy beams, and if this does not lead to undue complexity of operation it may be advantageous. However, it is perhaps worth noting that the director of at least one institute with a variable energy machine has indicated that after a few years' experience he would now prefer to have the equivalent (slightly higher) fixed energy cyclotron.

To the one remaining question - how big must the cyclotron be? - there is once again no straightforward answer. As has been stated above, the smallest available cyclotrons are designed to make only four radionuclides, whilst the biggest, with proton beams of up to 80 MeV, would be able to make good yields of any radionuclide which has so far found any application in nuclear medicine, and, of course, many more besides (27). As can be seen in Table 1, a fixed energy, multi-particle cyclotron capable of accelerating protons to only 26, 22 or even 15 MeV can sustain a useful and varied production programme, so this would certainly be true of today's "middle-range" of cyclotrons, giving proton beams of up to 35 or 40 MeV.

It could be, therefore, that as well as on cost, the final decision about size might rest on yet another consideration: whether the machine is to have any other function than radionuclide production. This is certainly the case in the U.K. where it was recently decided that a further cyclotron should be provided for fast neutron therapy and radionuclide production. For the fast neutrons to have a "depth dose" comparable to ^{60}Co therapy machines, a 40-MeV proton beam would suffice, but to compare treatment with those carried out on an 8-MV linac, a 70 MeV cyclotron would be necessary (28).

ACKNOWLEDGEMENTS

We wish to express our sincere thanks to the 42 individuals named in Table 1, for their generous response to our questionnaire, and to a further 5 who advised us that their cyclotrons in Bonn, College Station, Jyväskylä,

Maryland and Uppsala were not currently making any radionuclides for medical use. Our thanks are also due to the cyclotron manufacturers, listed in Table 3, for information received.

REFERENCES

1) Clinical and experimental applications of Krypton-81m. LAVENDER J.P., ed. Brit. J. Radiol, Special Report No.15, London 1978.

2) BLUE J.W., SODD V.J. : ^{123}I production from the β^+ decay of ^{123}Xe. In Uses of cyclotrons in chemistry, metallurgy and biology. C.B. Amphlett, ed. Butterworths, London, 1970, pp 138-148.

3) KONDO K., LAMBRECHT R.M., WOLF A.P. : ^{123}I production for Radiopharmaceuticals XX : Excitation functions of $^{124}Te(p,2n)^{123}I$ and $^{124}Te(p,n)^{124}I$ reactions and effect of target enrichment on radionuclidic purity. Int. J. Appl. Rad. Isotopes. 28 : 395-401, 1977.

4) ACERBI E., BIRATTI C., CASTIGLIONI M, et al: Production of ^{123}I for medical purposes at the Milan AVF cyclotron. Int. J. Appl. Rad. Isotopes 26 : 741-747, 1975.

5) GODART J., BARAT J.L., MENTHE A. et al : In beam collection of Xe-123 for carrier-free I-123 production. Int. J. Appl. Rad. Isotopes 28 : 967-969, 1977.

6) LAGUNAS-SOLAR M.C., JUNGERMAN J.A., PEEK N.F. et al : Large scale cyclotron production of high purity carrier-free Iodine-123. J. Labelled. Cmpds. Radiopharm: XVI : 224-225, 1979.

7) BIEVELEZ P., COGNEAU M., CHARLIER R. et al : Preparation of carrier-free ^{123}I with a high radionuclidic purity. J. Radioanal. Chem. 30 : 67-79, 1976.

8) SCHIMMEL A., KASPERSEN F.M., LINDNER L : Cyclotron production of $^{123}Xe(\rightarrow^{123}I)$ and ^{127}Xe using CH_2I_2 as target material. Int. J. Appl. Rad. Isotopes 30 : 63-64, 1979.

9) (a) CUNINGHAME J.G., MORRIS B., NICHOLS A.L. et al : Large scale production of ^{123}I from a flowing liquid target using the (p,5n) reaction. Int. J. Appl. Rad. Isotopes 27 : 597-603, 1976.
(b) CUNINGHAME J.G., HILL J.I.S., NICHOLS A.L. et al : The production of Iodine-123 on the Harwell variable energy cyclotron : AERE Report 9087, 1978.

10) WEINREICH R : In Iodine-123 in Western Europe, S.M. Qaim, G.Stöcklin, R.Weinreich eds., Jül-Conf-20, KFA Jülich, August 1976, pp 119-125.

11) (a) BLUE J.W. : US Patent 3971697, July 1976.
(b) BLUE J.W., LEONARD R., JHA S. et al : A cyclotron target for the production of radioxenons. Prog. Nucl. Med. 4 : 53-60, 1978. (Karger. Basel 1978).

12) LAMBRECHT R.M., NEIRINCKX R., WOLF A.P. : Cyclotron isotopes and radiopharmaceuticals XXIII. Novel anhydrous ^{18}F-fluorinating intermediates. Int. J. Appl. Rad. Isotopes. 29 : 175-183, 1978.

13) STRAATMAN G., WELCH M.J. : F-18-Labelled diethylamino sulphur trifluoride (DAST) : An F-for-OH fluorinating agent. J. Nucl. Med. 18 : 151-158, 1977.

14) TEWSON T.J., WELCH M.J., RAICHLE M. : (^{18}F) labelled 3-deoxy-3-fluoro-D-glucose : Synthesis and bio-distribution data. J. Nucl. Med. 19 : 1339-1345, 1978.

15) LAMBRECHT R.M., WOLF A.P. : Cyclotron production and use of inorganic radionuclides for positron emission tomography. Proc. 2nd Int. Symposium on Radiopharm. March 1979, Seattle, Washington, U.S.A.

16) HELUS F., BAUER H., MAIER-BORST W : Cyclotron production of short lived ^{21}Na. J. Labelled. Cmpds. Radiopharm. XVI : 107-108, 1979.

17) SAHAKUNDU S.M., QAIM S.M., STÖCKLIN G : Cyclotron production of short-lived ^{30}P. J. Labelled. Cmpds. Radiopharm. XVI : 106, 1979 and Int. J. Appl. Rad. Isotopes 30 : 3-5, 1979.

18) NICKLES R.J., GATLEY S.J., MADSEN M.T. et al : High yield synthesis of the sequence $^{13}N_2$, $^{13}N_2O$, $^{11}C_2H_2$, $H_2{}^{15}O$, $H_2{}^{14}O$ and F^- for rCBF studied. J. Labelled. Cmpds. Radiopharm. XVI : 90-91, 1979.

19) LAGUNAS-SOLAR M.C., JUNGERMAN J.A. : Cyclotron production of carrier-free Cobalt-55 : a new positron emitting label for Bleomycin. Int. J. App. Rad. and Isotopes 30 : 25-32, 1979.

20) KU T.H., RICHARDS P., STANG L.G.(Jnr.) et al : BLIP production of Fe-52 and the Mn-52m generator. Proc. 2nd Int. Symposium on Radiopharm. March 1979, Seattle, Washington U.S.A.

21) NEIRINCKX R.D., HOLMAN B.L., DAVIS M.A. : Tantalum-178 radiopharmaceuticals for lung and liver imaging. Ibid.

22) NEIRINCKX R.D., DAVIS M.A. : Development of a chromatographic ^{68}Ge-^{68}Ga generator yielding ionic gallium. Ibid.

23) FREMLIN J., STAMMERS K., STEWART F.R. : A new generator for Krypton-81m. Nucl. Inst. and Methods 156 : 369-373, 1978.

24) STANG L.G.(Jnr.) : The Brookhaven Linac Isotope Producer. Prog. Nucl. Med. 4 : 34-36, 1978 (Karger. Basel. 1978).

25) SUZUKI K., IWATA R : Multi-target assembly in an irradiation with high energy particles - Simultaneous production of ^{123}I, ^{62}Zn and ^{13}N. Int. J. Appl. Rad. and Isotopes. 28 : 663-665, 1977.

26) HARPER P.V: Potentials and problems of short-lived radionuclides in medical imaging applications. Int. J. Appl. Rad. Isotopes 28 : 5-11, 1977.

27) For useful information on this topic see : STÖCKLIN G., QAIM S.M., Recent trends in nuclear reaction data needs for nuclear medicine. Proc. Int. Conf. on Neutron Physics and Nuclear Data for Reactors and other Applied Purposes. Harwell, U.K. Sept. 1978 (to be published).

28) A Strategy for Future Research in Neutron Therapy. M.R.C. Cancer Therapy Committee : Working Party on Neutron Therapy, 1978. (unpublished).

TABLE 1. ACCELERATORS FOR RADIONUCLIDE PRODUCTION : DECEMBER 1978

A. NON-COMMERCIAL ACCELERATORS

Location Institute Person supplying information	Machine type Time spent on radio-nuclide production	Radionuclides produced Routinely at least 1/week; Irregularly; (in future)
BELGIUM		
Ghent University Institute for Nuclear Studies Dr. C. Vandecasteele	CGR-MeV 520 8 h/wk	Rb/Kr-81, N-13, (C-11)
Leige University Cyclotron Research Centre Dr. M.A. Guillaume	CGR-MeV 520 25 h/wk	C-11, N-13, O-15, F-18, K-43, Rb/Kr-81, Se-73, Ga-67, In-111
Louvain-la-Neuve University Nuclear Chemistry Lab., Dr. M. Cogneau	CGR-MeV 930 15 h/wk	Fe-52, I-123, Mg-28, At-211, (Tl-201)
CANADA		
Vancouver, B.C., TRIUMF Dr. B.D. Pate	TCC CP-42 and TRIUMF 500-MeV	I-123, (C-11, N-13, O-15, F-18, Fe-52, Se-77m, Ga-67, Ge-68, Cd-109, In-111, Xe-127, Tl-201)
Winnipeg University of Manitoba Dr. M.I. Gusdal	(50 MeV protons)	I-123, Rb/Kr-81, Rb-84
FINLAND		
Turku Abo Academy Dr. M. Brenner	Techsnabexport, USSR (20 MeV protons) 16 h/wk	C-11, F-18, Rb/Kr-81, O-15, K-43, Fe-52, Br-77, Tl-197-202, Re-181, Bi-206, (I-123)

FRANCE		
Orsay Service Hospitalier Fred. Joliot Dr. D.Comar	CGR-MeV 520 30 h/wk	C-11, N-13, O-15, F-18, K-43, Ru-97, Ge-68, Bi-206, (Rb/Kr-81)
GERMAN FEDERAL REPUBLIC		
Essen University Clinic Dr. H.J.Machulla	TCC CV-28 (50%)	C-11, F-18, I-123, (N-13, O-15)
Heidelberg Inst.f.Nuklearmedizin, DKFZ Dr. F.Helus	AEG - Compact	N-13, F-18, Rb/Kr-81, I-121, I-123, Hg-197m, O-15, Fe-52, Cu-61, Zn-62, Ru-97, Tl-201, (C-11)
Jülich Inst.f.Chemie 1, KFA Dr. G.Stöcklin	(1) TCC CV-28 24 h/wk	C-11, F-18, P-30, Cr-48, Br-77, I-123, Cl-34m, Tl-201 Pb-203, (more F-18 and Cr-48)
do	(2) "JULIC" (45 MeV protons) 12 h/wk	Mg-28, I-123 - every 2 wks., Br-76,77
Karlsruhe Kernforschungszentrum Dr. H.Schweickert	Isochronous (26 MeV protons) 5 h/wk	I-123 (500mCi/wk), (double I-123, Rb/Kr-81)
INDIA		
Calcutta Dr. R.S.Mani (Bhabha Atomic Res. Centre, Trombay)	V.E.C. (60 MeV protons)	(K-43, Fe-52, Ga-67, In-111, I-123, Tl-201)

ITALY		
Milan University Cyclotron Laboratory Dr. C.Birattari	A.V.F. (45 MeV protons) 10-15 h/wk	Rb/Kr-81, I-123, Tl-201, V-48, As-71-74, Cd-107-109, Hg-197, Pb-203, Bi-205-206
JAPAN		
Chiba Natl.Inst. of Radiol.Sciences Dr. T.Hiramoto	CGR-MeV 930 12 h/wk	C-11, N-13, O-15, F-18, Ti-45, Fe-52, Br-77, I-123
Saitama Inst.Phys.Chem.Research Dr. T.Nozaki	- (18 MeV protons) 5 h/wk	F-18, Mg-28, K-43, Fe-52, Br-77
Sendai Tohoku University Dr. S.Morita	CGR-MeV 680 30-40 h/wk (planned)	(F-18, Br-77)
Tokyo University Inst.of Medical Science Dr. Akira Ito	TCC CS-30 2 h/wk	F-18, Ti-44, Co-56, I-123
NETHERLANDS		
Eindhoven University of Technology Dr.R.L.P.van den Bosch	AVF (30 MeV protons) 8 h/wk	Br-77, I-123, Fe-55, Sr-87m, Cd-109, (Fe-52, Rb/Kr-81)
Groningen University Inst.of Nuclear Physics Dr. W.Vaalburg	AVF (65 MeV protons) 7 h/wk	C-11, N-13, Co-55, O-15, Fe-52, Rb/Kr-81, I-123, Hg-197

SWITZERLAND		
Villigen Natl.Inst.for Nucl.Research Dr. I.Huszar	VEC (72 MeV protons) and Ring accelerator (590 MeV protons) 3-4 h/wk	In-111, I-123, (Xe-127)
UNITED KINGDOM		
Birmingham University Dept. of Physics Prof. J.H.Fremlin	Fixed-frequency (10 MeV protons) 20 h/wk	Rb/Kr-81, I-123, Na-22, Po-208, (daily Rb/Kr-81)
Edinburgh M.R.C. Western General Hospital Mr. T.Saxton	TCC CS-30	(O-15, N-13, C-11, Rb/Kr-81)
Harwell Chem.Divn., AERE Mr.J.G.Cuninghame	VEC (58 MeV protons) 12 h/wk	I-123, S-38, Nb-92m, Rh-99m, Pu-236, Pu-237, (Tl-201)
London M.R.C.Hammersmith Hospital Mr. I.A.Watson	Classical 45" (8 MeV protons) 37 h/wk	C-11, N-13, O-15, F-18, K-43, Rb/Kr-81, Fe-52, I-123, Cu-61, Zn-62, Br-77, Ru-97, Cs-129, Pb-203 (K-38)
UNITED STATES OF AMERICA		
Bloomington Indiana Univ.Cyclotron Fac. Dr. D.L.Friesel	Separated sector VEC (200 MeV protons) (under 1%)	I-123
Boston Mass.Gen.Hospital Dr. G.L.Brownell	Allis-Chalmers (6 MeV deuterons) 40 h/wk	C-11, N-13, F-18, O-15,

Brookhaven Natl.Lab. Upton L.I., N.Y. Dr. A.P. Wolf	Isochronous 60" (36 MeV protons) 40-50 h/wk	C-11, N-13, O-15, F-18, K-38, Mn-51, Mn-52, Rb/Kr-81, I-123, At-211, Ne-19, Cl-34m, Fe-55, Br-76, Br-77, Se-97m, Tc-92, Os-191, Au-195m, Pb-203,
Brookhaven Natl.Lab. Upton L.I., N.Y. Dr. Louis Stang	BLIP (Lin.Acc.) (200 MeV protons) 145 h/wk	Mg-28, Fe/Mn-52, Ga-67, Rb/Kr-81, Ru-97, I-123, Ge-68 Cd-109, Sn-113, Te-118, Xe/I-122, Xe-127, W-178
Chicago University Franklin McLean Mem.Res.Inst. Dr. P.V. Harper	TCC CS-15 10 h/wk	C-11, N-13, O-15, F-18, Mn-51
Cleveland, Ohio NASA Dr. J.W. Blue	VEC (45 MeV protons) 8 h/wk	C-11, Tl-201
Davis, Calif. Crocker Nuclear Lab. Dr. M.C.Lagunas-Solar	VEC (66 MeV protons) 20-35 h/wk	I-123, Tl-201, N-13, Co-55, Rh-101m
Los Alamos, N.Mex. C.P.Anderson Meson Phys. Facility Dr. H.A. O'Brien	LAMPF (Lin.Acc.) (800 MeV protons) 168 h/wk	Fe/Mn-52, Br/Se-77, Sr/Rb-82, Hf/Lu-172, (Cu-67, Ge-68, Se-72, I-123, Xe-127)
Los Angeles, Calif. U.C.L.A.School of Medicine Dr. N.S. MacDonald	TCC CS-22 50 h/wk	C-11, N-13, F-18, Zn/Cu-62, Cu-64, Rb/Kr-81, O-15, Cd-107, In-111, I-123
Miami Beach, Fla. Mount Sinai Medical Centre	TCC CS-30	C-11, O-15, Ga-67, In-111, Tl-201

(information derived from Prog. nucl. Med. 4, 28 (Karger, Basel 1978)

Michigan State Univ.Cyclotron Lab. Dr. P.S.Miller	VEC (50 MeV protons) av. 6 h/wk	C-11, N-13
New York, N.Y. Sloan Kettering Inst. Dr. R.S.Tilbury	TCC CS-15 20 h/wk	C-11, N-13, O-15, F-18, K-38, Fe-52, Se-73
Oak Ridge, Tenn. O.R.N.L. Dr. S.W.Mosko	Isochronous Cyclotron (55 MeV protons) 2%	No data given
St.Louis, Mo. Washington University Dr. M.J.Welch	(1) Allis-Chalmers (6 MeV deuterons)	C-11, N-13, O-15, F-18
do	(2) 52" Cyclotron (30 MeV ^4He) 12 h/wk	Br-77, Kr-77, Rb/Kr-81
do	(3) TCC CS-15	(C-11, N-13, F-18)
Washington, D.C. Naval Research Lab. Dr. R.O.Bondelid	Sector-focusing cycl. (65 MeV protons) -	C-11, F-18, I-123

B. COMMERCIALLY OPERATED CYCLOTRONS

Firm Location Person supplying information	Machine type	Radionuclides produced
JAPAN		
Nihan Medi-Physics Takarazuka, Hyogo Dr. Masaaki Hazue	TCC CS-30	F-18, Ga-67, Rb-81, In-111, I-123, Tl-201
NETHERLANDS		
Mallinckrodt (formerly Duphar) Petten, N.Holland	Philips (28 MeV protons)	No information given
UNITED KINGDOM		
The Radiochemical Centre Amersham Dr. M. Finlan	Philips (27 MeV protons)	Ga-67, In-111, Co-57, As-73,74, Au-195, Be-7 Bi-206,207, Co-56, Fe-55, Mn-52,54, Ti-44, V-48
UNITED STATES		
Medi-Physics Dr. A. Fleischer		
(i) Arlington Heights, Ill.	Scanditronix MC-40	Rb-81, I-123, Tl-201
(ii) Emeryville, Calif.	TCC CS-22	Ga-67, Rb-81, In-111, I-123, Tl-201
(iii) South Plainfield, N.J.	TCC CS-22	Ga-67, Rb-81, I-123
New England Nuclear North Billerica, Mass. Dr. J. Need	(i) TCC CS-22	Co-57, Ga-67, Cd-109, In-111
	(ii) TCC CS-30	Ge-68, Au-195, Tl-201
	(iii) TCC CS-30	Ge-68, Au-195, Tl-201

TABLE 2. LIQUID-PHASE TARGET SYSTEMS USED TO PRODUCE I-123

Nuclear Reactions	Target Material	Incident Beam Energy (MeV)	I-123 yields (mCi/μAh)	Location	Ref.
$^{127}I(p,5n)^{123}Xe$	I_2	58	4	Grenoble	5
	NaI	65	21	Davis	6
	NaI/H_2O	60	4	Louvain	7
	CH_2I_2	52	4	Amsterdam	8
	CH_2I_2/I_2	58	6	Harwell	9
$^{127}I(d,6n)^{123}Xe$	NaI	64	8	Jülich	10
$^{133}Cs(p,p10n)^{123}Cs$	Cs	-	-	Vancouver	11

TABLE 3. COMMERCIALLY AVAILABLE CYCLOTRONS - FEBRUARY 1979

Country of Origin & Manufacturer	Model No.	Particle	Extracted beam Energy MeV	Extracted beam Current μA	Mass metric tons	Power UK	Approx. Cost US $	Location of example
FRANCE CGR-MeV **	*325	p	13.5	50	10		1.75M	-
		d	8	50				
	520	p	3-24	100	28	270	3M	Orsay
		d	3-15	100				Leige
		^{3}He	7-31	50				Ghent
		α	10-30	50				
	560	p	5-40	100	55		4.5M	-
		d	10-20	100				
		^{3}He	15-52	50				
		α	20-40	50				
	*560p	p	42	50	55		4M	-
	680	p	6-45	100	100	600	5.75M	Orleans
		d	12-30	100				Sendai
		^{3}He	12-70	50				
		α	12-60	50				
	*680p	p	50	50	100	600	4.35M	-
	930	p	10-80	40	200	1000	9M	Louvain
		d	10-50	40				Chiba
		^{3}He	20-130	30				
		α	20-100	30				

Country of Origin & Manufacturer	Model No.	Particle	Extracted beam Energy MeV	Extracted beam Current μA	Mass metric tons	Power UK	Approx. Cost US $	Location of example
JAPAN Japan Steel Works	*Baby	p d	10 7	30-50	15.5	80	700K F.O.B.	-
SWEDEN Scanditronix	*MC16F	p d	16 8		17		700- 900K	
	MC20	p d ^3He α	3-20 1.5-10 3-27 3-20	100 100 30 30	23	200	1.1M	Jyväskylä
	MC35	p d ^3He α	2-35 4-18 6-48 8-35	65 65 30 30	55	375	1.5M	Hannover
	MC40	p d ^3He α	9-40 5-20 7-53 9-40	65 65 30 30			1.6M	Arlington Heights
	*MC60F	p	60					
U.S.A. The Cyclotron Corporation ***	*CP16	p^- d^-	8-16 4-8	50 50	16.5	200		-
	*CS15	p d ^3He α	15 8 20 16	60 100 60 40	22	175		New York Chicago St. Louis

Country of Origin & Manufacturer	Model No.	Particle	Extracted beam Energy MeV	Extracted beam Current μA	Mass metric tons	Power UK	Approx. Cost US $	Location of example
The Cyclotron Corporation (Cont'd)	*CS30	p	26	60	29	250		Miami Beach
		d	15	100				Edinburgh
		^3He	38	60				Tokyo
		α	30	40				N. Billerica(2)
	CV28	p	2-24	60	29	250		Essen
		d	4-14	100				Jülich
		^3He	6-36	5-50				
		α	8-28	8-40				
	CP30	p^-	8-30	200	34			
		d^-	5-17	50				
	CP42	p^-	11-42	200	44	325		Vancouver
		d^-	6-24	50				
U.S.S.R. Techsnabexport		p	5-18	50	24	200		Turku
		d	3-10	50				
		^3He	8-24	25				
		α	6-20	25				

* fixed energy beams

** CGR-MeV prices include cost of installation

*** TCC prices range from $65K to $1.4M

BLIP PRODUCTION OF Fe-52 AND THE Mn-52m GENERATOR*

Thomas H. Ku, Powell Richards,
Louis G. Stang, and Thomas Prach

Brookhaven National Laboratory
Medical Department
Upton, New York 11973

ABSTRACT

Fe-52 ($t_{1/2}$=8.275 h) is the only radioisotope of iron with decay properties suitable for imaging applications (100% 169-keV and 112% annihilation gammas). It can also be used in a Fe-52/Mn-52m generator system as a source of Mn-52m ($t_{1/2}$=21.1m) which is a short-lived positron emitter (192.54% annihilation gammas). Clinically useful quantities of Fe-52 have been prepared at the Brookhaven Linac Isotope Producer (BLIP) by bombarding manganese or nickel targets with medium-energy protons. After chemical separation, Fe-52 is loaded onto a generator column from which carrier-free Mn-52m is eluted. The Mn-52m generator is potentially useful in positron emission computed tomography.

*This work was done under Contract No. EY-76-C-02-0016 with the UNITED STATES DEPARTMENT OF ENERGY.

INTRODUCTION

Iron-52 ($t_{1/2}$= 8.275 h) has decay characteristics suitable for detection on a gamma camera (169-keV gammas, 100% abundant), or a positron emission tomograph (511-keV positron annihilation gammas, 112% abundant). It is the only iron isotope useful in nuclear imaging applications. The major use for Fe-52 is in monitoring the distribution of hemopoietic marrow and the diseases of the blood-forming organs. It can also be used in a Fe-52/Mn-52m generator system (1) as a convenient source of short-lived Mn-52m which is well suited for computed emission tomography. Mn-52m ($t_{1/2}$=21.1 min) decays with a 192.54% abundance of annihilation photons and a 1.434-MeV photon in 97.8% abundance (2). The 21.1-min half life is sufficiently short to provide reasonably low radiation exposure and makes possible repeated studies at fairly short intervals. Yet the half-life is long enough to permit rapid synthesis of Mn-52m-labelled radiopharmaceuticals. Radiomanganese has been demonstrated to accumulate in the myocardium (3).

The widespread use of Fe-52 has been limited because of the difficulty in preparing sufficient quantities on the available cyclotrons. The previously reported production methods (4-8), including the bombardment of a chromium target with alpha or He-3 particles and of MnO_2 with protons, could produce Fe-52 batches limited to only several millicuries. This study was initiated to explore the possibility of producing substantial quantities at the BLIP (Brookhaven Linac Isotope Producer) (9,10), and to evaluate the Mn-52m generator.

PRODUCTION OF Fe-52

The method involves bombarding a manganese or a nickel target with a medium-energy proton accelerator. Manganese powder (99.99% pure, Atomergic Chemicals Corp., Plainview, NY), 2.62-g/cm^2 thick, is sealed between 0.25-mm copper windows and bombarded by a 70-MeV proton beam to form Fe-52 by the Mn-55(p,4n)Fe-52 reaction. With a 0.45-g/cm^2 (0.5-mm) nickel foil target (99.98% pure, Goodfellow Metals Cambridge Ltd., Cambridge, England), Fe-52 is produced mainly by the (p,3p4n) reaction between high abundance (68.3%) Ni-58 and 200-MeV protons. Other nickel isotopes also produce small additional contributions by the (p,3pXn) reactions. A typical bombardment lasts for 15 hours at an average beam current of 70 microamperes.

Following irradiation, the manganese target is dissolved by heating in 5N hydrochloric acid, and the nickel target in 12N nitric acid. For both targets, Fe-52 is separated from the target matrix and other reaction products by means of an anion-exchange method. After adjusting the target solution to 6N in HCl by mixing with concentrated hydrochloric acid, it is forced by air pressure through a pretreated ion exchange column (AG 1 X 8, 200-400 mesh resin, Bio-Rad Laboratories, Richmond, CA, 1.8 cm dia. X 2 cm long) at a flow rate of 25 ml/min at 50°C. The column is then washed with two portions of 100 ml of 6N HCl followed by two portions of 65 ml of 4N HCl at 4 ml/min at 80°C. Fe-52 is finally eluted with three 10-ml portions of 0.01N HCl at 80°C under gravity. The Fe-52 recovery is consistently higher than 90%.

The Fe-52 production parameters are summarized in Table 1. Yield determinations are performed using a calibrated Ge(Li) detector connected to a multichannel analyzer. For the target thicknesses used, the production rates are 0.20 mCi/μAh for the manganese target and 0.12 mCi/μAh for the nickel target; approximately 120 mCi and 70 mCi are produced from the manganese and nickel targets, respectively, in a typical bombardment of 15 hours at a 70 μA beam current. The gamma spectroscopy also reveals a trace of 44.6-d Fe-59 impurity in the Fe-52 from the nickel target (0.17% at end of bombardment). The 2.7-y Fe-55 impurity, which emits 5.9-keV (24.3%) and 6.5-keV (3.3%) X-rays and no gamma rays, is determined by means of a high-resolution Si(Li) X-ray detector. It amounts to 0.7% for a manganese target and 2.0% for a nickel target at end of bombardment. However, when used in a generator, these impurities would present no problem to the radionuclidic quality of Mn-52m since they are retained on the column and their daughters are nonradioactive.

Iron-52 produced in charged-particle accelerators is virtually carrier-free. However, stable iron could be introduced from target materials and reagents or through improper handling. In this work, high-purity manganese and nickel, containing less than 10 ppm iron impurity (according to suppliers' specifications), are utilized as target materials. Reagent-grade acids are used in the chemical processing. To remove the last traces of stable iron contaminants, the acids are further purified beforehand by passing the HCl through an anion exchange column and by distilling the HNO_3. Dilutions of the acids and solutions are done with ultrapure water ("Super Q water" filter system obtained from Millipore Corp., Bedford, MA, with resistance >10 Megohms). The amount of iron present in the final Fe-52 solution, as determined by atomic absorption, is of the order of 100-500 μg for both targets. The specific activity is higher than 0.14 mCi/μg at end of bombardment.

Because of the complexity of the Ni-58(p,3p4n)Fe-52 reaction, the excitation function is expected to show a monotonical rise from the threshold energy and then remain relatively flat once the maximum is reached. Consequently, an increase in nickel target thickness should raise the yield proportionately. The concentrations of Fe-55 and Fe-59 impurities in the product are also expected to remain unchanged for the same reason.

THE Fe-52/Mn-52m GENERATOR

The Fe-52 solution, or part of it, is transferred to a glass chamber and made 8N in HCl. The solution is subsequently forced by air pressure through the column of the generator in which Fe-52 is adsorbed as $FeCl_4^-$. An additional 5 ml of 8N HCl is run through the generator to wash the column.

The generator column consists of the lower two-thirds of a 1-ml plastic tuberculin syringe barrel with a cast male Leur-Lok connection (Burron Medical Products, Inc., Bethlehem, PA). It is press-fitted over a modified Teflon female Leur adaptor (Analtech Inc., Neward, DE) through which the outlet tubing is connected. The column is filled with 0.6 ml of pretreated anion exchange resin (AG 1 X 8, 200-400 mesh, Bio-Rad Laboratories, Richmond, CA) supported by a porous disc of inert polypropylene (Amicon Corp., Lexington, MA). The generator is housed inside a 2-cm-thick lead shield through the top of which the inlet (1.2-mm-i.d. X 60-cm-long Teflon tubing, Hamilton Co., Whittier, CA) and outlet (1.2-mm-i.d. X 60-cm-long PE50 Intramedic polyethylene tubing, Clay-Adams Inc., New York, NY) lines protrude. After loading Fe-52, the shielded assembly is

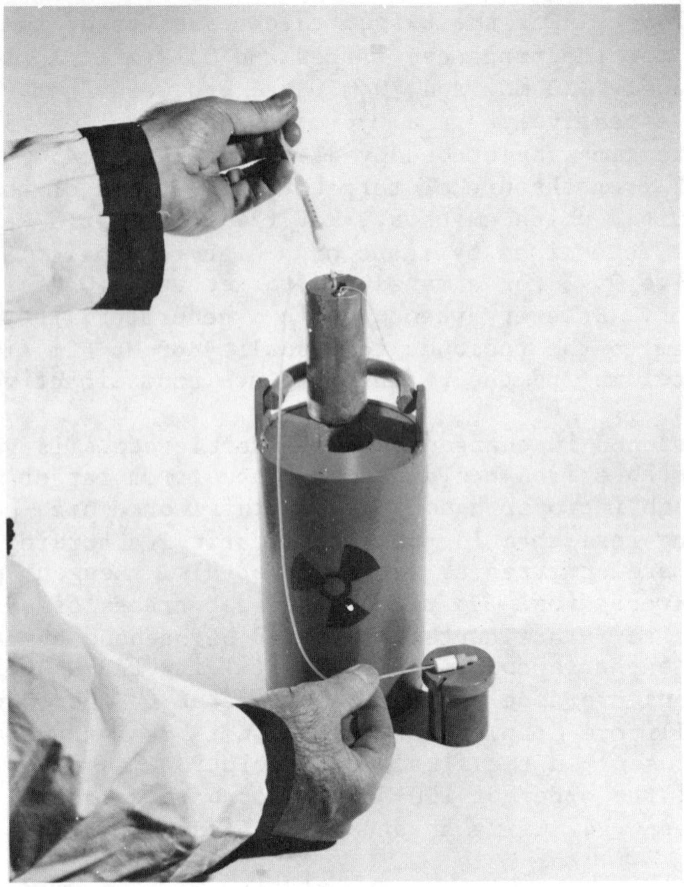

Fig. 1. The Mn-52m generator unit with lead shields.

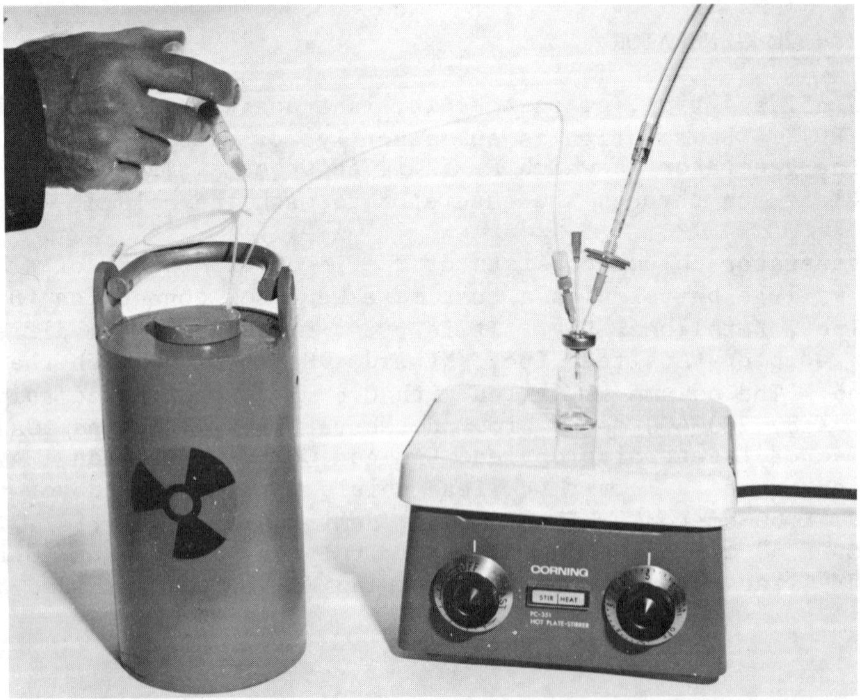

Fig. 2. Radionuclide generator system for Mn-52m.

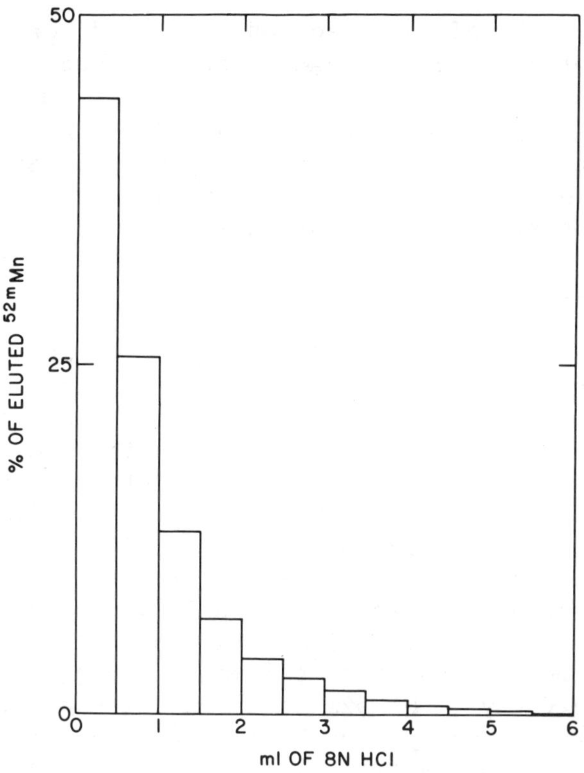

Fig. 3. Elution profile of the Mn-52m generator.

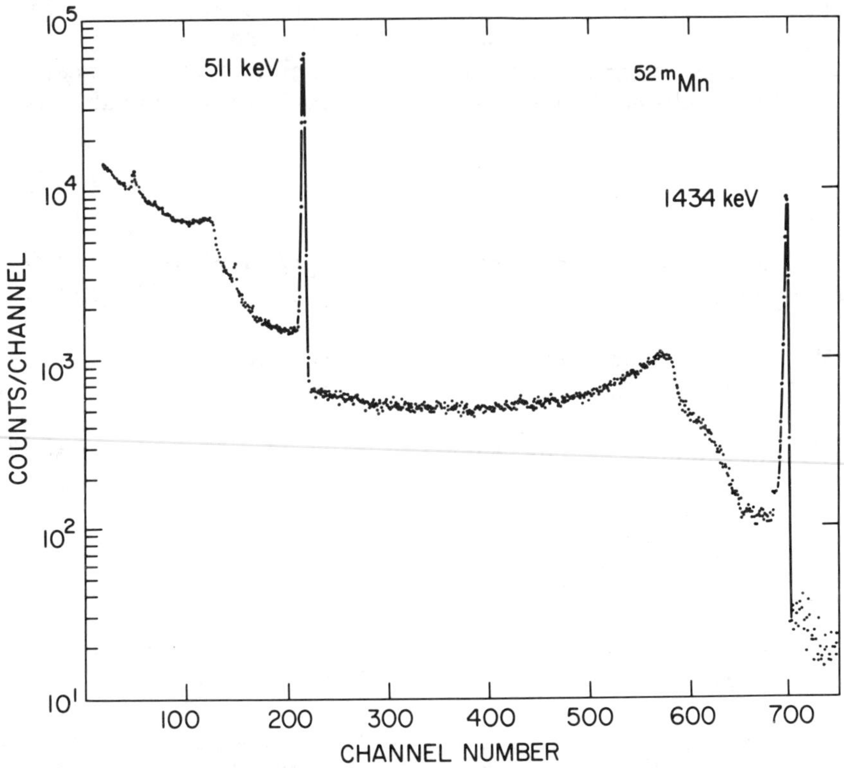

Fig. 4. Ge(Li) gamma spectrum of eluted Mn-52m.

taken out of the hot cell and inserted into a second 3.7-cm-thick lead shield. A generator unit together with the lead shields is shown in Fig. 1.

The generator is eluted with 2 ml of 8N HCl. The eluate is evaporated on a hot plate in a stream of filtered nitrogen gas until dry (Fig. 2). The Mn-52m is then redissolved in 3 ml of sodium acetate buffer (pH=5.6). The solution is passed through a disposable 0.45 μm bacterial filter (Millipore Corp., Bedford, MA) before injection to ensure sterility of the Mn-52m. The entire procedure takes less than 10 minutes.

A typical elution curve of the Fe-52/Mn-52m generator is shown in Fig. 3. With the column having a 0.6 ml resin bed, approximately 90% of the generated Mn-52m is recovered in 2 ml of 8N HCl. Sample elutions have been checked by independent laboratories (11,12) and found to be sterile and pyrogen-free. Fig. 4 is a gamma-ray spectrum of the eluted Mn-52m obtained using a Ge(Li) detector coupled to a multichannel analyzer. The radionuclidic impurities were determined after the short-lived Mn-52m had decayed. No Fe-52 breakthrough was detected. The only activity found was Mn-52 from the decay of Mn-52m. Mn-52m decays in two branches; 97.8% goes to stable Cr-52 and 2.2% goes to 5.67-d Mn-52. Decay-growth calculations (13) indicate that $3.2 \times 10^{-4}\%$ of Mn-52 (relative to Mn-52m) is formed in the generator column in a typical growth period of one hour. Mn-52 from the decay of Mn-52m at 118 minutes after elution would amount to $5.6 \times 10^{-3}\%$ of what the Mn-52m had been at time of elution. The presence of such small amounts would have a negligible impact on the total radiation dose to the patient.

The generator is being evaluated at Brookhaven National Laboratory in collaboration with the Department of Surgery, the State University of New York at Stony Brook, using the PETT III tomographic instrument. Preliminary studies for the detection and diagnosis of myocardial infarcts in a dog model have shown promising results (14). Rapid labelling techniques for the preparation of Mn-52m-labelled radiopharmaceuticals for other possible applications are also being developed.

REFERENCES

1. Atcher RW, Friedman AM, Huizenga JR, et al: Mn-52m -- A new short-lived, generator-produced radionuclide. J Nucl Med 19: 689, 1978 (abst)

2. Martin MJ, ed: Nuclear decay data for selected radionuclides. Oak Ridge National Laboratory Report 5114, 1976, pp 5-6

3. Schelbert H, Chauncey D, Halpern S, et al: Radioactive manganese: Its potential as a myocardial imaging agent. J Nucl Med 18: 642, 1977 (abst)

4. Yano Y, Anger HO: Production and chemical processing of Fe-52 for medical use. Int J Appl Radiat Isot 16: 153-156, 1965

5. Greene MW, Lebowitz E, Richards P, et al: Production of Fe-52 for medical use. Int J Appl Radiat Isot 21: 719-723, 1970

6. Saha GB, Farrer PA: Production of Fe-52 by the Mn-55(p,4n) Fe-52 reaction for medical use. Int J Appl Radiat Isot 22: 495-498, 1971

7. Dahl AS, Tilbury RS: The use of a compact, multi-particle cyclotron for the production of Fe-52, Ga-67, In-111 and I-123 for medical purposes. Int J Appl Radiat Isot 23: 431-438, 1972

8. Murakami Y, Akiha F, Ezawa O: Comparative studies on the advantage of He-4 or He-3 particles bombardment for the production of Fe-52 and I-123 for medical use. In Radiopharmaceuticals and Labeled Compounds, vol 1, Vienna, IAEA, 1973, pp 257-273

9. Richards P, Lebowitz E, Stang Jr LG: The Brookhaven Linac Isotope Producer (BLIP). In Radiopharmaceuticals and Labeled Compounds: vol 1, Vienna, IAEA, 1973, pp 325-341

10. Stang Jr LG: The Brookhaven Linac Isotope producer. Prog Nucl Med 4: 34-46, 1978

11. The South Shore Analytical and Research Laboratory, Inc., Islip, New York

12. Leberco Laboratories, Roselle Park, New Jersey

13. Friedlander G, Kennedy JW, Miller JM: Nuclear and Radiochemistry. New York, John Wiley and Sons, 2nd ed, 1964, pp 70-76

14. Hui J, Atkins HL, Som P, et al: Manganese-52m positron emission tomography for detecting myocardial infarction (to be published)

TABLE 1. Fe-52 PRODUCTION PARAMETERS

Target Material	Manganese (2.62 g/cm^2)	Nickel (0.45 g/cm^2)*
Nuclear Reaction	Mn-55(p,4n)Fe-52	Ni-58(p,3p4n)Fe-52
Incident Proton Energy	70 MeV	200 MeV
Exit Proton Energy	50 MeV	198.6 MeV*
Production Rate	0.20 mCi/μAh	0.12 mCi/μAh*
Irradiation	15 hours at 70 μA	15 hours at 70 μA
Total Yield (EOB)**	120 mCi	70 mCi*
Radioimpurities (EOB)**	0.7% Fe-55	2.0% Fe-55 and 0.17% Fe-59

*An increase in nickel target thickness would raise the yield proportionately without affecting the quality of Fe-52.

**End of bombardment.

POSITRON-EMITTING RADIONUCLIDES - PRESENT AND FUTURE STATUS

Richard M. Lambrecht

Chemistry Department
Brookhaven National Laboratory
Upton, New York 11973

ABSTRACT

A tabulation of 157 positron-emitting radionuclides which have the physical characteristics deemed appropriate for radiopharmaceutical use in conjunction with positron emission tomography is provided. The most promising radionuclides are within the production capabilities of a variable energy cyclotron accelerating protons to about 40 MeV and deuterons, helium-3, and helium-4 to compatable energies. To date only 27 positron emitting radionuclides have been subjected to radiopharmaceutical consideration, whereas only ^{11}C, ^{13}N, ^{15}O, ^{18}F, ^{38}K, and ^{68}Ga have proved to be especially promising.

INTRODUCTION

A combination of recent events (1-3) has resulted in a significant movement towards cyclotron and generator produced short-lived positron emitting radionuclides and radiopharmaceutical chemistry to be used in combination with positron emission tomography for the quantitative determination of metabolic function and disease state. Elsewhere in these proceedings Silvester (4) and Wolf (5) have discussed the production of medical radionuclides and organic radiopharmaceuticals labeled with cyclotron produced radionuclides. Demonstratable success with short-lived positron emitting radionuclides has been limited to ^{11}C, ^{13}N, ^{15}O, ^{18}F, ^{38}K, and ^{68}Ga although ^{14}O, ^{21}Na, ^{30}P, ^{34m}Cl, ^{45}Ti, ^{49}Cr, ^{51}Mn, ^{52m}Mn, ^{55}Co, ^{61}Cu, ^{62}Cu, ^{66}Ge, ^{75}Br, ^{76}Br, ^{77}Kr, $^{81,82m}Rb$, ^{82}Rb, ^{104}Ag, ^{121}I, ^{122}I, and ^{123}Xe have been reported to be in various stages of development. A complete bibliography of papers and reports published on all aspects of the fundamental and applied studies with positrons, positron emitters, positron emission tomography are compiled elsewhere (6,7) as part of a continuing effort. The intent here is to present the first comprehensive survey of the positron-emitting radionuclides which are deemed to have appropriate physical characteristics and be within the production capability of a cyclotron.

DISCUSSION

Table 1 is a compilation of published nuclear data for 157 nuclides listed in order of increasing atomic number and identifying the half life and the abundance of positrons per 100 decay events. An indication of the complexity of the positron energy spectra is summarized by listing the number of distinguishable positrons, and the median positron energy of the spectra calculated by Hogan (8) using the Fermi theory of beta decay. The Table lists 8 nuclides with a t 1/2 of 60 s or less, 12 with t 1/2 of 1-2 m, 26 with t 1/2 = 2-10 m, 24 with t 1/2 of 10-30 m, 40 with t 1/2 of 0.5-2 h, 27 with t 1/2 of 2-10 h, and 20 with t 1/2 > 10 h.

The status of the nuclear data of several of the radionuclides is not without question. The status of nuclear data, not only decay but also production, of some of the medical radionuclides has been reviewed (9) or compiled without evaluation (10,11). The prerequisite positron abundance necessary to make a nuclide useful is not yet defined due to rapid developments in instrumentation and the absence of extensive studies. Data on the median energy of the positron spectra is presented as a criteria indicative of the degree of influence the positron spectra and range-energy distribution will have on the resolution of the emission tomograph. The data are intended as a more practical aid. The interpretation of the effect of positron range on the resolution with a volume element viewed with a positron emission tomograph is subject to argument (12-21). Fig. 1 depicts an interpretation of the variation in the relative FWHM of one spectrometer (12) as a function of the relative path length (20) of the relatively uncomplicated positron spectra of ^{18}F, ^{11}C, ^{13}N, ^{15}O, and ^{68}Ga. Derenzo and Budinger (19) have suggested that it will not be possible to achieve a resolution for a positron emission tomograph of greater than the square root of (angular spread)^2 + (range spread)^2 which for ^{68}Ga is 4 mm.

The minimum accepted positron abundance for the purposes of Table 1 was 2.5%. Of the 157 nuclides, 86 decay with $\geq$ 50% abundance of positrons, i.e. the annihilation radiation is $\geq$ 100 γ's per 100 decay events.

Lambrecht et al. (23) have reported that $^{81,82m}Rb$ seems satisfactory for emission tomography of the myocardium. The result was initially surprising due to a report (24) that the positron abundance of ^{81}Rb was ∿ 12%. Recent decay scheme studies have shown the positron abundance of ^{81}Rb and ^{82m}Rb to be 27% and 26%, respectively (24).

The choice of a radionuclide for radiopharmaceutical applications depends on a combination of physical, chemical, and physiological considerations. For example, ^{38}K ($t\ 1/2 = 7.62$ m, $\beta^+ = 100\%$, $\gamma = 2.17$ MeV, 100%) has demonstrated (24) all the criteria required for an "ideal" radiopharmaceutical. The high energy gamma did not interfer with the tomographic quality obtainable with the PETT III. In fact, it has been suggested that the coincident annihilation radiation and the 2.17 MeV gamma should be considered (21,26) for dual tomographic and positron annihilation lifetime determinations *in vivo*.

The production and processing of simple inorganic species which have physiological properties to permit near-direct clinical use or which can be rapidly manipulated into synthetic intermediates for radiopharmaceutical preparation are key points in the widespread use of short-lived positron emitting radionuclides. These parameters will be discussed in detail elsewhere. The absorbed radiation dose becomes a more serious consideration with nuclides having a half life of > 3 h, except perhaps in therapeutic or high benefit-vs.-risk circumstances. Siegel (27) suggested ^{55}Co-labeled bleomycin may be one identifiable example, and research has progressed in this direction (28,29).

As the radiopharmaceutical sciences evolve in part into the direction of positron emission tomography, it is apparent that cyclotrons and accelerators (tandem, electron, and particle linacs) will be essential for production of radionuclides for near-direct clinical use, for on-line preparation of synthetic intermediates required in synthetic organic, metallo-organic and inorganic chemistry, and for production and distribution of longer-lived positron-emitters or their generators.

Justification of a medical cyclotron depends on the envisioned use. Initial economic factors argue for a single particle, low-energy machine devoted to the production of ^{11}C, ^{13}N, ^{15}O, and ^{18}F as the key positron-emitting radionuclides. Without doubt these radionuclides in combination with ^{68}Ge-^{68}Ga and possibly ^{82}Sr-^{82}Rb and ^{44}Ti-^{44}Sc generators will have a principle role in future radiopharmaceutical developments. However, scientific and philosophical arguments argue that too small a machine may be short-sighted.

Examples of other bioelements required for human metabolism are K, P, Se, Rb, and Mn. None of the appropriate radionuclides of these elements can be produced in acceptable yield and radionuclidic purity with a low energy single particle cyclotron. The $^{40}Ar(p,3n)^{38}K$ reaction with 32 MeV protons (26,30) and the $Cl(\alpha,n)^{38}K$ reaction with 15 MeV alphas (31) have been utilized for production of the ideal radionuclide. Sahakundu et al. (32) concluded the best route to ^{30}P was via the $^{27}Al(\alpha,n)^{30}P$ reactions with 28 MeV alphas. The production of ^{73}Se in acceptable radionuclidic purity and yield requires either ^{3}He or ^{4}He particles (33-35). Recently, $^{81,82m}Rb$ produced via the $^{82}Kr(p,xn)$ reactions on natural Kr was shown to be an appropriate radionuclidic pair for emission tomography of the myocardium (21) in instances where ^{38}K cannot be produced in-house. In addition, the production route with > 32 MeV protons is yet the most appropriate for routine production of ^{81}Rb-^{81m}Kr generators (36).

In many respects the ^{52}Fe-^{52m}Mn generator (37-39) is probably limited to research applications due to the fact that a high energy, high fluence accelerator is required to produce multi-mCi quantities of ^{52}Fe. The 8.27 h half life is a limitation since chemistry and shipment considerations will probably result in a generator of 1 day or less shelf life. By fortunate circumstance either of two Mn radionuclides can be produced in high yield and purity via the $^{50}Cr(p,2n)^{51}Mn$ or $^{50}Cr(p,n)^{52m}Mn$ nuclear reactions (30).

Positron emitting isotopes of presently used γ-emitting radionuclides which merit further consideration are listed in Table 2.

ACKNOWLEDGEMENTS

Research carried out at Brookhaven National Laboratory under contract with the U.S. Department of Energy and supported by its Office of Basic Energy Sciences.

REFERENCES

1. Phelps, M. E., Hoffman, E. J., Mullani, N. A. et al.: "Design considerations for a positron emission transaxial tomograph (PETT III)". IEEE Trans. Nucl Sci. NS-23: 516 (1976).

2. Brownell, G. L., Correia, J. A., Zamenhof, R. G.: "Positron instrumentation" in Recent Advances in Nuclear Medicine, Vol. 5 (Lawrence, J. N. and Budinger, T. F., eds), Grune and Stratton, Inc. (1978), pp. 1-48.

3. Budinger, T. F., Derenzo, S. E., Gullberg, G. T. et al.: "Emission computer assisted tomography with single photon and positron annihilation photon emitters". J. Comp. Assist. Tomogr. 1: 131 (1977).

4. Silvester, D. T.: "Radionuclide Production". 2nd Intern. Symposium on Radiopharmaceuticals, Seattle, Washington, March 18-23, 1979.

5. Wolf, A. P.: "Organic Radiopharmaceuticals. Recent Advances". 2nd Intern. Symposium on Radiopharmaceuticals, Seattle, Washington, March 18-23, 1979.

6. Lambrecht, R. M.: "Antimatter-matter interactions" I. Positron and Positronium, A Bibliography 1930-1974. BNL 50510, July, 1975, p.201.

7. Singru, R. M., Lal, K. B., Tao, S. J. et al.: "Positron annihilation data tables". Atomic Data and Nuclear Data Tables 20: 475-501 (1971).

8. Hogan, O. L.: "Beta spectra v. spectra of individual positron emitters". Report USNRDL-TR-1104, November 14, 1966, p.126.

9. Stocklin, G., Qaim, S. M.: "Recent trends in nuclear reaction data needs for nuclear medicine". Proc. Int. Conf. on Neutron Physics and Nuclear Data for Reactors and Other Applied Purposes, Harwell, September 25-29, 1978 (in press).

10. Burrows, T. W., Burt, J. S.: "The Bibliography of Integral Charged Particle Nuclear Data". BNL-NCS-50640, Second Edition, March, 1978, p.456.

11. Burrows, T. W., Holden, N. E.: "A source list of nuclear data bibliographies, compilations, and evaluations", BNL-NCS-50702, Second Edition, October, 1978, p.104.

12. Phelps, M. E., Hoffman, E. J., Huang, S. C. et al: "Effect of positron range on spatial resolution". J. Nucl. Med. 16: 649 (1975).

13. Muehllehner, G.: "Resolution limit of positron cameras". J. Nucl. Med. 17: 757 (1976).

14. Phelps, M. E., Hoffman, E. J.: "Reply No. 1". J. Nucl. Med. 17: 757 (1976).

15. Cho, Z. H.: "Reply No. 2". J. Nucl. Med. 17: 758 (1976).

16. Derenzo, S. E., Budinger, T. F., Cahoon, J. L. et al.: "High resolution computed tomography of positron emitters". IEEE Trans Nucl. Sci. NS-24 : 544 (1977).

17. Cho, Z. H., Chan, J. K., Ericksson, L. et al.: "Positron ranges obtained from biomedically important positron-emitting radionuclides". J. Nucl. Med. 16: 1174 (1975).

18. Muehllehner, G., Buchin, M. P., Dudek, J. H.: "Performance parameters of a positron imaging camera". IEEE Trans. Nucl. Sci. NS-23: 528 (1976).

19. Derenzo, S. E., Budinger, T. F.: "Resolution limit for positron-imaging devices". J. Nucl. Med. 18: 491 (1977).

20. Hoffman, E. J., Phelps, M. E.: "Reply". J. Nucl. Med. 18: 491 (1977).

21. Lambrecht, R. M.: "Antimatter-matter interactions. Positron annihilation, radiopharmaceuticals, and the positron camera" in The Chemistry of Radiopharmaceuticals, Heindel, M. D., Burns, H. D., Honda, T., and Brady, L. W. eds., New York, Masson Publishing, Inc., 1978, pp. 87-107.

22. Nelms, A. T.: "Energy loss and range of electrons and positrons". National Bureau of Standards Circular 577, U. S. Government Printing Office, Washington, D. C., July 26, 1956, p.30.

23. Lambrecht, R. M., Gallagher, B. M., Wolf, A. P. et al.: "Cyclotron Isotopes and Radiopharmaceuticals. XXIX. ^{81,82m}Rb for positron emission tomography". Intern. J. Appl. Radiat. Isot. (submitted).

24. Li-Scholz, A., Bakhru, H.: "Studies in the Decay of 4.7 h ^{81}Rb". Phys. Rev. 168: 1244 (1968).

25. Lederer, C. M., Shirley, V.S., eds: "Table of Isotopes", 7th Edition, New York, John Wiley and Sons, Inc., 1978.

26. Lambrecht, R. M., Hara, T., Gallagher, B. M. et al.: "Cyclotron Isotopes and Radiopharmaceuticals. XXVIII. Production of potassium-38 for myocardial perfusion studies". Intern. J. Appl. Radiat. Isot. 29: 667 (1978).

27. Siegel, B. A.: "Potential applications of emission tomography for tumor detection". Symp. Positron-Emitting Radionuclides and Tomography in Nuclear Medicine, Lake Ozark, MO, April 19-21, 1978 (unpublished).

28. Lagunas-Solar, M. C., Jungerman, J. A.: "Cyclotron Production of Carrier-free Cobalt-55, A new positron-emitting label for bleomycin". Intern. J. Appl. Radiat. Isot. 30: 25 (1979).

29. Watanabe, M., Horiuchi, K., Nakahara, H. et al.: "^{3}He bombardment of manganese for the production of ^{55}Co". J. Label Comp. and Radiopharm. 16: 216 (1979).

30. Lambrecht, R. M. and Wolf, A. P.: "The cyclotron production of ^{38}K, ^{51}Mn, ^{52m}Mn, and ^{77}Kr for positron emission tomography". J. Label Comp. and Radiopharm. 16: 129 (1979).

31. Tilbury, R. S., Dahl, J. R., Chandra, R. et al.: "Potassium-38 production with a cyclotron for medical use". J. Label Comp. and Radiopharm. 16: 127 (1979).

32. Sahakundu, S. M., Qaim, S. M., and Stocklin, G.: "Cyclotron production of short-lived ^{30}P". Intern. J. Appl. Radiat. Isot. 30: 3 (1979).

33. Guillaume, M., Lambrecht, R. M., Wolf, A. P.: "Cyclotron Isotopes and Radiopharmaceuticals. XXVII. ^{73}Se". Intern. J. Appl. Radiat. Isot. 29: 411 (1978).

34. Guillaume, M., Lambrecht, R. M., Christiaens, L. et al.: "Carrier free selenium-73 cyclotron production and seleno-radiopharmaceutical synthesis". J. Label Comp. and Radiopharm. 16: 126 (1979).

35. Nozaki, T., Itoh, Y., Iwamoto, M. et al.: "Yield of ^{73}Se and ^{77}Br for various reactions and their carrier- and/or salt free separation". J. Label. Comp. and Radiopharm. 16: 219 (1979).

36. Ruth, T. J., Lambrecht, R. M., Wolf, A. P. et al.: "Cyclotron Isotopes and Radiopharmaceuticals. XXX. Aspects of Production, Elution, and Automation of ^{81}Rb-^{81m}Kr Generators". Int. J. Appl. Radiat. Isot. (submitted).

37. Atcher, R. W., Friedman, A. M., Huizenga, J. R. et al.: "^{52m}Mn -- A New short-lived generator-produced radionuclide". J. Nucl. Med. 19: 689 (1979).

38. Grant, P. M., O'Brien, H. A., Jr., Bayhurst, B. P. et al.: "Spallation yields of ^{52}Fe, ^{67}Cu, ^{67}Ga, and ^{201}Tl for reactions of 800 MeV protons with Ni, As, Pb, and Bi Targets". J. Label Comp. and Radiopharm. 16: 212 (1979).

39. Waters, S. L., Kensett, M. J.: "Excitation functions and production yields for some radionuclides of ion from the alpha particle bombardment of natural chromium". J. Label Comp. and Radiopharm. 16: 218 (1978).

TABLE 1. CHARACTERISTICS OF POSITRON EMITTING RADIONUCLIDES HAVING POTENTIAL APPLICATIONS IN THE RADIOPHARMACEUTICAL SCIENCES

Nuclide[a]	Half life[b]	No. of Positrons[b,c]	Abundance[b,c]	E_{median}[c] MeV
^{11}C	20.38 m	1	0.998	0.375
^{13}N	9.96 m	1	0.998	0.475
^{14}O	70.6 s	3	1.000	0.768
^{15}O	122 s	1	0.99+	0.723
^{17}F	64.5 s	1	1.000	0.729
^{18}F	109.8 m	1	0.969	0.245
^{18}Ne	1.67 s	2	1.000	1.385
^{19}Ne	17.3 s	1	0.99+	*
^{21}Na	22.47 s	2	1.000	1.085
^{23}Mg	11.3 s	2	0.950	1.308
^{30}P	2.50 m	2	1.000	1.455
^{34m}Cl	32.0 m	3	0.53	0.712
^{38}K	7.61 m	1	0.995	1.113
$^{42m+g}Sc$	62.0 s	2	1.000	2.200
^{43}Sc	3.89 h	3	*	*
^{44}Sc	3.93 h	1	0.95	0.616
^{45}Ti	3.09 h	2	0.853	0.417
^{47}V	32.6 m	1	0.970	0.812
^{48}V	15.967d	1	0.496	0.283
^{49}Cr	41.9 m	3	0.932	0.627
^{50m}Mn	1.74 m	2	0.990	1.476
^{50}Mn	0.28 s	1	1.000	3.552
^{51}Mn	46.2 m	3	0.972	0.935

Nuclide[a]	Half life[b]	No. of Positrons[b,c]	Abundance[b,c]	E_{median}[c] MeV
^{52m}Mn	21.1 m	1	*	*
^{52}Mn	5.59 d	2	0.276	0.215
^{52}Fe	8.27 h	1	0.565	0.336
^{53}Fe	8.51 m	3	0.975	1.064
^{54m}Co	1.46 m	1	1.000	2.078
^{54}Co	0.19 s	1	1.000	3.438
^{55}Co	17.5 h	3	0.77	0.510
^{57}Ni	36.0 h	3	0.40	0.338
^{59}Cu	82 s	6	0.984	1.531
^{60}Cu	23.4 m	5	0.936	0.957
^{61}Cu	3.41 h	4	0.622	0.480
^{62}Cu	9.73 m	4	0.978	1.290
^{64}Cu	12.70 h	1	0.190	0.273
^{61}Zn	89.1 s	4	0.991	1.882
^{62}Zn	9.2 h	2	0.069	0.279
^{63}Zn	38.1 m	5	0.93	0.911
^{64}Ga	2.62 m	2	0.990	1.975
^{65}Ga	15.2 m	4	0.86	*
^{66}Ga	9.4 h	6	0.567	1.712
^{68}Ga	68.1 m	2	0.875	0.811
^{66}Ge	2.3 h	6	0.27	0.574
^{67}Ge	19.0 m	4	0.96	1.200
^{69}Ge	39.0 h	4	0.361	0.492
^{69}As	15 m	1	0.98	1.290

Nuclide[a]	Half life[b]	No. of Positrons[b,c]	Abundance[b,c]	E_{median}[c] MeV
^{70}As	53 m	2 (3)	0.84	0.680
^{71}As	61 h	2	0.32	00.343
^{72}As	26.0 h	6	0.777	1.147
^{74}As	17.78 d	1 (2)	0.309	0.406
^{71}Se	4.9 m	2	1.000	1.477
^{73}Se	7.1 h	2 (4)	0.65	0.566
^{74m}Br	41 m	2	*	*
^{74}Br	25.3 m	1	1.000	2.004
^{75}Br	98 m	4 (3)	0.755	0.376
^{76}Br	16.1 h	7	0.57	1.200
^{78}Br	6.46 m	2	0.932	1.076
^{74}Kr	11.5 m	1	1.000	1.385
^{77}Kr	75 m	3 (4)	0.80	0.741
^{79}Kr	35.0 h	2	0.07	0.254
^{79}Rb	23.0 m	3	0.84	1.063
^{80}Rb	34 s	2	0.977	1.827
^{81m}Rb	32 m	1	0.500	0.603
^{81}Rb	4.58 h	3	0.27	0.372
^{82m}Rb	6.2 h	1	0.256	0.332
^{82}Rb	1.25 m	3	0.962	1.382
^{81}Sr	26 m	2	0.87	1.244
^{83}Sr	32.4 h	(3)	0.24	(0.494)
^{84}Y	39 m	5	0.865	1.201
$^{85(g)}Y$	2.7 h	(4)	0.70	*

Nuclide[a]	Half life[b]	No. of Positrons[b,c]	Abundance[b,c]	E_{median}[c] MeV
$^{85(m)}Y$	4.9 h	(3)	0.55	(0.875)
^{86}Y	14.74 h	6	0.34	*
^{87}Zr	1.6 h	1	0.830	0.920
^{89}Zr	78.4 h	1	0.223	0.390
^{89}Nb	66 m	1	0.74	*
^{89}Nb	2.0 h	1	0.910	1.270
^{90}Nb	14.6 h	3	0.53	0.602
^{90}Mo	5.67 h	1	0.25	0.498
^{91m}Mo	65 s	3	0.280	1.210
^{91}Mo	15.49 m	1	0.941	*
^{92}Tc	4.4 m	1	1.000	1.857
^{93}Tc	2.7 h	3	0.13	0.248
^{94m}Tc	52 m	1 (3)	0.72	0.434
^{94}Tc	293 m	3 (1)	0.11	1.036
^{95}Ru	1.65 h	3	0.15	0.459
^{97m}Rh	44 m	1	(1.000)	(0.722)
^{98}Rh	8.7 m	1 (2)	(0.170)	(1.181)
^{100}Rh	20.8 h	5	0.049	0.919
^{98}Pd	18 m	*	*	*
^{99}Pd	21.4 m	4	*	*
^{102}Ag	13.0 m	2	0.68	*
^{103}Ag	1.10 h	3	0.42	*
^{104m}Ag	33 m	1	0.590	1.202
^{104}Ag	69 m	1	0.150	0.434

Nuclide[(a)]	Half life[(b)]	No. of Positrons[(b,c)]	Abundance[(b,c)]	E_{median}[(c)] MeV
^{106}Ag	24.0 m	2	0.610	0.826
^{105}Cd	56.0 m	(2)	*	*
^{106}In	6.3 m	(1)	*	*
^{107}In	32.4 m	1	0.35	0.882
^{108}In	39.6 m	(1)	0.110	1.512
^{108}In	58.0 m	1	0.20	0.585
^{109}In	4.2 h	(1)	0.06	*
^{110}In	69 m	1	0.72	1.006
^{109}Sn	18.0 m	(5)	*	*
^{111}Sn	35 m	1	0.290	0.640
^{114}Sb	3.5 m	2	1.000	0.144
^{115}Sb	31.8 m	1	0.33	0.665
^{116m}Sb	60.4 m	2	0.192	0.634
^{116}Sb	16 m	2	0.28	0.931
^{117}Sb	2.80 h	1	0.025	0.288
^{118}Sb	3.5 m	2	0.828	1.364
^{117}Te	62 m	(1)	0.30	*
^{120m}I	53 m	1	*	*
^{120}I	1.35 h	1	0.46	1.802
^{121}I	2.12 h	1	0.064	0.506
^{122}I	3.6 m	1 (3)	0.77	1.386
^{124}I	4.2 d	3	0.25	0.788
^{121}Xe	39 m	1	0.08	1.233
^{123}Xe	2.08 h	1	0.13	0.666

Nuclide[a]	Half life[b]	No. of Positrons[b,c]	Abundance[b,c]	E_{median}[c] MeV
^{125}Cs	45 m	2	0.39	*
^{126}Cs	1.64 m	2	0.82	1.636
^{127}Cs	6.2 h	4	0.035	0.364
^{128}Cs	3.6 m	4	0.61	1.138
^{130}Cs	29.9 m	1	*	*
^{130}La	8.7 m	(1)	*	*
^{131}La	61 m	3	0.245	0.603
^{132}La	4.8 h	1	0.810	1.637
^{133}La	3.91 h	1	1.000	0.534
^{134}La	6.67 m	2	0.618	1.186
^{136}La	9.87 m	2	0.332	0.855
^{131}Ce	10 m	1	0.113	*
^{133}Ce	5.4 h	1	0.150	0.578
^{135}Pr	25 m	1	0.25	1.111
^{137}Pr	1.28 m	1	0.255	0.754
^{138m}Pr	2.1 h	1	0.236	0.622
^{139}Pr	4.4 h	1	0.062	0.449
^{140}Pr	3.30 m	1	0.487	1.075
^{139}Nd	30 m	1	0.256	1.382
^{141}Pm	20.9 m	1	0.57	1.156
^{142}Pm	40.5 s	1	0.69	1.693
^{142}Sm	2.49 m	1	0.10	0.451
^{143}Sm	8.83 m	1	0.465	1.111
^{145}Gd	22 m	1	1.000	1.066

Nuclide[a]	Half life[b]	No. of Positrons[b,c]	Abundance[b,c]	E_{median}[c] MeV
^{152}Tb	17.5 h	3	0.145	0.938
^{155}Ho	49 m	(1)	(1.000)	(0.933)
^{156}Ho	56 m	(3)	(1.000)	(0.933)
^{158m}Ho	27 m	1.320	1.000	0.593
^{158}Er	2.4 h	1	1.000	0.362
^{164}Tm	2.0 m	2	0.39	*
^{165}Yb	10 m	(1)	*	*
^{168m}Lu	6.7 m	2	0.012	*
^{168}Lu	5.5 m	1	1.000	0.544
^{178}Re	13.2 m	1	0.113	1.372
^{180}Re	2.42 m	1	0.08	0.847
^{186}Ir	1.7 h	5	1.000	0.856
$^{190m2}Ir$	3.2 h	1	0.050	0.891
^{194}Au	39.5 h	2	0.03	0.604

(a) Arranged in order of increasing atomic number

(b) Lederer, C. M., Shirley, V. S. (eds.), Browne, E., Dairiki, J. M., Doebler, R. E., Shihab-Eldin, A. A., Jardine, L. J., Tuli, J. K., Buyrn, A. B. Table of Isotopes, 7th Ed., J. Wiley and Sons, New York, 1978.

(c) Hogan, O. L., "Beta Spectra V. Spectra of Individual Positron Emitters" Report USNRDL-TR-1101, 14 November 1966, 126 pp.

* Insufficient nuclear decay data available for (c) calculation

( ) Parenthesis indicates the value in table may be in error

TABLE 2

POSITRON-EMITTING ISOTOPES OF γ-EMITTING RADIONUCLIDES

Nuclide	t 1/2	β^+ Abundance	Production*
^{110}In	69 m	72 %	^{107}Ag(α,n) ^{112}Cd(p,n)
^{66}Ga	9.4 h	56.7%	^{66}Zn(α,4n)^{66}Ge $\xrightarrow[2.27\ h]{}$ ^{66}Ga ^{66}Zn(^{3}He,3n)
^{95}Ru	1.65 h	15 %	^{94,95}Ru(α,xn) ^{94,95}Ru(^{3}He,xn)

*isotopically enriched targets

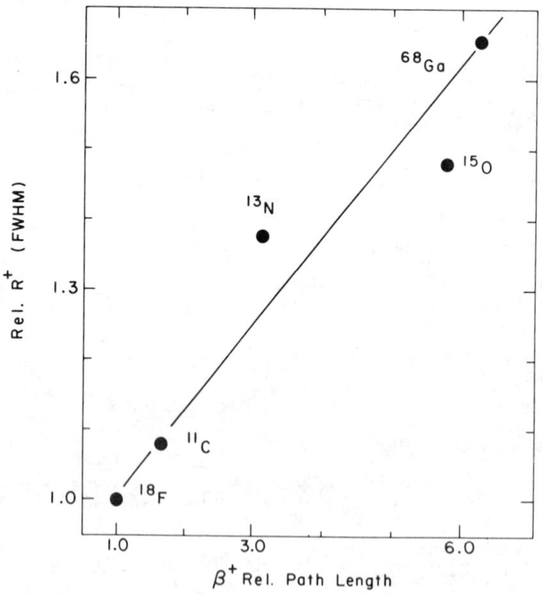

Figure 1.

Variation in the resolution function of a spectrometer as a function of the relative path length of positrons ^{11}C, ^{13}N, ^{15}O, ^{18}F, and ^{68}Ga in water.

HIGH-LEVEL PRODUCTION OF C-11-CARBOXYL-LABELED AMINO ACIDS

L.C. Washburn, T.T. Sun, B.L. Byrd, and R.L. Hayes

Medical and Health Sciences Division
Oak Ridge Associated Universities
Oak Ridge, Tennessee 37830

and

T.A. Butler and A.P. Callahan

Nuclear Medicine Technology Group
Health and Safety Research Division
Oak Ridge National Laboratory
Oak Ridge, Tennessee 37830

ABSTRACT

C-11-Labeled amino acids have significant potential as agents for positron tomographic functional imaging. We have developed a rapid, high-temperature, high-pressure modification of the Bücherer-Strecker amino acid synthesis and found it to be quite general for the production of C-11-carboxyl-labeled neutral amino acids. Production of C-11-carboxyl-labeled DL-tryptophan requires certain modifications in the procedure. Twelve different amino acids have been produced to date by this technique. Synthesis and chromatographic purification require approximately 40 min, and C-11-carboxyl-labeled amino acids have been produced in yields of up to 425 mCi. Two C-11-carboxyl-labeled amino acids are being investigated clinically for tumor scanning and two others for pancreatic imaging. Over 120 batches of the various agents have been produced for clinical use over a three-year period.

INTRODUCTION

C-11-Carboxyl-labeled amino acids ($T_{1/2}$ C-11 = 20.4 min) have recently been shown to have significant potential in nuclear medicine. [Carboxyl-^{11}C] DL-tryptophan (1,2) and DL-valine (2,3) are useful for pancreatic imaging, and [carboxyl-^{11}C] 1-aminocyclopentanecarboxylic acid (ACPC) (4,5) and 1-aminocyclobutanecarboxylic acid (ACBC) (6) are tumor-localizing agents. The great importance of amino acids in biochemical pathways suggests that C-11-carboxyl-labeled amino acids may have untapped potential for study of a multitude of physiological processes. Quantitative functional imaging with C-11-carboxyl-labeled amino acids is made possible by their suitability for positron computerized transaxial tomography (7).

We have developed a rapid, high-temperature, high-pressure modification of the Bücherer-Strecker amino acid synthesis and applied it to the synthesis and purification of a number of C-11-carboxyl-labeled amino acids (8). The objectives of this paper are to report some recent refinements in our high-level synthetic technique and to provide previously unreported experimental details.

TARGETRY AND PRODUCTION OF C-11 OXIDES

We use the Oak Ridge National Laboratory's 86-Inch Cyclotron for production of C-11. The combination of high beam currents (typically 150-175 μA) of 22 MeV protons with a 5 mm orbital separation distance permits the use of high-production internal-beam targets. A mixture of ^{11}CO and $^{11}CO_2$ is produced by the $^{11}B(p,n)^{11}C$ reaction. Figure 1 shows a cutaway view of our boron oxide (B_2O_3) target which we bombard with 22 MeV protons to produce C-11.

The B_2O_3 target is prepared as follows: B_2O_3 rods are prepared by heating B_2O_3 (m.p. 450°C) in a platinum crucible at 800°C for 30 min to remove any water present, cooling to ∿ 600°C, and drawing out strands of the material which rapidly harden upon cooling. Sections 60 mm long are then cut from those strands which have diameters of approximately 1 mm. These are stored in a closed glass tube containing Drierite until used for target preparation. A "boat" of 0.005-in.-thick Grafoil (Fiber Materials, Inc., Biddeford, ME) is formed in a mold having the same dimensions as the capsule tube (see below), to ensure a good fit of the completed target in the capsule tube. The "boat" is loaded with rods of B_2O_3, placed in a quartz tube, and heated at 800°C for 15 min in a tube furnace (Lindberg Type 123-1, 422-watt, Lindberg, Inc., Watertown, WI, or equivalent) regulated with a variable transformer. In order to minimize trapped air, the target is fused in an atmosphere of helium or nitrogen. Since B_2O_3 shrinks when fused, this procedure may be repeated several times to achieve the desired weight of B_2O_3 in the target, i.e., approximately 0.65 g.

The B_2O_3 and its support are contained in an aluminum capsule tube of 5 mm O.D. which is in turn contained in an aluminum jacket tube of 6.3 mm O.D. Between the two aluminum tubes is an annular space which carries 15 l/min of cooling water (Fig. 1). The 60 mm wide by 6 mm high proton beam is centered on the axis of the tube. The proton energy is attenuated from 22 MeV to 16 MeV at the target midplane in passing through the jacket wall, water annulus, and capsule wall. Air in the target and connecting tubing is removed by purging with helium and a static helium atmosphere is maintained during bombardment. The B_2O_3 is liquefied by the beam and the C-11 produced by the $^{11}B(p,n)^{11}C$ reaction combines with oxygen in the target. After bombardment it is essential that the molten B_2O_3 solidify (requiring about 30 sec in our case) to prevent the sweep gas from pushing it out of the beam's path. The resulting mixture of ^{11}CO and $^{11}CO_2$ is then swept from the target by 100 ml/min of helium to a remote control manipulator cell located adjacent to the cyclotron, where the bolus of C-11 activity is measured using an ion chamber (Fig. 2). Visual examination of several targets in a glove box after use has shown that, during bombardment, the molten B_2O_3 coats the inside of the capsule tube, facilitating good release of the mixture of C-11 oxides. Production of $^{11}CO/^{11}CO_2$ using this targetry has given a maximum yield of 5 Ci and a typical yield of 3-4 Ci for a 40-min bombardment at 150-175 μA.

No statistically significant advantage was observed in using B_2O_3 enriched in B-11 (98.4% B-11, made from enriched H_3BO_3 by heating at 800°C for 30 min) versus natural B_2O_3 (80.4% B-11) as a target material. The reason for this is unclear but may be related to the chemical form and/or chemical purity of the respective preparations. We use natural B_2O_3 (Ultrex, J. T. Baker Chemical Co., Phillipsburg, NJ) for our routine target fabrication.

CATALYTIC PRODUCTION OF $H^{11}CN$

$^{11}CH_4$ and $H^{11}CN$ are produced sequentially from $^{11}CO/^{11}CO_2$ by modifications of the catalytic methods of Bánfi et al (9). The 100 cc/min helium stream containing $^{11}CO/^{11}CO_2$ is mixed with 50 cc/min of hydrogen and passed through a quartz tube containing nickel on kieselguhr (Ni-3226, E 1/16, Harshaw Chemical Co., Cleveland, OH) (Fig. 2). The minimum path length of the catalyst is 5 cm, and it is held in place by a plug of nickel wire at each end. The quartz tube containing the nickel catalyst is first heated to 520°C in a hydrogen atmosphere using a regulated tube furnace (see above) in order to activate the catalyst, and the temperature is then lowered to 400°C and held at that level for the run. The nickel catalyst is no longer effective after it has cooled to room temperature and, therefore, is replaced after each day's run.

The output gas containing $^{11}CH_4$ is passed through a $CaSO_4$/NaOH trap to remove water and any unreacted $^{11}CO_2$. The gas stream is then mixed with 50 cc/min of anhydrous ammonia and passed through another quartz tube containing platinum catalyst held at 950°C using a tube furnace as above. The platinum catalyst is prepared from platinum screen by cleaning and etching with aqua regia, plating with platinum black, and crumpling into small wads to increase the surface area. The path length of the platinum catalyst is 5 cm, and it is prevented from moving downstream by a constriction in the quartz tube. The $H^{11}CN$-containing output gas from the platinum catalyst is finally bubbled into 1 ml of 0.005 N NaOH in a glass gas absorption column (0.7 cm x 25 cm) having a bulbed area near the top to enhance break-up of the bubbles (see Fig. 2).

We found that the catalytic conversion yield was always significantly lower for the first run than for subsequent runs on a given day. This was apparently due to release of water from the target which decreased the

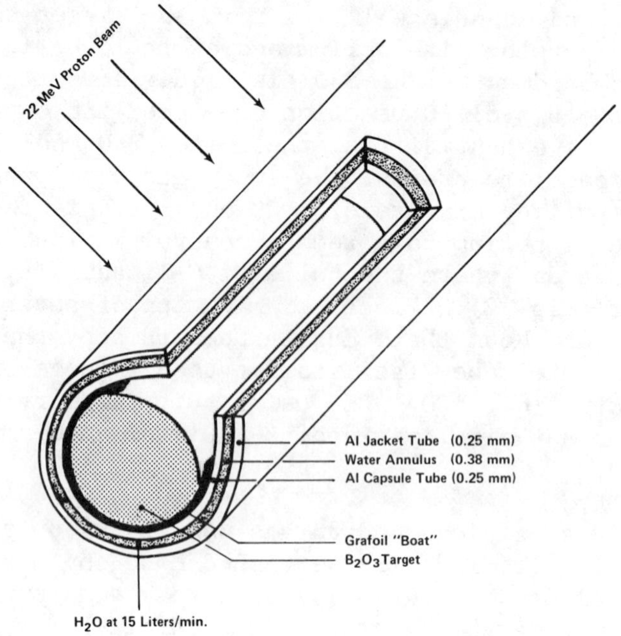

Figure 1.

Cutaway view of the B_2O_3 target assembly used for production of C-11 at the Oak Ridge National Laboratory's 86-Inch Cyclotron.

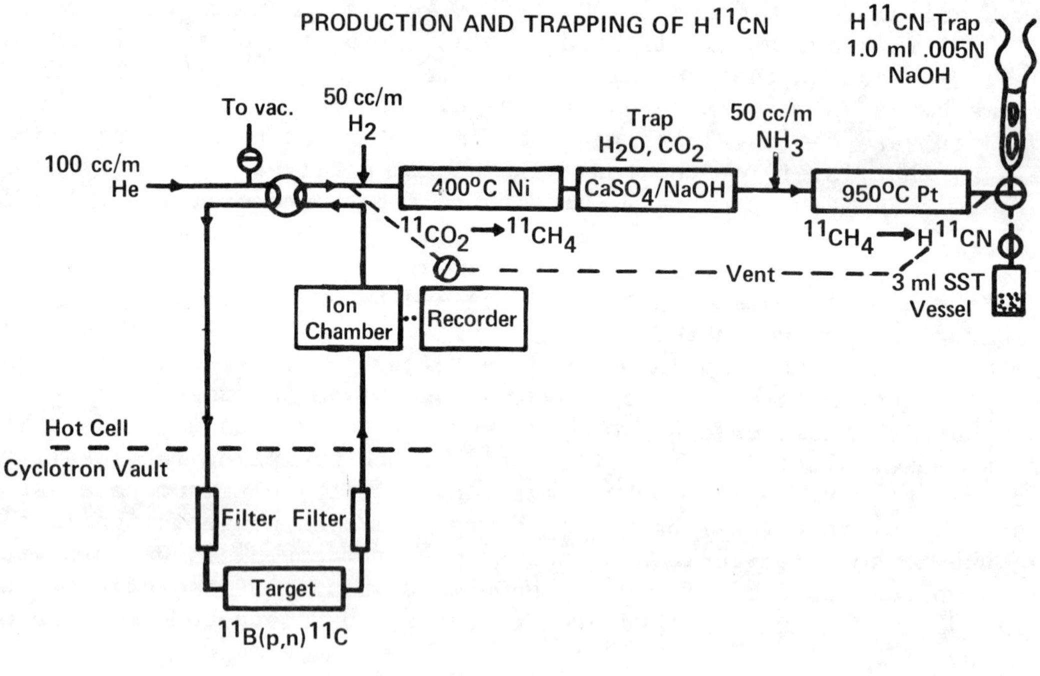

Figure 2.

Schematic of the system used for production of $^{11}CO/^{11}CO_2$, sequential catalytic conversion of $^{11}CO/^{11}CO_2$ to $H^{11}CN$, and trapping of $H^{11}CN$.

efficiency of the nickel catalyst. This problem can be circumvented by carrying out a 5-min bombardment and venting the resulting gas stream around the two catalysts. The trapped [^{11}C] cyanide is measured in an ion chamber after it is drained into a stainless steel reaction vessel (see below). The average combined efficiency for catalytic conversion from $^{11}CO/^{11}CO_2$ and trapping of $H^{11}CN$ is approximately 70%.

SYNTHESIS OF C-11-CARBOXYL-LABELED AMINO ACIDS

A stainless steel reaction vessel fitted with a needle valve (Fig. 3) is used for the high-temperature, high-pressure Bücherer-Strecker amino acid synthesis. The reaction vessel is fabricated from type 304 stainless steel using a 1 7/8-in. piece of standard 3/4-in. hexagonal stock. The vessel is formed by drilling a 7/16-in. diameter hole in the stock and fitting the top with standard 1/4-in. female N.P.T. threads. The reaction vessel has a concave bottom to facilitate complete removal of the reaction mixture. The internal volume beneath the threads is 3 ml.

The needle valve has 1/4-in. male N.P.T. threads which are wrapped with 1/4-in. Teflon tape before tightening to ensure a good seal with the reaction vessel. The valve's closure is based on metal-to-metal contact between a 6-32 stainless steel screw and the needle inlet, which has a depression to permit better seating of the screw. The screw has a 3/16-in. x 2-in. stainless steel rod welded to the top to enable the valve to be opened and closed using manipulators. The entire needle assembly is hinged to swing at a right angle with the body so that liquids can easily be added to or withdrawn from the reaction vessel using a needle and syringe.

Before starting a C-11-carboxyl-labeled amino acid production run, the reaction vessel is loaded with the required reagents. For neutral amino acids, the reagents are: the appropriate carbonyl compound, 0.5 mmole; $(NH_4)_2CO_3$, 0.75 mmole; NH_4Cl, 0.125 mmole; and KCN, 0.125 mmole. In the case of [carboxyl-^{11}C] DL-tryptophan, some modifications in the procedure are necessary. An aldehyde derivative, 3-indoleacetaldehyde bisulfite addition compound, is used because of the instability of the free aldehyde to either acid or base. In addition, tar formation caused processing difficulties when an excess of the carbonyl derivative was used and led to use of the following quantities of reagents for [carboxyl-^{11}C] DL-tryptophan: 3-indoleacetaldehyde bisulfite addition compound, 0.1 mmole; $(NH_4)_2CO_3$, 0.6 mmole; NH_4Cl, 0.1 mmole; and KCN, 0.1 mmole.

All further operations through chromatographic purification are carried out in a high-level hot cell using manipulators. The 1 ml of 0.005 N NaOH containing trapped [^{11}C] cyanide is drained from the gas absorption column into the reaction vessel. The column is rinsed with 0.5 ml of distilled water and the rinse also added to the reaction vessel. Opening and closing the vessel is assisted by transfer to a 3/4-in. hexagonal socket mounted to a lead brick. The closed reaction vessel is then heated in a modified hot plate assembly regulated by a variable transformer (Fig. 4). The heating device is an aluminum block affixed to a hot plate and has a hexagonal hole $\sim$ 2 in. deep and only slightly larger than the reaction vessel (to afford good heat transfer between the heating device and the reaction vessel). The

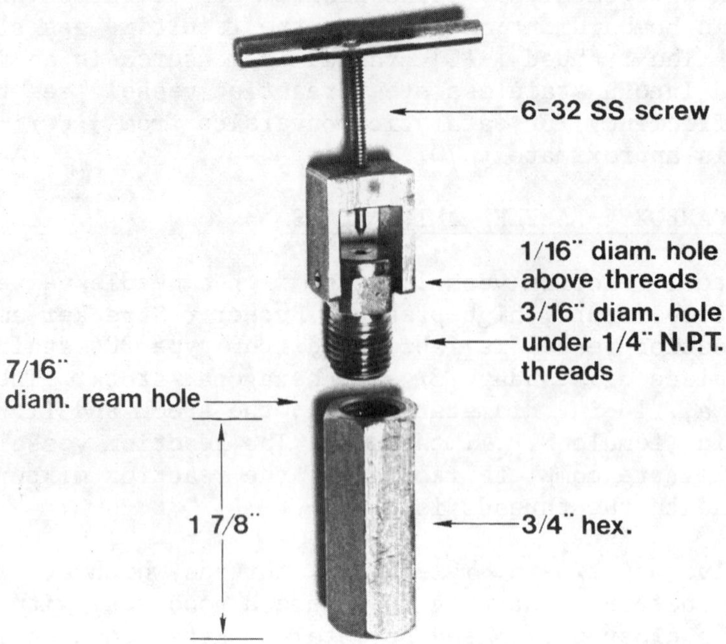

Figure 3.

Stainless steel reaction vessel and needle valve used in high-temperature, high-pressure modification of the Bücherer-Strecker amino acid synthesis.

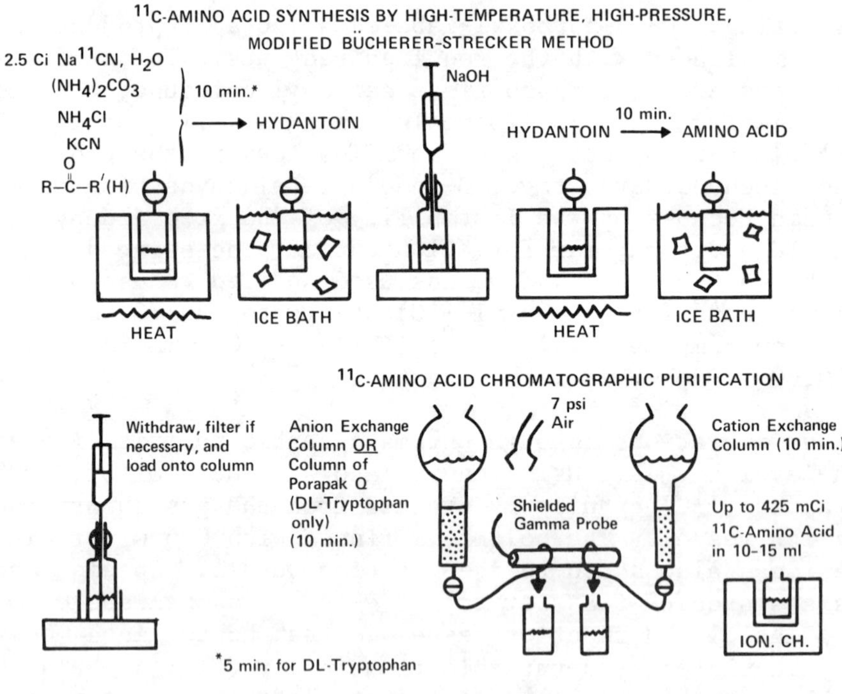

Figure 4.

Schematic of the method used for production of C-11-carboxyl-labeled amino acids by our high-temperature, high-pressure modification of the Bücherer-Strecker amino acid synthesis.

temperature is monitored using a mercury thermometer placed in an oil-filled hole in the block.

The synthesis time and temperature vary according to the C-11-carboxyl-labeled amino acid which is to be made. For neutral, aliphatic amino acids, the reaction time is 10 min for each of the two steps (see Fig. 4), and the reaction temperature is 210-220°C. For [carboxyl-^{11}C] DL-tryptophan the reaction times are 5 and 8 min for the first and second steps, respectively, and the reaction temperature is 240°C for both steps. We recently found that this change in the reaction parameters minimizes thermal decomposition of the intermediate hydantoin and produces higher and more reliable yields of [carboxyl-^{11}C] DL-tryptophan (1). For other aromatic amino acids, i.e., [carboxyl-^{11}C] DL-phenylalanine and DL-phenylglycine, the reaction time is 10 min for each step, and the reaction temperature is 190°C.

After the first heating step, the reaction vessel is cooled in an ice bath for ∿ 30 sec, opened, and 1 ml of 6.25 N NaOH is added. The second heating step then results in hydrolysis of the intermediate hydantoin to the amino acid. After cooling as before, the valve is opened and the reaction mixture is removed using a syringe and blunt 4-in., 18-gauge needle. The reaction vessel is rinsed twice with 1 ml of distilled water and the rinses added to the reaction mixture.

PURIFICATION OF C-11-CARBOXYL-LABELED AMINO ACIDS

C-11-Carboxyl-labeled amino acids are purified by rapid column chromatographic techniques (Fig. 4). Columns are fitted with female ball joints to permit increased column flow rates by pressurization with ∿ 7 psi of compressed air. In order to monitor the elution of radioactivity and thus minimize the elution volumes, the columns are eluted through tubing wrapped around a shielded gamma probe.

The purification procedure for neutral aliphatic and aromatic amino acids is as follows: In most cases, with the exception of C-11-carboxyl-labeled DL-valine, DL-phenylalanine, and ACBC, the reaction mixture is quite dark and should be filtered. This is accomplished by transferring the reaction mixture to a vertically-mounted 20-ml syringe body attached to a Millipore-type holder containing a coarse filter disc. The reaction mixture then is filtered by insertion of the plunger and expulsion of the liquid through the filter. The highly basic mixture is loaded directly onto a 1.5 cm x 5 cm, AG 1-X2, 100-200 mesh anion-exchange bed (Bio-Rad Laboratories, Rockville Center, NY), which has previously been converted to the hydroxide form by washing with 1 N NaOH. The column is rinsed twice with 30 ml of distilled water to remove neutral and cationic impurities and then eluted with 1 N HCl. The acidic eluate is loaded onto a 1.0 cm x 15 cm, AG 50W-X2, 50-100 mesh cation-exchange bed (Bio-Rad Laboratories) in the hydrogen form (previously washed with water and 1 N HCl). The column is rinsed twice with 30 ml of distilled water to remove anionic impurities, and the amino acid is eluted with 0.2 N NaOH. When activity is observed on the gamma probe, the column is allowed to flow slowly without pressurization, in order to elute the amino acid in a narrow band (∿ 15 ml). The C-11-carboxyl-labeled amino acid solution is collected in a sterile 25-ml bottle and the activity measured in an ion chamber.

For [carboxyl-^{11}C] DL-tryptophan a 1.0 cm x 3 cm column of 80-100 mesh Porapak Q (Waters Associates, Inc., Milford, MA) replaces the anion-exchange resin in the first purification step. The column is prepared beforehand by soaking the Porapak Q in acetone, loading the column, rinsing with 40 ml of water per gram of Porapak Q, and finally rinsing well with 0.1 N HCl. The cooled reaction mixture is filtered as above, acidified with 2 ml of 6 N HCl, and then loaded onto the column. The loaded column is washed twice with 30 ml of 0.1 N HCl and is then eluted with 50% aqueous ethanol. The eluate is acidified with 2 ml of 6 N HCl (already present in the receiver) and loaded directly onto a cation-exchange bed as in the purification of neutral amino acids.

A total of approximately 40 min is required for synthesis and purification of C-11-carboxyl-labeled amino acids by our method. The solution is then transported to the Medical and Health Sciences Division of Oak Ridge Associated Universities for final radiopharmaceutical processing. This consists of neutralization to pH 7.0 $\pm$ 0.2, filtration through a sterile 0.22 μ microfilter (Millipore Corporation, Bedford, MA) into a precapped, presterilized bottle, and a rapid (15 min) check for pyrogenic contamination using the Pyrotest Limulus amebocyte lysate technique (Difco Laboratories, Detroit, MI). The solution then is suitable for clinical use.

HIGH-LEVEL PRODUCTION OF C-11-CARBOXYL-LABELED AMINO ACIDS

Table 1 lists the amino acids which have been synthesized by our rapid, high-temperature, high-pressure Bücherer-Strecker technique. Most have been labeled with C-11, but a few were only labeled with C-14 using analogous procedures, because C-14 tissue distribution studies indicated that the C-11-labeled agents would not have significant nuclear medical potential. The C-11-carboxyl-labeled amino acids which are being investigated clinically (see Introduction) have all been produced many times; however, some of the other compounds have been made only once or a few times and the yield data may not be as indicative of the achievable yields. The chemical yields are based on the amount of stable cyanide used in the syntheses. The specific activity of C-11-carboxyl-labeled amino acids produced by this method generally is 15-40 mCi/mg at the completion of synthesis and purification. The low chemical yield for DL-α-methylvaline is probably due to steric hindrance.

In order to produce multiple batches of C-11-carboxyl-labeled amino acids in a given day, it is necessary to regenerate the chromatographic columns between batches. The anion exchange resin is regenerated by washing the column twice with 20 ml of water followed by washing with 40 ml of 3 N NaOH, and regeneration of the cation-exchange resin is analogous except that 40 ml of 3 N HCl is used in place of the NaOH. Regeneration of the Porapak Q is accomplished by rinsing the column with 20 ml of acetone and 60 ml of 0.1 N HCl.

Figure 5 shows how the various steps (cyclotron production of C-11 oxides, catalytic conversion and trapping of $H^{11}CN$, amino acid synthesis, amino acid purification, and column regeneration) are overlapped in order to produce multiple batches of various C-11-carboxyl-labeled amino acids at intervals of one hr. Further overlapping of the steps can result in production intervals as low as 45 min, but requirements regarding clinical turnover time

TABLE 1. PRODUCTION OF VARIOUS PURIFIED AMINO ACIDS USING A HIGH-TEMPERATURE, HIGH-PRESSURE MODIFICATION OF THE BÜCHERER-STRECKER TECHNIQUE

Amino acid	Precursor carbonyl compound	Batches C-11-labeled compound to date	Maximum chemical yield (%)	Average chemical yield* (%)	Maximum C-11 yield to date (mCi)
2-Aminobicyclo[2,2,1] heptane-2-carboxylic acid	Norcamphor	-	30		-
1-Aminocyclobutane carboxylic acid (ACBC)	Cyclobutanone	23	78	50	415
1-Aminocyclohexane-carboxylic acid	Cyclohexanone	1	55		75
1-Aminocyclopentane-carboxylic acid (ACPC)	Cyclopentanone	73	60	38	300
1-Amino-2-methylcyclo-pentanecarboxylic acid	2-Methylcyclopentanone	-	45		-
1-Amino-3-methylcyclo-pentanecarboxylic acid	3-Methylcyclopentanone	-	48		-
DL-Leucine	Isovaleraldehyde	-	45		-
DL-α-Methylvaline	3-Methyl-2-butanone	-	8		-
DL-Phenylalanine	Phenylacetaldehyde	4	68	55	425
DL-Phenylglycine	Benzaldehyde	1	31		150
DL-Tryptophan	3-Indoleacetaldehyde bisulfite addition compound	40	59	32	325
DL-Valine	Isobutyraldehyde	34	89	55	365

* Only for those amino acids which were synthesized four or more times.

CONTINUOUS LARGE-SCALE PRODUCTION OF ANIMO ACIDS

LABELED WITH CARBON-11

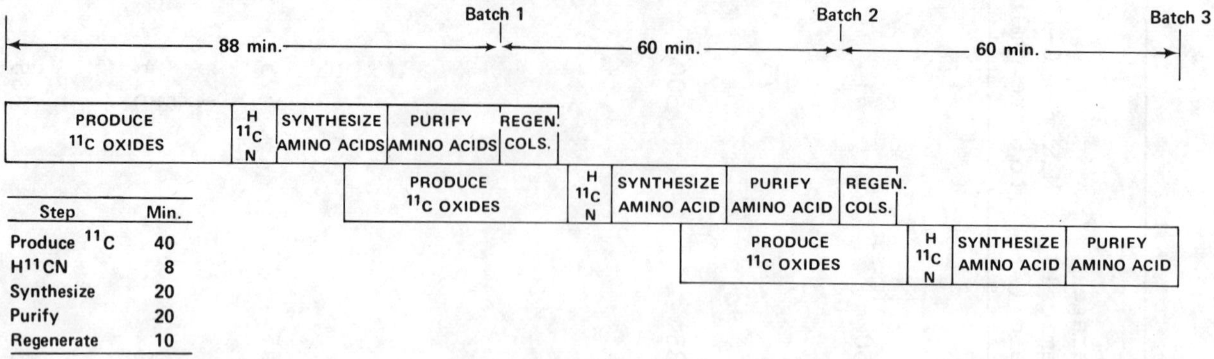

Figure 5.

Timing of production steps for multiple-batch preparation of C-11-carboxyl-labeled amino acids at one-hour intervals.

led to our routine production interval of approximately one hr. Over 120 batches of C-11-carboxyl-labeled amino acids have been produced for clinical tumor and pancreatic imaging studies over a three-year period.

ACKNOWLEDGEMENTS

This work was supported in part by USPHS Grant No. CA-14669 from the National Cancer Institute. Oak Ridge Associated Universities (ORAU) operates under contract number EY-76-C-05-0033 with the U.S. Department of Energy. The research at Oak Ridge National Laboratory (ORNL) was sponsored by the Office of Health and Environmental Research, U.S. Department of Energy, under contract number W-7405-eng-26 with the Union Carbide Corporation. The authors acknowledge the valuable contributions to this work made by T.J. Alford, J.E. Carlton, D.D. Ferguson, J.J. Rafter, and B.W. Wieland of ORAU, M.R. Skidmore of ORNL, and J.B. Anon and R.R. Turtle, formerly with ORAU.

REFERENCES

1. Washburn LC, Sun TT, Byrd BL, et al: DL-[Carboxyl-^{11}C] tryptophan, a potential agent for pancreatic imaging: Production and preclinical investigations. *J Nucl Med* (in press)

2. Hübner KF, Andrews GA, Buonocore E, et al: Imaging the human pancreas with carbon-11 labeled amino acids by rectilinear and positron tomographic scanning. *J Nucl Med* (in press)

3. Washburn LC, Wieland BW, Sun TT, et al: [1-^{11}C] DL-Valine, a potential pancreas-imaging agent. *J Nucl Med* 19:77-83, 1978

4. Hayes RL, Washburn LC, Wieland BW, et al: Carboxyl-labeled ^{11}C-1-aminocyclopentanecarboxylic acid, a potential agent for cancer detection. *J Nucl Med* 17:748-751, 1976

5. Hübner KF, Andrews GA, Washburn L, et al: Tumor location with 1-aminocyclopentane[^{11}C] carboxylic acid: Preliminary clinical trials with single-photon detection. *J Nucl Med* 18:1215-1221, 1977

6. Hayes RL, Washburn LC, Wieland BW, et al: Carboxyl-labeled C-11-ACBC, a possible new radiopharmaceutical for detection of cancer using positron tomography. *J Nucl Med* 18:639, 1977 (abstract)

7. Phelps ME: Emission computed tomography. *Semin Nucl Med* 7:337-365, 1977

8. Hayes RL, Washburn LC, Wieland BW, et al: Synthesis and purification of ^{11}C-carboxyl-labeled amino acids. *Int J Appl Radiat Isot* 29:186-187, 1978

9. Bánfi D, Mlinkó S, and Palágyi T: A new synthesis for the preparation of ^{14}C-labelled alkali cyanides. *J Labeled Comp* 7:221-223, 1971

CYCLOTRON PRODUCTION OF THALLIUM - 201 VIA THE $^{205}Tl(p,5n)^{201}Pb \dashrightarrow {}^{201}Tl$ REACTION

Manuel C. Lagunas-Solar, John A. Jungerman
and David W. Paulson

Crocker Nuclear Laboratory
University of California
Davis, CA 95616

ABSTRACT

The production of the clinically useful ^{201}Tl radioisotope was investigated based upon the utilization of the $^{205}Tl(p,5n)^{201}Pb \dashrightarrow {}^{201}Tl$ reaction. Thin-stacked metallic foils of 99.5% enriched ^{205}Tl were used as targets. Thick-target yields for the daughters ^{201}Tl, ^{200}Tl, and ^{202}Tl were calculated as a function of target thickness, bombarding time, target chemistry time, and incident proton energies. The results demonstrated the feasibility of utilizing the (p,5n) reaction to produce ^{201}Tl of similar radionuclidic composition and in yields comparable to the $^{203}Tl(p,3n)^{201}Pb \dashrightarrow {}^{201}Tl$ method.

INTRODUCTION

Thallium-201 for myocardial perfusion imaging studies, using the scintillation camera, is becoming a well-established clinical procedure for the evaluation and diagnosis of heart diseases (1). Its use in other areas of diagnosis is also being investigated. In order to satisfy its growing demand for clinical cardiology, a large research investment has been directed toward establishing the best method for its production by cyclotron technology. It is now generally accepted that the reaction $^{203}Tl(p,3n)^{201}Pb \dashrightarrow {}^{201}Tl$ constitutes the basis for the method most widely used for producing large quantities of ^{201}Tl (2). The (p,3n) reaction requires proton beam energies in the 28 to 30 MeV range and the use of enriched (and expensive) ^{203}Tl targets for yield optimization. The (p,3n) reaction is also the best suited for the characteristics of most presently operating commercial cyclotron facilities.

The utilization of higher proton beam energies (48 - 52 MeV) to produce ^{201}Tl via the $^{205}Tl(p,5n)^{201}Pb \dashrightarrow {}^{201}Tl$ reaction was suggested in 1976 (3). However, the lack of detailed information on the excitation functions and thin-target yields for the series of $^{205}Tl(p,xn)$ reactions prevented, at that time, a critical and documented evaluation of this alternate method.

A complete study of the $^{205}Tl(p,xn)$ reactions using 99.5% enriched ^{205}Tl and covering the 34 - 60 MeV energy range has been submitted recently by the authors (4). The study of the proton reactions induced on natural thallium targets in the 15 - 60 MeV range has already been reported (5). This information allowed us to evaluate and compare the (p,3n) and the (p,5n) methods. Several technical and practical advantages as well as some theoretical disadvantages of the (p,5n) method will be discussed. A new method is therefore proposed for production of ^{201}Tl. This (p,5n) method could be used solely or in tandem with a ^{203}Tl target to increase the production yield of ^{201}Tl per irradiation, if a 48 - 52 MeV proton accelerator is available.

MATERIALS AND METHODS

Enriched ^{205}Tl (99.5%) was obtained from Oak Ridge National Laboratory, as the thallic oxide, and reduced to metallic thallium in the presence of hydrogen. The metallic thallium was then used to prepare thin circular foils with thicknesses ranging from 0.16 to 0.20 g/cm^2, with an area of 1.98 cm^2. Target irradiations were carried out with an external proton beam with energies ranging from 45 to 60 MeV at the chemical target facility of the University of California's Crocker Nuclear Laboratory 76-inch isochronous cyclotron. The proton beam energy was calculated from cyclotron parameters and verified by neutron time-of-flight measurements. The estimated error in the energy of the proton beam amounted to $\pm$ 0.5 MeV. A well-collimated, defocussed proton beam was used in all the irradiations. The beam intensity was kept below 0.5 μA to avoid melting of the stacked foils. The total number of protons striking the stacked Tl foils target was measured by integrating the total charge (μC) collected by a Faraday cup. Special precautions were taken not to expose the foils for longer than 0.5 hr to avoid high counting rates and unnecessary exposure to the researchers. A calibrated Ge(Li) detector (30 - cm^3 coaxial) coupled to a 4096-channel pulse-height analyzer was used for counting each individual foil.

The number of foils to be handled in each experiment was kept to a minimum (< 15) in order to complete the counting of ^{200}Pb before a significant ingrowth of the daughter ^{201}Tl (from its 9.4 hr ^{201}Pb parent) became significant. Normally the first ^{200}Pb counting was completed within 60 min. EOB, therefore minimizing the interference of the 136 keV (2.65%) and 167 keV (10%) ^{201}Tl gamma rays, in the subsequent computer analysis of the 147.6 keV (28.4%) characteristic ^{200}Pb gamma ray. This precaution has proven to be most adequate in the 34 to 46 MeV energy region where the ^{200}Pb production is rather small, and the photopeak integration is further complicated by background and Compton scattering from other gamma rays. Detector dead-time losses were corrected by counting simultaneously a 60 Hz electronic pulser.

Apart from the low beam intensity experiments, a series of production runs were carried out using a thick-metallic 99.5% ^{205}Tl target. These production experiments were carried out using beam currents of up to 30 µA for an 8 hr irradiation with the purpose of testing target performance as well as the yield and radionuclidic purity of the final ^{201}Tl product. The methods for target/product radiochemical separation and the radionuclidic and radiochemical purity tests have been reported (2).

DATA ANALYSIS

For the evaluation of the counting data for each thin-foil and for the thick-target measurements, the values of the physical half-lives and gamma ray intensities were the same as previously reported (5). The method used for data reduction included uncertainties due to corrections for detector efficiencies (± 7%), dead-time losses (± 1%), decay time (± 1%), counting statistics (± 2%), beam current (± 5%), beam energy (± 10%), computer errors in the spectral integrations (± 4%), target thickness variations (± 6%), and foil gamma ray absorption corrections (± 2%). The overall uncertainty in the thin-target yield measurements can be obtained as a r.m.s. sum of the above uncertainties, for a total error of ± 15%. The overall uncertainty in the thick-target measurements, where the thick-foil targets were handled chemically, can be estimated by adding to the above uncertainties, the uncertainties due to gamma ray self-absorption corrections in solution (± 6%), and chemistry losses (± 3%). The total uncertainty can then be estimated as ± 17%.

RESULTS

Although the excitation functions and thin-target yields for the ^{205}Tl (p,xn) reactions were reported for 34 - 60 MeV protons (4), the relevant energy range for evaluating the ^{205}Tl(p,5n) reaction as a source of ^{201}Tl is limited by several factors. Table I gives the reaction Q-values, optimum proton energies for maximum thin-target yield, and the physical half lives of the parent-daughter radioisotopes induced by proton bombardment of ^{205}Tl targets. An inspection of the different Pb radioisotopes shown in Table I reveals that only ^{200}Pb (21.5 hr) and ^{202m}Pb (3.62 hr) should be considered as potential sources of radionuclidic contamination. They are inevitably generated during the production of ^{201}Tl via the (p,5n) reaction, and their decay produces ^{200}Tl (26.1 hr) and ^{202}Tl (293.5 hr), respectively. The decay of the ^{200}Tl and ^{202}Tl isotopes produce gamma ray emissions of sufficiently high energy (^{200}Tl, 88% 367 keV, 10% 579 keV, 8% 829 keV, 35% 1210 keV and others, ^{202}Tl, 95% 439 keV) to penetrate the lead collimator used in heart imaging. Septal penetration will then cause deterioration of the ^{201}Tl image quality.

TABLE I

REACTION Q-VALUES (MeV), OPTIMUM ENERGY, E_{op} (MeV), AND MAXIMUM THIN-TARGET YIELDS FOR PROTON-INDUCED NUCLEAR REACTIONS ON ^{205}Tl.

Reaction on ^{205}Tl	Product Isotope	$t_{\frac{1}{2}}$ (hr)	Daughter Isotope	$t_{\frac{1}{2}}$ (hr)	Q (MeV)	E (MeV)	Yield, A_o (mCi/μA-hr/g/cm^2)
(p,7n)	^{199m}Pb	0.20	^{199}Pb[1]	1.5	-47.6	---	---
(p,7n)	^{199}Pb	1.5	^{199}Tl	7.4	-47.6	---	---
(p,6n)	^{200}Pb	21.5	^{200}Tl	26.1	-39.1	58	8.5
(p,5n)	^{201}Pb	9.4	^{201}Tl	73.5	-31.9	45	31.8
(p,4n)	^{202m}Pb	3.62	^{202}Tl[2]	293.5	-23.3	36	25
(p,3n)	^{203}Pb	52.1	^{203}Tl	Stable	-16.5	26	6
(p,2n)	^{204m}Pb	1.115	^{204}Pb	Stable	- 6.8	20	38

1 ^{199m}Pb decays 100% to ^{199}Pb excited states (Lederer, C.M. et al., "Table of Isotopes", Sixth Ed., J. Wiley & Sons, 1967).

2 Only 9% of the ^{202m}Pb decays through electron capture to ^{202}Tl excited states. ^{202m}Pb decays 19% to ^{202}Pb (∿3 x 10^5y) excited states (McDonald, J.A. et al., Nucl. Phys. 3:513, 1957).

On the other hand, both the ^{203}Pb (52.1 hr) and the ^{204m}Pb (1.115 hr) radioisotopes decay to stable isotopes (^{203}Tl and ^{204}Pb, respectively) and are not relevant factors in selecting the best parameters to produce ^{201}Tl via the (p,5n) reaction. Furthermore, any ^{203}Pb and ^{204m}Pb radioactivity remaining at the end of processing of the ^{201}Tl can easily (and efficiently) be removed by chemical methods (precipitation and/or ion-exchange) (2). The presence of ^{199}Pb (1.5 hr) on the other hand, should be eliminated simply by limiting the incident proton energy on the ^{205}Tl target to a level below the energetic threshold for its formation (∿ 47.6 MeV). ^{199}Pb decays to the 7.4 hr ^{199}Tl radioisotope whose emissions (14% 455 keV, 12% 208 keV, 9% 247 keV, etc.) most likely would deteriorate the image quality of ^{201}Tl preparations.

In a separate study using a scintillation camera, a comparison was made between ^{201}Tl, made by the (p,3n) method, containing different levels of ^{200}Tl (1 - 15%) and ^{202}Tl (0.2 - 1.5%) as the only radionuclidic contaminants. Commercially available ^{201}Tl, both in the USA and Europe, contains 0 - 1% ^{200}Tl, 0.2 - 1% ^{202}Tl, and up to 0.5% ^{203}Pb at calibration times. Measurements of spatial resolution using a Rollo phantom indicated the significantly detrimental effect of ^{200}Tl. However, no differences were observed when the level of ^{200}Tl was $\leq 2\%$. The detailed methods and results of this work will be reported elsewhere (6). Therefore, and within the limitations imposed by the seemingly acceptable levels of ^{200}Tl and ^{202}Tl as radionuclidic impurities, and based upon the premise that a ^{200}Tl level of $\leq 2\%$ will not cause undesirable effects on image quality, the 34 - 52 MeV energy range seems the most adequate for a detailed exploration of different conditions for the production of ^{201}Tl via the $^{205}Tl(p,5n)$ reaction. Levels of up to 2% ^{202}Tl were found to have a lesser effect on image quality (6).

Figure 1 shows the proton energy dependence (34 - 52 MeV) of the thin-target yields (A_o) in mCi/μA-hr/g/cm^2, for the ^{201}Pb, ^{200}Pb, ^{202m}Pb, ^{203}Pb, and ^{204m}Pb isotopes formed via the series of $^{205}Tl(p,xn)$ reactions (see Table 1). The individual experimental points in Figure 1 are not shown. In the authors opinion, however, the yield curves represent the best computer-smoothed fittings to the experimental data. Therefore, these curves are used in subsequent thick-target yield calculations. The thin-target yields, shown in Figure 1, are corrected for decay to the end of bombardment and for decay during irradiation.

Table 2 gives the results of the thick-target yield calculations (in mCi/μA-hr) for ^{201}Pb, ^{200}Pb, and ^{202m}Pb. These calculations were carried out based upon the thin-target yields shown in Figure 1. The integration of the respective yields curves, with an integration step ΔE = 0.5 MeV, was performed by using the following formula:

$$\text{Thick-Target Yield (mCi/μA-hr)} = \sum_{E_i}^{E_e} A_o \, (dE/dx)^{-1} \, \Delta E \qquad (1)$$

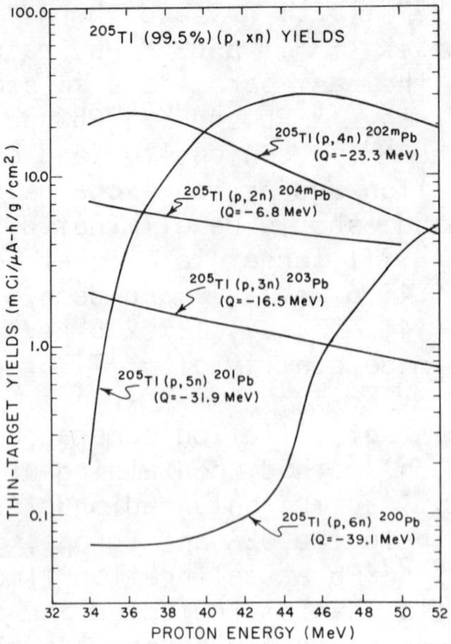

Figure 1

The thin-target yield, A_o (mCi/μA/hr/g/cm^2), is shown for ^{201}Pb, ^{200}Pb, ^{202m}Pb, ^{203}Pb, and ^{204m}Pb radionuclides for an enriched (99.5%) ^{205}Tl target as a function of proton energy.

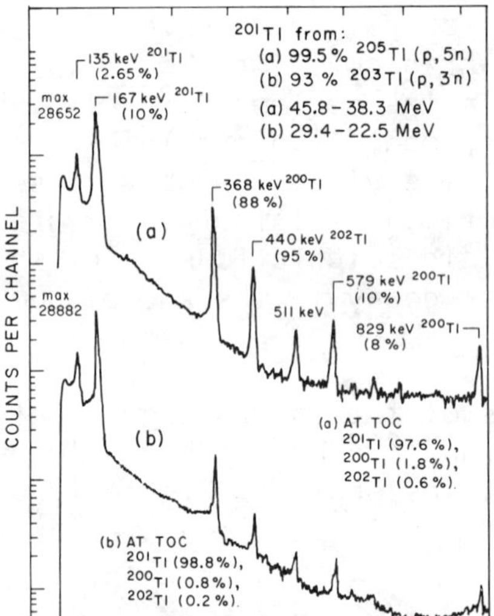

Figure 2

A comparison of Ge(Li) gamma ray spectra for ^{201}Tl made from ^{205}Tl(p,5n) and ^{203}Tl(p,3n) is shown in a semi-logarithmic scale. The spectra were obtained at time of calibration (TOC) for both products. Almost identical maximum for the ^{201}Tl 167 keV gamma ray facilitates a direct comparison of radionuclidic contamination levels.

TABLE 2

THICK-TARGET YIELDS FOR THE ^{201}Pb - ^{201}Tl, ^{200}Pb - ^{200}Tl, AND ^{202m}Pb - ^{202}Tl PARENT-DAUGHTER SYSTEMS IN THE 38 - 50 MeV ENERGY RANGE.[1]

E (MeV)	^{201}Pb	^{200}Pb (mCi/μA-hr)		^{202}Pb	^{201}Tl	^{200}Tl (mCi/μA-hr)		^{202}Tl	
49.8	2.05	297	$\times 10^{-3}$	0.480	0.194	98.6	$\times 10^{-3}$	4.99	$\times 10^{-4}$
49.3	2.08	243		0.503	0.197	80.7		5.23	
48.8	2.16	207		0.526	0.204	68.7		5.47	
48.3	2.26	171		0.547	0.246	56.8		5.69	
47.8	2.38	144		0.586	0.225	47.8		6.09	
47.3	2.44	122		0.614	0.231	40.5		6.39	
46.8	2.51	100		0.635	0.237	33.2		6.60	
46.3	2.57	81.3		0.655	0.243	27.0		6.81	
45.8	2.59	57.6		0.691	0.245	19.1		7.19	
45.3	2.61	39.2		0.718	0.247	13.0		7.47	
44.8	2.55	22.7		0.761	0.241	7.54		7.91	
44.3	2.49	15.3		0.795	0.236	5.08		8.27	
43.8	2.39	11.3		0.876	0.226	3.75		9.11	
43.3	2.29	9.89		0.910	0.217	3.28		9.46	
42.8	2.16	8.78		0.956	0.204	2.91		9.94	
42.3	2.02	7.93		0.995	0.191	2.63		10.3	
41.8	1.93	7.16		1.08	0.183	2.38		11.2	
41.3	1.76	6.80		1.14	0.166	2.26		11.9	
40.8	1.59	6.36		1.20	0.150	2.11		12.5	
40.3	1.43	6.16		1.28	0.135	2.04		13.3	
39.8	1.26	5.80		1.38	0.119	1.93		14.4	
39.3	1.18	5.52		1.47	0.112	1.83		15.3	
38.8	0.985	5.40		1.53	0.0932	1.79		15.9	
38.3	0.832	5.07		1.59	0.0787	1.68		16.5	

1. The proton energies (E) listed here are the average energy of the integration step (0.5 MeV) used for these calculations. Values for dE/dx were taken from Range and Stopping Power Tables. Rapport CEA-R 3042 (1966). Tl yields given here were calculated from parent-daughter relationships at 32-hrs (see text).

Also given in Table 2 are the thick-target yields for the Tl-daughter radioactivities (^{201}Tl, ^{200}Tl, ^{202}Tl). The yields for the Tl daughters were calculated for a 32-hr decay period from the end of processing (EOP, the end of Pb-Tl radiochemical separation). This decay period corresponds to the maximum ingrowth yield of the ^{201}Tl daughter product.

The optimum operational parameters for the production of ^{201}Tl via the (p,5n) reaction, is a rather complex combination of technical and experimental conditions many of which are unique to a particular cyclotron facility. Therefore, and for greater generality, the results given in Table 2 were analyzed only with consideration to the variations in the ^{201}Tl product as a function of incident proton energy, target thickness (or exit proton energy), bombardment time, and target-chemistry and decay times. Table 3 gives the results of this analysis. The Tl yields (mCi/µA-hr) given in Table 3 are corrected for the decay during the bombarding time (b_t) and for the initial decay losses during the time required for the initial Pb-Tl chemical separation (tc_t). For the U.C. Davis CNL's procedure, which is based on precipitation and/or ion-exchange, the time elapsed between EOB and EOP is 1.5 ± 0.2 hours. Furthermore, the Tl yields as given in Table 3 are effective yields at 32 hrs from EOP. (See Note 1, Table 2).

A comparison of gamma ray spectra for ^{201}Tl made by the (p,3n) and (p,5n) reactions is shown in Figure 2. The yield measured in a thick-target irradiation (8 hr, 30µA) was 2.1 mCi/µA-hr at 32 hr EOP (9% higher than calculated in Table 3, Mode H). As shown in Figure 2, the (p,5n) product has 2.25 and 3 times greater proportion of ^{200}Tl and ^{202}Tl, respectively. The significance of these higher proportions in the quality of images is being evaluated and will soon be reported (6). The increased patient dose as a result of the higher proportion of ^{200}Tl and ^{202}Tl has already been reported (< 15%) (7).

DISCUSSION

Several general conclusions can be reached by inspection of the results given in Table 3:

1) The incident proton energy is a dominant factor for the radionuclidic purity of ^{201}Tl, in particular for its proportion of ^{200}Tl.

2) By lowering the exit proton energy (see modes D - H, Table 3) the ^{201}Tl yield increases slightly, while the ^{200}Tl level decreases and the ^{202}Tl increases. Overall the ^{201}Tl purity level remains practically constant (± 8%).

3) As expected, longer bombardment times (see modes A - C, Table 3) will produce smaller ^{201}Tl yields, while decreasing (slightly) the proportion of ^{201}Tl and ^{202}Tl, but with a more marked increase in the ^{200}Tl level (30% increase). However, an irradiation of 8 hr is practical and was used as a preferred mode in our analyses. Among the factors which have been considered in analyzing the results given in Table 3, the target-chemistry time has remained constant in our calculations. Variations of this time will greatly influence the final ^{201}Tl yield because of decay of its ^{201}Pb parent. However, practical considerations (systems and facilities) may govern this parameter and limit its overall significance. It is possible to delay the completion of the initial Pb-Tl separation, at the expense of ^{201}Tl overall yield, to

TABLE 3

^{201}Tl YIELDS (mCi/μA hr) AND RADIONUCLIDIC PURITIES (%) AT 32 hr FROM END OF PROCESSING FOR DIFFERENT PRODUCTION MODES[1]

Modes	E_{in} (MeV)	E_{ex} (MeV)	b_t (hr)	tc_t (hr)	Yields (mCi/μA hr) and Purities (%) ^{201}Tl	(%)	^{200}Tl	(%)	^{202}Tl	(%)
(A)	45.3	38.3	4	1.5	2.25	(97.53)	0.048	(2.08)	0.0091	(0.39)
(B)	45.3	38.3	8	1.5	1.76	(97.15)	0.045	(2.48)	0.0067	(0.37)
(C)	45.3	38.3	12	1.5	1.55	(96.99)	0.043	(2.69)	0.0051	(0.32)
(D)	45.8	40.3	8	1.5	1.65	(96.63)	0.054	(3.16)	0.0035	(0.21)
(E)	45.8	39.8	8	1.5	1.73	(96.64)	0.055	(3.07)	0.0051	(0.29)
(F)	45.8	39.3	8	1.5	1.81	(96.65)	0.057	(3.04)	0.0057	(0.31)
(G)	45.8	38.8	8	1.5	1.87	(96.63)	0.059	(3.05)	0.0063	(0.32)
(H)	45.8	38.3	8	1.5	1.92	(96.63)	0.060	(3.02)	0.0069	(0.35)
			(TOC = 24 hr)		1.53	(97.55)	0.032	(2.04)	0.0065	(0.41)
(I)	44.8	38.8	8	1.5	1.54	(97.55)	0.033	(2.09)	0.0057	(0.36)
(J)	44.8	38.3	8	1.5	1.59	(97.52)	0.034	(2.09)	0.0064	(0.39)
(K)	44.3	38.8	8	1.5	1.37	(97.69)	0.027	(1.93)	0.0054	(0.38)
(L)	44.3	38.3	8	1.5	1.43	(97.61)	0.029	(1.98)	0.0061	(0.41)
(M)	44.3	41.3	8	1.5	0.96	(97.79)	0.019	(1.94)	0.0027	(0.27)
			(TOC = 24 hr)		.77	(98.40)	0.010	(1.28)	0.0026	(0.32)
(N)	49.8	38.3	8	1.5	3.10	(87.28)	0.443	(12.47)	0.0088	(0.25)
			(TOC = 120 hr)		1.00	(97.60)	0.018	(1.76)	0.0066	(0.64)

[1]The different production modes are shown here as a function of incident (E_{in}) and exit (E_{ex}) proton energies, bombardment (b_t) and target chemistry (tc_t) times. Tl yields are corrected for decay during bombardment, target-chemistry (1.5 hr) time, and calculated for the effective Tl daughter radioactivities at 32 hr from EOP.

decrease the final proportion of ^{202}Tl by allowing time for its short-lived precursor (^{202}Pb 3.62 hr) to decay. However, this delay will not significantly affect the proportions of ^{201}Tl and ^{200}Tl and, therefore, is not recommended.

The incident proton energy must be limited to $E_{in} < 46$ MeV so as to minimize the ^{200}Tl impurity level to about 3% (see Modes D - H, Table 3). This level will correspond to < 2% at time of calibration (∿ 24 hrs later), a level at which no significant deterioration of image quality has been observed (6). Lowering the incident proton energies to $E_{in} < 45.5$ MeV or to $E_{in} < 45$ MeV, will decrease the ^{200}Tl level to about 2% (∿ 1% at TOC) (see Modes I, J, Table 3) and to < 2% (< 1% at TOC)(see Modes K, L, Table 3), respectively. The effect of the incident proton energy on the ^{202}Tl level is smaller and practically constant (0.2 - 0.4%) for the production modes listed in Table 3.

The ^{201}Tl yields calculated for the different production modes is acceptable and better (by about 10%) than our observed yield using the ^{203}Tl(p,3n) reaction. Several production modes (see Modes D - L, Table 3) may be used to produce 330 - 460 mCi of ^{201}Tl (30 μA, 8 hr, at 32 hr EOP) per irradiation.

Another possibility, not yet explored, that also lends itself directly to allow a timely transportation of ^{201}Tl to distant manufacturing centers, is the use of higher incident proton energies. An example of this possibility is given in Table 3 (Mode N). A 49.8 to 38.3 MeV (11.5 MeV "thick" target) may be used to produce ^{201}Tl (87.28%) with yields of 3.10 mCi/μA hr. However, this mode will also produce 12.47% ^{200}Tl and 0.25% ^{202}Tl at 32 hr EOP. This material can then be transported (or simply stored) and finally after 5 days, processed to its final form. After this elapsed time, however, the yield is reduced to 1.00 mCi/μA hr of ^{201}Tl (97.60%) with only 1.76% ^{200}Tl and 0.64% ^{202}Tl. The alternatives of this strategy are certainly numerous, and with obvious implications on the reliability of ^{201}Tl supply to the medical community. On the other hand, the technical challenge of developing such a system is also of great magnitude.

In conclusion, we have demonstrated that the reaction ^{205}Tl(p,5n) can be used to produce ^{201}Tl in similar yields to those obtained from the ^{203}Tl(p,3n) reaction. The most serious limitation of this reaction is the unavoidable presence of ^{200}Tl in the final ^{201}Tl product. Although it is possible to minimize its presence, a final conclusion on the significance of the proposed ^{200}Tl levels (1 - 2%) will have to be demonstrated by imaging experimentation.

Finally, a tandem target (99.5% ^{205}Tl and 93% ^{203}Tl) under development at U.C. Davis, may generate amounts in excess of 900 mCi (32 hr EOP) per single irradiation. This is most certainly a judicious possibility considering the potential of ^{201}Tl is nuclear medicine.

ACKNOWLEDGMENTS

The authors wish to thank the operating staff at the Crocker Nuclear Laboratory for their considerable help during these experiments.

REFERENCES

(1) Ritchie J., Hamilton G.W., Wackers F.J.Th. (Eds.) "Thallium - 201 Myocardial Imaging". Raven Press, New York (1978).

(2) Williams D.L., Finn R.D., Campbell J.A., and Lagunas-Solar M.C., "Thallium-201: Physical Properties and Methods of Production", Chap. 10, in "Thallium-201 Myocardial Imaging", Raven Press, New York (1978).

(3) Lagunas-Solar M.C., Prog. Nucl. Med. 4:118-128 (Karger, Basel) 1978. Proc. 4th Medical Cyclotron Users Conference, Miami Beach, Fla., March 1976.

(4) Lagunas-Solar M.C., Jungerman J.A., and Paulson D.W., "Thallium-201 Yields and Excitation Functions for the Lead Radioactivities Produced by Irradiation of Thallium-205 with 34 - 60 MeV Protons". Int. J. Appl. Rad. and Isotopes (submitted Nov. 1978).

(5) Lagunas-Solar M.C., Jungerman J.A., Peek N.F., and Theus R.M. "Thallium-201 Yields and Excitation Functions for the Lead Radioactivities Produced by Irradiation of Natural Thallium with 15 - 60 MeV Protons". Int. J. Appl Rad. Isotopes 29: 159-165 (1978).

(6) Hines H.H., Lagunas-Solar M.C., and Glass E.C., "The Effect of Different Levels of ^{200}Tl and ^{202}Tl Radiocontamination on ^{201}Tl Imaging", (Work in Progress).

(7) Quaim S.M., Weinreich R., and Ollig H., "Production of ^{201}Tl and ^{203}Pb via Proton Induced Nuclear Reactions on Natural Thallium". Int. J. Appl Rad. and Isotopes (submitted).

DEVELOPMENT OF A CHROMATOGRAPHIC Ge-68-Ga-68 GENERATOR YIELDING IONIC GALLIUM

Rudi D. Neirinckx and Michael A. Davis

Joint Program in Nuclear Medicine, Harvard Medical School, and
Children's Hospital Medical Center
Boston, Massachusetts 02115

ABSTRACT

The behavior of germanium and gallium on various organic chromatographic supports was investigated. The aim was to develop a simple high-yield chromatographic system for the routine production of ionic Ga-68 from a Ge-68-Ga-68 generator. The most attractive systems developed are based on adsorption of Ge-68 on an anion-exchange resin (Bio-Rad AG1-X8) or on a new synthetic resin, developed in this laboratory specifically for adsorption of germanium. The generator based on AG1-X8 with 0.01N HF as the eluent yields 95% of the Ga-68 in a 3 ml volume. Up to 600 elutions were performed before Ge-68 breakthrough reached 10^{-2}%. The synthetic resin yielded more than 60% of available Ga-68 in 5 ml of 0.3N HCl. More than 600 collections were performed from the same generator with Ge-68 breakthrough levels of less than 10^{-3}%.

INTRODUCTION

Interest in the diagnostic use of positron-emitting radionuclides has greatly increased with the development of tomographic reconstruction methods. Most biologically important positron emitters are cyclotron-produced and because of their short half-lives unavailable in the typical hospital setting. Gallium-68, a radionuclide with a 68-min half-life, is accessible by means of a generator system based on its separation from the parent germanium-68 (287-day half-life). It is easily chelated so that a number of Ga-68 labeled compounds can be synthesized. The 68-min half-life is sufficiently short to provide reasonably low radiation exposure of the patient and yet adequately long to permit the synthesis of radiopharmaceuticals and the study of a variety of biologic functions.

In principle, a radioisotope generator system designed for use in nuclear medicine should be simple, and give reproducibly a high yield of the desired radionuclide with a very large separation factor from the parent radioisotope. Chromatographic systems offer the most convenient technique for generator design. However, the adsorption of germanium onto substrates has not been very successful, and is rarely used in analytical chemistry. The Ga-68 generator system that is commercially available (1) employs 5 millimolar EDTA (pH 7.0) to collect gallium from an aluminum oxide column, while germanium remains adsorbed (1). At this pH, Ga^{3+} is strongly complexed by EDTA. The main drawback of this generator is the necessity of additional chemical manipulations requiring a chemically competent staff if the Ga-68 is desired in a form other than the EDTA-complex. This separation of the Ga-EDTA-complex can be done by anion-exchange chromatography in hydrochloric acid medium (2), by extraction (3), or by pyrolysis (3).

Other workers have attempted to produce alternative column chromatographic methods for obtaining a simple ionic form of Ga-68 in solution. Kopecky at al (4) eluted an Al_2O_3 column (basic; Merck) with 0.2N HCl, collecting 48% of the Ga-68. Up to 200 elutions were performed and the aluminum and Ge-68 contamination of the eluent assayed. The Ge-68 level was less than $2 \cdot 10^{-3}$% of the Ga-68. In a Ga-68 labeled colloid for liver scanning this would increase the radiation dose by about 25% (5). Malyshev at al (6) used the hydroxides of zirconium, tin and titanium as adsorbents and either HCl, HNO_3 or CH_3COOH as eluents. The optimum conditions were found to be a zirconium oxide (ZrO_2) adsorbent eluted with 0.1N HCl. A 35% yield of Ga-68 was obtained with the Ge-68 breakthrough varying from $3 \cdot 10^{-2}$ to $6 \cdot 10^{-3}$%. This breakthrough would lead to dose-rate increases to the liver by Ga-68 labeled colloids of 400% to 75% respectively. Gallium can also be separated from germanium by extraction into diethyl acetone (7) or by extraction as a Ga-68-oxine complex (8).

Ion-exchange chromatographic procedures have relatively low selectivity for germanium because of the non-ionic character of germanium compounds. Anion-exchange behavior in HF on a strongly basic anion-exchange resin measured by Faris (9), shows that the heavier members of the IIIB series are only weakly complexed by fluoride, while the IVB series elements form very stable complexes that are strongly adsorbed. In the case of germanium this is due to the formation of germanium hexafluoride (GeF_6^{2-}) which could readily replace the anions on the anion-exchange resin. The complexation of Ga (III) with fluoride and the stability constants of the complexes were studied (10). There are no anionic gallium fluoride complexes for fluoride concentrations less than 10^{-3}

molar. A separation from germanium, which should still form a stable GeF_6^{2-} complex under these circumstances, therefore seems possible on an anion-exchange resin.

An alternative to the adsorption of germanium complexes on an anion-exchanger is the complexation of germanium by a chelate resin, an organic reagent, which is incorporated into a fairly inert matrix. Both germanium and gallium form complexes with polyhydroxy compounds (11,12,13) and the germanium compounds are more stable with polyphenols than with aliphatic polyalcohols. The condensation of phenols with formaldehyde has been extensively studied as the reaction products are of commercial importance. On such a resin germanium could be immobilized while under certain circumstances gallium might be eluted.

MATERIALS AND METHODS

Batch equilibration studies: Bio-Rad AG1-X8 200-400 mesh anion-exchange resin was used to measure the adsorption of both germanium and gallium from HF. Injectable Ga-68 labeled compounds, and therefore the eluate from the generator, should have ionic strengths close to that of blood. Therefore, the acidity of the eluents was limited to 0.1N HF or less.

A chelate resin was synthesized by the condensation of pyrogallol and formaldehyde using acid catalysis. The resulting polymer was ground to a particle size of approximately 2 to 50μ and used for batch equilibration studies with dilute HCl.

The adsorption of germanium and gallium onto the resins tested was determined by batch equilibration of 100-mg amounts of the dry resin with 5g of the eluents, containing Ge-68-Ga-68 equilibrium mixtures. The samples were shaken for ten minutes, centrifuged and 1 ml aliquots of the supernatants assayed for Ga-68. After correction for decay between the end of equilibration and the activity measurement, this activity was compared with the activity of the original eluent solution of the Ge-68-Ga-68 equilibrium mixture. Distribution coefficients (K_D) were calculated as

$$K_D = \frac{\text{Activity/gram of adsorbent}}{\text{Activity/gram of eluent}}$$

The K_D values for germanium were obtained by counting the same samples for Ga-68 24 hrs after equilibration.

Column experiments: Bio-Rad AG1-X8 200-400 mesh experimental columns were built with a diameter of 0.4 cm, and resin bed volumes of either 1.3 ml or 0.25 ml; these were loaded with a Ge-68-Ga-68 equilibrium mixture and eluted with dilute HF. The elution profile of Ga-68 from these experimental columns, the Ga-68 yield, and the breakthrough of Ge-68 as a function of the total eluted volume were measured. The breakthrough was expressed as percent of the total Ge-68 loaded onto the resin.

Chromatographic columns were loaded with the ground synthetic resin to bed volumes of 3 ml, and a Ge-68-Ga-68 equilibrium mixture adsorbed on the resin. The elution profile of Ga-68, the total yield for Ga-68 in an acceptably small collection volume and the breakthrough of Ge-68 were determined. The latter is carried out by counting the eluents for Ga-68 activity 24 hrs after collection.

The measurement of the elution profile, yield and breakthrough were repeated at regular intervals between continuous elutions of the resin.

RESULTS

Anion-exchange in dilute HF: The results of the batch equilibration studies are summarized in Table 1. The distribution coefficient (K_D) for germanium on Bio-Rad AG1-X8 in dilute HF was higher than 3,500 as long as the concentration of the acid was greater than 0.001N. The K_D values for gallium varied between 10 and 40 for concentrations between 0.1 and 0.001N.

Fig. 1 shows the elution profile of Ga-68 from 0.5 g of Bio-Rad AG1-X8 200-400 mesh anion-exchange resin in a 10 cm column, with a 0.4 cm diameter, washed with 0.01N HF. The cumulative Ga-68 collection yield shows that 95% of all Ga-68 on the column can be collected in 20 ml of 0.01N HF.

Measurements were made of the breakthrough of Ge-68 and the yield of Ga-68 as a function of the number of Ga-68 collections from a Bio-Rad AG1-X8 column (0.25 ml resin bed). For at least 600 elutions 90 ± 3% of the Ga-68 could be collected in 3 ml of 0.01N HF, while the Ge-68 breakthrough levels remained lower than 10^{-2}%.

Chelate formation with a pyrogallol-formaldehyde resin: The distribution constants (K_D) of Ge-68 and Ga-68 between the pyrogallol-formaldehyde resin and dilute HCl solutions are shown in Table 2. The sizes of the resin particles ranged from 2 to 50μ. On microscopic examination the larger particles looked like aggregates of particles of about 2μ diameter, but ultrasonication for five min did not break up these larger aggregates.

The increase in Ga-68 collection as a function of eluting volume of 0.3N HCl is depicted in Fig. 2. Typically 60-65% yields in volumes twice the bed volume of the generator were obtained. Higher yields were obtained with larger collection volumes. The elution of Ga-68 at the loading step is virtually complete (>99%) upon collection of the starting volume plus one column volume of additional 0.3N HCl.

The variation in the breakthrough of Ge-68 and of the elution yield for Ga-68 as a function of the number of collection volumes is shown in Table 3.

DISCUSSION

The high distribution coefficient for germanium between the anion-exchange resin and HF should allow hundreds of collections of Ga-68 from a generator without appreciable loss of germanium. Hydrolysis of the GeF_6^{2-} complex could lead to early losses of germanium, but this is unlikely in dilute HF. The total volume eluted through the 0.25 ml column represented 1,000 collection volumes for Ga-68 from the generator.

The choice of the HF concentration for the collection of Ga-68 depends on two conflicting considerations. The K_D for gallium is lowest for higher normalities of HF. However, higher HF concentrations mean more fluoride in the final sample. Lower HF concentrations necessitate the use of larger collection volumes for gallium. Although the overall fluoride content of the preparations would be lower, the longer elution times and the larger volume would make the generator less practical.

TABLE 1. DISTRIBUTION COEFFICIENTS OF ^{68}Ge AND ^{68}Ga BETWEEN BIO-RAD AG1-X8 200-400 MESH AND AQUEOUS SOLUTIONS OF DILUTE HF

HF Concentrations (N)	$K_{D_{Ge}}$	$K_{D_{Ga}}$
10^{-3}	$\geq$5,500*	40
$3 \cdot 10^{-3}$	$\geq$5,000	40
$4 \cdot 10^{-3}$		44
10^{-2}	12,000	27
$2 \cdot 10^{-2}$	6,000	21
$3 \cdot 10^{-2}$	>4,000	22
$6 \cdot 10^{-2}$		18
10^{-1}	>4,000	14

*The upper limit that can be determined under these experimental conditions

TABLE 2. DISTRIBUTION COEFFICIENTS OF GERMANIUM AND GALLIUM BETWEEN DILUTE HCl AND PYROGALLOL CONTAINING CHELATE RESIN

N_{HCl}	$K_{D_{Ge}}$	$K_{D_{Ga}}$
0 (0.9%NaCl)	≥ 5,000	50
0.1	≥ 5,000	1.8
0.2	≥ 5,000	1.6
0.3	≥ 5,000	0.5
0.5	≥ 5,000	0.3

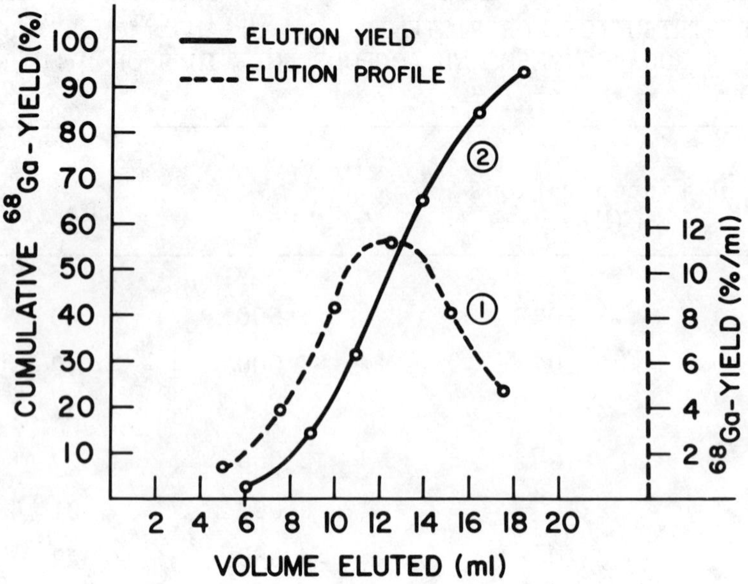

Figure 1

Elution profile (right-hand ordinate) and cumulative yield (left-hand ordinate) for Ga-68 collection from 0.5g Bio-Rad Ag1-X8 200-400 mesh anion-exchange resin by elution with 0.01N HF.

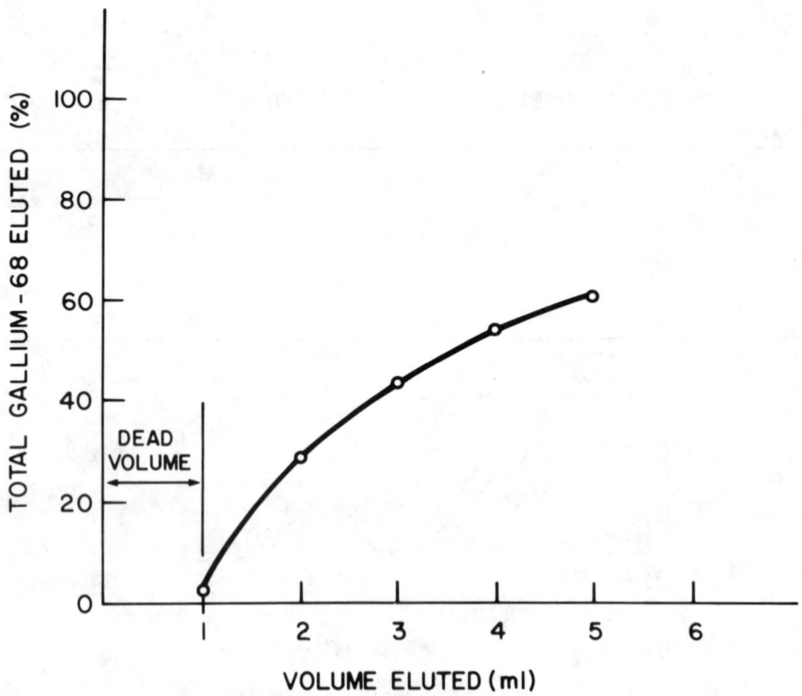

Figure 2

Cumulative yield for Ga-68 collection from pyrogallol-formaldehyde chelate resin by elution with 0.3N HCl.

Using 0.01N HF, the amount of sodium fluoride in a neutralized 3 ml eluent would be 3.10^{-5}mole or 1.3 mg. This would increase the fluoride in normal human whole blood by about the same amount as is already present (0.3 mg/ liter). This is equivalent to a typical prophylactic fluoride dose. The safety of these fluoride preparations will depend then on the speed with which fluoride is excreted or adsorbed. Since the gallium fluoride complex is very weak it is expected that it will not interfere with any preparation of Ga-68 labeled radiopharmaceuticals. If necessary it could easily be masked by the addition of a small amount of Al(III) ions, which complex fluoride ions very strongly.

The ratio of the distribution coefficients of germanium and gallium between Bio-Rad AG1-X8 and HF is about 450 with 0.01N HF. Using 0.001N HF the K_D for Ga(III) increases by a factor of only 1.5. Thus Ga-68 could be collected with 5 ml of 0.001N HF eluate which represents 0.2 mg NaF on neutralization. The ratio of the K_D's for germanium and gallium is probably higher with 0.001N HF than with 0.01N HF.

The capacity of the pyrogallol formaldehyde resin to adsorb germanium could be due to the reaction of germanium with the phenolic groups, as was described by Kurnevich at al (14) for the compound 3,4,5 (trihydroxy)-$C_6H_2COOCl_3$. In this compound the three hydroxyl groups participate in the formation of the germanium-complexes, with simultaneous splitting off of protons from two hydroxyl groups. The Ge-68 breakthrough, even after 600 collections of Ga-68, is lower than 10^{-3}% per collection volume. After 600 collections, the germanium is still retained on the top of the resin as could be expected considering the very high K_D value of this element under the conditions used.

It is thought that the Ge-68 migrates inside the resin particles, where the flow of the HCl solution may be low and the collection of the Ga-68 delayed. The 60% yield of Ga-68 can probably be improved by producing the pyrogallol-formaldehyde resin under more controlled conditions and grinding it to a narrower particle size range, or by condensing it on an inert substrate with a well-defined mesh size. As eluent for the chelate resin 0.3N HCl was chosen, because this will, upon neutralization with an equal volume of 0.3N NaOH yield 0.15M NaCl, which is isotonic and directly injectable, or usable for further labeling work.

ACKNOWLEDGMENTS

Thanks are due to Rebekah Taube for help in the preparation of the manuscript. This work was supported by DOE Contract E (11-1) - 4115

REFERENCES

1. Greene MW, Tucker WD: An improved gallium-68 cow. Int J Appl Radiat Isot 12:62, 1961

2. Hnatowich DJ: A method for the preparation and quality control of Ga-68-radiopharmaceuticals. J Nucl Med 16:764-768, 1975

3. Yano Y: Preparation and control of Ga-68 radiopharmaceuticals. In

Radiopharmaceuticals From Generator Produced Radionuclides. Vienna, IAEA, 1971, pp 117-125

4. Kopecky P, Mudrova B and Svoboda K: The study of conditions for the preparation and utilization of a Ge-68-Ga-68 generator. Int J Appl Radiat Isot 24:73-80, 1973

5. Barrall RC, Chakalian VM, Colombetti LG and Finston RA: Purity tests of a Ge-68-Ga-68 generator. Int J Appl Radiat Isot 22:149-154, 1971

6. Malyshev KV and Smirnov VV: Radiokhimiya. A generator of gallium-68 based on zirconium hydroxide. 17:137-140, 1957

7. Gleason GI: A Positron Cow. Int J Appl Radiat Isot 8:90-94, 1960

8. Ehrhardt GJ and Welch MJ: A new germanium-68/gallium-68 generator. J Nucl Med 19: 925-929, 1978

9. Faris JP: Adsorption of the elements from hydrofluoric acid by anion exchange. Anal Chem 32: 520-522, 1960

10. Babko AK and Shimadina LG: Stability of the fluoride complexes of certain metals. Zh Neorgan Khim 4:1060-1066, 1959

11. Antikainen PJ: The chelation of germanic acid with glycols and polyalcohols. Suomen Kem 31B:291-293, 1958

12. Antikainen PJ: A new germanium catecholate and its thermal decomposition. Suomen Kem 32:211-213, 1959

13. Antikainen PJ: Chelate formation between germanium acid and some glycols and polyhydric alcohols in aqueous solutions. Acta Chem Scand 13:312-322, 1959

14. Kurnevich GI, Vishnevskii VB and Loiko EM: Infrared spectra of mixed coordination compounds of germanium (IV). Zh Prikl Spektrosk 17:290-295, 1972

TABLE 3. BREAKTHROUGH OF GE-68 FROM PYROGALLOL-FORMALDEHYDE RESIN AND COLLECTION YIELD FOR GA-68, IN A TWO COLUMN VOLUMES, AS A FUNCTION OF NUMBER OF GA-68 COLLECTIONS

Collection #	Ge-68 Breakthrough/Collection (%)	Ga-68 Yield (%)
1	$< 10^{-3}$	99
50	$< 10^{-3}$	83
100	$< 10^{\ 3}$	63
200	$< 10^{-3}$	61
300	$< 10^{-3}$	-
500	$< 10^{-3}$	60
600	$< 10^{-3}$	60

TANTALUM-178 RADIOPHARMACEUTICALS FOR LUNG AND LIVER IMAGING

Rudi D. Neirinckx, B. Leonard Holman, and Michael A. Davis

Joint Program in Nuclear Medicine, Harvard Medical School, Children's Hospital Medical Center and Peter Bent Brigham Hospital, Boston, MA 02115

ABSTRACT

A W-178-Ta-178 generator system has been developed in our laboratory. The influence of the volume and HCl - H_2O_2 concentration of the eluent on Ta-178 yield was investigated and an electrophoretic study of the Ta-178 eluate was undertaken in order to provide useful information for synthesizing Ta-178 labeled radiopharmaceuticals. Tantalum-178 has been incorporated into particulate matter of discrete size range for imaging of the lungs (stannous human albumin microspheres and ferrous hydroxide macroaggregates) and liver (stannous human albumin minimicrospheres, tantalum hydroxide-gelatin colloid, microaggregated stannous albumin). Animal distribution in rats and scintigraphy studies in rabbits and monkeys suggest that these agents are comparable to the currently used Tc-99m lung and liver agents.

INTRODUCTION

Tantalum-178, the daughter of tungsten-178 ($T_{1/2}$=21 days), is a short-lived radionuclide ($T_{1/2}$= 9.3min) with possible clinical applications, particularly in pediatric and cardiovascular nuclear medicine. The predominant photon emissions are characteristic x-rays from hafnium (54-65 keV).

A practical W-178 - Ta-178 generator has been developed (1, 2) and procedures established for the production of the high activities of W-178 needed (3).

Because of its low energy and short half-life, the radionuclide was initially thought to have potential only for use with multiwire proportional chambers. Subsequent experience has shown that imaging is possible with an Anger scintillation camera and pinhole collimator. Tantalum-178-labeled radiopharmaceuticals would reduce patient radiation dose and facilitate repeat or multiple examinations. The ionic form of the Ta-178 eluted from the generator has been studied in order to elucidate the condition for optimum yields of Ta-178 and inhibition of W-178 breakthrough and to facilitate the labeling of compounds for lung and liver imaging.

MATERIALS AND METHODS

Ta-178 yield as a function of eluent volume and composition: the percent of Ta-178 that can be collected by elution from the anion-exchange resin (generator support) was measured by the batch equilibration of 0.25 grams of Bio-Rad AG1-X8 resin onto which W-178 had been previously adsorbed with a 10 or 40 fold excess of 0.05 to 0.50 N HCl. After equilibration the liquid phase was assayed for Ta-178 activity. The influence of H_2O_2 was studied by gradually increasing its concentration in 0.1 N HCl eluent from 0 to 3 %.

Electrophoresis of the Ta-178 eluate: The Ta-178 eluate neutralized with NaOH, phosphate buffer or citrate buffer was evaluated by electrophoresis at 275 V in a trisbarbital-sodium barbital buffer (pH 8.8).

Preparation of Lung Imaging Agents:

Ta-178 - Stannous human albumin microspheres (HAM).
One mg of stannous albumin microspheres (10-40 μ diameter)* was suspended in 100 lambda of a 5% Tween-80 solution in triply distilled water and 0.6 ml of W-178-Ta-178 generator eluate (0.1N HCl and 0.1% H_2O_2) added. The mixture was stirred electromagnetically for three min in a hot water bath, spun down at 2,500 rpm in a refrigerated centrifuge, and washed twice with saline. Supernatant, particles and washings were counted separately for Ta-178 radioactivity.

Ta 178 - Ferrous hydroxide macroaggregates (FHMA).
All the reagents used in this preparation were made up with water which had been purged with nitrogen gas overnight. Tantalum-178 was eluted from the generator with a solution of 0.1N HCl and 0.1% H_2O_2, evaporated to dryness, and redissolved in 0.5 ml of 0.1N HCl. This was added to 2 mg of ferrous ion dissolved in 0.5 ml of 0.1N HCl in an evacuated vial. After neutralization with NaOH and standing for two min, 1.5 ml of a 5% aqueous gelatin solution was added and the preparation shaken for one min prior to injection.

* NEN Pulmolite, New England Nuclear Corporation, North Billerica, MA

Preparation of Liver Imaging Agents:

Ta-178 - Stannous human albumin minimicrospheres.
Albumin minimicrospheres* with a diameter range of 0.5 to 2 μ were labeled with Ta-178 using the same procedure as described above for HAM. For each experiment, 0.5 mg of minimicrospheres was used.

[^{178} Ta] Tantalum hydroxide - gelatin colloid.
Tantalum-178 was eluted in 1 ml of 0.1N HCl and 0.1% H_2O_2. One ml of a 5% aqueous gelatin solution and 0.25 ml of phenol red were added after which the solution was neutralized with Na04 until an orange-red color appeared. The mixture was heated in a boiling water bath for five min.

Ta-178 - Microaggregated stannous albumin (MIAA).
Two mg microaggregated albumin**(80% of particles between 0.1 μ and 1 μ diameter) were suspended in 1 ml of isotonic saline and boiled for either 0, 2, 5 or 10 min with 1 ml of W-178-Ta-178 generator eluate. The particles were washed once and resuspended in isotonic saline.

Animal Distribution Studies: Wistar inbred albino rats from the Charles River Breeding Laboratories weighing 150 to 250 g were used for the animal distribution studies. One ml of the Ta-178 agents (10 to 100 μCi) was injected into the saphenous vein. The animals were sacrificed by ether asphyxiation five min (lung agents) or ten min (liver agents) after administration. Lungs, liver, spleen, bladder, kidneys, muscle, bone, fur and blood were assayed for radioactivity.

Scintigraphic Studies: To determine the feasibility of <u>in vivo</u> imaging with Ta-178, images were obtained in rabbits and monkeys, using a gamma camera (Searle LFOV) equipped with a pinhole collimator. The animals were typically injected intravenously with 1 ml solutions (5 mCi)of either a lung or liver scanning radiopharmaceutical. The electromagnetic radiation of Ta-178 between 54 keV and 65 keV was used and imaging commenced 2 to 5 min after injection. The camera was preset to collect 500,000 counts, and scintiphotos were taken. Repeat or additional studies were made 30 to 45 min after the preceding injection.

RESULTS

Ta-178 yield as a function of eluent volume and composition: The results and batch equilibration as a function of HCl concentration and eluent volume are summarized in Table 1. The yields of Ta-178 showed that increasing the volume of the mobile phase did not increase the Ta-178 yield in line with what is expected from the presence of only one Ta-178 species in the solution. The variation of the Ta-178 yield with 10 ml eluent, containing increasing concentrations of HCl is shown in Fig. 1, curve 1. Curve 2 represents the data by Babko et al (4) concerning the fraction of $Ta(OH)_4^+$ in an acidic solution of tantalic acid as a function of pH. The influence of H_2O_2 on the collection yield of Ta-178 in 10 ml of eluent is summarized in Table 1.

Electrophoresis of Ta-178 eluates: Electrophoresis of the neutralized Ta-178 eluates showed that the NaOH and phosphate buffered preparations contained only unchanged tantalum species, while the citrate-buffered preparations contained negatively changed tantalum citrate complexes.

* 3M Company, Minneapolis, Minnesota
** Microlite, New England Nuclear Corporation, North Billerica, MA

TABLE 1. Ta-178 COLLECTION FROM BIO-RAD AG 1-X8 ANION-EXCHANGE RESIN WITH HCl and H_2O_2

HCl (N)	H_2O_2(%)	% ^{178}Ta Collected in 2.5 ml	in 10 ml
0.05	0	12	32
0.10	0	43	55
	0.1		54
	0.2		53
	0.4		46
	1.0		37
	3.0		34
0.20	0	42	64
0.50	0	75	79

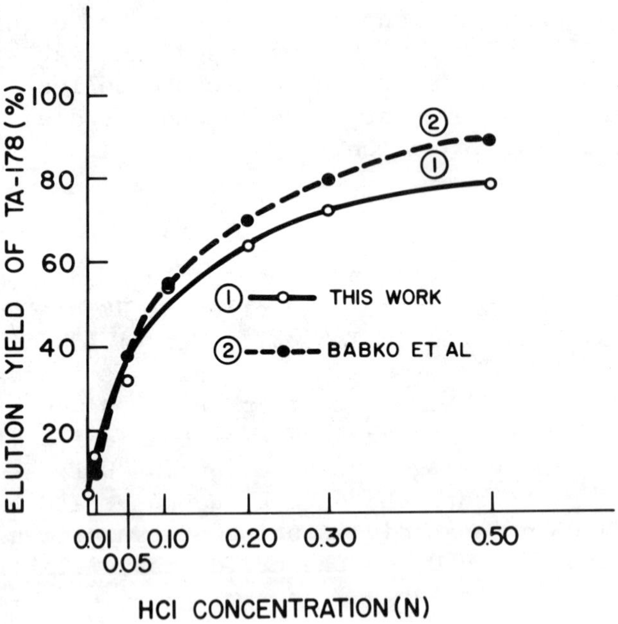

Figure 1

Elution yield for Ta-178 from a Bio Rad Agl-X8 200-400 mesh resin (Curve 1) and fraction of $Ta(OH)_4^+$ in an acid solution of tantalic acid (Curve 2) (4) as a function of HCl concentration.

Lung Imaging Agents:

Ta-178 - Stannous human albumin microspheres (HAM).

The yield for Ta-178 incorporation into stannous albumin microspheres was 98%. The *in vivo* distribution in rats five min after intravenous injection is shown in Table 2. There is a high concentration of radioactivity in the lungs.

Ta-178 - Ferrous hydroxide macroaggregates (FHMA).

The coprecipitation efficiency for Ta-178 with ferrous hydroxide was greater than 98%. The *in vivo* distribution in rats, five min after injection, is shown in Table 2.

Liver Imaging Agents:

Ta-178 - Stannous human albumin minimicrospheres.

Boiling the suspension of 0.5 ml minimicrospheres for three min resulted in a labeling efficiency of 98%. Less than 1% of the Ta-178 was collected in the supernatant upon washing twice with saline. The *in vivo* distribution in rats is shown in Table 2.

$[^{178}Ta]$ Tantalum hydroxide - gelatin colloid.

The labeling efficiency for the $[^{178}Ta]$ tantalum hydroxide-gelatin colloid preparation was 98%. The organ distribution in rats ten min after intravenous injection is shown in Table 2.

Ta-178 - Microaggregated stannous albumin.

The labeling efficiency for Mirolite increased only slowly withtime and was 26% after heating in a hot water bath for 10 min. Results of the animal distribution as a function of time showed the accumulation in the liver to be a slow process, reaching 60% after 20 min.

Scintigraphic Studies: The scintigraphic studies produced images comparable to those obtained with the corresponding Tc-99m radiopharmaceuticals. Figures 2 and 3 are examples of lung and liver scintigrams in the rabbit with Ta-178-HAM and Ta-178- minimicrospheres respectively.

DISCUSSION

The elution yields of Ta-178 as a function of the eluent volume show that more than one tantalum species is present in the solution. The species that can be collected from the anion-exchange resin with dilute HCl is most likely $Ta(OH)_4^+$, since the Ta-178 yield as a function of HCl concentration is in good agreement with the measurements by Babko et al (4) of the relative abundance of $Ta(OH)_4^+$ in an acid solution of tantalic acid. At higher pH values, the fraction of $Ta(OH)_5$ increases, as indicated by the electrophoresis of the neutralized Ta-178 eluates.

If the $Ta(OH)_4^+$ were in dynamic equilibrium with $Ta(OH)_5$, increasing the acidity of the eluent would lead to identical amounts of Ta-178 being eluted with decreasing volumes of eluent. Instead it is found that with increasing acidity higher yields of Ta-178 are obtained as sharp elution peaks in identical volumes. This can be explained by the absence of a

TABLE 2. IN VIVO DISTRIBUTION OF Ta-178 LABELED LUNG AND LIVER AGENTS IN RATS

	LUNG IMAGING AGENTS[a]		LIVER IMAGING AGENTS[b]	
Organ	Ta-178-HAM	Ta-178-FHMA	$Ta(OH)_5$ gelatin colloid	Ta-178-minimicrospheres
Liver	2.5±0.7[c,d]	1.8±0.1	80 ± 3	89 ± 2
Spleen	0.1±0.05	0.1±0.03	1.0±0.2	3.5±1.0
Lung	90 ± 2	95 ± 2	1.3±0.3	3.5±0.5
Kidneys	0.2±0.1	0.1±0.03	0.7±0.3	0.2±0.05
Bladder	<0.1	<0.1	0.6±0.5	<0.01
Blood	5 ± 1	<0.1	6.8±0.8	1.8±0.5
Muscle	<0.1	<0.1	3.3±0.6	1.4±0.3
Bone	<0.4	<0.1	3.9±0.4	1.0±0.3
Fat & Fur	<0.5	<0.1	3.0±0.1	0.4±0.1

a. Measured 5 min. after I.V. injection.

b. Measured 10 min. after I.V. injection.

c. Percent injected dose per organ.

d. Average of 6 rats ± S.E.M.

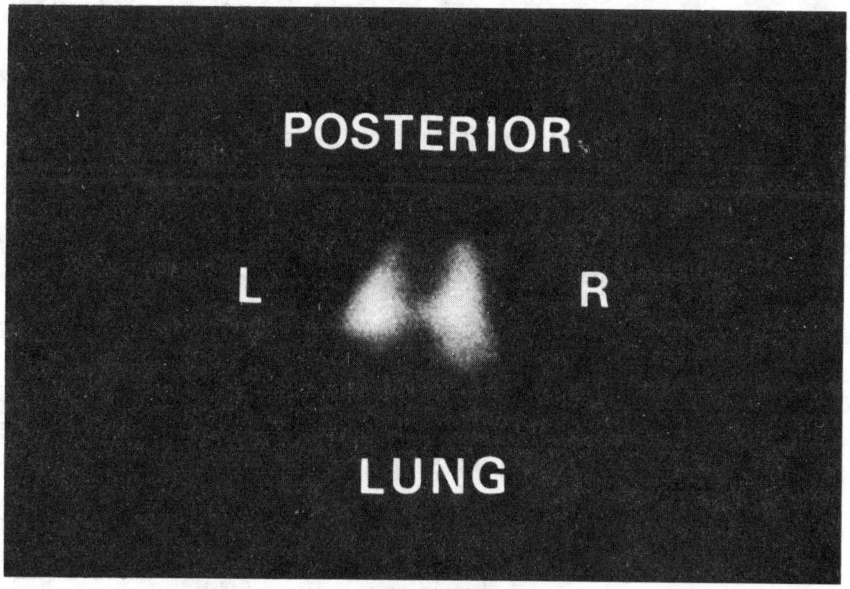

Figure 2

Scintiphoto of rabbit lungs obtained by collection of 500,000 counts on a Searle LFOV gamma camera, equipped with a pinhole collimator, 2 min after injection of 5mCi of Ta-178-HAM.

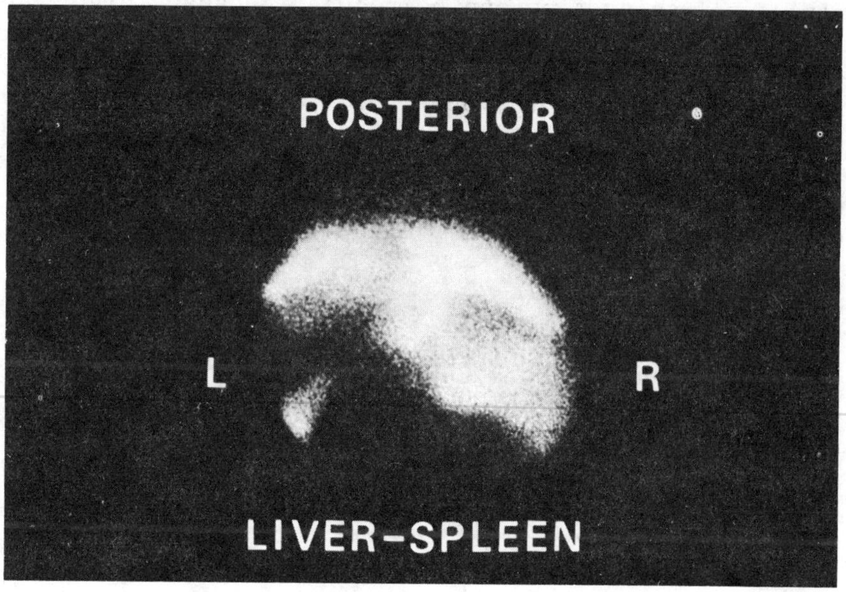

Figure 3

Scintiphoto of rabbit liver, obtained by collection of 500,000 counts on a Searle LFOV gamma camera, equipped with a pinhole collimator, 5 min after injection of 5mCi of Ta-178 - mini-micropheres.

dynamic equilibrium between $Ta(OH)_4^+$ and $Ta(OH)_5$. Experimentally, the relative amounts of each compound are pH dependent as can be expected for hydroxide formation.

The decrease of the elution yield of Ta-178 with increasing H_2O_2 concentration confirms the results of Bandi et al (5) who found that H_2O_2 caused tantalum to be more tightly bound by an anion-exchange resin. The 0.1% H_2O_2 used in an admixture with the dilute HCl for the generator operation has no appreciable influence on Ta-178 elution.

The Ta-178 in the unneutralized generator eluate is in the $Ta(OH)_4^+$ form and is incorporated easily into human albumin as shown by the labeling results. Upon neutralization, $Ta(OH)_5$ coprecipitates completely with $Fe(OH)_2$ to produce a Ta-178 labeled lung agent or can be coated with gelatin to form a Ta-178 labeled colloid with an acceptable size for use as a liver agent.

The procedures reported in this study for making Ta-178 labeled liver-spleen and lung agents can be completed within one half-life after elution of Ta-178 from the generator. Scintiphotos using these Ta-178 labeled radiopharmaceuticals are comparable to those obtained with Tc-99m labeled agents.

Among the liver scanning agents tested, the minimicrospheres collected in the liver to a greater degree than the [^{178}Ta] tantalum hydroxide-gelatin complex ten min after injection. This may be due to their more uniform size. Both agents however would be acceptable for clinical use.

Tantalum-178 offers an attractive alternative to Tc-99m which is the radionuclide of choice for most static and dynamic imaging procedures in nuclear medicine. Imaging time can be reduced because larger doses of Ta-178 can be administered. Even with larger doses the absorbed dose may be less because Ta-178 has a physical half-life 38.7 times shorter than Tc-99m, or equal photon yields 1.4 times as much Ta-178 agent as Tc-99m agent would be required. Thus, using Ta-178 a decrease in administered radiation dose by a factor of 25 can still be obtained while preserving identical photon yields. Furthermore, repeat studies can be carried out as soon as 30 min after a previous injection of a Ta-178 labeled radiopharmaceutical.

ACKNOWLEDGEMENTS

This work was supported by DOE Contract E(11-1)-4115. Thanks are due to Rebekah Taube for help in the preparation of the manuscript, and to Dr. Eli Livni and Alice Carmel for technical assistance.

REFERENCES

(1) Holman BL, Harris GI, Neirinckx RD, et al: Tantalum-178 - A short-lived nuclide for nuclear medicine: Production of the parent W-178. J Nucl Med 19: 510-513, 1978

(2) Neirinckx RD, Jones AG, Davis MA, et al: Tantalum-178: A short-lived nuclide for nuclear medicine: Development of a potential generator system. J Nucl Med 19: 514-519, 1978

(3) Neirinckx RD, Ku T, Holman BL, et al: Production and Purification of Tungsten-178. Int J Appl Radiat Isot, in press

(4) Babko AK, Lukachina VV, Nabivanets BL: Solubility and Acid-Base Properties of Tantalum and Niobium Hydroxides. Zh Neorgan Khim 8: 1839-1843, 1963; Engl transl 8:957-961

(5) Bandi WR, Buyok EG, Lewis LL, et al: Anion-exchange Separation of Zirconium, Titanium, Niobium, Tantalum, Tungsten and Molybdenum. Anal Chem 33: 1275-1278, 1961